COMPUTATIONAL MODELLING OF OBJECTS
REPRESENTED IN IMAGES:
FUNDAMENTALS, METHODS AND APPLICATIONS

BALKEMA – Proceedings and Monographs
in Engineering, Water and Earth Sciences

PROCEEDINGS OF THE INTERNATIONAL SYMPOSIUM CompIMAGE 2006, COIMBRA, PORTUGAL, 20–21 OCTOBER 2006

Computational Modelling of Objects Represented in Images

Fundamentals, Methods and Applications

Editors

João Manuel R.S. Tavares & R.M. Natal Jorge
Faculdade de Engenharia da Universidade do Porto, Porto, Portugal

CRC Press
Taylor & Francis Group
Boca Raton London New York

CRC Press is an imprint of the
Taylor & Francis Group, an **informa** business
A TAYLOR & FRANCIS BOOK

CRC Press
Taylor & Francis Group
6000 Broken Sound Parkway NW, Suite 300
Boca Raton, FL 33487-2742

First issued in hardback 2019

ISBN-13: 978-0-415-43349-5 (hbk)

Typeset by Charon Tec Ltd (A Macmillan Company), Chennai, India

Visit the Taylor & Francis Web site at
http://www.taylorandfrancis.com

and the CRC Press Web site at
http://www.crcpress.com

Computational Modelling of Objects Represented in Images –
João Manuel R.S. Tavares & R.M. Natal Jorge (eds)
© 2007 Taylor & Francis Group, London, ISBN 978-0-415-43349-5

Table of Contents

Computational Modelling of Objects Represented in Images –
João Manuel R.S. Tavares & R.M. Natal Jorge (eds)
© 2007 Taylor & Francis Group, London, ISBN 978-0-415-43349-5

Acknowledgements

The editors and the Symposium organizers acknowledge the support towards the publication of the Book of Proceedings and the organization of the International Symposium CompIMAGE to the following organizations:

– Centro Internacional de Matemática (CIM)
– Associação Portuguesa de Mecânica Teórica Aplicada e Computacional (APMTAC)
– Universidade do Porto (UP)
– Faculdade de Engenharia da Universidade do Porto (FEUP)
– Fundação para a Ciência e a Tecnologia (FCT)
– Instituto de Engenharia Mecânica – Pólo FEUP (IDMEC-Polo FEUP)
– Instituto de Engenharia Mecânica e Gestão Industrial (INEGI)
– Comissão de Coordenação e Desenvolvimento Regional do Norte (CCDRN)
– Centro de Estudos Euro Regionais Galiza-Norte de Portugal
– Portugal – Espanha Cooperação Transfronteiriça INTERREG III A
– Hotel D. Luis, Coimbra
– Sociedade Portuguesa de Neurorradiologia (SPN)
– Sociedade Portuguesa de Engenharia Biomédica (SPEB)
– Grupo Português de Computação Gráfica (GPCG)

Computational Modelling of Objects Represented in Images –
João Manuel R.S. Tavares & R.M. Natal Jorge (eds)
© 2007 Taylor & Francis Group, London, ISBN 978-0-415-43349-5

Preface

This book contains the keynote lectures and full papers presented at the International Symposium *Comp-IMAGE – Computational Modelling of Objects Represented in Images: Fundamentals, Methods and Applications*, held in Coimbra, Portugal, during the period 20–21 October 2006. The event had 8 invited lectures, 38 oral presentations distributed by ten sessions and 31 posters during two sessions. The contributions came from 19 countries: Algeria, Belgium, Brazil, Canada, Chile, China, Colombia, Czech Republic, France, Germany, Ireland, Italy, Japan, Netherlands, Portugal, Spain, Tunisia, United Kingdom, and United States of America.

In our days the research related with objects modelling has been a source of hard work in several distinct areas of science, such as mathematics, mechanical, physics and informatics. One major application of objects modelling is in medical area. For instance, it is possible to consider the use of statistical or physical procedures on medical images in order to model the represented objects. This computational modelling can have different goals; for instances, 3D shapes reconstruction, organs segmentation in 3D or 2D images, organs registration, etc. Others usual applications are: temporal tracking, analyses of deformation, recognition and simulation.

The main goal of *CompIMAGE* consisted in the provision of a comprehensive forum for discussion on the current state-of-the-art in the related fields. The Symposium covered: Image Processing and Analysis, Image Segmentation, Data Interpolation, Registration, Acquisition and Compression, 3D Reconstruction, Objects Tracking, Motion and Deformation Analysis, Objects Simulation, Medical Imaging, Computational Bioimaging and Visualization.

The Symposium brought together several researchers representing several fields related to Computational Vision, Computer Graphics, Computational Mechanical, Mathematics, Statistics, Medical Imaging, etc. The expertise spanned a broad range of techniques, such as finite element method, modal analyses, stochastic methods, principal components analyses, independent components analyses and distribution models.

The organizers would like to take this opportunity to thank to The International Center of Mathematics (CIM), to The Portuguese Association of Theoretical, Applied and Computational Mechanics (APMTAC), to all sponsors, to all members of the Scientific Committee, to all Invited Lecturers and to all Authors for submitting their contributions.

João Manuel R.S. Tavares
Renato M. Natal Jorge
(Symposium Organizers)

Computational Modelling of Objects Represented in Images –
João Manuel R.S. Tavares & R.M. Natal Jorge (eds)
© 2007 Taylor & Francis Group, London, ISBN 978-0-415-43349-5

Invited lectures

During the symposium were presented Invited Lectures by eight Expertises from five countries:

- Active and catadioptric vision systems for robotics applications, by Hélder Araújo – University of Coimbra, Portugal
- Digital geometry for image analysis and processing, by Valentin E. Brimkov – State University of New York, United States of America
- Multimodality in brain imaging: methodologic aspects and applications, by Sónia I. Gonçalves – Verheij – Vrije Universiteit Medical Centre, The Netherlands
- Perceptual users interfaces and their applications, by Francisco J. Perales – Universitat de les Illes Balears, Spain
- Contour detection by surround suppression of texture, by Nicolai Petkov – University of Groningen, The Netherlands
- Computer vision and digital inclusion of persons with special needs: Overview and state of art, by Hemerson Pistori – Dom Bosco Catholic University, Brasil
- MR diffusion tensor imaging, by M. Forjaz Secca – Universidade Nova de Lisboa, Portugal
- High Order Dual Methods for Staircase Reduction in Texture Extraction Problems, by Frederick E. Park – University of Michigan, United States of America

Computational Modelling of Objects Represented in Images –
João Manuel R.S. Tavares & R.M. Natal Jorge (eds)
© 2007 Taylor & Francis Group, London, ISBN 978-0-415-43349-5

Scientific committee

All works were submitted to an International Scientific Committee composed by sixty expert researchers from recognized institutions of fifteen countries:

Adelino F. Leite-Moreira	University of Porto, Portugal
Adérito Marcos	University of Minho, Portugal
Ana Mafalda	Pedro Hispano Hospital, Portugal
Augusto Goulão	Garcia de Orta Hospital, Portugal
Bernard Gosselin	Faculte Polytechnique de Mons, Belgium
Bogdan Raducanu	Computer Vision Center, Spain
David Steinman	University of Toronto, Canada
Demetri Terzopoulos	University of California, Los Angels, United States of America
Eduardo Borges Pires	Technical University of Lisbon, Portugal
Francisco Calheiros	University of Porto, Portugal
Francisco Perales	Universitat de les Illes Balears, Spain
Georgeta Oliveira	Pedro Hispano Hospital, Portugal
Gerhard A. Holzapfel	Graz University of Technology, Austria
Helcio R.B. Orlande	Federal University of Rio de Janeiro, Brasil
Hélder Araújo	University of Coimbra, Portugal
Hemerson Pistori	Dom Bosco Catholic University, Brasil
Hervé Delingette	The French National Institute for Research in Computer Science and Control, France
Ioannis Pitas	Aristotle University of Thessaloniki, Greece
Isabel M.A.P. Ramos	University of Porto, Portugal
Jan C. De Munck	Vrije Universiteit Amsterdam, Netherlands
Jan Kybic	Czech Technical University in Prague, Czech Republic
Javier Melenchón Maldonado	Universitat Ramon Llull, Barcelona, Spain
João A.C. Martins	Technical University of Lisbon, Portugal
João M.C.S. Abrantes	Technical University of Lisbon, Portugal
João Manuel R.S. Tavares	University of Porto, Portugal
João Martins Pisco	Universidade Nova de Lisboa, Portugal
João Paulo Vilas Boas	University of Porto, Portugal
Jorge Alves da Silva	University of Porto, Portugal
Jorge M.G. Barbosa	University of Porto, Portugal
Jorge S. Marques	Technical University of Lisbon, Portugal
José António Simões	University of Aveiro, Portugal
José Carlos Príncipe	University of Florida, United States of America
José Miguel Salles Dias	Microsoft, Portugal
Juan J. Villanueva	Autonomous University of Barcelona, Spain
Laurent Cohen	Universite Paris Dauphine, France
Leandro Machado	University of Porto, Portugal
Lionel Moisan	Université Paris V, France
Luís Alexandre	University of Beira Interior, Portugal
Manuel Doblaré	University of Zaragoza, Spain

Invited Keynotes/Lectures

Computational Modelling of Objects Represented in Images –
João Manuel R.S. Tavares & R.M. Natal Jorge (eds)
© 2007 Taylor & Francis Group, London, ISBN 978-0-415-43349-5

Active vision systems and catadioptric vision systems for robotic applications

Hélder J. Araújo

Institute of Systems and Robotics, Dept. of Electrical and Computer Engineering,
Polo II – University of Coimbra, Coimbra – Portugal

ABSTRACT: Many robotic applications benefit from a wide field of view. Wide fields of view facilitate tasks such as navigation and obstacle detection. A number of solutions have been proposed in the literature to enable the development of computer vision systems with panoramic fields of view. In most cases extended fields of view imply resolution tradeoffs. Most of these solutions also imply that the mathematical models of the systems are more complex and highly non-linear. Another important aspect for robotic applications is the ability to control and change the image acquisition parameters. In this paper we discuss aspects related to two of the commonly used solutions: active vision systems and catadioptric vision systems. We characterize several technical aspects relevant to robotic applications and discuss aspects related to their control and calibration.

1   INTRODUCTION

The issue of integrating panoramic and omnidirectional vision systems into robotic devices has to take into account requirements related to the relevant information to be extracted (and its usefulness for the tasks that have to be performed). In many robotic applications the difficulties that researchers faced when trying to extract information from images (or sequences of images) whose conditions of acquisition were not controlled, led to the development of solutions based on active systems. In these systems several camera parameters can be actively changed. This type of solutions was a result from the realization that the extraction of information from the images could be facilitated if the acquisition parameters could be changed to suit the specific nature of the data to be extracted. This kind of approaches was strongly influenced by the views of the psychophysicist J.J. Gibson (Gibson 1983; Gibson 1987). Gibson developed the concept of "ecological perception". According to his view it would be a mistake to consider the problems of visual perception from an abstract and static point of view. From his point of view perception is entirely dependent on a "matrix of stimuli" and perception is a direct consequence of the properties of the environment. Gibson's ideas require active organisms. Perception is a function of the interaction between the organism and the environment. These ideas were first adopted in computer vision by Ruzena Bajcsy who first proposed the concept of active perception in (Bajcsy 1985). In 1987, in a paper presented at the "*First International Conference on Computer Vision (ICCV)*" Yiannis Aloimonos and co-workers proposed the concept of active vision (Alioimonos et al. 1988) and formally proved the advantages of the approach. In particular they proved that an active observer can transform problems such as shape-from-shading, shape-from-contour and shape-from-texture from ill-posed problems into well-posed problems. Two other papers were important for the establishment of the solutions based on active vision namely (Bajcsy 1988) and (Ballard 1991). In both these papers both Bajcsy and Ballard defined active vision as being a sequential and interactive process for selecting and analyzing parts of a scene. Indeed, the main idea is that such a process can decrease the amount of computation required by the visual process, since it allows the reduction of the amount of information to be processed by selecting the features of the visual scene that are relevant to the task at stake. As a matter of fact active vision gets inspiration from the methods and processes that both mammals and insects use to extract information from the environment. The wide variety of the biological vision systems suggests that the solutions are highly dependent on the specific task or problem to be dealt with.

2   DEVELOPMENT OF ACTIVE VISION SYSTEMS

The first active vision systems that we developed were aimed at allowing the active tracking of people by a

Figure 1. Left: the initial configuration of the system; Center: final configuration.

Figure 3. Two images of the MDOF system.

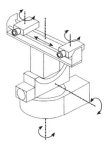

Figure 2. Degrees of freedom.

Figure 4. Ocular module of the MDOF system.

mobile platform. Those systems were based on a simple configuration and used stepper motors (Dias et al. 1993).

In figures 1 and 2 two configurations are displayed. The system has neck pan control, neck tilt control and eye pan control. The degrees of freedom are displayed on the right side of the same figure. This system was integrated into a mobile platform to allow the development of an application for tracking mobile targets (Dias et al. 1995; Dias et al. 1998). In this application the goal was to keep both the robot position and orientation (relative to the mobile target) approximately constant. The active vision system was essential for such a goal for two reasons:

- Target tracking requires that the target remains visible most of the time. It is easier to achieve such a goal if an active vision system or an omnidirectional vision system is used. Otherwise keeping the target in the field of view may require some complex maneuvering by the robot. Complex maneuvering by a non-holonomic robot may be slow.
- An active vision system allows the target to remain in the center of the image most of the time. It is therefore possible to use simpler segmentation methods, since for the regions close to the center of the image the simplification of the methods through their linearization do not result in large errors. On the other hand the active vision system itself allows the use of the system data, provided by the encoders, to estimate the target position and orientation relative to the robot.

The goal of obtaining a more effective target tracking led us to develop a more complex, faster and accurate active system. The previous system was slow and did not allow the optimization of the control aspects since the axes were actuated by stepper motors which did not provide neither position nor velocity feedback.

The technical limitations of the first system led us to develop a second system (MDOF) for which a much more demanding set of specifications was defined. Additional goals were also defined for the second system namely the study, development and real time implementation of visual behaviors. The mechanical degrees of freedom of the new system were the following (Batista et al. 1995):

- Neck pan, tilt and swing.
- Each eye had two mechanical degrees of freedom (pan and tilt) and three optical degrees of freedom (zoom, focus and aperture).

Figure 3 has two images of the MDOF system. In addition to the above mentioned degrees of freedom this system allows the mechanical adjustment of the distance between the cameras and the adjustment of the position of the optical center of the lens (so that it can be guaranteed that the pan axis of rotation of the eye intersects the optical center of the lens). The configuration of the ocular systems corresponds to the Fick geometry. The objective was to allow the independent tilt of both ocular systems. The ocular systems (figure 4) were developed using commercial lenses that were fitted with DC motors with

4

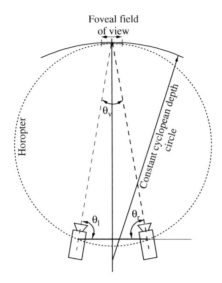

Figure 5. Symmetric configuration for tracking.

And the retinal velocities are:

$$
\begin{bmatrix} v_x^{l/r} \\ v_y^{l/r} \end{bmatrix} = \begin{bmatrix} \frac{f_x}{\bar{P}_{l/r}\cdot\hat{z}}\left(\bar{V}_{l/r}\cdot\hat{x}\right) - \frac{f_x}{(\bar{P}_{l/r}\cdot\hat{z})^2}\left(\bar{P}_{l/r}\cdot\hat{x}\right)\left(\bar{V}_{l/r}\cdot\hat{z}\right) \\ \frac{f_y}{\bar{P}_{l/r}\cdot\hat{z}}\left(\bar{V}_{l/r}\cdot\hat{y}\right) - \frac{f_y}{(\bar{P}_{l/r}\cdot\hat{z})^2}\left(\bar{P}_{l/r}\cdot\hat{y}\right)\left(\bar{V}_{l/r}\cdot\hat{z}\right) \end{bmatrix}
$$

On the other hand taking into account that:

$$
\bar{P}_{l/r} = R_{l/r}\bar{P} + T_{l/r}
$$

and since the symmetrical configuration implies that $\theta_l = -\theta_r = \theta$ (see figure 5 for the variables) it yields:

$$
\Delta_v = \begin{bmatrix} \Delta v_x \\ \Delta v_y \end{bmatrix} = \begin{bmatrix} \frac{t_z\, f\, sin(2\theta)}{Z} \\ 0 \end{bmatrix}
$$

And due to the symmetry it can also be shown that:

$$
\frac{\partial \theta}{\partial t} = \frac{\Delta v_x}{2f}
$$

This result simplifies the vergence control. Version movements are controlled based on the common component to both left and right optical flows. Pure version movements generate a null flow disparity. This type of motion occurs when the target moves along the horopter of the configuration. The pan rotation velocity (defined in the cyclopean coordinate system) is given by:

$$
\Omega^p = \frac{v_x^l + v_x^r}{2f\cos^2\theta}
$$

And the elevation/tilt rotation velocity (defined in the cyclopean coordinate system) is given by:

$$
\Omega^t = \frac{v_x^l + v_x^r}{2f\cos\theta}
$$

As a result of the configuration considered any motion outside the horopter with a non zero longitudinal velocity component projects into the retinas different motion fields. In this case (and for the symmetric configuration) the retinal velocities disparity is a function of the longitudinal component of the 3D velocity vector of the target.

This system was later applied to the development of surveillance applications (Batista et al. 1998). For that purpose a third (fixed) camera was integrated into the system. The third camera acquires a panoramic view of the environment. The aim is to enable tracking of multiple targets. If one of the targets has a suspicious behavior then its tracking is performed by the binocular system (Peixoto et al. 2000). This kind of operation requires a calibrated system. As a result the

harmonic drives. All the degrees of freedom are actuated by DC motors with harmonic drives. Each axis is also fitted with optical encoders allowing the feedback required for control. With this system several visual behaviors were implemented namely saccadic movements, smooth pursuit and vergence (Araújo et al. 1996; Batista et al. 1997).

To perform the binocular tracking of targets an original approach was applied to the vergence control (Batista et al. 1997). The required movements are decomposed into version movements and vergence movements. Tracking is performed based on the motion field/ optical flow. Optical flow from both left and right images are used as measurements. Considering a symmetric configuration for the fixation geometry it can be proved that vergence can be controlled using the flow retinal disparities (see figure 5 for the geometry and variables). The goal is to keep the target in the center of both images. The target moves in the vergence plane. Let us define a 3D coordinate system located in the middle point of the baseline, the cyclopean coordinate system. Let $\bar{P} = (X, Y, Z)$ be the coordinates of the target in the cyclopean coordinate system. In this coordinate system the target moves with velocity:

$$
\bar{V} = -\bar{\Omega} \times \bar{P} - \bar{t} \tag{1}
$$

In the coordinate systems of the left and right (l/r) cameras the velocity of the target is:

$$
\bar{V}_{l/r} = -R_{l/r}\left[\bar{\Omega} \times \bar{P} + \bar{T}\right]
$$

5

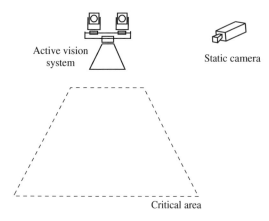

Figure 6.   Integration with a fixed camera.

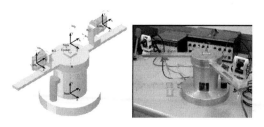

Figure 7.   Left: configuration of the new system; Right: a picture of the system.

target location estimated by the fixed camera can be converted into the coordinate frame of the binocular system (figure 6).

A third system was also built with the aim of simplifying the control and operation of the active system. In addition, and in order to enable the integration of such a system with a mobile platform, smaller and lighter ocular modules were designed and built. We also verified (Barreto et al. 1999) that it could be advantageous to have the possibility of controlling the relative orientation of both cameras. As a result we decided to develop the system of figure 7.

Active vision systems are biologically inspired. However they also have formal and analytical advantages from the point of view of extraction of information. Indeed, and in general, perception requires activity (Gibson 1983). On the other hand in biological systems there are a wide variety of eye configurations. As a result there was also research interest in evaluating the advantages of the diverse biological eye configurations. One of the simplest ways of building new types of eyes/cameras is by combining mirrors (specially curved mirrors) with cameras, creating catadioptric systems. These systems were first developed and used in Robotics (Yagi and Kawato 1990).

3   CATADIOPTRIC VISION SYSTEMS

An important contribution for the development and generalization of catadioptric vision systems were the results by Nayar (Baker and Nayar 1998). He proved that specific combinations of curved mirrors with cameras would generate images whose geometry complied with central projection. In particular the combination of a parabolic mirror with an orthographic camera and the combination of a hyperbolic mirror with a perspective camera (with the second focus of the hyperbola coinciding with the camera center of projection) would generate images whose geometry was central, i.e. with a single center projection. As a consequence all the results applicable to perspective images could be applied to the panoramic and omnidirectional images acquired with this type of systems. These systems have also interesting properties that we have also studied and addressed.

Another relevant contribution for the development and study of catadioptric systems were the results by Geyer and Daniilidis (Geyer and Daniilidis 2000; Geyer and Daniilidis 2001). They developed a general geometric model for image formation, applicable to all central catadioptric systems. Starting from that model we were able to clarify and detail some aspects of the image geometry in these systems (Barreto and Araujo 2001; Barreto and Araujo 2002), in particular in what concerns the projection of straight lines. Geometric image formation in these systems (mapping between 3D points and their images) can be decomposed into three steps:

- In the first step 3D points are mapped into an oriented projective plane. This is a linear mapping, described by a 3×4 matrix, similar to the matrix corresponding to a perspective camera;
- In the second step a non-linear mapping is applied to the oriented projective plane;
- In the third step a collineation is applied. The parameters of the collineation are a function of the intrinsic parameters of the camera, the mirror parameters and also of the camera pose relative to the mirror.

This is a general model that identifies the non-linear transformation that occurs in the geometric formation of central catadioptric images. This model allowed the analysis and study of the geometric properties of the projections of 3D straight lines. As a result a set of geometric properties of the conics that are images of 3D lines were proved. These properties are projective invariants and can be used for calibration and reconstruction. Some of the properties that were proved were:

- The absolute conic as well as the effective viewpoint can be estimated from the image of three lines in general position;

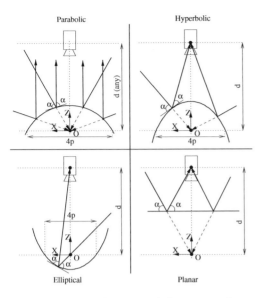

Figure 8. Configurations corresponding to central catadioptric vision systems.

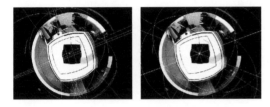

Figure 9. Central catadioptric images with conics.

- The position of the line at infinity in the catadioptric image plane can be estimated from the images of two lines;
- For the case of a hyperbolic or elliptical mirror the images of two lines are sufficient to calibrate the camera.

It was also proved that the images of three 3D straight lines are enough to calibrate any central catadioptric vision system. In addition several other results were proved establishing several useful relationships between the conics and the parameters of the projection system. Conic estimation is in general difficult since it is strongly affected by noise due to its nonlinear nature. Furthermore and in general the conics in the image are partially occluded. It is very difficult to estimate these conics with the available algorithms. However in these systems not all conics are images of 3D lines. Conics that result from the projection of 3D lines satisfy a set of necessary and sufficient conditions. In (Barreto and Araujo 2003) the necessary and sufficient conditions for a conic to be the image

of a line were proved for the case of a paracatadioptric system. These conditions lead to several relationships that must be satisfied by the parameters of the conics. Given a set of N conics these relationships define $3N - 5$ independent restrictions that must me satisfied. Since each image conic has 5 degrees of freedom the total number of parameters to be estimated is $5N$. These restrictions can be used in the process to estimate the conics leading to more robust and stable estimation algorithms.

Catadioptric systems and active systems allow bigger fields of view. Active systems also allow, in many applications, an easier and more robust way of extracting relevant information. A relevant issue when modelling the geometrical relationship between the 3D geometric elements and their projections is the definition of the coordinate system. This is specially important when dealing with motion estimation, since in that case the Jacobians for the transformations have to be computed. One can simplify the mathematical relationships between position and velocity in 3D and their image projections by selecting specific coordinate systems. In (Barreto and Araújo 2004) two restrictions that coordinate systems for active and catadioptric systems should meet (in order to simplify the mathematical relationships) were identified:

- The "compactness constraint". This restriction implies that for a coordinate system satisfying this constraint the image coordinates of a point depend only on two of the 3D coordinates. In this case the corresponding Jacobian matrix has a null column. This restriction is a necessary condition for having an invertible mapping between the 3D world and the projection;
- The "decoupling constraint". This restriction implies that each of the image coordinates depends only on one of the 3D coordinates.

In (Barreto and Araújo 2004) several examples of coordinate systems verifying both conditions are presented. One of the coordinate systems specified can be applied to perspective cameras and a coordinate system for central catadioptric systems is also specified. In many applications, and specially in tracking applications, the use of coordinate systems satisfying the restrictions described above has the advantage of allowing an invertible mapping of coordinates.

4 CONCLUSIONS

In this paper we have described the advantages of using active and catadioptric vision systems in Robotics applications. These advantages result from the ability of extending the fields of view and also of actively changing the system parameters. We also described specialized algorithms for controlling visual behaviors in active vision systems. Catadioptric systems were also characterized from the point of view of their geometric properties.

REFERENCES

Aloimonos, Y., I. Weiss, and A. Bandyopadhay (1988, January). Active vision. *International Journal of Computer Vision 1*(4), 333–356.

Araújo, H., J. Batista, P. Peixoto, and J. Dias (1996). Pursuit control in a binocular active vision system using optical flow. In *ICPR96 – Int. Conf. on Pattern Recognition*, Vienna, Austria.

Bajcsy, R. (1985, April). Active perception vs. passive perception. In *Third IEEE Workshop on Computer Vision*, Bellaire, MI. IEEE.

Bajcsy, R. (1988, August). Active perception. *IEEE Proceedings 76*(8), 996–1005.

Baker, S. and S. Nayar (1998). A theory of catadioptric image formation. In *Proc. of IEEE International Conference on Computer Vision*, Bombay, pp. 35–42.

Ballard, D. (1991, February). Animate vision. *Artificial Intelligence 48*(1), 57–86.

Barreto, J. and H. Araújo (2004). A general framework for the selection of world coordinate systems in perspective and catadioptric imaging applications. *Int. Journal of Computer Vision 57*(1), 23–47.

Barreto, J., P. Peixoto, J. Batista, and H. Araújo (1999, October 17–21). Tracking multiple objects in 3d. In *IROS'99 – IEEE/RSJ International Conference on Intelligent Robots and Systems*, Kyongju, Korea.

Barreto, J. P. and H. Araujo (2001, December). Issues on the geometry of central catadioptric imaging. In *Proc. of the IEEE Int. Conf. on Computer Vision and Pattern Recognition*, Kauai, Haway, USA.

Barreto, J. P. and H. Araujo (2002, May). Geometric properties of central catadioptric line images. In *Proc. of European Conference on Computer Vision*, Copenhagen, Denmark.

Barreto, J. P. and H. Araujo (2003, October). Paracatadioptric camera calibration using lines. In *Proc. of ICCV'2003 – IEEE Int. Conf. on Computer Vision*, Nice, France.

Batista, J., J. Dias, H. Araújo, and A. Almeida (1995, September). The isr multi-degrees-of-freedom active vision robot head: design and calibration. In *M2VIP'95 – Second International Conference on Mechatronics and Machine Vision in Practice*, Hong Kong.

Batista, J., P. P. Peixoto, and H. Araújo (1997, September). Real-time vergence and binocular gaze control. In *IROS97–IEEE/RSJ Int. Conf. on Intelligent Robots and Systems*, Grenoble, France.

Batista, J., P. Peixoto, and H. Araújo (1997). Visual behaviors for real-time control of a binocular active vision system. *IFAC Journal on Control Engineering Practice 5*(10), 1451–1461.

Batista, J., P. Peixoto, and H. Araújo (1998). Realtime active visual surveillance by integrating peripheral motion detection with foveated tracking. In *Proc. of the IEEE Workshop on Visual Surveillance*, pp. 18–25.

Dias, J., J. Batista, C. Simplício, H. Araújo, and A. Almeida (1993). Implementation of an active vision system. In *Proc. of the Intern. Workshop on Mechatronical Computer Systems for Perception and Action*, Halmstad, Sweden.

Dias, J., C. Paredes, I. Fonseca, H. Araújo, J. Batista, and A. Almeida (1998). Simulating pursuit with machines: Experiments with robots and artificial vision. *IEEE Trans. on Robot. and Automat. 14*(1), 1–18.

Dias, J., C. Paredes, I. Fonseca, H. A. J. Batista, and A. Almeida (1995, May). Simulating pursuit with machines: Experiments with robots and artificial vision. In *Proc. IEEE Int. Conf. on Robotics and Automation*, Nagoya, Japan.

Geyer, C. and K. Daniilidis (2000). An unifying theory for central panoramic systems and practical implications. In *in Proc. of European Conference on Computer Vision*, Dublin, pp. 445–461.

Geyer, C. and K. Daniilidis (2001). Catadioptric projective geometry. *International Journal of Computer Vision 43*, 223–243.

Gibson, J. J. (1983). *Senses Considered As Perceptual Systems*. Waveland Press.

Gibson, J. J. (1987). *The Ecological Approach to Visual Perception*. Lea.

Peixoto, P., J. Batista, and H. Araújo (2000). Integration of information from several vision sensors for a common task of surveillance. *Robotics and Autonomous Systems 31*, 99–108.

Yagi, Y. and S. Kawato (1990). Panoramic scene analysis with conic projection. In *Proc. of the International Conference on Robots and Systems*.

Computational Modelling of Objects Represented in Images –
João Manuel R.S. Tavares & R.M. Natal Jorge (eds)
© 2007 Taylor & Francis Group, London, ISBN 978-0-415-43349-5

Digital geometry for image analysis and processing

Valentin E. Brimkov

Mathematics Department, Buffalo State College, State University of New York, United States

ABSTRACT: Digital geometry is a discipline dealing with geometric properties of *digital objects* (also called *digital pictures*). These are usually modeled as sets of points with integer coordinates representing the pixels/voxels of the considered objects. Digital geometry is developed with the expectation that it would provide an adequate mathematical background for new advanced approaches and algorithms for various problems arising in image analysis and processing, computer graphics and related areas of visual computing. The present work provides a brief discussion on the motivation, basic directions, and achievements of digital geometry. A couple of typical examples of research problems and their solutions are also considered.

1 DIGITAL GEOMETRY: SUBJECT, MOTIVATION, AND GOALS

In this section we provide a short discussion on the motivation, basic directions, and goals of digital geometry.

1.1 *What is digital geometry?*

Digital geometry deals with geometric properties of *digital objects* (also called *digital pictures*). These are usually modeled as sets of points with integer coordinates representing the pixels/voxels of the considered digital objects. Digital geometry has established itself as an independent discipline comparatively recently, in the second half of the 20th century, with the initiation of research in visual computing, involving various applied areas such as image analysis and processing, computer vision, computer graphics and, more recently, multimedia technologies. The nature of the used research approaches and the obtained results puts digital geometry on the border of applied mathematics and theoretical computer science, as the framework of the performed research is determined by practical applications in mind.

Digital geometry is developed with the expectation to provide an adequate theoretical (mathematical) background for new advanced approaches to and algorithms for solving various problems arising in visual computing.

1.2 *Why digital geometry?*

In general, the development of digital geometry follows the one of classical geometry. The latter has appeared in the remote past as a collection of practical computation rules helpful for resolving certain everyday problems. Only much later, since Euclid, it starts turning to a rigorous mathematical subject. Over the centuries, new practical tasks motivate the rise of new geometries, such as analytical geometry, differential geometry, and, more recently, computational geometry, to mention a few. Most of these belong to the continuous domain of mathematics, with a few exceptions leaded by specific applications (e.g., combinatorial geometry and certain finite geometries).

In recent decades, the development of various branches of visual computing poses new challenges to the researchers. The objects operated in computer graphics, image analysis and processing are discrete sets of points. However, as a rule, continuous mathematics (in particular, classical geometry) is used for modeling and problem solving (for instance, in most works available in the SIGGRAPH volumes). Comparatively more rarely, ad-hoc algorithms are used for direct processing of discrete data.

As the computer images are discrete, it is quite natural the geometry involved to be discrete as well. Despite the presence of a lot of results, one should admit that a theory that could perfectly serve as a discrete analog of Euclidean geometry, is not completely developed yet. There is a simple reason for that: development of such a theory is hard, due to the discrete nature of the objects involved. In particular, this may cause ambiguity when one looks for the most reasonable definitions of even very basic discrete primitives, such as straight lines, circles, planes, etc. For instance, several definitions of a digital straight line are available in the literature (see, e.g., (Rosenfeld & Klette 2001)), each of which has advantages and disadvantages to the others. Certain paradoxes that do not exist in continuous spaces are also possible. See, e.g., examples in Section 1.1.4 of (Klette & Rosenfeld 2004).

It should also be mentioned that, as far as a theory of digital geometry exists, it is not very well-known and is rarely used by software developers. This is conditioned by a number of reasons. Pretty often, the insufficient mathematical background of the programmers does not allow them understand and apply advanced approaches involving more sophisticated technical machinery. It is surprising that the above can sometimes be observed even in computing laboratories of some very reputable organizations. Moreover, the industry usually requires to produce software in very short periods of time without serious theoretical research. Unfortunately, in many cases this leads to lowering the quality in terms of time and memory efficiency of the developed algorithms and of the accuracy and reliability of the obtained solutions. Sometimes this may indeed be an important shortcoming (e.g., in medical applications).

In view of the preceding discussion, digital geometry is aimed at becoming a rigorous theory that serves as a universal tool for modeling and resolving various problems. The theory should be easily applicable. It is also clear that digital geometry would be of little use if it remains unknown to those for whom is created. Thus an important task is to make it more popular by means of systematic education.

2 MATHEMATICS OF DIGITAL GEOMETRY

As already mentioned, digital geometry is a modern discipline that sets up itself as such in relation to its nowadays applications. However, it has its roots in a number of classical mathematical disciplines, such as number theory (since C.F. Gauss), geometry of numbers (since H. Minkowski), graph theory (since L. Euler), and combinatorial topology (since the middle of the 19th century). At present, research in digital geometry resorts to the above and some other mathematical disciplines. A more complete (although not exhaustive) list is given next.

- Number theory, geometry of numbers
- Classical Euclidean geometry, analytical geometry, affine geometry, projective geometry
- Algebraic geometry
- Vector spaces, metric spaces
- Combinatorial geometry, discrete geometry, tilings and patterns
- Computational geometry
- General topology, combinatorial topology
- Graph theory
- Linear programming, integer programming, Diophantine equations, polyhedral combinatorics, lattice polytopes

- Mathematical morphology
- Discrete dynamical systems, fractal theory
- Combinatorics on words
- Approximation theory, Diophantine approximations, continued fractions
- Probability theory and mathematical statistics
- Design and analysis of algorithms, complexity theory

Knowledge and approaches from the above-listed subjects are used to obtain theoretical results and design algorithms for solving various specific problems. Occasionally, results of digital geometry turn out to be known in different terms in the framework of earlier studies. Overall, however, digital geometry has provided a lot of new results, some of which are not only useful regarding practical applications, but also technically sound and deep from mathematical point of view.

3 MAIN DIRECTIONS

Digital geometry is germane with discrete geometry that deals with similar and some other related matters from a bit more general perspective (see the topics of Mathematical Subject Classification number 52Cxx). In particular, discrete geometry includes a number of subjects (e.g., ones related to matroid theory) that are not directly related to computer imagery, and tackles them from more abstract point of view. Instead, digital geometry is closely focused on problems arising from image analysis and processing, computer graphics, and related disciplines. Below we list some basic subjects of digital geometry, among others.

Digital topology

- Digital topologies (classification)
- Topology of digital objects (Basic topological invariants of curves and surfaces, Topology of digital curves and surfaces, Topology of linear digital objects)

Geometry of digital manifolds

- Geometry of digital curves and surfaces
- Digital straightness in 2D and 3D
- Digital planarity
- Length and curvature of digital arcs
- Area and curvature of digital surfaces
- Digital convexity

Transformations

- Axiomatic digital geometry
- Transformation groups and symmetries
- Neighborhood-preserving transformations
- Magnification and demagnification

Discrete tomography
Morphologic operations

- Dilation, erosion, simplification, segmentation, decomposition

Deformations

- Topology-preserving deformations, shrinking, thinning
- Deformations of curves, 3D pictures, and multivalued pictures

Picture properties

- Moments
- Operations on pictures
- Invariant properties
- Spatial relations

For detailed presentation of these and other areas of digital geometry the reader is referred to the recent monograph (Klette & Rosenfeld 2004). The next section illustrates typical research in digital geometry by considering some important problems related to digital planarity.

4 EXAMPLES: SOME PROBLEMS OF DIGITAL PLANARITY

Similar to classical geometry, linear objects play a central role in digital geometry. Theoretical research on digital planarity is naturally driven by important practical applications in image analysis, pattern recognition and volume modeling. In this section we review some basic algorithms for digital plane recognition, digital surface segmentation, and digital polyhedra generation. Before that, let us recall one of the available several equivalent definitions of a digital plane.

Definition 1 *A set* $D_{a,b,c,\mu,\omega} = \{(i,j,k) \in \mathbb{Z}^3 : \mu \leq ai + bj + ck < \mu + \omega\}$ *is called a* digital plane *with* normal $\mathbf{n} = (a,b,c)$, *intercept* μ, *and thickness* ω.

If $\omega = \max\{|a|, |b|, |c|\}$, then $D_{a,b,c,\mu,\omega}$ is called a *naive plane*, that is the thinnest hole-free digital plane. A digital plane with $\omega = |a| + |b| + |c|$ is called *standard*. A *digital plane segment* (DPS) is a connected portion of a digital plane. One can define lower (resp. upper) supporting points that determine the lower (resp. upper) supporting continuous planes defining a digital plane (see Figure 1). The preimage of a DPS, S, is the set of planes whose digitizations contain S. It appears to be the solution of a system of linear inequalities with unknowns α_1, α_2, and β. Thus, it is a convex polyhedron (possibly empty).

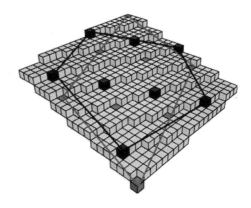

Figure 1. Illustration of a subset of a digital plane $D_{7,17,57,0,57}$ with its lower and upper convex hulls on the supporting planes.

4.1 *DPS recognition and digital surface segmentation*

DPS recognition and digital surface segmentation are fundamental problems in image analysis. Table 1 lists different algorithms and their computational costs. All complexity bounds are given with respect to the number n of grid points in S.

In (Stojmenović & Tošić 1991) a DPS recognition algorithm based on convex hull separability is proposed. The recognition of DPSs in grid adjacency models is discussed in (Veelaert 1994; Klette et al. 1996) (recognition by least-square optimization), and (Megiddo 1984, Preparata & Shamos 1985; Vittone & Chassery 2000; Buzer 2002) (linear programming when the dimension is fixed). (Debled-Renesson & Reveillès 1994) proposes an approach based on tests for existence of lower and upper supporting planes for a given set of points.

(Françon et al. 1996) suggests a recognition method for DPSs by converting the problem into a system of n^2 linear inequalities, where n is the cardinality of the given set of points. The system is solved by the Fourier elimination algorithm. One can also apply the CDD algorithm[1] for solving systems of linear inequalities by successive intersection of half-spaces defined by inequalities (Fukuda & Prodon 1996). A very efficient incremental algorithm based on a similar approach is proposed in (Klette & Sun 2001). It also provides a polyhedrization of a given digital surface. The main motivation for digital surface polyhedrization comes from medical imaging and other visualization problems where discrete volumes of voxels result from

---

[1] **C** implementation of the **D**ouble **D**escription (CDD) Method of Motzkin et al., see http://www.ifor.math. ethz.ch/~fukuda/cdd_home/cdd.html

Table 1. Algorithms for DPS recognition.

Reference	Description	Complexity	Comments
(Kim 1984)	Detection of a supporting plane	$O(n^4)$	Based on an incorrect theorem
(Megiddo 1984)	Linear programming	$O(n)$	
(Preparata & Shamos 1985)	Linear programming	$O(n \log n)$	Provides the complete preimage
(Kim 1991)	Detection of a supporting plane	$O(n^2 \log n)$	Optimized (Kim 1984), also based on an incorrect theorem
(Stojmenović & Tosić 1991)	Convex hull separability	$O(n \log n)$	
(Veelaert 1994)	Evenness property	$O(n^2)$	Rectang. DPS
(Debled-Renesson & Reveillès 1994)	Arithmetic structure	n.a.	Rectang. DPS
(Reveillès 1995)	Arithmetic geometry	$O(n)$	Rectang. DPS
(Vittone & Chassery 2000)	Linear programming and Farey series	$O(n^3 \log n)$	Preimage computation with arithmetic solutions
(Klette & Sun 2001)	Comb. procedure	n.a.	
(Buzer 2002)	Linear programming for DPS recognition	$O(n)$	On-line algorithm
(Gérard et al. 2005)	Convex hull analysis	$O(n^7)$	fast algorithm in practice

scanning and MRI techniques. Since digital medical images involve a huge number of points, it is quite problematic to apply traditional rendering or texture algorithms to obtain satisfactory visualization. Moreover, one can face difficulties in storing or transmitting data of that size. There are multiple sources of data being transmitted for many diverse uses, such as telemedicine, mine detection, tele-maintenance, ATR, visual display, cueing, and others. In all these applications the coding compression methodology used is paramount. For this, one can try to transform a discrete data set to a polyhedron, such that the number of its 2-facets is as small as possible. Such polyhedrizations are also searched for the purposes of geometric approximation of surfaces as well as for surface area and volume estimation. Note that the optimization version of this last problem is NP-hard (the first proof is available in another paper by the author in the present volume (Brimkov 2006)).

Typical timing results for the last three algorithms are shown in Figure 2, using a polyhedrized digital ellipsoid at grid resolutions ranging from 10 to 100. Specifically, (Klette & Sun 2001) takes advantage of certain geometric properties of digital planes and repeatedly updates a list of supporting planes. The set of points is accepted as a DPS iff the final list of planes is non-empty. The updating step is time-efficient.

One can perform a breadth-first search of the face graph to agglomerate the faces into DPSs. Figure 3 illustrates results of the agglomeration process for a digitized sphere and for an ellipsoid with semi-axes 20, 16, and 12. Faces that have the same gray level belong to the same DPS. The respective numbers of faces of the digital surfaces of the sphere and ellipsoid are 7,584 and 4,744, respectively. The numbers of DPSs are 285 and 197; the average sizes of these DPSs are 27 and 24 faces.

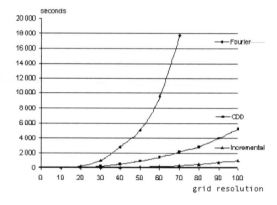

Figure 2. Running times of three DPS recognition algorithms on a PIII 450 running Linux.

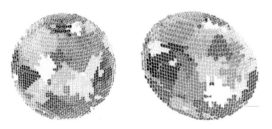

Figure 3. Agglomeration into DPSs of the faces of a sphere and an ellipsoid (grid resolution $h = 40$).

To complete the polyhedrization process, one sets all the face vertices that are incident to at least three of the DPSs to be vertices of the polyhedron. Figure 4 shows the final polyhedra for the sphere and ellipsoid. Note that these polyhedra are not simple; their surfaces are not hole-free.

Figure 4. A polyhedrized sphere and ellipsoid.

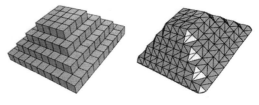

Figure 6. A {0, 1}-binary object and a Marching-Cubes surface obtained with an iso-level in [0, 1].

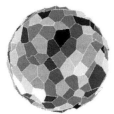

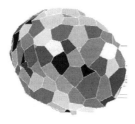

Figure 5. The polyhedrized sphere and ellipsoid where the breadth-first search depth is restricted to 7.

Figure 7. *Left:* Final result on the object of Figure 6. *Right:* Result on a sphere of radius 25.

Restricting the depth of the breadth-first search changes the polyhedrization from global to local and results in "more uniform" polyhedra. Figure 5 shows results when the depth is restricted to 7. The number of small DPSs is reduced and the sizes of the DPSs are more evenly distributed. The respective numbers of DPSs are 282 and 180 and their average sizes are 27 and 26; note that these are nearly the same as in the unrestricted case.

The output of Klette-Sun's algorithm is not, in general, a valid polyhedron but like a *patchwork* of planar segments. It is desirable to obtain a polyhedron with the following reversibility property: the polyhedron digitization coincides with the originally given set of grid points. An algorithm from (Coeurjolly et al. 2004) addresses the problem of such a reversible polyhedrization.

The main idea is to simplify the polyhedron obtained by a Marching-Cubes (MC) algorithm (Lorensen & Cline 1987), using information about the digital surface segmentation. The MC algorithm is a widely used isosurface generation algorithm in 3D volume data. This method considers local grid point configurations to replace them by small triangles composing the global isosurface. With a reference to (Lachaud & Montanvert 2000), the triangulated surface obtained by the MC algorithm is a combinatorial manifold. In other words, the surface is closed, hole-free and without self-crossing. Furthermore, the object boundary quantization of this polyhedron is exactly the input binary object. See Figure 6.

The output of the algorithm is a digital polyhedron such that a large facet is associated to each recognized

DPS. The facets of the polyhedron are stitched together by strips of triangles. These triangles are called *non-homogeneous* in (Coeurjolly et al. 2004) because their three vertices do not belong to the same digital plane. The obtained polyhedron is a combinatorial manifold and possesses the reversibility property. See Figure 7.

For more details on the presented problems and algorithms we refer to the recent survey on digital planarity (Brimkov et al. 2003).

4.2 *Digital polyhedra generation*

In this section we briefly consider certain problems that are in a sense reverse to those of the previous section. One of these is DPS generation. Usually straightforward methods for its solution follow directly from the particular definition of a digital plane. See, e.g., (Debled-Renesson & Reveillès 1994) for an algorithm based on Reveillès definition of arithmetic planes. A related problem is the digitization (scan-conversion) of a given space polygon. An efficient practical algorithm has been proposed in (Kaufman 1987). Algorithms involving "supercovers" (i.e., "thick" digitizations including all voxels intersected by the given polygon) have been proposed in (Andres et al. 1997a). Discrete linear manifolds within a "standard model" (i.e., based on standard planes) have been defined in (Andres 2003).

For various applications in surface modeling it is reasonable to work with an appropriate polyhedral approximation of a given surface rather than with the surface itself. Often this is the only possibility since the surface may not be available in an explicit form. Thus having suitable algorithms for digitizing a polyhedral surface is of significant practical importance. The above-mentioned supercover approach has been applied to polyhedra digitization (Andres et al. 1997b). The faces of the obtained digital polyhedra admit analytical description. They are portions of planes' supercovers that are thicker than the (naive) digital planes. As discussed in the literature, the optimal ground for polyhedra digitization is naturally provided by the naive digital planes. However, it has been unclear for a long time how to define a "naive" digital polygon and especially its edges, so that the overall digitization admits no holes along the edges of the resulting digital polyhedron. These theoretical obstacles have been recently overcome by employing relevant mathematical approaches. Specifically, three different algorithms have been proposed. The first one (Barneva et al. 2000) is based on reducing the 3D problem to a 2D one by projecting the surface polygons on suitable coordinate planes, next digitizing the obtained 2D polygons, and then restoring the 3D digital polygons. The generated digital polygons are portions of the naive planes associated with the facets of the surface (see Figure 8). Another algorithm (Brimkov & Barneva 2002) is based on introducing new classes of 3D lines and planes (called *graceful*) which are used to approximate the surface polygons and their edges, respectively. The algorithm from (Brimkov et al. 2000) approximates directly every space polygon by a digital one, which is again the thinnest possible, while the polygons' edges are approximated by the thinnest possible naive 3D straight lines defined algorithmically in (Kaufman & Shimony 1986) and analytically in (Figueiredo & Reveillès 1995) and (Brimkov et al. 2000). This algorithm provides an "optimal solution" while being optimally fast and using memory space of optimal order. In fact, the obtained discretization

appears to be *minimally thin*, in a sense that removing an arbitrary voxel from the digital surface leads to occurrence of a hole in it.

All these algorithms assure hole-free discretizations. They run in time that is linear in the number of the generated voxels, which are stored in a 2D array. Moreover, the generated 3D digital polygons admit analytical description.

5 CONCLUDING REMARKS

The purpose of this paper was to introduce the reader to the subject of digital geometry. In the last two decades the interest to the latter in the scientific community is constantly increasing. A number of related conferences provide researchers in the field with opportunities for regular meetings. Industrial interest to the subject is increasing, as well. The author believes that digital geometry is setting itself as a valuable theoretical foundation for research and software development in image analysis, image processing, and in other related areas.

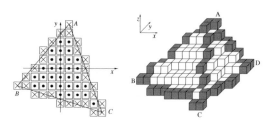

Figure 8. *Left:* A 2D digital triangle determined by the points $A(1, 5)$, $B(-5, -2)$, and $C(5, -5)$. The triangle border consists of the three digital line segments determined by pairs of vertices of the continuous triangle. *Right:* Digital triangular mesh with vertices $A(1, 5, 4)$, $B(-5, -2, 0)$, $C(5, -5, -2)$, $D(8, 2, 0)$. The projection of the left one onto the coordinate plane Oxy is the 2D digital triangle on the left. The border voxels corresponding to the border pixels of the 2D digital triangle are in dark gray.

REFERENCES

Andres, E. 2003. Discrete linear objects in dimension n: the standard model. *Graphical Models* 65: 92–111.

Andres, E., Nehlig, Ph. & Françon, J. 1997. Supercover of straight lines, planes and triangles. In E. Ahronovitz & C. Fiorio (eds). Proc. *Discrete Geometry for Computer Imagery*: 87–98. LNCS 1347, Berlin: Springer.

Andres, E., Nehlig, Ph. & Françon, J. 1997. Tunnel-free supercover 3D polygons and polyhedra. In Proc. *Eurographics*: C3–C13.

Barneva, R.P., Brimkov, V.E. & Nehlig, Ph. 2000. Thin discrete triangular meshes. *Theoretical Computer Science* 246: 73–105.

Brimkov, V.E. & Barneva, R.P. 2002. Graceful planes and lines. *Theoretical Computer Science* 283: 151–170.

Brimkov, V.E. 2006. Discrete volume polyhedrization: complexity and bounds on performance. In Proc. *CompIMAGE – Computational Modelling of Objects Represented in Images: Fundamentals, Methods and Applications*. October 21–22, Coimbra, Portugal. Taylor & Francis Group Pu.

Brimkov, V.E., Barneva, R.P. & Nehlig, Ph. 2000. Minimally thin discrete triangulations. In A. Kaufman, R. Yagel & M. Chen (eds), *Volume Graphics*: 51–70 (Chapter 3). London: Springer.

Brimkov, V.E., Coeurjolly, D. & Klette, R. 2003. Digital planarity – a review. CITR-TR-142, University of Auckland, NZ, January 2003, http://www.citr.auckland.ac.nz/techreports/. To appear in *Discrete Applied Mathematics*.

Buzer, L. 2002. An incremental linear time algorithm for digital line and plane recognition using a linear incremental feasibility problem. In A. J.-P. Braquelaire, J.-O. Lachaud & A. Vialard (eds), Proc. *Discrete Geometry for Computer Imagery*: 372–381. LNCS 2301, Berlin: Springer.

Coeurjolly, D., Guillaume, A. & Sivignon, I. 2004. Reversible discrete volume polyhedrization using Marching Cubes simplification. In Proc. *Vision Geometry XII*, SPIE 5300: 1–11.

Debled-Rennesson, I. & Reveillès, J.-P. 1994. A new approach to digital planes. In Proc. *Vision Geometry III*, SPIE 2356: 12–21.

Dorst, L. & Smeulders, A.W.M. 1984. Discrete representation of straight lines. *IEEE Trans. Pattern Analysis Machine Intelligence* 6: 450–463.

Figueiredo, O. & Reveillès, J.-P. 1995. A contribution to 3D digital lines. In Proc. *Discrete Geometry for Computer Imagery*: 187–198.

Françon, J., Schramm, J.M. & Tajine, M. 1996. Recognizing arithmetic straight lines and planes. In S. Miguet, A. Montanvert & S. Ubéda (eds), Proc. *Discrete Geometry for Computer Imagery*: 141–150. LNCS 1176, Berlin: Springer.

Fukuda, K. & Prodon, A. 1996. Double description method revisited. In Proc. *Combinatorics and Computer Science*: 91–111. LNCS 1120, Berlin: Springer.

Kaufman, A. 1987. An algorithm for 3D scanconversion of polygons. In Proc. *Eurographics*: 197–208.

Kaufman, A. & Shimony, E. 1986. 3D scan-conversion algorithms for voxel-based graphics. In Proc. *Workshop on Interactive 3D Graphics*: 45–75. New York: ACM.

Kim, C.E. 1984. Three-dimensional digital planes. *IEEE Trans. Pattern Analysis Machine Intelligence* 6: 639–645.

Klette, R., Stojmenović, I. & Žunić, J. 1996. A parametrization of digital planes by least square fits and generalizations. *Graphical Models Image Processing* 58: 295–300.

Klette, R. & Rosenfeld, A. 2004. *Digital Geometry – Geometric Methods for Digital Picture Analysis*. San Francisco: Morgan Kaufmann.

Klette, R. & Sun, H.-J. 2001. Digital planar segment based polyhedrization for surface area estimation. In C. Arcelli, L.P. Cordella, & G. Sanniti di Baja (eds), *Visual Form*: 356–366. Berlin: Springer.

Lachaud, J.-O. & Montanvert, A. 2000. Continuous analogs of digital boundaries: A topological approach to iso-surfaces. *Graphical Models and Image Processing* 62: 129–164.

Lindenbaum, M. & Bruckstein, A.M. 1993. On recursive, $O(n)$ partitioning of a digitized curve into digital straight segments. *IEEE Trans. Pattern Analysis Machine Intelligence* 15: 949–953.

Lorensen, W.E. & Cline, H.E. 1987. Marching cubes: a high resolution 3D surface construction algorithm. *Computer Graphics* 21: 163–170.

McIlroy, M.D. 1985. A note on discrete representation of lines. *AT&T Technical J.* 64: 481–490.

Megiddo, N. 1984. Linear programming in linear time when the dimension is fixed. *J. ACM* 31: 114–127.

Preparata, F.P. & Shamos, M.I. 1985. *Computational Geometry: An Introduction*. New York: Springer.

Rosenfeld, A. & Klette, R. 2001. Digital straightness. In *Electronic Notes in Theoretical Computer Science* 46.

Stojmenović, I. & Tošić, I. 1991. Digitization schemes and the recognition of digital straight lines, hyperplanes and flats in arbitrary dimensions. *Vision Geometry, Contemporary Mathematics Series* 119: 197–212.

Veelaert, P. 1994. Digital planarity of rectangular surface segments. *IEEE Trans. Pattern Analysis Machine Intelligence* 16: 647–652.

Vittone, J. & Chassery, J.-M. 2000. Recognition of digital naive planes and polyhedrization. In G. Borgefors, I. Nyström, and G. Sanniti di Baja (eds), Proc. *Discrete Geometry for Computer Imagery* 296–307. LNCS 1953, Berlin: Springer.

Computational Modelling of Objects Represented in Images –
João Manuel R.S. Tavares & R.M. Natal Jorge (eds)
© 2007 Taylor & Francis Group, London, ISBN 978-0-415-43349-5

Multimodality in brain imaging: Methodological aspects and applications

S.I. Gonçalves
Institute of Biophysics and Biomedical Engineering, University of Lisbon, Lisbon, Portugal
Vrije University Medical Center, Amsterdam, The Netherlands

J.C. Munck
Vrije University Medical Center, Amsterdam, The Netherlands

F.H. Lopes Silva
Swammerdam Institute for Life Sciences, University of Amsterdam, Amsterdam, The Netherlands

ABSTRACT: This paper illustrates the concept of Multimodality in Brain Imaging by giving two examples of multimodal techniques. The first one solves a classical problem of the EEG (Electroencephalogram) Inverse Problem by computing the resistivities of brain, skull and scalp "in vivo" and individually for each subject, thus considerably decreasing the systematic errors associated to the EEG Inverse Problem solution. The second example consists of the co-registration of EEG and fMRI (functional magnetic resonance imaging), a novel technique that consists of recording the EEG inside the MR (magnetic resonance) scanner during the acquisition of the fMRI sequence. It poses several questions of technical and theoretical nature that will be described. Furthermore, the adopted solutions will be explained as well as the consequences the use of this technique has for the future of Brain Imaging.

1 INTRODUCTION

The study of the brain using technical rather than empirical methods dates back from the beginning of the twentieth century when Hans Berger (1929) recorded for the first time, on the scalp, the electric signals of the brain using a technique named as Electroencephalography (EEG). He recorded the temporal fluctuations of the brain electric activity during relaxed wakefulness and at the time, he observed that these temporal fluctuations were dominated by a signal at a frequency of approximately 10 Hz. It was mainly recorded from scalp locations over the occipital area and was enhanced by eye closure being almost blocked by eye opening. He named this signal as the "alpha rhythm" (Lopes da Silva, 1991) and connected it with the state of relaxed wakefulness.

Since then, many technical developments were responsible not only for the development of EEG but also for the invention of other more modern techniques such as fMRI (functional magnetic resonance imaging) (Ogawa et al., 1990). Furthermore, the acknowledgment that in order to obtain a better understanding of the brain processes it was better to combine different techniques, thus taking advantage of the benefits of each, opened the way to multimodality in brain

imaging which is illustrated by the two applications described in this paper.

Contrary to the beginning, nowadays EEG systems acquire data in a digital format therefore allowing for post-processing methods to be applied. The EEG is characterized by a high temporal resolution, which is of the order of the ms. Thus, it is able to follow the brain processes at the time scale of the neuronal activity. The spatial resolution however, is conditioned by the limitations of the EEG Inverse Problem (IP) (Zoles, 1998). The IP is a method by which it is possible to localize the neuronal sources that are responsible for a certain feature of the EEG signal. A mathematical model is applied where the measured electric potentials are the known variables and the neuronal source is the unknown. A geometrical model is adopted for the head, usually described by three compartments corresponding to the brain, skull and scalp. The electric resistivities are also assumed to be known. Since the IP is, by nature ill-posed (Zoles, 1998), i.e. several source distributions originate the same electric potential distribution, a source model has to be assumed in order to constrain the IP. The dipole model is often used to describe sources of brain electric activity. A classic problem of the EEG IP is the uncertainty with which the electric resistivities are known (Faes et al.,

1999). Although classically the resistivity values were taken from literature, those contained large systematic errors due to several factors such as measurements "in vitro" instead of "in vivo" or the variability of experimental procedures. In this paper, two methods will be described to compute the electric resistivities "in vivo" for each subject. The first one is based on the principles of EIT (Electric Impedance Tomography) and the second is based on the combined analysis of MEG and EEG data corresponding to somatosensory evoked fields (SEF) and somatosensory evoked potentials (SEP).

Functional magnetic resonance imaging (Ogawa et al., 1990) is a brain imaging technique which is able to measure small changes in blood oxygenation level associated with brain activity (e.g. due to the activation of a certain brain area, functionally related to the task performed by the subject), which cause slight variations in the MR signal. Thus, by comparing the signal in MR images acquired during the period when the subject is performing some task (task period) to that acquired during the period when the subject is at rest (rest period), it is possible to determine which brain areas were activated and thus involved in processing a given task. The high spatial resolution of MRI (of the order of mm) makes it the preferred technique to use for accurate localization of a certain brain process. However, the temporal resolution is only of the order of 1 second. Recently, the possibility of co-registering EEG and fMRI has become available (Lemieux et al., 1997; Lemieux et al., 1999). This technique associates the high temporal resolution of EEG with the high spatial resolution of fMRI, thus having the potential to give new perspectives to source localization (Bénar et al., 2003). However it has several problems due to the combination of two techniques that are both based on electromagnetism. Thus, the fMRI sequence perturbs the EEG and the EEG influences the quality of the fMRI images. Furthermore, the pulse artifact, resulting from the cardiac activity also appears on the EEG.

In this paper, the methodology applied in co-registered EEG-fMRI will be described and its application to the study of the alpha rhythm will be illustrated.

2   METHODS

2.1   The estimation of brain resistivities using EIT and the combined analysis of SEF/SEP data

2.1.1   Theory
The application of the EIT method to compute equivalent electrical resistivities using spherical models for the head has already been demonstrated in (Gonçalves et al., 2000). In this paper the same method is applied

using realistic models for the head and the Boundary Element Method (BEM). The mathematical details of the algorithm are given in (Gonçalves et al., 2003).

In the SEF/SEP method the analysis is performed in 2 steps:

1. The sources, which best explain the recorded magnetic N20 data in the time samples around the latency of the response are computed with the single dipole model. Therefore both the positions and the tangential components become known parameters;
2. EEG data is used to compute the radial components of the dipoles as well as the brain electrical conductivity and the brain to skull conductivity ratio. Since both the dipole's positions and tangential components are already known the parameters aforementioned remain the only unknowns to be determined.

The potential Ψ_i measured on a sensor i located at $\vec{r}_i$ at time instant j can be written as:

$$\begin{aligned}
\Psi_{ij} = {}& M_j^R \rho_{brain} \Theta_{ij}^R \left(\vec{r}_i, \vec{r}_{d,j}, \frac{\rho_{brain}}{\rho_{skull}} \right) \\
& + M_j^{T1} \rho_{brain} \Theta_{ij}^{T1} \left(\vec{r}_i, \vec{r}_{d,j}, \frac{\rho_{brain}}{\rho_{skull}} \right) + \\
& + M_j^{T2} \rho_{brain} \Theta_{ij}^{T2} \left(\vec{r}_i, \vec{r}_{d,j}, \frac{\rho_{brain}}{\rho_{skull}} \right)
\end{aligned} \tag{1}$$

where $\vec{r}_{d,j} =$ dipole position at time instant j; $\rho_{brain} =$ resistivity of the brain; $\rho_{skull} =$ resistivity of the skull; $M_j^R =$ radial component of the dipole moment at time instant j; $M_j^T =$ tangential components of the dipole at time instant j: M_j^{T1}, M_j^{T2}; Θ_{ij}^R and Θ_{ij}^T are respectively the potential generated on sensor i at time instant j by the radial and tangential components of the dipole on a sphere of normalized conductivities: $(1, \rho_{brain}/\rho_{skull}, 1)$.

The IP is defined as the computation of the parameters M_j^R, ρ_{brain} and $\rho_{brain}/\rho_{skull}$ that best explain the electrical potentials measured on the scalp. If the potential measured on a certain sensor i at time instant j is considered, then the difference between measured (Ψ_{ij}) and predicted ($\tilde{\Psi}_{ij}$) potential is expressed as:

$$\cos t = \frac{\sum_{ii'j} \left(\tilde{\Psi}_{ij} - \Psi_{ij} \right) C_{ii'}^{-1} \left(\tilde{\Psi}_{i'j} - \Psi_{i'j} \right)}{\sum_{ii'j} \Psi_{ij} C_{ii'}^{-1} \Psi_{i'j}} \tag{2}$$

where i and i' run over the total number of measuring sensors, j runs over the number of time samples analysed and $C_{ii'}$ is the spatial covariance (de Munck et al., 2002) between EEG channels i and i'. Although $\rho_{brain}/\rho_{skull}$ must be estimated using non-linear minimisation procedures, M_j^R and ρ_{brain} are, in principle,

18

determined by exact expressions because they are linear parameters.

2.1.2 Head models
The spherical model is defined in two different ways. In the *default model*, a sphere is fitted to the grid of electrodes used in each subject and taken as the description of the skin. The radii of the two spheres describing the skull and the brain are respectively equal to 0.92 R_{skin} and 0.87 R_{skin}, where R_{skin} is the radius of the sphere describing the skin. The same relative skull thickness (r.s.t), equal to 0.05, is taken for all subjects.

In the *individual model* instead of taking the factors 0.92 and 0.87 for all subjects, these parameters are adjusted for each subject by fitting a set of concentric spheres to their realistic model, therefore adjusting the r.s.t. for each subject. The approximation $\rho_{brain} = \rho_{skin}$ is made.

The realistic model is derived from the individual MR scans of the subjects using BEM (Gonçalves et al., 2003).

2.1.3 Data acquisition
Data from 6 normal subjects was acquired using the Omega MEG/EEG system (CTF Systems Inc.). Data for the SEF/SEP analysis was obtained from the stimulation of the median nerve using an electrical current. The frequency of the stimulus was set at 2 Hz and its duration was 0.2 ms. Current intensity ranged from 2.5 to 12 mA. After the onset of the stimulus, MEG and EEG data were recorded simultaneously at a rate of 1250 Hz, using 151 MEG channels and 64 EEG channels, with the electrodes positioned according to the extended 10–20 system. Data was filtered on-line with an anti-aliasing low-pass filter at 400 Hz and off-line using a low-pass filter at 300 Hz and a high-pass filter at 5 Hz. In order to improve the signal-to-noise-ratio, 500 to 700 artifact-free epochs of 0.4 seconds had to be averaged.

The acquisition of data for EIT analysis was made using the aforementioned EEG system with the electrodes positioned according to the extended 10–20 system. Current was injected on a pair of electrodes measuring the potential distribution on the remaining sensors, this procedure being repeated for several injection pairs. The injection electrodes were positioned with a maximum separation between them, and the reference electrode located approximately halfway between injection and extracting electrodes to decrease the effects of local variations of the skull conductivity on the results. The injection-extraction electrode pairs were chosen to cover the whole perimeter of the head.

The current generator produced a 60 Hz sinusoidal electrical current of 10 μA rms. Data was acquired at a rate of 1250 Hz, using on-line high and low-pass filters at 0.16 Hz and 300 Hz respectively. For each injection pair, epochs of 105 seconds were recorded.

2.2 Co-registered EEG-fMRI

2.2.1 Algorithm to correct EEG data for gradient and pulse artifacts
A method that is a variant of the average subtraction method (Allen et al., 2000) is used. This method, contrary to previous studies, is based on simple procedures to correct the EEG data for both gradient and pulse artifacts and where most effort and complexity is directed towards the temporal alignment of EEG and fMRI data which is a crucial point for correction. The features of the algorithm include the automatic computation of MR sequence timing parameters to be used in the temporal alignment of EEG and fMRI data with sub-sample precision and the use of both slice and volume artifact templates to correct the data. Furthermore, the information about timing parameters required a priori is highly decreased and only the number of fMRI volumes and slices as well as the name of the volume marker on the EEG are needed as input for the gradient artifact correction algorithm. The pulse artifact correction follows the same approach proposed by Allen et al. (1998) but introduces two innovative elements. Firstly, the pulse artifact markers are derived from the pulse signal recorded by the MR scanner and not from the ECG. This has several advantages, one of the most important being the fact that the pulse signal is not corrupted by gradient artifacts. Secondly, the problem of the non-stationarity of the pulse artifact is approached in an innovative way by applying a clustering algorithm (Van't Ent et al., 2003). Details of the algorithm can be found in (Gonçalves et al., 2006b).

2.2.2 Study of the alpha rhythm
In this study and similarly to what others have done, the spontaneous variations of alpha rhythm are studied using continuous EEG/fMRI. However, instead of taking an approach that focuses primarily on group results, individual results are analyzed in order to investigate inter-subject variations and to determine whether these can be explained by the state of wakefulness of the subject.

2.2.2.1 Experimental procedure
Data were recorded from eight healthy subjects (four females, four males, mean age 34, ± 8 years) during rest. In order to record spontaneous variations of the alpha rhythm, the subjects, who were kept in the dark, were instructed to lie still inside the scanner, keeping their eyes closed.

The EEG was acquired using an MR compatible EEG amplifier (SD MRI, Micromed, Treviso, Italy) and a cap providing 19 Ag/AgCl electrodes positioned according to the 10/20 system. For patient safety reasons the wires were carefully arranged such that loops and physical contact with the subject were avoided. EEG data were acquired at a rate of 1024 Hz using the

Clinic-Acquisition software package (Micromed, Treviso, Italy). An anti-aliasing hardware low-pass filter at 268.8 Hz was applied. The EEG amplifier had a resolution of 16 bits, an input impedance larger than 10 MΩ and a CMRR = 105 dB at 50 Hz. Each channel had differential inputs and used one sigma-delta AD converter. For RF protection, a low-pass filter at 200 Hz (20 dB/decade) was applied and a 12 KΩ current limiting resistor (Lemieux et al., 1997) was attached to each electrode lead.

Functional images were acquired on a 1.5 T MR scanner (Magnetom Sonata, Siemens, Erlangen, Germany) using a T2* weighted EPI sequence (TR = 3000 ms, TE = 60 ms, 64 $\times$ 64 matrix, FOV = 211 $\times$ 211 mm, slice thickness = 3 mm (10% gap), voxel size = 3.3 $\times$ 3.3 $\times$ 3 mm^3) with 24 transversal slices covering the complete occipital lobe and most of the parietal and frontal lobes. In the protocol 400 volumes (i.e. 20 minutes of data) were acquired for each subject. In general, the acquisition was made continuously. For some subjects however, the acquisition was split in two series of 10 minutes in order to alleviate the discomfort caused by the electrode pressure on the back of the head. In addition, and prior to the functional scans, a high resolution MPRAGE sequence (TR = 2700 ms, TI = 950 ms, TE = 5.18 ms, 256 $\times$ 192 $\times$ 160 matrix, FOV = 256 $\times$ 192 $\times$ 240 mm, voxel size = 1.0 $\times$ 1.0 $\times$ 1.5 mm^3), was acquired for each subject, in order to provide the anatomical reference for the functional scan. The functional images were aligned to the anatomical ones by assuming that there is no motion between the anatomical scan and the first EPI scan.

The electrocardiogram (ECG) was recorded during the fMRI sequence using the electrodes provided with the scanner.

2.2.2.2 Analysis of EEG and fMRI data

In order to determine the spectral characteristics of interest of the corrected EEG data the fluctuations in power of the alpha band were computed. An advanced FFT algorithm was applied in order to compute a spectrogram sampled at the same frequency as the fMRI data, i.e. 0.33 Hz. Thus, the FFT was computed using a rectangular window of 3 s without overlap. Subsequently, the power time series were averaged over all alpha band frequencies (8–12 Hz), thus obtaining a single power time series per channel.

In this study, a classic bipolar montage was used in order to emphasize local variations. For the analysis, the following derivations were considered: C3-P3, P3-O1, O1-T5, T5-T3, C4-P4, P4-O2, O2-T6 and T6-T4. The power time series, averaged over these channels, was considered for further calculations. However, for some subjects, this procedure did not yield the most significant statistical parametric maps and, therefore, a subset of derivations was taken instead.

The MR data were motion corrected and spatially smoothed using a 4 mm radius spatial filter in the software package AFNI (Cox, 1996). Next, the BOLD signal in each voxel was correlated to the average power time series using the General Linear Model (GLM) formalism (Cox et al., 1995), which was implemented in a software package developed in-house and which was validated by comparison with AFNI. In the context of the GLM, the average power time series was taken as the regressor and both an offset and a linear trend were considered as covariates. Furthermore, the averaged power time series was first convolved with a canonical hemodynamic response function (HRF) prior to using it as the regressor in the GLM. This is a standard procedure in the field of fMRI analysis to take into account the delay in the hemodynamic response.

The correction for multiple comparisons was made by means of controlling the false-discovery rate (FDR) according to Benjamini and Hochberg (1995).

3   RESULTS

3.1   The estimation of brain resistivities using EIT and the combined analysis of SEF/SEP data

For each subject, the equivalent electrical resistivities for brain (skin) and skull were computed using the EIT method and the SEF/SEP analysis using spherical models for the head. The results are presented on Table 1. It shows the values obtained for the equivalent electrical properties of brain (skin) and skull, expressed in terms of electrical resistivities, with both EIT and SEF/SEP methods. The results obtained with both methods suggest an agreement, especially for the values of $\rho_{skull}/\rho_{brain}$, between both methods. For subjects 1 and 6, the agreement between the absolute values of the resistivities is not as good as for the other subjects. However for these two subjects, the values of $\rho_{skull}/\rho_{brain}$ given by EIT and SEF/SEP are not as similar as for the other subjects and therefore larger differences in the absolute values of the resistivities are more likely to appear. In figure 1 a plot of ρ_{brain} as a function of ρ_{skull}, using the EIT method, is shown using both spherical and realistic models for the head. It can be seen that there is a clear trend of decrease in the variation (ratio between maximum and minimum values) of ρ_{brain} and ρ_{skull} when using realistic models.

3.2   Co-registered EEG-fMRI: study of the alpha rhythm

In figures 2, 3 and 4 results of the statistical parametric maps (SPM's) corresponding to subjects 1, 5 and 6 and resulting from the correlation between the temporal variations of the alpha power and the BOLD signal, are shown since they illustrate the results obtained for

20

Table 1. Results obtained with EIT and SEF/SEP, expressed in terms of electrical resistivities. In addition the average and standard deviation (SD) values associated to the average of ρ_{brain} and ρ_{skull} are also presented.

	EIT METHOD			SEF/SEP METHOD		
Subject	ρ_{brain} (Ω.cm)	ρ_{skull} (Ω.cm)	$\rho_{skull}/$ ρ_{brain}	ρ_{brain} (Ω.cm)	ρ_{skull} (Ω.cm)	$\rho_{skull}/$ ρ_{brain}
1 (f, 27 y)	440	13300	30	175	7441	43
2 (m, 41 y)	245	30800	127	–	–	–
3 (f, 34 y)	280	26900	96	280	24000	86
4 (m, 40 y)	295	20000	68	250	16300	65
5 (m, 27 y)	245	16100	66	250	18600	74
6 (f, 25 y)	330	15000	45	215	14600	68
Average	305	20355		234	16185	
SD(%)	24	35		17	37	

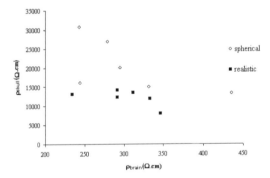

Figure 1. Plot of the resistivity of the skull as a function of the resistivity of the brain, obtained with the EIT method using both spherical and realistic models. The results with the spherical model were obtained using the same relative skull thickness for all subjects.

all subjects. The color scales identify the correlation thresholds.

For subject 1 the areas where BOLD is *negatively* correlated to the alpha rhythm are mainly located within the occipital gyri, precuneus, inferior temporal gyri, pre- and post-central gyrus. The positive correlations are located in the thalamus.

For subject 5, only positive correlations were found in the superior frontal gyrus. In addition, smaller regions *negatively* correlated to alpha are observed in the vicinity of the pre- and post-central gyrus and also in the superior and middle temporal gyrus. However, in the case of subject 6, for whom only *positive* correlations were found, these were located in the cortex,

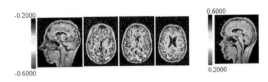

Figure 2. Individual correlation maps represented in sagital slices and various consecutive axial slices for subject 1. (FDR thresh = 0.01, all derivations.)

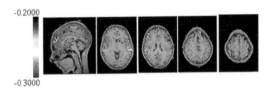

Figure 3. Individual correlation maps represented in sagital slices and various consecutive axial slices for subject 5. (FDR thresh = 0.01, subset of derivations: C3-P3, C4-P4, T5-T3, T6-T4.)

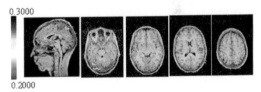

Figure 4. Individual correlation maps represented in sagital slices and various consecutive axial slices for subject 5. (FDR thresh = 0.01, subset of derivations: C3-P3, C4-P4, P3-O1, P4-O2).

namely in the orbitofrontal gyri, precuneus and post-central gyrus and superior and middle temporal gyrus.

All results were thresholded at a false detection rate (FDR) of 0.01 (Gonçalves et al., 2006a).

4 DISCUSSION AND CONCLUSIONS

In this paper, the concept of multimodality in brain imaging is illustrated using two different methodologies. The first methodology is applied to compute "in vivo" and individually for each subject the equivalent resistivities of brain, skull and scalp. For that, two techniques are used: EIT and the combined analysis of SEF/SEP data, the latter illustrating multimodality through the use of both EEG and MEG data simultaneously.

The second methodology consists of co-registering EEG and fMRI. This new multimodal technique is applied to study the sources of the alpha rhythm and its relationship to BOLD.

The results obtained in the study of the resistivities suggest the existence of variability in the values of $\rho_{skull}/\rho_{brain}$ among subjects that may be important to take into account to decrease the sources of error in EEG modelling. Due to the limited size of the surveyed population, it is not possible to make a stronger generalization of this statement. However, the results obtained with the realistic model show that the variability in resistivities and the resisitivities' ratio decreases. This suggests that the observed variability with spherical models can partially be explained by geometric errors associated with head model. The results with the realistic model also show even with head geometry correction, there are still variations to be accounted for, thus pointing to the necessity of calibrating the values of $\rho_{skull}/\rho_{brain}$, ρ_{skull} and ρ_{brain} by measuring them "in vivo" for each subject.

The results presented in this paper regarding the analysis of the alpha rhythm using co-registered EEG-fMRI showed evidence that the "resting state" is neither a static phenomenon nor reproducible in time. Rather it may differ considerably between subjects and even within one subject (results not shown). This intra- and inter-subject variability, which is known within the EEG framework, is probably caused by fluctuations in processes such as attention, memory load or spontaneous cognitive processes and is likely to be overlooked in group analysis. We are aware of the limited number of subjects that was included in the present study. However the acknowledgment of such striking intra- and inter-subject differences in the correlation between spontaneous alpha rhythm variations and the BOLD signal, mostly overlooked in previous studies, is considered of great interest as the "resting state" remains the cornerstone of fMRI. In this line of reasoning, the question of inter-subject variability should be addressed when comparing fMRI results from different subjects. The concomitant recording of the EEG may help to define the "resting state" in a more objective way in general fMRI studies. A possible method to do this would be to use the average alpha power time series, computed from the simultaneously recorded EEG, as a covariate in the fMRI analysis.

REFERENCES

Allen, P. J., Polizzi, G., Krakow, K., Fish, D. R., Lemieux, L., 1998. Identification of EEG events in the MR scanner: the problem of pulse artifact and a method for its subtraction. NeuroImage, **8**, 229–239.

Allen, P. J., Josephs, O., Turner, R., 2000. A method for removing imaging artifact from continuous EEG recorded during functional MRI. NeuroImage, **12**, 230–239.

Bénar, C. G., Aghakhani, Y., Wang, Y., Izenberg, A., Al-Asmi, A., Dubeau, F., Gotman, J., 2003. Quality of EEG simultaneous EEG-fMRI for epilepsy. Clin Neurophysiol, **114**, 569–580.

Benjamini, Y., Hochberg, Y., 1995. Controlling the false discovery rate: A practical and powerful approach to multiple testing. J. R. Statist. Soc. B, **57(1)**, 289–300.

Cox, R. W., Jesmanowicz, A., Hyde, J. S., 1995. Real-Time functional magnetic resonance imaging, MRM, **33**, 230–236.

Cox, R. W., 1996. AFNI: Software for analysis and visualization of functional magnetic resonance neuroimages, Comput Biomed Res, **29**, 162–173.

Faes T. J. C., Van der Meij H., Heethaar R. M., De Munck J. C., 1999. The Resistivity of Human Tissue (100 Hz–10 MHz): a meta-analysis of review studies, Physiological Measurement 20(4): R1–R10.

Gonçalves, S., de Munck, J. C., Heethaar, R. M., Lopes da Silva, F. H., van Dijk, B. W., 2000. The Application of Electric Impedance Tomography to reduce systematic errors in the EEG Inverse Problem – A Simulation Study. Physiol. Meas., **21**, 379–393.

Gonçalves, S. I., de Munck, J. C., Verbunt, J. P. A., Bijma, F., Heethaar, R. M., Lopes da Silva, F. H., 2003. In vivo measurement of the Brain and Skull resistivities using an EIT based method and realistic models for the head. IEEE Trans. Biomed. Eng., 50(6), 754–767.

Gonçalves, S. I., de Munck, J. C., Pouwels, P. J. W., Schoonhoven, R., Kuijer, J. P. A., Maurits, N. M., Hoogduin, J. M., Van Someren, E. J. W., Heethaar, R. M., Lopes da Silva, F. H., 2006a. Correlating the alpha rhythm to BOLD using simultaneous EEG/fMRI: inter-subject variability. NeuroImage, 30(1), 203–213.

Gonçalves, S. I., de Munck, J. C., Pouwels, P. J. W., Kuijer, J. P. A., Heethaar, R. M., 2006b. Automatic artifact removal in co-registered EEG/fMRI, NeuroImage, under review.

Koles, Z. J., 1998. Trends in EEG source localization. Electroenceph. Clin. Neurophysiol., **106**, 127–137.

Lemieux, L., Allen, P. J., Franconi, F., Symms, M. R., Fish, D. R., 1997. Recording of EEG during fMRI experiments: patient safety. MRM, **38**, 943–952.

Lemieux, L., Allen, P. J., Krakow, K., Symms, M. R., Fish, D. R., 1999. Methodological issues in EEG-correlated functional MRI experiments. IJBEM, **1(1)**, 87–95.

Lopes da Silva, F. H., 1991. Neural mechanisms underlying brain waves: from neural membranes to networks. Electroencephal Clin Neurophysiol, **79**, 81–93.

de Munck, J. C., Huizenga, H. M., Waldorp, L. J., Heethaar, R.M., 2003. Estimating stationary dipoles from MEG/EEG data contaminated with spatially and temporally correlated background noise, IEEE Trans. Signal Proc, **50**, 1565–1572.

Ogawa, S., Lee, T. M., Kay, A. R., Tank, D. W., 1990. Tank brain magnetic resonance imaging with contrast dependent on blood oxygenation. Proc. Natl. Acad. Scie. USA, **87**, 9868–9872.

Van't Ent, D., Manshanden, I., Ossenblok, P., Velis, D.N., de Munck, J. C., Verbunt, J. P. A., Lopes da Silva, F.H., 2003. Spike cluster analysis in neocortical localization related epilepsy yields clinically significant equivalent source localization results in magnetoencephalogram (MEG). Clin Neurophysiol **114**, 1948–1962.

Computational Modelling of Objects Represented in Images –
João Manuel R.S. Tavares & R.M. Natal Jorge (eds)
© 2007 Taylor & Francis Group, London, ISBN 978-0-415-43349-5

Perceptual users interfaces and their applications

Francisco Jose Perales López

Unidad de Gráficos y Visión por Ordenador, UGyVpO, Departamento de Matemáticas e Informática de la UIB,
Palma de Mallorca, Baleares, Spain

ABSTRACT: In this paper we would like to introduce the main ideas about the news new paradigms for communications between computers and persons. In the late 1800s, Marey and Muybridge conducted independent studies of human and animal motion by shooting multiple photographs of moving subjects over a short period of time (Muybridge, 1955), (Muybridge, 1957). Actually at the present time, the study of human movement using a computer is very useful and can apply be applied to many areas. One such of these applications is the advanced perceptual three-dimensional interfaces.

For achieve these advanced and natural interaction new types of devices and interfaces must be designed. The process to capture human gestures and motion analysis in front using a commercial camera is needed to be able in order to allow computers to understand our actions and react in a intelligent manner to interact together in a multimodal application context.

This problem is a challenging topic in many areas. In computer graphics, is very complex to animate realistic articulated figures like humans or animals. In computer vision, is very interesting to design new algorithms to detect, track, and recover articulate motion; for example an avatar making complex actions in front a digital video camera. And finally, the intelligent agents must be carefully designed to understand the natural and incomplete information given by the end-user, and driven to drive correctly the adequate actions to reach the final objectives in the practical application domain.

We would like to present in the next sections basic concepts and definitions in multimodal and perceptual user interfaces, a brief state of the art and some examples of virtual and augmented systems applied to games and disable people.

Keywords: Human motion capture, humanoid, real and synthetic images, matching, calibration, biomechanical & graphic model, computer vision techniques, perceptual & multimodal interfaces, artificial intelligence, agents.

1   INTRODUCTION

1.1   *Basic concepts & definitions*

The research of new human-computer interfaces has become a growing field in computer science, which aims to attain the development of more natural, intuitive, unobtrusive, and efficient interfaces. This objective has come up with the concept of Perceptual User Interfaces (PUI), that are turning out to be very popular as they seek intend to make the user interface more natural and compelling by taking advantage of the ways in which people naturally interact with each other and with the world. PUIs can use speech and sound recognition and generation (ARS & TTS), computer vision, graphical animation and visualization, language understanding, touch-based sensing and feedback (haptics), learning, user modeling and dialog management.

These new interfaces can be used in different scenarios (offices, cars, domotics houses, hospitals). But more important are the potential users of the system. PUIs offer assistive technology for people with physical disabilities, which can help them to lead have more independent lives. They also contribute to new and more powerful interaction experiences for any kind of audience.

Of all the communication channels through where interface information can travel, computer vision provides a lot of information that can be used for detection and recognition of human's actions and gestures, which can be analyzed and applied to interaction purposes.

When sitting in front of a computer and with the use of standards web cameras, very common devices nowadays, heads and faces can be presumed to be visible. Therefore, systems based in head or face feature detection and tracking, and face gesture or expression recognition can become very effective human-computer interfaces.

Of course, difficulties can arise from in-plane (tilted head, upside down) and out-of-plane (frontal view, side view) rotations of the head, facial hair, glasses, lighting variations and cluttered background.

To answer some important questions, first we must know about the basic concepts related with human computer interaction. What is a HCI?

- The study of people, computer technology, and the ways in which these influence each other.
- The discipline that aims to make computer technology more usable by people.
- The design, evaluation and implementation of interactive computing systems for human use.

Also the User interface is the portion of computer program which the user interacts with.

In general, in the new actual environments we handle several types of information so we can consider interfaces with multimodality inputs. So, a Multimodal User Interface is a system that combines two or more input modalities in a coordinated manner (Turk et al., 2005).

- Modality: particular sense (sight, hearing, touch, taste and smell). For example, command-line uses sight and touch.
- Channel: pathway through which information is transmitted. For example, GUIs use mouse and keyboard.
- The Multimodal User Interfaces use multiples modalities to implement multiple channels of communication. The evolution of these types of multimodality provides a new paradigm that is the Perceptual User Interfaces.
- Highly interactive, multimodal user interfaces modeled after natural human-to-human interaction.
- For people to be able to interact with computers in a similar fashion to how they interact which each other and with the physical world.

In general we assume in PUI a high degree of natural communication with computer, in the same manner or degree with the person to person communications. The perceptual modalities can be summarized in the main sequent areas:

- Speech recognition.
- Language understanding.
- Pen-based gesture.
- Sensors for body tracking.
- Non-speech sound.
- Haptic input and force feedback.
- Computer vision.

Specific cases are the systems where the visual information is the main input to the system, specifically information added to the agent or intelligent system. The Visual Based Interfaces (VBI) use computer vision to perceive these cues in order to support interactive systems. The visual cues are important in communication process. The most useful visual cues are:

- Presence and location (detection & tracking).
- Identity (recognition).
- Expression (facial feature tracking, expression recognition).
- Focus of attention (head/face/gaze tracking).
- Body posture and movement (body modeling and tracking).
- Gesture (gesture recognition, hand tracking).
- Activity (analysis of body movement).

The VBI are applicable to several areas but video and computer games play an integral part in the lives of many people, in particular children. In (Höysniemi et al., 2005) we can discover a novel computer vision based game technologies that aim to give players more immersive and physically challenging games experiences.

1.2 State of the art in PUI and human motion understanding

In this section we would like to introduce the more usually techniques used in human motion capturing tracking and understanding and their application in the Perceptual User Interfaces. The main objective is to give the guidelines and references in a schematic manner. In the References section we can know more about these basic ideas and projects.

In a general sense, motion capture is the process of recording a live motion event and translating it into usable mathematical terms. This is done by tracking a number of key points or regions/segments in space over time and combining them to obtain a three-dimensional representation of the performance. Then, using this technology we can translate a live movement or performance into a digital performance. The captured object could be anything that exist in the real world and make some motion. In the case that we use makers, these key points are the areas that best represent the motion of the subject's different moving part of the person.

Performance animation is not the same as motion capture, although many people use the terms interchangeably. Whereas motion capture pertains to the technology used to collect motion, the term performance animation means the actual performance that the animation designers use to bring a character to life, independently of the technology used. The performance animation is the final product of a character driven by a performer; a motion capture is only the collection of data that represent the motion.

There are different ways of capturing motion. Some systems use cameras that digitize different views of the movement, which are then used to put together the position of key points or markers normally reflective. Others use electromagnetic fields or ultrasound to track a group of sensors. Also mechanical systems based on linked structures or armatures that use potentiometers to determine the rotation of each link are also available. Finally we can design combinations of two or more of these technologies to reduce the inherent limitations. But new technologies are also aiming the possibility to make a real-time tracking of an unlimited number of key points or all the segments of the person with no space limitations at the highest frequency possible with the smallest margin of error. We are thinking in non invasive systems that use computer vision procedures and can recover the movement from a non invasive optical data using sophisticate motion models. These systems are at the moment in early state but in near future will be use by commercial purposes. For more precise historical and technical information you can read (Badler et al., 1993), (Menache, 2006).

Briefly, we would like to present the types of motion capture systems using a classification criterion as outside-in, inside-out, and inside-in depending of where the captures sources and sensors are placed.

– An outside-in system uses external sensors to collect data from sources placed on the body.
– Inside-out systems have sensors placed on the body that collect external sources.
– Inside-in systems have their sources and sensors placed on the body.

Examples of the first kind of systems are camera-based systems, where the cameras are the sensors, and the reflective markers are the sources. Electromagnetic systems are examples of inside-out systems, whose sensors move in an externally generated electromagnetic field. Finally the inside-in systems are electromechanical suits, in which the sensors are potentiometers and the sources are the actual joints inside the body. So the principal technologies used today for representing these categories are optical, electromagnetic, and electromechanical human tracking systems.

In the area of face and hands interaction we can also classify the different approaches that have been used for non invasive face/head-based interfaces. For the control of the position some systems analyze facial cues such as colour distributions, head geometry or motion. Other works track facial features or gaze including infrared lighting. To simulate the user's events it is possible to use facial gesture recognition. There are several papers dealing with theses topics and is impossible to review all the techniques in a short paper so only representative papers are commented here and in bibliography. In the case of the hands gestures we concentrate to solve the complexity in capturing and analyzing the articulated hand motion. Using biomechanical hand model we can reduce the highly dimension joint space and the joint correlations constrains help to solve this problem (Wu et al., 2005).

The facial analysis and synthesis is also a key area in the future applications using perceptual advance user interfaces. The final objective is to understand the expression or complex actions that the person is doing in front of a camera with the facial features. The automatic analysis of facial expression is very complex and includes several problems: detection of an image segmented as a face, extraction of the facial expression information, and classification of the expression (emotion categories). A system that performs these operations accurately and in real time would form a very big step in achieving a human-like interaction between users and machines. The main techniques and prototypes are in the survey (Pantic et al., 2000), (Hjelmas et al., 2001).

In this short paper we consider some examples of VBI where facial gestures are atomic facial feature motions such as eye blinking, winks or mouth's opening. Other systems contemplate the head gesture recognition that implies overall head motions or facial expression recognition that combines changes of the mentioned facial features to express an emotion.

2 SOME EXAMPLES OF VBI

In this section we would like to describe some classical examples of VBI at our laboratory. In particular we present:

– Hand gestures as game inputs.
– Face-based for accessibility
– Posture recovery for interaction in virtual worlds.

2.1 An example of hand gesture game

The VBI presented here is based of skin color information and trends to classify game commands for a children gesture interaction. The main visual cues uses are:

– Color segmentation (Gaussian model in RGB).
– Region tracking (blob based).
– Shape analysis for command recognition.

The next picture is an example of the one basic state of the hand in front of a webcam (Manresa-Yee et al., 2006). The systems is to be able to recognize some basics states (close, open, left, right, front & back) and the application can drives the navigation of the children inside a farm to rescue animals. The system is adapted to use in a 2d environment or in 3D if is needed. Only standard hardware are needed and no external constrains are imposed. The real time is reached and the testing procedures done guarantee the robust criteria adopted.

Figure 1. Basic hand state.

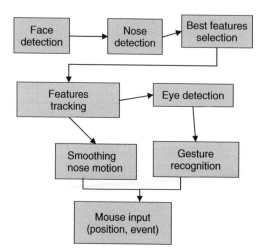

Figure 2. Global flow diagram.

Figure 3. Facial gesture user interface.

2.2 Face-based for accessibility

In this case the VBI considered is oriented to accessibility for disable people. We assume that these disable people cant move the hands and arms but has a good control of the face without spasmodic movements. Also, we assume no invasive markers or painted features and low cost devices. In this case, the visual cues are: (see above figure 2)

In the next picture we can see the interface and the features tracked. In particular the nose is detection is the most stable are to give the mouse & the eyes

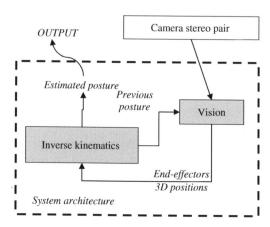

Figure 4. General flow diagram.

Figure 5. Example of virtual workshop.

are considered to simulate mouse actions (Manresa-Yee et al., 2005), (Manresa-Yee et al., 2006). At the moment, the eye tracking and actions classification must be improve to avoid no robust actions detection due to image quality and user variability.

2.3 Posture recovery for interaction in virtual worlds

In this application, is used a biomechanical representation of human body to improve the analysis of human movement in front of a virtual environment.

We use a stereo workbench with two digital color cameras as we can see in the pictures. In general the systems propose criteria based in:

− Evaluate the collision of one body part with a 3D obstacle primitive (plane, cylinder or sphere) from the environment.
− The body part must be different from the hand to show the interest of full body tracking.
− The body part must move in 3D when the hand moves toward its own goal to show the necessity of full body tracking.
− The (visual) feedback we provide when a collision occur must help the user to adapt its

Figure 6.   Example of gesture game environment.

posture/movement to avoid the collision with the virtual obstacle.

The flow diagram for track the end effectors is based in partial input vision of hands and recover the arm using the inverse kinematics models (Boulic et al., 2006) .

Also this system is able to guide a 3D stereo game in front the workbench in real time. We can compute the degree of freedoms of every segmented of the arm and control the 3D interaction with virtual world (Jaume-i-Capó et al., 2006).

3   CONCLUSIONS

In this paper, we have presented the main ideas and future trends in Advanced User Interfaces. Up to our knowledge we consider that the future interfaces will be multimodal and using natural communications procedures in the sense of human to human communications systems. The visual information and the integration in a high level domain of agents will be the most challenge areas to reach intelligent interfaces. The wide number of applications and the increase of multimedia computer capabilities give us a solid and good background to develop multidisciplinary applications managing multimodal information. We are specially oriented to disable and elderly people because the integration in communications technologies must be guaranteed, especially with these groups.

4   ACKNOWLEDGEMENTS

This work has been subsidized by the national project MCYT TIN2004-07926 of the Spanish Government. Also we would like to give many thanks to Dr. X. Varona, C. Manresa, A. Jaume, D. Arellano by pictures and previous work.

REFERENCES

Badler, N., Phillips, C., Webber, B. Simulating Humans. Computer Graphics Animation and Control. Oxford University Press, 1993.

Boulic, R., Varona, J., Unzueta, L., Peinado, M., Suescun, A., Perales, F.J.: "Evaluation of on-line analytic and numeric inverse kinematics approaches driven by partial vision input". *Virtual Reality* 10(1): 48–61, 2006.

Hjelmas, E., Kee Low, B. Face detection: A Survey, Computer Vision & Image Understanding 83, 236–274, 2001.

Höysniemi, J., Hämäläinen, P., Turkki, L., Rouvi, T. Children's Intuitive Gestures in Vision-Based Action Games, Communications of the ACM, pp. 45–50, January 2005, Vol. 48, N. 1.

Jaume-i-Capó, A., Varona, J., Perales, F.J.: "Interactive applications driven by human gestures", in SIACG 2006 Ibero-American Symposium on Computer Graphics, Santiago de Compostela, Spain, July 2006.

Manresa-Yee, C., Varona, J., Mas, R., Perales, F.J.: "Hand Tracking and Gesture Recognition for Human-Computer Interaction". Electronic Letters on Computer Vision and Image Analysis 5(3): 96–104, 2005.

Manresa-Yee, C., Varona, J., Perales, F.J.: "Towards Hands-Free Interfaces Based on Real-Time Robust Facial Gesture Recognition". *AMDO'06, Lecture Notes in Computer Science* 4069: 504–513, 2006.

Menache, A. "Understanding Motion Capture for Computer Animation and Video Games", Morgan-Kaufman, 2000.

Muybridge, E. (1957). "Animals in Motion". Dover Publications.

Pantic, M., Rothkrantz, J. M. Automatic Analysis of Facial Expressions: The state of the art. IEEE Transactions PAMI, pp. 1324–1922, Vol. 22, N. 12, December 2000.

Turk, M., Kölsch, M.: Perceptual Interfaces. In Medioni, G., Kang S.B. (eds): Emerging Topics in Computer Vision, Prentice Hall, 2005.

Turk, M., Robertson, G.: Perceptual User Interfaces. Communications of the ACM 43: 32–34, 2000.

Wu Y., Lin J., Huang T, IEEE Transactions PAMI, pp 1910–1922, IEEE, 2005.

Computational Modelling of Objects Represented in Images –
João Manuel R.S. Tavares & R.M. Natal Jorge (eds)
© 2007 Taylor & Francis Group, London, ISBN 978-0-415-43349-5

Contour detection by surround suppression of texture

Nicolai Petkov

Institute of Mathematics and Computing Science, University of Groningen, The Netherlands

ABSTRACT: Based on a keynote lecture at CompImage 2006, Coimbra, Oct. 20–21, 2006, an overview is given of our activities in modelling and using surround inhibition for contour detection. The effect of suppression of a line or edge stimulus by similar surrounding stimuli is known from visual perception studies. It can be related to non-classical receptive field (non-CRF) inhibition that is found in 80% of the orientation selective neurons in the primary visual cortex. A computational model of surround suppression is presented. It acts as a feature contrast computation for oriented stimuli: the response to an edge at a given position is suppressed by other edge responses in the surround. Consequently, the responses to texture edges are strongly reduced while the responses to contours are scarcely affected. The model gives results that are in line with perception. A surround suppression step is added to a Gabor energy filter and to the Canny edge detector. In either case it improves considerably the detection of contours. The biological utility of the neural mechanism of surround inhibition might be that of quick pre-attentive detection of object contours in natural environments rich in texture. In computer vision, a surround suppression step can be added to virtually any edge detector with limited local support in order to improve its contour detection performance.

1 INTRODUCTION

The visual perception of an edge or line can be influenced by other such stimuli in the surroundings. Fig. 1 illustrates the effect of band limited texture noise on the visibility contrast of letters (Solomon and Pelli 1994). The effect is, however, not specific of letters and is due to the interaction of the texture noise with the contours (Petkov and Westenberg 2003). Fig. 2 demonstrates a similar effect caused by 'distractor' stimuli surrounding letters and Fig. 3 shows two classical examples for the role of the context in which a visual stimulus is presented.

Oriented visual stimuli are involved in all the examples given and one may wish to relate these phenomena to the properties of orientation selective visual neurons. The classical theory of such neurons includes the concept of a receptive field that is local and limited in space (Hubel and Wiesel 1962; Daugman 1985). Such a receptive field can cover, for instance, only one line segment and a small area around it in the images shown in Fig. 2. Then it should give the same output for a line segment that is a part of a letter in both

Figure 2. The higher the number of line segments that surround the letters, the more difficult to see the letters.

Figure 3. The triangle shown in the first image is present in the second image, too, but there it is perceived as an incomplete triangle with only two legs (Galli and Zama 1931). The third and fourth image illustrate the orientation-contrast pop out effect (Noth-durft 1991).

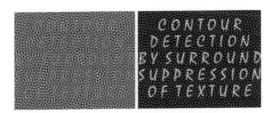

Figure 1. The recognition of the low contrast letters in the left image is made more difficult or even impossible by the texture. At high contrast (right image) the letters are well visible.

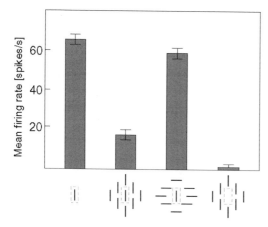

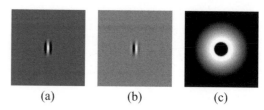

Figure 5. (a) Symmetric and (b) antisymmetric Gabor function. (c) Surround.

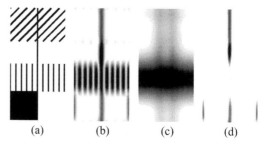

Figure 6. (a) Input image. (b) Gabor energy. (c) Suppression term. (d) Surround suppressed Gabor energy.

Figure 4. The response of an orientation selective neuron to different stimuli (shown below the diagram). (first bar) The neuron responds strongly to an optimal stimulus in its CRF (schematically shown as a dashed line rectangle). (second bar) The response is suppressed by the addition of other similarly oriented stimuli to the surroundings. (third bar) Orthogonal stimuli have only slight effect. (fourth bar) There is no response if the central stimulus is missing. (Adapted from (Nothdurft et al. 1999)).

images shown in that figure and this cannot explain the difference in saliency of the letters in the two images.

Recent neurophysiological research has shown that the concept of a classical receptive field is not sufficient to adequately describe the responses of orientation selective neurons in the primary visual cortex to complex visual stimuli and that 80% of these neurons exhibit surround suppression of various types (Nothdurft et al. 1999). The type of neural response shown in Fig. 4 can, for instance, be used to explain the orientation-contrast pop out effect.

In the following I give a brief overview of our previous and current work on modelling surround suppression and using it in computer vision for improving the performance of contour detection operators. Mathematical formulae are left out of this presentation, detailed explanations are avoided and references are mostly limited to our work on this topic.

2 COMPUTATIONAL MODEL

Figs. 5a and 5b show a pair of Gabor functions that can be used in a Gabor energy filter which is defined as follows: an input image is convolved with each of these functions and the results are squared and summed up pixel-wise; then the pixel-wise square root is computed. Such a filter is considered to be an adequate first order approximation of the function of complex cells. Figs. 6a and 6b show an input image and the

corresponding output of a Gabor energy filter with preferred vertical orientation.

As can be seen in Fig. 5ab the Gabor functions have significant values only in a limited region that can be thought of as the CRF of a corresponding complex cell. Fig. 5c shows the surround of such a CRF, modelled by a weighting function that is a difference of two concentric Gaussian functions with different spatial extents. We model surround inhibition by convolving the output of the Gabor energy filter (Fig. 6b) with the surround weighting function (Fig. 5c) in order to obtain an inhibition term for each pixel (Fig. 6c). This inhibition term is multiplied by a coefficient, that we call inhibition strength, and then subtracted from the Gabor energy filter and the result is half-wave rectified. The output of this surround suppressed Gabor filter is shown in Fig. 6d. For a mathematical formulation of a Gabor energy filter and surround suppression the reader is referred to (Petkov and Westenberg 2003; Grigorescu et al. 2003).

While the original Gabor energy filter enhances all lines and edges of its preferred (in this case vertical) orientation, the surround suppressed filter enhances only lines and edges that are isolated, i.e. not surrounded by other lines and edges with the same orientation.

Fig. 7 shows a pair of input images and the corresponding output images for a Gabor energy filter and a surround suppressed Gabor energy filter computed as superpositions of filters with different preferred orientations. One can see that the surround suppressed

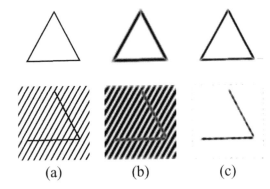

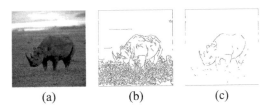

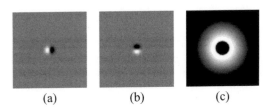

Figure 8. (a) Input image. Binarized output of (b) a Gabor energy filter and (c) a surround suppressed Gabor energy filter.

Figure 7. (a) Input images. (b) Gabor energy. (c) Surround suppressed Gabor energy.

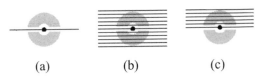

Figure 9. (a-b) x- and y-derivatives of a Gaussian function used in the Canny edge detector. (c) Weighting function used for surround suppression.

Gabor energy filter enhances only two legs of the triangle, the ones that are not surrounded by lines of the same orientation, and suppresses the line grating. This filter thus reproduces the saliency of these legs as perceived by human observers. Further examples can be found in (Petkov and Westenberg 2003).

3 COMPUTER VISION APPLICATION

The output of a (surround suppressed) Gabor energy filter can be binarized by non-maxima suppression and thresholding, Fig. 8. The binarized Gabor energy filter detects the contours of an object of interest but at the same time it detects all the texture edges present in the scene. This is a major disadvantage when the output of an edge detector is to be used for object recognition. Modern contour based shape analysis methods (Grigorescu and Petkov 2003) rely on clean object contours and texture noise has a devastating effect on their performance. In that respect, the suppressed Gabor energy operator gives results that are much more useful for shape analysis.

The idea of applying surround suppression is, however, not limited to the Gabor filter only. It can be applied to any operator with local support. Figs. 9a and 9b show the x- and y-derivatives of a Gaussian function. They can be used for scale-dependent gradient computation and subsequent binarization for edge detection (Canny 1986). Fig. 9c shows a surround weighting function that can be convolved with the gradient field in order to compute an inhibition term that, multiplied with a certain coefficient, is subtracted from the gradient (magnitude) field. After half-wave rectification the result can be binarized using the standard procedure. Thus, by inserting a surround suppression step between the gradient computation and the binarization of the Canny edge detector, the performance

Figure 10. (a) A band region is excluded from the surround to prevent self-inhibition of lines. (b) The split surround suppresses edge texture in homogeneous areas but (c) leaves texture boundaries unaffected because in the latter case the inhibition term is determined by the half-ring that collects less edge response.

of that operator in contour detection tasks can be improved very substantially (Grigorescu et al. 2004).

The effectiveness of surround suppression can be improved by using a surround that is not circularly symmetric. One can prevent the side effect of self-inhibition of lines by excluding the area that lies in the continuation of the central optimal stimulus from the inhibition surround, Fig. 10a. In this way the surround is split into two half-rings flanking the CRF. A further improvement can be achieved by computing a separate inhibition term for each of these half-rings and using the lesser of the two values for inhibition. This leads to improved detection of texture boundaries, Fig. 10b,c.

Finally, one can apply surround suppressed contour detection in a multi-scale scheme and combine the results from the different resolutions. Fig. 11c illustrates the results obtained with such a muli-scale operator (see (Papari et al. 2006) for details).

Figure 11. (a) Input image. (b) Output of the Canny edge detector. (c) Output of a multi-scale surround suppressed contour operator.

Quantitative evaluation of the contribution of surround inhibition to texture suppression and contour detection and comparisons with other operators have been done in (Grigorescu et al. 2003; Grigorescu et al. 2004; Papari et al. 2006). The algorithms sketched above can be found as internet enabled executable programs on the site matlabserver.cs.rug.nl and a data set of natural images with corresponding hand-drawn contour ground truth images can be downloaded from www.cs.rug.nl\~imaging.

4 CONCLUSIONS

By means of a computational model and computational experiments we have shown that surround suppression is the most likely origin of certain perceptual effects that concern the saliency of contours in the presence of surrounding texture. Of course, the creation of puzzling effects cannot be the purpose of a neural mechanism. The biological role of surround suppression should rather be searched for in the structure of natural images. Typically, such images are rich in texture due to grass, foliage, pebbles, sand, water, etc. Suppression of texture edges leads to improved visibility of object contours and thus makes objects easier to detect. The quick pre-attentive detection of contours in a natural environment can be a matter of survival for both pray and predator. Indeed, camouflage strategies in both natural and artificial systems typically involve the use of texture patterns (on the skin of an animal or on military clothes or equipment) that are similar to the texture patterns in the surroundings, making object contours difficult to detect or even invisible.

We have demonstrated that surround suppression can be deployed in virtually any edge detection operator with limited local support. It helps remove or reduce the amount of texture edges that should be considered as noise in contour-based object recognition algorithms.

REFERENCES

Canny, J. F. (1986). A computational approach to edge detection. *IEEE Trans. Pattern Analysis and Machine Intelligence 8*(6), 679–698.

Daugman, J. G. (1985). Uncertainty relations for resolution in space, spatial frequency, and orientation optimized by two-dimensional visual cortical filters. *J. Optical Society of America A 2*, 1160–1169.

Galli, A. and A. Zama (1931). Untersuchungen über die Wahrnehmung ebener geometrischer Figuren, die ganz oder teilweise von anderen geometrischen Figuren verdeckt sind. *Zeitschrift für Psychologie 31*, 308–348.

Grigorescu, C. and N. Petkov (2003). Distance sets for shape filters and shape recognition. *IEEE Trans. Image Processing 12*(10), 1274–1286.

Grigorescu, C., N. Petkov, and M. A. Westenberg (2003). Contour detection based on nonclassical receptive field inhibition. *IEEE Trans. Image Processing 12*(7), 729–739.

Grigorescu, C., N. Petkov, and M. A. Westenberg (2004). Contour and boundary detection improved by surround suppression of texture edges. *Image and Vision Computing 22*(8), 609–622.

Hubel, D. H. and T. N. Wiesel (1962). Receptive fields, binocular interaction, and functional architecture in the cat's visual cortex. *J. Physiology (London) 160*, 106–154.

Nothdurft, H. C. (1991). Texture segmentation and pop-out from orientation contrast. *Vision Research 31*, 1073–1078.

Nothdurft, H. C., J. Gallant, and D. C. van Essen (1999). Response modulation by texture surround in primate area V1: Correlates of "popout" under anesthesia. *Visual Neuroscience 16*(1), 15–34.

Papari, G., P. Campisi, N. Petkov, and A. Neri (2006). Contour detection by multiresolution surround inhibition. In *Proc. Int. Conf. on Image Processing ICIP 2006, Atlanta, GA, October 8–11 2006*, Volume CD-ROM, pp. 4 pages. IEEE.

Petkov, N. and M. A. Westenberg (2003). Suppression of contour perception by band-limited noise and its relation to non-classical receptive field inhibition. *Biological Cybernetics 88*(10), 236–246.

Solomon, J. A. and D. G. Pelli (1994). The visual filter mediating letter identification. *Nature 369*, 395–397.

Computational Modelling of Objects Represented in Images –
João Manuel R.S. Tavares & R.M. Natal Jorge (eds)
© 2007 Taylor & Francis Group, London, ISBN 978-0-415-43349-5

Computer vision and digital inclusion of persons with special needs: Overview and state of art

Hemerson Pistori

UCDB – Dom Bosco Catholic University, Campo Grande, Brazil
GPEC – Research Group in Engineering and Computing

ABSTRACT: This survey paper addresses some issues related to the application of computer vision techniques to improve the welfare of people with special needs. The main problems and current work on topics like sign language processing and wheelchair control will be presented. The paper also introduces an ongoing project that aims at creating a free software environment that will include implementations of a large amount of computer vision, pattern recognition and machine learning techniques, tuned to the problems related to the digital inclusion of people with special needs. The software will also serve as an experimental environment, where new techniques will be implemented, tested and compared.

1 INTRODUCTION

Unfortunately, worldwide, up to date statistics on the quality of life of people with disabilities are not available. However, a recent comprehensive survey on disabled people situation in the United States revealed some very important issues. According to this survey, only 35 percent of adults with special needs are employed, 26 percent live in poverty and 21 percent do not complete high school studies. All these numbers are two or three times worse than for people without disabilities (Krane and Hanson 2004), and probably represent an optimistic upper bound on the situation for most of the countries, particularly the poorest ones.

The United States census also estimated, in 2002, that 11.5 percent of their citizens had severe disability, while 18.1 percent presented some level of disability (Steinmetz 2006). The projection of these numbers to the world population would give more than one billion of people with some level of disability. Computer based assistive technologies are emerging as an important mean to improve communication, mobility and self care abilities for people with special needs. This paper focuses on computer vision based assistive technologies and present current work with high relevance to the problem of deafs and quadriplegics social inclusion.

Besides the social justification for the researches on computer vision based assistive technologies, the problems faced in this area are specially challenging and serve as testbeds for many computer vision techniques. Usually, the systems must implement multiple deformable objects tracking, in complex and moving background, with noise and occlusions. In this paper, two prominent problems will be surveyed: (1) the automatic translation of sign language to written language and (2) the control of a wheelchair using face expressions and movements. Works on each of this problems are presented in the next two sections. A software specially devised to serve as a complete environment for the development of human-machine interaction systems based on computer vision is described in Section 4. The last section is reserved for discussions, conclusions and comments on emerging research topics.

2 SIGN LANGUAGE RECOGNITION

Two important misconceptions are usually associated to sign languages. First, sign languages are not universal. Estimates on the number of sign languages around the world vary from 4000 to 20000 (Woll, Sutton-Spence, and Elton 2001), and as it happens with oral languages, they also present regional dialects and accents. Signs in sign languages are not simple mappings from words or letters in oral languages. Sign languages have their own grammatical structures, many of them without an oral language analog, and which evolve naturally in a very sophisticated, and not completely controlled way, in order to optimize the use of body parts shape, movement and spatial relation. Signs that correspond to alphabetical digits do exist, however, they are only used in some special cases, like in proper names finger-spelling.

As the mechanical movements of hands, face and torso are much slower than the vocal tract control, sign languages are highly dependent on context sensitivity and a sort of multi-channel information transmission in order to achieve the same speed, in communication, as oral languages. Long sentences, in spoken language, can be translated to short sentences in sign language, by the parallel use of facial expressions and different shapes and movements from the right and left hand. Even the written representation of sign languages is an open and challenging problem, as current formal language theory is not appropriate in this context (Huenerfauth 2005).

A sign language recognition system presents almost all of the most difficult and interesting problems in computer vision, as it requires real-time segmentation, tracking and classification of multiple, deformable, self-occluding and non-linear moving objects. The first works on sign language recognition focused on static hand shape recognition (Pistori and Neto 2004), however, in the last five year, many important results on the recognition of complete gestures and face expressions have been reported.

As it happens in speech recognition, hidden Markov models are the most cited technique in gesture recognition. In (Holden, Lee, and Owens 2005), correct classification rates of 97% at the sentence level, and of 99% at the word level have been reported. The proposed system is based on hidden Markov models, trained with 379 utterances, and tested on 163 sentences from the Australian Sign Language. A two stage classifier, combining hidden Markov models and support vector machines were tested on a very large vocabulary, with 4942 different signs and 59304 samples of Chinese Sign Language. An 89.40% correct classification rate was achieved, with the help of two cyber gloves and 3 space-position tracker, to circumvent the hands tracking problem (Ye, Yao, and Jiang 2004).

An attempt to achieve good generalization, with fewer samples, has been reported in (Bowden, Windridge, Kadir, Zisserman, and Brady 2004). In Bowden work, Independent Component Analysis, hidden Markov models and high-level linguist knowledge were combined to build a system that correctly recognised 96.7% of 43 words from the British sign language, using just 1 sample of each word for training and 200 for testing. This work also builds on the assumption that the dominant hand convey most of the information necessary to recognize the 43 words tested.

Another dominant topic in sign language recognition is skin color based segmentation. Several different color spaces, including HSV, normalized RGB, YCbCr have been investigated in the construction of both parametric and non-parametric skin color models. Real time performance, with a position error from 0.8 to 1.8 pixels, for tracking hands in a Thai Sign Language recognition system, has been recently reported in (Soontranon, Aramvith, and Chalidabhongse 2005).

The combination of image features and information captured by accelerometers and data gloves is also being explored, with increments of up to 94.26% in tracking precision (Culver 2004).

In order to deal with accents from different signers, without recurrent to very large training sets, a recent work has evaluated the application of some adaptation techniques already in use in speech recognition. In (von Agris, Schneider, Zieren, and Kraiss 2006), Maximum Likelihood Linear Regression has been used to adapt the parameters of the underlying hidden Markov model. Experiments using a test corpus consisting of 153 isolated signs from British Sign Language, performed by four signers, in a control environment, showed a relative improvement of up to 41.6% compared to the independent user baseline (no adaptation).

Two of the most cited annotated test corpus, for sign language investigation, are the Purdue's RVL-SLLL (Martinez, Wilbur, Shay, and Kak 2002) and *signstream*, one of the most comprehensive collection of annotated videos of American sign language utterances, including images captured from different angles and signers. An extensive survey of 156 papers on sign language analysis can be found in (Ong and Ranganath 2005).

3 WHEELCHAIR CONTROL

Many people, including paraplegics, quadriplegics and elders, depend, permanently or temporarily, on wheelchair for locomotion. The control of an electric wheelchair when joysticks are not an option, includes devices that can be driven by tongue (Struijk 2006) and forehead muscle contractions (Felzer and Freisleben 2002). Computer vision, however, is being explored in the development of wheelchair control mechanisms based on gesture, facial expression, head movements and eye gaze. In this case, the users are free of any mechanical contact and images captured through cameras attached to the wheelchair are interpreted and used to produce the driving commands. Computer vision techniques, specially robotic navigation, are also being explored in the construction of the so called intelligent wheelchairs, which can recognize the environment to avoid collision and even drive, autonomously, the user to some predefined places (Ono, Uchiyama, and Potter 2004). This survey is focused on the works that use images from the user face and hand to generate commands to a wheelchair.

In a recent paper, the control of a wheelchair by face movements has been explored. Multilinear discriminant classifiers were trained to achieve 100% accuracy in face detection. Head pose (turn left and turn right) were estimated by support vector machines, with linear kernels, achieving a 95.7% correct pose identification rate (Bauckhage, Kaster, and Rotenstein 2006). Matsumoto combined head motion, gaze direction, blinks and lip motion information, captured from a standard PC camera pair, to build a wheelchair controller that has been tested in real situations. Face detection has been implemented by template matching of 3d facial model that included eyebrows, eyes and mouth (Matsumoto, Ido, Takemura, Koeda, and Ogasawara 2004).

The system proposed by Jia, in (P. Jia and Yuan 2006), uses nose pose estimation, by template matching, to control the moving direction of the wheelchair. Five face expressions, discriminated using an algorithm that combines Adaboost and Camshift, working on Haar-like features extracted from the face image, were used to supply additional controls to the wheelchair, like stop and start moving. Tracking was implemented using the mean shift algorithm in the hue space. Only one, out-of-the-shell webcamera, were used as the image capture device. The problem of wrong command interpretation, when the user makes fast head movements in response to some external stimulus, has been tackled in (Kuno, Murakami, and Shimada 2001). The system distinguishes among slow and fast head movements and includes pedestrian recognition to avoid collision.

4   SIGUS

No general method can solve all the problems discussed in the last sections and the success of some algorithms and techniques are usually task dependent and demands a hard work in parameter fine tuning. The SIGUS platform[1] is a software environment intended to aid computer vision investigators in building and testing systems tuned to the problem of machine-human interaction using corporal expressions, including hand gestures, face expression and eye-gaze.

The platform is built upon two mature open source libraries, the ImageJ (Abramoff, Magelhaes, and Ram 2004; Rusband 2006), for digital image processing and the Weka (Witten and Frank 2005), for machine learning. These two packages were extended with new features and integrated to some other libraries, like randomj, for random number generation (used in Monte Carlo based approaches), mical, for grammar induction and jama, for linear algebra routines. Integration of the SIGUS platform to the SciLab package (Pires and Rogers 2002; Gomez 2006) is also under development.

Training datasets with thousands of images and tools to help annotation and ground truth generation are also part of SIGUS platform. The datasets will include images from hundreds of different gestures and sentences of Brazilian sign language, taken from persons with different skin colors, ages, signaling skill levels, in different environments (office, classroom, outdoor, etc) and illumination setup (natural, incandescent lamps, fluorescent lamps, etc). A face expressions and head movements dataset is also being created, with images taken from cameras attached to wheelchairs, in a large variety of environments and with different persons.

Ground truth sets are being generated to test segmentation, feature extraction, dimension reduction, tracking and classification algorithms in real situations. Different metrics and dissimilarity measures to be used both in performance comparison and by distance based algorithms, are being implemented, including Hausdorff, Chebyshev, Bhattacharrya, Chi-squared, Kullback-Leibler Divergence, Mahalanobis and Earth mover distance.

Several skin segmentation techniques are already available in the platform, including the ones based on mixture models, Gaussian models and some non-parametrics models. The platform also implements adaptive background segmentation, particle filter based tracking and via Weka, a great variety of supervised learning techniques, including state-of-art support vector machines and radial basis networks. Hidden Markov models, for sign language recognition, are also available. Most of the implemented algorithms presents graphical counterparts that can be used to monitor its execution in real time and to adjust parameters. For instance, in particle filters tracking, it is possible to monitor, in a intuitive graphical manner, the number and weights of the particle set.

Figure 1 illustrates a software created using SIGUS platform. In this software, the user can associate different visual sign (face expressions, face movements

Figure 1.   Example of an application created using SIGUS platform: a drum, for quadriplegic people, that can be played by face movements.

[1] http://www.gpec.ucdb.br/sigus

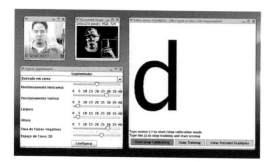

Figure 2. Example of an application created using SIGUS platform: a drum, for quadriplegic persons, that can be played by face movements.

or hands configuration, for instance), to each piece of a virtual drum. After a calibration and training phase, the virtual drum can be played without any direct contact to a mouse or keyboard, just by reproducing face or hand configurations in front of a standard webcam, positioned in front of the user. An alphabetical editor, that translates sign language hand configurations to letters and digits, is another example of a software prototype available with the SIGUS platform (Figure 2).

5 DISCUSSIONS AND CONCLUSIONS

This paper presented an overview of some recent work related to the use of computer vision techniques to improve human-machine interaction for people with special needs. A new free software platform, not intended to be just another general purpose computer vision package, but a development environment tailored to the programming and testing of systems for people with special needs were also presented.

The two problems presented in this paper, namely the sign language translation and wheelchair control have been investigated in other areas than computer vision, in systems based on another kind of sensors and devices. An important future work is the creation and application of new methods to compare the performance of systems based on very different technologies, like computer vision, electroculography (William and Kirsch 2005) or electroencefalogram (Kauhanen, Nykopp, Lehtonen, Jylnki, Heikkonen, Rantanen, Alaranta, and Sams 2006) based wheelchair control. There is also a great need for benchmark datasets. Authors usually report correct classification rates that cannot be mutually compared, as the testing assumptions are too different.

An important criticism to the work of many computer scientists that deals with systems for people with special needs is the lack of a more consistent and methodical interaction and participation of the final

users during research, development and testing. Fehr (Fehr, Langbein, and Skaar 2000) experiments with some systems prototypes showed that many of the proposed human-machine interfaces, supposed to be aimed at people with special needs, presented serious problems of usability.

Another problem, not explored in this paper, that is being tackled using computer vision techniques, is vision-tactile substitution. The problem consists in translating the images captured from a camera, usually installed in sun-glasses, to a low-resolution matrix of tactile stimulators. In the most common approach, a matrix of electrodes is attached to the body of a blind person, usually in a place with high tactile sensibility, like the tongue, the belly, the fingertips or the forehead (Tachi 2006; Tyler 2000). The processed image or some high level information, extracted from the scene (like a text or a number), is "printed" on the skin of the blind, who can be trained to associate the "printed" signs to visual information.

ACKNOWLEDGMENTS

This work has been funded by the Dom Bosco Catholic University, UCDB, the Foundation of Teaching, Science and Technology Development of Mato Grosso do Sul State, FUNDECT, and the Brazilian Studies and Projects Funding Body, FINEP. The author of this paper and some of his advisees hold scholarships from the Brazilian National Counsel of Technological and Scientific Development, CNPQ.

REFERENCES

Abramoff, M. D., P. J. Magelhaes, and S. J. Ram (2004). Image processing with ImageJ. *Biophotonics International 11*(7), 36–42.

Bauckhage, C., T. Kaster, and A. M. Rotenstein (2006). Fast learning for customizable head pose recognition in robotic wheelchair control. In *FGR '06: Proceedings of the 7th International Conference on Automatic Face and Gesture Recognition (FGR06)*, Los Alamitos, CA, USA, pp. 311–316. IEEE Computer Society.

Bowden, R., D. Windridge, T. Kadir, A. Zisserman, and M. Brady (2004, May 11–14). A linguistic feature vector for the visual interpretation of sign language. In *ECCV 2004, 8th European Conference on Computer Vision*, pp. 390–401.

Culver, V. R. (2004). A hybrid sign language recognition system. In *Eighth IEEE International Symposium on Wearable Computers (ISWC'04)*, Los Alamitos, CA, USA, pp. 30–33. IEEE Computer Society.

Fehr, L., W. E. Langbein, and S. B. Skaar (2000). Adequacy of power wheelchair control interfaces for persons with severe disabilities: A clinical survey. *Journal of Rehabilitation Research and Development 37*(3), 353–360.

Felzer, T. and B. Freisleben (2002). Hawcos: the "hands-free" wheelchair control system. In *Proceedings of the 5th international ACM conference on Assistive technologies (ASSETS)*, pp. 127–134.

Gomez, C. (2006, Junho). Scilab. *Available at http://www. scilab.org/*.

Holden, E.-J., G. Lee, and R. Owens (2005). Automatic recognition of colloquial australian sign language. In *WACV-MOTION '05: Proceedings of the IEEE Workshop on Motion and Video Computing (WACV/MOTION'05) –* Volume 2, Washington, DC, USA, pp. 183–188. IEEE Computer Society.

Huenerfauth, M. (2005). Representing coordination and non-coordination in an american sign language animation. In *Assets '05: Proceedings of the 7th International ACM SIGACCESS Conference on Computers and Accessibility,* New York, NY, USA, pp. 44–51. ACM Press.

Kauhanen, L., T. Nykopp, J. Lehtonen, P. Jylnki, J. Heikkonen, P. Rantanen, H. Alaranta, and M. Sams (2006). EEG and MEG brain-computer interface for tetraplegic patients. *IEEE Transactions on Neural Systems and Rehabilitation Engineering 14*(2).

Krane, D. and K. Hanson (2004). *National Organization on Disability and Harris 2004 Survey of Americans with Disabilities.* New Work: Harris Interactive.

Kuno, Y., Y. Murakami, and N. Shimada (2001). User and social interfaces by observing human faces for intelligent wheelchairs. In *PUI '01: Proceedings of the 2001 Workshop on Perceptive User Interfaces*, New York, NY, USA, pp. 1–4. ACM Press.

Martnez, A., R. Wilbur, R. Shay, and A. Kak (2002). Purdue RVL-SLLL ASL database for automatic recognition of american sign language. In *Proc. IEEE International Conference on Multimodal Interfaces.*

Matsumoto, Y., J. Ido, K. Takemura, M. Koeda, and T. Ogasawara (2004, May 17–19). Portable facial information measurement system and its application to human modeling and human interfaces. In *6th IEEE International Conference on Automatic Face and Gesture Recognition (FGR 2004)*, pp. 475–480.

Ong, S. C. W. and S. Ranganath (2005). Automatic sign language analysis: A survey and the future beyond lexical meaning. *IEEE Trans. Pattern Anal. Mach. Intell. 27*(6), 873–891.

Ono, Y., H. Uchiyama, and W. Potter (2004). A mobile robot for corridor navigation: a multi-agent approach. In *ACM-SE 42: Proceedings of the 42nd Annual Southeast Regional Conference*, New York, NY, USA, pp. 379–384. ACM Press.

P. Jia, H. Hu, T. L. and K. Yuan (2006, October). Head gesture recognition for hands-free control of an intelligent wheelchair. *Journal of Industrial Robot 33*(6).

Pires, P. S. M. and D. A. Rogers (2002, November). Free/open source software: An alternative for engineering students. In *Proceed. of FIE – Frontiers in Education.*

Pistori, H. and J. J. Neto (2004, September 29). An experiment on handshape sign recognition using adaptive technology: Preliminary results. *Lecture Notes in Artificial Intelligence. XVII Brazilian Symposium on Artificial Intelligence – SBIA'04 3171*, 464–473.

Rusband, W. (2006, Junho). Imagej. *Available at http://rsb.info. nih.gov/ij/*.

Soontranon, N., S. Aramvith, and T. H. Chalidabhongse (2005). Improved face and hand tracking for sign language recognition. In *ITCC '05: Proceedings of the International Conference on Information Technology: Coding and Computing (ITCC'05)*, Washington, DC, USA, pp. 141–146. IEEE Computer Society.

Steinmetz, E. (2006, May). Americans With Disabilities: 2002 – Household Economic Studies. Technical Report Current Population Reports, Bureau of The Census – U.S. Department of Commerce, Washington, DC.

Struijk, L. N. S. A. (2006). A tongue based control for disabled people. *Lecture Notes in Artificial Intelligence. Computers Helping People with Special Needs. 4061*, 913–918.

Tachi, H. K. Y. K. S. (2006). Forehead electrotactile display for vision substitution. In *Euro-Haptics.*

Tyler, K. A. K. M. E. (2000). Effect of electrode geometry and intensity control method on comfort of electrotactile stimulation on the tongue. In *Proc. of the ASME, Dynamic Systems and Control Division*, Volume 2, pp. 1239–43.

von Agris, U., D. Schneider, J. Zieren, and K.-F. Kraiss (2006, June 17–22). Rapid signer adaptation for isolated sign language recognition. Los Alamitos, CA, USA, pp. 159–164. IEEE Computer Society.

Williams, M. and R. Kirsch (2005, July 5–8). Feasibility of electroculography as a command interface for a high tetraplegia neural prosthesis. In *10th Annual Conference of the International Functional Electrical Stimulation Society*, Montreal, Canada.

Witten, I. H. and E. Frank (2005). *Data Mining: Practical Machine Learning Tools and Techniques.* San Francisco: Morgan Kaufmann.

Woll, B., R. Sutton-Spence, and F. Elton (2001, October). *The Sociolinguistics of Sign Languages*, Chapter Multilingualism – the global approach to sign languages. Cambridge, UK: Cambridge University Press.

Ye, J., H. Yao, and F. Jiang (2004). Based on HMM and SVM multilayer architecture classifier for chinese sign language recognition with large vocabulary. In *3rd International Conference on Image and Graphics (ICIG'04)*, pp. 377–380.

Computational Modelling of Objects Represented in Images –
João Manuel R.S. Tavares & R.M. Natal Jorge (eds)
© 2007 Taylor & Francis Group, London, ISBN 978-0-415-43349-5

MR diffusion tensor imaging

M. Forjaz Secca

Cefitec, Physics Dep., Univ. Nova de Lisboa, Caparica, Portugal
Ressonância Magnética, Caselas, Lisboa, Portugal

ABSTRACT: Magnetic Resonance Imaging is arguably the most versatile and complete imaging modality in Medicine nowadays. Its performance and techniques are totally dependent on complex computational procedures relying on powerful computers. Diffusion Imaging is one of the newest and most exciting of the advanced MRI techniques, providing structural and microscopic functional information about the mobility of water in the intra and extra cellular spaces and the shape of cellular spaces where water is present. The most dramatic of these techniques is the one called White Matter Fiber Tracking, which allows us to see the structure of the axon fibers within the brain.

1 INTRODUCTION

Nuclear Magnetic Resonance (NMR) is an established scientific technique that started as a field of Physics but later extended to Chemistry, Biochemistry and Medicine. Its importance is so acknowledged that several Nobel prizes were won in this field: two in Physics (Rabi in 1944 and Bloch and Purcell in 1952), two in Chemistry (Ernst in 1991 and Wüthrich in 2002) and one in Physiology and Medicine (Lauterbur and Mansfield in 2003). The 2003 award was a recognition of the breakthrough in medical diagnostics and research that the development of magnetic resonance imaging represents.

The Nuclear in NMR disappeared because of the negative connections people made with nuclear radiotherapy and nuclear medicine, thus the technique become know simply as MR Imaging (MRI).

One of the reasons MRI represents a breakthrough in medical diagnostics is its extreme versatility and wealth of information, making possible the existence of several advanced imaging techniques, like brain functional imaging, cardiac functional imaging, angiographic imaging, perfusion imaging, spectroscopic imaging and diffusion imaging.

2 DIGITAL IMAGING

An image is a representation of an object or a person. This representation can be either two or three dimensional. For an image to be seen, for it to exist, we need light, we need to see it.

A medical image is an attempt to see the human body, transform it into an image. The simplest medical image is an external or surface image, like a simple photograph. This concept also covers endoscopic images, since they are still surface images. However, in Medicine we want to see more than the surface, we want to see inside the body, or an internal image, allowing us to "see" the invisible. Obviously the images that interest us are *in vivo* images and preferably non, or minimally, invasive. With the discovery of X-rays imaging entered another world, leaving the exterior and making it possible to see the interior of the body.

Simple X-ray imaging was then followed by ultrasound, CT, MRI, Nuclear medicine, PET and a few other more specific techniques like OCT.

Since the advent of ultrasound, all new medical images became reconstructed images, totally dependent on computers to be reconstructed. What computers do is allow us to convert a set of physical measurements into two-dimensional data that can be represented on a screen as a medical image.

The images are not really two-dimensional, because to each point in a two-dimensional space corresponds either a value that gives us the grey level, producing in fact three-dimensional information, or two values that give us colour information, corresponding to each point not only an intensity level but also the frequency of the colour, producing in this way four-dimensional information. Traditional X-ray images are analog images, however most modern medical images are digital images. Both in TC and MRI the pixels in the image contain more than planar information, they incorporate information from a certain depth of the imaging slice, in reality to each pixel there is a volume element of tissue associated with it. This volume element is called a voxel.

It is essential to remember that digital medical images are reconstructed images, they are not the real body; as we manipulate these images we are creating another reality, that can or cannot correspond to the truth; therefore medical images should be used with care and with a careful knowledge of the causes.

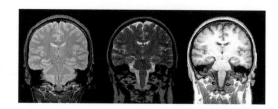

Figure 1. Example of a DP weighted image, a T2 weighted image and a T1 weighted image acquired on the same person.

3 MAGNETIC RESONANCE IMAGING (MRI)

3.1 How does MRI work?

About 70% of our body weight is water, which contains Oxygen and Hydrogen. The nuclei of Hydrogen consist of protons, which behave like little magnets, because they have a magnetic moment. When placed in a magnetic field the protons align with the field and precess around its axis, in a way similar to a spinning top.

The frequency of precession is given by

$$\omega = \gamma B \tag{1}$$

This means that the higher the magnetic field the faster the protons turn.

If radiofrequency electromagnetic radiation is emitted around the protons at a frequency in resonance with their precession frequency the protons can absorb this radiation increasing their energy.

Once the external radiofrequency is stopped the protons re-emit the radiation absorbed at a rate that differs from tissue to tissue and depends on the relaxation rates characteristic of each tissue.

The medical image is constructed because the external field does not remain homogeneous, but has pulsed magnetic field gradients that modulate the field in space and provide a spatial codification that allow us to know from where in space each signal is coming.

3.2 Main parameters of MRI

The main parameters of MRI are: the proton density, DP, the longitudinal relaxation time, T1, and the transverse relaxation time, T2, which are characteristics of each tissue that depend on their physical and chemical constitution.

By adjusting the correct acquisition variables for the MRI sequences, all MRI images can have their contrasts weighted by any of these parameters, or a mixture of them.

3.3 Other MRI techniques

Apart from the standard techniques, MR images can be made sensitive to a wealth of other physical, chemical and physiological parameters, like blood oxygenation level, blood velocity, blood perfusion, chemical constitution and molecular diffusion.

These so called advanced techniques provide a wide range of imaging modalities that have given MRI its reputation for versatility and wealth of information.

4 DIFFUSION IMAGING

4.1 What is diffusion?

Diffusion is the random translational motion of molecules in a medium, due to the random collisions provoked by their thermal motion. Every fluid has a characteristic diffusion constant, D, related to the mobility of its molecules.

Einstein showed that the mean square of the distance traveled by the molecules in a three-dimensional liquid is given by:

$$\langle r^2 \rangle = 6Dt \tag{2}$$

The MR signal can be made sensitive to displacements of water molecules between 10^{-8} and 10^{-4} m in a timescale of a few milliseconds to a few seconds.

So MRI is capable of measuring the microscopic mobility of the water molecules in tissues. This mobility can be a very important parameter because it is conditioned by the cellular structural restrictions and it can increase if there is destruction of biological barriers by disease.

4.2 How to measure diffusion?

In the presence of a magnetic field gradient, the MR signal is dependent on T2 and the strength of the gradient.

$$M(t) = M(0)\exp\left(-\frac{TE}{T_2}\right)\exp(-bD) \tag{3}$$

where TE is the acquisition echo time, D is the diffusion coefficient and b is a value that quantifies the magnetic field gradient.

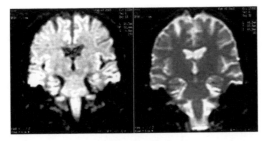

Figure 2. Example of two images acquired with a high gradient and with no gradient (T2), used to calculate the DC.

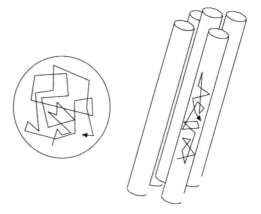

Figure 4. Diagram showing the difference between isotropic and anisotropic diffusion.

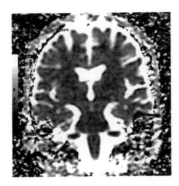

Figure 3. ADC map calculated from the images in Figure 2.

The factor b, known as the b value, has units of s/mm² and can vary from 0 up to the order of 10,000 s/mm². It is given by:

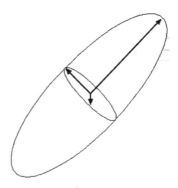

Figure 5. Example of a diffusion tensor ellipsoid.

$$b = \gamma^2 \delta^2 g^2 \left(\Delta - \frac{1}{3}\delta \right) \qquad (4)$$

where γ, δ and Δ are parameters that give the intensity, duration and spacing of the magnetic field diffusion gradients.

In practice TE is kept short and equation 3 is used to calculate the diffusion coefficient from two different images acquired on the same location: one with a high gradient and one with no gradient.

From the exponential fit a value for the diffusion coefficient is calculated for each voxel and a map is created. Because the diffusion within each voxel also incorporates the flow of blood in all the capillaries present in the voxel, the value obtained is called the apparent diffusion coefficient (ADC).

With isotropic diffusion, diffusion is described by the single scalar parameter, D, as shown above, but with anisotropic diffusion we require a diffusion tensor, **D**, to fully describe the mobility of the molecules in the different directions and the correlations between the different directions.

The diffusion tensor can be represented by a 3 × 3 matrix:

$$\mathbf{D} = \begin{pmatrix} D_{xx} & D_{xy} & D_{xz} \\ D_{yx} & D_{yy} & D_{yz} \\ D_{zx} & D_{zy} & D_{zz} \end{pmatrix} \qquad (5)$$

The diffusion tensor can also be represented by an ellipsoid, whose three axis correspond to the 3 eigenvectors of the tensor.

4.3 *Main parameters*

With the use of an appropriate set of gradients it is possible to determine not only the magnitude of the local ADC's, but also their directions.

The number of gradients can range from 6 to more than one hundred, which generates a large amount of quantitative data that can be analyzed in many different ways.

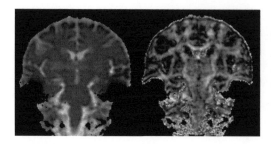

Figure 6. Examples of sagittal mean diffusivity and FA maps.

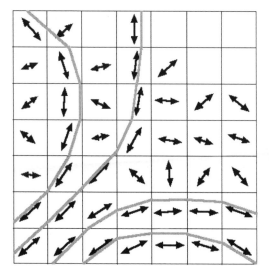

Figure 7. Diagram of the algorithm used for tracking, showing the direction of greatest diffusion.

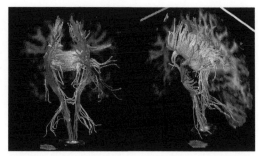

Figure 8. Images of fiber tracts showing the corpus callosum (Courtesy of Derek Jones).

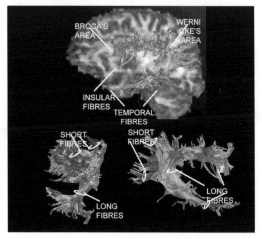

Figure 9. Images of fiber tracts showing the connections between Broca's area and Wernicke's area (Courtesy of Derek Jones).

Although there are many more, the main parameters are: the mean diffusivity and fractional anisotropy (FA).

The mean diffusivity is the ADC averaged in all directions. It is an important parameter because an increase in mean diffusivity could result from an increase in extracellular space due to a defect of neurogenesis or cell loss.

Fractional Anisotropy (FA) indicates how directional the diffusion is and can be visualized by the ratio between the longer and the shorter axis of the ellipsoid. A reduced FA can be attributed to increased or abnormally located grey matter or pathological white matter with abnormal myelination or ectopic neurons and with tissue destruction.

4.4 *3D representation*

Using the calculated diffusion tensor parameters we obtain not only information about the anisotropy in diffusion but also about the actual direction of the ellipsoid, which gives us a three-dimensional representation of the main direction of diffusion.

This is extremely important if we consider that in white matter in the brain the cell shapes are tubular and limit diffusion mainly to one direction, implying a very high directional anisotropy.

If we could follow the main diffusion directions from voxel to voxel we would be able to build the tracts of the white matter fibers within the brain.

This is what is called white matter fiber tracking or diffusion tensor tractography.

4.5 *Fiber tracking*

In order to track and estimate the fiber path several deterministic methods have been developed and applied, like steering or propagation methods (streamlines, tensor deflection or tensorlines) or tract propagation methods (FACT, Euler or Runge Kutta).

All these algorithms compare the direction of maximum diffusion within one voxel with the directions of diffusion in neighboring voxels and look for the most probable neighboring voxel to connect with it, thus building up a continuous tract along the volume acquired, Figure 7.

There are many different ways of presenting the calculated fiber tracts, from two-dimensional colour maps to three-dimensional rotating structures, representing lines, tubes or strips. In Figures 8 and 9 two of these techniques are shown.

One of the main difficulties encountered in fiber tracking is the solution of the problem of the crossing of fibers, which occurs in several regions of the brain and can lead to huge errors.

In white matter tractography errors are cumulative, therefore even small errors can produce catastrophic results. There are several methods to estimate tract error, like Probabilistic tractography and Bootstrap tractography, which involve complex computational calculations and produce more reliable tract data.

5 CONCLUSIONS

MR Diffusion Tensor Imaging is a very promising and powerful technique to estimate white matter fiber trajectories *in vivo*. It is a technique that is highly dependent on complex algorithms and heavy computing power. It has many applications in medicine, both in the fundamental and clinical aspects, like establishing the fiber connections between different eloquent parts of the brain, determining whether a tumour has displaced the fibers within a certain area of the brain or is encapsulating them, or determining if particular fiber connections have been destroyed.

There are still many aspects to be developed and the future looks promising for diffusion imaging, and we can be sure that all these aspects will rely heavily on the power of computational imaging.

REFERENCES

Basser, P.J. & Jones, D.K. 2002. Diffusion-tensor MRI: theory, experimental design and data analysis – a technical review. *NMR Biomed* 15: 456–467.

Le Bihan, D. & Breton, E. 1985. Imagerie de diffusion *in vivo* par resonance magnétique nucléaire. *CR Acad Sci Paris* 301: 1109–1112.

LeBihan, D. & Mangin, J.F. & Poupon, C. & Clark, C.A. & Pappata, S. & Molko, N. & Chabriat, H. 2001. Diffusion Tensor Imaging: Concepts and applications. *JMRI* 13: 534–546.

Pierpaoli, C. & Jezzard, P. & Basser, P.J. & Barnett, A. & DiChiro, G. 1996. Diffusion tensor imaging of the human brain. *Radiology*, 201: 637–648.

Watts, R. & Liston, C. & Niogi, S. & Ulug, A.M. 2003. Fiber tracking using magnetic resonance diffusion tensor imaging and its applications to human brain development. *Mental Retard and Develop Disab Res Rev* 9(3): 168–177.

Contributed papers

Computational Modelling of Objects Represented in Images –
João Manuel R.S. Tavares & R.M. Natal Jorge (eds)
© 2007 Taylor & Francis Group, London, ISBN 978-0-415-43349-5

Object contour tracking in videos by using adaptive mixture models and shape priors

Mohand Saïd Allili & Djemel Ziou

University of Sherbrooke, Department of Computer Science, Faculty of Science, Sherbrooke (QC), Canada

ABSTRACT: In this paper, we propose a novel object tracking algorithm in video sequences. The method is based on object mixture matching between successive frames of the sequence by using active contours. Only the segmentation of the objects in the first frame is required for initialization. The evolution of the object contour on a current frame aims to find the maximum fidelity of the mixture likelihood for the same object between successive frames while having the best fit of the mixture parameters to the homogenous parts of the objects. To permit for a precise and robust tracking, region, boundary and shape information are coupled in the model. The method permits for tracking multi-class objects on cluttered and non-static backgrounds. We validate our approach on examples of tracking performed on real video sequences.

1 INTRODUCTION

Object detection and tracking has a wide range of applications in computer vision and image processing, like video surveillance, video object retrieval, robotics, etc. In the past, using foreground (object) distribution matching has been substantially investigated for tracking. In this approach, the object is tracked on each frame of the sequence based on its color appearance. In (5), the authors proposed to use the mean-shift analysis to track the distribution of a tracked object. However, the object shape is constrained to be an ellipse, which restricts the applicability of the approach when dealing with deformable objects. In (13), the tracking is performed by minimizing a distance between the histograms of the reference and target objects. However, the objects may undergo variations due to noise, illuminations changes and self-shadowing that may decrease the efficiency of histogram matching. Besides, histogram representation is sparse and less effective for high-dimensional data. Recently, the authors in (11) proposed to track objects by competing their statistics and those of the background. However, when there is a similarity between the statistics of the object and the background or illumination changes, the object contour may be diverted and fail to capture the real boundaries. To alleviate the problem of illumination changes, an approach based on adaptive mixture models was proposed in (9). Indeed, using mixture models provides a flexibility and power to represent

accurately multi-dimensional data (8). Furthermore, they are less sensible to noise than mere histogram representation. However, the object is constrained to be a box which does not allow for tracking the boundaries of deformable objects.

In the present paper, we propose an object tracking approach based on matching finite mixture models between successive frames of the sequence and active contours. Having the reference mixture model as a prior, the aim is to deform a contour on a current frame for each object in a fashion that fits a mixture model to its homogenous parts and constrains the mixture to be similar to the one of the reference object. This permits to track deformable objects with slight variations of color appearance. To have a robust tracking to cluttered backgrounds, region, boundary and shape information are combined to steer the object contours evolution. The method is implemented by using level set contours that allow for automatic topology changes and stable numerical schemes (10).

This paper is organized as follows: In section (2), we present the details of the proposed model for tracking. In section (3), we show a validation of the proposed approach on examples of object tracking on real video sequences.

2 FORMULATION OF THE TRACKING MODEL

2.1 *Object mixture modelling*

Let: $\Omega \subset \Re^2$ be the domain of the image and R_{obj} the region of the object to be tracked through an image sequence, which is constituted of the frames $I^{(n)}$ where

This work was supported by the Natural Sciences and Engineering Research Council (NSERC Canada).

$(0 \leq n < \infty)$. Let $M(\Theta^{(n)})$ and $M(\Theta^{(n+1)})$ be the two mixture models that characterize the object in two consequent frames $I^{(n)}$ and $I^{(n+1)}$, where $\Theta^{(n)}$ and $\Theta^{(n+1)}$ designate the mixture parameters respectively. In the first frame, we suppose the mixture model is calculated by designating manually the area of the object and then estimate the parameters by using the Maximum Likelihood method. The data $\mathcal{D}$ of the object could be real valued, like image intensity, or vector-valued, like color and texture features. In our case, we use a feature vector $U(\mathbf{x}) = (u_1(\mathbf{x}), \ldots, u_d(\mathbf{x}))$ that combines color and texture characteristics where $\mathbf{x}$ represents the pixel coordinates (x, y). The probability of the feature vector $U(\mathbf{x})$ in the object is given by:

$$p(U(\mathbf{x})/\Theta) = \sum_{k=1}^{K} \vartheta_k \cdot p(U(\mathbf{x})/\theta_k) \tag{1}$$

where K is the number of components in the mixture. Θ represents the set of mixture parameters which include the mixing parameters $\vartheta = (\vartheta_1, \ldots, \vartheta_K)$ and the parameters of the mixture densities $(\theta_1, \ldots, \theta_K)$. The likelihood function is given as follows:

$$\mathcal{L}(\mathcal{D}, \Theta) = \prod_{\mathbf{X} \in R_{obj}} p(U(\mathbf{x})/\Theta) \tag{2}$$

The mixture parameters are obtained by minimizing the minus log-likelihood function $E(\Theta) = -\log(\mathcal{L}(\mathcal{D}, \Theta))$ according to the parameters Θ:

$$\hat{\Theta}$$
$$= \text{argmin}_{\Theta} \left\{ -\sum_{\mathbf{X} \in R_{obj}} \log \left(\sum_{k=1}^{K} \vartheta_k \cdot p(U(\mathbf{x})/\theta_k) \right) \right\} \tag{3}$$

Let us denote the region of the object in the consecutive frames $I^{(n)}$ and $I^{(n+1)}$ respectively by $R_{obj}^{(n)}$ and $R_{obj}^{(n+1)}$. We formulate our tracking model by minimizing the distance between the Log-likelihoods $E(\Theta^{(n)})$ and $E(\Theta^{(n+1)})$ of the object in the frames $I^{(n)}$ and $I^{(n+1)}$. Having $\vec{\gamma}$ as a planar curve that represents the object contour in the frame $I^{(n+1)}$, the tracking model seeks for the optimal curve $\hat{\vec{\gamma}}$ and mixture parameters $\hat{\Theta}^{(n+1)}$ which maximize on the one hand the likelihood of the object data to the parameters and minimize, on the other hand, the distance between the object mixture likelihoods respectively in the frames $I^{(n)}$ and $I^{(n+1)}$. To match the object mixture likelihoods, we propose to minimize the following functional:

$$J(\vec{\gamma}, \Theta^{(n+1)}) = \left\{ E(\vec{\gamma}, \Theta^{(n+1)}) - E(\Theta^{(n)}) \right\} \tag{4}$$

Remark that without the contour $\vec{\gamma}$, the functional J is the EM algorithm that resolves iteratively the

maximum likelihood estimation of the mixture parameters, where the convergence of the algorithm is sought by minimizing the distance between $E(\Theta^{(n+1)})$ and $E(\Theta^{(n)})$ (3). Note that the functional (4) can rewritten as follows:

$$J(\vec{\gamma}, \Theta^{(n+1)})$$

$$= \iint_{R_{obj}^{(n+1)}} -\log \left(\frac{p(U(\mathbf{x})/\Theta^{(n+1)})}{p(U(\mathbf{x})/\Theta^{(n)})} \right) d\mathbf{x} \tag{5}$$

$$= \iint_{R_{obj}^{(n+1)}} -\log \left(\sum_{k=1}^{K} \frac{\vartheta_k^{(n+1)} p(U(\mathbf{x})/\theta_k^{(n+1)})}{p(U(\mathbf{x})/\Theta^{(n)})} \right.$$

$$\left. \frac{p(\theta_k^{(n)}/U(\mathbf{x}))}{p(\theta_k^{(n)}/U(\mathbf{x}))} \right) d\mathbf{x} \tag{6}$$

In the functional (6), we multiplied both the numerator and the dominator of the fraction of likelihoods by the posterior the term $p(\theta_k^{(n)}/U(\mathbf{x}))$. By using the property of Jensen's inequality to a concave function (3), we obtain the following inequality:

$$-\log \left(\sum_{k=1}^{K} \frac{\vartheta_k^{(n+1)} p(U(\mathbf{x})/\theta_k^{(n+1)})}{p(U(\mathbf{x})/\Theta^{(n)})} \frac{p(\theta_k^{(n)}/U(\mathbf{x}))}{p(\theta_k^{(n)}/U(\mathbf{x}))} \right) \leq$$

$$\underbrace{-\sum_{k=1}^{K} p(\theta_k^{(n)}/U(\mathbf{x})) \log \left(\frac{\vartheta_k^{(n+1)} p(U(\mathbf{x})/\theta_k^{(n+1)})}{p(U(\mathbf{x})/\Theta^{(n)}) p(\theta_k^{(n)}/U(\mathbf{x}))} \right)}_{Q(\mathbf{x}, \Theta^{(n+1)})} \tag{7}$$

Then, from equations (4) and (7), it follows that:

$$E(\vec{\gamma}, \Theta^{(n+1)}) \leq E(\Theta^{(n)}) + \underbrace{\iint_{R_{obj}^{(n+1)}} Q(\mathbf{x}, \Theta^{(n+1)}) d\mathbf{x}}_{Q(\vec{\gamma}, \Theta^{(n+1)})} \tag{8}$$

As we seek to match the likelihoods of the object in the frames $I^{(n)}$ and $I^{(n+1)}$, minimizing the functional (4) according to the new mixture parameters: $\Theta^{(n+1)}$ and the contour: $\vec{\gamma}$ will lead to decrease the value of the energy $E^{(n+1)}(\vec{\gamma}, \Theta^{(n+1)})$ that is lower-bounded by $E^{(n)}$ and upper-bounded by $E^{(n)} + Q(\vec{\gamma}, \Theta^{(n+1)})$. Thereby, minimizing (4) amounts to minimize the term $Q(\vec{\gamma}, \Theta^{(n+1)})$. Note that this term can be rewritten as a sum of two terms: Q_1 that depends on the contour and the mixture parameters $(\vec{\gamma}, \Theta^{(n+1)})$ and Q_2 that depends only on the object contour $\vec{\gamma}$. Having this, and adding the constraint that the mixing parameters $\vartheta_k^{(n+1)}$ should satisfy the condition: $\sum_{k=1}^{K} \vartheta_k^{(n+1)} = 1$, we obtain the following new energy minimization:

$$(\hat{\vec{\gamma}}, \hat{\Theta}^{(n+1)}) = \text{argmin}_{\vec{\gamma}, \Theta^{(n+1)}} \left\{ Q(\vec{\gamma}, \Theta^{(n+1)}) \right\} \quad (9)$$

with:

$$Q(\vec{\gamma}, \Theta^{(n+1)}) =$$

$$\underbrace{\iint_{R_{obj}^{(n+1)}} -\sum_{k=1}^{K} p(\theta_k^{(n)}/U(\mathbf{x})) \log \left(\vartheta_k^{(n+1)} p(U(\mathbf{x})/\theta_k^{(n+1)}) \right) d\mathbf{x}}_{Q_1(\vec{\gamma}, \Theta^{(n+1)})}$$

$$+ \underbrace{\iint_{R_{obj}^{(n+1)}} \sum_{k=1}^{K} p(\theta_k^{(n)}/U(\mathbf{x})) \log \left(\vartheta_k^{(n)} p(U(\mathbf{x})/\theta^{(n)}) \right) d\mathbf{x}}_{Q_2(\vec{\gamma})}$$

$$+ \lambda \underbrace{\left(\sum_{k=1}^{K} \vartheta_k^{(n+1)} - 1 \right)}_{Q_3(\lambda, \vartheta^{(n+1)})}$$

where λ is a Lagrange multiplier to add a constraint to the model for obtaining mixing parameters ϑ_k that sum to 1. The interpretation of the different terms in the functional (9) is given as follows: (1) The minimization of the first term Q_1 according to the new mixture parameters leads to the best fit of the data inside the contour $\vec{\gamma}$ to the homogenous parts of the object. Having the mixture potability multiplied by the posterior probability of the old mixture parameters is a regularization that constrains the new parameters of the mixture to not deviate from the old ones. (2) The term Q_2 does not contribute when minimizing according to the new parameters since it does not depend on the latter. The term Q_3 depends only on the mixing parameters and constrains the summation of the latter to be 1. On the other hand, minimizing the energy according to the contour $\vec{\gamma}$ moves the contour to capture the data that fit to the new parameters of the mixture (by using the term Q_1) and assures the proximity between the data likelihood of the object in the first and second frames (by using the term Q_2).

2.2 Adding boundary and shape priors

As the tracking is achieved by the means of competing the statistics of the object and the background, conflicts may arise when the object is in contact with parts of the background that have the same statistical properties as the object. In this case, the contour may deviate from the true boundary of the object. To cope with this problem, we use boundary and shape information to reinforce the model.

To keep the contour smooth and ensure a good alignment with the object boundaries, we add a regularization term to the model that contains the boundary information. Indeed, there may be a strong correlation between the boundary of the tracked object and

the image edges. To detect the edges in the multivalued image $U(\mathbf{x})$, we use the method proposed in (7). This method estimates the boundary plausibility of the object, that we denote by $\|\nabla U(\vec{\gamma}(s))\|$, by calculating the strongest first order directional derivative of the image. The boundary energy is added to the model in a similar fashion to the approaches in (4).

We use a prior shape information for the object to constrain the object contour to keep its shape in the presence of conflicts with the background. For the moment, we assume the energy resulting from the prior shape information is denoted by: $Q_s(\vec{\gamma})$. In section (2.4), we discuss how to formalize the shape energy. Finally, we obtain the following energy functional:

$$J^*(\vec{\gamma}, \Theta^{(n+1)})$$

$$= \alpha(Q_1(\vec{\gamma}, \Theta^{(n+1)}) + Q_2(\vec{\gamma}) + Q_3(\lambda, \vartheta))$$

$$+ \beta \oint_0^{L(\vec{\gamma})} \xi \left(\|\nabla U(\vec{\gamma}(s))\| \right) ds + \delta Q_s(\vec{\gamma}) \quad (10)$$

with s denoting the arc-length parameter, $L(\vec{\gamma})$ is the length of the curve and ξ is a strictly decreasing function of the boundary plausibility $\|\nabla U(\vec{\gamma}(s))\|$. The constants α, β and δ determine the contribution of the region, boundary and shape information in the above functional.

2.3 Region tracking by curve and statistics evolution

Here, we assume the data in the object are represented by a mixture of multivariate Gaussian distributions. Each density distribution in the mixture is given by:

$$p(U(\mathbf{x})/\theta_k) =$$

$$\frac{1}{2\pi|\Sigma_k|^{1/2}} \exp \left\{ -\frac{1}{2}(U(\mathbf{x}) - \mu_k)^T \Sigma_k^{-1}(U(\mathbf{x}) - \mu_k) \right\} \quad (11)$$

We estimate the number of components of the mixture model by using the AIC criterion (2). After introducing the Gaussian formulation in the functional (9), the term $Q_1(\vec{\gamma}, \Theta^{(n+1)})$ becomes as follows:

$$Q_1(\vec{\gamma}, \Theta^{(n+1)}) = \quad (12)$$

$$-\iint_{R_{obj}^{(n+1)}} \sum_{k=1}^{K} p(\theta_k^{(n)}/U(\mathbf{x})) \left\{ \log \left(\frac{\vartheta_k^{(n+1)}}{|\Sigma_k^{(n+1)}|} \right) \right.$$

$$\left. -\frac{1}{2}(U(\mathbf{x}) - \mu_k^{(n+1)})^T \Sigma_k^{(n+1)^{-1}} (U(\mathbf{x}) - \mu_k^{(n+1)}) \right\} d\mathbf{x}$$

$$(13)$$

Here, we have two alternatives for solving the minimization of the functional $Q(\vec{\gamma}, \Theta^{(n+1)}) = Q_1(\vec{\gamma}, \Theta^{(n+1)}) + Q_2(\vec{\gamma}) + Q_3(\lambda, \vartheta)$. In the first alternative, we minimize the functional (9) alternatively between the contour $\vec{\gamma}$ and the mixture parameters $\Theta^{(n+1)}$. In the second alternative, we use shape derivatives by writing the parameters of the mixture as functions of the domain of the region of the object. For the sake of simplicity, we use the first approach, where we minimize alternatingly J^* on the object contour $\vec{\gamma}$ and the mixture parameters $\Theta^{(n+1)}$. By using Euler-Lagrange equations, the minimization according to $\vec{\gamma}$ leads to the following motion equation of the contour:

$$\frac{\partial \vec{\gamma}(s,t)}{\partial t} = -\Big(\alpha [Q_1(\vec{\gamma}(s,t)) + Q_2(\vec{\gamma}(s,t))]$$

$$+ \ \beta Q_b(\vec{\gamma}(s,t)) + \delta Q_s(\vec{\gamma}(s,t)) \Big) \cdot \vec{\mathcal{N}} \quad (14)$$

with:

$Q_1(\vec{\gamma}(s,t))$

$$= \ \sum_{k=1}^{K} p(\theta_k^{(n)}/U(\vec{\gamma}(s,t))) \left[\log\left(\frac{\vartheta_k^{(n+1)}}{|\Sigma_k^{(n+1)}|} \right) - \frac{1}{2} \right.$$

$$\left. (U(\vec{\gamma}(s,t)) - \mu_k^{(n+1)})^T \Sigma_k^{(n+1)^{-1}} (U(\vec{\gamma}(s,t)) - \mu_k^{(n+1)}) \right]$$

$Q_2(\vec{\gamma}(s,t))$

$$= \ \sum_{k=1}^{K} p(\theta_k^{(n)}/U(\vec{\gamma}(s,t))) \left[\log\left(\frac{\vartheta_k^{(n)}}{|\Sigma_k^{(n)}|} \right) - \frac{1}{2} \right.$$

$$\left. (U(\vec{\gamma}(s,t)) - \mu_k^{(n)})^T \Sigma_k^{(n)^{-1}} (U(\vec{\gamma}(s,t)) - \mu_k^{(n)}) \right]$$

$Q_b(\vec{\gamma}(s,t))$

$$= \ \xi(\vec{\gamma}(s,t))\kappa + \nabla \xi(\vec{\gamma}(s,t)) \cdot \vec{\mathcal{N}}$$

where κ is the curvature of the contour $\vec{\gamma}$ and $\vec{\mathcal{N}}$ is its inward normal vector. To permit for the curve to change automatically its topology during its evolution, we use the level formalism proposed in (10). Having $\Phi: \Re^2 \to \Re$ as a level set distance function and $\vec{\gamma}$ is represented implicitly by its zero level set, we obtain the following motion equation for the level set function Φ:

$$\frac{\partial \Phi}{\partial t} = -\Big(\alpha [Q_1(\Phi) + Q_2(\Phi)]$$

$$+ \ \beta Q_b(\Phi) + \delta Q_s(\Phi) \Big) \cdot \|\nabla \Phi\| \quad (15)$$

The minimization of the functional (10) according to the mixture parameters leads to the following updating equations for the parameters $(\vartheta_k, \mu_k, \Sigma_k), k \in \{1, \ldots, K\}$:

$$\mu_k^{(n+1)} = \frac{\iint_{R_{obj}^{(n+1)}} t_k(\mathbf{x}) U(\mathbf{x}) d\mathbf{x}}{\iint_{R_{obj}} t_k(\mathbf{x}) d\mathbf{x}} \quad (16)$$

$$\Sigma_k^{(n+1)} =$$

$$\frac{\iint_{R_{obj}^{(n+1)}} t_k(\mathbf{x}) \left[U(\mathbf{x}) - \mu_k^{(n+1)})(U(\mathbf{x}) - \mu_k^{(n+1)})^T \right] d\mathbf{x}}{\iint_{R_{obj}^{(n+1)}} p(\theta_k^{(n)}/U(\mathbf{x})) d\mathbf{x}} \quad (17)$$

$$\vartheta_k^{(n+1)} = \frac{\iint_{R_{obj}^{(n+1)}} t_k(\mathbf{x}) d\mathbf{x}}{\iint_{R_{obj}^{(n+1)}} d\mathbf{x}} \quad (18)$$

with $t_k(\mathbf{x}) = p(\theta_k^{(n)}/U(\mathbf{x}))$.

2.4 Shape information

To calculate the distance between two level set functions which represent the shape of the objects, we use an improved version of the approach proposed in (6). To minimize the distance between a reference level set function Φ_r and a current level set function Φ, the authors minimized the following energy functional:

$$Q_s = \iint_{R_{obj}} (\Phi_r - \Phi)^2 \frac{h(\Phi_r) + h(\Phi)}{2} d\mathbf{x} \quad (19)$$

with $h(\Phi) = \frac{H(\Phi)}{\iint_{R_{obj}} H(\Phi) d\mathbf{x}}$ and H is the heaviside function. Several good properties of the above distance can be found in (6).

In our approach, the reference level set Φ_r is the one that was produced from the latest tracking result. However, since an object contour may translate and rotate between the frames $I^{(n)}$ and $I^{(n+1)}$, we should take into account this by using euclidian transformations. To compare the two functions Φ_r and Φ, we should firstly align them in $I^{(n+1)}$, as shown on fig. (1). The new reference shape Φ_r' is obtained translating and rotating Φ_r. The translation vector T is approximated by the difference between the centers of gravity of Φ_r and Φ that are c and c'. For the rotation matrix, we use the approach proposed in (12). The expected scale change of the object is determined by having the previous frames results. Let $S_k^{(n)}$ represent the size of the object k in the frame $I^{(n)}$. The scale change of the object is determined by: $\gamma = S_k^{(n)}/S_k^{(n-1)}$. In our model, the contribution of the shape information aims to resolve the conflicts with the background where the contour may divert away from the true object. Suppose that by evolving the object contour Φ_k on a current frame $I^{(n+1)}$, we have the scale of the object changes by $\gamma' = S_k^{(n+1)}/S_k^{(n)}$. Then the shape term (19) is added

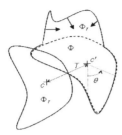

Figure 1. Shape transformation between successive frames of the sequence.

to the energy functional (10) weighted by the factor $\delta = \gamma' / \gamma$.

3 EXPERIMENTS AND DISCUSSION

In the experiments that we conducted, we used a combination of color and texture features for region information. We use CIE-$L^*a^*b^*$ color space which is perceptually uniform. For texture, we used features calculated from the correlogram of the pixel neighborhood (1). An element of the correlogram matrix $C^{\upsilon,\theta}(c_i; c_j)$ should give the probability that, given a pixel $\mathbf{x}_1$ of color c_i, a pixel $\mathbf{x}_2$ at distance υ and orientation θ from $\mathbf{x}_1$ is of color c_j. We calculate the correlogram for 4 orientations $(\upsilon, 0), (\upsilon, \frac{\pi}{4}), (\upsilon, \frac{\pi}{2})$ and $(\upsilon, \frac{3\pi}{4})$. Let Υ be the total number of displacements υ. We derive from each correlogram three typical characteristics that are namely: *Inverse-Difference-Moment* (*IDM*), *Energy* (*E*) and *Correlation* (*C*). *E* and *C* measure respectively the homogeneity of the texture. *IDM*, however, measures the coarseness of the texture. By having $\upsilon = 2$, we end up with 9 dimensions for the feature vector $U(\mathbf{x})$. We set $\alpha = \beta = 0.5$ for the level set curve evolution equation (15).

In fig. (2), we show an example of "car" tracking moving on a textured background. The video sequence contains 339 frames. To compare our method with the state of the art, we implemented the method proposed in (11) which performs the tracking by using active contours and region competition between the object and the background. We show the tracking results for the frames 55, 65, 80, 100, 120 and 140 of the sequence. As the moving object was in contact with parts of the background that have similar region information, the contour was diverted from the true object boundary for the method in (11). Our method succeeded in tracking the true boundary of the car for the same frames of the sequence by matching the mixture models of the target object with the reference object under the boundary and shape constraints. The tracking took about one to two seconds in each frame of the sequence, which is obviously slow for doing real-time

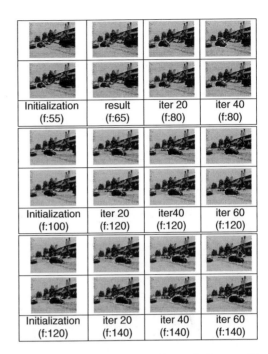

Initialization (f:55)	result (f:65)	iter 20 (f:80)	iter 40 (f:80)
Initialization (f:100)	iter 20 (f:120)	iter40 (f:120)	iter 60 (f:120)
Initialization (f:120)	iter 20 (f:140)	iter 40 (f:140)	iter 60 (f:140)

Figure 2. Comparison of the tracking method proposed in (11) (first row of each array) to our tracking method (second row of each array). We show the tracking results in the frames f:55, f:65, f:80, f:100, f:120 and f:140.

tracking. In the future, we will seek an optimization of the computation time.

In fig. (3), we show another example that compares our method to the method proposed in (11) and the method proposed (13). The last method uses histogram matching to track a single object in a video sequence. On the example a man is walking in a cluttered background, and the whole scene is taken with a mobile camera (the sequence was downloaded here: http://people.csail.mit.edu/torralba/images/). In the figure, we show the result of tracking for two successive frames performed on the above methods and ours. In the first row, we show respectively: the previous position of the object with the contour delineating it, the current frame, the boundary plausibility and the contour initialization on the current frame. The second and third rows show respectively the tracking results by using (11) and (13), while the last row shows the results of our method. The results of (11) can be explained by the fact that since the object and background share some region characteristics, the object contour shrunk to concede some pixels to the background. The method proposed in (13) uses histogram matching to track the object and, since the aim is to have the same color histogram as the reference object, the contour stuck in a local minima by replacing a part of the object body by the background (see row 3).

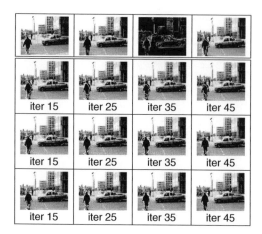

Figure 3. Comparison of tracking methods. The second, third and fourth rows show the evolution of the tracking by using respectively the method proposed in (11), in (13) and our method.

Figure 4. Object tracking with illumination changes in frames f:1, f:75, f:122 and f:129. The first, second and third rows show respectively the tracking results by using: the method in (13), the method in (11) and our method.

Since no additional boundary information is used to move the contour, it was not able to correct its position afterwards. In the last row of the figure, our method did well by capturing exactly the boundaries of the object thanks to the combination of region, boundary and shape information.

Finally, we show on fig. (4) an example of tracking under an illumination change. The video sequence contains 143 frames and the tracked object is the face. We show respectively the frames number 1, 75, 122 and 129. The method proposed in (13) lost the object in the frame 122 when the change of illumination started. In the method proposed in (11), the contour lost the object in the frame 129 where there is a severe illumination change. Our model lost a part of the object but kept the most part of it even with a severe illumination change, which demonstrates the performance of the approach. We note that since the object in this sequence moves

slowly, the tracking took less than one second for each frame in this sequence.

4 CONCLUSIONS

We proposed in this work a novel tracking method based on matching mixture models of the objects between successive frames of a video sequence. The tracking is performed by combining region, boundary and shape information for accurate object localization. The method showed promising results in comparison to recent tracking methods in that it permits for achieving an accurate tracking under cluttered backgrounds and illumination changes. In future work, we should investigate the use of more region features to better model the object and combine tracking and prediction to enhance the speed and robustness of the method.

REFERENCES

[1] M.S. Allili and D. Ziou. Automatic Color-Texture Image Segmentation by Using Active Contours. *In proc. of IEEE IWICPAS, LNCS 4153*, 495–504, 2006.

[2] H. Akaike. A New Look at the Statistical Model Identification. *IEEE Trans. AC*, 9:716–723, 1974.

[3] C. Bishops. Neural Networks for Pattern Recognition. *Clarendon Press, Oxford*, 1995.

[4] V. Caselles, R. Kimmel and G. Shapiro. Geodesic Active Contours. *IJCV*, 22:61–79, 1997.

[5] D. Comaniciu, V. Ramesh and P. Meer. Kernel-based Object Tracking. *IEEE Trans. PAMI*, 25(5):564–577, 2003.

[6] D. Cremers and S. Soatto. A Pseudo-Distance for Shape Priors in Level Set Segmentation. *IEEE Workshop on VLSM*, 169–176, 2003.

[7] C. Drewniok. Multispectral Edge Detection: Some Experiments on Data From Landsat-TM. *IJ. on Remote Sensing*, 15(18):3743–3765, 1994.

[8] J. McLachlan and G.Peel. Finite Mixture Models. *John Wiley & sons*, 2000.

[9] S.J. McKenna, Y. Raja and S.Gong. Tracking Colour Objects Using Adaptive Mixture Models. *Image and Vision Computing*, 17:225–231, 1999.

[10] S. Osher and J. Sethian. Fronts Propagating With Curvature-Dependant Speed: Algorithms Based on Hammilton-Jacobi Formulations, J. of Comp. Physics, 22:12–49, 1988.

[11] A. Yilmaz, X. Li, and M. Shah. Contour-based object tracking with occlusion handling in video acquired using mobile cameras. *IEEE Trans. PAMI*, 26(11):16, 2004.

[12] T. Zhang and D. Freedman. Tracking objects using density matching and shape priors. *IEEE ICCV*, 1950–1954, 2003.

[13] T. Zhang and D. Freedman. Improving Performance of Distribution Tracking through Background Mismatch. *IEEE Trans. PAMI*, 27(2):282–287, 2005.

Computational Modelling of Objects Represented in Images –
João Manuel R.S. Tavares & R.M. Natal Jorge (eds)
© 2007 Taylor & Francis Group, London, ISBN 978-0-415-43349-5

Image analysis methodology to study the evolution of chromium penetration in chromium tannage

A.L. Amaral, M. Mota & E.C. Ferreira
Chemical and Biological Technology Department, Polytechnic Institute of Bragança, Bragança, Portugal
Centre of Biological Engineering, University of Minho, Braga, Portugal

A. Crispim
Department of Chemical Engineering, Inst. Sup. Eng. do Porto, Porto, Portugal

ABSTRACT: The chrome tanning process is the main method employed to improve the hide stabilization, and chrome tanning in organic solvent medium has been tried by some authors, namely for ovine skins, but the process is not yet clear and fully understood. In recent years the application of image analysis methodologies to bioreactors is widely growing mainly on biomass and materials characterization. In the present work, the use of image analysis was employed to examine chromium penetration during bovine hide tannage in the presence of terpentine. Results showed that the overall chromium penetration rate was considerably faster in the first two minutes, becoming then much slower and linear until completion. Furthermore in the initial stages of penetration the chromium diffused much faster on the flesh that on the grain side of the hide. A second study performed on the flesh side only allowed to determine the chromium flux in the hide.

1 INTRODUCTION

1.1 *Tanning industry background*

The conversion of animal hides into leather is a classical method to dispose of organic waste products from the meat industry and, at the same time, to create commodities for the consumer. The essential step in leather manufacturing is the tanning process. Before the tannage process, hides are subjected to various preliminary cleaning processes. After the removal of different non-collagenous constituents by soaking and unhairing in alkaline medium the connective tissue is mechanically removed by fleshing, and some interstitial proteins are extracted by bating with mixtures of proteolytic enzymes.

Tanning with chromium is a process used since the end of the XIX century (Julien 1981). In this tannage process, the cleaned hide material, practically pure collagen, is brought into an acid medium by pickling and then stabilized with complex basic sulphates of trivalent chromium. The process is done in a water medium and is explained in detail by many authors (Martignone 1997, Adzet 1985, Heidemann 1993, Bienckiewicz 1983, Gratacós *et al.* 1962, Grasser 1934). Some processes have been tested and developed to increase the up-take of chromium by the hide with success, but there is still an important chromium discharge to the wastewater. The chrome tanning in an organic solvent medium is a possibility to reduce the chromium discharge, tried by some authors (Chagne *et al.* 1996), namely with ovine skins. The process is not well understood, mainly for bovine hides, and needs more investigation to have conclusive results. The aim of this work is to study the chromium penetration during the chromium tanning of bovine hides in the presence of terpentine

1.2 *Image processing and analysis*

Image processing and analysis have become nowadays a very important tool with a large field of applications. The image analysis systems strength resides on the ability to remove the subjectiveness of human analysis, the possibility to extract quantitative data that would be very difficult or impossible to obtain by other means and avoid tedious and time-consuming tasks to human researchers (Russ 1995, Gonzalez & Woods 1992). Furthermore, with the exponential increase in computer processing capabilities and affordability as well as better imaging systems, image analysis has became a standard routine in many day-to-day applications and scientific studies (Amaral *et al.* 2005, 2004a, b).

Image processing and analysis of grey scale images seems, therefore, as a quite appropriate methodology to allow the chromium penetration assessment. The image processing method must however take the highest care in the determination of the chromium penetration area within the hide from the grey scale images. Therefore, the segmentation step emerges as a key stage in the correct chromium area assessment. Although many segmentation procedures could be considered it is recommended to use an automated or semi-automated algorithm. The resulting binary image may still require some cleaning such as debris removal, filling, etc, before the determination of the chromium penetration area.

2 MATERIALS AND METHODS

2.1 *Tanning procedure*

The hides were obtained from a lot of salted hides of 20–30 kg from Monteiro Ribas Indústrias (a Portuguese leather company from Oporto), at the lime splitted state with a thickness of 4.5 mm. All the used hide pieces, always from the same zone, near the backbone, were first delimed and pickled by a conventional process: 400 g of pickled hide were weighed, sammed, cut in eight similar pieces, put into a small drum, with temperature control at a speed of 24 rpm. The hide pieces run at 30°C with 200% (w/w) of terpentine and 6% (w/w) of chromium salt and 0.8% (w/w) of sodium formiate.

2.2 *Image acquisition*

Each hide piece was immediately dried at 50°C in an oven, for 2 hours, and the chromium penetration was studied by image analysis: the image acquisition was performed by the visualization of a transversal cut of the hide piece in an Olympus SZ4045TR-CTV stereomicroscope (Olympus, Tokyo) at 40× magnification linked to a Sony AVCD5CE camera (Sony, Tokyo) and a DT3155 frame grabber (Data Translation, Marlboro). The images were digitalized with a size of 768 × 576 pixels and 256 grey levels by the software Global Lab Image 3.21 (Data Translation, Marlboro). An example of a sequence of images obtained from the hide pieces taken during the chromium penetration trials is presented in Figure 2.

2.3 *Image analysis software*

An image processing and analysis software was developed in Matlab 5.3 (The Mathworks, Natick) in order to correlate the grey level intensities with chromium concentrations in the hide piece and assess the chromium concentration with time through the analysis of the chromium concentration gradient both in the grain and in the flesh side.

The software is divided in six stages: Image acquisition; image pre-treatment; image cropping; identification of chromium penetration areas; determination of the intensity values and gradient; binary image and data recording.

Figure 1. Small drums.

Table 1. Penetration assays.

Sample	1	2	3	4	5	6	7	8
Time (min)	2	7	15	25	40	60	90	120

In the first stage the software allows the acquisition of the hide pieces in 8 bit format (256 colors) in a number of supported file types. Furthermore, the acquired image range is then set a minimum of zero and a maximum of one.

The second stage is the image pre-treatment where is first applied to the image a gray-scale opening with a 20 × 20 pixels mask. This procedure allows for the elimination of small peaks within both the chromium penetrated and the non-penetrated hide zones. Furthermore, an average filter of 20 × 20 pixels is applied to reduce the noise present in the image, mainly on the chromium gradient.

The next stage is the image cropping to establish the grain side and the flesh side. This is performed by simply applying a vertical image cut-off between the two zones and establishing a grain side sub-image and a flesh side sub-image. Care must be taken, therefore, on the image acquisition so that the hide is acquired in the upright position and centered with regard to the horizontal span of the image.

Next the chromium penetration areas must be identified to further determine their intensity values and gradient for both the grain and flesh side. In order to do so the image is analyzed row by row in the up-down direction. First the limits of the hide are determined based on their intensity differences from the homogeneous darker background. Then the limit of the chromium penetration is established given the pixels neighborhood differences between the changing intensity penetrated area and the constant intensity non-penetrated area.

Subsequently the intensity values and gradient of the chromium penetrated zones is determined for both the grain and flesh side. Upon the determination of the chromium penetration limits in the preceding stage, all the pixels between those two limits for each row are used with a twofold purpose: determine the average intensity of the penetrated zone and the chromium gradient. For the determination of the chromium gradient the intensity pixel values and position are fed to a linear regression algorithm and the slope is calculated. Once this procedure has been performed for all the rows of the image an overall intensity gradient proportional to the chromium gradient can be determined.

Finally a binary image of the chromium penetrated zones for both the grain and flesh side is created and saved. The data of the average intensity values as well as the intensity gradient are also saved in text format.

3 RESULTS AND DISCUSSION

3.1 Grain and flesh side penetration assay

A study was performed on the hide chromium penetration from two opposite sides: Grain side and flesh side. An example of the images obtained in this chromium penetration study is presented in Figure 2.

The obtained results in terms of normalized grayscale intensities are presented in Table 2 and their evolution with time in Figure 3.

The obtained results allowed to determine that the overall chromium penetration rate was considerably faster in the first 2 minutes and then much slower and linear until the complete penetration (normalized penetration of 1). The amount of time needed to the complete dissemination of the chromium in the hide was found to be 60 minutes in this assy. Furthermore, it could also be shown that, in the initial stage of penetration the chromium diffused much faster on the flesh side that on the grain side of the hide.

3.2 Flesh side penetration assay

Another study was performed on the chromium penetration by the flesh side and an example of the images obtained is presented in Figure 4.

This study was based on two fundamental assumptions:

The first assumption is that the average grayscale intensity found in the hide zone penetrated by the chromium, (C_β) is inversely proportional to the average of the chromium concentration in the hide, corrected by the hide average grey intensity.

The second assumption is that the relation between the gradient of the grayscale intensities in the hide near the periphery and in the penetration front, is proportional to the gradient of chromium concentration, but with opposite signs.

Table 3 presents the dependence of the penetrated hide grayscale intensity with time and the respective penetrated thickness (G_β) determined directly from the images.

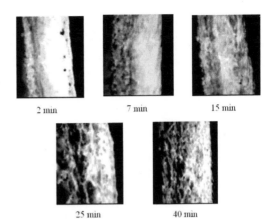

2 min   7 min   15 min

25 min   40 min

Figure 2. Images of the chromium penetration study from both the grain and flesh sides.

Table 2. Normalized grayscale intensities with time.

Time (min)	Normalized grayscale intensity		
	Grain side	Flesh side	Total
2	0.02	0.25	0.27
7	0.02	0.30	0.32
15	0.21	0.20	0.41
25	0.42	0.12	0.54
40	0.33	0.37	0.70
60	0.50	0.50	1.00
90	0.50	0.50	1.00
120	0.50	0.50	1.00

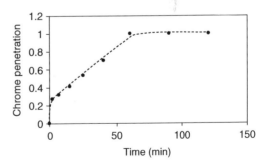

Figure 3. Evolution of the chromium penetration with time.

Figure 5 presents the correlation between the grayscale intensity and the assay time. From the analysis of this figure it was clear that the grayscale intensity (C_β) of the non-penetrated hide was, for this assay 0.7039, being hence the correcting intensity factor for the establishment of the intensity versus chromium concentration correlation.

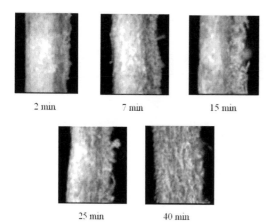

2 min   7 min   15 min

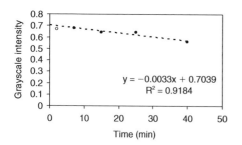

25 min   40 min

Figure 4.   Images of the chromium penetration study from the flesh side.

Table 3.   Normalized grayscale intensities and penetrated thickness with time.

Time (min)	C_β	G_β
2	0.671	0.00233
7	0.680	0.00263
15	0.643	0.00293
25	0.641	0.00238
40	0.563	0.00190

$$y = -0.0033x + 0.7039$$
$$R^2 = 0.9184$$

Figure 5.   Grayscale intensity as a function of time.

Taking into account the second assumption, the chromium flux (chromium quantity penetrated by time) is proportional to the slope of the grayscale intensity correlation with time, and therefore proportional to $0.0033\,min^{-1}$.

4   CONCLUSIONS

The study performed on the hide chromium penetration from the grain and flesh sides allowed to establish that the overall chromium penetration rate was considerably faster in the first 2 minutes and then much slower and linear until complete penetration. It was also shown that, in the initial stage of penetration the chromium diffused much faster on the flesh side that on the grain side of the hide.

A second study was performed on the chromium penetration by the flesh side in order to determine chromium quantity penetrated by time (chromium flux) into the hide. The results have showed that the chromium flux, proportional to the slope of the grayscale intensity correlation with time, was found to be proportional to $0.0033\,min^{-1}$.

ACKNOWLEDGEMENTS

The authors would like to acknowledge Monteiro Ribas Indústrias, a Portuguese tannery, for its collaboration in the drum trials, for the raw materials it provided and for encouraging research in the leather industry.

REFERENCES

Adzet, J. M. 1985. *Quimica – Tecnica de teneria*. Barcelona.
Amaral, A.L., Pereira, M.A., da Motta, M., Pons, M.N., Mota, M., Ferreira, E.C., Alves, M.M. 2004a. Development of Image Analysis Techniques as a Tool to Detect and Quantify Morphological Changes in Anaerobic Sludge: II. Application to a Granule Deterioration Process Trigered by the Contact with Oleic Acid. *Biotechnology and Bioengineering* 87: 2: 194–199.
Amaral, A.L., da Motta, M., Pons, M.N., Vivier, H., Mota, M., Ferreira, E.C. 2004b. Survey of Protozoa and Metazoa Populations in Wastewater Treatment Plants by Image Analysis and Discriminant Analysis. *Environmetrics* 15: 4: 381–390.
Amaral, A.L. & Ferreira, E.C. 2005. Activated Sludge Monitoring of a Wastewater Treatment Plant using Image Analysis and Partial Least Squares. *Analytica Chimica Acta* 544: 246–253.
Bienkiewicz, K. 1983. *Physical chemistry of leather making*. Florida: Robert E. Krieger Publishing company, inc.
Chagne, V., Silvestre, F., Gaset, A. 1986. Review of chrome tanning: Part 2. *Leather* April: 77–84.
Gonzalez, R.C., Woods, R.E., 1992. *Digital image processing*. Reading, Addison-Wesley Publishing Company.
Gratacós, E., Boleda, J., Portvella, M., Adzet, J. M., Lluch, G. 1962. *Tecnología quimica del cuero*. Barcelona.
Grasser, G. 1934. *Petit traité pratique et théorique du tannage au chrome*. Paris.
Heidemann, E. 1993. *Fundaments of leather manufacture*. Dramstadt: Eduard Roether KG.
Jullien, I. 1981. *Le tannage au chrome*. Lyon: Centre technique du cuir.
Martignone, G. 1997. *Manuale di pratica conciaria*. Milan: EDITMA.
Russ, C.R. 1995. *The image processing handbook*. Boca raton: CRC Press.

Computational Modelling of Objects Represented in Images –
João Manuel R.S. Tavares & R.M. Natal Jorge (eds)
© *2007 Taylor & Francis Group, London, ISBN 978-0-415-43349-5*

Surveillance and tracking in feature point region with predictive filter of variable state

D.A. Aracena-Pizarro
University School of Industry, Informatics and Systems Engineering, Area of Computer and Informatics Engineering, University of Tarapaca Arica (UTA), Chile

C.L. Tozzi
Department of Computer Engineering and Industrial Automation, School of Electrical and Computer Engineering, State University of Campinas (UNICAMP), São Paulo, Brazil

ABSTRACT: Surveillance in today's world is a very common issue in computational vision. This activity is present in literature in two different ways: first, as having both camera and objects in motion (Behrad et al. 2000); second, having detection of moving objects by means of one static camera (Lipton et al. 1998). This paper is centered in the last approach, where the interest is to find the movement of objects in images by detecting temporal differences and to define the movement region, which is analyzed by growing region, selecting one region and tracking the object.

Once the region is selected, the interest points are determined through a modified corner detector of Harris et al. (1988). A reference data bank is created, to be used in the matching process and determining the characteristic of corresponding points. With these corresponding points, the movement parameters of the region can be estimated and the prediction filter (VSDF) in the tracking cycle initialized. The method that is developed here consists in considering the tracking cycle a matching process by normalized correlation with the help of the prediction filter to adjust the estimated measurements.

Thus a method that allows tracking of points of interest in a surveillance region, in a stream of images with significative results to implement appropriate real time algorithms. In this stage of our research Matlab and regular digital cameras were used for prototype design of tools and experimenting.

Keywords: Surveillance, computer vision, motion detection, matching, tracking, variable status dimension filters.

1 INTRODUCTION

The tracking and surveillance is one of the main challenges of interest the field of computational vision. There are many applications in this field which cover a wide expectrum, for instance: the security service or monitoring, cars tracking or any kind of vehicle, target tracking or aims, space applications among others. The motion detection and tracking in this area (e.g. rigid mobiles in general) is a complex problem which demands sophisticated solutions, where normally the initial detection method and the motion estimated method are present.

Historically, the classification of regions was taken from unique images or fixed ones. Nowadays video images are analyzed to detect regions in motion (Haritaoglu et al. 1998; Wren et al. 1997). The philosophy behind the segmentations techniques based on motion is to segment an image in motion objects' regions vs. motionless objects through temporal difference techniques. From the motion objects' regions were considered those with the highest level of information with the purpose of doing the object tracking, that is to say, in the detection phase, an appropriated number of points of interest; for the matching process, a very large amount of corresponded points and, for the motion estimation, a sufficient amount of corresponding points to determine in an appropriate way the fundamental matrix and the motion's parameters.

A tracking system based on points of interest from a motion objects in a region, it is mainly composed by two big modules: detection and tracking cycle. The detection module consists of feature detection phases in an automatic way, in this case to detect the interested point in the motion object, there are several proposals to solve them in specific literature, the ones which were analyzed and studied such as (Schmid et al. 2000, Berger et al. 1999, Forstner 1994, Harris et al. 1988 and Trucco et al. 1998), chosing the modified Harris' detector. Inside the initial detection module, the matching process allows us to establish the correspondence of the two following images making possible the estimation in motion process, which determines the initial data for the tracking cycle. The tracking cycle's module considers a standard matching process to verify and recheck the correspondence of the estimated points and to determine its presence. The estimation of it is done using a predictive filter that allows us to estimate the motion and the position of the points in the following image. Normally, the vision tracking systems assume a known motion, even though what is traditional it is assume a related motion (Kanade et al. 1991) or in the use of a predictive filter, Kalman (KF), Kalman Extended (EKF) or Implicit (IKF), it is assumed like a known motion with velocity or constant acceleration (Bar-Shalon et al. 1993). In this case the use of a predictive filter was adapted to make sure the feature tracking (points of interest) variables of the in motion objects, from easy implementation and from the quick update of the states and matrices of information or covariances, which participate in the recursive filter (McLauchlan et al. 1999).

It describes a tracking system based on points of interests taken from an object or region motion, from an outdoor video image. The tracking system is based in three basic principles: a) the temporal consistency which gives a way of detecting the object and defining regions in motion, while it rejects the motionless background pixels; b) the combination of the detection of the motion with the matching algorithm with centered templates in the points of interest of estimated interest, mechanism which offers more strength to the tracking method and; c) the use of a specialized predictive filter, which allows us to estimate the tracking of the points of interest of selected objects even counting on a minimum number of corresponding points in the region of interest.

1.1 Works relative to tracking and surveillance

There is an appreciable number of proposed methods for the tracking of regions, but just a few of them consider the tracking of points of an object in movement inside an specific region of an image. In general the tracking is treated considering a model, that is to say, with a camera in movement and an static scene or the objects are moving and an static camera is watching the scene or the environment. Shi et al. (1994), Jaynes et al. (1999), Fusiello et al. (1998), Soatto et al. (1994), Ravela et al. (1996) and Wren et al. (1997), show tracking of vehicles in highways, to estimate the movement. Koller et al. (1998), uses the tracking to estimate the speed of movement and the position of a mobile in particular to mark its contours in a virtual way. Lipton et al. (1998) and Collins et al. (2000), performs the tracking

of regions through an algorithm of matching by normalized correlation SSD (Square Sum of Difference) to estimate the movement through the calculation of the optical flow, in that way we can determine the tracking of people and vehicles according to the areas of the occupying of bodies. Haritaoglu et al. (1998), makes a tracking of regions based on normalized correlation SSD and a predictive Extended Kalman Filter EKF, being the first work of tracking that operates in real time, his proposal allow to estimate the position of the region to determine a window search in a small part of the total of the image, reducing the time of execution of the correlation process. Azarbayejani et al. (1995), performs the tracking of the characteristic points or interest in prepared internal environments (indoor), considering a affine movement and EKF, to get the goal he uses in a more practical way the development of the Kalman Filter defined in Broida et al. (1990). The tracking without considering filters of prediction in the present time are more frequent and they are based on matching algorithm by normalized correlation NCC (additions of the products in proportion with the standard deflections of the windows that participate) or SSD, but they are mainly tracking of regions (Collins et al. 2000).

Its nature VSDF allows the changing of the points in the tracking, allowing in an easy way the elimination and the incorporation of new points that appear in the observation. The conformation of the filter considers an estimation of the initial movement, it is made by applying the algorithm of two views or the eight points. (Hartley 1997; Zhang et al. 1996) To determine the fundamental matrix and the parameters recovery of movements, rotation, and translation, through the factorization process, show a good behaviour and speed in the tracking of the interest points in the region. The calculation of the fundamental matrix and the factorization requires doing the process of calibration of cameras to determine the intrinsic parameters, which are considered without modification in the tracking cycle. The calibration method proposed by Turceryan et al. (1996) was used, adapting the refining process proposed by Weng et al. (1992). This combination of vision techniques allow to have a system of tracking able to follow an object in movement at any moment and with more information of interest leaving the application opened.

2 SYSTEM ARCHITECTURE

The proposed system consists on two main modules, called of Detection and Tracking Cycle. In the first module, the objects in movement are detected by temporal difference of the initial images, characterized by regions in movement, selecting the region that has an appropriate number of characteristic points, that allow to conform a data base that facilitate the way to operate with these points in the following phases-matching, to determine the correspondence, correlation, parameters estimation of movements and the tracking cycle. In the second module, the points of interest of the selected region are tracking in the sequence of images. The search of the points is made in the entrance of each image, detected in the previous image, by normalized correlation NCC or SSD. The windows of search for the correlation are centred in the resultant points of the prediction phase. In the case that the number of correspondent points are significant or the user changes the region to track, the process comes back to the first module.

The objects in movements are detected in the sequence of images using temporal difference. The regions are detected by regions growing. These selected regions can be tracked using a combination of movement and image correlation.

3 DETECTION MODULE

In the initial detection module it is necessary to determine the regions in movement in a known place, surveillance by an static camera watching the environment. The areas are detected and then analysed by the occupation area because a person or a vehicle uses a bigger area, which facilitates to determine the point of operation to eliminate the ghosts movements, considering a minimum area of connected region, we can discard regions that are not of interest to analyse applying techniques of image morphology.

3.1 Temporal difference

There are many variants of the methods of temporal difference between images, a simple and appropriate method is its use by absolute difference in the sequence of images. A value of threshold cut is selected from the studio of the observed environment; this value is used to determine the changes of the grey levels with respect to an image. If I_n is the intensity of the nth image, so the difference function Δ_n is

$$\Delta_n = |I_n - I_{n-1}| \tag{1}$$

and the objects in movements in the difference image are extracted by the use of the thresholding T.

$$R_n(u, v) = \begin{cases} 1 & , \quad \Delta_n(u, v) \geq T \\ 0 & , \quad \Delta_n(u, v) < T \end{cases} \tag{2}$$

After that the resultant image that has the objects in movement are filtrated through an morphologic algorithm of detection by area open, in the case that are more than one detected object, in this application, it is selected the one that has the bigger morphologic area,

(a)   (b)

(c)

                          (d)

Figure 1. a) y b) A vehicle or person rectangular region. c) y d) the pixels in movement.

to create, at the end, a rectangular region that considers the bottom pixels and the region in movement, as the figure 1 shows.

The rectangular region is used as a base to the extraction of the interest points that belong to the pixels in movement. Once detected the movement of a particular object, the present image and the next are used for the process of matching and determination of the correspondent points, which in the cycle of tracking are tracked in the following images. This characteristic defers from the traditional process of the of tracking cycle that supports the process in matching by normalized correlation in each region, related to this, there are proposal in the literature that shows a poor behaviour, when exists occlusion of one part of the object in process (Lipton et al. 1998; Behrad et al. 2000) while the process of tracking by points of interest allows to do the process of tracking even part of the object is hidden or stops its movement.

3.2 Detection, matching and movement estimation modules

In detection module, our concern is centred in the use of the method to determine clue points or points of interest, through the taking advantage of the information of luminosity and levels of grey. These points of interest are extracted through the use of algorithm of Harris et al. (1988), who uses the function of self-correlation of the image that calculates the measure where a point of interest or a point of relevant and characteristic features are present.

In the present time latest works are natural referent to Kanade et al. (1991) proposals. For this work the criterion function is defined in Trucco et al. (1998), which selects the points that have a higher value to the minor of the value of auto correlation, eliminating the next points through an algorithm of *non maximum suppression.*

Pilu (1997), presents a proposal that allow to determine the correspondence of the points between two successive images, for that, he uses as a local measure a value of confidence determine by the normalized correlation NCC. This result is integrated to an algorithm of approximation, where it is established as an objective function to conform the matrix of similarity. The correspondence establishment process is also used in the cycle of tracking.

Movement estimations had a larger attention from the introduction of the concept Essential Matrix proposed in the initial works of Longuet-Higgings. Nowadays this problem is seen mainly by lineal algebra techniques, in a simple computer way (see, Zhang et al. 1996, Hartley 1995–1997).

4 MODULE OF TRACKING

This module involves the processes of matching by correlation, that shows a better performance regarding the ones proposed that are existing in the literature (see, Shapiro et al. 2001 and Trucco et al. 1998), where the estimated points of the next image are centered in a mask of fixed size. Once the points are refined the new position of these ones is defined in the region of the next image, through the predictive filter VSDF. The algorithm is switched to the Detection module, when there is not a proper number of corresponding points.

A new improvement introduced before executing the algorithm of correlation is to achieve a process of updating the masks or templates. This is justified by the variations of the radiometric properties that make the appearance of the features of interest change. If we consider that the appearance of the vicinity of one feature slowly changes in time (not abruptly), then the new mask can be derived from the linear combination of the pixels of the image, like the template of the current vicinity and the previous ones for a particular feature. In this case

$$\tau_t = \lambda \tau_t + (1 - \lambda)\tau_{t-1} \qquad (3)$$

where τ_t is the new mask that will be utilized in the search of the feature in the image t, τ_{t-1} is the mask of the used feature in time $t-1$ and λ is a probabilistic parameter from 0 to 1 utilized as weight of the contribution of the current image and the new mask in the next images. The tests were done with λ from 0.8 to 0.9, showing that variations that go from 10 to 15%.

The module of tracking consists of four steps, described as follow:

1. The current masks are adjusted for each point of interest and the method of normalized correlation.
2. The vectors of disparity are calculated and it is based on these vectors the refinement of the points of interest.
3. If the number of points is appropriate, the new location of the refined corresponding points is

updated. Otherwise it returns to the module of initial detection.
4. Through the predictive filter it estimates the position of the investigated points in the next image.

4.1 Predictive filter

For the tracking scheme with recursive filter, in this application a Variable Status Dimension Filter is used (VSDF) (Bar-Shalom 1993, Welch et al. 2001, Brown 1997, McLauchlan 1995), that consists in determine a group of global statuses as a vector x, and a group of local statuses to be estimated $y_i, i = 1 \ldots n$. The whole group of global and local parameters forms the status vector and they're related by the observations $z_i(j)$ in time j by the equation of measurement

$$z_i(j) = h_i(j; x; y_i) + w_i(j), \qquad (4)$$

where $w_i(j)$ is the vector of Gaussian distribution with mean value zero and covariant $R(j)$. Like any other filter of Kalman algorithm, the rules of updating are derived from

$$P^{-1} \leftarrow P^{-1} + H^T R^{-1} H, \qquad \hat{x} \leftarrow \hat{x} + xd, \qquad (5)$$

where P is the matrix of covariant (or of information) and $xd = PH^T R^{-1}v$. Here H is the Jacobian of the matrix $\partial h / \partial x$. Both H and h are evaluated in the previous value of $\hat{x}$.

For the recursive filter, the next definitions for its implementation are considered: the global statuses that represent the parameters of movement $x = (\theta, \omega, \kappa, T_x, T_y, T_z)$, where the angles θ, ω, κ are the angles of Euler that represent the rotation matrix R, and (T_x, T_y, T_z) represent the translation of the axis x, y, z; the local statuses correspond to the 3D points $P = (X, Y, Z)$; $y = P$, they are the projection of the measure in the image plane; and the equation of measure corresponds to the perspective equation $z = (u_i, v_i)^T$.

To implement the recursive update of the filter VSDF it needs to calculate the Jacobians of h according to x and according to y, which are easily obtained by the differentiation, details in McLauchlan (1995–1999).

5 RESULTS

The algorithms were implemented in a computer PC Pentium III EC/P600 MHz using Matlab and the libraries ("toolbox") of morphology (http://www.mathworks.com). These were tested with sequences of images captured in an external environment of the Saucache Campus, in the Tarapacá University with different vehicles and landscapes. The images were

Figure 2.   Interest points detection and first tracking.

(b)

Figure 3.   Sequence of tracking with mobiles of rigid structure. a) Sequence with a Jeep, it shows in the images 4, 7 and 12. b) Sequence of a Sedan car in the images 4, 12 y 17.

captured with a Javeline NTSC color camera, with lenses and filters that are manually adjusted for distance and luminosity of the day; with a 320 × 240 pixels resolution. Figure 2 shows the results of the process of detection and the points of interest of a Luv Van in the field of observation.

In figure 3 it shows the results of the tracking of a Jeep and a car, with a number of characteristic points, which allowed establishing the correspondence of the selected points between two consecutive images, to extract the parameters of initial movement of the mobiles and initiate the recursive VSDF filter, to be used in the next sequences. The image sequence allows visualizing the precision of tracking, which depends primarily on the initial movement parameters, the quality of the process of matching and the determination of correspondence of the considered points in the images. In a natural way, the vehicle being tracked on the road keeps changing size and the pixels of the points of interest are changing while it is getting close or going away, this problem gets complicated, when fixed masks are considered (normally the initial), that's why the annexed lineal combination method offers capacity to the algorithm to do the tracking in a satisfying way.

These sequences were processed considering the normalized correlation NCC (Hou et al. 2003). The method has an appropriate behavior when the conditions of luminosity don't produce shadows of relevance, allowing to follow, most of the time, the body in movement, but when the shadow is significant, for example, at the end of the afternoon, the shadows make difficult to determine the area or object in movement and it can consider points of tracking that vary significantly in the sequence. Like it is shown in figure 3.

The process of correlation is the neuralgic part of the method, because its result influences in the answers of the next stages.

6   CONCLUSIONS Y COMMENTS

In the scheme of implemented tracking were considered quick methods that present an automatic appropriate operation in each stage of the model, assuming that the sequence is manifested as unknown and random movement. This fact makes a condition of the selection of the algorithms in all stages (detection, matching, initial estimation of movement, tracking cycle). The corners and/or points of interest were considered, because they have relevant characteristics such as distinction, unity, invariability and stability, that in this case is desirable, because the tracking is possible even, when the mobile stops or has hidden parts. It's also possible by this kind of feature, to amplify the spectrum of applications to add or estimate more parameters.

The tracking algorithm, the recursive predictive filter, shows a proper behavior, in the points that by effect of the luminosity they vary, because it allows to be updated in a fast way. The relevant characteristics of this kind of tool is that allows to count on the information of the past events and in a very simple way to handle incorporations and eliminations of the statuses, keeping a stable operation level and the algorithm presents more accurate predictions. This differs from the logic of the Kalman filters, linear or non linear (Brown et al. 1997, Bar-Shalom et al. 1993, Welch 2001).

To construct a repetitive tracking algorithm in the most automatic way, without any manual election (the opposite happens in most of the works written in literature about the subject or the points, correspondence, movement, are known or there are intentional marks put in the environment, etc.).

As future work, it is expected to apply this model of tracking to make applications of Augmented Reality in external environments, as also utilize this method in surveillance applications that add information to the objects and regions that allow helping the observers in security, guide, historical reconstruction, etc. Another aspect that needs work is how to eliminate the effect of the shadows, in a way that allows only to consider the real points of the object and to do its tracking. The solution of this last problem would allow us to do superposition's and annotations with better record quality.

REFERENCES

Azarbayejani, A. and Pentland, A. 1995. Recursive estimation of motion, structure, and focal length, *IEEE Transactions on Pattern Analysis and Machine Intelligence*, 17(6): 562–575.

Bar-Shalom, Y. and Li, X. 1993. *Estimation and Tracking: principles, techniques and software.* Artech House. Boston.

Behrad, A. Shahrokni, A. Motamedi, S.A. and Madani, K. 2000. A Robust Vision-Based Moving Target Detection and Tracking System. In *madani@univ-paris12.fr*

Berger, M-O, Wrobel-Dautcourt, B. Petitjean, S. and Simon, G. 1999. Mixing synthetic and video images of an outdoor urban environment. *Machine Vision and Applications.* Vol. (1999), (11), pp. 145–159, Nov.

Broida, T. Chandrashekhar, S. and Chellappa, R. 1990. Recursive 3D Motion Estimation from a Monocular Image Sequence, *IEEE Transactions on Aerospace and Electronic Systems* Vol. 26, 4, July. pp. 639–656.

Brown, R. G. and Hwang, Y. C. 1997. *Introduction to Random Signals and Applied Kalman Filtering*, 3rd Edition, John Wiley & Sins, Inc.

Collins, Lipton, Kanade, Fujiyoshi, Duggins, Tsin, Tolliver, Enomoto, and Hasegawa, 2000. A System for Video Surveillance and Monitoring: VSAM Final Report, *Technical report CMU-RI-TR-00-12*, Robotics Institute, Carnegie Mellon University, May.

Forstner, W. 1994. A Framework for Low Level Feature Extraction. *Lecture Notes in Computer Science*, Vol. 802, Computer Vision – ECCV'94. pp. 383–394.

Fusiello, A. Tommasini, T. Trucco, E. and Roberto, V. 1998. Making Good Features to Track Better, *in Proceeding of the IEEE Conference on Computer Vision and Pattern Recog.* pp. 178–183, Santa Barbara , CA.

Haritaoglu, I. Harwood, D. and Davis, L.S. 1998. W^4 Who? When? Where? What? A Real Time System for Detecting and Tracking People. *In 3rd International Conference on Face and Gesture Recognition.* Nara. Japan, pp: 222–227.

Harris, C. and Stephens, M. 1988. A Combined Corner and Edge Detector. *In Proceeding 4th Alvey Vision Conference* (AVC88) pp. 147–151.

Hartley, R. I. 1995. In Defence of the 8-point Algorithm, in *Proceeding of the IEEE International Conference on Computer Vision.*

Hartley, R. I. 1997. Kruppa's Equations Derived from Fundamental Matrix, *IEEE Transactions on Pattern Analysis and Machine Intelligence*, Vol. 19, (2), pp. 133–135, Feb.

Jaynes, C. 1999. Fast Feature Extraction of Mechanical Parts in Motion, *Technical Report*, Department of Computer Science. University of Kentucky.

Kanade, T. and Tomassi, C. 1991. Detection and Tracking of point features. *Technical Report CMU-CS-91-132.* Carnegie Mellon University, April.

Koller, D. Klinker, G. Rose, E. Breen, D. Whitaker, R. Tuceryan M. · 1998. Real-time Vision-Based Camera Tracking for Augmented Reality Applications. *Tec. Rep. California Inst. Of Technology.* e-mail dieter. koller@autodesk.com

Lipton, A. J. Fujiyoshi, H. and Patil, R. S. 1998. Moving Target Classification and Tracking from Real-Time Video. In *proc. IEEE Workshop on Applications of Computer Vision (WACV).* Princeton NJ, October, pp. 8–14.

McLauchlan, P. and Murray, D. 1995. A Unifying Framework for Structure and Motion recovery From Image Sequence, *In Proc. 5th Int'l Conf. On Computer Vision*, Boston, pp. 314–320, June.

McLauchlan, P. 1999. The Variable State Dimension Filter. *Technical Report VSSP 4/99.* University of Surrey, Dept. of Electrical Engineering, Nov.

Pilu, M. 1997. A Direct Method for Stereo Correspondence based on Singular Value Decomsition, *in Proc. CVPR'97*, pp. 261–266.

Ravela, S. Draper, B. Lim, J. and Weiss, R. 1996. Tracking Object Motion Across Aspect Changes for Augmented Reality, *in Proc. ARPA Image Understanding Workshop*, Palm Spring, CA.

Schmid, C. Mohr, R. and Bauckhage, C. 2000. Evaluation of Interest Point Detectors. *International Journal of Computer Vision.* Vol. 37, (2), pp. 151–172.

Shapiro, Linda, and Stockman, George, 2000. Computer Vision.

Soatto, S. Frezza, R. Perona, P. 1994. Motion Estimation via Dynamic Vision, *Technical Report CIT-CDS 94-004*, California Institute of Technology.

Trucco, E. and Verri, A. 1998. *Introductory techniques for 3D Computer Vision.* Prentice Hall. NJ.

Tuceryan, M. Greer, D. Whitaker, R. Breen, D. Crapton, C. and Ahlers, K. 1995. Calibration Requirements and Procedures for a Monitor-Based Augmented Reality System, *IEEE Trans. On Visualization and Computer Graphics*, Vol. 1, 3, Sep. pp. 255–273.

Welch, G. and Bishop, G. 2001. Kalman Filter, *SIGGRAPH 2001*, Los Angeles, CA, August.

Weng, J. Cohen, P. and Herniou, 1992. Camera Calibration with Distortion Models Accuracy Evaluation. *IEEE Transactions on Pattern Analysis and Machine Intelligence*, Vol. 14, (10), pp. 965–980.

Wren, C. Azarbayejani, A. Darrell, T. and Pentland, A. 1997 Pfinder: Real-time tracking of the human body, *IEEE Transactions on Pattern Analysis and Machine Intelligence*, Vol. 19, (7), pp. 780–785.

Zhang, Z. Deriche, R. Faugeras, O. Luong, Q. 1996. A Robust Technique for Matching Two Uncalibrated Images Through the Recovery of the Unknown Epipolar Geometry, *INRIA Research Report 2273*, May.

Computational Modelling of Objects Represented in Images –
João Manuel R.S. Tavares & R.M. Natal Jorge (eds)
© 2007 Taylor & Francis Group, London, ISBN 978-0-415-43349-5

Terrestrial laser technology in sporting craft 3D modelling

P. Arias, J. Armesto, H. Lorenzo & C. Ordóñez
Department of Natural Resources Engineering & Environment, University of Vigo, ETSE Minas,
Campus Marcosende, Vigo, Spain

ABSTRACT: The urgent need to increase productivity and competitively demands from sporting craft builders the incorporation of newer design and manufacturing technologies, as CAD/CAM and CNC machining systems. This paper describes the sporting boats' 3D surface modelling, needed for automated manufacturing processes, through terrestrial laser scanner technology. The methodology followed for data collection and data processing is described in detail, advantages and limitations of this technique are displayed and the accuracy of the obtained results is also estimated. According to obtained results, this technology proves to provide accurate results and to be cost and time effective.

1 INTRODUCTION

The demand for fishing, sporting and pleasure craft in developed countries has undergone a steady increase in recent years and all medium-term forecasts point towards the continuity of this trend, due to growing purchasing power and greater leisure time of citizens in these countries. This fact, together with the gradual decrease within the European shipping market (COM 2003), are the likely causes of the increasing number of companies entering the sporting boats sector.

Nowadays, most sporting boat manufacturing processes are largely manual and rely on the expertise of individual operatives; resulting boats are therefore usually unique models and differ from any other. In addition, rigorous quality control programmes and assembly flow diagrams are rarely implemented or applied during their construction. As a result, parts are frequently wasted or re-worked due to production errors, causing production delays and cost increases. This situation combined with the urgent need to increase productivity and competitively, is placing pressure on shipbuilders to improve production processes by incorporating new design and manufacturing technologies. Automation is one major step towards achieving the quality, cost and delivery targets and the coordination of design and manufacturing processes.

Due cost effectiveness and high quality results, CAD/CAM and CNC machining is one of the largest areas of expansion and renovation in the automation and coordination of design and manufacturing processes in a wide range of manufacturing processes (Karunakaran et al. 2005). In order to program the machine's operations, this manufacturing process requires 3D digital models of the objects that are to be manufactured. Reverse engineering, which concerns with the generation of 3D CAD models of physical objects, may be employed to obtain digital models of boats in situations wherein the serial automated replication of an existing boat model is needed (because of satisfactory aesthetic design and successful navigability performance).

Apart from its usefulness in the automation of manufacturing, having a 3D digital model of the product is vital for the following concerns: a) the reuse of the information for re-engineering or repair processes (when analysis and modifications are required to construct a new and improved product); b) the navigability simulation field; c) the product documentation for virtual advertising and merchandising systems; d) the possibility of sending an electronic format of the model to allow the replication of the boat in different, remote sites (Bordegoni & Filippi 2001); e) quality control processes (through shape plans, symmetry analysis, etc).

Terrestrial laser scanner is a recently developed reverse engineering technology for data collection and subsequent 3D modelling whose applications in industrial and civil construction fields are increasing day by day. The outlines for effective terrestrial laser scanning in capturing point clouds and measuring distances are described in Lichti et al.(2002). According to Gruen & Akca (2005) laser scanning technology might be considered valid and precise for creating high-resolution polygon meshes, surfaces or solid models of objects. Nevertheless achieving smooth surface representations (needed for reasons of manufacturability) from point clouds is the key task for the efficient application of this technology to reverse engineering processes (Yau et al. 2005).

The problem of obtaining smooth surface models from laser point clouds can be stated as follows: given a set V of n points in the 3D space, a surface S might be found that approximates or interpolates V. A huge variety of procedures have been developed in the last years concerning this task. When unstructured data are focused, generally a first triangulation of the points is performed, in other words, a piecewise linear function is obtained that interpolates V. This introduces a first level of organization as it defines the basic topological structure of the point cloud. The process of remeshing to obtain a regular polygon mesh can be considered a second level of organization that leads to a better smooth surface approximation (Floater 1997). The next step is segmentation into smaller point regions, in such a way that later each region can be well-approximated by a single surface. Finally the individual surfaces are trimmed and joined together (Benko et al. 2001).

For the segmentation three main options might be accomplished: iterative approaches or user assistance for segmentation or holding strong assumptions related to the object topology (Benko et al. 2001). Automatic segmentation procedures that have been developed until the date, besides to be sensitive or computationally complex or both, can only be applied to simple topology data (Liu et al. 2003); consequently they are feasible in a limited number of applications since several industrial environments, as boats structure, can not be represented as a set of plans, cylinders, spheres, and other simple surfaces.

In this paper a boat 3D modelling procedure from unstructured laser scanned data is described. It is based on an automated polygonation, user guided segmentation and semi-automated surface fitting. Advantages and limitations of this technique are displayed and the accuracy of the obtained results is also estimated.

This paper is part of a broader project whose final aim consists of building a pleasure boat prototype with reinforced polyester with fibreglass on the basis of the structure of an existing one. The 3D digital models of the deck and the hull are needed to carry out reengineering and redesigning processes, navigability simulations, etc. The size of the existing boat measures 12 m in overall length, with a beam of 4.5 m and the draft measuring 2.4 m.

2 INSTRUMENTATION

- Terrestrial laser scanner. A Leica Cyrax 2500 laser was used. This is a pulsed time-of-flight scanner that records up to 1000 points/second with a precision of 6 mm with a resolution capacity of less than 1 millimetre (Leica 2005). For maximum distances up to 50 metres the precision achieved is around 2 millimetres. Each scanning operation requires approximately 10 minutes and each scan covers a 40° vertical × 40° horizontal field-of-view.
- Magnetic targets.
- Laptop computer: Pentium IV, 2.4 GHz, 1GB RAM, 40 GB hard disk. Terrestrial laser applications require the scanner to be connected to a computer in which the point clusters recorded by the laser are stored in real time.
- Tripod. A tripod provides the necessary support to ensure the terrestrial laser scanner is set at suitable height and kept stable and immobile during scanning operations.
- Software application for point cloud matching.
- Software for point deleting and point cloud filtering.
- Software for generating 3D surface models from the pre-processed point clouds.
- Software for creating hull and deck cross-sections from the 3D surface models.

Resulting models are based on standard file formats (e.g. IGES) compatible with most of the commercial software.

3 METHOD

3.1 Data collection design

This step focuses on the design of the data collection phase in order to optimize reflectivity conditions and to minimize measuring time and distorting factors during data capture:

- Estimation of the number of scans needed to scan the whole object, avoiding shadow areas and maintaining overlap between consecutive scans, whilst keeping the number of scans to a minimum.
- Estimation of the number of scanner stations needed to perform the estimated number of scans.
- Environmental factors. Surrounding environment is analyzed in order to detect those elements in or around the scanning area that could interfere in data collection: sources of vibrations, high noise, shadowing elements, etc.
- Craft support. The stability of the craft support is verified in order to avoid any kind of vibrations or movement during the scanning process.
- Correct lighting conditions. This is an essential parameter, as the point clouds assembly is based on magnetic target recognition, and they can only be recognised when light is diffused and at the right intense.

3.2 Data collection

1. Magnetic target distribution. These targets were employed because they are automatically

recognised by the software system, so subsequent assembly of point clouds is faster and easier. Ten magnetic targets were distributed over and around the craft, in such a way that at least five were registered with each scan.

2. Scanner stationing and data collection. Any vibration, strong noise or any movement of the scanner or the craft itself are avoided during the scanning process.

3. Point cloud assembly. Each registered cloud of points was added to the others after scanning in order to detect incomplete data, corrupted files, etc. Prior to assembly, magnetic targets were identified and marked in each point cloud. Later these magnetic targets were taken as a reference for point clouds matching. At the end of this phase all the collected point clouds were aligned in a common global coordinates system.

3.3 Data pre-processing

Deck and hull data are processed independently of each other:

1. Point deletion. Due to field of vision of the scanner, several elements apart from the craft were recorded: walls, ceiling and floors of the premise, other boats, work tools and other materials, etc. In this stage the scans are analysed in order to detect and remove points that do not belong to the objects that are to be modelled. Moreover, points that were registered with erroneous values and that are clearly outside the area of the craft surface were deleted. The 52% of the entire hull assembled point cloud and the 66% of the deck point cloud were deleted in this step. The resulting point cloud of the hull is shown in figure 1.

2. Redundancy filtering. As a consequence of the data capture design (overlap between different scans is needed in order to reassemble), there is a significant amount of redundant information in the assembled point cloud. In this phase this redundant information was removed from the cloud of points in order to reduce the volume of data and to simplify subsequent operations (see figure 2 for resulting hull point cloud). The filtering intensity is reliant on the available hardware and software for data processing. Nevertheless, filtering might be light in areas with complex curves and surfaces. It can be more intense in areas with simple shapes, as lots of middling points in these areas do not show any relevant information for the construction of the 3D models. Filtering is performed through deletion of points whose distance with others is within a previously established global mean distance. Only the 6% of the entire hull point cloud left after redundancy filtering. For the deck the 19% was preserved.

3.4 Data processing: craft 3D modelling

The 3D polygon mesh models were constructed either for the hull (see figure 3) or the deck of the craft from the filtered point clouds. Then different procedures were performed for the hull and the deck.

The hull modelling was performed as follows:

1. Autosurfacing. This is a semi-automated process where NURBS surfaces (Non-Uniform Rational B-Spline) are fitted to the polygon mesh. Primary, user guided segmentation is performed. Boundaries

Figure 2. Assembled point clouds for the craft hull after redundancy filtering.

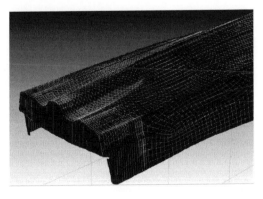

Figure 3. Detail view of the hull polygon mesh model.

Figure 1. Assembled point clouds for the craft hull after deleting points.

of regions are selected from the polygon mesh lines. Then, the number of surfaces needed for the description of the topology of the craft is optimized automatically through decimation of the initial polygon mesh. For each region, a NURBS surface is fitted. Resulting surfaces are joined to obtain the whole model. Some joins and edges are manually smoothed in a stronger or lighter degree depending on the kind of surface and the characteristics of the element to be modelled.

2. Identification of incomplete surfaces and manual completion. Those areas where the system was unable to automatically generate the corresponding surfaces were manually completed until the whole three-dimensional model of the craft was obtained (see figure 4 for result).

Autosurfacing could not be performed for the deck modelling due to the abundance of elements with complex curves and surfaces. In this case it was performed as follows:

1. Sectioning. The polygon mesh model of the craft deck was sectioned into a series of transversal and longitudinal planes, obtaining a set of section slides. For each section slide appropriate curves where derived: straight lines, regular curves and NURBS were used.

2. Surface generation. The transversal and longitudinal planes defined in the previous step divide the deck into sections. The corresponding surface was obtained for each section taking its boundary edges (resulting of the previous step) as reference and defining backup rail edges when polysurfaces (considered as surfaces formed of NURBS (Yin & Jiang 2004)) should be fitted to the polygon mesh model of the deck.

3.5 Accuracy assessment of the modelling process

No method for laser data accuracy analysis has yet emerged that is generally accepted (Geng et al. 2004;

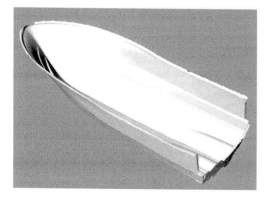

Figure 4. Resulting hull surface 3D model.

(Boehnen & Flynn 2005); the accuracy of the resulting 3D models is especially difficult to estimate when the real shape of the object is unknown and the measuring equipment has not been previously calibrated. As these should be the expected circumstances in most cases where laser scanning would be applied to boat modelling, the thickness of the non filtered point clouds have been considered the most adequate indicator of the quality of the data collection process and the point positioning accuracy. To assess the thickness of the point clouds, the scans aligned in a global coordinates system are sectioned by perpendicular planes to the stern-bow symmetry axis. Twenty homogeneously distributed sections were analyzed.

4 RESULTS

Data collection of the targeted boat with Leica laser scanning needed eight scans: 4 for the hull, 4 for the deck; these were taken from seven different stations. Over six million points were measured, but only the 6% and 19% of the point clouds of the hull and deck respectively (see table 1 for quantitative details) were processed for the polygon mesh generation. It has been observed that laser data processing requires experienced and skilful operators, especially for redundancy filtering step, as there is a risk of filtering out too much data, with the consequent loss of information, or too little data filtering, causing subsequent problems due to excessive information and overly large files.

Table 1. Quantitative details of a sporting boat 3D modelling process (deck and hull separately) through terrestrial laser scanner. The time consumes of each step is indicated as a percentage value of the whole processing time.

Hull		Time (%)
Data collection	2,454,784 points	4
Points deleting	1,169,237 points	3
Redundancy filtering	141,841 points	2
Polygon mesh model	282,600 faces	2
Autosurfacing/sectioning	92 surfaces	1
Manual completion/ surfaces generation	24 surfaces	6

Deck		Time (%)
Data collection	1,178,197 points	4
Points deleting	520,952 points	3
Redundancy filtering	217,976 points	2
Polygon mesh model	469,950 faces	2
Autosurfacing/ sectioning	3,424 section slides	2
Manual completion/ surfaces generation	277 polysurfaces, 95 regular surfaces	70

The time needed either for data collection or for each step of data processing and 3D modelling is indicated in table 1 as a percent value of the whole time taken up by the whole process.

Note that though a great amount of points is measured, data collection is the less time consuming step (8%; see table 1), while completing surfaces consumes most processing time. In this last step strong differences between hull and deck modelling appear. The craft hull, which is a surface with few frequently smooth discontinuities, was automatically modelled by autosurfacing in the 80% of its surface. On the other side, autosurfacing could not be performed for the deck modelling, mainly due to plenty of features with different sizes and shapes, and especially of parts with complex curves and surfaces. In this case regular and irregular curves were obtained for a set of 120 section slides of the deck polygon mesh, and then the corresponding surfaces could be obtained. This step required the 70% of the whole time (see table 1).

According to these results, terrestrial laser scanning might be considered a cost and time efficient technology for craft hull 3D modelling. Moreover, although the deck modelling has proved to be a slow hard manual process, terrestrial laser scanning should be considered a competitive technology for deck craft 3D modelling, in relation to traditional modelling techniques, either manual or topographic methods.

The precision analysis of the modelling process data through Leica terrestrial laser scanner, estimated by the thickness of the point clouds, showed that accuracy values decrease in proportion to the distance between the laser and the object being measured. In this case it was confirmed that thickness values are less than 5 mm for the analyzed sections of the boat.

Although the positioning accuracy of points measured through laser scanner technology is highly dependent on signal reflected from the surface, and this factor itself depends on varying parameters such as the colour and the roughness of the surface, or the angle at which the surface is scanned, this maximum value of thickness might serve as a reference for the accuracy to expect when sporting boats with similar size are measured.

5 CONCLUSIONS

This study clearly shows the reliability of terrestrial laser scanning in the generation 3D surface models of pleasure craft. These models, as well as being essential for automated production processes, also allows for

engineering processes, adjustments to be made to the deck and hull, navigability simulations, quality control concerning asymmetry etc. Many of these processes cannot be carried out using traditional measuring techniques (tape measures, levels, etc.) or with topographic methods and in any case the results obtained using these methods are undoubtedly less reliable.

REFERENCES

Benko, P., Martin, R.R. & Várady, T. 2001. Algorithms for reverse engineering boundary representation models. *Computer Aided Design* 33 (11): 839–851.

Boehnen, C. & Flynn, P. 2005. Accuracy of 3D Scanning Technologies in Face Scanning Scenario. In *Proc. of the 5th International Conference on 3D Digital Imaging and Modelling*, CD-ROM.

Bordegoni, M. & Filippi, S. 2001. Reverse Engineering for Molding", In *Proc. of ADM 2001 – 12th International Conference on Design Tools and Methods in Industrial Engineering*.

COM 2003, 232. (Commission of the European Communities, 232). 2003 *Seventh report on the situation in world shipbuilding* http://europa.eu.int/ (last access: July 8th, 2006).

Karunakaran, K.P., Agrawal, S., Vengurlekar, P.D., Sahasrabudhe, O.S., Pushpa, V. & Ely, R. 2005. Segmented object manufacturing. *IIE Transactions* 39: 291–302.

Floater, M.S. 1997. Parameterization and smooth approximation of surface triangulation. *Computer Aided Geometric Design* 14 (3): 231–250.

Geng, J., Zhuang, P., May, P., Yi, S. & Tunnell, D. 2004. 3DFaceCam – A Fast and Accurate 3D Facial Imaging Device for Biometrics Applications. In A.K. Jain & N.K. Ratha (eds), *Proc. SPIE, Biometric Technology for Human Identification*, 5404: 316–327.

Gruen, A. & Akca, D. 2005. Least squares 3D surface and curve matching. *ISPRS Journal of Photogrammetry and Remote Sensing* 59 (3): 151–174.

Leica Geosystems AG. 2005. http://www.leica-geosystems. com (last access: November, 6th, 2005).

Lichti, D.D., Gordon, S.J. & Stewar, M.P. 2002. Ground-Based Laser Scanners: operation, systems and applications. *Geomatica* 56 (1): 21–33.

Liu, G.H., Wong, Y.S., Zhang, Y.F. & Loh, H.T. 2003. Modelling cloud data for prototype manufacturing. *Journal of Materials Processing Technology* 138: 53–57.

Yau, H.T., Lee, R.K., Chuang, C.M. & Hsu, C.Y. 2005. NC Tool-Path Generation Based on Surfel Models Constructed from Random Scanned Data. *Computer-Aided Design & Applications* 2: 1–4.

Yin, Z. & Jiang, S. 2004. Iso-phote based adaptive surface fitting to digitized points and its applications in region-based tool path generation, slicing and surface triangulation. *Computers in Industry* 55: 15–28.

Computational Modelling of Objects Represented in Images –
João Manuel R.S. Tavares & R.M. Natal Jorge (eds)
© 2007 Taylor & Francis Group, London, ISBN 978-0-415-43349-5

External anatomical shapes reconstruction from turntable image sequences using a single off-the-shelf camera

Teresa C.S. Azevedo
INEGI – Inst. de Eng. Mecânica e Gestão Industrial, LOME – Lab. Óptica e Mecânica Experimental
FEUP – Faculdade de Engenharia da Universidade do Porto, Portugal

João Manuel R.S. Tavares & Mário A.P. Vaz
INEGI, LOME
DEMEGI – Departamento de Engenharia Mecânica e Gestão Industrial, FEUP, Portugal

ABSTRACT: Three-dimensional (3D) human body reconstruction and modeling from multiple two-dimensional (2D) images is a topic of great interest and research but still remains a difficult problem. Volumetric based methods are commonly used to solve this type of problem. With those methods, from a turntable image sequence of an object, a 3D model can be built. The final goal of the work presented in this paper is to reconstruct all-round 3D models of the human body, using a single off-the-shelf camera and volumetric techniques, with the best accuracy and photorealistic appearance.

1 HUMAN BODY 3D RECONSTRUCTION

3D imaging of physical objects has become a strong topic of research, mostly due to the increasing improvements in computational resources. Particularly, 3D reconstruction of the human body promise to open a very wide variety of applications: medicine, virtual reality, ergonomics applications, etc.

Classical devices for external anatomical shape reconstruction are 3D scanners ((Levoy, Pulli et al., 2000), (Rocchini, Cignoni et al., 2001)). They are expensive but simple to use and fast. The technology involved can vary from laser light, structured light or time of flight, with an accuracy of millions of points ((Boehler, Heinz et al., 2001), (Remondino, 2004)).

There are other techniques called image-based. 3D human body models are built from images, using approaches such as anthropometric statistics ((Barrón and Kakadiaris, 2000), (Mori and Malik, 2006)), multi-view geometry ((Remondino, 2004)) or occluding contours (Boyer and Berger, 1997).

1.1 Volumetric methods

Traditional methods, like stereo matching or disparity maps, fail to capture shapes with complicated topology, like the human body because of its smoothness (Zeng, Lhuillier et al., 2005).

Volumetric or voxel-based approaches have been quite popular for some time ((Seitz and Dyer, 1997), (Slabaugh, Culbertson et al., 2001)). They assume that

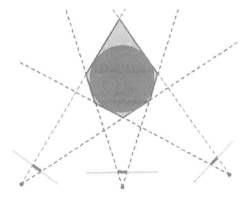

Figure 1. Visual hull from three viewpoints.

there is a bounded volume in which the object of interest is placed.

Shape from silhouettes ((Laurentini, 1994), (Srivastava and Ahuja, 1990)) is one of the most used volumetric methods. It combines silhouette images of an object with calibration information to set the visual rays in scene space for all silhouette points, which define a generalized cone within the object must be placed. The intersection of these cones defines a volume of scene space (*visual hull*) in which the object is guaranteed to be (Figure 1). Accuracy depends on the number of views used, the positions of each viewpoint, the calibration quality and the object's complexity.

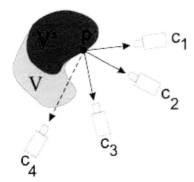

Figure 2. Non photo-consistency lemma: if p isn't photo-consistent with some camera(s), it is non consistent with the entire set of cameras.

$$2D\ point \longleftarrow maps\ to \longleftarrow 3D\ point$$

$$s \begin{bmatrix} x \\ y \\ 1 \end{bmatrix} = \begin{bmatrix} f_x & \gamma & p_x \\ 0 & f_y & p_y \\ 0 & 0 & 1 \end{bmatrix} \begin{bmatrix} r_{11} & r_{21} & r_{31} & t_1 \\ r_{12} & r_{22} & r_{32} & t_2 \\ r_{13} & r_{23} & r_{33} & t_3 \end{bmatrix} \begin{bmatrix} X \\ Y \\ Z \\ 1 \end{bmatrix}$$

Intrinsics   Extrinsics

Figure 3. Camera's intrinsic and extrinsic parameters, mapping 3D world point coordinates into 2D image point coordinates.

Much refined voxel models can be obtained using space carving ((Kutulatos and Steiz, 1998), (Sainz, Bagherzadeh et al., 2002)). The reconstruction is initialized with a bounding box of voxels containing the true 3D object. The 3D shape of the object is constructed by removing (*carving*) voxels that are not photo-consistent with the reference views (Figure 2).

From a reconstructed volumetric model, a polygonal surface approximation can be constructed using, for example, the Marching Cubes algorithm ((Heckbert and Garland, 1997), (Lorensen and Cline, 1987)).

Table 1. Pseudo-code for GVC algorithm.

```
initialize surface voxel list (SVL)
for every voxel V
  carved(V) = false
loop {
  visibilityChanged = false
  compute item buffers by rendering voxels on SVL
  for every voxel V in SVL {
    compute vis(V)
    if (consist(vis(V)) = false) {
      visibilityChanged = true
      carved(V) = true
      remove V from SVL
      for all voxels N that are adjacent to V
        if (carved(n) = false and N is not in SVL)
          add N to SVL
    }
  }
  if (visibilityChanged = false) {
    save voxel space
    quit
  }
}
```

2 OUR METHODOLOGY

In this paper, we present a volumetric method – Space Carving – for object 3D reconstruction.

First step is camera calibration, which means finding the map between the 3D world and the 2D image space (Figure 3).

We based our approach in Zhang's algorithm (Zhang, 2000). Several views of a chessboard pattern were used, acquired with a single off-the-shelf camera. Taken as inputs the correspondences between 2D image points and 3D scene points, Zhang's algorithm outputs the intrinsic camera matrix (focal length and principal point; skew γ is admitted as zero), distortion coefficients and extrinsic parameters.

For each calibration image, keeping the camera untouched, the object to reconstruct is placed on top of the pattern and another image is taken. So, for each turntable position we have a pattern image (for calibration) and an object image (for 3D reconstruction).

To avoid all calibration images from having the pattern on the same plane, some extra images of the pattern in different orientations were taken.

After calibration is done, for each object image of the turntable sequence, background/foreground segmentation is performed using some basic image processing tools.

After the segmentation process, the object's model is built using a *Generalized Voxel Coloring* (GVC) technique (Loper, 2002). It allows arbitrary camera placement and generally provides better results than some other color-consistency algorithms (Slabaugh, Culbertson et al., 1999). Specifics of the algorithm are on Table 1.

Finally, the volumetric model is polygonized (smoothed) using the Marching Cubes algorithm.

3 EXPERIMENTAL RESULTS

Our method was tested on two objects, very different on size and shape complexity: an human upper-torso (Figure 4) and an hand model (Figure 5).

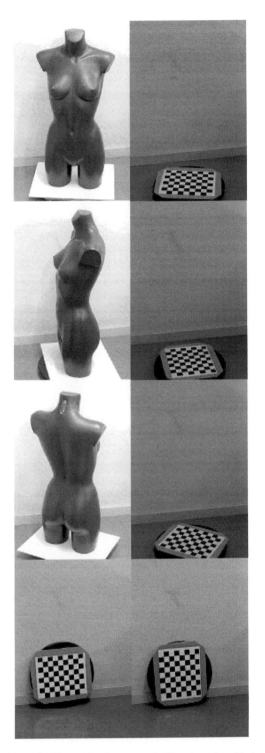

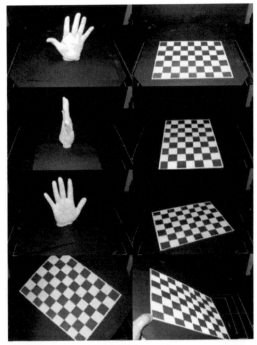

Figure 5. Hand object image sample. Three top pairs: object and pattern pair of turntable sequence; last pair: extra images used in calibration.

Table 2. Intrinsic camera calibration results for torso model.

Intrinsic camera matrix		Distortion coefficients	
f_x	4908.4560 ± 185.6349	k_1	-0.2031 ± 0.2887
f_y	5029.6398 ± 226.8837	k_2	7.0357 ± 23.0532
c_x	681.3248 ± 207.6812	p_1	0.0509 ± 0.0145
c_y	1640.2373 ± 347.0643	p_2	0.0035 ± 0.0099

Table 3. Intrinsic camera calibration results for hand model.

Intrinsic camera matrix		Distortion coefficients	
f_x	2294.0405 ± 16.5807	k_1	-0.1218 ± 0.0139
f_y	2297.0050 ± 15.3810	k_2	0.0000 ± 0.0000
c_x	988.4635 ± 26.0595	p_1	0.0083 ± 0.0021
c_y	687.3008 ± 29.1624	p_2	0.0032 ± 0.0027

3.1 Calibration results

The camera to be calibrated was a NIKON D200 CCD camera. The chessboard plane contained 7×9 squares, so there where 48 inner corners, each square measuring 30×30 mm.

On Table 2 and Table 3 are the intrinsic results for the torso and hand model, respectively.

Regarding extrinsic calibration results, the standard deviation of the reprojection error (in pixel), in both

Figure 4. Torso object image sample. Three top pairs: object and pattern pair of turntable sequence; last pair: extra images used in calibration.

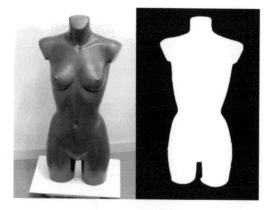

Figure 6. Torso image and respective silhouette.

Figure 7. Hand image and respective silhouette.

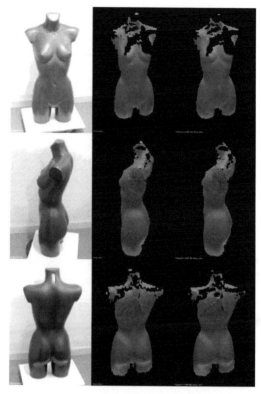

x and y directions, was [0.3584; 0.3648] for the upper-torso object and [0.5513; 0.8851] for the hand object.

It can be seen that, although the reprojection error is slightly higher for the hand model, the intrinsic parameters are much more accurate.

Figure 8. 3D model of the upper-torso model: left, original image; middle, voxel model; right, polygonized model.

3.2 Segmentation results

Background/foreground segmentation means to binarize an image, extracting the object of interest (foreground) from the rest of the image (background).

Using some image processing tools, like Sobel operator for edge detection, and different morphology functions in combination, as flood-fill operation to fill separated regions, it was possible to obtain reasonably good silhouettes for both objects turntable image sequence (Figure 6 and Figure 7).

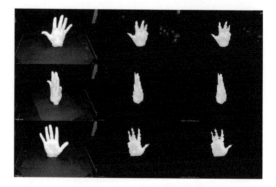

Figure 9. 3D model of the hand model: left, original image; middle, voxel model; right, polygonized model.

3.3 Carving results

Some snapshots of the 3D models obtained for both objects can be seen in Figures 8 and Figure 9.

Although not very good for any of the objects, three-dimensional modelling had better results for the upper-torso object, because its shape has no abrupt changes comparing to the hand object and also because the higher distance between camera and object minimizes the effects of calibration errors.

4 CONCLUSIONS AND FUTURE WORK

We can easily conclude that to have a complete and accurate 3D model of an object is very difficult. It gets even harder when the objects to modelize have strong shape variations, like the fingers in our hand model. Also, good calibration results are essential to get exact information about the object placement in 3D space.

Further work will concentrate on improving the calibration method, specially on dividing into intrinsic and extrinsic calibration. Some chessboard pattern images will be used for intrinsic parameters estimation and the rotation and translation of the object will be estimated directly by the object images.

Other improvements will be on Space Carving, mainly on the study of the influency of voxel coloring (since our objects have smooth colors).

ACKNOWLEDGMENTS

This work was partially done in the scope of the project "Segmentation, Tracking and Motion Analysis of Deformable (2D/3D) Objects using Physical Principles", with reference POSC/EEA-SRI/55386/2004, financially supported by *FCT – Fundação para a Ciência e a Tecnologia* from Portugal.

REFERENCES

C. Barrón, I. A. Kakadiaris, *Estimating Anthropometry and Pose from a Single Image*, IEEE Computer Society Conference on Computer Vision and Pattern Recognition Proceedings, New York:669–676, 2000.

W. Boehler, G. Heinz, A. Marbs, *The potential of non-contact close range laser scanners for cultural heritage recording*, XVIII International Symposium of CIPA Proceedings, Working Group VI, Potsdam, Germany, 2001.

E. Boyer, M.-O. Berger, *3D Surface Reconstruction Using Occluding Contours*, International Journal of Computer Vision, 22 (3):219–233, 1997.

P. S. Heckbert, M. Garland, *Survey of Polygonal Simplification Algorithms*, Multiresolution Surface Modeling Course, Siggraph Course Notes, ACM Press, New York (25), 1997.

K. N. Kutulatos, S. M. Steiz, *A Theory of Shape by Space Carving*, Technical Report TR692, Computer Science Department, University of Rochester, New York, USA, 1998.

A. Laurentini, *The visual hull concept for silhouette-based image understanding*, IEEE Transactions on Pattern Analysis and Machine Intelligence, 16 (2):150–162, 1994.

M. Levoy, K. Pulli, B. Curless, et al., *The Digital Michelangelo Project: 3D scanning of large statues*, Siggraph 2000, Computer Graphics Proceedings, ACM Press, ACM SIGGRAPH, Addison Wesley:131–144, 2000.

M. Loper, *Archimedes – A Generalized Voxel Coloring Implementation*, http://www.loper.org/~matt/Archimedes/ Archimedes_docs/html/main.html, 2002.

W. E. Lorensen, H. E. Cline, *Marching cubes: A high resolution 3D surface construction algorithm*, International Conference on Computer Graphics and Interactive Techniques Proceedings, ACM Press, New York, USA, 21 (4):163–169, 1987.

G. Mori, J. Malik, *Recovering 3D Human Body Configurations Using Shape Contexts*, IEEE Transactions on Pattern Analysis and Machine Intelligence, 28 (7):1052–1062, 2006.

F. Remondino, *3-D Reconstruction of Static Human Body Shape from Image Sequence*, Computer Vision and Image Understanding, 93 (1):65–85, 2004.

C. Rocchini, P. Cignoni, C. Montani, et al., *A low cost 3D scanner based on structured light*, EUROGRAPHICS Proceedings, Interlaken, Switzerland, 20 (3):299–308, 2001.

M. Sainz, N. Bagherzadeh, A. Susin, *Carving 3D Models from Uncalibrated Views*, Proceedings of the 5th IASTED International Conference Computer Graphics and Imaging, Hawaii, USA:144–149, 2002.

S. N. Seitz, C. R. Dyer, *Photorealistic Scene Reconstruction by Voxel Coloring*, Proceedings of the Computer Vision and Pattern Recognition Conference, San Juan, Puerto Rico:1067–1073, 1997.

G. Slabaugh, W. B. Culbertson, T. Malzbender, et al., *A survey of methods for volumetric scene reconstruction from photographs*, International Workshop on Volume Graphics Proceedings, New York, USA:21–22, 2001.

G. G. Slabaugh, W. B. Culbertson, T. Malzbender, *Generalized Voxel Coloring*, Workshop on Vision Algorithms Proceedings, Corfu, Greece:100–115, 1999.

S. K. Srivastava, N. Ahuja, *Octree generation from object silhouettes in perspective views*, Computer Vision, Graphics and Image Processing, 49:68–84, 1990.

G. Zeng, M. Lhuillier, L. Quan, *Recent Methods for Reconstructing Surfaces from Multiple Images*, Lecture Notes in Computer Science, Springer Verlag, 3519:429–447, 2005.

Z. Zhang, *A Flexible New Technique for Camera Calibration*, IEEE Transactions on Pattern Analysis and Machine Intelligence, 22 (11):1330–1334, 2000.

Computational Modelling of Objects Represented in Images –
João Manuel R.S. Tavares & R.M. Natal Jorge (eds)
© 2007 Taylor & Francis Group, London, ISBN 978-0-415-43349-5

Patient specific heart models from high resolution CT

Chandrajit Bajaj, Samrat Goswami, Zeyun Yu & Yongjie Zhang

Department of Computer Sciences, Computational Visualization Center, Institute of Computational Engineering and Sciences, University of Texas at Austin, Austin, Texas, USA

Yuri Bazilevs & Thomas J.R. Hughes

Department of Aerospace Engineering and Engineering Mechanics, Institute of Computational Engineering and Sciences, University of Texas at Austin, Austin, Texas, USA

ABSTRACT: Computer Tomography (CT) and in particular super fast, 64 and 256 detector CT has rapidly advanced over recent years, such that high resolution cardiac imaging has become a reality. In this paper, we provide a solution to the problem of automatically constructing three dimensional (3D) finite-element mesh models (FEM) of the human heart directly from high resolution CT. Our overall computational pipeline from 3D imaging to FEM models has five main steps, namely, (i) discrete voxel segmentation of the CT (ii) discrete topological noise filtering to remove non-regularized, and small geometric measure artifacts (iii) a reconstruction of the inner and outer surface boundaries of the human heart and its chambers (iv) computation of the medial axis of the heart boundaries and a volumetric decomposition of the heart into tubular, planar and chunky regions, (v) a flexible match and fit of each of the decomposed volumetric regions using segmented anatomical volumetric templates obtained from a 3D model heart.

1 INTRODUCTION

Computer aided diagnosis and treatment of cardiovascular disease, in particular atherosclerosis, left ventricular hypertrophy, valvular dysfunction, increasingly rely on faithful patient specific heart FEM (finite element mesh) models that can be used in full-cycle simulation of pulsatile blood flow through the heart. An emerging methodology to construct spatially realistic human heart models is via super fast, 64 and 256 detector (high resolution) Computer Tomographic (CT) imaging (Toshiba Medical Systems – 64 Slice CT 2006).

Volume rendering of one such CT64 dataset is shown in Figure 2. Although state-of-the-art, the imaging is only the first step of a significant computational sequence of image and geometry processing steps, that are necessary for generating a robust and spatially realistic FEM model. In this paper, we present such a computational pipeline (see Figure 1) that processes the imaging data, and additionally uses an anatomically correct template heart 3D model, to construct an anatomically correct patient specific FEM model.

Our overall computational pipeline from 3D imaging to FEM models has five main steps, namely, (i) discrete voxel segmentation of the CT (ii) discrete topological noise filtering to remove non-regularized, and small geometric measure artifacts (iii) a reconstruction of the inner and outer surface boundaries of the human heart and its chambers

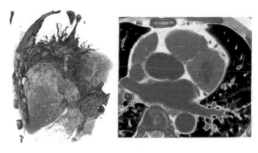

Figure 2. Left subfigure shows the volume rendering of a subvolume of the input CT imaging data. Right subfigure shows a portion of the cross-section of the input. One full slice is shown in the two leftmost subfigures in Figure 3.

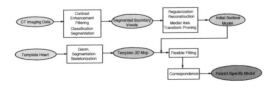

Figure 1. The Computational pipeline.

(iv) computation of the medial axis of the heart boundaries and a volumetric decomposition of the heart into tubular, planar and chunky regions, (v) a flexible match and fit of each of the decomposed volumetric regions using segmented anatomical volumetric templates obtained from a 3D model heart. The pipeline has two major components. One path works on the imaging data, filters the noise, segments the main components of the heart and builds an initial surface mesh. As discussed in Section 2, this initial model is not always correct because of missing information and topological inconsistency. To circumvent that, we create a segmented 3D map from the template heart model and annotate it properly that helps build a correspondence with the initial and possibly incomplete model created from the imaging data. Further we match and fit the patient specific model with the segmented 3D map and inherit the information encoded there to fill up the gap of the missing information as well as remove the noisy spurious components that could not be avoided due to ambiguity in the imaging data. Figure 6(a) shows the template model used in this paper.

Section 2 discusses the steps necessary to build the initial heart model from the imaging data. Section 3 discusses the steps necessary to build the template segmented 3D map. Finally, in Section 4, we discuss the tasks that are needed to accomplish to finally obtain the patient specific model of heart.

2 PATIENT SPECIFIC DATA PROCESSING

In this section, we discuss the issues that need to be tackled in order to process the imaging data effectively. In the following subsections, we discuss the four main steps, namely *Segmentation or Classification*, *Regularization*, *Reconstruction* and *Pruning*. Additionally, in each of these subsections, we show the results of each of these steps on an imaging dataset. The dataset is courtesy Dr. Charlie Walvaert of Austin Heart Hospital. The imaging dataset is of dimension $512 \times 512 \times 432$ and the spacing in x, y, z directions are respectively 0.390625 mm, 0.390625 mm, 0.3 mm.

2.1 *Image segmentation*

Segmentation is a way to dissect the features of interest from their surroundings. In case of the heart dataset, we aim to extract the four chambers automatically. To this end, we have developed a computational procedure based on the fast marching method (Sethian 1996; Malladi and Sethian 1998; Sethian 1999). In this method, a contour is initialized from a pre-chosen seed point, and the contour is allowed to grow until a certain stopping condition is reached. The traditional fast marching method is designed for single object segmentation. In order to segment multiple objects, like the chambers in the heart data, a seed for each of the

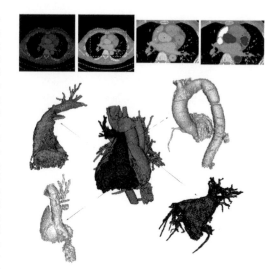

Figure 3. Top row shows the slice of the input image, the result after contrast enhancement and anisotropic filtering, seed selection for segmentation and finally the result of segmentation process on that single slice. The second row shows the overall segmentation of the imaging data into four subvolumes. The dissociation is enhanced by coloring it differently. The subfigures show the exploded view of the individual subvolumes.

components must be chosen. When contours from different seed points, they should stop each other on their boundaries. This multi-seeded fast marching method (Yu and Bajaj 2005; Sifakis and Tziritas 2001) simultaneously segment all the components and hence is extremely useful to separate multiple features that are too close to segment sequentially.

Since the given patient-specific heart data is very noisy and has a low contrast, it is always useful to filter the noise and enhance the contrast before we segment the features of interest. We have developed a fast and adaptive method for contrast enhancement (Yu and Bajaj 2004), which was applied to the given heart data. Additionally, an anisotropic noise reduction approach (Perona and Malik 1990; Bajaj, Wu, and Xu 2003) was employed to smooth out the noise as seen in the original data. Compared to the isotropic (e.g., Gaussian) filtering methods, the anisotropic approaches can preserve sharp features much better while the noise is reduced. Figure 3 shows the result of image segmentation on the imaging data.

2.2 *Regularization*

This step is optionally employed to remove the artifacts of the voxel-based classification. The level-set based segmentation can involuntarily produce voxels which are, although connected, can be thought of as dangling components. Such spurious subset of voxels pose

Figure 4. Reconstruction of four segmented boundaries of the patient specific heart. With each surface mesh the corresponding pointset is shown in the same color.

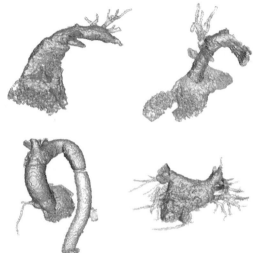

Figure 5. Result of pruning: The green portions in each of the four extracted components contains reliable information that can be used to match the patient specific data with the template model. The missing or spurious portions of each component are drawn in white.

problem in reconstructing a surface out of the boundary voxels. Therefore a careful selection/removal of a subset of voxels output by the segmentation step is crucial. The result of this step is a set of boundary voxels whose centers lie on the surface of the segmented boundary with all dangling components removed. In our experiments, we have not encountered such cases with the dataset that we dealt with. Nevertheless, this step should optionally be included in the pipeline for curation of the segmented region boundaries.

2.3 Reconstruction

From the regularized voxel centers, we reconstruct the surface that faithfully depict the surface triangle mesh of the boundary in question. There are number of reconstruction technique available for this purpose (Bajaj, Bernardini, and Xu 1995; Amenta and Bern 1999). For our purpose we use the *TightCocone* algorithm by (Dey and Goswami 2003). Sometimes, the noise present in the data is a major challenge to deal with and to circumvent that problem we employ the version of that algorithm that deals with noisy point cloud – *RobustCocone* (Dey and Goswami 2004).

This step results in the surface meshes which are good candidates for further fitting operation. The reconstructed surfaces of the components for the dataset is shown in Figure 4. As apparent from the pictures, the reconstructed surface contains several portions which are noisy and also some blood vessels which should not be used for matching the patient specific data with the template atlas. In the next subsection we describe that step.

2.4 Pruning

The quality of the imaging data is the main bottleneck in the modeling procedure and therefore it is necessary to clean-up the data before building the correspondence with the template atlas. The goal of this step is to identify the portions of the patient heart data which are missing and also the spurious components erroneously classified in the segmentation process because of weak intensity variation. We employ the geometry segmentation approach on the regularized pointset to achieve this goal. The segmentation approach, as described in (Dey, Giesen, and Goswami 2003), relies on a parameter which determines if two adjacent maxima should be clustered together and form a bigger segment. Choosing this parameter carefully removes the small spurious components as well as the noisy incomplete portions of the patient heart models. Figure 5 shows the result of this step.

After this step we have a weaker annotation of the parts of the patient specific heart models – the ones which are stable and can be used for correlation with the atlas; and the ones which are either spurious or incomplete. The segments belonging to the second class can not be used for building correspondence with the template parts.

3 TEMPLATE PREPARATION

The patient specific imaging data is often incomplete and contains topologically inconsistent and spurious components. We rectify such anomalies by inheritance of topology from the template heart model. In order to do so, we first process the template geometry and annotate it following the heart anatomy. We call the template geometry T, which has two distinct components – T_{out} and T_{in}. The analog of T_{in} in the patient data is the inner wall of the heart which interfaces with the blood being circulated and the analog of T_{out} is the outer boundary where the heart is embedded among soft tissues and muscles. Below we describe the major steps in processing the solid bounded by T_{in}, and construct a segmented 3D map of the template geometry which is key to inherit the topological and anatomical information into the patient specific imaging data.

3.1 Geometry segmentation

Given T_{in}, we decompose it into 4 connected components -

1. Left Atrium – T_{LA}
2. Left Ventricle – T_{LV}
3. Right Atrium – T_{RA}
4. Right Ventricle – T_{RV}

Left and right ventricles are additionally segmented into the valves and aortic arches.

The key ingredient in this segmentation process is the careful analysis of the critical points of the distance function induced by each T_*.

Given any shape S, one can define a distance function $h_S : \mathbb{R}^3 \to \mathbb{R}$ which assigns to every point in the three dimensional space its distance to the nearest point on the object S. The function h_S can be approximated by a similar function h_P when S is known only via a finite set of points P sampled from S. This function, which is popularly known as *distance function* has a rich history of application and especially the critical point structure of this function encodes a lot of information about the shape attributes of S. For a list of prior work, and especially on the topological invariants of the critical point structure, see (Bajaj, Bernardini, and Xu 1995; Edelsbrunner 2002; Giesen and John 2003).

For the purpose of segmentation, we use the partition of space by gradient uniformity which is otherwise known as the *stable manifold* of the critical points. The stable manifolds are computed efficiently via the Voronoi-Delaunay diagram of the pointset P. Details are given in (Dey, Giesen, and Goswami 2003).

3.2 Annotation

In the context of shape attributes, it is often required to annotate the decomposed parts as *tubular* or *flat* or *blobby*. Such annotation also can be performed via careful analysis of the critical point structure of h_P. The key ingredient to achieve this is a construct analogous to stable manifold, *unstable manifold*. These are

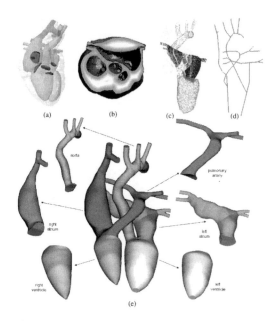

Figure 6. (a) Template model of human heart. The inner boundary is shown inside the transparent outer boundary of heart. (b) A cross section through the middle that is colored according to the value of the Signed Distance Function. (c) The skeleton of all four chambers (colored cyan). (d) Only the skeleton to help visualize the connectivity structure. (e) Complete segmented template 3D map.

partitions of space in accordance with negated gradient uniformity. It was shown in (Goswami, Dey, and Bajaj 2006), that unstable manifold of the index 1 and 2 saddle points reveal the *flat* and *tubular* features respectively. For our purpose, we apply the annotation process on every decomposed part.

As the Figure 6 (c,d) shows, the unstable manifold of the index 2 saddle points additionally produces the skeleton of the object. For tubular regions, the skeletons are particularly useful as they can be used to fit a NURBS model as was done previously by (Zhang, Bazilevs, Goswami, Bajaj, and Hughes 2006).

3.3 Template Segmented 3D Map creation

The process of segmentation and feature annotation create a complete description of the template which we call a *Template Segmented 3D Map*. This segmented 3D Map has different components of the model heart properly decomposed and tagged with domain knowledge of heart anatomy as to which component corresponds to which ventricle or aortic arch or atrium etc. We show the prepared atlas in Figure 6.

Once the template map is created, we build a correspondence table as to which part of the segmented patient data should be matched with which

Figure 7. Correspondence between the informative components of the patient specific heart (green) and the template model (red). Every green part from the initial patient model is matched with the red part of the template.

part of the segmented template. Figure 7 shows the correspondence.

4 CONCLUSION AND FUTURE WORK

In this paper, we have presented our current status of ongoing work on creating a patient specific model of heart from high resolution CT imaging data. We have developed a pipeline and described the steps that constitute the pipeline.

The remaining step is to fit the solids from the template atlas to the pruned components of the patient heart flexibly without violating the topological invariants of the template that conform with the heart anatomy. The template provides the invariant which, after the flexible fitting is performed, shall help fill up the missing information in the image and also remove the extraneous components from the model faithfully.

ACKNOWLEDGMENTS

We would like to thank New York University for the initial heart model, which we modified to suit our purpose as a heart template model. We also thank Dr. Charlie Walvaert of the Austin Heart Hospital for providing us with the CT64 thoracic scan. Thanks are also due to Joe Rivera and Jasun Sun of CVC, for their immense help in data processing. Lastly, we would also like to thank Jyamiti group of The Ohio State University for the reconstruction and segmentation software, namely *TightCococne*, *RobustCocone* and *SegMatch*. This research is supported in part by NSF grants ITR-EIA-0325550, CNS-0540033 and NIH grants P20 RR020647, R01 GM074258-021 and R01-GM073087.

REFERENCES

Amenta, N. and M. Bern (1999). Surface reconstruction by voronoi filtering. *Discr. Comput. Geom. 22*(4), 481–504.

Bajaj, C., F. Bernardini, and G. Xu (1995). Automatic reconstruction of surfaces and scalar fields from 3D scans. In *ACM SIGGRAPH*, pp. 109–118.

Bajaj, C., Q. Wu, and G. Xu (2003). Level-set based volumetric anisotropic diffusion for 3d image denoising. In *ICES Technical Report, University of Texas at Austin.*

Dey, T. K., J. Giesen, and S. Goswami (2003). Shape segmentation and matching with flow discretization. In F. Dehne, J.-R. Sack, and M. Smid (Eds), *Proc. Workshop Algorithms Data Strucutres (WADS 03)*, LNCS 2748, Berlin, Germany, pp. 25–36.

Dey, T. K. and S. Goswami (2003). Tight Cocone: A watertight surface reconstructor. In *Proc. 8th ACM Sympos. Solid Modeling and Applications*, pp. 127–134.

Dey, T. K. and S. Goswami (2004). Provable surface reconstruction from noisy samples. In *Proc. 20th ACM-SIAM Sympos. Comput. Geom.*, pp. 330–339.

Edelsbrunner, H. (2002). Surface reconstruction by wrapping finite point sets in space. In B. Aronov, S. Basu, J. Pach, and M. Sharir (Eds.), *Ricky Pollack and Eli Goodman Festschrift*, pp. 379–404. Springer-Verlag.

Giesen, J. and M. John (2003). The flow complex: a data structure for geometric modeling. In *Proc. 14th ACM-SIAM Sympos. Discrete Algorithms*, pp. 285–294.

Goswami, S., T. K. Dey, and C. L. Bajaj (2006). Identifying flat and tubular regions of a shape by unstable manifolds. In *Proc. 11th Sympos. Solid and Physical Modeling*, pp. 27–37.

Malladi, R. and J. Sethian (1998). A real-time algorithm for medical shape recovery. In *IEEE International Conference on Computer Vision*, pp. 304–310.

Perona, P. and J. Malik (1990). Scale-space and edge detection using anisotropic diffusion. *IEEE Trans. on Pattern Analysis and Machine Intelligence 12*(7), 629–639.

Sethian, J. (1996). A marching level set method for monotonically advancing fronts. *Proc. Natl. Acad. Sci. 93*(4), 1591–1595.

Sethian, J. (1999). *Level Set Methods and Fast Marching Methods, 2nd edition*. Cambridge University pPress.

Sifakis, E. and G. Tziritas (2001). Moving object localization using a multi-label fast marching algorithm. *Signal Processing: Image Communication 16*(10), 963–976.

Toshiba Medical Systems – 64 Slice CT (2006, July). Clinical advancement in volumetric CT.

Yu, Z. and C. Bajaj (2004). A fast and adaptive algorithm for image contrast enhancement. In *Proc. IEEE International Conference on Image Processing*, pp. 1001–1004.

Yu, Z. and C. Bajaj (2005). Automatic ultra-structure segmentation of reconstructed cryo-em maps of icosahedral viruses. *IEEE Transactions on Image Processing: Special Issue on Molecular and Cellular Bioimaging 14*(9), 1324–1337.

Zhang, Y., Y. Bazilevs, S. Goswami, C. L. Bajaj, and T. J. R. Hughes (2006). Patient-specific vascular nurbs modeling for isogeometric analysis of blood flow. In *Proc. 15th Int. Meshing Roundtable*, pp. 73–92.

Computational Modelling of Objects Represented in Images –
João Manuel R.S. Tavares & R.M. Natal Jorge (eds)
© 2007 Taylor & Francis Group, London, ISBN 978-0-415-43349-5

Exploring non-Gaussian behaviour in EEG data using random field methods

Shangqi Bao & J.F. Ralph
Department of Electrical Engineering & Electronics, University of Liverpool, Liverpool, UK

G. Meyer & S. Wuerger
Centre for Cognitive Neuroscience, School of Psychology, University of Liverpool, Liverpool, UK

ABSTRACT: This paper investigates the use of thresholded random fields for the detection of significant activation in the evoked potentials in Electroencephalography (EEG) data. Three different activation states are investigated: two uni-modal signals (audio (A) and video (V) response) and one cross-modal signal (using simultaneous audio and video stimuli, A + V). The data is assessed using standard Gaussian-based hypothesis testing methods and more general methods based on non-Gaussian random fields.

Keywords: Signal detection, EEG processing, non-Gaussian random fields.

1 INTRODUCTION

Electroencephalography (EEG) is a noninvasive tool that records the electric activity of the brain, using extremely sensitive electrodes placed on scalp. The resulting activity is a set of raw EEG potentials which map the electrical activity of the brain. Particular activity associated with the processing of a specific event is called an '*event-related potential*' (ERP), whose morphological structure (shape) is consistent over a number stimuli of the same type. Unfortunately, ERPs are of relatively small amplitude, ranging from less than 1 microvolt to as many as 10 microvolts. This is in comparison to background EEG values that can range from 10 to 100 microvolts. As a result of this size disparity ERPs cannot be seen easily in a raw EEG tracing. Another inherent property of these records is that there is a large amount of noise caused by internal or external sources such as muscle movement, blood flow, distance between neural sources and electrodes. These may lead to a poor signal to noise ratio and limit the ability to extract and interpret the signals received.

When processing EEG signals the challenge is to determine the underlying neural sources that generate the measured activities. The challenge for signal processing is two-fold: to reduce the noise level (without corrupting the underlying data) and to detect significant changes in electrical activity when the stimulus is changed. This paper concentrates on the second of these problems – detecting changes in the evoked potentials due to different types of stimuli. There are basically two kinds of tests for whether a response is present: one based on the similarity of a measurement across replications (the T-test or phase coherence) and the other based on the difference between a measurement at the frequency of stimulation (signal) and other measurements (noise) in the spectrum (the F-test).

Both of the aforementioned ways to detect significant changes in a signal use confidence intervals and measures of statistical significance. However, most statistical significance tests rely on parametric methods and are based on the Gaussian distribution. This paper compares the results of a non-Gaussian method based on thresholding random fields and a standard Gaussian significance test (the F-test). Due to the strongly non-Gaussian nature of the EEG data, we would expect that the F-test method is unreliable, since a signal which is purely noise is likely to fail tests by exceeding the number of false positives allowed by the Gaussian assumption. The Random Field Theory (RTF) has been developed to give adjusted p-values for existing confidence test results for brain activation [1]. Its use in generating a more general threshold method, which is based on non-Gaussian random fields, is assessed.

2 DATA & METHODS

2.1 Stimuli

Nine subjects participated in this study. The stimuli consist of three different signals: two uni-modal stimuli (a pulsed audio signal and a video pulse from an LED – both at fixed frequencies) and a cross-modal signal (simultaneous audio and video stimulation). The experiment was repeated 20 times for each subject and a total of 180 trials were generated for each state. The frequencies of the audio and video stimuli were 20 Hz and 13 Hz respectively. The response signals/evoked potentials are recorded by a 129-channel, whole-head axial gradiometer system (using 1 reference channel placed in the middle of the calvariae – i.e. the dome of the skull).

The records to each stimulus were pre-selected to discard the obviously distorted trials and off-work channels. The rest of the data were taken from 400 to 5000 ms with 3.6 ms resolution, which are believed as post-stimulus response. (The first 400 ms were discarded to remove the comparatively large 'P300' signal generated by the onset of the stimuli, which persists for approximately 300 ms). Each of the channel records were de-noised by a zero-phase forward and reverse digital filter. Then, the Fourier transform was taken of all the remaining trials. The results were then plotted as the localized frequency responses on a headplot image (see Figure 1 – generated using EEGLAB [2]). The headplots show the evoked potentials at particular frequencies, referenced to the position of the electrodes during the experiment.

2.2 Significant test protocols

These Fourier spectrums consisting of 3 kinds of stimuli were initially accessed by looking at the amplitude

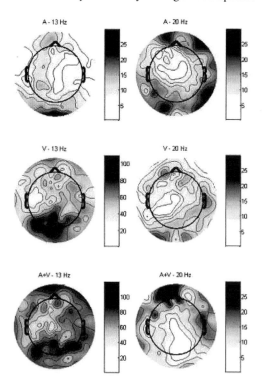

Figure 1. Example headplot images for the three different stimuli for the two frequencies of interest.

spectra of the de-noised potentials. These spectra showed the responses at the frequencies equal to the modulation rates of the carrier frequencies (A [20 Hz] and/or V [13 Hz]). However, no obvious signal was found any of the frequencies where some type of mixing might be expected – if some low level processing was generating any harmonic mixing (e.g. a signal might be expected at around 7 Hz in the A + V cases due to nonlinear mixing of the known frequencies). To investigate whether any underlying (or hidden) periodicity was present the flowing significance test methods were used.

2.2.1 F-test for hidden periodicity

This test derives from the initial description of Schuster and further work by Fisher [3]. To apply these tests, the significance of fluctuations needs to be compared to the normal background fluctuations, where no known signal is present. Any significant deviation from normal background fluctuations can be characterized by a significance level or probability, which indicates the likelihood of the signal being due to an underlying process and not the background noise. An estimate of the background noise can be obtained from frequencies where no stimulus occurred. We estimated the signal-to-noise ratio by comparing the power at each stimulus-frequency (a_s^2), equivalent to the sum of the squares of the real and imaginary parts of the FFT, to the average power at 120 nearby frequencies (60 above and 60 below the stimulus frequency), excluding the frequencies where there were other stimuli [4]:

$$R_F = \frac{120 \cdot |a_s|^2}{\sum_{i=s-60,\, i \neq s}^{i=s+60} |a_i|^2}$$

Since the spectra were derived from a sweep lasting 4.6 s, power measurements were available at a resolution of 1/4.6 or 0.217 Hz. The noise estimates therefore came from 13.0 Hz (i.e. 0.217 × 60) above and below the frequency of interest. The significance of this ratio can be evaluated using an F distribution table with (2 and 240) degrees of freedom.

2.2.2 Random field theory

Random field theory was devised to solve multiple comparison problems in functional imaging [5]. It uses results that give the expected Euler characteristic (EC) for a smooth statistical map that has been thresholded. As with the F-test, the method generates a distribution that represents the underlying properties of the background noise (in this case the Euler characteristic of a thresholded random field with no signal), and then compares the received signals with this model for the background noise. The deviation between the received signal response and the underlying noise model give an indication of the likelihood that a true signal is present.

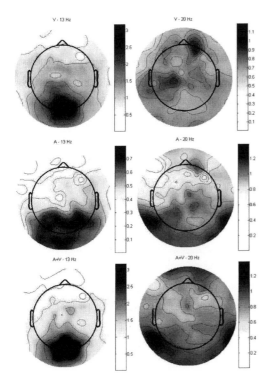

Figure 2. Example headplot images of F-value for the three different stimuli for the two frequencies of interest.

The application of RFT is in stages. First is the estimation of the smoothness (spatial correlation) of the selected statistical map. Then, the statistical map is thresholded at different levels to map changes in the number for regions above the thresholds. The number of regions is called the *Euler Characteristic*. The by mapping the expected EC as a function of the threshold value, it is possible to obtain a good approximation for the probability of observing one or more distinct regions at that threshold. Methods to estimate the expected EC are discussed in detail in reference [6].

3 RESULTS

The example results for implementation of F-test on the known carrier frequencies are shown in Figure 2. Unfortunately, there is no significant response at other frequencies of interest (such as the difference $20\,\text{Hz} - 13\,\text{Hz} = 7\,\text{Hz}$). The question then turns to what is the minimum amplitude can be detected as significant by an F-test? In other words, how sensitive is the F-test to periodic signals in the presence of so much noise? (This should allow us to set an upper limit on any undetected signal that was present). To solve this problem, a sinusoidal signal was added in to a sensor from the front part of brain where no neural response should appear. Figure 3 illustrates the difference between the original brain activity and the adjusted one. From hundreds of tests, the minimum amplitude is around 8% of the background amplitude.

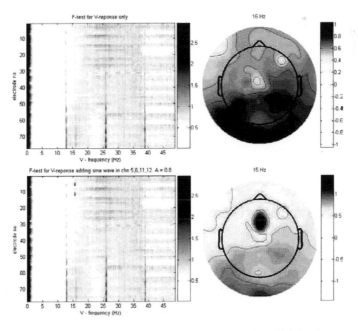

Figure 3. Frequency spectrum and headplot of original response and added in sinusoidal signal.

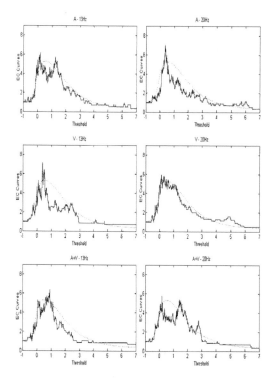

Figure 4. Example Euler Characteristic (EC) curves for the three different stimuli for the two frequencies of interest.

This further indicated that the F-test is unreliable since most of the background noise is larger than the minimum amplitude which was detected as significant.

Due to the failure of the F-test, RTF was applied to the EEG records. Before starting to threshold the headplot images, it is necessary to scale the entire raw images to be zero mean and unit variance. The re-adjusted images were then thresholded – that is, setting all the pixels to zero with their value less than or equal to the threshold and the rest of pixels above the threshold to be one. The variation in the number of regions of connected pixels appearing in the thresholded image, as the threshold varies, were recorded as curves. By comparing the features of the EC curves for each frequency, differences between the interesting frequencies and the others can be found.

To verify the nature of the headplot images, thousands of under-Gaussian-distributed random images were generated, using the appropriate scaling, image means and Gaussian smoothing. A comparison of the averaged EC curves for the sample images and the experimental images is plotted in Figure 4. It is clear that the EC curves of the experimental images are significantly different to the simulated Gaussian data, which indicates the non-Gaussian nature of the experimental data. This is also borne out by comparisons of the corresponding F-test data.

By comparing the averaged EC curves for the different frequencies (i.e. using only the experimental data), it is possible to determine which frequencies contain more (or less) significant areas of activation compared to the mean response. The region of high activation towards the back of the head shown in the V – 13 Hz headplot in Figure 1 corresponds to the pronounced 'plateau' between threshold values 1–3 in the corresponding EC curve in Figure 4. This method further proved the absence of low-level integrated brain activity on the expected frequency (7 Hz) as its EC curve is very similar than the baseline curves from generated random images.

4 CONCLUSIONS

This paper has applied two significance test methods (the F-test and Random field methods) to detect the significance level of brain evoked potentials. The presence of highly-correlated non-Gaussian noise in the evoked potentials recorded by EEG sensors means that standard Gaussian-based significance tests may be unreliable – with more false positive results than expected. The use of thresholded random fields can be used to assist in the rejection of false positives and improve the overall detection of significant areas of electrical activity within the brain. The results indicate that there is no significant activation in the brain at the frequencies where low-level neural processing might be expected – i.e. frequencies corresponding to nonlinear mixing between the applied signals (such as the frequency difference 20 Hz – 13 Hz = 7 Hz in this case).

REFERENCES

[1] Worsley, K.J., Marrett, S., Neelin, P., Vandal, A.C., Friston, K.J., and Evans, A.C. (1996). 'A unified statistical approach for determining significant signals in images of cerebral activation'. Human Brain Mapping, 4: 58–73.

[2] Delorme, A., Makeig, S. (2004). 'EEGLAB: an open source toolbox for analysis of single-trial EEG dynamics', Journal of Neuroscience Methods 134: 9–21.

[3] Schuster A. (1898) 'On the investigation of hidden periodicities with application to a supposed 26 day period of meteorological phenomena'. Terrestr. Magnet. Atmos. Electr. 3: 13–41.

[4] Picton, T.W., Dimitrijevic, A., John, M.S., and Van Roon, P. (2001). 'The use of phase in the detection of auditory steady-state responses'. Clin. Neurophysiol., 112, 1698–1711.

[5] Brett, M., Penny, W., and Kiebel, S., (2003) 'An introduction to random field theory, in: Human Brain Function II', Academic Press, London, UK, Chapter 14.

[6] Worsley, K.J., Evans A.C., Marrett, S., and Neelin, P. (1992) 'A three dimensional statistical analysis for CBF activation studies in the human brain'. Journal of Cerebral Blood Flow and Metabolism, 12: 900–918.

Computational Modelling of Objects Represented in Images –
João Manuel R.S. Tavares & R.M. Natal Jorge (eds)
© 2007 Taylor & Francis Group, London, ISBN 978-0-415-43349-5

Multimodal macula mapping by deformable image registration

P. Baptista
Center of New Technologies for Medicine (CNTM) from the Association for Innovation and Biomedical Research on Light and Image (AIBILI), Coimbra, Portugal

J. Ferreira
Institute of Systems and Robotics (ISR), Coimbra, Portugal

R. Bernardes
Center of New Technologies for Medicine (CNTM) from the Association for Innovation and Biomedical Research on Light and Image (AIBILI), Coimbra, Portugal

J. Dias
Institute of Systems and Robotics (ISR), Coimbra, Portugal
Faculty of Sciences and Technology, University of Coimbra, Coimbra, Portugal

J. Cunha-Vaz
Center of New Technologies for Medicine (CNTM) from the Association for Innovation and Biomedical Research on Light and Image (AIBILI), Coimbra, Portugal
Institute of Biomedical Research on Light and Image (IBILI), Faculty of Medicine, University of Coimbra, Coimbra, Portugal

ABSTRACT: This work aims to present the applied methods for the registration of different imaging modalities of the human macula. Multimodal image registration is herewith considered the process of overlaying two or more images of the same eye, taken from different viewpoints, at different times, with different sizes and fields-of-view and using different sensors, thus imaging different retinal structures. Additionally, the registration of retinal images from *in vivo* patients using scanning systems has to deal with intrinsic image distortions due to saccadic eye movements. In order to correct for differences in eye fundus scanned angles, image sizes and feature expressions in different modalities, a *feature vector* was built for each modality embodying complementary information from the image alone.

Keywords: Multimodal, Registration, Deformable Registration.

1 INTRODUCTION

Multimodal macula mapping is the combination of a variety of diagnostic imaging modalities to examine the macular region in order to obtain information on its structure and function. The macula is located in the posterior pole of the retina and is responsible for detailed and color vision, thus constituting an important area for human vision, as any macular alteration will, sooner or later, affect visual acuity (Bernardes, Lobo, and Cunha-Vaz 2002). Currently, a wealth of retina imaging modalities exists, either on daily clinical practice or research environments, such as color fundus photography, red-free fundus photography, fluorescein angiography, indocyanine green angiography, optical coherence tomography, retina

flowmeter, fundus autofluorescence, leakage analysis, multifocal electroretinography, etc. The potential for multimodal macular mapping was demonstrated in (Bernardes, Lobo, and Cunha-Vaz 2002).

Laliberté and Gagnon stated in (Laliberté and Gagnon 2003) that ophthalmology is one of the medical areas where diagnosis implies the analysis of a large number of images. This is true even for a single eye followed during a short period of time due to the number of complementary examination procedures in use for pathologies like diabetic retinopathy (DR) or age-related macular degeneration (AMD), to name but a few. Duncan and Ayache, in (Duncan and Ayache 2000), gave an overview of medical image analysis discussing, among other topics, methodologies for image matching/registration. On their comments on multimodal image registration we can find the problems due to images being formed by various imaging modalities producing different types of information and often at different spatial resolutions.

Published work on multimodal retinal image registration deals mostly with images acquired using area sensors, with which the whole image is acquired at once (Laliberté and Gagnon 2003; Zana and Klein 1999; Matsopoulos, Asvestas, Mouravliansky, and Delibasis 2004), while available retinal imaging modalities encompass a number of different sensors and sensor types.

As the use of confocal scanning laser ophthalmoscopes (CSLO) (Webb, Hughes, and Delori 1987) becomes more commonplace, a new problem arises for the registration of this sort of images, more precisely the intrinsic image distortions due to saccadic eye movements, which in turn can be of two kinds: voluntary (e.g., when the eyes move from one fixation point to another as in reading) and involuntary (e.g., due to following the flying spot that illuminates the eye fundus). These effects in the image will obviously depend on the amount of time for the full scan, e.g., 400 ms for the Zeiss CSLO used in (Bernardes, Lobo, and Cunha-Vaz 2002) to 1.6 s for the CSLO used in this work. This suggests the need for deformable image registration procedures to correct for local deformations as opposed to the statement of global deformations on ophthalmic images by Laliberté and Gagnon in (Laliberté and Gagnon 2003) and the statement of Zana and Klein in (Zana and Klein 1999), where the authors restricted their work to central images containing the macula, papila and temporal vessels to limit deformations. In (Pinz, Bernögger, Datlinger, and Kruger 1998), Pinz *et al.* published a multimodal imaging system based on a scanning laser ophthalmoscope without taking into consideration the distortions herein considered. Nevertheless, the imaging device used in their work was not specified besides image size and scanned angle, 760×430 pixels and $40°$, respectively. The system may, therefore, be fast enough to produce results similar to an area sensor and thus not showing local distortions.

In our work we will focus on two different types of light detectors – area sensors, either of color or monochromatic type, and single-point sensors (e.g., photodiodes) – to acquire *in vivo* 2D morphology and/or functional information of the human retina. Moreover, we will address the problem of registering different modalities based on different sensors and covering different fields-of-view (FOV), e.g., $20°$ high-speed fluorescein angiography centered on the fovea (performed by a CSLO) to a $50°$ color fundus photograph. Another major difference to published literature is related to the fact that being centered on the foveal area and using small FOV, we are faced with a low number of retinal structures. Furthermore,

the modalities herewith considered gather a different level of details for this particular region of the human eye, therefore not being possible to apply the methods presented in (Laliberté and Gagnon 2003; Zana and Klein 1999; Matsopoulos, Asvestas, Mouravliansky, and Delibasis 2004).

In the remaining of this text we will present a set of methodologies that allow automatic registration of images from different imaging modalities (e.g., retinographies and fluorescein angiographies) and using different sensor types (area and single-point sensors).

2 METHODS

2.1 *Modalities*

One of the most widely used retinal imaging modality is the retinography (color fundus photograph), either on film or digital format. Currently, a large number of digital color fundus photography devices exist, producing images ranging in size from thousands – to mega–pixels. In this work, a Zeiss Fundus Camera system, model FF450, was used to acquire 768×576 pixels RGB images of $50°$ FOV, centered on the fovea, thus imaging the whole macula, optic disc and temporal retinal arcades.

The other modality herewith considered is the fluorescein angiography, i.e., a fundus photograph taken after the administration of a dye (sodium fluorescein, NaFl, $C_{20}H_{10}O_5Na_2$). Fluorescein angiographies may be performed using either a regular camera, similar to the ones used for retinographies, or using a CSLO, thus being named high-speed fluorescein angiographies or SLO.FA. Using a series of fluorescein angiographies taken with a CSLO in different confocal planes, it is possible to compute the leakage of fluorescein into the vitreous and/or the permeability of the blood–retinal barrier to fluorescein (Lobo, Bernardes, Santos, and Cunha-Vaz 1999; Bernardes, Dias, and Cunha-Vaz 2005). In this work, a Heidelberg Retina Angiograph was used to acquire $20°$ SLO.FA of 256×256 pixels (maximum).

2.2 *Features*

The features available for each modality herein considered are the fovea, the optic disc and the vascular network for retinographies, and the fovea and a small set of the vascular network for the SLO.FA.

The retinography is pre-processed to be converted into a gray-scale image, where the fovea appears as a dark region, the optic disc appears as a bright area and the vessels, veins and arteries, are darker than the surrounding tissue, i.e., the background.

Vascular network expression in fluorescein angiographies is dependent on the time after injection, i.e., the amount of time elapsed since the dye was injected into the patient blood-stream. After injection, arteries become brighter as fluorescein starts appearing in the retinal circulation and veins are still dark. Thereafter, also veins end up getting brighter as fluorescein is collected after microcirculation. The foveal area is normally kept darker than background. For the time frame of interest for our application, 10 to 20 minutes after dye injection, the fovea is characterized by a dark region and the vascular network by bright vessels.

As different modalities may represent the same feature in different ways and cover different areas of the eye fundus using different image sizes, a feature vector of the type $f_{modality} = \langle modality, color\ model, field-of-view, image\ size, optic\ disc\ diameter, time-after-injection, ...\rangle$ is created for each modality, encompassing additional information from the image alone.

As modalities herein considered cover different areas of the eye fundus, 20° and 50°, a coarse registration approach is mandatory by computing the location of common features for both modalities. Therefore, for each modality, the location of the fovea is determined by computing their cross-correlation with an inverted Gaussian (1), with σ experimentally determined ($k = 1$).

$$g(i,j) = k\left[1 - \frac{1}{2}exp\left(\frac{-(i^2 + j^2)}{2\sigma^2}\right)\right] \ . \tag{1}$$

For each modality, the vascular network is detected through skeletonization after vascular tree segmentation. This segmentation was achieved resorting to contourlets (Do 2003), an efficient directional multiresolution image representation which provides the usual advantages of wavelet analysis, adding anisotropy and directionality properties which the latter lacks, presenting the basis to a more robust method for smooth contour segmentation. In other words, it provides the same level of performance of wavelets using fewer coefficients for this type of segmentation (Do 2003) – this fact was confirmed in our experimental studies.

On the other hand, comparatively to other techniques based on radically different approaches for segmentation, such as morphology or differential geometry, the contourlet method has proved in our experimental studies to be more robust in the presence of noise in images, or for low resolution/small FOV angle modalities (e.g. 20° SLO.FA images) where even apparently high signal-to-noise ratios greatly influence the vessel detection procedure.

2.3 Global rigid registration

Since modalities have different image sizes and FOV and the registration approach will not be based on

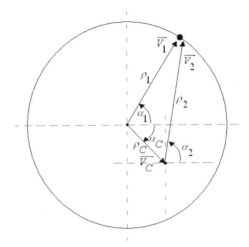

Figure 1. Vectorial representation of the translation of the centre of the fovea.

computing a transformation matrix (affine or perspective) based on a set of corresponding landmarks, it is mandatory to have all modalities at the same scale, with each pixel representing the same FOV. In order to avoid the creation of artifacts, instead of interpolation we followed the approach of decimating the higher resolution modality to the resolution of the one having the lower number of pixels per degree of the eye fundus. Therefore, as the retinography presents a resolution of 14.5 pixels/degree and the SLO.FA presents 12.8 pixels/degree, the retinography was decimated to the latter resolution at the pre-processing stage.

Having extracted both the location of the fovea and the retinal vascular tree for each modality, it is possible to compute the rotation between modalities by computing the translation in polar coordinates of the extracted retinal vascular network.

As the location of the fovea in one modality may be estimated with some displacement relatively to the other modality, this displacement influences the rigid registration both on the translational part as well as in the rotational part, as seen in Figure 1. This relative displacement produces a distortion in the representation of the vascular network in polar coordinates, both in the angle and distance to the reference, i.e., $\alpha_1 \rightarrow \alpha_2$ and $\rho_1 \rightarrow \rho_2$ (Fig. 1), being these effects dependent on the relative distances between references (relative positions of detected foveas) and the point to be mapped into polar coordinates.

By computing block translations in polar coordinates, it is possible to compute both α_C and ρ_C (Fig. 1), since for α_C and $\alpha_C + \pi/2$ there is a zero shift in the angle and the maximal shift in ρ_C, being it equal to half the difference between ρ_{α_C} and $\rho_{\alpha_C + \pi/2}$. Both ρ_C and

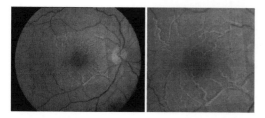

Figure 2. Result of rigid registration showing room for improvement.

α_C can be determined by a minimization algorithm as $\rho_2 = f(\rho_1, \rho_C, \alpha_1, \alpha_C)$ (2).

$$\rho_2 = \sqrt{\rho_1^2 + \rho_C^2 - 2\rho_1\rho_C \cos(\alpha_1 - \alpha_C)} \ . \tag{2}$$

This procedure allows to simultaneously correct differences in the computed fovea between images (modalities) as well as correct the respective distortions in polar coordinates, thus making it possible to compute the rotation between modalities. It is now possible to perform global rigid registration between modalities which represents the first step for deformable registration, as seen in Figure 2, i.e. to compute a 3×3 transformation matrix $\mathbf{P}$,

$$P = T(x_{retino}, y_{retino}) \times R(\theta) \times T(-x_{slo.fa}, -y_{slo.fa}) \tag{3}$$

where (x_{retino}, y_{retino}) and $(x_{slo.fa}, y_{slo.fa})$ (translation parameters) are the homogenous fovea coordinates on both modalities (after correction) and θ is the estimated rotation angle between modalities.

2.4 Deformable registration

After the initial approach using global rigid registration, it is possible to fine-tune the registration by considering the SLO.FA as a set of overlapping windows ($W_{SLO.FA}$) of size $M \times M$ centered on pixel (x_i, y_i), corresponding to pixel (x_f, y_f) on the retinography, i.e. $\mathbf{X_f} = \mathbf{P}\mathbf{X_i}$, with $\mathbf{X_f} = [x_f, y_f, 1]^T$, $\mathbf{X_i} = [x_i, y_i, 1]^T$ and $\mathbf{P}$ as computed in (3).

A W_{RETINO} window of size $N \times N$ centered on (x_f, y_f) (on the retinography) constitutes the search space where a match will be sought for the $W_{SLO.FA}$ window. This search is made by computing a similarity measurement using the partitioned intensity uniformity (PIU) defined by (4). For details see (Hill and Batchelor 2001).

$$PIU_A = \sum_b \frac{n_b}{N} \frac{\sigma_A(b)}{\mu_A(b)} \ . \tag{4}$$

For rotation angles below a given threshold (θ_{thres}), the matching is performed by ignoring the rotation,

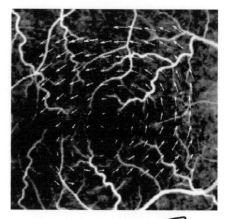

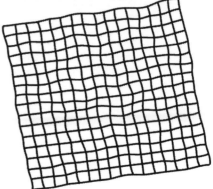

Figure 3. The deformation field calculated using equations above (scaled for better clarity) over the SLO.FA (top) and the grid deformation (bottom).

while for angles above θ_{thres} the SLO.FA is previously rigidly registered.

For each (W_{RETINO}, $W_{SLO.FA}$) window pair a translation is therefore computed on top of the previous transformation $\mathbf{P}$, thus being computed a transformation $\mathbf{P}'$ for each $W_{SLO.FA}(i, j)$ window position as given by

$$\mathbf{P}'_{i,j} = \begin{bmatrix} 1 & 0 & \Delta x(i,j) \\ 0 & 1 & \Delta y(i,j) \\ 0 & 0 & 1 \end{bmatrix} \mathbf{P} \ , \tag{5}$$

where i, j represents the $W_{SLO.FA}$ window index.

A set of control points, computed by (5), allows to establish two thin-plate splines (TPS), one for the displacement in the x-direction and the other for the y-direction. Figure 3 shows the computed deformation field.

3 RESULTS AND CONCLUSIONS

The final result achieved with the developed methodology as explained in this text is shown on Figure 4. The quality of the registration was qualitatively

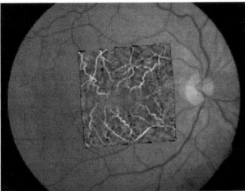

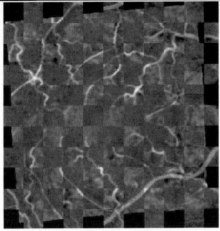

Figure 4. Top: The SLO.FA (gray-scale portion) shown over the full retinography (color). Bottom: Gray-scale chessboard-like representation of both modalities (only the common portion).

assessed as no quantitative algorithm was implemented yet. Nevertheless, it is now clear the need for the deformable registration for multimodal macula mapping when considering SLO–like modalities, and that the approach followed produces registration with the necessary accuracy for the current purposes.

REFERENCES

Bernardes, R., J. Dias, and J. Cunha-Vaz (2005, January). Mapping the human blood-retinal barrier function. *IEEE Trans. Biomed. Eng. 52*(1), 106–116.

Bernardes, R., C. Lobo, and J. Cunha-Vaz (2002, November–December). Multimodal macula mapping: A new approach to study diseases of the macula. *Surv Ophthalmol 47*(6), 580–589.

Do, M. (2003).Contourlets and sparse image expansions. In *Wavelets: Applications in Signal and Image Processing X. Proceedings of the SPIE*, pp. 560–570.

Duncan, J. and N. Ayache (2000, January). Medical image analysis: Progress over two decades and the challenges ahead. *IEEE Transactions on Pattern Analysis an Machine Intelligence 22*(1), 85–106.

Hill, D. and P. Batchelor (2001). Registration methodology: Concepts and algorithms. In J. Hajnal, D. Hill, and D. Hawkes (Eds.), *Medical Image Registration.*, The Biomedical Engineering Series, pp. 39–70. CRC Press.

Laliberté, F. and L. Gagnon (2003, May). Registration and fusion of retinal images – an evaluation study. *IEEE T Med Imaging 22*(5), 661–673.

Lobo, C., R. Bernardes, F. Santos, and J. Cunha-Vaz (1999, May). Mapping retinal fluorescein leakage with confocal scanning laser fluorometry of the human vitreous. *Arch Ophthalmol 117*(5), 631–637.

Matsopoulos, G., P. Asvestas, N. Mouravliansky, and K. Delibasis (2004, December). Multimodal registration of retinal images using self organizing maps. *IEEE T Med Imaging 23*(12), 1557–1563.

Pinz, A., S. Bernögger, P. Datlinger, and A. Kruger (1998, August). Mapping the human retina. *IEEE T Med Imaging 17*(4), 606–619.

Webb, R., G. Hughes, and F. Delori (1987, April). Confocal scanning laser ophthalmoscope. *Appl. Opt. 26*(8), 1492–1499.

Zana, F. and J. C. Klein (1999, May). A multimodal registration algorithm of eye fundus images using vessels detection and hough transform. *IEEE T Med Imaging 18*(5), 419–428.

Computational Modelling of Objects Represented in Images –
João Manuel R.S. Tavares & R.M. Natal Jorge (eds)
© 2007 Taylor & Francis Group, London, ISBN 978-0-415-43349-5

Isogeometric analysis of blood flow: A NURBS-based approach

Yuri Bazilevs, Yongjie Zhang, Victor M. Calo, Samrat Goswami,
Chandrajit L. Bajaj, & Thomas J.R. Hughes
Institute for Computational Engineering and Sciences, The University of Texas at Austin, United States

ABSTRACT: We describe a new approach for constructing patient-specific vascular geometries suitable for isogeometric fluid and fluid-structure interaction analysis of blood flow in arteries. We use solid NURBS (non-uniform rational B-splines) to define vascular geometries as well to perform analysis. It is argued in this paper that this new approach is a viable alternative to the finite element method, which is a standard tool for analysis of vascular systems. Advantages of the new approach are discussed and the technique is demonstrated on a variety of patient-specific arterial models.

1 INTRODUCTION

Patient-specific modeling and simulation-based medical planning has recently become an attractive avenue of research. This research is aimed at providing physicians with tools to construct and analyze combined anatomical/physiological models in order to predict the outcome of alternative treatment plans for an individual patient using techniques from image processing and computational mechanics. The finite element method is considered a standard tool for cardiovascular simulation. In the pioneering paper on the subject (Taylor, Hughes, and Zarins 1998) used real-life geometries to simulate blood flow thus opening the door for designing predictive technologies for vascular modeling and treatment planning.

In the finite element method complex geometrical objects are represented by a mesh of finite elements, typically piece-wise linear tetrahedra or hexahedra. Figure 1 illustrates examples of unstructured tetrahedral and hexahedral mesh of a patient-specific abdominal aorta built using the techniques outlined in (Zhang, Bajaj, and Sohn 2005) and (Zhang and Bajaj 2006). Basis functions, usually of low order, are constructed on the finite element partition, thus generating discrete spaces for approximate solutions of the underlying partial differential equations (see, e.g., (Hughes 2000)).

Isogeomeric Analysis is a new computational technique that improves on and generalizes the standard finite element method. It was first introduced in (Hughes, Cottrell, and Bazilevs 2005), and expanded on in (Cottrell, Reali, Bazilevs, and Hughes 2006). In an effort to instantiate the concept of isogeometric analysis, an analysis framework based on NURBS was built. Mathematical theory of the NURBS-based

approach was put forth in (Bazilevs, da Veiga, Cottrell, Hughes, and Sangalli 2006).

In this work we advocate the use of the NURBS-based isogeometric analysis for vascular applications as an alternative to the standard finite element approach. Some of the motivating factors are: (1) NURBS are able to compactly and accurately represent smooth exact geometries, that are natural for arterial systems, but unattainable in the faceted finite-element representation (2) NURBS-based isogeometric analysis is inherently a higher-order technique with approximation properties superior to low-order finite elements. Both factors, as well as some additional ones discussed later in this paper, should render fluid and structural computations more physiologically realistic.

A skeleton-based sweeping method was developed to construct hexahedral solid NURBS meshes for patient-specific models from imaging data. Templates are designed for various branching configurations to decompose the geometry into mapped meshable regions. Piece-wise linear semi-structured hexahedral meshes can also be constructed using this approach. Figure 1 shows examples of semi-structured hexahedral and solid NURBS meshes of a patient-specific abdominal aorta.

The remainder of the paper is organized as follows. In Section 2 we give a brief review of isogeometric analysis based on NURBS. In Section 3 we present a detailed discussion of the NURBS-based arterial cross-section construction and its implications on the analysis procedures from the stand point of accuracy and implementational convenience. In Section 4 we present examples of patient-specific models. In Section 5 we draw conclusions and outline future research directions.

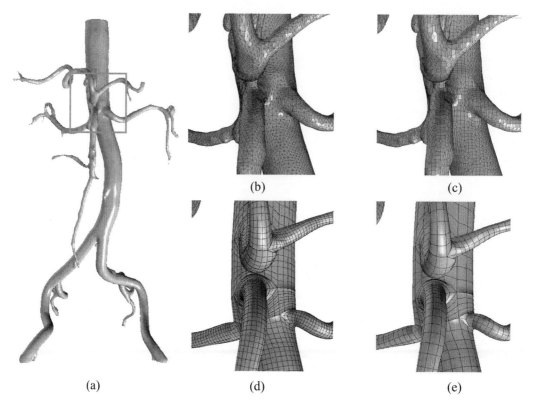

Figure 1. (a) – Geometrical model of a patient-specific abdominal aorta; (b) – Tetrahedral mesh; (c) – Unstructured hexahedral mesh; (d) – Semi-structured hexahedral mesh; (e) – Solid NURBS mesh. All meshes show a zoom on the branching area and are analysis suitable.

2 ISOGEOMETRIC ANALYSIS USING NURBS

In a NURBS-based isogeometric analysis a physical domain in $\mathbb{R}^3$ is defined as a union of patches. A patch, denoted by Ω, is an image under a NURBS mapping of a parametric domain $(0, 1)^3$

$$\Omega =$$

$$\{\mathbf{x} = (x, y, z) \in \mathbb{R}^3 \mid \mathbf{x} = \mathbf{F}(\xi, \eta, \zeta), \ 0 < \xi, \eta, \zeta < 1\}, (1)$$

where

$$\mathbf{F}(\xi, \eta, \zeta) = \sum_{i=1}^{n} \sum_{j=1}^{m} \sum_{k=1}^{l} R_{i,j,k}^{p,q,r}(\xi, \eta, \zeta) \mathbf{C}_{i,j,k}, \quad (2)$$

$$R_{i,j,k}^{p,q,r} = \frac{N_{i,p}(\xi) M_{j,q}(\eta) L_{k,r}(\zeta) w_{i,j,k}}{\sum_{\hat{i}=1}^{n} \sum_{\hat{j}=1}^{m} \sum_{\hat{k}=1}^{l} N_{\hat{i},p}(\xi) M_{\hat{j},q}(\eta) L_{\hat{k},r}(\zeta) w_{\hat{i},\hat{j},\hat{k}}} \quad (3)$$

In the above, $R_{i,j,k}^{p,q,r}(\xi, \eta, \zeta)$'s are the rational basis functions, and $\mathbf{C}_{i,j,k}$'s $\in \mathbb{R}^3$ are the control points. In the

definition of the rational basis, $N_{i,p}(\xi)$'s, $M_{j,q}(\eta)$'s, and $L_{k,r}(\zeta)$'s, are the univariate B-spline basis functions of polynomial degree p, q, and r; $w_{i,j,k}$'s, strictly positive, are the weights.

In isogeometric analysis the geometry generation step involves construction of a control mesh, which is a piecewise multi-linear interpolation of control points, and the corresponding rational basis functions. The initial mesh encapsulates the 'exact geometry' and, in fact, defines it parametrically.

For the purposes of analysis, the isoparametric concept is invoked (see (Hughes 2000)). The basis for the solution space in the physical domain is defined through a push forward of the rational basis functions defined in (2) (see (Bazilevs, da Veiga, cottrell, Hughes, and Sangalli 2006) for details). Coefficients of the basis functions, defining the solution fields in question (e.g., displacement, velocity, etc.), are called control variables.

As a consequence of the parametric definition of the 'exact' geometry at the coarsest level of discretization, mesh refinement can be performed automatically without further communication with the original description. This is an enormous benefit. There are NURBS analogues of finite element *h*- and *p*-refinement, and there is also a variant of *p*-refinement, which is termed *k*-refinement, in which the continuity of functions is systematically increased. This seems to have no analogue in traditional finite element analysis but is a feature shared by some meshless methods. For the details of the refinement algorithms see (Hughes, cottrell, and Bazilevs 2005).

3 CONSTRUCTION OF THE ARTERIAL CROSS-SECTION

Blood vessels are tubular objects, therefore we choose the sweeping method to construct meshes for isogeometric analysis. A central feature of the approach is the construction of the arterial cross-section template, which is based on the NURBS definition of the circle. A solid NURBS description of an arterial branch is then obtained by extrusion of a circular surface along the vessel path, projection of the control points onto the true surface, and filling the volume radially inward. Arterial systems also engender various branchings and intersections, which are also handled with a template-based approach described in detail in (Zhang, Bazilevs, Goswami, Bajaj, and Hughes 2006). The above procedure generates a multi-patch, tri-variate description of a patient-specific arterial geometry that is also analysis suitable.

In this section we focus on the construction of the cross-section template specific to fluid-structure interaction analysis. We identify the area occupied by the blood, or the fluid region, and the arterial wall, or the solid region. These two subdomains are separated by the luminal surface, or the fluid-solid boundary. Figure 2, which shows a schematic of the NURBS mesh

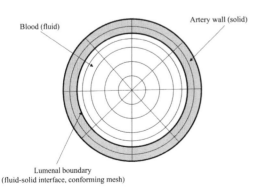

Blood (fluid)

Artery wall (solid)

Lumenal boundary
(fluid-solid interface, conforming mesh)

Figure 2. Arterial cross-section template based on a NURBS mesh of a circle. Fluid and solid regions are identified and separated by an interface. For analysis purposes basis functions are made C^0-continuous at the fluid-solid boundary.

of a circular cross-section, gives an illustration of the above decomposition. NURBS elements are defined as areas enclosed between isoparametric lines. Note that the isoparametric lines correspond to the radial and circumferential directions, and both engender linear parameterization. For computational purposes we isolate the fluid and solid regions by a C^0 line as the solution is not expected to have regularity beyond C^0 at the multi-physics interface. It is important to note that this does not introduce any changes in the geometry of the object or its parameterization.

Human arteries are not circular, hence projection of the template onto the true surface is necessary. Only control points that govern the cross-section geometry are involved in the projection process, while the underlying parametric description of the cross-section stays unchanged. The end result of this construction is shown in Figure 3 which shows the mapping of the template cross-section onto the patient-specific geometry. Here the isoparametric lines are somewhat distorted so as to conform to the true geometry, while the topology of the fluid and solid subdomains is preserved along with their interface.

As compared to the standard finite element method, the above approach has significant benefits for analysis, both in terms of accuracy and implementational convenience:

1. In the case of a flow in a straight circular pipe driven by a constant pressure gradient, NURBS basis of quadratic order gives rise to a point-wise exact solution to the incompressible Navier-Stokes equations. This also has implications on the overall accuracy of the approach.
2. Parametric definition of the NURBS mesh in the fluid region allows one to refine the boundary layer region near the arterial wall. This is crucial for overall accuracy as well as for obtaining accurate wall quantities, such as wall shear stress, which plays an important role in predicting the onset of vascular disease. It is well known that unstructured finite element boundary layer meshes lead to much less accurate solutions for a comparable number of degrees of freedom. In order to circumvent this

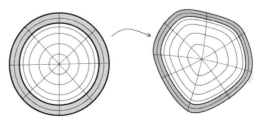

Figure 3. Arterial cross-section template is mapped onto a subject-specific geometry in a way that the topology of the fluid and solid subdomains remains unchanged.

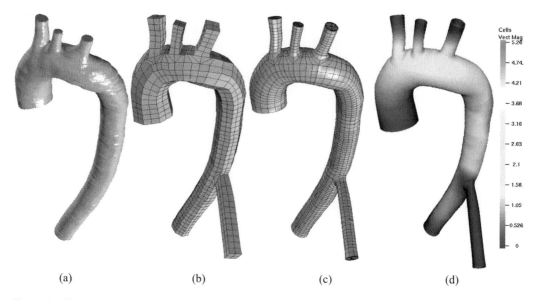

Figure 4. Thoracic aorta. (a) – surface geometry; (b) – control mesh; (c) – solid NURBS (41,526 elements); (d) – fluid-structure interaction simulation results: contours of the arterial wall velocity (cm/s) during late diastole plotted on the current configuration. Computational model contains an LVAD branch, which is assumed rigid.

shortcoming, adaptive boundary layer meshing is required, which is not an easy task, especially for unsteady flows. For recent work in this direction see (Sahni, Muller, Jansen, Shephard, and Taylor 2006).

3. Parametric definition of the NURBS mesh in the solid region allows one to easily define material anisotropy which is present in the arterial wall. See (Holzapfel 2004) for arterial wall material modeling which accounts for anisotropic behavior.

4. Fluid structure interaction applications involve motion of the fluid region. This is typically done by solving an auxiliary linear elastic boundary value problem for mesh movement (see, e.g., (Bazilevs, Calo, Zhang, and Hughes 2006)). Parametric mesh definition in the fluid region allows for a straight forward specification of these elastic mesh parameters. For example, we "stiffen" the mesh in the radial direction so as to preserve boundary layer elements during mesh motion.

4 NUMERICAL EXAMPLES

In this section we present applications of the new methodology to two patient-specific vascular models: a model of the thoracic aorta and a model of the abdominal aorta. Isogeometric analysis is then used to compute blood flow in the models. In all cases, time-dependent, viscous, incompressible Navier-Stokes equations were used as the blood model. The fluid density and dynamic viscosity were chosen to be representative of blood flow. Both models are subjected to a time-periodic inflow boundary condition, which simulates the input from a beating heart. All examples present fluid-structure interaction calculations using the isogeometric approach. The wall is assumed to be nonlinear elastic (see Bazilevs (Bazilevs, Calo, Zhang, and Hughes 2006) for the details of the mathematical formulation). All simulations were run in parallel.

Thoracic aorta model: Data for this model was obtained from CT Angiography imaging data of a healthy male over-30 volunteer. An extra branch, representing a left ventricular assist device (LVAD), was added to the arterial model. Evaluation of LVADs, as well as other electromechanical devices used to support proper blood circulation, is of great interest to the cardiovascular community. The surface geometry, the control mesh, and the solid NURBS model are shown in Figures 4a–4c. Figure 4d shows a result of the fluid-structure interaction simulation. Note that the inlet and the three smaller outlet branches were extended for the purposes of analysis.

Abdominal aorta: Data for this model was obtained from 64-slice CT angiography of a healthy male over 55 years of age. The surface geometry, the control mesh, and the solid NURBS model are shown in Figures 5a–5c. Figure 5d shows a result of the fluid-structure simulation. A computational study using a truncated geometrical model of this aorta was performed in (Bazilevs, Calo, Zhang, and Hughes 2006).

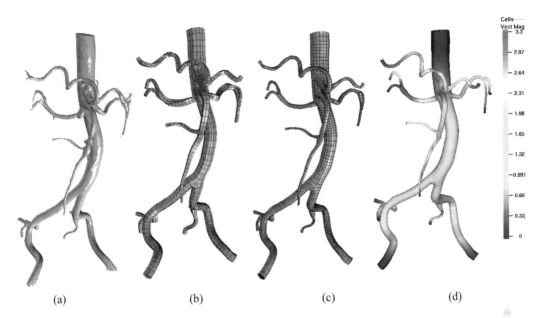

Figure 5. Abdominal aorta. (a) – surface geometry; (b) – control mesh; (c) – solid NURBS mesh (73,314 elements); (d) – fluid-structure interaction simulation results: contours of the arterial wall velocity (cm/s) during late systole plotted on the current configuration. Only major branches are kept in (b–d).

5 CONCLUSIONS AND FUTURE WORK

We have developed a NURBS-based modeling and simulation approach for fluid and fluid-structure interaction analysis of blood flow in arteries. Our technique is a viable alternative to the standard finite element method. It possesses a set of attractive features from the view point of accuracy and implementational convenience not engendered by standard finite element discretizations. Rigorous comparison with the standard finite element computations are still necessary in order to strengthen our conjecture.

We have focused on NURBS modeling. Techniques such as A-patches (Bajaj, Chen, and Xu 1995), T-splines (Sederberg, Cardon, Finnigan, North, Zheng, and Lyche 2004), and subdivision (Cirak, Scott, Antonsson, Ortiz, and Schröder 2002) are currently under investigation.

We have successfully applied our method to two patient-specific examples, which involve a thoracic aorta model, and an abdominal aorta model. As part of future work, we would like to apply the techniques described here to modeling and analysis of the human heart.

ACKNOWLEDGEMENTS

Y. Bazilevs and Y. Zhang were partially supported by the J. T. Oden ICES Postdoctoral Fellowship at the Institute for Computational Engineering and Sciences. This research of Y. Zhang and C. Bajaj was supported in part by NSF grants EIA-0325550, CNS-0540033, and NIH grants P20-RR020647, R01-GM074258, R01-GM073087. This support is gratefully acknowledged. We would also like to thank Fred Nugen, Bob Moser, and Jeff Gohean for providing us with the data for the thoracic aorta model.

REFERENCES

Bajaj, C., J. Chen, and G. Xu (1995). Modeling with cubic A-patches. *ACM Transactions on Graphics 14*, 103–133.

Bazilevs, Y., V. M. Calo, Y. Zhang, and T. J. R. Hughes (2006). Isogeometric fluid-structure interaction analysis with applications to arterial blood flow. *Computational Mechanics 38*, 310–322.

Bazilevs, Y., L. B. da Veiga, J. A. Cottrell, T. J. R. Hughes, and G. Sangalli (2006). Isogeometric analysis: Approximation, stability and error estimates for *h*-refined meshes. *Mathematical Models and Methods in Applied Sciences 16*, 1031–1090.

Cirak, F., M. J. Scott, E. K. Antonsson, M. Ortiz, and P. Schröder (2002). Integrated modeling, finite-element analysis, and engineering design for thin-shell structures using subdivision. *Computer-Aided Design 34*, 137–148.

Cottrell, J. A., A. Reali, Y. Bazilevs, and T. J. R. Hughes (2006). Isogeometric analysis of structural vibrations. *Computer Methods in Applied Mechanics and Engineering 195*, 5257–5296.

Holzapfel, G. A. (2004). Computational biomechanics of soft biological tissue. In E. Stein, R. D. Borst, and

T. J. R. Hughes (Eds.), *Encyclopedia of Computational Mechanics, Vol. 2, Solids and Structures*, Chapter 18. Wiley.

Hughes, T. J. R. (2000). *The Finite Element Method: Linear Static and Dynamic Finite Element Analysis*. Mineola, NY: Dover Publications.

Hughes, T. J. R., J. A. Cottrell, and Y. Bazilevs (2005). Isogeometric analysis: CAD, finite elements, NURBS, exact geometry, and mesh refinement. *Computer Methods in Applied Mechanics and Engineering 194*, 4135–4195.

Sahni, O., J. Muller, K. E. Jansen, M. S. Shephard, and C. A. Taylor (2006). Efficient anisotropic adaptive discretization of the cardiovascular system. *Computer Methods in Applied Mechanics and Engineering 195*, 5634–5655.

Sederberg, T. W., D. L. Cardon, G. T. Finnigan, N. S. North, J. Zheng, and T. Lyche (2004). T-Spline Simplification and Local Refinement. *ACM Transactions on Graphics (TOG), SIGGRAPH 23*, 276–283.

Taylor, C. A., T. J. Hughes, and C. K. Zarins (1998). Finite element modeling of blood flow in arteries. *Computer Methods in Applied Mechanics and Engineering 158*, 155–196.

Zhang, Y. and C. Bajaj (2006). Adaptive and quality quadrilateral/hexahedral meshing from volumetric data. *Computer Methods in Applied Mechanics and Engineering 195*, 942–960.

Zhang, Y., C. Bajaj, and B. S. Sohn (2005). 3D finite element meshing from imaging data. *Computer Methods in Applied Mechanics and Engineering 194*, 5083–5106.

Zhang, Y., Y. Bazilevs, S. Goswami, C. L. Bajaj, and T. J. R. Hughes (2006). Patient-specific vascular NURBS modeling for isogeometric analysis of blood flow. In *Proceedings of the International Meshing Roundtable Conference*.

Computational Modelling of Objects Represented in Images –
João Manuel R.S. Tavares & R.M. Natal Jorge (eds)
© 2007 Taylor & Francis Group, London, ISBN 978-0-415-43349-5

Computing statistics from a graph representation of road networks in satellite images for indexing and retrieval

Avik Bhattacharya, Ian H. Jermyn, Xavier Descombes & Josiane Zerubia
Ariana (joint research group INRIA/I3S), INRIA, Sophia Antipolis, Cedex France

ABSTRACT: Retrieval from remote sensing image archives relies on the extraction of pertinent information from the data about the entity of interest (*e.g.* land cover type), and on the robustness of this extraction to nuisance variables (*e.g.* illumination). Most image-based characterizations are not invariant to such variables. However, other semantic entities in the image may be strongly correlated with the entity of interest and their properties can therefore be used to characterize this entity. Road networks are one example: their properties vary considerably, for example, from urban to rural areas. This paper takes the first steps towards classification (and hence retrieval) based on this idea. We study the dependence of a number of network features on the class of the image ('urban' or 'rural'). The chosen features include measures of the network density, connectedness, and 'curviness'. The feature distributions of the two classes are well separated in feature space, thus providing a basis for retrieval. Classification using kernel k-means confirms this conclusion.

1 INTRODUCTION

The retrieval of images from large remote sensing image archives relies on the ability to extract pertinent information from the data, and on the robustness of this extraction (Daschiel and Datcu 2005). In particular, most queries will not concern, for example, the atmospheric conditions or illumination present when the images were acquired, but instead information that is invariant to these quantities, for instance land cover type. Most image-based characterizations are, however, far from invariant to changes sin such nuisance variables, and thus fail to be robust when dealing with a large variety of images acquired under different conditions. Characterizations based on semantic entities detected in the scene, in contrast, are invariant to such changes, and inferences based on such entities can thus be used to retrieve images in a robust way. For this to work, the properties of the entity in question must be strongly dependent on the query. Road networks provide one example of such an entity: their topological and geometrical properties vary considerably, for example, from urban to rural areas. Features computed from an extracted road network can therefore in principle be used to characterize images, or parts of images,

as belonging to one of these classes. Our work differs from much previous work, for example (Wilson and Hancock 1997; Luo and Hancock 2001), in that we are not interested in identifying the same network in different images, or in a map and an image, and producing a detailed correspondence, but rather in using more general network properties to characterize other entities in an image. The work to be described in this paper takes the first steps towards classification (and hence retrieval) based on such inter-semantic dependencies. Specifically, we study the dependence of a number of road network features on the class of the image, which for the moment we restrict to be either 'urban' or 'rural'. The chosen features when taken in isolation measure the density of the network and its 'curviness' in various ways. Taken together they also measure its connectedness. We find that the feature distributions of the two classes are well separated in feature space, and thus provide a basis for retrieval given an appropriate feature metric. Classification based on kernel k-means confirms that this is the case. We also discover an unexpected relation between two of the features that is consistent across images and classes.

2 NETWORK EXTRACTION AND REPRESENTATION

In order to compute topological and geometrical features of the network, we first need to extract the road

This work was partially funded by the French Space Agency (CNES), ACI QuerySat, the STIC INRIA-Tunisia programme, and EU NoE Muscle (FP6-507752). The data was kindly provided by CNES and by Sup'com, Tunis.

network from an image, and then convert the output to an appropriate representation. This representation should be independent of the output of the extraction algorithm, since we do not want to be committed to any single such method. We consider two network extraction methods (Lacoste et al. 2005; Rochery et al. 2006). The method of (Rochery et al. 2006) is based on 'higher-order active contours'. Higher-order active contours are a generalization of standard active contours that use long-range interactions between contour points to include non-trivial prior information about region shape, in this case that the region should be network-like, that is composed of arms with roughly parallel sides meeting at junctions. The output of this method is a distance function defining the region corresponding to the road network. The method of (Lacoste et al. 2005) models the line network as an object process, where the objects are interacting line segments. The output of (Lacoste et al. 2005) is a set of line segments of varying length, orientation, and position. This output is converted to the output of (Rochery et al. 2006) by performing a dilation and then a distance function computation on the resulting binary image.

The distance function resulting from these methods is then converted to a graph representation of the road network for feature computation purposes. The graph itself captures the network topology, while the network geometry is encoded by decorating vertices and edges with geometric information. The conversion is performed by computing the shock locus of the distance function using the method of (Dimitrov et al. 2000; Siddiqi et al. 2002), extended to deal with multiple, multiply-connected, components. The method identifies shock points by examining the limiting behaviour of the average outward flux of the distance function as the region enclosing the shock point shrinks to zero. A threshold on this flux yields an approximation to the shock locus. The graph is then constructed by taking triple points and end points as vertices, corresponding to junctions and termini, while the edges are composed of all other points, and correspond to road segments between junctions and termini. Figure 1 shows an example of the representation graph. The road network (top right) is first extracted from the input image (top left). The methods cited above are then used to generate the shock locus (bottom left), which is then converted to the graph representation (bottom right).

The vertices and edges are decorated with geometric quantities computed from the shock locus. The features are then computed from the graph and its decorations. They are described in the next section.

2.1 Feature vectors

We focus on five features, summarized in table 1. They fall into three groups: two measures of 'density', two

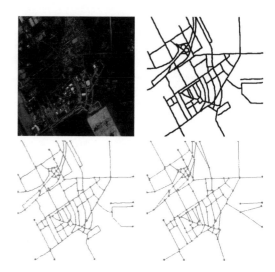

Figure 1. An example of a graph representation. Top left: original image ©CNES; top right: extracted road network; bottom left: shock locus of road network; graph representation.

Table 1. Summary of features.

Notation	Description
m	Number of edges in graph
Ω	Area of image
a	Quadrant label
l_e	Length of road segment corresponding to edge e
m_v	Number of edges at a vertex $\sum_{e:v\in e} 1$
N_J	Number of junction vertices $\sum_{v:m_v>2} 1$
$\tilde{N}_J$	Junction density $\Omega^{-1} N_J$
$\tilde{L}$	Length density $\Omega^{-1} \sum_e l_e$
d_e	Euclidean distance between vertices in an edge
p_e	Ratio of lengths l_e/d_e
var(p)	Ratio of lengths variance
	$m^{-1} \sum_e p_e^2 - (m^{-1} \sum_e p_e)^2$
k_e	Average curvature $l_e^{-1} \int_e ds \, |k_e(s)|$
var(k)	Average curvature variance
	$m^{-1} \sum_e k_e^2 - (m^{-1} \sum_e k_e)^2$
$M_{J,a}$	Number of junction edges per quadrant
	$\sum_{v\in a: m_v>2} m_v$
$\tilde{M}_{J,a}$	Density of junction edges per quadrant $\Omega_a^{-1} M_{J,a}$
var($\tilde{M}_J$)	Variance of density of junction edges
	$(1/4) \sum_a \tilde{M}_{J,a}^2 - ((1/4) \sum_a \tilde{M}_{J,a})^2$

measures of 'curviness', and one measure of 'homogeneity'. Let v be a vertex, and e be an edge. Let l_e be the length of the road segment corresponding to e, and let d_e be the length of e, that is the Euclidean distance between its two vertices. Let $m_v = \sum_{e:v\in e} 1$ be the number of edges at a vertex. Then $N_J = \sum_{v:m_v>2} 1$

is the number of junction vertices. Let Ω be the area of the image in pixels. We define the 'junction density' to be $\tilde{N}_J = \Omega^{-1} N_J$. This is intuitively a useful measure to separate urban and rural areas: we expect urban areas to have a higher value of $\tilde{N}_J$ than rural areas. Similarly, we define the 'length density' to be $\tilde{L} = \Omega^{-1} \sum_e l_e$. Again, we expect urban areas to have a higher value of $\tilde{L}$ than rural areas. Note than one can have a high value of $\tilde{L}$ and a low value of $\tilde{N}_J$ if junctions are complex and the road segments are 'space-filling'.

Let $p_e = l_e/d_e$, and $k_e = l_e^{-1} \int_e ds |k_e(s)|$, *i.e.* the absolute curvature per unit length of the road segment corresponding to e. Although it may seem natural to characterize the network using the average values per edge of these quantities, in practice we have found that more useful features are obtained by using their variances. We thus define the 'ratio of lengths variance' to be the variance of p_e over edges, var(p), and the 'average curvature variance' to be the variance of k_e over edges, var(k). Note that it is quite possible to have a large value of p_e for an edge while having a small value of k_e if the road segment is composed of long straight segments, and vice-versa, if the road 'wiggles' rapidly around the straight line joining the two vertices in the edge. We expect rural areas to have high values of one of these two quantities, while urban areas will probably have low values, although this is less obvious than for the density measures.

To measure network homogeneity, we divide each image into four quadrants, labelled a. Subscript a indicates quantities evaluated for quadrant a rather than the whole image. Let $M_{J,a} = \sum_{v \in a: m_v > 2} m_v$ be the number of edges emanating from junctions in quadrant a. This is very nearly twice the number of edges in a, but it is convenient to restrict ourselves to junctions to avoid spurious termini at the boundary of the image. Let $\tilde{M}_{J,a} = \Omega_a^{-1} M_{J,a}$ be the density of such edges in quadrant a. Then we define the 'network inhomogeneity' to be the variance of $\tilde{M}_{J,a}$ over quadrants, var($\tilde{M}_J$).

In the experiments reported in the next section, all the images have the same resolution. However, more generally we need to consider the scaling of the above quantities with image resolution. We assume that changing the resolution of the image does not change the extracted road network. This can happen, for example, if the network extracted from a lower resolution image lacks certain roads contained in the network extracted from a higher resolution image because they are less than one pixel wide. This effectively limits the range of the resolutions that we can consider simultaneously. Having assumed this, invariance to image resolution is easily accomplished by converting quantities in pixel units to physical units using the image resolution.

Figure 2 shows scatter plots of selected pairs of the features described above as computed from a database

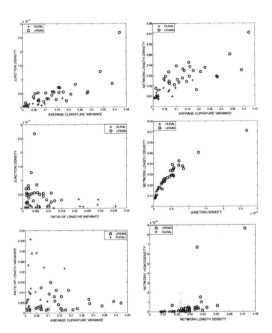

Figure 2. Scatter plots of selected pairs of features. Red stars correspond to rural areas, blue circles to urban areas. From left to right, top to bottom: $\tilde{N}_J$ versus var(k); $\tilde{L}$ versus var(k); $\tilde{N}_J$ versus var(p); $\tilde{L}$ versus $\tilde{N}_J$; var(p) versus var(k); var($\tilde{M}_J$) versus $\tilde{L}$.

of 52 SPOT5, 5 m resolution images, 26 images of each class, representing various types of urban and rural landscapes. The plots show, from left to right, top to bottom: $\tilde{N}_J$ versus var(k); $\tilde{L}$ versus var(k); $\tilde{N}_J$ versus var(p); $\tilde{L}$ versus $\tilde{N}_J$; var(p) versus var(k); var($\tilde{M}_J$) versus $\tilde{L}$. Blue circles correspond to images of urban areas, red stars from images of rural areas.

As can be seen from the plots, the junction densities, $\tilde{N}_J$, for urban areas are for the most part higher and more varied than those for rural areas, where the values are small. The network length density, $\tilde{L}$, behaves similarly. The behaviour of the average curvature variance, var$k$, is perhaps less expected. Urban areas show generally higher values of this feature, and there is also a wide spread of values, while rural areas demonstrate, with a few exceptions, very little curvature variance. The ratio of lengths variance, var(p) is also interesting. Both classes cluster around low values, again with a few exceptions in the case of rural areas. The average curvature variance varies widely w.r.t ratio of length variance for urban areas, whereas the ratio of length variance varies widely w.r.t average curvature variance for rural areas. The network inhomogeneity var($\tilde{M}_J$) for rural areas is low and does not vary with with the network length density, whereas for urban areas the network inhomogeneity is low but varies widely with the network length density. Perhaps the most intriguing plot is length density against junction density, in which both rural and urban data points follow a well defined curve, well approximated by $\tilde{L} = 1.4\tilde{N}_J^{1/2}$. Naively, if there is on average one junction for every a^2 pixels, the junctions will be separated by a distance $O(a)$. If each junction has r edges, there will be on average $r/2$ segments of length $O(a)$ for every a^2 pixels, and thus $\tilde{L} \simeq (r/2)\tilde{N}_J^{1/2}$. For a square lattice $\tilde{L} = 2\tilde{N}_J^{1/2}$, so that in some sense road networks are 'less connected' than a square lattice. However, this analysis effectively assumes a uniform distribution of junctions and no symmetry-breaking 'clustering' effects due to dependencies between different junction positions. In general, there seems to be no reason *a priori* why even the exponent $1/2$ should be consistent across images and classes, let alone the pre-factor. It remains to be seen whether this relation is preserved in a larger data set. Finally, and most importantly, note that the points from the two classes are quite well separated in many of the plots, making it reasonable to use these features for classification.

3 CLUSTERING

The above results indicate that the selected features represent suitable choices for classification based on road network properties given an appropriate feature metric. In order to classify the images from the two classes, we use the kernel k-means algorithm since the feature data are not linearly separable. We use a Gaussian kernel,

$$\psi(X_1, X_2) = e^{-\frac{\|X_1 - X_2\|^2}{2\sigma^2}},$$

where X_1 and X_2 are two feature vectors. The clustering result, displayed in table 2, shows that the two

Table 2. Kernel k-means clustering result with $\sigma = 0.5$.

	Urban	Rural
Class 1	1	19
Class 2	25	7

classes can be well partitioned using the above five features. 19 and 25 images from 'rural' and 'urban' classes respectively were correctly classified, while 1 and 7 images from 'urban' and 'rural' classes respectively were incorrectly classified.

4 CONCLUSION

The preliminary studies reported above indicate that certain features of road networks can serve as characterizations for various image classes that are robust to nuisance parameters and in principle also to imaging modality. Future work will involve extracting road networks from many more images and studying the features described above and others (*e.g.* area density, number and angles of roads at junctions, network connectivity) on the resulting database of graphs. Probabilistic models of selected features will be developed, both for the above coarse classes, and refinements of them. These models will then enable classification and retrieval based on road networks.

REFERENCES

Daschiel, H. and M. Datcu (2005). Image information mining system evaluation using information-theoretic measures. *EURASIP Journal on Applied Signal Processing 14*, 2153–2163.

Dimitrov, P., C. Phillips, and K. Siddiqi (2000). Robust and efficient skeletal graphs. In *Proc. IEEE Computer Vision and Pattern Recognition (CVPR)*, Hilton Head Island, USA, pp. 1417–1423.

Lacoste, C., X. Descombes, and J. Zerubia (2005). Point processes for unsupervised line network extraction in remote sensing. *IEEE Trans. Pattern Analysis and Machine Intelligence 27*(10), 1568–1579.

Luo, B. and E. R. Hancock (2001, October). Structural graph matching using the EM algorithm and singular value decomposition. *IEEE Trans. Pattern Analysis and Machine Intelligence 23*(10), 1120–1136.

Rochery, M., I. H. Jermyn, and J. Zerubia (2006). Higher-order active contours. *International Journal of Computer Vision 69*(1), 27–42.

Siddiqi, K., S. Bouix, A. Tannenbaum, and S. W. Zucker (2002). Hamilton-jacobi skeleton. *International Journal of Computer Vision 48*(3), 215–231.

Wilson, R. C. and E. R. Hancock (1997). Structural matching by discrete relaxation. *IEEE Trans. Pattern Analysis and Machine Intelligence 19*(6), 634–648.

Computational Modelling of Objects Represented in Images –
João Manuel R.S. Tavares & R.M. Natal Jorge (eds)
© 2007 Taylor & Francis Group, London, ISBN 978-0-415-43349-5

3D visualization of radiosurgical treatment plans – experience with Java3D and VTK

O. Blanck, A. Schlaefer & A. Schweikard
University of Luebeck, Institute for Robotics and Congnitive Systems, Germany

ABSTRACT: In radiosurgery, high energy beams are used to irradiate a tumor region and to deliver a therapeutic dose. A robotic radiosurgery system can use a large number of beams starting at different points around the patient and oriented towards the tumor. Treatment planning consists of finding a set of beams that efficiently delivers the dose subject to clinical and physical constraints. Conventionally, the evaluation of treatment plans is based on two-dimensional graphics. To assess the plan quality and to understand what makes a beam efficient under a certain clinical objective, we have developed a three-dimensional visualization tool. In this paper we will briefly present the tool and our experience with using Java3D and VTK for its implementation.

Keywords: Computer Aided Therapy and Treatment, Robotic Radiosurgery, Telemedicine.

1 INTRODUCTION

A common form of cancer radiation therapy is *percutaneous irradiation*, where high energy beams are used to irradiate a tumor region from outside the patient's body. The beams typically intersect in the target region to deliver a therapeutic dose while sparing critical structures in its proximity.

In *radiosurgery*, beams from a wide range of positions around the patient are used. Hence the overlap of beams outside the tumor is small, and the high dose region is precisely covering the tumor. As a result it is safe to deliver higher doses to the tumor, destroying its cells more rapidly, much similar to a surgical tool.

Early examples for radiosurgical systems include the Gamma Knife (Elekta AB, Stockholm) and conventional gantry based linear accelerators (linac) equipped with cylinder collimators. The Gamma Knife was designed for the treatment of intracranial targets. Using a stereotactical frame attached to the patient's head, beams from up to 201 Cobalt 60 sources intersect at an iso-center, thus forming a small spherical region of high dose. Arbitrarily shaped tumors can be treated by placing a number of spheres with different radii accordingly. A similar iso-centric treatment can be performed with a gantry based system by moving the linac accordingly.

A relatively new development is *robotic radiosurgery* where a robot arm places a miniature linac at different positions around the patient. Given the six degrees of freedom of the robot, the linac can be placed at virtually any position around the patient, and beams can be oriented towards arbitrary points

in the target region. While for practical treatment the number of potential linac positions is restricted to approximately 100, there is no limitation on a beam's orientation. Hence, a comparatively large number of non-iso-centric and non-coplanar beams can be used for treatment.

An example for a robotic radiosurgery system is the CyberKnife (Accuray Inc., Sunnyvale, California), which is also capable of compensating for the tumor's motion due to breathing, see Figure 1.

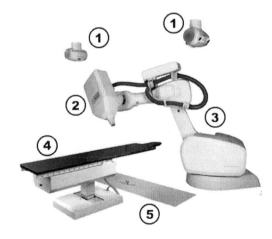

Figure 1. CyberKnife System – 1. X-ray sources for system correction during treatment, 2. Miniature LINAC, 3. Robot arm, 4. Patient couch, 5. X-ray cameras.

Treatment planning is typically based on an *computed tomography* (CT). The relevant structures, specifically the *planning target volume* (PTV) and the *organs at risk* (OAR), are outlined as contours on the CT slices and are referred to as volumes of interest (VOI). The treatment plan is generated with inverse planning [1]. A two-dimensional graphical visualization of the dose distribution is an important tool to evaluate treatment plans. Drawing iso-dose lines for different dose levels on the axial, sagittal, or coronal CT-slices it is possible to assess how well the high-doses cover the PTV and spare the OAR.

However, when a large number of non-coplanar beams is used, it becomes more difficult to assess a single beam's effect to the PTV on two-dimensional slices. A three-dimensional possibly autostereoscopic display of the PTV, the OAR, and the beams can be helpful in this case. For the Gamma Knife the selection of beams and spheres is performed manually, and Olofsson et al. [2] describe the use of haptic feedback and stereoscopic visualization to guide the process. A 3D scene containing the PTV, the OAR, and the iso-dose surfaces is generated and displayed via shutter glasses to provide depth information. The human planner can place and replace spheres until the dose distribution is satisfactory. Since the dose is shown as an iso-surface, the target boundaries and areas of particular high dose (hot spots) may be invisible. Haptic feedback is used to guide the planner towards points inside the target and to avoid hot spots. Three-dimensional stereoscopic visualization has also been proposed in [3] and [4].

A widely used toolkit for three-dimensional visualization is the Java3D (Java.net/Sun Microsystems, California) extension to the Java programming language. Java3D is a freely available and provides the programmer with a set of 3D graphical tools based on OpenGL (or DirectX) for building virtual 3D environments. In Java3D the entire scene is described as a virtual universe using a tree data structure of the particular components, and Java3D code can easily be embedded in web applications. It has been used in a variety of medical projects, too [5–7].

Another toolkit specifically designed for the visualization of medical image data is the *Visualization ToolKit* (VTK, Kitware Inc., New York) [8]. VTK is an open source toolkit providing the programmer with a set of libraries and utilities for constructing advanced 3D environments based on OpenGL. It consists of a C++ class library, and several interpreted interface layers including Tcl/Tk, Java, and Python, and supports a wide variety of visualization algorithms and volumetric methods. In VTK the entire scene is described by a visualization pipeline containing the particular components. Examples for the use of VTK include the medical imaging interaction

toolkit and the open-source software for navigating in multidimensional DICOM images [9–11].

When planning a radiosurgical treatment, computerized optimization methods are used to find the most suitable beams. However, selecting beam orientations is an intricate task, and given certain a clinical objective, e.g. treatment time or target dose homogeneity, it is often difficult to understand what makes a beam efficient. We have developed a tool to facilitate an enhanced analysis based on three-dimensional representations of organs and beams, and both, Java3D and VTK have been used for the implementation. We will briefly present the rationale behind our system, and discuss our experiences using Java3D and VTK for three-dimensional visualization and autostereoscopic display of medical data.

2 MATERIAL AND METHODS

The reason for experimenting with both, VTK and Java3D, lies in their respective advantages within the context of our project. VTK offers a wealth of functions and methods for displaying image data, including a built in support for autostereoscopic displays (ASD). Java3D on the other hand can be easily integrated in heterogeneous environments due to its platform independence. It can also be embedded in web applications.

To visualize a treatment plan we first have to import the CT images and the plan data including the contours of the PTV and OAR. Using the physics data for the CyberKnife and the CT data of the patient we create a DoseModel to calculate the dose deposition by the beams in the patient's body. A treatment plan containing PTV, OAR, and beams is shown by either VTK

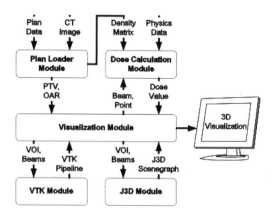

Figure 2. System overview containing the three main modules plan loader, dose calculation, and visualization, the VTK Module, the Java3D Module, and their dependencies.

or Java3D. An overview of the system modules is given in Figure 2.

A patient's anatomy with the different volumes of interest, the beams and the resulting dose distribution is essentially of three dimensional nature. Hence, we discretize the VOI into a grid-like structure of voxels to facilitate an accurate visual understanding of the dose distribution and its three-dimensional extent. As the VOI are of different importance, the resolution of the grid can vary. In the three-dimensional scene, each voxel is represented by a sphere, with the color defined according to the voxel's dose. Due to the overlapping of beams inside the tumor hot spots can arise that are invisible to the viewer. To identify such hot spots we can vary the size of the spheres, which allows a look 'inside' the volume.

Depending on the type of analysis, beams are visualized in different ways. For a first impression and to compare a large number of beams, each beam is represented by a small cylinder along its central axis. When assessing the dose effect of a beam, a pencil shaped surface is shown which denotes the area where the dose coefficient is above a defined threshold. To study beam placement, a beam's central axis and two perpendicular disks representing the outer limit of the beam and the areas where the dose coefficient exceeds a threshold are shown. The visualization of beams and volumes in our tool can be seen in Figure 3.

In Java3D, a scene graph with a tree structure forms the basis for the visualization. Its root object is a VirtualUniverse with a Canvas3D for drawing the three-dimensional scene. Volumes of interest and beams are represented by BranchGroup nodes, and each single voxel is defined by a TransformGroup node containing a colored sphere. Similarly, a TransformGroup and cone or QuadArray objects are used to represent beams, see Figure 4.

In VTK, the 3D visualization is generated as the primitives forming the scene pass a rendering pipeline. At the beginning of the pipeline stands the raw data or the geometry of the visualized object, e.g. a vtkSphereSource or a vtkConeSource, in case of a voxel or beam, respectively. For the beams pencil shaped surface we use a vtkCellArray. Subsequently, a vtkMapper and a vtkActor define the objects shape, appearance and position. The scene is then rendered (vtkRenderer) and shown on the screen (vtkRendererWindow). User interaction is implemented on top of the vtkRendererWindow, and the VTK rendering pipeline can be seen in Figure 5. A mapping of the objects used in our tool to VTK and Java3D can be seen in Table 1.

To visualize the patient's body surface with VTK we used vtkImageData and windowing with smoothing filter as described in Schroeder at al. [8]. Similarly we followed Schroeder et al. using a the vtkImageData and a vtkVolumeTextureMapper3D to display the CT. To create a more realistic three-dimensional visualization with VTK we used an auto stereoscopic display (SeeReal Technologies GmbH, Dresden, Germany) as shown in Figure 6.

Our tests were performed on Standard PC Hardware including a 2,53 GHz Intel Pentium 4 with 1 GB RAM and an ATI Mobility Radeon 9000 graphics card running Windows XP, and a 2,41 GHz 64 Bit AMD Dual

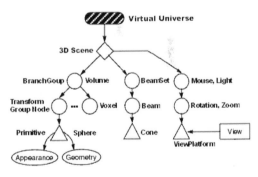

Figure 4. Java3D SceneGraph from Universe to Primitives, as used in the visualization tool.

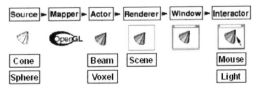

Figure 5. VTK rendering pipeline from Source to Window, as used in the visualization tool.

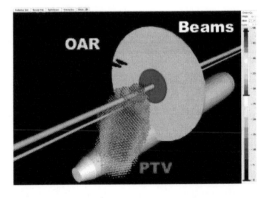

Figure 3. Visualization of beams and volumes. The voxels of PTV (front) and OAR are represented by hypsometrically colored spheres. Beams are shown as either the iso-dose-coefficient surface (pencil beam), as the beam's central axis, or by disks showing the outer limit (yellow) and the high-dose area (red).

VTO	Java3D	VTK
Scene	VirtualUniverse	vtkRendererWindow
	Canvas3D	vtkRenderer
View	TransformGroup	vtkCamera
	Light	vtkLight
Volume	BranchGroup	Vector of vtkActor
BeamSet		
Voxel	TransformGroup	vtkSphereSource
	Sphere	vtkMapper
	Appearance	vtkActor
Beam	TransformGroup	vtkConeSource
(cone)	Cone	vtkMapper
	Appearance	vtkActor
Beam	TransformGroup	vtkPoints
(pencil)	QuadArray	vtkCellArray
		vtkActor

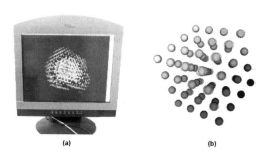

Figure 6.   (a) 3D Monitor/(b) 3D Software rendering.

Core with 4 GB RAM and a NVIDIA GeForce 7800 GTX graphics card running Linux.

3   RESULTS

The tool was used to evaluate existing treatment plans with respect to the identification of cold and hot spots in the different volumes of interest and to assess the effect of beams. The coloring of the different voxels according to their dose and the shading of the primitives provides a good visual understanding of the treatment plan and its properties in both implementations with Java3D and with VTK. The spatial representation of the beams facilitates a much better understanding of the beams chosen by the optimization process and the interplay of beams forming the dose distribution, compare Figures 7–9. Using the autostereoscopic display, the three-dimensional impression of the treatment plan was much improved.

Figure 7.   Visualization of a spine tumor with two pencil beams. With the pencil shape it is possible to study the beams' spatial properties and relate them to the weights assigned.

Figure 8.   Visualization of the spine tumor from Figure 7 with all treatment beams. The beams overlap in the PTV and avoid the spinal cord and the esophagus.

Java3D provides a stable and robust API with a good documentation, and can be easily used on different platforms. It also integrates well with java applets making it possible to deploy and share the software via a network. This is an advantage in a medical context with its often inhomogeneous hard- and software environments. We used our tool with scenes of up to 30000 spheres and obtained a stable visualization. However, setting the values for light and materials properly was sometimes difficult but crucial to get good results. Java3D does not provide built in support for the ASD we use, although stereo rendering could be implemented by the user.

VTK features a rich API to create 3D scenes and visualization algorithms including for example polygon reduction, image convolution and Delaunay

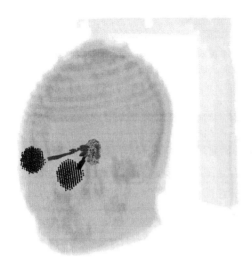

Figure 9. CT Volume generated using VTKTextureMapper3D and VOI. The top figure displays the dose distribution in PTV and OAR, while in the bottom figure the treatment beams with their different activation times measured in monitor units (MU) (white = high MU/black = low MU) are shown.

triangulation. The rendering model also supports 2D, volumetric, and texture-based approaches that can be used in any combination. VTK also supports autostereoscopic rendering for commercial ASDs. While VTK is available for different operating systems, the installation process can be cumbersome. The 3D impression using the standard settings for light and materials was very realistic, especially in combination

Table 2. Rendering time of Java3D and VTK (object generation and calculation time plus real rendering time).

Rendering time	Java3D	VTK
1153 Voxel	1504.7 ms	496 ms
6501 Voxel	9992.5 ms	9979 ms
16382 Voxel	29534 ms	48135 ms
930 Beams	28981 ms	25878 ms

with the ASD. However, we had occasional problems rendering multiple objects with different transparency.

For both, VTK and Java3D we measured the rendering time from within a Java application. We considered the complete rendering process including object generation, setting color and shape, and rendering the scene. For a small number of objects VTK was faster, while for large scenes Java3D performed better, see Table 2.

4 CONCLUSION

The developed tool and its three-dimensional visualization proved to be useful for research on beam placement heuristics, and the autostereoscopic display improves the three-dimensional effect. It is difficult to compare VTK and Java3D, as both have different merits. VTK presents an interesting and powerful alternative for 3D visualization and more advanced tasks. However, we plan to focus the development of our three-dimensional visualization tool on Java3D, primarily due to its platform independence and web integration, the faster rendering of large scenes and the simpler API.

REFERENCES

[1] A. Schweikard, M. Bodduluri, J. R. Adler: Planning for camera-guided robotic radiosurgery. IEEE transactions on robotics and automation, 14(6):951–962, 1998

[2] I. Olofsson, K. Lundin, M. Cooper, P. Kjaell, A. Ynnerman: A haptic interface for dose planning in stereotactic radio-surgery. Pro. 8th Int. Con. Inf. Vis. (2004) 200–205

[3] R. Hubbold, D. Hancock, C. Moore: Autostereoscopic display for radiotherapy planning. Con. Stereo. Dis. and V.R. Sys. IV (1997)

[4] A. Schlaefer, O. Blanck, A. Schweikard: Autostereoscopic display of the 3D dose distribution to assess beam placement for robotic radiosurgery. Medical Physics, 32, 2122, 2005

[5] G. Guttmann: Spilling the Beans on Java 3D: A Tool for the Virtual Anatomist. The Anatomical Record, 257:73–79, 1999, 73

[6] T. Can, Y. Wang, Y. Wang, J. Su: FPV: fast protein visualization using Java 3D. Bioinformatics, 19/8, 2003, 913–922

[7] S. Huang, R. Baimouratov, P. Xiao, A. Ananthasubramaniam, W. Nowinski: A Medical Imaging and Visualization Toolkit in Java. Journal of Digital Imaging, Vol. 19, No 1, 2006, 17–29

[8] W. Schroeder, L. Avila, W. Hoffman: Visualizing with VTK: A Tutorial. IEEE Computer Graphics and Applications, Sep/Oct 2000, 0272–1716

[9] D. Gobbi, T. Peters: Generalized 3D nonlinear transformations for medical imaging: an object-oriented implementation in VTK. Computerized Medical Imaging and Graphics 27 (2003) 255–265

[10] A. Rosset, L. Spadola, O. Ratib: OsiriX: An Open-Source Software for Navigating in Multidimensional DICOM Images. Journal of Digital Imaging, Vol 17, No 3 (September), 2004: pp 205–216

[11] I. Wolf, M. Vetter, I. Wegner, T. Bttger, M. Nolden, M. Scbinger, M. Hastenteufel, T. Kunert, H. Meinzer: The Medical Imaging Interaction Toolkit. Medical Image Analysis, Vol 9, 2005, 594–604

Computational Modelling of Objects Represented in Images –
João Manuel R.S. Tavares & R.M. Natal Jorge (eds)
© 2007 Taylor & Francis Group, London, ISBN 978-0-415-43349-5

A 3-D mechanical model for the pelvic surgery

M. Boukerrou, C. Rubod & M. Cosson
Clinique de chirurgie gynécologique, Hôpital Jeanne de Flandre, CHRU, Lille, France

M. Brieu
Laboratoire de Mécanique de Lille, CNRS, Ecole Centrale de Lille, France

M. Vermandel & P. Dubois
Institut de Technologie Médicale, Inserm, CHRU, Lille, France

ABSTRACT: Authors present a new method for building a 3-D mechanical model of the pelvic cavity. This model makes use of the data both from medical imaging (MRI) to recover the anatomical sets of the patient and from experimental measurements of the mechanical behaviour of regarded soft tissues. The 3-D geometry of the organs is reconstructed from 3 sets of orthogonal MRI slices. Uniaxial tension tests are carried out on organic samples of tissues harvested in per-operative conditions or on cadavers. The final aim of this study is to deliver an efficient and customised help to the surgical decision.

1   INTRODUCTION

Pelvi-genital prolapses consist in protuberance of all or part of the vaginal wall. These disorders are common, as they affect one woman in three for all ages combined [Samuelson 1999]. Their treatments are mainly surgical, involving anatomical restoration and functional surgery. Despite the rise of new techniques, the frequency of the prolapse's relapses still remains high (up to 60% of patients over the age of 60 years) [Swift 2000].

Most of the offered treatments are still empirical and poorly valued. That's why some recent works rose, tempting to provide assistance to the surgical decision [Bercley 1999, Singh 2001, Socrate 2005]. We present in this paper an original methodological approach, dealing with the achievement of a mechanical model of the pelvic female cavity: it makes a combined use of medical imaging and of mechanical tests on organic tissues. The final aim is to deliver a new efficient and customised tool for helping, by means of numerical simulation, the preoperative decision as well as the postoperative check-up.

2   METHODOLOGY

Such a model requires 2 kinds of data:

- the geometry of the pelvic cavity
- the mechanical behaviour of the various organic elements included in this cavity.

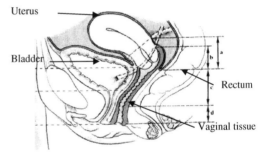

Figure 1.   Simplified anatomy of the pelvic cavity.

2.1   *Anatomical recalls*

The pelvic cavity includes various components (soft organs, bones, ligaments) (Figure 1). The vagina makes a supporting hammock for the pelvic viscera [Kamina 1995]. The vaginal tissue acts then as an interface where forces are transmitted between every pelvic organ (specially the bladder and the rectum): it is therefore highly involved in the prolapse process and will be the main topic of our works. In this initial step of our study, we will highly simplify our model of the pelvic cavity by restricting it only to the bladder, the rectum and the vagina (as deformable organs) linked to the sacrum and pubic bone (rigid bodies) by ligaments.

2.2   *The cavity imaging*

We made use of the standard preoperative protocol which includes for every patient a MRI acquisition

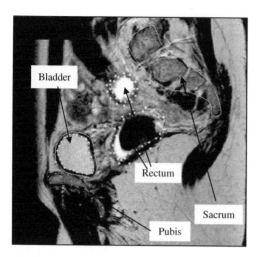

Figure 2. Example of operator defined contours of the various organs.

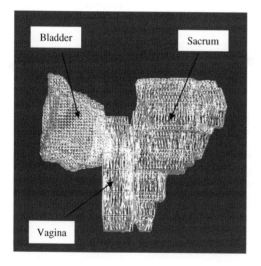

Figure 3. 3D geometrical model of the 3 soft organs.

according to 3 orientations (axial, sagittal and coronal) in supine position. The organic volumes are outlined by means of a contrast product. We can thus take benefit from the obtained images (3 mm thick contiguous slices) on which the contours of the 3 bodies are manually outlined (Figure 2). The meshes of the 3-D surfacic shapes are then computed on each incidence and the reconstructed volumes are then merged from the 3 incidences (Figure 3) by fusion in fuzzy logic using an original algorithm [Vial 2001].

The total number of meshes ranges from about 6000 to 10000 for each volume.

A 3-D F.E. model is built by use of the ABAQUS™ software from the MRI imaging data of the 3-D organic shapes.

Figure 4. Tensile test on an organic (vaginal) soft tissue sample.

2.3 *The mechanical behaviour of organic tissues*

The mechanical properties of the human soft tissues are still very poorly known [Ettema 1998, Goh 2003]. Indeed it is not possible to isolate living organs which are deeply inserted inside the human body. So we make uni-axial tension tests:

– in situ: for ligaments on fresh cadavers: we use a surgical forceps that we specially designed for mechanical measurements [Dubois 2002]
– ex-vivo: for organic tissues samples (preoperative vaginal tissues) or ex-situ (from fresh cadavers) for rectal and vesical tissues: we use in this case a conventional tension test machine (Instron 4302™) as shown in Figure 4.

3 RESULTS

3.1 *3-D organic geometry*

The geometrical model includes the 3 deformable (soft) retained organs (bladder, rectum and vagina) and the 2 rigid structures (sacrum and pubis) as shown in the Figure 5. The considered mechanical model differs from shell structures and requires manually outline of both inner and outer shapes of the soft organs (thickness about 1 mm); a contact surface between neighboring organs has also to be warranted. Rigid bone structures (as sacrum and pubis) are also marked. It must be noticed that MRI imaging does not show the ligaments.

3.2 *Mechanical behavior of the organic tissues*

An example of various curves recorded on the 3 soft organs (bladder, vagina and rectum) is given Figure 6. They show a good differentiation between the behavior

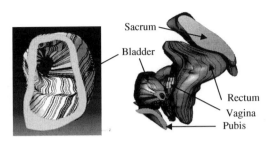

Figure 5. The reconstructed organic volumes.

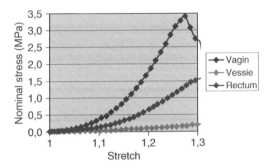

Figure 6. Tension-test experimental curves on various organic tissues.

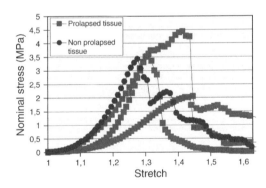

Figure 7. Tension-test experimental curves on pathologic (prolapsed) and healthy (non prolapsed) vaginal tissues.

of each organ, the bladder proving – as expected – to be the more elastic one.

Ex-situ measurements made on fresh cadavers exhibit (Figure 7) some differentiation between pathologic (prolapsed) and healthy (non prolapsed) vaginal tissues.

3.3 3-D mechanical model of the pelvic cavity

The three organic shapes are inserted in the 3-D finite element software. Several assumptions are made to establish the boundary conditions, as well for the bone attachment's points, as for the geometry and for the contact properties of (and between) the different neighboring organs. The experimental data of the various tissues are applied to the meshes of the organic surfaces.

We shall now use this model to evaluate the stretch 3D mapping on shapes obtained from MRI acquisitions. The applied pressures will have been previously recorded in vivo on the corresponding women, by means of a specific multisensor intravaginal probe during efforts (cough, testing).

4 CONCLUSION

All these mechanical tests confirm a good reproducibility of measurements and good differentiation between healthy and pathologic patients. This first study very clearly showed that the tissues exhibit a damageable anisotropic hyper elastic behaviour. We shall next introduce the ligaments' strength and interactions between organs in the 3D mechanical simulation to deduce the constitutive model from their characteristic parameters. We shall also make confrontation between dynamic M.R.I. and numerical simulation.

IRM imaging delivers a satisfactory geometry of the organic shapes. But the realistic definition of surfaces of contact is of prime importance to know the interactions between the bodies. We are working at the improvement of our protocol.

The determination of a reliable model of the pelvic cavity might help in our understanding of the physiopathology of prolapse. In the longer term, a customized mechanical model might allow to simulate, for every patient, the actual constraints and the results of the planned surgeries. This prospect for a predictive mechanical model of the vaginal cavity is interesting both for the simulation of surgical strategies and for the pre- and post-operative diagnosis: it could bring an efficient help in the choice of the best suited technique to each patient. The mechanical characteristics of synthetic intra-vaginal prosthetic tissues could also be valued with the same manner.

REFERENCES

Bercley J., Weghorst S., Glastone H., Raugi G., Berg D. M., 1999. Fast Finite Element Modeling for Surgical Simulation. *Stud. Health Technol. Inform.*, 62: 55–61.

Dubois P., Thommen Q., Jambon A. C. 2002., In vivo measurement of surgical gestures, *IEEE Trans. on Biomed. Eng.*, 49(1): 49–54.

Ettema G., Goh J., Forwood M. 1998. A new method to measure elastic properties of plastic-viscoelastic connective tissue. *Med. Eng. Phys.*, 20(4): 308–314.

Goh J. 2003. Biomechanical and biochemical assessments for pelvic organ prolapse. *Curr Opin Obst Gynecol.*, 15: 391–394.

Kamina P. 1995. *Anatomie clinique du petit bassin et perinée.* Maloine Ed.

Samuelsson E., Victor F., Tibblin G., Svardsudd K. 1999. Signs of genital prolapse in a Swedish population of women 20 to 59 years of age and possible related factors. *Am J. Obstet. Gynecol.* 180(2): 299–305.

Singh K., W. Reid M., L. Berger A. 2001. Assessment and Grading of Pelvic Organ Prolapse by Use of Dynamic magnetic Resonance Imaging. *Am. J. Obstet. Gynecol.*, 185(1): 71–77.

Socrate S., Paskaleva A.P., Myers K.M., House M. 2005. Connection between uterine contractions and cervical dilation: a biomechanical theory of cervical deformation, 1st *Intern. Conf. on Mechanics of Biomaterials & Tissues, Hawai'i, 11–15 dec. 2005.*

Swift S. 2000. The distribution of pelvis organ support in a population of female subjects seen for routine gynecologic health care. *Am J. Obstet Gynecol*, 183(2): 277–85.

Vial S., Gibon D., Vasseur C., Rousseau J. 2001. Volume delineation by fusion of fuzzy sets obtained from multiplanar tomographic images. *IEEE Trans. Med. Imaging*, 20(12): 1362–1372.

Computational Modelling of Objects Represented in Images –
João Manuel R.S. Tavares & R.M. Natal Jorge (eds)
© 2007 Taylor & Francis Group, London, ISBN 978-0-415-43349-5

Monte Carlo simulations studies in small animal PET

S. Branco & P. Almeida

Instituto de Biofísica e Engenharia Biomédica, Faculdade de Ciências da Universidade de Lisboa, Lisboa, Portugal

S. Jan

Service Hospitalier Frédéric Joliot, CEA/DSV/DRM, Orsay, France

ABSTRACT: This work is based on the use of an implemented PET simulation system dedicated for small animal PET imaging using. GATE, a Monte Carlo simulation platform based on the Geant4 libraries, is well suited for modeling the microPET® FOCUS system and to implement realistic phantoms, as the MOBY phantom and imaging maps of real exams. The use of the microPET® FOCUS model with GATE was validated for spatial resolution, counting rates performances, imaging contrast recovery and quantitative analysis. Results of realistic simulations for static and dynamic studies are presented. These simulations include realistic injected doses in the animal and realistic time framing.

1 INTRODUCTION

The rapid growth in genetics and molecular biology combined with the development of techniques for genetically engineering had led to an increased interest *in vivo* small animal imaging. In this context, Positron Emission Tomography (PET) is an extremely powerful tool. PET is a non-invasive nuclear medicine technique which provides spatial and temporal distribution of radiotracers allowing the understudying of physiological and metabolic functions of the body (Wernick, M. N. & Aarsvold, J. N. 2004).

Monte Carlo simulations studies are essential tools in modeling of imaging systems, developing and assessing of tomographic reconstruction algorithms or evaluating correction methods for improved image quantification. We used, in this study, the Geant4 Application for Tomographic Emission (GATE) Monte Carlo platform (Jan, S. et al. 2004).

The objective in this work is to produce realistic simulations exams for the fluoride and the FDG radiotracers, using the validated microPET® FOCUS system for the GATE platform and real mouse phantoms descriptions.

2 MATERIAL AND METHODS

2.1 *The microPET® FOCUS 220 system*

The microPET® FOCUS 220 system is one of the latest generation of commercial scanners dedicated to high resolution animal imaging such as rodents (mice and rats) and the primates (macaque and small baboon). The scanner consists in 4 detectors rings: each ring is made of 42 detector blocks – with a total of 168 blocks. Each detector block is composed by a matrix of 12×12 crystals LSO with dimensions $1.5 \times 1.5 \times 10\,mm^3$. Its axial Field of View (FOV) have 7.6 cm and a diameter of 26.0 cm. The FOCUS system have a volume resolution of 2.5 µL and an absolute sensitivity of 3.4%, both at the center of the FOV (Tai, Y. et al. 2005).

2.2 *The GATE platform*

GATE is a generic Monte Carlo simulation platform, designed to answer the specific needs of nuclear medicine imaging, based on Geant4 libraries a well established code for radiation transport. In addition to the potentialities provided by Geant4, GATE includes specific modules necessary to perform realistic simulations, including the management of time dependent processes (detector and source movements, radioactive decay, dynamic acquisitions), and complex source distributions.

The geometry description of the microPET® scanner is illustrated by the Figure 1. The geometry description of the microPET® FOCUS system and simulations have, for the both of them, been validated for quantitative analysis.

2.3 *Mouse phantoms descriptions*

In order to reproduce realistic images we used real mouse phantoms descriptions. We used, in these

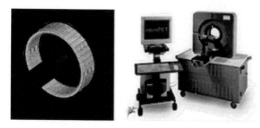

Figure 1. MicroPET® FOCUS geometry performed by GATE (left) and the microPET® FOCUS system (right).

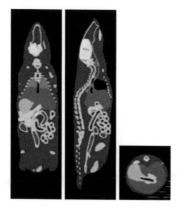

Figure 2. Coronal, sagittal and transaxial slices generated by the MOBY program.

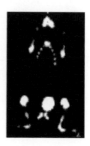

Figure 3. Coronal slices of the ⁻F mouse phantom: attenuation phantom slice (left) and emission phantom slice (right).

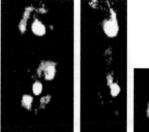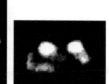

Figure 4. Coronal, sagital and transaxial slices for the activity distribution of the FDG mouse phantom.

studies, the MOBY phantom (Segars, W. P. et al. 2004) which combines the realism of a voxelized phantom, with the flexibility of a mathematical phantom, based on non-uniform rational B-splines. The default whole body MOBY phantom consists of a $128 \times 128 \times 448$ matrix with 0.25 mm side cubic voxels, which is illustrated in the Figure 2. We used a matrix size of $40 \times 40 \times 124$ with a 0.5 mm side cubic voxels.

To generate a more realistic mouse phantom, for a dedicated exam, we used a real whole body mouse acquisition, performed at the CEA/SHFJ. For this exam, we injected 400 μCi of fluoride marked with ^{18}F (labeled ⁻F). The acquisition started 20 minutes post injection with a duration of 20 minutes. The binning of data produced 4 frames of 900 s. As shown in the Figure 3 the phantom consists in an emission map composed by $101 \times 55 \times 95$ voxels of 0.474519 mm in the x side and y side, and 0.796000 mm in the z side, and an attenuation map composed by $25 \times 14 \times 24$ voxels of 1.917050 mm in the x side, 1.864180 mm in the y side and 3.150830 mm in the z side.

In addition, we used a real dynamic whole body mouse [^{18}F]FDG exam to generate an emission map that consist in a matrix of $104 \times 61 \times 95$ voxels with the x, y and z sizes of 0.454719 mm, 0.454719 mm and 0.796000 mm respectively. The real exam was performed with a microPET® FOCUS 220 system at CEA/SHFJ. The mouse was injected with an activity of 220 μCi and scanned during 90 minutes. The image was reconstructed using FORE + OSEM2D algorithms (16 subsets and 4 iterations), data normalization and corrections for dead time, scatter and attenuation was done. The description of the phantom is illustrated by the Figure 4.

2.4 Validations of the microPET® FOCUS using GATE

Sets of simulations were compared against experimental data and showed a very good data agreement. The spatial resolution was measured using a glass capillary filled with ^{18}F aqueous source placed at various positions in the radial FOV. The differences between GATE simulations and experimental data for radial and tangential resolutions showed a discrepancy lower than 3%. To measure the counting rate performances was used a 34 mL and a 3 cm cylindrical phantom filled with ^{11}C aqueous source with an initial activity of 12 mCi. The comparison between simulated and experimental data for prompt and delay coincidences are in very good agreement for a large range of activity in the FOV. The complete validation results is given by (Jan, S. et al. 2005).

2.5 Phantoms implementation and cluster computing

The phantoms emission maps were used on GATE to distribute the activity in different structures and the attenuation maps to replicate the mouse body absorption.

The complete simulation platform was installed on a cluster of 512 CPUs with 64 bits architecture and a Linux operating system. To speed up the simulations we used the compression phantom algorithm available in the last GATE release (v3.0.0).

2.6 Simulation and reconstruction protocol

The simulations were performed under realistic acquisitions parameters. The energy window was set to 350 keV – 750 keV, and an energy resolution of 26% at 511 keV was applied. A coincidence time window of 6 ns was set and delayed coincidences sorter was included. A non paralysable dead time model was applied on the single events, and also on the coincidence processor. The data transfer from the scanner to the hard disk was also simulated. Background noise due to the LSO decay was included.

Two different protocols were used. In the first we defined an activity map close to the $^-$F distribution, based on the real whole body mouse exam for the MOBY phantom and the $^-$F mouse phantom. For the second protocol we used the Time Activity Curves (TACs) related with the real [^{18}F]FDG biodistribution with whole body FDG exam, also used to generate the FDG mouse phantom. With the TACs we defined an activity distribution close to the real FDG biodistribution to reproduce metabolic mouse images for the MOBY phantom and the FDG mouse phantom.

In both protocols we defined a gamma/gamma emission source to mimic a full correction of positron range and gamma accolinearity. In addition, these physical effects are not taking into account to speed up the computing time and to reproduce gold standard simulation results. All the data results were reconstructed using FORE+OSEM2D algorithms (16 subsets and 4 iterations).

2.7 [^{18}F]FDG biodistribution

The 2-Deoxy-[^{18}F]fluoro-D-glucose (FDG) is one of the commonly radiotracer used in cancer research. FDG is an analogue of glucose and is taken up by living cells through the normal glucose pathway.

We used the dynamic FDG mouse exam described in the section 2.3 to compute the FDG biodistribution. Regions of Interest (ROIs) were drawn around the bladder, heart, liver, kidneys and whole body. The activity in each ROI was normalized by the total body activity in order to obtain a relative concentration in

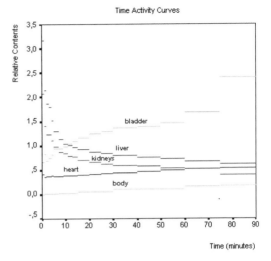

Figure 5. TAC for each ROI over a period of 90 minutes.

each organ at each time point. Figure 5 shows the TAC for each organ at each time frame, used as input in our simulation studies.

3 RESULTS

3.1 Simulation studies for the $^-$F radiotracer

We simulated an activity map close to the mouse body ^{18}F-fluoride distribution, with a direct gamma/gamma emission at the latest acquisition time frame. This simulation set-up was applied in the $^-$F mouse phantom and the MOBY phantom. The quantitative and qualitative results from simulations and experimental data are quite similar. Figure 6 shows the results after a full simulation of a gamma/gamma emission source for 1.53×10^{10} particles, with the $^-$F mouse phantom. We simulated $830\,\mu$Ci in the whole body mouse for 500 s acquisition time.

The resolution differences between the real data and the simulation images are due to the non normalization and attenuation corrections and also because of the resolution degradation produced by the scanner on simulated data.

We defined a similar activity distribution on the bones, spleen, skull and the bladder for the MOBY phantom to reproduce the $^-$F distribution. The ^{18}F-fluoride tracer normally accumulates at the bones and highlights skeletal system of the animal. Figure 7 illustrates the accumulated activity after a 4.0×10^{10} particles generated which is equivalent of 2.21 mCi injected dose with 500 s for the acquisition time.

The results are in a very good agreement with the expected.

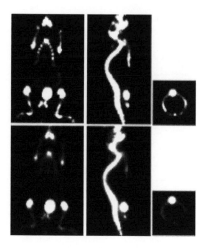

Figure 6. Coronal, sagital and transaxial slices of the real exam (top) and the simulation (bottom) for the ⁻F mouse phantom (bottom).

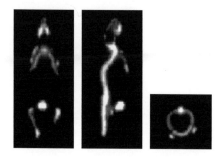

Figure 7. Slices for the MOBY phantom after a full simulation with an activity distribution close to the ⁻F distribution.

3.2 Simulation studies for the FDG radiotracer

We simulated a realistic FDG exam with the MOBY phantom. The activity map introduced in the phantom was based in the FDG distribution, as previous stated, at the first and the last time frames. For the first frame we simulated $219\,\mu Ci$ for an acquisition time of $60\,s$ which produced 2.9×10^9 particles. We simulated $131\,\mu Ci$ for 4.2×10^{10} particles with 15 minutes acquisition time, which corresponds to the last time frame. The results are illustrated by Figure 8.

As we can see in the Figure 8, in the last simulated frame the differences between the activity distribution for the liver and the heart are very similar, disabling a clear separation between the two organs. The other factor that contributes for that reason is the specific description of the organs inside the MOBY phantom. In this frame is possible to see the higher accumulated activity inside the bladder, compared with the first frame. These results are strongly in agreement with the computed FDG biodistribution.

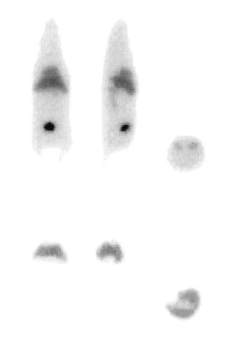

Figure 8. Coronal, sagital and transaxial slices for the simulation of the full MOBY phantom with an activity distribution map close to the FDG distribution: top – at the last activity frame; bottom – at the first activity frame.

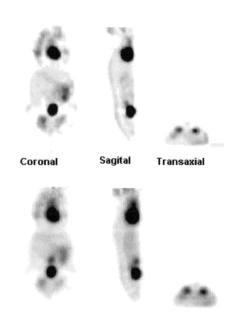

Coronal   **Sagital**   **Transaxial**

Figure 9. Slices of the real FDG exam (top) and the simulated exam (bottom) for the FDG mouse phantom, with an activity map distribution close to the last activity frame.

114

We used the FDG mouse phantom to compute also a realistic FDG exam. The defined activity map distribution is close to the last activity time frame. We defined an acquisition simulated time of 15 minutes which is equivalent to an activity of $112\,\mu$Ci resulting in 3.5×10^{10} tracking particles. Figure 9 shows the reconstructed results.

The qualitative FDG distribution inside the body is clearly in a good agreement with the real exam.

4 CONCLUSIONS AND PERSPECTIVES

The use of the microPET® FOCUS system shows that GATE is well suited to model the FOCUS system for quantitative analysis and to implement realistic voxelized mouse body phantoms. We showed that the GATE platform can simulate realistic small animal PET acquisition under realistic conditions to improve the quantitative analysis in mouse body studies.

We are developing dynamic simulations, with the MOBY phantom, including the breath motion and optimization protocols for the FDG uptake on the lung including the tumor motion.

In the near future, realistic dynamic simulations with different radiotracers, as the FLT, for whole body exams will be done. Biological kinetics will be implemented inside the GATE platform.

REFERENCES

Jan, S. et al. 2004. GATE: a simulation toolkit for PET and SPECT. *Physics in Medicine and Biology* 49(19): 4543–4561.

Jan, S. et al. 2005. Monte Carlo Simulations of the microPET FOCUS system for small rodents imaging applications. *IEEE Nuclear Science Symposium Conference Record* 3: 1653–1657.

Segars, W.P. et al. 2004. Development of a 4-D Digital Mouse Phantom for Molecular Imaging Research. *Molecular Imaging and Biology* 6(3): 149–159.

Tai, Y. et al. 2005. Performance Evaluation of the microPET FOCUS : A Third-Generation microPET Scanner Dedicated to Animal Imaging. *Journal of Nuclear Medicine* 46: 455–463.

Wernick, M. N. & Aarsvold, J. N. 2004. *Emission Tomography: The Fundamentals of PET and SPECT*. California: Elsevier Academic Press.

Computational Modelling of Objects Represented in Images –
João Manuel R.S. Tavares & R.M. Natal Jorge (eds)
© 2007 Taylor & Francis Group, London, ISBN 978-0-415-43349-5

Discrete volume polyhedrization: Complexity and bounds on performance

Valentin E. Brimkov

Mathematics Department, Buffalo State College, State University of New York, USA

ABSTRACT: Given a set $M \subset \mathbb{Z}^3$, an enclosing polyhedron for M is any polyhedron P such that the set of integer points contained in P is precisely M. Representing a discrete volume by an enclosing polyhedron is a fundamental problem in visualization. In this paper we propose the first proof of the long-standing conjecture that the problem of finding an enclosing polyhedron with a minimal number of 2-facets is strongly NP-hard. We also provide a lower bound for that number.

1 INTRODUCTION

This paper deals with a problem usually known as "discrete volume polyhedrization" and further abbreviated DVP. In its most general form, it is the following:

Given a set $M \subset \mathbb{Z}^3$, we look for a (possibly non-convex) polyhedron P, such that the set $P_{\mathbb{Z}}$ of integer points contained in P is precisely M.[1] Such a polyhedron will be called *enclosing* for M. Usually, the 2-dimensional facets of P (2-facets, for short) are required to be convex polygons, as two adjacent polygons may be co-planar. Their number $f_2(P)$ is desired to be as small as possible. Define *polyhedral complexity* of M as $PC(M) = \min_P \{f_2(P) : P$ is an enclosing polyhedron for $M\}$. If $f_2(P) = PC(M)$, P is *minimally enclosing* for M.

Our goal is to study the computational complexity of finding minimally enclosing polyhedron for a given integer set M, as well as to estimate the polyhedral complexity of such a set.

1.1 Motivation

The main motivation for the above problem comes from medical imaging and other visualization problems where discrete volumes of voxels result from scanning and MRI techniques. Since digital medical images involve a huge number of points, it is quite problematic to apply traditional rendering or texture algorithms to obtain satisfactory visualization. Moreover, one can face difficulties in storing or

transmitting data of that size. There are multiple sources of data being transmitted for many diverse uses, such as telemedicine, mine detection, telemaintenance, ATR, visual display, cueing, and others. In all these applications the coding compression methodology used is paramount. For this, one can try to transform a discrete data set to a polyhedron P such that the number of its 2-facets is as small as possible. Such polyhedrizations are also searched for the purposes of geometric approximation of surfaces as well as for surface area and volume estimation.

1.2 Related work and open questions

In recent years DVP is attracting an increasing interest, mainly driven by its rich set of applications. A substantial body of literature impossible to report here is developed on the subject that has been considered from diverse points of view. For instance, a lot of related work is available in any volume of ACM SIGGRAPH series. The most widely used algorithm is the Marching cube method (Lorensen & Cline 1987) that generates a triangulated polyhedral surface in which small triangles model local configurations of voxels. The shortcoming of that method is that the number of triangular facets of the surface may be comparable with the number of points of M. Moreover, the obtained polyhedron is not always hole-free. For other related studies see, e.g. (Borianne & Françon 1994; Burguet & Malgouyres 2000; Debled-Renesson & Reveillès 1994; Kim & Rosenfeld 1982; Kim & Stoimenović 1991; Klette & Sun 2001; Papier 1995; Papier & Françon 1999; Sivignon 2004; Sivignon et al. 2004; Sivignon et al. 2004; Vittone & Chassery 2000; Yu & Klette 2002). The intrinsic complexity of the problem explains the overwhelming usage of

---

[1] In practice, M is often obtained through "digitization" of some (usually unknown) set $S \subset \mathbb{R}^3$ of full dimension.

greedy algorithms, that, as a rule, are not accompanied with a rigorous performance estimation. Most importantly, the question about the computational complexity of DVP is still open. The lack of deep theoretical results is somewhat surprising in view of the forcible practical importance of the problem.

DVP is germane with some well-known problems in theory of lattice polytopes. For instance, tight bounds exist for the number of the vertices, edges, and facets of a lattice polytope obtained as a convex hull of the integer points in a large ball (or a convex body satisfying certain conditions (Bárány & Larman 1998)). We rely on this and other results in obtaining ours. It is not hard to realize (see Section 4) that if in the DVP formulation M is an arbitrary set, then a bound $PC(M) = \Theta(|M|)$ holds where $|M|$ is the cardinality of M. The problem of estimating the polyhedral complexity of a discrete set M becomes nontrivial when $M = S_{\mathbb{Z}}$ for some convex body S, therefore this is the case for which we consider it.

1.3 Results

The main results of the paper are the following.

1. The first proof of the long-standing conjecture that the optimization version of DVP is strongly NP-hard (Section 3). The demonstration is based on the concept of "well-sized lattice polygons" that feature certain optimal properties. This theory, briefly presented in Section 2, may be of certain independent interest.
2. A lower bound on the polyhedral complexity of a given set $M = S_{\mathbb{Z}} \subset \mathbb{Z}^n$ where S is a large (hyper) ball (Section 4). This result implies a bound on the performance of the convex hull algorithm applied to DVP.

1.4 Some notations

Let S be a *body* in $\mathbb{R}^n$, i.e., a subset of $\mathbb{R}^n$ of full dimension $dim(S) = n$. Denote $S_{\mathbb{Z}} = S \cap \mathbb{Z}^n$. For a set $A \subseteq \mathbb{R}^n$, $diam(A) = \max_{x,y \in A} \|x - y\|$ is its *diameter* where $\|.\|$ is the Euclidean norm. By kA we denote the homothetic image of A under a homothety with center $\mathbf{0}$ and a constant of proportionality $k \in \mathbb{R}_+$. By $conv(A)$ we denote the convex hull of A. Given two sets $A, B \subset \mathbb{R}^n$, $\rho(A, B)$ is the Euclidean distance between them. Given a polyhedron $P \subset \mathbb{R}^n$, the number of its *i*-facets is denoted by $f_i(P)$, $0 \leq i \leq n$.

2 WELL-SIZED LATTICE POLYGONS

Let $P \subset \mathbb{R}^2$ be a polygon. By $\nu(P)$ we denote the minimal number of convex polygons into which P can be decomposed. We now define *decomposition complexity* of a set $M \subset \mathbb{Z}^2$ as $DC(M) = \min_P \{\nu(P) : P$

is an enclosing polygon for $M\}$. We will say that P is *minimally decomposable* with respect to $M = P_{\mathbb{Z}}$ if $\nu(P) = DC(M)$.

Now let $P \subset \mathbb{R}^2$ be a lattice polygon. We call P *irreducible* if there is no real number $\alpha < 1$ for which αP is still a lattice polygon. Otherwise P is *reducible*. Informally, by *size* of P we will mean its diameter.

Definition 1 *A lattice polygon P is* well-sized *if the following conditions are met:*

(i) *P is minimally enclosing for $P_{\mathbb{Z}}$;*
(ii) *P is minimally decomposable with respect to $P_{\mathbb{Z}}$.*

We call P a *least well-sized* polygon if there is no $\alpha < 1$ for which αP is well-sized. If a well-sized polygon is an irreducible lattice polygon, then clearly it is also a least well-sized polygon. Note that a least well-sized polygon may be reducible (see Fig. 2a).

Obviously, an arbitrary lattice polygon P is enclosing for the set $P_{\mathbb{Z}}$ but not necessarily well-sized, as one or both of conditions (i), (ii) may not be satisfied, both for irreducible (Fig. 1) or reducible lattice polygons (Fig. 2).

Finding $\nu(P)$ is a strongly NP-hard problem (Lingas 1982), so testing if a polygon is well-sized or not is hard, in general. For our pseudopolynomial reduction in the next section, however, we will need an arbitrary lattice polygon P that appears to be well-sized. To this

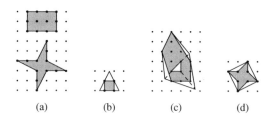

(a)   (b)   (c)   (d)

Figure 1. Irreducible lattice polygons (in gray): (a) two minimally enclosing minimally decomposable, (b) non-minimally enclosing minimally decomposable, (c) minimally enclosing non-minimally decomposable, (d) non-minimally enclosing non-minimally decomposable.

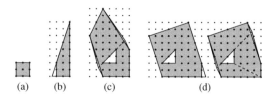

(a)   (b)   (c)   (d)

Figure 2. Reducible lattice polygons (in gray): (a) minimally enclosing minimally decomposable, (b) non-minimally enclosing minimally decomposable, (c) minimally enclosing non-minimally decomposable, (d) non-minimally enclosing non-minimally decomposable.

end, we will show that for any lattice polygon P there is a number $\delta_P \in \mathbb{Z}_+$, so that $\delta_P P$ is well-sized. δ_P is called a *sizing factor* for P. The procedure of computing δ_P and $\delta_P P$ is called *sizing* of P and is described next.

Let P be a lattice polygon. In what follows, w.l.o.g. we assume that P contains the origin $\mathbf{0}$. Let $W = \{v^1, v^2, \ldots, v^m\}$ be the set of vertices of P where $v^i = (v_1^i, v_2^i) \in \mathbb{Z}^2$, $1 \leq i \leq m$. Let

$$\phi(P) = \max_{v^j, v^k \in W} \{3, \|v^i - v^j\|\}, \ 1 \leq j, k \leq m.$$

For any $v^i \in W$, we define

$$\psi_i(P) = \max_{j, k \neq i} \{3/\rho(v^i, \overline{v^j v^k})\}$$

where $\overline{v^j v^k}$ is the straight line through v^j and v^k. Then we set $\psi(P) = \max_i \psi_i(P)$ and eventually define

$$\delta_P = \max\{\phi(P), \psi(P)\}. \tag{1}$$

We have the following plain fact.

Lemma 1 *Given a lattice polygon P with $d = diam(P)$ and $0 \in P$, then:*

1. *There is a polynomial $p(x)$ such that $\delta_P = O(p(d))$.*
2. *δ_P can be computed in time $q(m, \log d)$ where $q(x,y)$ is a polynomial in two variables.*

Next we list a series of lemmas that reveal certain properties of well-sized polygons. These will be later used in the proof of strong NP-hardness of the optimization DVP problem, but may be of independent interest as well.

Let P be a lattice polygon.

Lemma 2 *For any integer $k \geq 2\delta_P^2$, the polygon kP is minimally enclosing for $(kP)_{\mathbb{Z}}$.*

Lemma 3 *For any integer $k \geq 2\delta_P^2$, $v(kP) \leq v(P')$ for any polygon P' (not necessarily a lattice one) that is enclosing for $(kP)_{\mathbb{Z}}$.*

Corollary 1 *For any integer $k \geq 2\delta_P^2$, the polygon kP is well-sized.*

Lemma 4 *Let $k \geq 2\delta_P^2$. By Lemma 2, kP is minimally enclosing for $P_{\mathbb{Z}}$. Let P' be another minimally enclosing polygon for $P_{\mathbb{Z}}$. Then there is a one-to-one correspondence between the vertices of kP and P', such that a vertex v of kP is corresponding to a vertex v' of P' if and only if $\|v - v'\| \leq c\delta_P$ for some positive constant c.*

Lemma 5 *Let kP and P' be as in Lemma 4. Let u, v, w be vertices of P and u', v', w' their corresponding vertices of P'. Then w is in the left/right half-plane determined by the line $\overline{uv}$ iff w' is in the left/right half-plane determined by the line $\overline{u'v'}$.*

We call a *line set* any set $\{v^1, v^2, \ldots, v^l\}$ of $l \geq 3$ vertices of P that belong to the straight line segment $\overline{u^{(1)} u^{(l)}} \subset P$.

Lemma 6 *Let kP and P' be as in Lemma 4 and let one of the following conditions be met:*

1. *Both kP and P' do not contain line sets.*
2. *kP contains a line set $\{v^1, v^2, \ldots, v^l\}$ if and only if P' contains a line set $\{\bar{v}^1, \bar{v}^2, \ldots, \bar{v}^l\}$ where a vertex v^i of kP is corresponding to a vertex $\bar{v}^i$ of P' for $i = 1, 2, \ldots, l$.*

Then $v(kP) = v(P')$.

Lemma 7 *Let kP and P' be as in Lemma 4 and let the following conditions be met:*

1. *If $\{\bar{v}^1, \bar{v}^2, \ldots, \bar{v}^l\}$ is a line set in P', then the set of their corresponding vertices $\{v^1, v^2, \ldots, v^l\}$ in kP is a line set in kP.*
2. *There exists a line set $\{v^1, v^2, \ldots, v^l\}$ in kP such that the corresponding set of vertices $\{\bar{v}^1, \bar{v}^2, \ldots, \bar{v}^l\}$ in P' is not a line set in P'.*

Then $v(kP) \leq v(P')$.

The proofs of the above lemmas are based on combinatorial-geometric reasoning. They are not hard but somewhat lengthy and therefore omitted.

3 OPTIMIZATION DVP IS STRONGLY NP-HARD

Consider the optimization version of DVP.

Optimization Discrete Surface Polyhedrization (OptDVP):

Instance: A set $M \subset \mathbb{Z}^3$ and a bound $\beta \in \mathbb{Z}_+$.
Problem: Decide if there is a polyhedron P, such that $M = P_{\mathbb{Z}}$ and with no more than β facets that are all convex polygons some of which may be co-planar.

We prove that OptDVP is strongly NP-hard by exhibiting a pseudopolynomial reduction to it from the following problem known to be strongly NP-hard (Lingas 1982).

Minimal Number Convex Partition (MNCP):

Instance: A simple polygon P (given by a sequence of pairs of integer-coordinate points in the plane) with non-rectilinear holes and a bound $\alpha \in \mathbb{Z}_+$.
Problem: Decide if P can be decomposed into no more than α convex polygons with vertices among those of P.

Note that MNCP is in P if holes are not allowed (Chaselle & Dobkin 1979).

For getting acquainted with the notion of strong NP-hardness, pseudopolynomial reduction, and related matters the reader is referred to (Garey & Johnson

1979). Here we only recall some basic points. Let $\Pi = (D_\Pi, Y_\Pi)$ and $\Pi' = (D_{\Pi'}, Y_{\Pi'})$ be decision problems with instance sets D_Π and $D_{\Pi'}$, respectively, and sets of instances with answer "yes" Y_Π and $Y_{\Pi'}$, respectively. Denote by $Max[I]$, $Length[I]$, $Max'[I']$, $Length'[I']$ the maximal number and the input length of the instances $I \in D_\Pi$ and $I' \in D_{\Pi'}$, respectively. A *pseudopolynomial reduction* from Π to Π' is a function $f : D_\Pi \to D_{\Pi'}$ such that:

(a) for all $I \in D_\Pi$, $I \in Y_\Pi$ iff $f(I) \in Y_{\Pi'}$,
(b) f can be computed in time polynomial in two variables: $Max[I]$ and $Length[I]$,
(c) there exists a polynomial q_1 such that, for all $I \in D_\Pi$, $q_1(Length'[f(I)]) \geq Length[I]$,
(d) there exists a two-variable polynomial q_2 such that, for all $I \in D_\Pi$, $Max'[f(I)] \leq q_2(Max[I], Length[I])$.

It is well-known (Garey & Johnson 1979) that if Π is strongly NP-hard and there is a pseudopolynomial reduction from Π to Π', then Π' is strongly NP-hard. We now sketch a proof of the following theorem.

Theorem 1 *OptDVP is strongly NP-hard.*

Proof It is not hard to see that by Lemma 1 the pseudopolynomial reduction conditions (b),(c), and (d) are all satisfied. We show next that condition (a) is met as well, i.e., given an instance $I = (X, \alpha)$ of MNCP, it is possible to construct an instance $I' = (M, \beta)$ of DVP such that the solution of I' is "yes" if and only if the solution of I is "yes."

We construct a polytope $P = X \times \tau \subset \mathbb{R}^2 \times \mathbb{R}^1$, i.e., $P \subset \mathbb{R}^3$ is the Cartesian product of X and an interval $\tau = [0, t]$ where t is an integer constant greater than two (see Fig. 3). Then let $M = P \cap \mathbb{Z}^n$. Finally, we set $\beta = 2\alpha + k$.

The so-constructed instance $I' = (P, 2\alpha + k)$ of DVP has solution "yes" iff the instance $I = (X, \alpha)$ of

MNCP has solution "yes." In the one direction the proof is trivial: if the solution for I is "yes" then obviously so is the solution for I' since P can be decomposed into no more than $2\alpha + k$ convex polygons. To demonstrate the other direction, we have to make sure that if the set M can be represented as $M = P' \cap \mathbb{Z}^n$ for some polyhedron P' with no more than $2\alpha + k$ facets, then X can be partitioned into no more than α convex polygons. For this, it is enough to show that any such polyhedron $P' \neq P$ cannot have smaller number of convex facets than P. It is not hard to realize that the latter is implied by Lemmas 2 to 7.

4 LOWER POLYHEDRAL COMPLEXITY BOUND

If in the DVP formulation the set S (and thus also M) is arbitrary, then we have the bound $f_i(P) = \Theta(|M|)$. (See Fig. 4 for an example in dimension two.) In what follows we assume that $M = S_\mathbb{Z}$ where S is a ball with a sufficiently large diameter and $\mathbf{0} \in S$. Our bound will also hold if $S \in C(D)$ where $C(D)$ is the family of convex bodies with C^2 boundary and radius of curvature at every point and every direction between $1/D$ and D, $D \geq 1$. Our considerations apply to an arbitrary dimension n.

Let $P = conv(S_\mathbb{Z})$. We have $P_\mathbb{Z} = S_\mathbb{Z}$ and $conv(P_\mathbb{Z}) = conv(S_\mathbb{Z}) = P$. Let P^* be a convex polytope with a minimal number of $(n-1)$-facets, such that $P_\mathbb{Z}^* = P_\mathbb{Z} = S_\mathbb{Z}$. Then, $conv(P_\mathbb{Z}^*) = P$. Denote $d_P = diam(P)$, $d_{P^*} = diam(P^*)$, and $d_S = diam(S)$. Clearly, $diam(P) \leq diam(P^*)$ and $diam(P) \leq diam(S)$. It is easy to see that any of the relations $diam(P^*) < diam(S)$, $diam(P^*) = diam(S)$, and $diam(P^*) > diam(S)$ is possible. For S being a sphere we have

$$d_P \leq d_{P^*}, d_S \leq 2d_P, d_S \leq 2d_{P^*}, d_{P^*} \leq 2d_S. \qquad (2)$$

In what follows, we will suppose that these diameters are sufficiently large.

An upper bound $O\left(d_S^{\frac{n(n-1)}{(n+1)}}\right)$ on the number of facets of P^* follows from (Bárány & Larman 1998) (see Fact 2 below). A lower bound is provided by the following theorem.

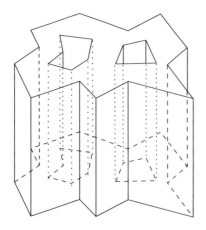

Figure 3. Illustration to the proof of Theorem 1: the polytope P.

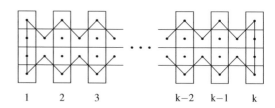

1   2   3 $\quad$ $k-2$ $k-1$ k

Figure 4. Example in 2D with $|M| = 6k - 2$ and $f_i(P) = 4k - 2$ for any $k \in \mathbb{Z}_+$.

Theorem 2 *For every $n \geq 2$ there is a constant $c(n)$ depending only on n such that*

$$f_{n-1}(P^*) \geq c(n) \frac{d_S^{\frac{n(n-1)}{(n+1)\lfloor n/2 \rfloor}}}{\log^{\frac{n-1}{\lfloor n/2 \rfloor}} d_S}.$$

Proof We use the following well-known results.

Fact 1 *Let P and P' be bounded polytopes with vertices $v_1, v_2, \ldots, v_k$ and $v_1', v_2', \ldots, v_k'$, respectively. If $\|v_i - v_i'\| < \varepsilon$, where ε is an enough small positive real number, then P and P' have the same number of k-facets $(0 \leq k \leq n-1)$.*

This last fact, that can be classified as belonging to the mathematical folklore, easily follows from theory of polyhedra (see, e.g., (Papadimitriou & Steiglitz 1982)). It implies that, given a polytope P, there is a rational polytope P', to which the conditions of Fact 1 apply. Thus in what follows, one can assume that both convex polytopes P and P^* are rational.

Fact 2 (Bárány & Larman 1998) *Let $K \in C(D)$ and $\bar{K} = conv(K \cap \mathbb{Z}^n)$. Then for every $n \geq 2$ there are constants $c_1(n)$ and $c_2(n)$ such that for all $k \in \{0, 1, \ldots, n-1\}$,*

$$c_1(n)d^{n\frac{n-1}{n+1}} \leq f_k(\bar{K}) \leq c_2(n)d^{n\frac{n-1}{n+1}} \quad (3)$$

where $d = diam(K)$ is sufficiently large.

Fact 3 (Chirkov 1997) *Let $Q = \{x \in \mathbb{R}^n \ : \ Ax \leq \mathbf{b}, A \in \mathbb{Z}^{m \times n}, \mathbf{b} \in \mathbb{Z}^m\}$. Consider the polytope $conv(Q_{\mathbb{Z}})$ and denote by $V(A, \mathbf{b})$ the set of its vertices. Then*

$$V(A, \mathbf{b}) \leq c(n)m^{\lfloor n/2 \rfloor}\log^{n-1}(1 + d) \quad (4)$$

where $d = \max_{x,y \in Q} \max_{j=1,2,\ldots,n} |x_j - y_j|$.

Obviously, Fact 3 holds also if d is defined by the Euclidean metric, i.e., if $d = diam(Q) = \max_{x,y \in Q} (\sum_{j=1}^n (x_j - y_j)^2)^{1/2}$.

From (3) we have that there exist $c_1(n)$ and $c_2(n)$ with $c_1(n)d^{n\frac{n-1}{n+1}} \leq f_k(P) \leq c_2(n)d^{n\frac{n-1}{n+1}}$. Denote for short $m = f_{n-1}(P)$, $m_* = f_{n-1}(P^*)$, $v = f_0(P)$, and $v_* = f_0(conv(P_{\mathbb{Z}}^*))$. Clearly, $v = v_*$. Then

$$m^* \leq m \leq c_2(n)d_S^{n\frac{n-1}{n+1}}. \quad (5)$$

Multiplying by $\frac{c_1(n)}{c_2(n)}$ both sides of the last inequality in (5) and using once again Fact 2, we get

$$\frac{c_1(n)}{c_2(n)}m \leq c_1(n)d_S^{n\frac{n-1}{n+1}} \leq v.$$

Now, keeping Fact 1 in mind, from (4) we have

$$v = v^* \leq c'(n)m_*^{\lfloor n/2 \rfloor}\log^{n-1}(1 + d_{P^*})$$

and thus

$$c_1(n)d_S^{n\frac{n-1}{n+1}} \leq c'(n)m_*^{\lfloor n/2 \rfloor}\log^{n-1}(1 + d_{P^*})$$

where $c'(n)$ is a constant depending only on n. Then, bearing (2) in mind, we obtain consecutively $m_*^{\lfloor n/2 \rfloor} \geq c''(n)\frac{d_S^{\frac{n(n-1)}{n+1}}}{\log^{n-1}(1+d_{P^*})}$ and $m_* = f_{n-1}(P^*) \geq$

$$c(n)\frac{d_S^{\frac{n(n-1)}{(n+1)\lfloor n/2 \rfloor}}}{\log^{\frac{n-1}{\lfloor n/2 \rfloor}} d_{P^*}} \geq \frac{d_P^{\frac{n(n-1)}{(n+1)\lfloor n/2 \rfloor}}}{\log^{\frac{n-1}{\lfloor n/2 \rfloor}} d_{P^*}} \quad \text{for some constants } c''(n)$$

and $c(n)$, respectively.

Corollary 2 *Let S be a disc in $\mathbb{R}^2$ with a radius r. Then there is a constant $\beta_1 \in \mathbb{Z}_+$ such that $f_1(P^*) \geq \beta_1 \frac{r^{2/3}}{\log(r+1)}$.*

Let S be a sphere in $\mathbb{R}^3$ with a radius r. Then there is a constant $\beta_2 \in \mathbb{Z}_+$ such that $f_2(P^*) \geq \beta_2 \frac{r^{3/2}}{\log^2(r+1)}$.

The first of the above bounds has been recently obtained in (De Vieilleville et al. 2005).

Theorem 2 implies that is exists a constant $\alpha(n)$ depending only on n such that $\frac{f_{n-1}(P)}{f_{n-1}(P^*)} \leq \Delta(d_S, n) = \alpha(n)d_P^{(1-\frac{1}{\lfloor n/2 \rfloor})n\frac{n-1}{n+1}}\log^{\frac{n-1}{\lfloor n/2 \rfloor}}(1 + d_S)$, i.e., the number of facets of the solution provided by the convex hull algorithm is within a factor $\Delta(d_S, n)$ from the optimal one. In particular, for $n = 2$ and $n = 3$ we have $\frac{f_1(P)}{f_1(P^*)} \leq \alpha_1 \log d_S$ and $\frac{f_2(P)}{f_2(P^*)} \leq \alpha_2 \log^2 d_S$, respectively, for some constants $\alpha_1, \alpha_2 \in \mathbb{Z}_+$.

5 CONCLUDING REMARKS

In this paper we showed that the general discrete volume polyhedrization problem is strongly NP-hard. We also gave a lower bound on its polyhedral complexity. Here are some questions that are still open and subject of future interest.

1. Is OptDSP in NP?
2. Is OptDSP NP-hard if non-convex polygonal facets are allowed?
3. Is OptDSP NP-hard when S is a convex body in $\mathbb{R}^3$? Is OptDSP NP-hard when S is a sphere?
4. Find a tight bound on $PC(M)$ when M is the set of integer points in a sphere or a circle.

REFERENCES

Bárány, I. & Larman, D.G. 1998. The convex hull of the integer points in a large ball. *Math. Annalen* 312: 167–181.

Borianne, Ph. & Françon, J. 1994. Reversible polyhedrization algorithm of discrete volumes. In J.-M. Chassery & A. Montanvert (eds). *Discrete Geometry for Computer Imagery*: 157–168.

Burguet, J. & Malgouyres, R. 2000. Strong thinning and polyhedrization of the surface of a voxel object. In G. Borgefors, I. Nyström, and G. Sanniti di Baja (eds). *Discrete Geometry for Computer Imagery*: 222–234. LNCS 1953, Berlin: Springer.

Chaselle, B. & Dobkin, D.P. 1979. Decomposing a polygon into its convex parts. In *Proc. 11th Annual ACM Sympos. on Theory Comput.*: 38–48.

Chirkov, A.Y. 1979. On the relation of upper bounds on the number of vertices of the convex hull of the integer points of a polyhedron with its metric characteristics (in Russian). In V.E. Alekseev, M.A. Antonets & V.N. Shevchenko (eds). *Proc. 2nd Internat. Conf. "Mathematical Algorithms,"*: 169–174. N. Novgorod: Nizhegorod University Press.

Debled-Renesson, I. & Reveillès, J.-P. 1994. An incremental algorithm for digital plane recognition. In J.-M. Chassery & A. Montanvert (eds). *Discrete Geometry for Computer Imagery*: 157–168.

De Vieilleville, F., Lachaud, J.-O. & Feschet, F. 2005. Maximal digital straight segments and convergence of discrete geometric estimators. In Proc. *SCIA'05*: 988–997.

Garey, M. & Johnson, D. 1979. *Computers and Intractability*. San Francisco: W.H. Freeman & Company.

Kim, C.E. & Rosenfeld, A. 1982. Convex digital solids. *IEEE Transactions on Pattern Analysis and Machine Intelligence* 4(6): 612–618.

Kim, C.E. & Stoimenović, I. 1991. On the recognition of digital planes in three dimensional space. *Pattern Recognition Letters* 32: 612–618.

Klette, R. & Sun, H.-J. 2001. Digital planar segment based polyhedrization for surface area estimation. In C. Arcelli, L.P. Cordella & G. Sanniti di Baja (eds). *Visual Form 2001*: 356–366. Berlin:, Springer.

Lingas, A. 1982. The power of non-rectilinear holes. In *Proc. 9th Internat. Colloc. on Automata, Languages, and Programming*: 369–383. LNCS 140, Springer.

Lorensen, W.E. & Cline, H.E. 1987. Marching cubes: a high resolution 3d surface construction algorithm. *Computer Graphics* 21(4): 163–169.

Papadimitriou, Ch. & Steiglitz, K. 1982. Combinatorial Optimization. New Jersey: Prentice-Hall.

Papier, L. 1995. Polyhédrisation et visualisation d'objets discrets tridimensionnels. Ph.D. thesis, Université Louis Pasteur, Strasbourg, France.

Papier, L. & Françon, J. 1999. Polyhedrization of the boundary of a voxel object. In C. Bertrand, M. Couprie & L. Perroton (eds). *Discrete Geometry for Computer Imagery*: 425–434. LNCS 1568, Springer.

Sivignon, I. 2004. De la caractérisation des primitives à la reconstruction polyédrique de surfaces en géométrie descrète. Ph.D. thesis, Institut National Polytechnique de Grenoble, France.

Sivignon, I., Dupont, F. & Chassery, J.-M. 2004. Discrete surface segmentation into discrete planes, In R. Klette & J. Žunić (eds). *International Workshop on Combinatorial Image Analysis*: 458–473. LNCS 3322, Springer.

Sivignon, I., Dupont, F. & Chassery, J.-M. 2004. Decomposition of three-dimensional discrete objects surface into discrete plane pieces. *Algorithmica* 38: 25–43.

Vittone, J. & Chassery, J.-M. 2000. Recognition of digital naive planes and polyhedrization. In G. Borgefors, I. Nyström & G. Sanniti di Baja (eds), *Discrete Geometry for Computer Imagery*: 296–309. LNCS 1953, Springer.

Yu, L. & Klette, R. 2002. An approximation calculation of relative convex hulls for surface area estimation of 3D objects. In R. Kasturi, D. Laurenderau & C. Suen (eds). *IAPR International Conference on Pattern Recognition* Vol. 1: 131–134. Québec (Canada): IEEE Computer Society.

Computational Modelling of Objects Represented in Images –
João Manuel R.S. Tavares & R.M. Natal Jorge (eds)
© 2007 Taylor & Francis Group, London, ISBN 978-0-415-43349-5

Contextual classification of remotely sensed images with integer linear programming

Manuel Lameiras Campagnolo

Dept. Matemática, Instituto Superior de Agronomia, Univ. Técnica de Lisboa and SQIG – IT, Portugal

Jorge Orestes Cerdeira

Dept. Matemática, Instituto Superior de Agronomia, Univ. Técnica de Lisboa, Portugal

ABSTRACT: Supervised classification techniques are commonly used to assign pixels of multispectral satellite imagery to a predefined set of classes in order to generate or update land use or land cover maps from remote sensed data. These techniques have a limited ability in expressing spatial relationships among pixels. We propose a new contextual approach to address this issue. In particular, we present an integer linear programming formulation which restricts the number of distinct objects in the classified image and we propose a heuristic for the resulting problem. We test it on one generated data set and one real data set.

1 INTRODUCTION

In a multispectral remote sensed image with n bands, each pixel of the image is described by a n-dimensional vector called the pixel's spectral signature. In image classification, one considers k distinct classes and looks for the "best" assignment of each pixel to one and only one class. Many techniques have been proposed to solve this problem (Landgrebe 2003) but the achieved solutions are frequently not satisfactory. This usually happens when classes are hard to discriminate in the space of spectral signatures and when existing spatial relations among pixels in the remotely sensed images are not fully explored.

Formally, an assignment of pixels is a function y such that $y_i^c = 1$ if pixel i is assigned to class c and is 0 otherwise. For simplicity, we say that pixel i is of class c if $y_i^c = 1$. Supervised classification techniques typically look for a partition $R_1, \ldots, R_k \subset \mathbb{R}^n$ of the space of spectral signatures such that the best decision rule is given by $y_i^c = 1$ if and only if $x^{(i)} \in R_c$ and $x^{(i)}$ is the spectral signature of pixel i. If d is an appropriate distance between the pixels' signatures and classes, this is equivalent to minimize the global function $D = \sum_{i,c} d_i^c y_i^c$, where d_i^c is the distance from pixel i to class c. Classifiers differ in the choice of function d. For instance, quadratic discriminant classifiers (usually known as "maximum likelihood classifiers" in the remote sensing literature) use the Mahalanobis distance, which is a function of the spectral signature of each pixel and of the mean vector and variance-covariance matrix of each class. Alternatively, the m

nearest neighbors non-parametric classifier estimates the probability of pixel i belonging to class c as the proportion of the closest m neighbors of i (in the spectral space) that belong to class c. Then, the class which is closest to pixel i is the one with the highest probability. Other classifiers like neural networks, support vector machines and decision trees can be described in the same framework.

However, something is missing in the approaches outlined above. Since d depends uniquely on the spectral signature of each individual pixel and on the description of the classes, then the spatial relations among pixels in the image are not accounted for. This per-pixel *non contextual classification* exhibits the well known "salt and pepper" effect. In opposition, classification techniques which take into account the spatial neighborhood relations among pixels are called "contextual". As far as we know, currently used contextual classifiers only incorporate in the decision rule information about the *local* neighbors of each pixel. The most straightforward approach for local contextual classification is to revise pixel-to-class assignment *a posteriori*. A simple example is the application of a 3×3 mode filter to the image after classification. A popular and more realistic model of local spatial context is given by Markov random fields in the framework of statistical Bayesian classification (Tso and Mather 1999).

While local decision rules may be suitable for certain cases, they may not apply to all spatial objects that can be identified in satellite imagery. For instance, if a patch of pixels of the same class is crossed by a one-pixel-wide feature of a different class (for instance, a road in a satellite image), which may not be clearly discriminated in the spectral space from the surrounding pixels, then a local rule would tend to remove this feature from the classified image.

2   THE CONCEPTUAL MODEL

We propose here a *global* contextual classifier. As before, and given any distance d between pixels and classes in the spectral space, we look for the assignment y that minimizes $D = \sum_{i,c} d_i^c y_i^c$. However, we add a restriction to the problem in order to prevent the "salt and pepper" effect derived from excessive number of small size patches of pixels. A patch is a maximal set of contiguous pixels belonging to the same class. One can view a patch as a spatial object in the classified image. Since objects in the image are somewhat spatially continuous, the key idea is to restrict the number of patches in the image, while preserving meaningful "narrow" objects. To turn this more precise consider the graph $G = (P, E)$, where P is the set of pixels and E is the set of pairs of pixels which are neighbors in the image. If y is any assignment of pixels to classes, a patch is a connected component (i.e., a maximal connected subgraph) of the subgraph of G induced by the set of pixels in a same class under y. Thus, different patches may be from the same class, but two neighbor patches (i.e., with a pair of neighbor pixels not both in the same patch) have to belong to different classes.

In this setting, the optimal assignment is given by the solution of

$$\min D = \sum_{i,c} d_i^c y_i^c \tag{1}$$

$$\sum_c y_i^c = 1, \text{ for every } i \tag{2}$$

$$y_i^c \in \{0,1\}, \text{ for every } i, c \tag{3}$$

number of patches $\leq N$ (4)

where N is a suitable number of patches for the image to be processed.

3   ANALYSIS

We now show how to turn condition (4) into a linear description and we propose a heuristic for the resulting problem.

3.1   *A linear model*

The key idea is to define a spanning forest T (i.e., an acyclic subgraph spanning all vertices) of G with N components, and assign to the same class all pixels within the same component of T.

Let us therefore consider decision variables x_{ij} on every edge $[i,j]$ of G, and define $T = \{[i,j] \in E: x_{ij} = 1\}$ with

$$\sum_{i,j} x_{ij} = |P| - N \tag{5}$$

$$x_{ij} \in \{0,1\}, \text{ for every } i, j \tag{6}$$

$$\sum_{i,j \in E(S)} x_{ij} \leq |S| - 1, \text{ for every } S \subset P \tag{7}$$

where $E(S)$ is the set of edges of G with both ends in S. Equation (5) ensures that forest T has $|P| - N$ edges, and inequalities (7) are the usual subtour elimination constraints preventing T from having cycles. Together, (5) and (7) imply that the forest has exactly N components.

We now have to guarantee that pixels within the same component of T will be assigned to the same class. The following simple observation allows us to guarantee this just from the knowledge of the edges of T. If two pixels from the same component of T are in different classes, then there will be some edge $[i,j]$ of T such that i and j are in different classes. Thus, the valid inequalities

$$1 - x_{ij} \geq y_i^c - y_j^c, \text{ for every } i, j, c \tag{8}$$

$$1 - x_{ij} \geq y_j^c - y_i^c, \text{ for every } i, j, c \tag{9}$$

ensures that whenever $[i,j]$ is in T, i.e., $x_{ij} = 1$, then i and j belong to the same class.

Note that distinct neighbor components of the N-component forest T may belong to the same class. However, if (y^*, x^*) is an optimal solution of (1)–(3),(5)–(9), then y^* is an optimal solution of (1)–(4), where the number of patches can be strictly smaller than N.

The use of an integer linear programming solver to deal with the formulation (1)–(3), (5)–(9) has to handle the huge number of constraints (7). One possible approach is to start with only (1)–(3), (5),(6), (8),(9), and successively add inequalities (7) that are violated by the current solution until a feasible solution is reached, which would be optimal. However, the number of constraints (7) to be added may make this procedure unpractical, even for a small data set (say, an image with 400 pixels like the one described in subsection 4.1).

To speed up this procedure classes of strong valid inequalities (i.e., cuts) for this classification problem have to be devised. This is the subject of polyhedral combinatorics theory: see for example (Pulleyblank 1983) or (Schrijver 1995). Nevertheless, any such integer cutting algorithm will certainly suffer from serious limitations regarding the size of the data sets. Hence, the development of accurate heuristics is a reasonable option for the problem in hand.

3.2 A heuristic approach

We design a heuristic for (1)–(3), (5)–(9) that can be viewed as a dynamic version of Kruskal's algorithm for minimum spanning trees (Kruskal 1956).

The algorithm performs $|P| - N + 1$ iterations. In each iteration $t = 0, 1, \ldots, |P| - N$ a spanning forest T_t with $|P| - t$ components is (implicity) defined, and all the pixels within each component I of T_t are assigned to the same class c. Thus, $D_I = \sum_{v \in I} d_v^c$ is the contribution of the component I for the objective function (1), and $\sum_I D_I$ is the value of the current solution.

At iteration 0 the spanning forest T_0 has $|P|$ components, each consisting of a single pixel which is assigned to its closest class. This is the non contextual classification.

In iteration $t \geq 1$, for each pair of adjacent components I, J of T_{t-1}, the value

$$c_{IJ} = \min_{\{c\}} \left\{ \sum_{v \in I} d_v^c + \sum_{v \in J} d_v^c \right\} - (D_I + D_J) \qquad (10)$$

is computed. The weight c_{IJ} on "edge" (I, J) is the minimum increment on the objective function (1) derived from merging components I and J within the same class. The "edge" (I^*, J^*) for which the minimum weight

$$c_{I^*J^*} = \min_{\{I,J\}} \{c_{IJ}\}$$

is attained is added to T_{t-1}, meaning that components I^* and J^* will be contracted into a single component of T_t, with every pixel of $I^* \cup J^*$ assigned to the same class (that one which was used to settle $c_{I^*J^*}$ in (10)).

The algorithm can be implemented with time and space complexities $O(k|E| + k|P|(|P| - N))$ and $O(k|P| + |E|)$, respectively.

4 EXAMPLES

To assess the quality of the solutions produced by the above heuristic algorithm, we use two data sets. The first is a small generated data set with 400 pixels and 2 classes. The second is a data set with 12513 pixels and 5 classes derived from SPOT-XS satellite imagery.

4.1 Generated data

This data set has two classes. Class 1 represents some linear features (e.g. roads) and class 2 represents an homogeneous background. The image has 20 rows and 20 columns. It was generated from the image in Figure 1, adding Gaussian noise to each class. Specifically, the amount of noise is such that the

Figure 1. Reference image with two classes.

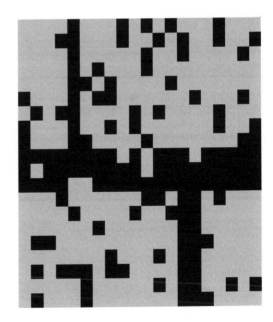

Figure 2. Non contextual classification after adding noise to Figure 1.

probability of misclassification for a pixel extracted from class 1 is 10%. For class 2, the corresponding probability is 20%. The resulting non contextual (per pixel) classification is shown in Figure 2.

Figure 3. Contextual classification with 5 patches (heuristic solution).

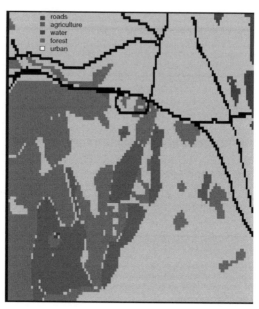

Figure 4. Land cover map obtained by photo-interpretation.

Since the number of patches in the reference image is 5, we run our algorithm up to that number of components. The resulting classification is shown in Figure 3.

4.2 *Real data*

The image covers the Monsanto area, in Lisbon, Portugal, with SPOT-XS data from the summer of 1990. The spatial resolution is 20 meters. It is a small area of 2580 meters by 1940 meters covered by 12513 pixels. We have only considered five classes: water, roads, urban areas, forested areas and agricultural areas. We used a random sample of 500 pixels (approximately 4% of the total) for estimating the spectral signatures of the classes. The distances d_i^c were computed using the k-nearest neighbors technique. The land cover map of that area is known from the photo-interpretation of aerial photographs of the same year (see Figure 4).

The resulting per pixel classification is shown in Figure 5. In this map, approximately 75% of the pixels are well classified according to the land cover map. We computed the value of the objective function (1) for every iteration of the heuristic algorithm. The values are plotted in Figure 6. The objective function (for the per pixel classification) stays constant until the iteration $t = 12181$, meaning that the non contextual classification has $12513 - 12181 = 332$ patches. Note that for a number of components less than 48,

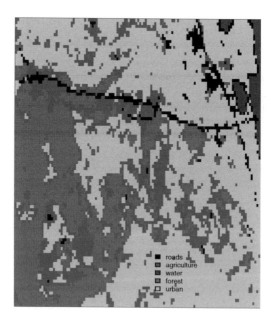

Figure 5. Non contextual classification of the SPOT-XS image.

i.e. for iteration $t > 12513 - 48$, the objective function is strictly increasing, which means that for less than 48 components all components of the solution are patches. We show in Figure 7 the resulting map with 15

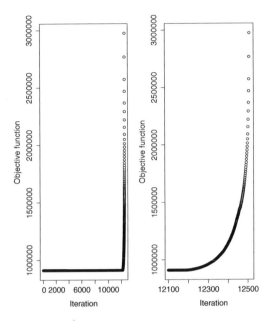

Figure 6.   Objective function values for iterations 0 through 12508 of the algorithm. The figure on the right highlights iterations 12100 to 12508.

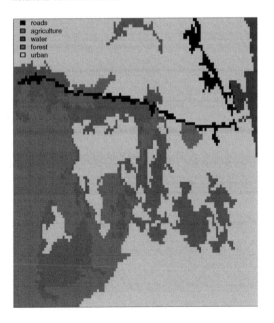

Figure 7.   Contextual classification of the SPOT-XS image with 15 patches (heuristic solution).

patches. For this map, the percentage of pixels misclassified by the non contextual classifier that are correctly classified by the algorithm is 24.62%. In opposition, the percentage of pixels correctly assigned in the non-contextual classification that are re-assigned to a different class is 4.17%.

5   CONCLUSIONS

We propose a mathematical model to deal with global contextual classification of multispectral images. Furthermore we were able to devise a 0–1 linear formulation (1)–(3), (5)–(9), and developed a heuristic algorithm for the problem.

The computational tests carried out allow us to conclude that the heuristic approach deals with some of the problems of the non contextual classification such as the "salt and pepper" effect. However, the tests we ran show that the algorithm behaves poorly regarding two kinds of features in the image. Firstly, it does not remove spurious "peninsulas" attached to relevant objects in the image (see Figure 3). Secondly, it is not successful in restoring linear features (e.g. roads) of the original image (compare Figures 5 and 7).

The first limitation is inherent to the model (1)–(4). We discuss below a possible way to overcome this limitation. The second is due to the particular heuristic we used. Since components defined by the algorithm are never broken apart, pixels in the background that are merged together in early iterations cannot be re-assigned to the linear features. However, the optimal solution of (1)–(4) is not contrived in this manner.

As we said above, we think that the current formalization can be further improved. We believe that the restrictions (2)–(4) are effective in accounting for the spatial relations among pixels of the image. However, the objective function (1) suffers from the lack of accuracy on the distances d_i^c, which are estimated from the data. For pixels whose spectral signatures are clearly related to one class, the estimates are reliable. However, for the remainder pixels, the random variability on the estimated values of d_i^c will affect significantly the results. A possible approach to overcome this could be to "randomize" (1) to take into account that phenomena. One possibility is to repeatedly run our algorithm on slightly perturbed values of the distances between pixels and classes. Then, one of the possible outcomes could be selected (for example, by an expert or by comparing their accuracy) as the solution of the problem. We intend to develop these ideas in a forthcoming paper.

The present work describes a new approach for contextual classification of remotely sensed images. Its global nature distinguishes it from usual techniques based on local post-classification filters or local random fields. Our formulation of the problem in a combinatorial optimization setting opens the way to explore many tools and techniques which are seldom applied to image classification. This paper is just a first step in that direction.

127

ACKNOWLEDGMENTS

This work was partially supported by FCT and EU FEDER, namely by CLC, grant SFRH/BPD/21012/2004 (MLC) and by POCTI Program from FCT (JOC). MLC's contribution was partly done while visiting the Department of Geography at UMD. We would like to thank Chris Justice from UMD and Leonor Pinto from ISEG for useful discussions on the subject of this paper.

REFERENCES

Kruskal, J. B. (1956). On the shortest spanning-subtree of a graph and the traveling salesman problem. *Proceedings of the American Mathematical Society 7*, 48–50.

Landgrebe, D. (2003). *Signal Theory Methods in Multispectral Remote Sensing*. John Wiley and Sons.

Pulleyblank, W. R. (1983). Polyhedral combinatorics. In *Mathematical Programming – The State of the Art*, pp. 312–345. Springer, Berlin.

Schrijver, A. (1995). Polyhedral combinatorics. In *Handbook of Combinatorics*, Volume II, pp. 1649–1704. Elsevier, Amsterdam.

Tso, B. and P. Mather (1999). Classification of multisource remote sensing images using a Markov random field. *IEEE Transactions on Geoscience and Remote Sensing 37*, 1255–1260.

Computational Modelling of Objects Represented in Images –
João Manuel R.S. Tavares & R.M. Natal Jorge (eds)
© *2007 Taylor & Francis Group, London, ISBN 978-0-415-43349-5*

Two methodologies for iris detection and location in face images

Fernando Jorge Soares Carvalho
Departamento de Matemática, Instituto Superior de Engenharia do Porto
R. Dr. Bernardino de Almeida, Porto, Portugal

João Manuel R.S. Tavares
Lab. de Óptica e Mecânica Experimental, Instituto de Eng. Mecânica e Gestão Industrial Dep. de Engenharia
Mecânica e Gestão Industrial, Faculdade de Engenharia da Universidade do Porto
Rua Dr. Roberto Frias, Porto, Portugal

ABSTRACT: In this paper two methodologies to detect and locate the iris of the eye in static images, are presented. One uses the Hough's transform and other is based on deformable templates. In both methodologies, the contour of the iris is represented geometrically by a circumference and its radius and centre are considered as its control parameters. The dynamic update of these parameters allows the determination of its final values, enabling the fully definition of the iris in the image. After a description of the considered methodologies, the advantages and disadvantages between them are presented, as well as some experimental results, conclusions and perspectives of future work.

1 INTRODUCTION

In Computational Vision, one of the main development areas is the detection and location of faces, either considering static images or image sequences. In this area, the face can be detected as a whole, using, for example, a geometric contour that defines it in a global way, or can be accomplished by identifying certain facial features, as the eye, the mouth, etc.

The methodologies for face detection and location in an input image are usually limited by the manner in which the face appears in that image. Thus, some existent methodologies consider, for example, that the face should be without facial hair and glasses, and in a front position, without partial occlusion, etc.

In (Yang et al. 2002), the existent methodologies for face detection and location, are divided into four different categories: based on knowledge; based on invariant features; based on appearance; and based on template matching.

The methodologies based on knowledge, are based in a set of rules coming from the knowledge of the human face to be considerer and they lead to the detection process, as the example of the work described by (Yang & Huang 1994). Some of those rules are: the existent of symmetry between facial features, as for instance, the eyes; the intensity values that appear in the centre of the face, usually uniform; the significant difference among the intensity values in the top of the face and in its centre; etc.

The methodologies based on invariant features are intended to detect facial features that exist independently from the variation of the face position or of the vision point used and the brightness conditions, like: skin texture (Dai & Nakano 1996); geometric position of facial features (Leung et al. 1995); skin colour (Rademacher 2001); etc.

In the other hand, the methodologies based on the appearance, use models that are learned from a set of training images and capture the representative variability of the appearance; examples are: eigenfaces (Turk & Pentland 1991); neuronal networks (Rowley et al. 1998); Hidden Markov Models (Samaria & Young 1994); etc.

Finally, the methodologies based on template matching, use templates that define a face or a just single feature of it. The detection is then lead by a dynamic process of comparison between the used template and the input image; examples are: cross correlation (Carvalho & Tavares 2005); deformable templates (Yuille et al. 1992); circular templates using the Hough's transform (Toennies et al. 1998); etc.

In this work, we use two methodologies, one that uses the Hough's transform and another one based in template matching, both sharing the same purpose: the detection and location of the iris in input images.

This paper is organized as follows: in section 2, the iris model and the energy fields used in the methodologies considered are presented; in the next section, both methodologies used for iris detection and location are described, pointing some of their advantages and disadvantages; in section 4, some experimental results are introduced; and finally, in the last section, we present some conclusions and perspectives of future work.

2 IRIS MODEL AND ENERGY FIELDS

2.1 Iris model

The iris of the human's eye can be geometrically represented by a circumference, Figure 1. The dynamic changing of the circumference's parameters allows: its displacement, by changing and update its centre; and its rigid deformation by updating its radius.

2.2 Energy fields

The methodologies that we describe in the next section use image energy fields that result from the application of specific morphologic operators, on the input image, that enhance important features of the eye, Figure 2. In this work, we considered three energy fields: intensity edges; intensity valleys; and intensity levels.

The edges energy field was obtained in this work using the Canny's edge detector for enhancing areas of the input image with high contrast, which means detecting the contours presented in the input image. On the other hand, the valleys energy field enhance areas of the input image with reduced intensities, and the input image intensity contains generic information about the intensities distribution.

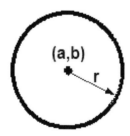

Figure 1. Iris model, represented by a circumference with radius r and centre with coordinates (a, b).

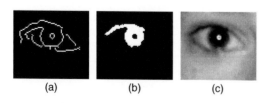

(a)   (b)   (c)

Figure 2. Representative image of the energy fields: intensity edges (a); intensity valleys (b); intensity levels (c).

3 USED METHODOLOGIES

In this section, is present a description of the two methodologies used in this work for the iris detection and location.

3.1 Iris detection and location using the Hough's transform

Basically, the Hough's transform consists on the determination of the circumference's control parameters, by searching in the Hough's space for a set of parameters that better describes it.

The algorithm considers the energy fields obtained from the edges intensity in binary levels. Then, in that energy field, are located all pixels with unitary intensity; being considered, in this case, circumference's centres with coordinates (x, y):

$$(x - a)^2 + (y - b)^2 = r^2.$$ (1)

Keeping the radius value constant and considering its angular magnitude θ between 0 and 2π, we determine the potentials coordinates of the iris' centre (a, b), using the parametric circumference equations:

$$\begin{cases} a = x - r \times \cos(\theta) \\ b = y - r \times sen(\theta) \end{cases}.$$ (2)

In Figure 3, an example of the parameters space is shown, considering the radius equal to 5 pixels.

Next, the coordinates of the iris' centre as well as the radius' used, are recorded in a parameter accumulator, the maximum number of occurrences of a certain parameter is then analyzed; that is, the maximum number of intercessions among all obtained circumferences is determined.

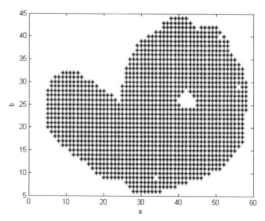

Figure 3. An example of the parameters' space.

130

It is considered that the algorithm always converges, whenever an only set of parameters of the accumulator contains the largest number of intercessions in the parameters' space.

3.2 Iris detection and location using a deformable template

The iris detection using a circumference as a deformable template consists on the dynamic and iterative update of its parameters; lead by an energy function that characterizes the energy cost of its deformation. Thus, low energy cost assures that the matching between the template and the facial feature in analysis is well done and so a correct detection was obtained. Otherwise, if the deformation energy cost is high then the template is different from the facial feature in analysis and the matching is not adequate.

The total energy E, is a result of several primitives using the energy fields considered in section 2.

The dynamic update of the template's parameters is processed by an optimal method: the steepest descendent gradient. As an example, during the iris' radius update are used the Equations 3 and 4:

$$r\_new = r\_old + \Delta t \times r'_t, \qquad (3)$$

where r_new is the radius of the new iteration, r_old the radius of previous iteration, Δt the space step, r'_t the variation rate of the radius in time, translated by the symmetrical variation rate of the total energy considering a small radius variation:

$$r'_t = \frac{dr}{dt} = -\frac{\partial E}{\partial r}. \qquad (4)$$

The algorithm used was divided into three processing phases: the first one, uses energy from the intensity valleys, allowing that the template have a correct position in the iris' centre; in the second phase, the previous energy field is used and the field resulting from the image of intensities allow the reach of the iris' radius, giving its correct scale; in the third and last phase, the energy from the intensity edges is used, allowing the fine adjustment of the iris' radius and centre. In section 4, is shown an example of this processing.

3.3 Comparison between the two methodologies

Although we can expect similar results while using these methodologies in a detection application, they both present some advantages and disadvantages.

The methodology based on the Hough's transform has the advantage of finding every circular shape presented in the input image, independently of its location, obtaining a good global detection. Its main disadvantage is related with the high demand in terms of computing resources.

On the other hand, the methodology that uses a deformable template has the advantage of requiring reduced computing resources to converge. Its disadvantage lies in the necessity of defining the initial values for the control parameters of the template used. The initial parameters should be properly chosen and closer to the final ones. Then, we can conclude that this methodology is a good local detector and a weak global detector. When considering the presence of several circular shapes in an input image, the template usually follows the circular shape that is closer to the initial position.

4 EXPERIMENTAL RESULTS

4.1 Hough's transform

In Figure 4, the sequence of the parameters' space is represented graphically, resulting from the consideration of different radius, between 5 and

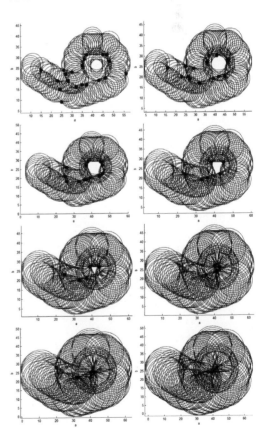

Figure 4. Representative image of the sequence parameters' space, considering different radius values, from left to the right and top to the bottom, the radius are respectively: 5, 6, 7, 8, 9, 10, 11 and 12 pixels.

12 pixels. In table 1, are presented the used radius, the record accumulate values, related to the maximum number of currencies with the same number of intersections, and the coordinates of the iris' centre.

It is possible to observe that the maximum number of the accumulator occurrences is going to become smaller as the radius is getting bigger, indicating that the algorithm converges. Thus, from Table 1, we can see that the iris' radius is assumed to be 12 pixels, and its centre has coordinates (41, 25) pixels, as it is shown in Figure 5, considering the left top corner of image as the origin of the referential used.

4.2 Deformable template

In Figure 6, we can observe an input image and the results of the matching process using the three phases processing describe in section 3.

Table 1. Results obtained using the methodology based on the Hough's transform.

Radius (Pixels)	Nº of intersections	Centre (Pixels)
5	20	–
6	14	–
7	13	–
8	6	–
9	5	–
10	5	–
11	4	–
12	1	(41,25)

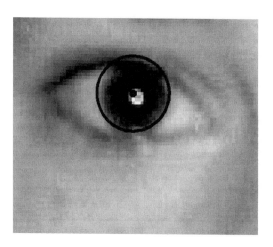

Figure 5. Input image, with a representation of a circumference with 12 pixels of radius centred on the coordinates point (41,25) pixels found using the methodology based on the Hough's transform.

In Table 2, we have the results from the following matching strategy: in the first phase, it was set of radius of the circumference used and calculated the coordinates of the iris' centre; in the second phase, based on the knowledge of the iris' centre previously determined, the value of the iris' radius was obtained; and in the third and last phase took place the fine adjustment of the template's control parameters.

In the example being considered, the obtained results were: the iris's centre is located in the point with coordinates (42, 25) pixels and the radius is defined as equal to 11.3 pixels.

5 CONCLUSIONS AND PERSPECTIVES OF FUTURE WORK

Having in mind the obtained results, we can consider that both methodologies considered in this paper allow

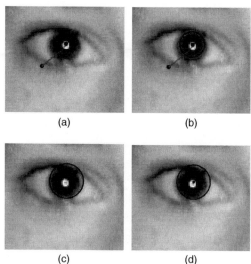

Figure 6. Detection and location of an iris in an input image using the methodology based on template matching: results of each phase (a to c), circumference obtained (d).

Table 2. Results obtained using the template matching methodology during its three phases.

Phase	Initial radius (Pixels)	Final radius	Initial centre	Final centre
1	10	–	(26,40)	(42,25)
2	6	8,5	–	–
3	8,5	11,3	(42,25)	(42,25)

the detection and location of the human iris in input images. It was also verified that exist a great similarity of the parameters' values, which define the iris' location and radius in the input image, obtained by both methodologies considered.

As a future work, he hope to apply the methodologies considered in this paper in the tracking of the human eye in image sequences, considering the deformable template proposed in (Yuille et al. 1992).

ACKNOWLEDGMENTS

This work was partially done in the scope of the project "Segmentation, Tracking and Motion Analysis of Deformable (2D/3D) Objects using Physical Principles", with reference POSC/EEA-SRI/55386/2004, financially supported by FCT – Fundação para a Ciência e a Tecnologia from Portugal.

REFERENCES

Carvalho, Fernando Jorge Soares & Tavares, João Manuel, R. S. 2005. Detecção e Extracção de Características do Olho Humano a partir de um Modelo Protótipo Deformável. At Encontro Nacional de Visualização Científica, Espinho, Portugal.

Dai, Y. & Nakano, Y. 1996. Face-Texture Model Based on SGLD and Its Application in Face Detection in a Color Scene. Pattern Recognition 29(6), pp. 1007–1017.

Leung, T.K., Burl, M.C. & Perona, P. 1995. Finding Faces in Cluttered Scenes Using Random Labeled Graph Matching. Proc. 5th IEEE International Conference Computer Vision, Cambridge, MA, USA: pp. 637–644.

Rowley, H., Baluja, S. & Kanade, T. 1998. Neural Network-Based Face Detection. IEEE Transactions Pattern Analysis and Machine Intelligence 208(1), pp. 23–38.

Rademacher, David 2001. *Face detection*. Colorado School of Mines, Introduction to computer Vision and Image.

Samaria, F. & Young, S. 1994. HMM Based Architecture for Face Identification. Image and Vision Computing 12, pp. 537–543.

Turk, M. & Pentland, A. 1991. Eigenfaces for Recognition. Journal of Cognitive Neuroscience 3(1), pp. 71–86.

Toennies, K., Behrens, F. & Aurnhammer, M. 1998. Feasibility of hough-transform based iris localisation for real-time-application. In *Proceedings of the 16th International Conference on Pattern Recognition (ICPR'02)*, Quebec, vol. 2, pp. 299–305.

Yuille, A., Hallinan, P. & Cohen, D. 1992. Feature Extraction from Faces Using Deformable Templates. *International Journal of Computer Vision* 8, pp. 99–111.

Yang, G. & Huang, T. S. 1994. Human Face Detection in Complex Background. Pattern Recognition 27(1), pp. 53–63.

Yang, Ming, Kriegman, David, J. & Ahuja, Narendra 2002. Detecting Faces in Images: A Survey. IEEE Transactions on Pattern Analysis and Machine Intelligence 248(1), pp. 34–58.

Computational Modelling of Objects Represented in Images –
João Manuel R.S. Tavares & R.M. Natal Jorge (eds)
© 2007 Taylor & Francis Group, London, ISBN 978-0-415-43349-5

Accessing earth observation data using JPEG2000

Hélder Carvalho, Carlos Serrão, António Serra & Miguel Dias
ISCTE/Adetti, Ed. ISCTE, Av. Das Forças Armadas, Lisboa, Portugal

ABSTRACT: Applications like, change detection, global monitoring, disaster detection and management have emerging requirements that need the availability of large amounts of data. This data is currently being captured by a multiplicity of instruments and Earth Observation (EO) sensors originating large volumes of data that needs to be stored, processed and accessed in order to be useful. The authors of this paper have been involved on an ESA-founded project, called HICOD2000 to study the applicability of the new image encoding standard – JPEG2000 – to EO products. This paper presents and describes the system that was developed for HICOD2000 project, which allows, not only the encoding and decoding of several EO products, but also supports some of the security requirements identified previously that allows ESA to define and apply efficient EO data access security policies and even to exploit some EO products electronic commerce over the Internet. This system was integrated with the existing ESA Ground Segment systems specifically the Services Support Environment (SSE).

Keywords: DRM, JPEG2000, Earth Observation, encryption, protection tool.

1 INTRODUCTION

Change detection, global monitoring, disaster detection and management applications have emerging requirements that need the availability of large amounts of data. This data is currently being captured by a multiplicity of instruments and Earth Observation (EO) sensors producing large amounts of data that needs to be stored, processed and accessed in order to be useful – as an example, ENVISAT accumulates, in a yearly basis, several hundred terabytes of data.

The authors of this paper were involved on an ESA-sponsored project, called HICOD2000 to study the applicability of the new image encoding standard – JPEG2000 – to Earth Observation (EO) products. This paper presents and describes the set of applications that were developed in the HICOD2000 project which allows, not only the encoding and decoding of several EO products, but also supports some of the security requirements identified previously that allows ESA to define and apply efficient EO data access security policies and even to exploit some EO products electronic commerce over the Internet.

During the project development, the work was mostly performed on ENVISAT and SPOT5 satellites EO data. The ENVISAT satellite contains a payload of eight measuring instruments and each of the resulting products has been examined during the project to determine the JPEG2000 coding suitability. As a result, several products have been identified: MERIS, ASAR, GOMOS, ATTSR, RA2 and SCIAMACHY.

Each of the products is currently coded in ESA's proprietary format called Payload Data Segment (PDS) which is a non compressed data format which usually results in quite large product files, not very suitable for Internet transfers. SPOT5 vegetation products were also considered in the HICOD2000 system. These products are stored in the NCSA's Hierarchical Data Format (HDF).

The JPEG2000 encoding process that was adopted in the project had into consideration the fact that no information loss should occur in the resulting JPEG2000-flavoured version of the original EO products. JPEG2000 is used in lossless coding operation to allow the exact recovery of the original product (bit by bit). This represents one of the big results of the proposed system – the possibility to compress the ENVISAT and SPOT5 EO products to lossless JPEG2000 format, and later recover the original format without information loss – this was a requirement due to the fact that there were already in place a great set of legacy PDS data manipulation tools that would not be affected by the encoding scheme. In fact, the HICOD2000 system could be plugged in these legacy systems.

Security was also a critical issue on the system. The possibility to directly protect the EO data and to manage the access to that protected data was something that the system also had to support. Since the project was using JPEG2000 technology to have a newer representation of the EO products, it was also developed a mechanism to allow these JPEG2000 EO products to be protected. The approach was to protect the content itself and not only the data channel – like what is easily found on SSL/TLS network connections. This opened new possibilities to ESA in terms of new business models and the definition of new EO data access policies – for instance it could be possible to "lock" the access to some sensitive data on a specific EO product and to allow the access to the remaining data (this parameterization could also be done on a user basis).

The solution adopted was based on one of the parts of JPEG2000 standard, called JPSEC, which deals with code-stream protection. Therefore it was developed a method to protect the EO product according to its resolution levels, and it was integrated a DRM solution to control the user access granularity to each of the EO product resolutions. Using this solution ESA could profit in terms of flexibility of its traditional business models, and allow the price differentiation of its own EO products according to the resolution level selected by the client. On the client side a visualization tool was improved to control the resolution level access according with the one the client purchased.

Finally, this system was integrated with the existing ESA Ground Segment systems specifically the Services Support Environment (SSE) and is now on a test phase.

2 JPEG2000, JPSEC AND DRM SOLUTION

Remote sensing applications handle large size images. An efficient compression algorithm would save significant storage space. Also image manipulation, such as zoom, pan and metadata edition may be valuable features. Content security is also a relevant aspect to be considered, those images that have intrinsic value and must be protected both in terms of privacy protection, conditional access and B2B and B2C e-commerce business processes. The study of these problems and the creation of a system that could combine a solution to solve them, were key objectives of this work.

2.1 JPEG2000 image compression standard

The image compression standard JPEG2000 (Boliek, Martin 2002) brings not only powerful compression performance but also new functionality unavailable in previous standards (such as region of interest, scalability and random access to image data, through flexible code stream description of the image). The compression performance using JPEG2000 overcomes JPEG, with the drawback of a much higher complexity. For software and hardware platforms, the complexity of JPEG2000 is believed to be around ten times higher than JPEG (Taubman, David S. & Marcellin, Michael W. 2001). This is a crucial problem for its adoption for real-time systems such as earth observation satellites.

Figure 1. JPEG (left) and JPEG2000 (right) low bit-rate comparison.

JPEG2000 introduces flexibility in the transmission of images with a progressive improvement in image quality. The standard supports the capability of transmitting arbitrary regions of interest inside the image, with greater fidelity than others. It is also capable of providing lower resolution and/or low quality representations of an image for quick viewing, by only decoding a selected portion of the total transmitted image. JPEG2000 is able to code a wide range of images, from black and white, to grayscale, full-color (24 bit/pixel) images, to hyper-spectral space and planetary images that typically contain several dozen-color bands (Taubman, David S. & Marcellin, Michael W. 2001). Another important feature is the ability to provide error-resilience during transmission; that is to say, an error during progressive transmission will only affect a small portion of the final image rather than the whole image. As it stands currently, an image than has been compressed to the same final size with JPEG2000 and JPEG, will show much less visible artifacts with the first, due to the increased ability of wavelets to represent an image at low resolution than with the later, where the individual 8x8 blocks become noticeable for large compression ratios (Fig. 1). The two formats behave similarly for high quality reproductions but JPEG2000 increasingly outperforms JPEG as more and more compression is introduced, while maintaining the same visual fidelity (Taubman, David S. & Marcellin, Michael W. 2001).

2.2 JPEG2000 security

The issue of secure access to satellite imagery, is sensitive, since they contain value that image producers and providers, think that requires protection. Part 8 of the JPEG2000 standard, called JPSEC (Conan, Vania 2005) provides solutions to this problem by specifying a set of standardized tools and solutions that ensures security of transaction, protection of contents, and protection of technologies, allowing applications to generate, consume, and exchange JPEG2000 secured bit-streams. JPSEC addresses, among other applications: encryption (flexible mechanism to allow for encryption of image content and metadata), source authentication (verification of authenticity of the source), data integrity (data integrity verification), conditional access (conditional access to portions of an image or its associated metadata), ownership protection (protection of the content owner rights). In order to protect the content techniques such as digital signatures, watermarking, encryption, and key generation and management, are used.

2.3 Digital Rights Management

An effective implementation of the JPSEC was combined with a Digital Rights Management (DRM) solution, which deploys protection and security, and controlled access to content and which is particularly relevant in the digital content e-commerce. This DRM solution, called OpenSDRM (Serrão, C. et al. 2003), which is composed of several optional elements covering the content distribution value chain, from content production to content usage. It covers several major aspects of the content distribution and trading: content production, preparation and registration, content, interactive content distribution, content negotiation and acquisition, strong actors and user's authentication and conditional visualization and playback.

3 EO PRODUCTS COMPRESSION AND PROTECTION

3.1 The Payload Data Segment

The ENVISAT satellite contains a payload of eight measuring instruments. Each instrument is tuned to a specific range in the electromagnetic spectrum, mostly in the infrared, the microwave and the visible spectrum. Some instruments are pointed directly at the Earth's surface while others point obliquely at the atmosphere. Some are intended to image the earth under different viewing conditions while others are intended to measure physical quantities or concentrations of chemical substances. The instruments store the measurement data in raw format, and send it back to ground stations at specific locations in their orbit path. This raw measurement data is afterwards combined to form different EO product levels.

The resulting products are stored in the Payload Data Segment file format (PDS) which contains a Main Product Header (MPH) and a Specific Product Header (SPH) followed by product dependent data. Some products have Measurement Data Segments (MDS) where the measurement data is stored either in integer or floating point format, depending on the product. This data format does not feature any compression and so can usually result in quite large product files, not very suitable for Internet transfer. Knowing that the current JPEG2000 standard does not specify the manipulation of floating point data, each of the ENVISAT instruments products was examined to determine it's suitability for JPEG2000 coding (Boliek, Martin 2002, Christopoulos C., Skodras A. &

Ebrahimi T. 2000), and as a result several have been identified: MERIS (ESA 2002e), ASAR (ESA 2002b), GOMOS (ESA 2002d), ATTSR (ESA 2002a), RA2 (ESA 2002c) and SCIAMACHY (ESA 1998, ESA 1999).

3.2 The PDS parser

The PDS parser is a tool able to assemble and disassemble PDS files. The parser starts the parsing process by identifying the product through is MPH header present in all products. After that it reads the SPH header, which is specific to each product. From this point forward, the parser is able do read all product specific data and separates it in two classes: JPEG2000 compressible data and metadata. The JPEG2000 compressible data, the image samples present on the MDS, are stored in raw files, one per band. The metadata, all data not suitable for JPEG2000 compression, is stored in a XML kind file. The reconstruction process (unparsing) start by reading the metadata file to identify the product and after that reads all metadata and raw data to the reconstructed file into its original order.

3.3 The Hierarchical Data Format

The SPOT5 satellite products, considered in the project, are about vegetation. The vegetation products are stored in the Hierarchical Data Format (HDF). The HDF format is a self describing and extensible file format which uses tagged objects that have standard meanings. The purpose is to store both known format description and data in the same file (NSA 2003). Shortly, this format consists in data structures organized in a hierarchy. No compression algorithm is applied to the data structures resulting in quite large files.

3.4 The HDF parser

The HDF parser parsing process separates the data in two different classes: JPEG2000 compressible data and metadata. As in the case of the PDS format, also the HDF format contains data which is not suitable for JPEG2000 compression. The suitable data structures for JPEG2000 encoding are the Scientific Data Sets, which contain the image samples. Other structures are considered as metadata and will be stored in an XML kind file.

3.5 The protection tool

The implemented protection tool is JPSEC compliant. The JPEG2000 syntax requires that any two consecutive bytes in the encrypted packet body must not have a value greater than $0 \times FF8F$ (Conan, Vania 2005). The protection tool security relies in the security of the Advanced Encryption Standard (AES) in output feedback mode (OFB). The algorithm checks

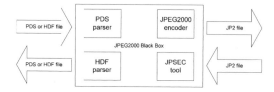

Figure 2.   JPEG2000 black box.

the presence of bytes with value $0 \times FF$ living them in clear text, to the others ciphering is applied. Although some bytes kept in clear text this algorithm protects 99.15% of the coded image data (Hongjun, Wu & Di, Ma 2004).

3.6   *Putting all together: the HICOD2000 black box*

The conversion process of EO products, either PDS or HDF, to JPEG2000, and vice versa, is automatic. A tool named HICOD2000 Black Box (H2KBB) was developed to perform the conversion process in an unattended way. The H2KBB (Fig. 2) is a set of tools combined in such a way that the input file type is detected and the compression or decompression process is automatically started.

The H2KBB identifies the input file type from the file signature. If the file is a PDS or HDF file it will invoke the appropriate parsing tool and generate the correct compression and decompression command lines to be handed to the JPEG2000 codec. The codec used was the Kakadu that can be found in www.kakadusoftware.com. To be able to identify the original EO product from the JPEG2000 file, the H2KBB inserts a tag in the decoding command line with the parser identification. After this the compression phase is started. If required the compressed file will be protected by the JPSEC protection tool.

On detecting a JPEG2000 input file the reverse path is made. If the JPEG2000 file has been protected it must be handed to the JPSEC protection tool to remove the protection. Next it will be decompressed using the embedded command line in the JPEG2000 file. The last step is invoking the correct parser to reconstruct the original file.

4   SSE INTEGRATION

The Service Support Environment (SSE) project was developed to be the foundation of a future EO services related marketplace. Prior to the introduction of the SSE, each service provider needed to develop its own portal, including all the non-service related infrastructure and then it needed to make the availability of the service known to the potential clients, while on the other hand, the potential clients do not easily find the available services, and when they do,

these usually supply only a fraction of the user needs, forcing him to use several providers. The SSE portal (services.eoportal.org) was developed to address these problems, while at the same time developing the EO economy through the development of the already mentioned "marketplace".

The two main components resulting of this activity are:

- The SSE portal, where users access the available services, order the products they are interested in, pay for them in a secure manner and with a large selection of payment methods, and track the progress of their order.
- The MASS toolbox, which service providers can use to chain their services into the MASS catalogue, eventually combining with other services to produce added-value services.

From the user point of view, access to the SSE portal is like the access to any other Web site, where he can register, browse the list of available products, create an order, pay for it and track its status. The most noticeable difference should be the wide range of available services.

The real difference however is from the service provider's point of view. A crucial factor in the design of the SSE system is the possibility of interoperability and chaining of services. This means that a service provider does not need to build its services from scratch, starting by processing the raw satellite (and ancillary) data – it can request the parts it needs from other service providers and just perform the steps required to create his products. This is done in several ways, first through the use of standard technologies such as XML, SOAP or WSDL which facilitates the development, and through the provision of a toolbox that simplifies the integration of the services within the SSE portal.

In order to make our solution available on the SSE portal, as a data conversion service, the service needed to be converted into SOAP based services.

On the SSE portal side this was done by configuring the service using three files, a WSDL file, a XSL style Sheet and a Schema file. These files are used to design the user interface (Fig. 3), catch and validate the service parameters, and link to the web service provider Toolbox server, were the service should be installed.

On the service server side, the Toolbox functionalities, allow the specification of the operation that has to be performed when a SOAP message is received. When a message is sent, from the SSE portal to the web service provider Toolbox server, the web service, previously installed and configured, through the use of XML Schema Definition (XSD) files, triggers the solution application. The result of this application is caught by the Toolbox and then sent to the SSE Portal in

Figure 3. SSE service order page.

HICOD2000 Client Application

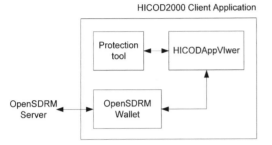

Figure 4. Client architecture.

a compatible SOAP message. This message states the unsuccess or success of the service with the resulting link to the product.

5 IMAGE ACCESS SOLUTION

In order to display JPEG2000 EO products and to control the access to such products there was the need to create a client module that integrated these functionalities.

The client module (Fig. 4) is composed of three main components, the JPEG2000 viewer application, the protection tool (described in section 3.5) and the OpenSDRMWallet.

5.1 JPEG2000 viewer

The function of the JPEG2000 viewer component is to visualize and interact with EO JPEG2000 products. This component can open both JPEG2000 regular files and JPEG2000 EO products protected or not.

The most relevant user requests will be to access images, and to navigate through those images. Therefore, when the user uses this component to open an image it checks if the image is protected or not. If the image is unprotected then it is automatically rendered on the viewer allowing the user to navigate freely to any of the image available resolutions.

If the viewer detects that the image is protected (Fig. 5), the viewer has to contact the OpenSDRMWallet to obtain the appropriate keys to decipher the image at each of the possible resolutions the user is trying to access and has acquired the right to do so. If those

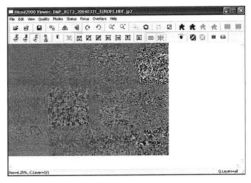

Figure 5. Protected image rendering.

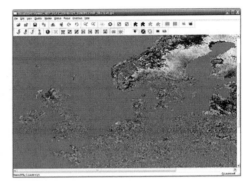

Figure 6. Image rendering after protection removal.

keys are returned with success to the viewer, they are handed to the JPSEC tool to decipher the image so it can be rendered correctly on the viewer. The viewer is signaled to render the image up to the requested resolution level (Fig. 6).

5.2 OpenSDRMWallet

The OpenSDRMWallet component is used to control the access to content licensing information as well as to cryptographic keys needed to decipher image data. This component determines how a given content can be used by a given user. Being a license repository the OpenSDRMWallet can be invoked by the local machine to retrieve the keys to a given license. If the license is not found on the repository the OpenSDRMWallet connects to the OpenSDRM license server, and tries to download a license. The returned license is then stored, securely, on the user-side.

6 CONCLUSIONS

Satellite imaging is one of the areas that can take advantage of the new image coding standard

JPEG2000. Not only in terms of image compression, a quite relevant aspect, but also from the perspective of other parts in the standard, namely security.

The efficient compression scheme of JPEG2000 can handle images with higher quality and dimensions and provide satellites with increased storage capability. The standard can also provide more flexibility due to the compression options it offers. This could help satellite providers to reduce time-to-market in solutions for satellite imaging e-commerce, which can be turned into an important competitive advantage.

In general, the described solution was developed as part of a set of exploratory actions to find new technological opportunities to integrate in ESA processes. In this specific case, HICOD2000 proved to ESA that JPEG2000 was an interesting technology that could be used to encode and distribute some EO products. The outcomes of the project have proved that some HCOD2000 technologies may be applied in EO systems. One of the major opportunities, for the future exploitation of the solution presented, is the application of JPEG2000 for lossless compression of a range of ENVISAT products, as well as further research following the JP3D JPEG2000 extension activities, namely in coping with floating point data sets, in order to be able to comply with all the possible EO products conversion.

REFERENCES

Boliek, Martin (ed.) 2001. *JPEG 2000 Image Coding System: Extensions*. ISO/IEC.

Boliek, Martin (ed.) 2002. *JPEG 2000 Image Coding System: Core Coding System*. ISO/IEC.

Christopoulos C., Skodras A. & Ebrahimi T. 2000. The JPEG2000 Still Image Coding System: An Overview. In *IEEE Transactions on Consumer Electronics*, Vol. 46, No. 4, pp. 1103–1127.

Conan, Vania (ed.) 2005. *JPEG 2000 Image Coding System: Secure JPEG 2000*. ISO/IEC.

ESA 1998, *SCIAMACHY Level 1b to 2 Off-Line Processing: Algorithm Theoretical Basis Document, Issue 1B*. European Space Agency.

ESA 1999, *SCIAMACHY Level 0 to 1c Processing: Algorithm Theoretical Basis Document, Issue 1*. European Space Agency.

ESA 2002a. *The AATSR Handbook, Issue 1.0*. European Space Agency.

ESA 2002b. *ASAR Product Handbook, Issue 1.0*. European Space Agency.

ESA 2002c. *ENVISAT RA2/MWR Product Handbook, Issue 1.0*. European Space Agency.

ESA 2002d. *GOMOS Product Handbook, Issue 1.0*. European Space Agency.

ESA 2002e. *MERIS Product Handbook, Issue 1.0*. European Space Agency.

Hongjun, Wu & Di, Ma 2004. *Efficient and secure encryption schemes for JPEG2000. IEEE International Conference on Acoustics, Speech, and Signal Processing*, Vol. 5, pp. 869–872.

NSA 2003. *HDF4 User's Guide – HDF4 Release 2.0*. National Center for Supercomputing Applications.

Serrão C., Neves D., Kudumakis P., Barker T. & Balestri M. 2003. *OpenSDRM – An Open and Secure Digital Rights Management solution, IADIS International Conference e-Society 2003*, Vol. 2, pp. 647–650.

Taubman, David S. & Marcellin, Michael W. 2002. *JPEG2000: Image Compression Fundamentals, Standards and Practice*. Kluwer Academic Publishers.

Computational Modelling of Objects Represented in Images –
João Manuel R.S. Tavares & R.M. Natal Jorge (eds)
© 2007 Taylor & Francis Group, London, ISBN 978-0-415-43349-5

Generating segmented tetrahedral meshes from regular volume data for simulation and visualization applications

Alex J. Cuadros-Vargas & Luis G. Nonato
University of São Paulo, Brazil

Eduardo Tejada & Thomas Ertl
University of Stuttgart, Germany

ABSTRACT: This paper describes a method to generate Delaunay tetrahedral meshes directly from regular volume data by vertex insertion driven by an error criterion. By adapting the mesh to the input data, the method produces meshes suitable for fast and precise volume ray-casting. A partitioning method that splits the mesh into homogeneous sub-meshes is also presented. Such mesh segmentation strategy turns out very useful in applications such as numerical simulation and focus-and-context rendering, where well defined structures are of paramount importance.

1 INTRODUCTION

Computational simulation in domains defined from regular volume data has stimulated the technological development in many branches of science, e.g. flow simulation (Steinman 2002) and virtual surgery (Bro-Nielsen and Cotin 1996). One of the main challenges in this context is to decompose the domain in a mesh suitable for simulation as well as visualization, as most algorithms proposed to perform this task are not aimed at generating meshes respecting the features of the whole volume, turning it difficult a realistic volume rendering of the mesh. Another drawback regarding mesh generation from regular volumes concerns the automation of the process. In general, the algorithms strongly rely on extensive pre-processing steps, which typically consist in segmenting regions of interest, using the boundary surfaces of such regions as input to a mesh generator (Cebral and Lohner 1999).

Techniques that act directly on the input volumetric data (Hale 2001), in general, line up the mesh with the volume features, but do not worry about segmenting structures contained in the volumes. The mesh segmentation is essential in applications such as physically-based deformation of volumetric bodies where different stiffness constants must be assigned to different materials within the mesh. Segmented meshes can also be exploited for rendering purposes, such as using focus-and-context techniques and combining surface and volume rendering to highlight the boundaries between different regions.

In this work we show a method that can be used to generate segmented Delaunay tetrahedral meshes, directly from the input volume, aiming at obtaining useful characteristics for volume visualization. These properties have been exploited and incorporated in our GPU-based single-pass ray-caster for tetrahedral meshes (Tejada and Ertl 2005).

2 RELATED WORK

One of the earliest works on generating meshes from images was proposed by Terzopulos and Vasilescu (1992), who approach the problem using adaptive meshes assembled as a set of nodal masses connected by springs. Their method moves the mesh nodes by gradient and curvature information so as to concentrate nodes in regions where a rapid variation of the data is present. A similar approach has been proposed by Hale (2001), which makes use of a potential energy function to line up points of a regular lattice with image features and generates the final mesh by Delaunay tessellation. A pre-processing step is employed to reduce noise and highlight sharp regions.

Aiming at minimizing the interpolation error between the mesh and the original image, Garland and Heckbert (1995) proposed an iterative method that inserts, in a Delaunay tessellation, the points of maximal interpolation error. Grosso et al. (1997) use an error measure derived from a least-square approximation to guide an iterative tetrahedral subdivision. The subdivision follows a set of rules applied over tetrahedra and octahedra marked for refinement. Roxborough and Nielson () also employ least-square approximation to estimate the interpolation error, but, differently to Grosso et al., only a single subdivision rule that splits the longest edge of a tetrahedron is applied. Takahashi et al. (2004) proposed an adaptive tetrahedralization method for computing topological volume skeletons, using topological and geometrical criteria to iteratively sub-divide the tetrahedra.

The iterative method proposed by Garcia et al. (1999) controls the maximum root-mean-square error (RMS) by choosing the vertices of the mesh from a curvature image, that is, more vertices are placed in areas with high curvatures. The mesh model is built by generating the Delaunay tessellation from the chosen vertices. Regions with high RMS error are resampled and the Delaunay tessellation updated. An interesting non-iterative algorithm has been presented by Yang et al. (2003). Their method makes use of zero-crossing jointly with the Floyd-Steinberg error-diffusion algorithm to choose a set of points from which the Delaunay tessellation is calculated.

A common characteristic of the strategies described above is that a mesh is generated directly from the original dataset, not demanding a pre-segmentation. Some algorithms, however, strongly rely on a pre-segmentation process to define regions of interest from which a mesh is built. An example is the algorithm proposed by Cebral and Lohner (1999) which makes use of a heavy user-assisted pre-processing to binarize the original image. The mesh is generated by an advancing-front technique that starts from the surface defined by the binary data set. Zhang et al. (2003) and Berti (2004) make use of thresholding and an octree decomposition to generate an adaptive tetrahedral mesh inside the region stipulated by the threshold. Mesh improvement techniques are used by both methods in order to smooth and fit the mesh into the thresholded region.

Our method presents characteristics from both classes of algorithms described above. The mesh is generated directly from the regular volume data, thus avoiding the stressful pre-processing step. Since the generated tetrahedra tend to be fitted within homogeneous regions of the volume, the mesh can be segmented according to the volume data.

3 BASIC CONCEPTS

Let S be a set of points in $\mathbb{R}^3$. A *tetrahedral mesh* of S is a three-dimensional simplicial complex M whose vertices are the points in S, and any k-simplex of M, $k = 0, 1, 2$, is contained in at least a 3-simplex (tetrahedron) of M. If the union of all simplices in M makes up the convex hull of S and the circumsphere of each tetrahedron in M does not contain any point of S in its interior then M is called a *Delaunay mesh*. Delaunay meshes are closely related to Voronoi diagrams, which can be obtained by associating to each k-simplex of M a $3 - k$ cell in the diagram. In fact, each vertex of the diagram is the center of the empty sphere circumscribing a Delaunay tetrahedron and each vertex v_i of the Delaunay mesh is contained in the three-dimensional cell of the diagram that contains the points in $\mathbb{R}^3$ closest to v_i (Fortune 1992).

Let S be a set of points and M be a tetrahedral mesh of S. If M can be decomposed as $M = M_1 \cup M_2 \cup \cdots \cup M_k$, where each M_i is also a mesh and $M_i \cap M_j, i \neq j$ is either empty or a k-dimensional simplicial complex, $0 \leq k \leq 2$, then $\{M_1, M_2, \cdots, M_k\}$ is said to be a *k-partitioning* of M in *sub-meshes* M_i, $i = 1, \ldots, k$.

A $X \times Y \times Z$ *regular volume* is a function $I : [0, \ldots, X] \times [0, \ldots, Y] \times [0, \ldots, Z] \to \mathbb{R}^+$ that assigns to each point $p \in [0, \ldots, X] \times [0, \ldots, Y] \times [0, \ldots, Z] \subset \mathbb{Z}^3$ a non-negative scalar $I(p)$. The pair $(p, I(p))$ is called *voxel*.

4 GENERATING DELAUNAY TETRAHEDRAL MESHES FROM STRUCTURED VOLUMETRIC DATA

As mentioned above, our method first generates a tetrahedralization that respects a desired feature of the volumetric data. In our case, we intend to generate a mesh whose tetrahedra do not cross different regions of the volume, that is, each tetrahedron should be contained in only one homogeneous region of the volume. Then, a partition of this mesh into homogeneous sub-meshes. We describe below these two processes in detail.

4.1 Generating the mesh

Let T be the set of tetrahedra of a Delaunay mesh M whose vertices are points of a regular volume I and $E : T \to \mathbb{R}^+$ be a function that associates an error measure to each tetrahedron in T. In fact, function E measures how good a tetrahedron is regarding a specific property, that is, E enables to decide whether or not a tetrahedron must belong to the tetrahedralization. Different strategies have been proposed to define the function E, which usually rely on evaluating E by traversing all voxels inside a tetrahedron t so as to decide, based on some characteristic of the image, whether or not t is a suitable tetrahedron. In general, when E indicates that t is a bad tetrahedron, the mesh is updated by inserting new points within t (Garland and Heckbert 1995). Although widely employed, this strategy presents two main drawbacks. Traversing all voxels within a tetrahedron may demand a high computational effort, as every time the mesh is updated all new tetrahedra must be scanned in order to re-evaluate E. Another problem arises during the insertion of new points, which, when not handled properly, can result in an accumulation of points around already existing vertices. To avoid this we adopt an strategy based on the medians of the tetrahedra to evaluate E. Traversing only medians one can reduce the computational effort while being effective in detecting tetrahedra that cross different regions of the volume.

Let h_1, h_2, h_3, h_4 be the four medians of a tetrahedron $t \in M$, $\mathcal{E}$ be a high-pass filter and $c_{\mathcal{E}}$ be an user defined scalar. Consider the set of points

$$P_j^t = \{p \in h_j \mid \mathcal{E}(p) \geq c_{\mathcal{E}}\} \text{ for } j = 1, \ldots, 4.$$

and let $P^t = \bigcup_{j=1}^4 P_j^t$. Then, P^t is the set of high frequency points (detected by $\mathcal{E}$) of a tetrahedron t. In our experiments we used the Sobel 3D operator as the filter $\mathcal{E}$.

Let $A \subset \mathbb{R}^3$ be a finite and non empty set. We define the distance between some point $p \in \mathbb{R}^3$ and A as $D(p, A) := \min\{d^2(p, a) \mid a \in A\}$, where $d(\cdot, \cdot)$ is the Euclidean distance. Note that, since A is a finite set, $\exists a_* \in A$ such that $d^2(p, a_*) = D(p, A)$.

Let w^t be the circumcenter of $t \in M$. Consider $P^t \neq \emptyset$. As P^t is finite, we have that, there always exists some $p_*^t \in P^t$ such that $d^2(w^t, p_*^t) = D(w^t, P^t)$. Let V be de set of vertices of M. From these definitions de error function E can be stated as:

$$E(t) = \begin{cases} 0 & \text{if } P^t = \emptyset \\ D(p_*^t, V) & \text{if } P^t \neq \emptyset \end{cases}$$

As p_*^t is a high frequency point, it is supposed to be on the boundary of different regions of the volume I. Therefore, $D(p_*^t, V)$ measures how much a tetrahedron t invades a region of the volume. Thus, values of $E(t)$ close to zero indicate that t is well fitted within a region contained in the volume. Therefore, a tetrahedron t is considered unsuitable if $E(t) > c_E$, where c_E is an user defined scalar. An unsuitable tetrahedron t is eliminated by inserting p_*^t in M.

It is a well known fact that circumcenters are distant points of the vertices of a Delaunay tetrahedralization. Since the point p_*^t, in a tetrahedron t, is chosen to be the closest point to a circumcenter w^t, then p_*^t is, in general, a distant point of the vertices of a Delaunay mesh M. Therefore, the problem of accumulating points around already existing vertices is also reduced.

4.2 Partitioning

Suppose that a mesh M has already been partitioned in n sub-meshes $M_1, M_2, \ldots, M_n$, $n > k$, where k is the target number of partitions. Thus, $n - k$ sub-meshes must be merged in order to get a k-partitioning. The

merge operation aims at gluing the most similar sub-meshes, where similarity is measured from a set of properties extracted from the sub-meshes.

Voxel information as well as topological and geometrical properties can be considered when comparing sub-meshes. Let $S_I(M_i, M_j)$ and $S_{TG}(M_i, M_j)$ be functions that measure similarity regarding voxel information and topological/geometrical properties of M_i and M_j respectively. From S_I and S_{TG} we can define a general similarity function as $S(M_i, M_j) = w_1 S_I(M_i, M_j) + w_2 S_{TG}(M_i, M_j)$, $w_1 + w_2 = 1$; $0 \leq w_1, w_2 \leq 1$, where the weights w_1 and w_2 are used to adjust the relevancy of voxel information and topological/geometric properties in the similarity measure.

Function S_I compares an array of characteristics extracted from the voxels in M_i and M_j, assigning a value between 0 and 1. In our context, values close to zero mean highly similar sub-meshes whereas values close to one indicate not similar sub-meshes. A simple way to define S_I, which has been adopted in our implementation, is compare the average of voxels values from each sub-mesh, that is, $S_I(M_i, M_j) = |H(M_i) - H(M_j)|/|H(M_i) + H(M_j)|$, where $H(M_k)$ is a function that associates to each sub-mesh M_k, $k = i, j$ the average of voxel values. In order to avoid to scan every voxel in a new sub-mesh generated from the merge operation, we compute $H(t) = \frac{1}{n_t}\Sigma_{p \in t}I(p)$ for each tetrahedron t, where n_t is the number of voxels in t, before starting the mesh segmentation process. From the value of H in each tetrahedron, we can estimate $H(M_k)$ as the weighted average of H in the tetrahedra, that is, $H(M_k) = \Sigma_{t \in M_k} Vol(t)H(t)/\Sigma_{t \in M_k} Vol(t)$, where $Vol(t)$ is the volume of tetrahedron t. It is worth noting that at the beginning each tetrahedron is considered as a sub-mesh.

S_{TG} is concerned with topological and geometrical properties to compare two sub-meshes. Aiming at favoring the merging of "small" neighbor sub-meshes into the largest ones, we define S_{TG} as follows:

$$S_{TG}(M_i, M_j) = \begin{cases} \frac{\min\{Vol(M_i), Vol(M_j)\}}{\max\{Vol(M_i), Vol(M_j)\}} & \text{if } M_i \cap M_j \neq \emptyset \\ 1 & \text{if } M_i \cap M_j = \emptyset \end{cases}$$

Therefore, $S_{TG}(M_i, M_j)$ assumes smaller values when M_i and M_j are neighbor sub-meshes ($M_i \cap M_j \neq \emptyset$) and the volumes are very different.

Using a priority queue sorted by S we can recover the most similar sub-meshes in a very efficient way. Thus, the sub-meshes with highest priority (the smallest value of $S(M_i, M_j)$) are merged up to a k-partitioning is reached.

Figure 1 (top-left) shows an example of a segmented mesh generated with this process. Notice that multiple structures are meshed simultaneously without pre-processing the original data.

5 RENDERING OF SEGMENTED TETRAHEDRAL MESHES

We adapted our single-pass ray-caster for tetrahedral grids (Tejada and Ertl 2005) to exploit the characteristics of the meshes generated with the approach described in Section 4, namely the segmentation and the fact that the scalar value is constant within a tetrahedron. Also, since the mesh is convex, no convexification pre-processing or re-entry handling must be performed.

Since the scalar is constant within a tetrahedron the accumulated associated color $\widetilde{C}_k$ and opacity α_k for the ray segment between the entry point x_k^i and exit point x_k^o of the k-th tetrahedron $t^{(k)}$ intersected by the ray can be calculated as $\widetilde{C}_k = \tilde{c}(H(t^{(k)}))\alpha_k$ and $\alpha_k = 1 - \exp(-\tau(H(t^{(k)})) d(x_k^i, x_k^o))$ respectively, where $H(t^{(k)})$ is the scalar value at tetrahedron $t^{(k)}$ and $\tilde{c}$ and τ are transfer functions for color and extinction densities respectively. Therefore, approximation improvement approaches such as pre-integrated volume rendering (Engel, Kraus, and Ertl 2001) are not necessary. The ray-integral is then accumulated as usual as $\widetilde{C}_{acc} = \widetilde{C}'_{k-1} + (1 - \alpha'_{k-1})\widetilde{C}_k$ and $\alpha_{acc} = \alpha'_{k-1} + (1 - \alpha_{k-1})\alpha_k$, where $\widetilde{C}'_{k-1} = \widetilde{C}_{acc}$ and $\alpha'_{k-1} = \alpha_{acc}$ are the accumulated associated color and opacity after the $(k-1)$-th intersected tetrahedron respectively. Renderings obtained using this modified ray-caster are shown in Figure 1.

Another consequence of having constant scalars within the tetrahedra is that the gradient within it becomes zero. However a rough way to obtain a non-zero gradient within each tetrahedron is setting the scalar value $s(v_i)$ at vertex v_i as $s(v_i) = \Sigma_{t_i \in St(v_i)} H(t_i)Vol(t_i)/\Sigma_{t_i \in St(v_i)} Vol(t_i)$, where $St(v_i)$ is the star of v_i. Thus, a constant gradient per tetrahedron is computed as done in previous work (Tejada and Ertl 2005). Since the gradient is also constant within each tetrahedron, two-dimensional transfer functions indexed by scalar and gradient magnitude can be used (Fig. 2: bottom-left).

The segmentation of the grid can be exploited by focus-and-context techniques. Also, the boundaries between the sub-meshes can be highlighted by rendering them as translucent surfaces during ray-casting (Fig. 2: bottom-right). This is accomplished by simply storing a further scalar for each face of the tetrahedra and fetching it during ray-traversal for the current tetrahedron. This scalar is the identifier of the boundary to which the face belongs. If the face is not a boundary face, a negative value is set. In order to interactively support color and opacity picking for the boundaries, the boundary identifier is used to fetch the corresponding color and opacity from a one-dimensional color table. The fetched color c_k^s and opacity α_k^s are then accumulated as $\widetilde{C}'_k = \widetilde{C}'_{acc} + (1 - \alpha'_{acc})\alpha_k^s c_k^s$ and $\alpha'_k = \alpha'_{acc} + (1 - \alpha'_{acc})\alpha_k^s$ (assuming that c_k^s is a non-associated color).

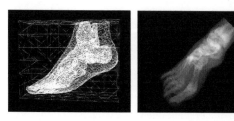

Figure 1. Partitioned mesh obtained for the Knee dataset and its respective volume rendering (top). Volume rendering of the meshes obtained for the Foot and Engine datasets (bottom).

6 RESULTS

The tests performed were carried out on a standard PC with a 3.20 GHz 64-bits processor and 2 GB of RAM and an NVidia 7800 GTX graphics board.

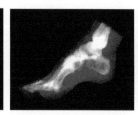
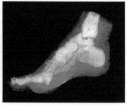

Figure 2. Mesh generated for the Visible Male Foot dataset (left). Volume rendering of the mesh (middle-left). Visualization using two-dimensional transfer functions (middle-right) and combined volume and surface rendering using the segmented mesh (right).

In Table 1 performance results are shown. In the table we also present ratios between the size in bytes of the tetrahedral mesh stored in a format we defined (detailed below) and the original volume stored as a stack of PGM files. The format we used for storing the tetrahedral mesh is as follows:

```
KeyWord_Sub_meshes #Sub-meshes
List of Sub-meshes IDs

KeyWord_Points #Points
X-Coord. Y-Coord. Z-Coord. Scalar Type*
...

Keyword_Tetrahedra #Tetrahedra
ID-Vertex0 ID-Vertex1 ID-Vertex2
ID-Vertex3
Scalar ID-Sub-mesh
...
```

```
* Type is used to classify the vertex as boundary
vertex or internal vertex.
```

Results concerning the performance of the hardware-implementation of the ray-caster are also presented in Table 1. It is important to mention that the ray-traversal included the boundaries surface check performed for each tetrahedron intersected by the ray and that two-dimensional transfer functions were used in all cases.

As discussed before, the segmentation can be exploited, not only for rendering purposes, but for applications where different materials present in the mesh must be identified. An example of this is volume deformation, where tetrahedral grids have proven to be more suitable than regular grids and different material properties, e.g. stiffness parameter, are desirable for different parts of the mesh (See Fig. 4 extracted from the work by Tejada and Ertl (2005)).

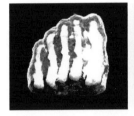

Figure 4. Mesh generated with our algorithm from the Foot data set suffering a physically-based deformation.

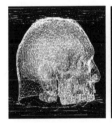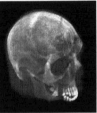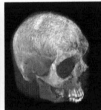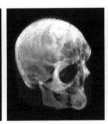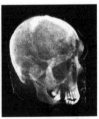

Figure 3. Renderings of a Delaunay tetrahedral mesh extracted from the Visible Human Head regular volume with our approach. From left to right: wireframe of the extracted mesh, volume rendering, combined volume rendering and surface rendering (for the boundary between two sub-meshes), rendering with one-dimensional transfer function and rendering with two-dimensional transfer function.

Table 1. Results obtained for the mesh generation algorithm and modified single-pass ray-caster. Processing times in seconds, input volume size, generated mesh size and compression ratio in bytes are shown for the mesh generation method. Performance results for the ray-caster modified for the tetrahedral meshes generated for two different viewport sizes are shown in frames per second.

Dataset	Processing time [sec]				Volume size [voxels]	Mesh size [tetrahedra]	Size ratio [bytes]	Rendering time [fps]	
	Mesh generation	Scalar values	Segmentation	Total				640 × 480	1600 × 1200
Knee	802.80	22.45	4.48	829.72	26471255	258600	0.218	3.76	0.80
Visible male head	961.14	28.79	3.75	993.68	77070336	389060	0.122	2.92	0.63
Visible male foot	218.32	6.08	1.24	225.65	4944354	101932	0.297	4.46	0.93
Foot	84.47	16.06	3.16	103.69	16777216	258600	0.256	3.41	0.74
Engine	264.37	7.40	1.38	273.15	8388608	122910	0.275	4.24	0.89

7 CONCLUSION

We have shown that the meshes generated by our strategy are suitable for volume visualization. In fact, the storage of a scalar representing the average of the voxel values in each tetrahedron allows the usage of multi-dimensional transfer functions in the hardware implementation of the ray-caster whilst avoiding pre-integration. The quality of the volume rendering carried out on the tetrahedral mesh generated by our method is comparable with the traditional regular grid volume visualization. However, the generated tetrahedral meshes can also be used in applications other than visualization, such as physically-based deformation.

However, it is necessary to point out that our approach does not generate tetrahedral meshes suitable for all kinds of numerical simulation, since the tetrahedra do not present a high quality shape. A technique devoted to improve the quality of the tetrahedra whilst preserving the alignment of the mesh with the volume is currently being implemented. This technique should turn the meshes suitable for a larger set of numerical applications.

Another point that deserves some comments is the mesh segmentation strategy. Although very simple, the mesh segmentation scheme presented good results in most of the tests we carried out, but its performance is not satisfactory in some cases. Aiming at improving the results obtained, we are adapting a powerful statistical segmentation strategy to work with tetrahedral meshes. We see such statistical approach as a promising tool.

Finally, concerning the rendering algorithm, the most critical problem that must be addressed arises due to the constant gradient within each tetrahedron, which affects the rendering for large tetrahedra when two-dimensional transfer functions dependent of the gradient are used or when non-polygonal isosurfaces are rendered. Interpolation techniques are been investigated in order to sort out this problem.

ACKNOWLEDGMENTS

We acknowledge the financial support of FAPESP – the State of Sao Paulo Research Funding Agency – (Grants #03/02815-0 and #02/05243-4), CNPq – the Brazilian National Research Council – (Grant #307268/2003-9), and DAAD – the German Academic Exchange Service – (Grant #A/04/08711).

REFERENCES

Berti, G. (2004). Image-based unstructured 3D mesh generation for medical applications. In *ECCOMAS*.

Bro-Nielsen, M. and S. Cotin (1996). Real-time volumetric deformable models for surgery simulation using finite elements and condensation. *Computer Graphics Forum 15*(3), 57–66.

Cebral, J. and R. Lohner (1999). From medical images to CFD meshes. In *International Meshing Roundtable*, pp. 321–331.

Engel, K., M. Kraus, and T. Ertl (2001). High-Quality Pre-Integrated Volume Rendering Using Hardware-Accelerated Pixel Shading. In *Workshop on Graphics Hardware*, pp. 9–16.

Fortune, S. (1992). Voronoi diagrams and Delaunay triangulation. In *Computing in Euclidean Geometry*, Volume 1 of *Lecture Notes Series on Computing*, pp. 193–233.

García, M., A. Sappa, and B. Vintimilla (1999). Efficient approximation of gray-scale images through bounded error triangular meshes. In *IEEE Image Processing*, pp. 168–170.

Garland, M. and P. Heckbert (1995). Fast polygonal approximation of terrains and height fields. Technical Report CMU-CS-95-181, Carnegie Mellon University.

Grosso, R., C. Lürig, and T. Ertl (1997). The Multilevel Finite Element Method for Adaptive Mesh Optimization and Visualization of Volume Data. In *IEEE Visualization*.

Hale, D. (2001). Atomic Images – a method for meshing digital images. In *10th International Meshing Roundtable*, pp. 185–196.

Roxborough, T. and G. M. Nielson. Tetrahedron based, least squares, progressive volume models with application to freehand ultrasound data. In *IEEE Visualization*.

Steinman, D. (2002). Image-based computational fluid dynamics modeling in realistic arterial geometries. *Ann. Biomed. Eng. 30*, 483–497.

Takahashi, S., Y. Takeshima, G. Nielson, and I. Fujishiro (2004). Topological volume skeletonization using adaptive tetrahedralization. In *Geometric Modeling and Processing*, pp. 227–236.

Tejada, E. and T. Ertl (2005). Large Steps in GPU-based Deformable Bodies Simulation. *Simulation Practice and Theory. Special Issue on Programmable Graphics Hardware 13*(9), 703–715.

Terzopoulos, D. and M. Vasilescu (1992). Sampling and reconstruction with adaptive meshes. In *IEEE Int. Conf. Comp. Vision, Pattern Recog.*, pp. 829–831.

Yang, Y., M. Wernick, and J. Brankov (2003). A fast approach for accurate content-adaptive mesh generation. *IEEE Trans. on Image Processing 12*(8), 866–881.

Zhang, Y., C. Bajaj, and B.-S. Sohn (2003). Adaptive and quality 3D meshing from imaging data. In *ACM Symposium on Solid modeling and applications*, pp. 286–291.

Computational Modelling of Objects Represented in Images –
João Manuel R.S. Tavares & R.M. Natal Jorge (eds)
© 2007 Taylor & Francis Group, London, ISBN 978-0-415-43349-5

Microstructure features identification in ferritic-paerlitic ductile irons

A. De Santis
Dipartimento di Informatica e Sistemistica "A.Ruberti" – Università di Roma "La Sapienza", Italy

O. Di Bartolomeo
Dipartimento di Meccanica, Strutture, Ambiente e Territorio – Università di Cassino, Italy

D. Iacoviello
Dipartimento di Informatica e Sistemistica "A.Ruberti" – Università di Roma "La Sapienza", Italy

F. Iacoviello
Dipartimento di Meccanica, Strutture, Ambiente e Territorio – Università di Cassino, Italy

ABSTRACT: Ductile irons offer a wide range of mechanical properties at a lower cost than the older malleable iron. These properties mainly depend on the shape characteristics of the metal matrix microstructure and on the graphite elements morphology; these geometrical features are currently evaluated by the experts visual inspection. This work provides an automatic procedure for a reliable standard estimation of the material microstructure morphology based on a novel image segmentation technique. The procedure has been successfully tested on specimens of different kinds of ductile irons of a typical production.

1 INTRODUCTION

Ductile cast irons are characterized by a wide range of mechanical properties, mainly depending on microstructural factors, as phases (characterized by volume fraction, size and distribution), graphite particles (characterized by number, size and shape) and defects (as porosity, inclusions, segregated elements etc.). Ductile iron advantages that have led to its success are numerous, and they can be summarized easily: versatility, and higher performances at lower cost. This versatility is especially evident in the area of mechanical properties where ductile iron offers the designer the option of choosing high ductility (up to 18% elongation), or high strength, with tensile strengths exceeding 825 MPa. Austempered ductile iron offers even greater mechanical properties and wear resistance, providing tensile strengths exceeding 1600 MPa.

Ductile irons are characterized by a tensile resistance that is similar to that of steel and by an interesting fatigue crack propagation resistance. Mechanical properties are strongly affected by chemical composition, matrix microstructure and graphite nodule characteristics. Matrix could range from fully ferritic, to fully paerlitic, from martensitic to bainitic depending on the chemical composition and on the heat treatment. Ferritic-paerlitic ductile irons are widely used because they are able to summarize both a

high castability and good mechanical properties. The best combination is obtained with similar ferrite and paerlite volume fraction (50%–50%). Concerning the graphite elements shape, a very high *nodularity* is strongly recommended. The peculiar morphology of graphite elements in ductile irons is responsible of their good ductility and toughness. Characterized by a rough spherical shape, graphite particles contained in ductile irons are also known as "nodules". They act as "crack arresters", with a consequent increase of toughness, ductility and crack propagation resistance Iacoviello & Di Cocco (2003). Three typical shapes of graphite nodules are shown in Fig. 1 Li et al. (2000), and they range from a spheroidal graphite with smooth (Fig. 1a) or rough (Fig. 1b) surface to deformed

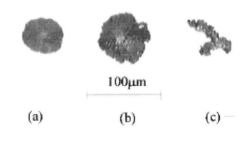

100μm

(a)   (b)   (c)

Figure 1.   Some typical graphite nodules.

graphite (Fig. 1.c): the highest are the differences of the graphite particles shape from perfect spheres and the worst are the mechanical performances of the ductile iron, especially considering the toughness and the crack propagation resistance. As a consequence, it is strongly recommended the control of the graphite particles shape. The analysis of the graphite particles shape is possible both directly (by means of metallographic measurements), and indirectly (e.g. by means of ultrasonic measurements). Direct method implies a metallographic preparation procedure (specimen cut followed by a surface polishing up to 1 μm diamond powder) and a surface light optical microscope (LOM) observation, usually performed considering a ×100 magnification. If it is also necessary a microstructure observation, a chemical etching is also performed before the LOM analysis (usually by means of Nital 1% or 3% solution for some seconds).

No standardized approach is defined neither for research activity nor for the routine quality control; some quantitative approaches are proposed, based on the definitions of a shape factor, defined as nodule section area/circumscribed circle area, Li et al. (2000), Imasogie & Wendt (2004).

Images obtained by means of LOM, despite a good visual appearance, are represented by a quite irregular signal due to various kind of degradations stemming from the very acquisition process: additive noise, albedos due to dust and specimen oxidation, artifacts coming from scratches occurring during the specimen preparation. A high performance image analysis procedure to robustly evaluate the nodules shape characteristics and the ferrite/paerlite fractions, can be obtained within the framework of image *segmentation*: the original image is partitioned into disjoint domains where the signal has homogeneous characteristics, and passing from one domain to another these characteristics vary significantly Mumford & Shah (1989), Chan & Vese (2001). Within each domain a best minimum square fit to the data is also provided so that a *cartoon* image is obtained; this is the closest approximant of the original picture in the class of piecewise constant functions. The segmented image preserves all the graphic information relevant to our analysis but with a lower number of different gray levels: for the considered applications up to 4 levels proved to be sufficient, as compared to 256 levels (8-bit data) of the original LOM data. The segmentation problem can be solved by various techniques, Morel & Solimini (1995) with different pros and cons. In this paper a region based method in De Santis & Iacoviello (2005) was preferred since it can deal with the complex topology of the graphite nodules without compromising the numerical complexity: as a consequence the ductile iron metallographies can be reliably segmented and evaluated in real time. As opposite to standard methods that are based on a image continuum model, the novelty of the proposed method consists in defining the segmentation problem and the related algorithm directly in the discrete domain.

The paper is organized as follows. In section 2 the proposed image segmentation procedure is outlined. In section 3 real data experiments are provided while concluding remarks along with some possible further developments are presented in section 4.

2 THE IMAGE SEGMENTATION PROCEDURE

Real data consist in the samples $\{I_{i,j}\}$ of the original image I over a grid of points

$$D = \{(i,j), i = 1,\ldots,N; j = 1,\ldots,M\}$$

Standard instrumentation provides a 8-bit data measurement, therefore 256 gray levels are available within the conventional range of [0–1]. In the considered application an image representation that preserves the relevant information content but with a much lower number of gray levels is advisable. The shape of the image sub-regions with homogeneous gray level gives the information of interest since defines different objects over a common background. Such a simpler representation can be obtained by a piece-wise constant image segmentation I_s defined as follows

$$I_s = \sum_{k=1}^{K} c_k \chi_{D_k}, \qquad B = \left(\bigcup_{k=1}^{K} \partial D_k \right) \qquad (1)$$

where $\{D_k\}, k = 1, 2, \ldots, K$ is a finite disjoint partition of the image domain D with boundary ∂D, and the the c_k's are the constant values assigned to each sub region D_k; moreover

$$\chi_{D_k}(i,j) = \begin{cases} 1 & (i,j) \in D_k \\ 0 & otherwise \end{cases}$$

The morphology of the ductile cast iron metallographic planes can be adequately represented by segmentation with either 2 (binarization) or 4 gray levels. A 4 levels segmentation can be efficiently obtained by a sequence of two binarizations: first the image is binarized and then each of the sub regions obtained is further binarized. Therefore the proposed segmentation algorithm will be outlined just for the binarization case; a detailed version can be found in De Santis & Iacoviello (2005). Let us then consider the case that $D = D_1 U D_2$; the two disjoint (not necessarily connected) components D_1 and D_2 can be defined by means of a *level set function*. Let:

$$\phi = \{\phi_{i,j}\} : D \to \mathbb{R}$$

We can write:

$$D_1 = \{(i,j):\phi_{i,j} \geq 0\}, \quad D_2 = \{(i,j):\phi_{i,j} < 0\}$$

The boundary B of the segmentation is known once the boundary points of D_1 or D_2 are defined. In the pixels adjacent to any point of ∂D_1 or ∂D_2 function ϕ has at least one sign change, while in the interior points it has none. Let $H(\cdot)$ and $\delta(\cdot)$ denote the Heaviside and the Dirac function respectively. The following function

$$\rho(z_{i,j}) = \sum_{\ell=-1}^{1}\left[\left(H(z_{i+\ell,j}) - H(z_{i,j})\right)^2 + \left(H(z_{i,j+\ell}) - H(z_{i,j})\right)^2\right]$$

counts the number of changes of sign of $z_{i,j}$ in a coordinate neighborhood of pixel (i,j); therefore we can check a boundary point with the following detector:

$$\gamma(\phi_{i,j}) = H\left(3 - \rho(\phi_{i,j})\right)\left[1 - \delta(\rho(\phi_{i,j}))\right]$$

Function γ is zero both on interior points ($\rho = 0$) and on isolated points ($\rho = 4$), and is one on the boundary points ($\rho = 1, 2, 3$). Now, given any real data I, segmentation (1) can be obtained by minimizing the following functional

$$E(c_1, c_2; \phi) = \lambda \sum_{i,j} H(\phi_{i,j})(I_{i,j} - c_1)^2$$

$$+ \lambda \sum_{i,j}\left(1 - H(\phi_{i,j})\right)(I_{i,j} - c_2)^2 + \frac{\alpha}{2}\sum_{i,j}\phi_{i,j}^2$$

$$+ \mu\sum_{i,j} H(\phi_{i,j}) + \nu\sum_{i,j}\gamma(\phi_{i,j}) \qquad (2)$$

This is a discrete version of the early Mumford and Shah variational formulation of the segmentation problem Mumford & Shah (1989); the first two terms are just the fit error within D_1 and D_2; the last two terms evaluate the "area" of D_1 and D_2 and the "length" of the segmentation boundary. Parameters λ, μ and ν are weights that can be used to enhance the contribution of one term with respect to the others. The third term is peculiar of the proposed formulation and is responsible of the functional convexity so that necessary and sufficient conditions for global minimum are available. Function E is not smooth with respect to ϕ because of the presence of the generalized functions $H(\cdot)$ and $\delta(\cdot)$. As suggested in Chan & Vese (2001) we relax the definition of (2) by considering smooth approximants of the generalized functions

$$H(z) \simeq H_\varepsilon(z) = \frac{1}{2}\left(1 + \frac{2}{\pi}\tan^{-1}\left(\frac{z}{\varepsilon}\right)\right)$$

$$\delta(z) \simeq \delta_\varepsilon(z) = \frac{1}{\pi}\frac{\varepsilon}{\varepsilon^2 + z^2}$$

We therefore modify the definitions of ρ and γ by using H_ε, δ_ε in place of H, δ, and define the following cost function

$$E_\varepsilon(c_1, c_2; \phi) = \lambda\sum_{i,j} H_\varepsilon(\phi_{i,j})(I_{i,j} - c_1)^2$$

$$+ \lambda\sum_{i,j}\left(1 - H_\varepsilon(\phi_{i,j})\right)(I_{i,j} - c_2)^2$$

$$+ \frac{\alpha}{2}\sum_{i,j}\phi_{i,j}^2 + \mu\sum_{i,j} H_\varepsilon(\phi_{i,j}) + \nu\sum_{i,j}\gamma_\varepsilon(\phi_{i,j}) \quad (3)$$

Necessary and sufficient conditions for the existence of a unique optimal solution can be found in De Santis & Iacoviello (2005). The optimal segmentation is given by

$$c_1 = \sum_{i,j} H_\varepsilon(\phi_{i,j}) I_{i,j} \bigg/ \sum_{i,j} H_\varepsilon(\phi_{i,j}),$$

$$c_2 = \sum_{i,j}\left(1 - H_\varepsilon(\phi_{i,j})\right) I_{i,j} \bigg/ \sum_{i,j}\left(1 - H_\varepsilon(\phi_{i,j})\right)$$

$$\phi_{i,j} = -\frac{1}{\alpha}\Bigg(\lambda p_{i,j} + \mu + \nu\sum_{\ell=-1}^{1}\Big[\left(H_\varepsilon(\phi_{i+\ell,j}) - H_\varepsilon(\phi_{i,j})\right) +$$

$$+ \left(H_\varepsilon(\phi_{i,j+\ell}) - H_\varepsilon(\phi_{i,j})\right)\Big]q_{i,j}\Bigg)\delta_\varepsilon(\phi_{i,j}) \qquad (4)$$

that is the Euler-Lagrange equation for (3). Functions $p_{i,j}$ and $q_{i,j}$ in (4) have the following definitions

$$p_{i,j} = (I_{i,j} - c_1)^2 - (I_{i,j} - c_2)^2$$

$$q_{i,j} = 2\Big[\delta_\varepsilon\left(3 - \rho_\varepsilon(\phi_{i,j})\right)\left(1 - \delta_\varepsilon\left(\rho_\varepsilon(\phi_{i,j})\right)\right)$$

$$- H_\varepsilon\left(3 - \rho_\varepsilon(\phi_{i,j})\right)\frac{2\rho_\varepsilon(\phi_{i,j})}{\varepsilon^2 + \rho_\varepsilon^2(\phi_{i,j})}\delta_\varepsilon\left(\rho_\varepsilon(\phi_{i,j})\right)\Big]$$

Function ϕ that solves equation (4) is obtained as the limit of a sequence $\{\phi^n\}$ by introducing a fictitious time evolution according to the following evolution equation

$$\phi_{i,j}^{n+1} = -\frac{1}{\alpha}\Bigg[\lambda p_{i,j}^n + \mu + \nu q_{i,j}^n\Big(\left(H_\varepsilon(\phi_{i-1,j}^n) - H_\varepsilon(\phi_{i,j}^n)\right)$$

$$+ \left(H_\varepsilon(\phi_{i+1,j}^n) - H_\varepsilon(\phi_{i,j}^n)\right) + \left(H_\varepsilon(\phi_{i,j-1}^n) - H_\varepsilon(\phi_{i,j}^n)\right)$$

$$+ \left(H_\varepsilon(\phi_{i,j+1}^n) - H_\varepsilon(\phi_{i,j}^n)\right)\Big)\Bigg]\delta_\varepsilon(\phi_{i,j}^n) \qquad (5)$$

where $p_{i,j}^n$ and $q_{i,j}^n$ are functions $p_{i,j}$ and $q_{i,j}$ computed by replacing in their expression $\phi_{i,j}$ by $\phi_{i,j}^n$. The solution of (5) converges to the solution of (4), De Santis & Iacoviello (2005).

A first level of analysis is aimed to the characterization of the graphite nodules shape. As discussed in the Introduction this is responsible of good ductility and toughness of the material. We are interested in parameters such as: the degree of "granularity", intended as the number of nodules per unit area; the nodules "solidity" as the ratio between the nodule area and the area of its convex hull; the nodules "eccentricity" that measures how much is the shape of a nodule close to a circle. The best situation is represented by a material with nodules featuring a solidity close to 1 and eccentricity close to 0, Fig. 1a. As opposite Fig. 1c represents a deformed nodule with very low solidity and surely not round. In Fig. 1b we have an intermediate situation where, due to the nodule contour roughness, solidity decreases with respect to Fig 1a, while we still have a good degree of roundness.

Our discussion about Fig. 1 specimens is just what an expert does by visual inspection. It is therefore of paramount importance to give a standard quantitative evaluation of the nodules shape characteristics, especially in real situation where, on the same specimen, nodules with characteristics shifting between the cases shown in Fig. 1 are present (see Fig. 2a). To this aim a specimen binarization can be performed to separate the nodules from the background; this is not as simple task as it appears since, due to the specimen preparation, the picture is corrupted by various degradation such as scratches, albedos, noise, coming from oxidation, dust and mechanical abrasion. All of this rules out simple processing techniques based on thresholding. Therefore we used the procedure described in section 2 with the following choice of the parameters: $\varepsilon = 1$, $\lambda = 10^3$, $\mu = 1$, $\nu = 1$, and $\alpha = 1$. For binarized images, the Matlab Image Processing Toolbox provides functions to label all the objects over the background and to evaluate their morphological properties.

In Fig. 2b we displayed just two sets of the shape parameters estimates among the $n = 63$ nodules. To provide an overall evaluation of the quality of the material specimen of Fig. 2a we report in Fig. 3 the histograms of the nodules distribution over the values of parameters of interest that is area, solidity and eccentricity. Hence, by computing standard sample averages by means of the distributions of Fig. 3, the operator can obtain a specimen signature to classify the quality of the current ductile cast iron production.

A second level of analysis consists in determining the characteristics of the metallic matrix, that depends on phases volume fractions and their morphology. It could range from fully ferritic (with a body cubic centered lattice), to fully paerlitic (a lamaellar mix of ferrite and cementite), from martensitic (with a tetragonal body centered lattice) to bainitic (a very fine mix of ferrite and cementite) depending on the chemical

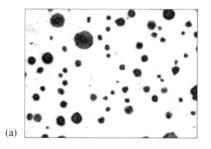

(a)

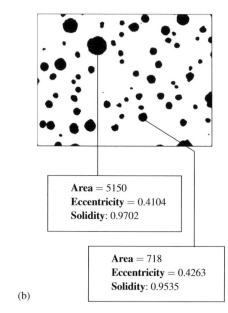

(b)

| Area $= 5150$ |
| Eccentricity $= 0.4104$ |
| Solidity: 0.9702 |

| Area $= 718$ |
| Eccentricity $= 0.4263$ |
| Solidity: 0.9535 |

Figure 2. (a) Original specimen. (b) Optimal binarization with shape parameters values for two different nodules.

composition and on the heat treatment. Microstructure analysis performed by means of optical microscopes depends on the specimens preparing procedure, and by the chemical or electrochemical etch. Different etching conditions (e.g. solution chemical composition, temperature or etching duration) imply different phases evidences. Light areas correspond to ferrite, gray areas correspond to paerlite (lamaellas are not evident because the magnification is too low) and black nodules correspond to graphite. In order to evaluate ferrite/paerlite ratio it is necessary to measure the ratio between the light/gray area to the total available area (light plus gray area).

The specimen of Fig. 4a represents a material with an estimated 70% of ferrite. From the binarized image of Fig. 4b we measured the area of the light zone and computed the ratio obtaining 55%. Nevertheless

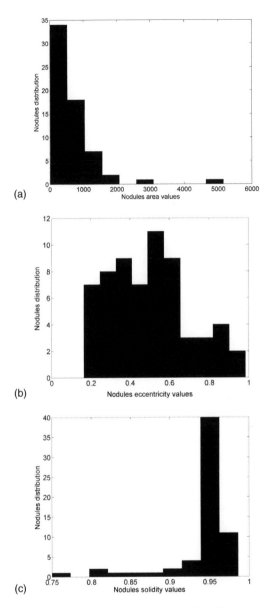

(a)

(b)

(c)

Figure 3. (a) Area histogram. (b) Solidity histogram. (c) Eccentricity histogram.

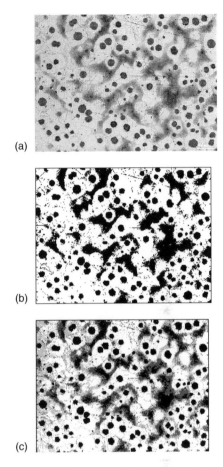

(a)

(b)

(c)

Figure 4. (a) Original data with 70% ferrite (estimation by visual inspection of an expert); (b) 2-levels segmentation; (c) 4-levels segmentation.

the 4-levels segmentation the measured ferrite percentage was 90%; a lower value was obtained since the 4-levels segmentation allows identifying some residual of the heat treatment that can be easily missed by visual inspection.

this evaluation suffers of a bias due to the area occupied by the nodules that should be subtracted from the total area to obtain the net area available for the heat treatment. To this aim we need to perform a 4-levels segmentation to identify the nodules as the darkest elements in the picture. In this case the measured ferrite percentage rises up to 80%.

The same evaluation was performed on the image of Fig. 5, which was estimated at 100% of ferrite. After

4 CONCLUSIONS

In this work a novel discrete set up to the image segmentation problem was proposed and applied to the analysis of the geometry of the ductile cast irons metallographic specimens. A robust estimation procedure was devised for a quantitative evaluation of parameters describing the morphology of the material microstructure and nodules shape. These estimates provide a rich set of standard geometrical characteristics that may support the experts in the quality evaluation of ductile cast iron productions. As a further development,

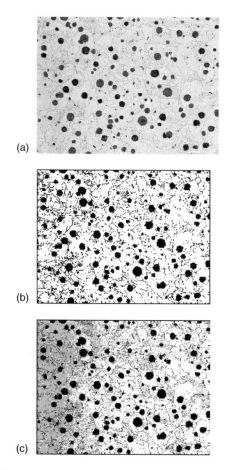

(a)

(b)

(c)

Figure 5. (a) Original data with 100% ferrite (estimation by visual inspection of an expert); (b) 2-levels segmentation; (c) 4-levels segmentation.

a neural network classifier can be trained by using the geometrical properties of the cast iron nodules and microstructure to evaluate the material quality directly in terms of its mechanical characteristics.

REFERENCES

Iacoviello, F., Di Cocco, V. 2003. Proceedings of International Conference on Fatigue Crack Paths, Parma, n. 116, Italy.
Jianming Li, Li Lu, Man On Lai. 2000. Materials Characterisation. 45: 83–88.
Imasogie, B.I., Wendt, U. 2004. Characterization of Graphite Particle Shape in Spheroidal Graphite Iron Using a Computer Based Image Image Analyzer. *Journal of Minerals and Materials Characterization & Engineering,* 3 (1): 1–12.
Mumford, D., Shah, J. 1989. Optimal approximations by piecewise smooth functions and associated variational problems. *Comm. Pure Appl. Math.* 42 (4).
Chan, T., Vese, L. 2001. Active Contours Without Edges. *IEEE Trans. on Image Processing,* 10 (2).
Morel, J.M., Solimini, S. 1995. Progress in Nonlinear Differential Equations and Their Applications, 14 Boston MA., Birkhauser .
De Santis, A., Iacoviello, D. 2005. Discrete level set approach to image segmentation. Technical Report n.13, ARACNE, available at: http://www.dis.uniroma1.it/~iacoviel/Articoli/DeSantis_new_2006.pdf.

Computational Modelling of Objects Represented in Images –
João Manuel R.S. Tavares & R.M. Natal Jorge (eds)
© *2007 Taylor & Francis Group, London, ISBN 978-0-415-43349-5*

Discrete level set segmentation for pupil morphology characterization

A. De Santis & D. Iacoviello

Dept. of Computer and Systems Science, University of Rome "La Sapienza" – Italy

ABSTRACT: The pupil morphological characteristics are of great interest for non invasive early diagnosis of the central nervous system response to environmental stimuli of different nature. Their evaluation in subjects suffering some typical diseases such as diabetes, Alzheimer disease, schizophrenia, drug and alcohol addiction is of concern. In this paper geometrical pupil features such as area, centroid coordinates, eccentricity, major and minor axes lengths are estimated by a procedure based on an image segmentation algorithm. It exploits the level set formulation of the related variational problem. A discrete set up of this problem is proposed: an arbitrary initial curve is evolved towards the unique optimal segmentation boundary by a difference equation. Numerical tests are performed on real pupillometry data taken in different illumination conditions showing a high degree of robustness of the shape parameters estimates.

1 INTRODUCTION

The importance of studying the human pupil shape and fluctuations is well established Wyatt (1995), Kristek (1965); it relies on the correlation of the pupil morphological properties with the changes of some physiological parameters in subjects suffering different diseases. The human pupil is characterized by an autonomic innervations; indeed the pupil diameter reflects the balance between the sympathetic and parasympathetic tone, that is the balance between the central and peripheral autonomic activity. The pupil size determination is a key point in important fields such as eye surgery and early diagnosis of diseases characterized by an unbalance of the nervous system. What is usually investigated is the pupil response to light, acoustic and cognitive stimuli in order to provide a non invasive diagnosis of illness such as: depression, Siegle et al. (2004), schizophrenia Steinhauer & Hakerem (1992), narcolepsy O'Neill & Trick (2001), Alzheimer and Parkinson disease Granholm (2003), diabetes Pittasch (2002), alcohol and drug abuse Kloeden & McClean (1995). Another important field of application of the eye movement measurement is the design of man-machine interfaces; for instance an eye-tracking device for eye's activity rehabilitation is proposed in Lin et al. (2004).

The existing procedures to measure the pupil diameter and its fluctuations range from totally heuristics methods to model based estimation algorithms Twa et al. (2004), Wingate et al. (1998): the pupil size is estimated by manual procedures with the help of a ruler or other graders on the pupillometer pictures;

best fit of circular templates and optimal thresholding are also adopted. An active contour segmentation technique is proposed in Tilmant et al. (2003).

In all these examples, it is clear that the crucial point consists in isolating the pupil pixels from the other elements on the eye picture; this is a simple instance of what is called *image segmentation*. The problem of segmentation is well established in image processing; it simply consists in separating different objects on a scene while providing a smoothed replica of the original data. One of the main research line lies within the variational approach dated back to Mumford & Shah (1989). In this formulation an *energy functional*, in general non-convex, is defined over a suitable 2D function sample space F and its global minimum provides the optimal segmentation.

To the aim of estimating the pupil morphology parameters a region based optimal segmentation procedure is proposed in the framework of the level set formulation of the Munford and Shah model, Chan & Vese (2001). As opposite to the standard formulation based on a continuum image model, we design our procedure directly in the discrete domain by considering an image model that is a more realistic representation of the real data which are the samples of the original image over a grid of points De Santis & Iacoviello (2005). Moreover, a modification of the energy functional that makes the problem convex is considered; as a consequence, necessary and sufficient conditions of a unique global solution are obtained in the space of piece-wise constant functions. The optimal solution is recursively computed by a nonlinear finite difference equation, which describes the evolution of the level set function from an initial configuration; the steady state zero level set defines the sought segmentation. As opposite to continuum models, no approximation scheme is required, no reinitialization during the level set function evolution is needed. The complex topology of the typical biomedical images can be easily dealt with, starting from any initial configuration of the level set function.

In Section 2 the discrete set up for the segmentation problem is presented and the theory is applied in Section 3 to pupillometer data; here morphological parameters of the pupil are estimated and some further development are also discussed. Some conclusions are provided in Section 4.

2 DISCRETE LEVEL SET ANALYSIS

The level set formulation of the variational segmentation problem is well established in the continuum set up, Mumford & Shah (1989), Chan & Vese (2001). In this section a brief outline of the discrete formulation of this problem is provided and the qualifying points are discussed omitting the proofs; a detailed derivation of the proposed method can be found in De Santis & Iacoviello (2005). In accordance to the available 8-bit real data, a monochromatic image I is defined as an array of real values $\{I_{i,j}\}$ on a grid of points

$$D = \left\{(i,j), i = 1,\dots,N; \ j = 1,\dots,M\right\}.$$

In most of the applications a real image can be satisfactorily approximated by a piecewise constant 2D function. Let $\{D_k\}, k = 1, 2, \dots, K$ be a finite disjoint partition of the image domain D with boundary ∂D, a piecewise constant segmentation I_s of an image I is defined as follows:

$$I_s = \sum_{k=1}^{K} c_k \chi_{D_k}, \quad \chi_{D_k}(i,j) = \begin{cases} 1 & for \ (i,j) \in D_k \\ 0 & otherwise \end{cases} \quad (1)$$

where the c_k's are the constant values assigned to each sub region D_k and the union of the ∂D_k's constitutes the boundary of the segmentation.

For simplicity, but without loss of generality, from now on let us consider $K = 2$; hence we have a segmentation with two different components D_1 and $D_2 = D \backslash D_1$, with two grey levels c_1 and c_2. Following the approach proposed in Chan & Vese (2001), De Santis & Iacoviello (2005), the two components can be suitably represented by the level sets of a function

$$\phi = \left\{\phi_{i,j}\right\} : D \to \mathbb{R}$$

$$D_1 = \left\{(i,j) : \phi_{i,j} \geq 0\right\}, \quad D_2 = \left\{(i,j) : \phi_{i,j} < 0\right\}$$

The boundary of the segmentation $B = \partial D_1 \cup \partial D_2$ is known once the boundary points of D_1 or D_2 are

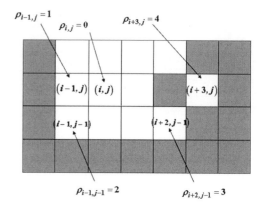

Figure 1. Internal, boundary and isolated points.

defined. As opposite to the continuum model, the points of ∂D_1 can not be simply defined as the zero level set of ϕ since, on a grid, the level set function needs not assume the zero value when passing from D_1 to D_2. Indeed, it is the number of sign changes of ϕ in the neighbor of a pixel that determines if the point is an interior point or a boundary point. With $H(\cdot)$ denoting the Heaviside function, the following function

$$\rho(z_{i,j}) = \sum_{\ell=-1}^{1} \left[\left(H(z_{i+\ell,j}) - H(z_{i,j})\right)^2 + \left(H(z_{i,j+\ell}) - H(z_{i,j})\right)^2\right]$$

counts the number of changes of sign of $z_{i,j}$ in a neighborhood of pixel (i,j); therefore we can check a boundary point with the following detector

$$\gamma(\phi_{i,j}) = H\left(3 - \rho(\phi_{i,j})\right)\left[1 - \delta(\rho(\phi_{i,j}))\right]$$

where δ is the Dirac function.

Function γ is zero both on interior points ($\rho = 0$) and on isolated points ($\rho = 4$); it is one on the boundary points ($\rho = 1, 2, 3$), see Fig.1.

Now, given any real data I, the segmentation (1) can be obtained by minimizing the following functional

$$E(c_1, c_2; \phi) = \lambda \sum_{i,j} H(\phi_{i,j})(I_{i,j} - c_1)^2$$

$$+ \lambda \sum_{i,j} \left(1 - H(\phi_{i,j})\right)(I_{i,j} - c_2)^2$$

$$+ \frac{\alpha}{2} \sum_{i,j} \phi_{i,j}^2 + \mu \sum_{i,j} H(\phi_{i,j}) + \nu \sum_{i,j} \gamma(\phi_{i,j}) \quad (2)$$

The first two terms are just the fit error within the regions where $\phi \geq 0$ or $\phi < 0$; the last two terms evaluate the "area" of D_1 and D_2 and the "length" of segmentation boundary. They are actually counting measures and therefore the "area" is the number of pixels within a region and the length is the number of boundary pixels. The third term is peculiar of the proposed formulation and was not present in the original formulation of Mumford & Shah (1989); while it has no physical meaning in terms of the segmentation properties, it is responsible of the functional convexity so that necessary and sufficient conditions for global minimum are available. Parameters λ, μ, ν and α are weights that can be used to enhance the contribution of one term with respect to the others in (2).

Function E is not smooth with respect to ϕ because of the presence of the generalized functions. To avoid complex analysis, we prefer, as suggested in Chan & Vese (2001), to relax its definition considering that the generalized functions are the weak limit ($w - \lim$) of sequence of smooth functions in the space of 2-D integrable functions

$$H(z) = w - \lim_{\varepsilon \to 0} \frac{1}{2}\left(1 + \frac{2}{\pi}\tan^{-1}\left(\frac{z}{\varepsilon}\right)\right) = w - \lim_{\varepsilon \to 0} H_\varepsilon(z)$$

$$\delta(z) = w - \lim_{\varepsilon \to 0} \frac{1}{\pi}\frac{\varepsilon}{\varepsilon^2 + z^2} = w - \lim_{\varepsilon \to 0} \delta_\varepsilon(z)$$

We therefore modify the definitions of ρ and γ by using $H_\varepsilon, \delta_\varepsilon$ in place of H, δ, and define the following cost function

$$E_\varepsilon(c_1, c_2; \phi) = \lambda \sum_{i,j} H_\varepsilon(\phi_{i,j})(I_{i,j} - c_1)^2$$

$$+ \lambda \sum_{i,j}(1 - H_\varepsilon(\phi_{i,j}))(I_{i,j} - c_2)^2$$

$$+ \frac{\alpha}{2}\sum_{i,j}\phi_{i,j}^2 + \mu\sum_{i,j}H_\varepsilon(\phi_{i,j}) + \nu\sum_{i,j}\gamma_\varepsilon(\phi_{i,j}) \quad (3)$$

Necessary and sufficient conditions for the existence of a unique optimal solution can be found in De Santis & Iacoviello (2005). The optimal segmentation is given by

$$c_1 = \frac{\sum_{i,j} H_\varepsilon(\phi_{i,j}) I_{i,j}}{\sum_{i,j} H_\varepsilon(\phi_{i,j})}, \quad c_2 = \frac{\sum_{i,j}(1 - H_\varepsilon(\phi_{i,j})) I_{i,j}}{\sum_{i,j}(1 - H_\varepsilon(\phi_{i,j}))} \quad (4)$$

$$0 = \alpha\phi_{i,j} + \{\lambda p_{i,j} + \mu$$

$$+ \nu\sum_{\ell=-1}^{1}\left[\left(H_\varepsilon(\phi_{i+\ell,j}) - H_\varepsilon(\phi_{i,j})\right)\right. \quad (5)$$

$$\left. + \left(H_\varepsilon(\phi_{i,j+\ell}) - H_\varepsilon(\phi_{i,j})\right)\right] q_{i,j}\} \delta_\varepsilon(\phi_{i,j})$$

Formula (5) is the Euler-Lagrange equation for (3). Functions $p_{i,j}$ and $q_{i,j}$ in (5) have the following definitions

$$p_{i,j} = (I_{i,j} - c_1)^2 - (I_{i,j} - c_2)^2$$

$$q_{i,j} = 2\left[\delta_\varepsilon(3 - \rho_\varepsilon(\phi_{i,j}))(1 - \delta_\varepsilon(\rho_\varepsilon(\phi_{i,j})))\right.$$

$$\left. - H_\varepsilon(3 - \rho_\varepsilon(\phi_{i,j}))\frac{2\rho_\varepsilon(\phi_{i,j})}{\varepsilon^2 + \rho_\varepsilon^2(\phi_{i,j})}\delta_\varepsilon(\rho_\varepsilon(\phi_{i,j}))\right]$$

Solution of (5) is usually obtained as steady state solution of a fictitious nonlinear finite difference evolution equation; the following evolution equation is proposed

$$\phi_{i,j}^{n+1} = -\frac{1}{\alpha}\left[\lambda p_{i,j}^n + \mu + \nu q_{i,j}^n\left(\left(H_\varepsilon(\phi_{i-1,j}^n) - H_\varepsilon(\phi_{i,j}^n)\right)\right.\right.$$

$$+ \left(H_\varepsilon(\phi_{i+1,j}^n) - H_\varepsilon(\phi_{i,j}^n)\right) + \left(H_\varepsilon(\phi_{i,j-1}^n) - H_\varepsilon(\phi_{i,j}^n)\right)$$

$$\left.\left. + \left(H_\varepsilon(\phi_{i,j+1}^n) - H_\varepsilon(\phi_{i,j}^n)\right)\right)\right]\delta_\varepsilon(\phi_{i,j}^n) \quad (6)$$

where $p_{i,j}^n$ and $q_{i,j}^n$ are functions $p_{i,j}$ and $q_{i,j}$ computed by replacing in their expression $\phi_{i,j}$ by $\phi_{i,j}^n$. The solution of (6) converges to the solution of (5), De Santis & Iacoviello (2005).

In case of multi level segmentation the energy functional (2) and the optimality conditions can be straightforward adapted to the presence of more than one level set function. Nevertheless in the experiments we will consider, the 4-levels segmentation is obtained by a hierarchical binarization processing; this consists in a first binarization and the two regions obtained are separately further binarized.

3 EXPERIMENTAL DESIGN AND NUMERICAL RESULTS

Our experimental setup consisted of the pupillometer shown in Fig. 2.

There is a chin and forehead rest, and a micro infrared CCD camera is positioned at 20 cm from the eyes. The subject was side illuminated by a gas discharge lamp of 100 w, and was asked to look into the camera lens. The camera was connected to a PC by a commercial video capture board with a frame rate of 10 frames/sec at a resolution of 274 × 376, with a 8-bit intensity level. Data were collected under different illumination conditions corresponding to daylight and darkened room.

The theory of Section 2 is applied to the images captured by the pupillometer. Data were preprocessed to obtain a signal with better intensity level distribution (gamma correction with $g = 0.3$) and a smoother

155

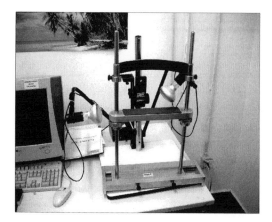

Figure 2. Pupillometer set up.

behavior (gaussian filtering with $\sigma = 4$); this avoids artifacts in the segmented images due to the fine iris texture. Indeed, this is not necessarily required, as in experiments with good room illumination and patients with light color iris.

To the aim of measuring different morphological features of the pupil, we need to label the different eye elements on the pictures, select the pupil and then evaluate some of its shape parameters. A four levels segmentation has proved to be suitable for a simplified eye picture representation, preserving all the information relevant to our analysis. The four levels segmentation is obtained by a recursive binary segmentation: this is a hierarchical process where, at the second step, each of the regions obtained by the first binarization are further binarized.

Fig. 3 shows the algorithm outcome for a typical pupillometer data in a daylight condition with the following choice

$$\varepsilon = 1, \ \lambda = 10^5, \ \mu = 1, \ \nu = 1, \alpha = 1.$$

From Fig. 3(c) we see that the pupil belongs to the set of darker elements; indeed this is always the case in the eye, due also to the chosen experimental setup. All the darker elements can be easily separated by the others by a simple logical operation; then an image with these elements as white objects over a black background is obtained, Fig. 3(d). The Matlab image processing toolbox provides useful instructions to label every object on the background and to measure its shape parameters: the instruction *bwlabel* can label all the white objects and subsequently the instruction *regionprops* evaluates some morphological characteristics of any object labeled by the previous instruction, such as: area, centroid coordinates, equivalent diameter, eccentricity. The area is just the number of pixels that build up the object. The equivalent diameter is the diameter of a circle that has the same area as the object.

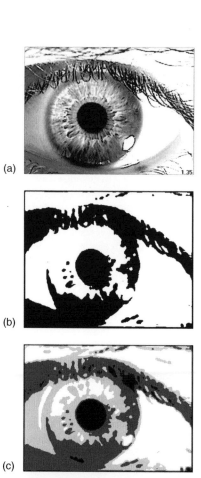

(a)

(b)

(c)

(d)

Area: 199
Centroid: [119.5025 248.2211]
Eccentricity: 0.9723
EquivDiameter: 15.9177

Area: 3440
Centroid: [149.7172 151.7017]
Eccentricity: 0.2604
EquivDiameter: 66.1811

Figure 3. Algorithm outcome on a typical pupil picture. (a) Original data; (b) early binarization; (c) 4-levels segmentation; (d) pupil localization and example of shape parameters evaluation of two elements.

The eccentricity is within [0–1] and quantifies how much the object is far from being circular: the value zero corresponds to a perfectly circular shape.

The set of shape parameters values corresponding to the pupil can be easily extracted since the pupil has surely the lowest eccentricity value within the labeled objects, Fig. 3(d).

The proposed procedure therefore provides a robust optimal estimation of those parameters such as the centroid coordinates and the equivalent diameter, which was the task of the various methods found in the relevant literature referenced in the Introduction. Other parameters are also available that may be considered for a more complete characterization of the pupil shape, especially in problems involving diseases that imply the pupil deformation. It is worth noting that the detection of the pupil is straightforward from the segmentation, no manual picture zoning is performed.

As a concluding remark we stress that dynamic measures of the pupil size are obtained by recording the estimation of the pupil diameter, the centroid position, the orientation and so forth, for any picture of the sequence captured by the pupillometer. These data can be further analyzed to study the pupil size dynamics supporting the response of the nervous system to given stimuli in controls and patients with specific pathologies. It is therefore of paramount importance that the proposed pupil estimation procedure does not require any parameters update on different frames of the same sequence.

4 CONCLUSIONS

The segmentation of biomedical images is a key point in the analysis of data for non-invasive diagnosis. In this paper we focus on pupillometer data since the pupil morphology and dynamics support the investigation of the reaction of the autonomous nervous system to various inputs in healthy individuals and in patients suffering some specific diseases. We propose a fully automated procedure for image segmentation based on the level set theory. We provide a novel discrete set up, which yields the level set evolution equation directly as a non-linear difference equation, whose steady state solution gives the unique optimal segmentation. For the chosen application a 4-levels segmentation is considered; it is suitable to extract the pupil from the eye picture and measure various morphological parameters, such as pupil diameter, centroid, area and others. The segmentation procedure is robust with respect to changes in illumination and does not require any tuning within the same frames sequence.

REFERENCES

Wyatt, H.J. 1995. The form of human pupil. *Vision Res.*, 35: 2021–2036.

Kristek, A. 1965. The physiological pupillogram of the human eye. *Czechoslovakian J. Ophthalmol.*, 21: 28–35.

Siegle, G.J., Steinhauer, S.R., Thase, M.E. 2004. Pupillary assessment and computational modelling of the Stroop task in depression. *International Journal of Psychophysiology*, 52: 63–76.

Steinhauer, S.R., Hakerem, G., 1992. The pupillary response in cognitive psychophysiology and schizophrenia. *Psychophysiology and experimental psychopathology: a tribute to Samuel Sutton.* New York: NY Academy of Sciences, 658: 182–204.

O'Neill, W., Trick, K. 2001. The narcoleptic cognitive papillary response. *IEEE Trans. on Biomedical Engineering*, 48 (9) 963–968.

Granholm, E., Morris, S., Galasko, D., Shults, C., Rogers, E., Vukov, B. 2003. Tropicamide effects on pupil size and pupillary light reflexes in Alzheimer's and Parkinson's disease. *International Journal of Psychophysiology.* 47: 95–115.

Pittasch, D., Lobmann, R., Behrens-Baumann, W., Lehnert, H. 2002. Pupil signs of sympathetic autonomic neuropathy in patients with type 1 diabets. *Diabets Care* 25 (9) 1545–1551.

C.N. Kloeden and A.J. McClean (Eds) 1995. Alcohol, Drugs and Traffic Safety *Oculomotor and Pupil Tests to Identify Alcohol Impairment.* 2: 887–880. Adelaide: NHMRC Road Accident Research Unit, University of Adelaide.

Lin, C.S., Huan, C.C., Chan, C.N., Yeh, M.S., Chiu, C.C. 2004. Design of a computer game using an eye-tracking device for eye's activity rehabilitation. *Optics and Lasers in Engineering* 42: 91–108.

Twa, M.D., Bay, M.D., Hayes, J., Bullimore, M. 2004. Estimation of pupil size by digital photography. *J.Cataract Refract Surg.* 30: 381–389.

Wingate, M., Hood, B., Phan, P. 1998. Automated pupil size determination for evaluating fluctuations in physiological arousal. *Proc. of 4th International Conference on Signal Processing,* 1666–1669.

Tilmant, C., Sarry, L., Gindre, G., Boire, J.Y. 2003. Monitoring and modeling of pupillary dynamics. *Proc. of the 25th annual International Conference of the IEEE EMBS,* Cancun, 678–681.

Mumford, D., Shah, J. 1989. Optimal approximations by piecewise smooth functions and associated variational problems. *Comm. Pure Appl. Math.* 42 (4).

Chan, T., Vese, L. 2001. Active Contours Without Edges. *IEEE Trans. on Image Processing* 10 (2).

De Santis, A., Iacoviello, D. 2005. Discrete level set approach to image segmentation, Technical Report 13, ARACNE, 2005, available at: http://www.dis.uniroma1.it/~iacoviel/Articoli/DeSantis_new_2006.pdf.

Computational Modelling of Objects Represented in Images –
João Manuel R.S. Tavares & R.M. Natal Jorge (eds)
© 2007 Taylor & Francis Group, London, ISBN 978-0-415-43349-5

3D watershed transformation based on connected faces structure

Sébastien Delest, Romuald Boné & Hubert Cardot
Université François Rabelais de Tours, Laboratoire Informatique, Tours, France

ABSTRACT: This paper describes a new approach for watershed segmentation on triangular mesh. A common approach for 3D patch-type segmentation is to use a growing or flooding process on a connected vertices structure. Watershed transformation is independent of the structure because only elements with neighbors and height criterion are required. It is widely used on 2D images, 3D meshes and 3D images. Here, we implement a connected faces structure and adapt the segmentation process to it. Connected faces structure offers different curvature information and neighborhood. We use the two structures on the hierarchical queue watershed transformation with a hierarchical merging process and we present a comparison of the two approaches.

1 INTRODUCTION

Polygonal meshes offer an efficient representation of 3D surfaces, in particular triangular meshes, which are used in many applications. In this paper, we deal with triangular meshes, however, connected faces structure can be used with other types of polygonal meshes. Mesh segmentation is used in many applications as texture atlas, shape analysis, recognition, matching, shape modeling and retrieval. The shape of models is important and can lead to different segmentation approaches depending on whether it deals with natural shapes or mechanical parts. The mesh segmentation methods are principally classified to patch-type and part-type where patch-type is related to surface segmentation of the model and part-type corresponds to shape and parts segmentation.

In the following, we propose a patch-type segmentation based on watershed transformation. This method, suggested by (Digabel and Lantuéjoul 1978), is a morphological tool commonly used for 2D image segmentation. (Vincent and Soille 1991) and (Meyer 1991) give a solid approach of the flooding principle. (Beucher and Meyer 1992) introduced new techniques based on hierarchical queues and (Beucher 2004) improved these methods with an unbiased watershed. The 3D mesh segmentation is closed to the 2D approach; pixels are replaced by vertex and connectivity is no more fixed. The hierarchical queue has been adapted to triangular meshes in (Betser, Delest, and Boné 2005) and we proposed a new approach where pixels are replaced by triangular faces. The main idea is to take advantage of a new information: the face curvature.

The remainder of the paper proceeds as follows: the next section deals with the curvature calculation

from connected vertex and connected faces structures. Section 3 presents the 2D hierarchical queue unbiased watershed transformation algorithm and then shows how to adapt it to 3D. Section 4 explains the methods to limit over-segmentation and section 5 presents results.

2 CURVATURE CALCULATION

To perform the watershed transformation, we need a structure where each element must be associated to a neighborhood and a criterion. In 2D, a pixel has 8 neighbor pixels. In case of polygonal meshes, vertex and faces have not a fixed connectivity, as shows in Figure 1.

For 2D images, the criterion corresponds usually to gray level or gradient magnitude; for polygonal meshes, this criterion is the vertex or face curvature. Several methods are proposed to calculate the vertex curvature. Mangan and Whitaker have pointed out the covariance matrix efficiency in (Mangan and Whitaker 1999), Gaussian, mean and principal curvatures have been detailed in (Meyer, Desbrun, Schröder, and Barr 2003) and a recent approach (Yamauchi, Gumhold, Zayer, and Seidel 2005) has been proposed to estimate Gaussian curvature from Gauss area on the Gauss map. In order to compare the vertex and face connection

Figure 1. Neighborhood relationships for pixels, vertices and triangles.

approaches, we use the covariance matrix, which offers the most relevant curvature information. The covariance matrix is given by the variance and covariance in all the three directions:

$$\sigma_{uu}^2 = \frac{1}{N}\sum_{i=0}^{N}(u_i + \bar{u})^2$$

$$\sigma_{uv}^2 = \frac{1}{N}\sum_{i=0}^{N}(u_i + \bar{u})(v_i + \bar{v})$$

$$C = \|M\| \text{ with } M = \begin{bmatrix} \sigma_{xx} & \sigma_{xy} & \sigma_{xz} \\ \sigma_{yx} & \sigma_{yy} & \sigma_{yz} \\ \sigma_{zx} & \sigma_{zy} & \sigma_{zz} \end{bmatrix}$$

where σ_{uu} corresponds to the u coordinates standard deviation in the element neighborhood, and σ_{uv} corresponds to the square root of the covariance between u and v coordinates. N is the number of triangles associated with this element and $\{x_t\, y_t\, z_t\}$ are the components of the normal from triangle t. The curvature C is defined by the norm of the covariance matrix M.

In connected faces structure case, there are more neighbors than in connected vertices structure case; that allows to easily differentiate the significant curvature, in a noise reduction way. Neighbor triangles do not have the same type of connection; some are connected to the face by an edge and others by a vertex as we can see in Figure 2. Keeping all connected triangles gives the best curvature information and watershed computation. While keeping only 3 maximum triangles connected by edges assures the best time calculation but a less accurate vertex curvature and watershed (Figure 3).

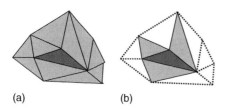

(a)   (b)

Figure 2. (a) Triangles neighborhood with all connected triangles; (b) triangles neighborhood with triangles connected by edges.

Figure 3. Curvature computation based on vertices, triangles and triangles connected by edge structures. High curvatures are whites.

3 WATERSHED TRANSFORMATION

This section describes the approach used in this watershed algorithm and the features of the implementation. On 2D images, the watershed transformation is a segmentation technique which simulates water rising on the gray level (Figure 4) or gradient magnitude of the input image from the local minima.

The flooding starts from local minima and incrementally fill the basins until it connects to its neighbors. A watershed is generated where basin collisions occur (Figure 5).

The use of hierarchical queue is one of the best solutions to build quickly a watershed on 2D images (Beucher 2004). The hierarchical queue is made up of several FIFO queue and each queue corresponds to a level (gray level or gradient magnitude). Queues are sorted by level and a queue can be unstacked only if previous queues are empty (Figure 6).

First, the minima are stacked into their corresponding queue. Each pixel or plateau minimum gets a label. Then, the cycle of extraction and stacking

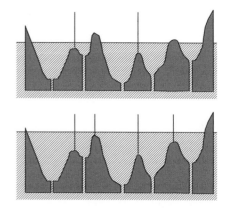

Figure 4. Lena image map generated from its gray levels intensities.

Figure 5. Two levels of flooding. Minima have an infinite depth and basin collisions generate a watershed.

up can begin: the first pixel of the first queue is unstacked, its non conflicting neighbors are stacked in their corresponding queues and received their root pixel label. The conflicting pixels are labelled as watershed but this watershed transformation is biased because this approach induces the pixel processing order accordingly with the queue stacking order. This arbitrary means of spilling water onto plateau pixels generates watershed positioning errors which, by accumulation, can cause a significant shift. To avoid this problem, (Beucher 2004) proposes an unbiased watershed.

To adapt the hierarchical queue watershed transformation to polygonal meshes, we use vertex or face structure connection instead of 2D image structure; the height function is now the vertex or face curvature. In Figure 7, the results are obtained by calculating the hierarchical queue unbiased watershed algorithm of 3D model with the two approaches. The section 5 offers a comparison of these methods.

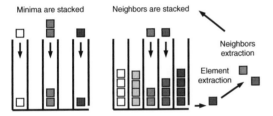

Figure 6. Watershed transformation based on hierarchical queue.

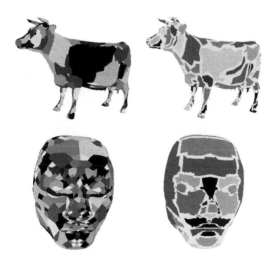

Figure 7. Hierarchical queue watershed transformation on connected vertices structure (left models) and connected faces structure (right models, watershed faces are kept).

4 HIERARCHICAL SEGMENTATION

As it can be seen in Figure 7, the output of the hierarchical queue watershed algorithm suffers quite badly from over-segmentation. To avoid this problem, the use of markers allows forcing the number of regions; that supposes the characteristics of the interesting objects are known. Another possibility is to use hierarchical segmentation. We are interested in this last method which offers a solid structure to reach the best level of segmentation.

The result of the watershed is used as an input partition of the hierarchical segmentation algorithm. This partition is composed of regions separated by watershed. We use here a simple hierarchical algorithm based on graphs; on 2D images, Marcotegui and Beucher (Marcotegui and Beucher 2005) went further with the use of the minimum spanning tree and the waterfalls. To compute the hierarchical segmentation, we first considered two neighbor catchment basins as an edge with a depth valuation (Figure 8).

The depth P of the watershed saddle point S between regions A and B corresponds to the difference between the saddle point curvature and the minimum curvature of the regions A and B:

$$P(S_i(R_a, R_b)) = C(S_i) - \min(R_a, R_b) .$$
$$= \max(h_a, h_b)$$

This edge also gets the elements (vertex or face) list which composed the watershed. We give to each region its elements list, its minimum curvature and its edge watershed list. To represent the region merging, we use nodes which can have children regions or nodes. For each region, we create a node which is made up of this region as children, the region label and the minimum curvature of this region. Then, we can compute the merging. This process needs to first sort the edge watershed by depth. From the minimum depth, we can successively make the merging; each merging generates a node which gets a node children list, a new label,

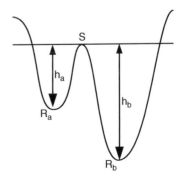

Figure 8. Watershed depth.

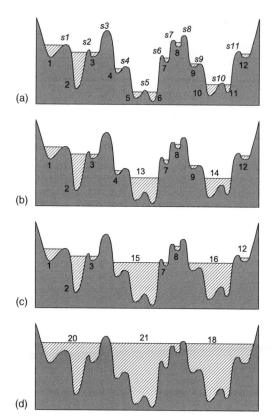

(a)

(b)

(c)

(d)

Figure 9. Several merging levels. (a) Regions and saddle points created from watershed transformation; (b) at the 2nd, the 4th and the 9th merging iteration, the region R14, R16 and R21 respectively are created.

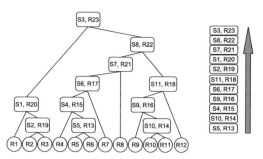

Figure 10. Regions merging tree. Node numbers correspond to the merging order from node R13.

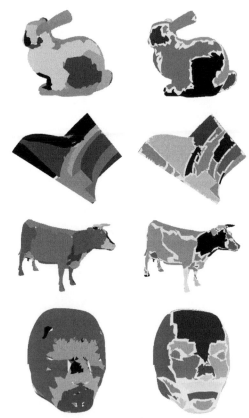

Figure 11. Comparison of the two approaches (connected vertices structure on the left, connected faces structure on the right).

an external edge watershed list and an internal edge watershed list; it is necessary to update the sorted list at each merging. When nodes are merged, theirs shared watersheds are removed, however, they must be store in the node as internal watershed because they exist at an inferior level of the tree (internal watershed belong to only one node). That allows to quickly label elements from the merging level. Figure 9a shows regions and saddle points created from watershed transformation; Figure 9b, 9c and 9d correspond to the 2nd, the 4th and the 9th merging iteration. At the first iteration, the region R13 is created, at the second, R14 is generated, etc. until the last iteration where R23 is created and all regions are merged.

Figure 10 shows the regions merging tree, each node corresponds to a segmentation level and when a level is chosen, all nodes which have an inferior level and a superior parent level are used for the labelling purpose. For example, at level 21, there are 3 regions and levels 18, 20 and 21 can be used for the labelling as shown in the Figure 9d.

5 ANALYSIS

Mechanical parts have generally high edge curvatures and in this case, vertex structure is better adapted as we can see on Figure 11. This is not the case for natural shape where connected faces structure gives

Table 1. Summary of the relative features of the different approaches applied to the models of Figure 11.

Models	Type	Merging level/Max	Computation time (ms)
Stanford Bunny	Vertex	123/155	128
(2503 vertices, 5804 faces)	Face	76/96	208
Fandisk	Vertex	258/280	376
(6475 vertices, 12946 faces)	Face	63/85	1010
Cow	Vertex	146/210	155
(2903 vertices, 5804 faces)	Face	92/143	265
Face	Vertex	10/31	109
(2363 vertices, 4640 faces)	Face	334/385	131

a patch-type segmentation more interesting; for example, the face is difficult to segment with a connected vertices structure while connected faces structure gives a very good segmentation.

Statistically, there are more faces than vertices and faces have a greater neighborhood than vertices; then, connected faces structure gives a computation time proportionately important. In a noise filtering way, a greater number of faces allows to easily differentiate the significant curvature. Connected faces structures leads to a lower number of regions but also to a longer watershed computation; hierarchical segmentation process and search for the best segmentation level of merging are faster.

Table 1 shows the number of regions for each model and each type of structure after the watershed transformation; it appears the last number level (where all regions are merged) and the chosen segmentation level of merging. Pre-processing and post-processing are not addressed here and they constitute complementary approaches to improve the segmentation. Before watershed transformation, we could use a noise curvature reduction and mean shift filtering (Yamauchi, Lee, Lee, Ohtake, Belyaev, and Seidel 2005).

6 CONCLUSIONS

Our approach uses the face curvature information to compute the 3D segmentation. We have presented a comparison of segmentation with different structures and showed that connected faces structure allows an interesting segmentation on natural shapes as the face and gives the smallest merging tree that allows a fast search of the best level of merging. These two structures bring different information so that will be interesting to use them simultaneously to improve the watershed transformation.

ACKNOWLEDGEMENTS

We are grateful to Stanford Computer Graphics Laboratory and the web site 3D Cafe (www.3dcafe.com) for making their models available online.

REFERENCES

Betser, J., S. Delest, and R. Boné (2005, September). Unbiased watershed hierarchical 3d segmentation. In *Proceedings of the Fifth IASTED International Conference on Visualization, Imaging and Image Processing,* Benidorm, Spain, pp. 412–417.

Beucher, S. (2004, Avril). Algorithmes sans biais de ligne de partage des eaux. Technical report, Centre de Morphologie Mathématique de l'École des Mines de Paris.

Beucher, S. and F. Meyer (1992). The morphological approach of segmentation: the watershed transformation. In E. Dougherty (Ed.), *Mathematical Morphology in Image Processing,* New York, pp. 433–481.

Digabel, H. and C. Lantuéjoul (1978). Iterative algorithms. In *2nd European Symp. Quantitative Analysis of Microstructures in Material Science,* Caen, France, pp. 85–99. Biology and Medicine.

Mangan, A. P. and R. T. Whitaker (1999). Partitioning 3d surface meshes using watershed segmentation. *IEEE Transactions On Visualization And Computer Graphics 5*(4), 308–321.

Marcotegui, B. and S. Beucher (2005). Fast implementation of waterfall based on graphs. Volume 30 of *Computational Imaging and Vision,* pp. 177–186. Dordrecht: Springer-Verlag.

Meyer, F. (1991). Un algorithme optimal de ligne de partage des eaux. In *Actes du 8ième Congrès AFCET,* Lyon-Villeurbanne, France, pp. 847–859.

Meyer, M., M. Desbrun, P. Schröder, and A. H. Barr (2003). Discrete differential-geometry operators for triangulated 2-manifolds. In H.- C. Hege and K. Polthier (Eds.), *Visualization and Mathematics III,* pp. 35–57. Heidelberg: Springer-Verlag.

Vincent, L. and P. Soille (1991).Watersheds in digital spaces: An efficient algorithm based on immersion simulations. *IEEE Transactions Pattern Analysis and Machine Intelligence 13*(6), 583–598.

Yamauchi, H., S. Gumhold, R. Zayer, and H.-P. Seidel (2005, September). Mesh segmentation driven by gaussian curvature. *Visual Computer 21*(8–10), 649–658.

Yamauchi, H., S. Lee, Y. Lee, Y. Ohtake, A. Belyaev, and H.-P. Seidel (2005, June). Feature sensitive mesh segmentation with mean shift. In *Shape Modeling International 2005,* Cambridge, MA, USA, pp. 236–243. IEEE.

Computational Modelling of Objects Represented in Images –
João Manuel R.S. Tavares & R.M. Natal Jorge (eds)
© 2007 Taylor & Francis Group, London, ISBN 978-0-415-43349-5

Computer-aided detection of pulmonary nodules in low-dose CT

P. Delogu & M.E. Fantacci
Dipartimento di Fisica dell'Università & INFN, Pisa, Italy

I. Gori
Bracco Imaging S.p.A., Milano, Italy

A. Preite Martinez
Centro Studi e Ricerche Enrico Fermi, Roma, Italy

A. Retico
Istituto Nazionale di Fisica Nucleare, Sezione di Pisa, Italy

ABSTRACT: A computer-aided detection (CAD) system for the identification of pulmonary nodules in low-dose multi-detector helical CT images with 1.25-mm slice thickness is being developed in the framework of the INFN-supported MAGIC-5 Italian project. The basic modules of our lung-CAD system, a dot-enhancement filter for nodule candidate selection and a voxel-based neural classifier for false-positive finding reduction, are described. Preliminary results obtained on the so-far collected database of lung CT scans are discussed.

1 INTRODUCTION

Lung cancer is one of the most relevant public health problems. Despite significant research efforts and advances in the understanding of tumor biology, there has been no reduction of the mortality over the last decades. Lung cancer most commonly manifests itself with the formation of non-calcified pulmonary nodules. Helical Computed Tomography (CT) is recognized as the best imaging modality for the detection of pulmonary nodules (Diederich et al. 2001). The amount of data that need to be interpreted in CT examinations can be very large, especially when thin collimation settings are used, thus generating up to about 300 2-dimensional images per scan, corresponding to about 150 MB of data.

In order to support radiologists in the identification of early-stage pathological objects, researchers have recently begun to explore computer-aided detection (CAD) methods in this area.

The First Italian Randomized Controlled Trial aiming at studying the potential impact of screening on a high-risk population using low-dose helical CT has recently started (Italung-CT trial). In this framework and in the framework of the MAGIC-V (INFN CSN-V) project, we are developing a CAD system for pulmonary nodule identification, intended to be used as a supportive tool in screening protocols. The system is based on a dot-enhancement filter and a neural classifier for the reduction of the number of false-positive (FP) findings. This paper describes a comparison between the performances of two different voxel-based neural approaches originally developed for the classification of the selected region of interest (ROI) by the dot-enhancement filter.

2 THE CAD SYSTEM ARCHITECTURE

Pulmonary nodules may be characterized by very low CT values and/or low contrast, they may have CT values similar to those of blood vessels and airway walls to which they could also be strongly connected. Examples of lung internal nodules are shown in Fig. 1.

An important and difficult task in the automated nodule detection is the initial selection of the nodule candidates within the lung volume. In the identification process, lung nodules are modeled as spherical objects and a dot-enhancement filter is applied to

Figure 1. Some examples of pulmonary nodules.

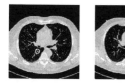

Figure 2. Some examples of false positive findings generated by the dot-enhancement filter.

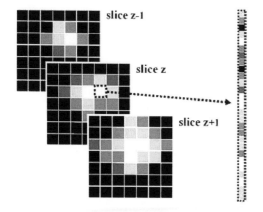

Figure 3. Voxel-based neural approach to false-positive reduction.

the 3D matrix of voxel data. The filter attempts to determine the local geometrical characteristics of each voxel, by computing the eigenvalues of the Hessian matrix and evaluating a likelihood function that was specifically built to discriminate between the local morphology of linear, planar and spherical objects, the latest modeled as having 3D Gaussian sections (Li et al. 2003; Delogu et al. 2005). A simple peak-detection algorithm (i.e. a local maximum detector) is then applied to the filter output to detect the filtered-signal peaks.

Since most FP findings are provided by crossings between blood vessels (see Fig. 2), we attempted to reduce the amount of FP/scan by developing a procedure which we called voxel-based neural approach (VBNA). According to that method, each voxel of a region of interest (ROI) is characterized by the grey level intensity values of its neighbors (see Fig. 3). We developed, implemented and compared two different VBNA procedures. In the first, the CT values of the voxels in a 3D neighborhood of each voxel of a ROI are rolled down into vectors of features (147 features) to be analyzed by a neural classifier. In the second procedure (Gori, I. & Mattiuzzi, M. 2005), 6 additional features constituted by the eigenvalues of the gradient and the Hessian matrices are computed for each voxel and encoded to the feature vectors (153 features). A feed-forward neural network is implemented

at this stage to assign each voxel either to the nodule or normal tissue target class.

A candidate nodule is then characterized as "CAD nodule" if the number of pixels within its ROI tagged as "nodule" by the neural classifier is above some relative threshold. A free response receiver operating characteristic (FROC) curve for our CAD system can therefore be evaluated at different threshold levels.

3 DATA ANALYSIS AND RESULTS

The CAD system was developed and tested on a dataset of low-dose (screening setting: 140 kV, 70 ÷ 80 mA) CT scans with reconstructed slice thickness of 1.25 mm. The scans were collected and annotated by experienced radiologists in the framework of the screening trial being conducted in Italy (Italung-CT). The database available for this study consists of 14 scans, containing 24 internal nodules. Each scan is a sequence of about 300 slices stored in the DICOM (Digital Imaging and Communications in Medicine) format.

First of all, the lung volume is segmented out of the whole 3D data array by means of a purposely built segmentation algorithm that identifies the internal region of the lung (Antonelli et al. 2005).

The 3D dot-enhancement filter applied to the selected lung regions shows a very high sensitivity. In particular, the lists generated by the peak-detector algorithm for all CT are empirically truncated so to include all annotated nodules. According to this procedure, a 100% sensitivity to internal nodules is obtained at a maximum (average) number of 54 (52.3) FP/scan.

With respect to the VBNA procedure for FP reduction, the dataset was randomly partitioned into train and test sets; the performances of the trained neural networks were evaluated both on the test sets and on the whole dataset.

In the first VBNA approach, 147 features, derived from a 2D region of 7×7 voxels for 3 consecutive slices with the voxel to be classified in the center, constitute each vector of the feature dataset. Two three-layer feed-forward neural networks with 147 input, were trained on two different random partitions of the dataset into train and test sets. The performances achieved in each trial for the correct classification of individual pixels are reported in Table 1, where the sensitivity and the specificity values obtained on the test sets, on the whole datasets and the average values on the two trials are shown.

Also in case of the second VBNA approach, when 6 additional features are encoded to each feature vector, two different neural networks were trained, obtaining the performances shown in Table 2.

The comparison between the average sensitivity and specificity obtained in the first and second approach

Table 1. VBNA with 147 features.

Test			Train + test	
Sens %	Spec %		Sens %	Spec %
71.7	82.7		81.5	87.6
73.4	78.1		78.1	80.6
		average	79.8	84.0

Table 2. VBNA with 153 features.

Test			Train + test	
Sens %	Spec %		Sens %	Spec %
75.3	78.6		85.5	83.3
79.9	84.1		84.6	88.1
		average	85.1	85.7

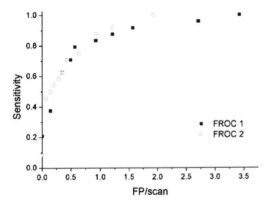

Figure 4. FROC curves obtained on a dataset on 14 CD scans containing 24 internal nodules.

proves that the implementation of some features exploiting the morphology of the voxel neighborhood in addition to the textural features directly derived by the sequence of the voxel intensity values can improve the system discriminating capability.

Once the VBNA approach with 153 features has been applied to each ROI, the FROC curves have been evaluated for the two trained neural networks. They are shown in Fig. 4. Both curves show the 100% sensitivity at a very low level of false positives (1.5–3.5). The sensitivity value remains very high (85% range) even at less than 1 false positive per scan.

4 CONCLUSIONS

In this work we compared two different voxel-based neural approaches (VBNA) to the FP reduction in the framework of the development of a CAD system for the identification of pulmonary internal nodules in low-dose CT scans. The VBNA approach where for each voxel 6 morphological features have been added to the vector of rolled-down voxel neighborhood, has shown the best FP reduction capability.

In conclusion, the dot-enhancement pre-processing algorithm provides a good sensitivity in the identification of nodule candidates and the VBNA with 153 features is an effective approach to the problem of false positives reduction. In particular, the VBNA approach allows reducing the average rate of 52.3 FP findings per scan generated by the dot-enhancement filter to 1.9 FP/scan for a sensitivity of 100% (24 nodules detected out of 24). If the sensitivity value is decreased to 87.5% (21 nodules detected out of 24), a rate of 0.9 FP/scan is obtained. These preliminary results are promising, albeit a validation against an independent validation database is required.

ACKNOWLEDGMENTS

We thank the *MAGIC-5 Collaboration (INFN, CSN-V)* for contributing to this research. We acknowledge Dr. L. Battolla, Dr. F. Falaschi and Dr. C. Spinelli of the U.O. Radiodiagnostica 2 dell'Azienda Ospedaliera Universitaria Pisana and Prof. D. Caramella and Dr. T. Tarantino of the Divisione di Radiologia Diagnostica e Interventistica del Dipartimento di Oncologia, Trapianti e Nuove Tecnologie in Medicina dell'Università di Pisa for providing the annotated database of CT scans. We are grateful to Dr M. Mattiuzzi from Bracco Imaging S.p.A. for useful discussions.

REFERENCES

Antonelli, M. et al. 2005, Segmentation and Reconstruction of the Lung Volume in CT Images, 20th *Annual ACM Symposium on Applied Computing, Santa Fe, New Mexico, March 13–17*, Vol I 255–259.

Delogu, P. et al 2005. Preprocessing methods for nodule detection in lung CT. *Computer Assisted Radiology and Surgery; Proc.* 19th *International Congress and Exhibition, Berlin, Germany. International Congress Series* 1281 1099–1104.

Diederich, S. et al. 2001. Detection of pulmonary nodules at spiral CT: comparison of maximum intensity projection sliding slabs and single-image reporting. *Eur. Radiol.* 11(8): 1345–50.

Gori, I. & Mattiuzzi, M. 2005. A method for coding pixels or voxels of a digital or digitalized image : Pixel Matrix Theory (PMT). *European Patent Application* 05425316.6.

Li, Q. et al. 2003. Selective enhancement filters for nodules, vessels, and airway walls in two- and three-dimensional CT scans. *Med. Phys.* 30(8): 2040–51.

Italung-CT trial. *http://www.med.unifi.it*

Computational Modelling of Objects Represented in Images –
João Manuel R.S. Tavares & R.M. Natal Jorge (eds)
© 2007 Taylor & Francis Group, London, ISBN 978-0-415-43349-5

Analyzing objects in images for estimating the delamination influence on load carrying capacity of composite laminates

L.M. Durão & A.G. Magalhães
ISEP – Instituto Superior de Engenharia do Porto, Portugal

João Manuel R.S. Tavres & A.T. Marques
FEUP – Faculdade de Engenharia da Universidade do Porto, Portugal

ABSTRACT: The use of fiber reinforced plastics has increased in the last decades due to their unique properties. Advantages of their use are related with low weight, high strength and stiffness. Drilling of composite plates can be carried out in conventional machinery with some adaptations. However, the presence of typical defects like delamination can affect mechanical properties of produced parts. In this paper delamination influence in bearing stress of drilled hybrid carbon+glass/epoxy quasi-isotropic plates is studied by analyzing objects in images. Conclusions show that damage minimization is an important mean to improve mechanical properties of the joint area of the plate and the appropriateness of the image processing techniques used.

1 INTRODUCTION

1.1 *Composites drilling*

The use of composite laminates in structures has enabled a considerable weight reduction and, consequently, an improvement in their dynamic characteristics. Although the early development of these materials has been related with aeronautical and aerospace usage, recent years have seen the spread of their application in many other industries like automotive, railway, naval, sporting goods and many others.

Their widespread use is yet limited by the cost and by the difficulties found during machining and joining of parts. Although composites components are produced to near-net shape, machining is often needed, as it turns out to fulfill requirements related with tolerances or assembly needs. As these parts need to be connected to others, drilling is often used for producing holes for bolts, rivets or screws.

Drilling is a machining operation that can be characterized by the existence of two effects: an extrusion or indentation caused the drill chisel edge that has null or very small linear speed and orthogonal cut exerted by the rotating cutting edges at a certain linear speed that is the result of tool diameter and rotational rate. In fact, the cutting action is more efficient at the outer regions of the cutting lips than near drill centre, Langella et al (2005).

As composites are neither homogeneous nor isotropic this operation raises specific problems that can be related with subsequent damage in the region around the hole. The most frequent defects caused by drilling are delamination, fiber pull-out, interlaminar cracking or thermal damages. Rapid tool wear, as a result of material abrasiveness, can also be an important factor in damage occurrence. These damages cannot be disregarded as they can result in a loss of static or fatigue strength of the part, Persson et al (1997). However, delamination has been considered as the most severe problem as it reduces the load carrying capacity of composite parts and so it needs to be avoided, Abrate (1997).

Delamination is a damage characterized by the separation of adjacent plies caused by an external action. It depends not only on fibre nature but also on resin type and respective adjacent properties. In drilling operations, delamination is a consequence of the indentation force exerted by the drill chisel edge stationary centre that acts more as a pierce than as a drill. The laminate under the drill tends to be drawn away from the upper plies, breaking the interlaminar bond in the region around the hole. As the drill approaches the end of the laminate, the uncut thickness becomes smaller and the resistance to deformation decreases. At some point, the loading exceeds the interlaminar bond strength and delamination occurs – figure 1.

Push-out delamination caused by this piercing action can be reduced if thrust force during drilling is minimized, Hocheng & Dharan (1990). The reduction of force can be accomplished by different ways

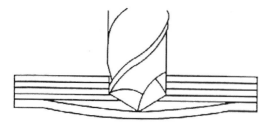

Figure 1. Delamination mechanism (Hocheng & Dharan, 1990).

and so several studies had been published with the aim to suggest alternative drilling methods to avoid delamination onset. Piquet et al (2000) listed some basic rules to be observed in drill geometry design. Tungsten carbide should be used as material, a rake angle of 6° will reduce peel-up delamination, a great number of cutting edges, up to six, will facilitate heat removal by increasing tool/part contact length. Park et al (1995) applied the helical-feed method to avoid fuzzing and delamination. Dharan & Won (2000) proposed an intelligent machining scheme to avoid delamination. Stone & Krishnamurthy (1996) studied the implementation of a neural thrust force controller that updates feed rate every three spindle revolutions. Tsao & Hocheng (2004) compared three different drills with varying parameters and concluded for the major importance of feed rate and drill diameter on delamination. The importance of feed rate was also established by Davim et al. (2003). A comprehensive summary of the steps towards free-delamination holes can be found in Hocheng & Tsao (2005).

Another possible approach is the execution of a pilot hole in order to reduce thrust force during drilling. The pilot hole reduces the chisel edge effect on the drill thrust force and, consequently, delamination hazard. Reduction of thrust force can reach 50%, according to the work of Won & Dharan (2002). Recently, Tsao & Hocheng (2003) also studied the effect of chisel length and pilot hole on delamination. According to these authors the pilot hole diameter should be around 15 to 20% of final drill diameter to minimize delamination risk.

Finally, the use of a sacrificial backup plate, if possible, can reduce delamination by providing a support for the uncut plies of a laminate.

1.2 Damage criteria

After hole completion it is necessary to define delamination criteria that allow the comparison of the damage caused by different drilling parameters. Remind that, due to the unique nature of composites, such comparison is only valid to plates fabricated according to

identical stacking sequence, same type of reinforcement fibre and identical fibre fraction. Two ratios were established for damage evaluation.

The first is *Delamination Factor* (F_d), Chen (1997), proposed as a quotient between the maximum delaminated diameter (D_{max}) and hole nominal diameter (D),

$$F_d = D_{max} / D \qquad (1)$$

The second is *Damage Ratio* (D_{RAT}), Mehta et al (1992). It was defined as the ratio of hole peripheral damage area (D_{MAR}) to nominal drilled hole area (A_{AVG}),

$$D_{RAT} = D_{MAR} / A_{AVG} \qquad (2)$$

Both criteria are based on the existence of digitized damage images obtained by radiography, ultrasound inspection or computerized tomography that must be analyzed using suitable image processing techniques, Tsao & Hocheng (2005). Those techniques can be pixel counting of digitized areas or direct measurement on the images or even on the part, if material is not opaque.

1.3 Bearing test

The main purpose of hole making in a plate is the possibility to assemble it to different parts in a structure. As parts will be subjected to efforts during service that will cause stress at the hole surrounding area, it is important for design engineers to know the load carrying capacity of the connection. That can be analyzed by a bearing test according to *ASTM D5961-01, procedure A*, which determines the load bearing response of multidirectional polymer composite laminates reinforced by high-modulus fibers. The results from this test, whose are believed to be affected by machined hole quality, will enable to compare different drilling options and determine the relative influence of delamination.

1.4 Hybrid laminates

The term hybrid refers to a composite that has more than one type of fibre or matrix in its construction. One of the main attractive when using hybrids is their synergy effect or 'hybrid effect'. A certain property, like tensile strength, will have a final value higher than the one predicted by the rule of mixtures. Hybrids present unique features that can be used to meet design requirements in a cost-effective way. Some of these advantages are balanced strength and stiffness, reduced weight or cost, improved fatigue or impact resistance and others, Schwartz (1988). Sometimes, a negative effect may also occur, Bader (1994).

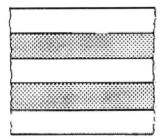

Figure 2. Interply hybrid (Bader, 1994).

Among the several type of hybrid composites, this work will deal with interply hybrids (figure 2), that consist of plies from two or more different unidirectional composites (carbon/epoxy, glass/epoxy) stacked in a specific sequence. This combination of carbon and glass fibres is referred as a good mix, as price and mechanical properties can be balanced according to designer needs.

2 EXPERIMENTAL WORK

2.1 Materials and tools

In order to perform the experimental work, a batch of hybrid composite plates using carbon/epoxy and glass/epoxy pre-preg materials was made. The stacking sequence – $[(0/-45/90/45)_C/(0/-45/90/45)_{3V}]_s$ – has the purpose of obtaining quasi-isotropic properties in the plates to be drilled and an optimization of characteristics combining glass fiber reinforced plies with carbon fiber reinforced plies. In fact, when comparing properties of plain glass/epoxy plates with these hybrid plates – table 1 – it is possible to observe that a percentage of 25% of carbon/epoxy plies has enabled an analogous increase of tensile strength and modulus and a great increment of flexural modulus, which is the most interesting feature. In order to have this increase in flexural modulus it is important that the material with higher flexural strength, carbon/epoxy in this case, is located at the outer plies of the laminate. The opposite option, that is to say with carbon/epoxy plies in the middle of the laminate, will be better for delamination reduction, keeping the same tensile properties but the increase of flexural modulus will be merely 10%.

All the plates used for comparison had a stacking sequence with the purpose of obtaining quasi-isotropic properties, so identical as the one described above but using one only type of reinforcement fibre. The laminates were cured under a pressure of 3 kPa and a temperature of 140°C for one hour in a hot plate press and air cooled. The final thickness of all plates was 4 mm.

Drilling experiments were executed in a vertical machining centre with 6 mm diameter drills. Initial cutting parameters were selected according to tools

Table 1. Properties of quasi-isotropic plates.

Material	R_m MPa	E GPa	ν	E_{fl} GPa
Gl/Ep	394	21.8	0.53	25.7
Hybrid	454	27.8	0.49	43.2
Carbon/Ep	771	49.9	0.51	130

Table 2. Machining parameters.

Tool	Cutting speed m/min	Feed rate mm/rev
HSS twist	20	0.09
Carbide twist	100	0.025
Carbide Brad	80	0.025
Carbide Dagger	38	0.05
Carbide special step	53	0.025

manufacturer advice and later confirmed or changed according to experimental results. Finally a 'best set' was selected for each tool used. Five types of drills and two tool materials were experimented for comparison: twist drill in HSS and tungsten carbide, brad, dagger and special step drill all in tungsten carbide only. Machining parameters are shown in table 2.

Twist drill is a standard geometry tool. Associated with the use of the twist drill, a pilot hole of 1.1 mm diameter was performed. The intention of pilot hole was to reduce the maximum thrust force achieved and decrease delamination around the hole by cancelling the chisel edge effect of the drill. The use of HSS has the purpose of confirming the general idea that this material is not adequate for fibre reinforced composites drilling. Brad drill is a special edged drill firstly designed for cutting wood, with edges in scythe shape, that causes the tensioning of the fibres prior to cut, thus enabling a 'clean cut' and a smooth machined surface. Dagger drills are designed for fibre reinforced laminates and had a very sharp tip – 30° – in order to reduce the indentation effect. Due to its particular geometry it needs to have enough space at the exit side of the plate, which sometimes is not possible, limiting its application in field work. The special step drill follows a suggestion from Dharan (2000) and was developed in the aim of a dissertation, Durão (2005). This special tool has two drilling diameters – 1.25 and 6 mm – dividing the drilling operation, and consequently the thrust force, in two stages. This division of the drilling operation also cancels the chisel edge effect for the final hole diameter drilling. The diameter transition has a conical shape, for a soft transition. The reduction of delamination risk by reducing the maximum thrust force is also looked for. Another advantage of

Figure 3. A radiography of a drilled part.

this tool is the possibility of executing the hole in one operation only.

Axial thrust force (F_z) and torque (M_t) during drilling were monitored with a *Kistler 4782* dynamometer associated to a multichannel amplifier and a personal computer for data collection. All the plates were drilled without the use of a sacrificial plate.

2.2 Delamination measurement

Delamination extension is not possible to be measured by a visual inspection as carbon/epoxy plates are opaque. So, the plates need to be inspected by enhanced radiography. Delamination can be detected using a contrasting fluid, like di-iodomethane. For that it is necessary to immerse the plate in the contrasting fluid for one and a half hour in a dark chamber. After time elapsed, plates are cleaned and radiographed. The final result of a developed film can be seen at figure 3.

The resulting images were then processed using a previously developed image analysis and processing platform, with the help of Computational Vision techniques. The use of Computational Vision has the purpose to reckon from the images obtained by radiography, information regarding damaged area or diameter. This process has the advantage of reducing operator dependence to measure the dimensions wanted, thus increasing results reliability. An existing processing and image analysis platform was used, Tavares (2000), Tavares (2002). This platform turned possible the use of some standard Computational Vision techniques, Awcock (1995), Jain (1995), Schalkoff (1989) of image processing like filtering, segmentation and region analysis.

For radiographic images, the first step is the manual selection of the interest zone, in order to reduce computer processing time. Then it is necessary to pre-process the sub image by using smoothing filters to reduce sudden changes of intensity, thus noise. Then, the smoothed image is segmented in interest areas. As this step can result in a large number of segmented areas it is followed by the elimination of those who are nor relevant. Noise areas are eliminated by the application of erosion and dilation morphologic filters. After this step, each image is composed of the damaged area and its background. The final step is the use of a region processing and analysis algorithm to

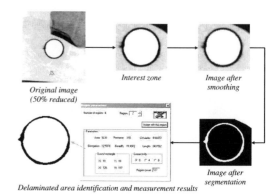

Original image (50% reduced) *Interest zone* *Image after smoothing*

Delaminated area identification and measurement results *Image after segmentation*

Figure 4. Example of the determination of damage measurements in an radiography image.

distinguish the several regions of each image. For each region the required measurements are made.

In the end it was possible to obtain the dimensions judged as necessary in order to have a damage evaluation according to equations (1) or (2). Results considering the two criteria mentioned were determined using the results given by this technique. In figure 4 some steps of the processing of a radiographic image are presented.

2.3 Bearing test

The last phase of experimental work was the execution of the bearing test. The main purpose of hole making in a plate is to assemble it to other parts in a structure. It is important to know what is the load carrying capacity of a plate. This test is designed to determine the bearing response of multidirectional polymer matrix reinforced composite laminates and it follows the orientations determined in *ASTM D5961-01* standard test method. This procedure provides data for compare the effect of several machining methods used in this work. The tests were carried out at an *INSTRON 4208* with a speed of 2 mm/min. Deformation data was collected by two independent systems: one connected to the test machine and other from two LVDT installed at the to test jig and connected to the plate. In order to have a valid test, a maximum load followed by a 30% decrease should be noticed, before shear out of the plate. Valid test results were then plotted against data regarding delamination by the two criteria referred.

3 RESULTS

The results considered in this work were the maximum thrust force and torque during drilling, evaluation of delamination based on the criteria referred as *Delamination Factor*, equations (1), and *Damage Ratio*,

Table 3. Drill geometry comparison results.

Tool	Maximum thrust force N	Maximum torque Nm	Bearing stress MPa
HSS twist	104	0.22	709
Carbide twist	28	0.34	748
Carbide Brad	42	0.16	733
Carbide Dagger	52	0.31	728
Special step	43	0.49	764

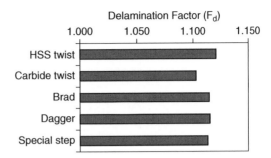

Figure 5. Comparison of delamination factor for several tools.

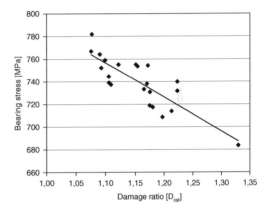

Figure 6. Bearing stress to damage ratio correlation.

equation (2) and finally by the bearing stress test values.

Thrust force and torque data obtained during drilling are presented in table 3. In the same table the results of the bearing stress for each tool are shown. These values are the average of five tests under identical experimental conditions for each tool geometry involved in this study.

From the results presented it is possible to observe that there is a substantial difference when changing from HSS to carbide twist drills. The thrust force has a strong reduction and the bearing stress of the plate notably increases. Although some of these effects can be explained by the use of different cutting parameters, it must be reminded that those were optimized for each drill geometry and material prior to the experimental work here described. Comparing the results of thrust force with bearing stress, it can be said that a reduction of thrust force during drilling, reducing the risk of delamination onset, has resulted in higher values of bearing stress. Clearly, drill geometry has a definitive influence on machined hole quality, affecting properties of plates to assemble in a structure. It was evident that the higher values of bearing stress correspond to the plates with lower delamination. A 14% difference was found when comparing the better individual value of bearing stress with the lower one. When considering average values, this difference comes to 7% between special step and HSS twist drill, that has always the worst results.

Torque values are always low for all drills considered and no influence on delamination or other damage was identified as relevant.

After drilling, the plates were radiographed and the images processed as described in 2.2. The results from the application of Computational Vision techniques are presented in figure 5, considering *Delamination Factor* criteria and a nominal hole diameter of 6 mm.

A relation connecting damage ratio and bearing stress was established from the data collected during experimental work. This relation can be seen in figure 6, with a correlation factor higher than 0.8. A second relation linking delamination factor to bearing stress was also set up, although the correlation factor was lower (around 0.5). In both cases a linear

correlation was tried, returning a line with negative slope. This slope confirms that delamination is an important factor in load carrying capacity of a laminate plate with assembly by screws, rivets or bolts.

From the data in figure 6, and the results of table 3, it is possible to say that higher thrust forces during drilling lead to higher delamination around the hole and, consequently, a decrease in plate mechanical strength, given by bearing stress test results.

4 CONCLUSIONS

Five different tools for drilling hybrid laminates were compared, one of them in HSS and the others in tungsten carbide. These four carbide drills had different geometries: twist, brad, dagger and a special step design. With such purpose a laminate with two types of reinforcement – carbon and glass – fibres in an epoxy matrix was drilled using the tools mentioned. Each drill was used at the cutting parameters considered as 'best set'. Forces were monitored during drilling, delamination evaluated through radiography associated with an analysis and processing platform.

173

Finally, drilled laminate strength was measured by a bearing test.

Based on the experimental work presented it is possible to draw some conclusions.

A correct choice of cutting speed and feed rate will reduce delamination. This damage can be evaluated by a non-destructive test like enhanced radiography.

From the drills experimented, HSS twist drill had the lower results, showing the inadequacy of the use of this material in drilling tools for fiber reinforced plastics.

Carbide twist drill had the best results for thrust force and *Delamination Factor*, but bearing stress test results were lower by about 2% from the best.

Special step drilled plates had the higher results of bearing stress, although these plates do not have the lower delamination factor value.

The image processing techniques used showed to be adequate to analyze the image objects involved. Their use can be easily extended to other materials that are suitable to be radiographed, C-scanned or examined by any other imaging technique.

Delamination extension can be correlated with bearing stress showing that higher delamination has a correspondence with lower bearing stresses.

The relationships vary if a different composite laminate – material or stacking sequence – is used. This has to be present when designing a structure using fibre reinforced laminates.

REFERENCES

Abrate, S., 1997, Machining of Composite Materials, *Composites Engineering Handbook*, Ed. P. K. Mallick, Marcel Dekker, New York, pp. 777–809.

Awcock, G. W., Thomas, R., 1995, *Applied image processing*, McGRAW-HILL International Editions, New York.

Bader, M.G., 1994, Hybrid effect, *Handbook of Polymer-fibre Composites*, Longman Scientific Technical, 225–230.

Chen, W. C., 1997, Some experimental investigations in the drilling of carbon fibre-reinforced plastic (CFRP) composite laminates, *Int. J. of Machine Tools and Manufacture*, 37, 1097–1108.

Davim, J. P., Reis, P., 2003, Furação de laminados epóxidos reforçados a fibras de carbono (CFRP): relação entre as forças de corte e o factor de delaminagem, *VI Congr. Ibero-Americano de Engenharia Mecânica*, Coimbra, pp. 1079–1084.

Dharan, C. K. H., 2000, *Conference on composites machining*, INEGI-PORTO.

Dharan, C. H. K., Won, M. S., 2000, Machining parameters for an intelligent machining system for composite laminates, *Int. J. of Machine Tools and Manufacture*, 39, pp. 415–426.

Durão, L. M. P., 2005, *Machining of Hybrid Composites*, PhD. thesis, FEUP, Porto.

Hocheng, H., Dharan, C. K. H., 1990, Delamination during drilling in composite laminates, *J. of Engineering for Industry*, 112, 236–239.

Hocheng, H., Tsao, C.C., 2005, The path towards delamination-free drilling of composite materials, *J. of Materials Processing Technology*, 167, p. 251–264.

Jain, R., Kasturi, R., Schunck, B. G., 1995, *Machine Vision*, McGRAW-HILL International Editions, New York.

Langella, A., Nele, L., Maio, A., 2005, A torque and thrust prediction model for drilling of composite materials, *Composites A*, 36, p. 83–93.

Mehta, M., Reinhart, T. J., Soni, A. H., 1992, Effect of fastener hole drilling anomalies on structural integrity of PMR-15/Gr composite laminates, *Proc. of the Machining Composite Materials Symposium, ASM Materials Week*, 113–126.

Park, K. Y., Choi, J. H., Lee, D. G., 1995, Delamination-free and high efficiency drilling of carbon fiber reinforced plastics, *J. of Composite Materials*, 29, pp. 1988–2002.

Persson, E., Eriksson, I., Zackrisson, L., 1997 Effects of hole machining defects on strength and fatigue life of composite laminates, *Composites A*, 28, 141–151.

Piquet, R., Ferret, B., Lachaud, F., Swider, P., 2000, Experimental analysis of drilling damage in thin carbon/epoxy plate using special drills, *Composites A*, 31, p. 1107–1115.

Schalkoff, R. J., 1989, *Digital image processing and computer vision*, John Willey & Sons, Inc.

Schwartz, M. M., 1988, *Composite Materials Handbook*, McGraw Hill.

Stone, R., Krishnamurthy, K., 1996, A Neural Network Thrust Force Controller to Minimize Delamination During Drilling of Graphite-Epoxy Composites, *Int. J. Machine Tools and Manufacture*, 36,pp. 985–1003.

Tavares, J. M. R. S., 2000, Tese de Doutoramento: *"Análise de Movimento de Corpos Deformáveis usando Visão Computacional"*, FEUP.

Tavares, J. M. R. S., Barbosa, J. G., Padilha, A. J., 2002, Apresentação de um Banco de Desenvolvimento e Ensaio para Objectos Deformáveis, *RESI – Revista Electrónica de Sistemas de Informação*, vol. 1.

Tsao, C. C., Hocheng, H., 2003, The effect of chisel length and associated pilot hole on delamination when drilling composite materials, *Int. J. of Machine Tools and Manufacture* 43, pp. 1087–1092.

Tsao, C. C., Hocheng, H., 2004, Taguchi analysis of delamination associated with various drill bits in drilling of composite material, *Int. J. of Machine Tools and Manufacture*, 44, 1085–1090.

Tsao, C. C., Hocheng, H., 2005, Computerized tomography and C-Scan for measuring delamination in the drilling of composite materials using various drills, *Int. J. of Machine Tools and Manufacture*, 45, 1282–1287.

Won, M. S., Dharan, C. H. K., 2002, Chisel edge and pilot hole effects in drilling composite laminates, *Trans. of ASME J. of Manufacturing Science and Engineering*, 124, pp. 242–247.

Computational Modelling of Objects Represented in Images –
João Manuel R.S. Tavares & R.M. Natal Jorge (eds)
© 2007 Taylor & Francis Group, London, ISBN 978-0-415-43349-5

Decimation and smoothing of triangular meshes based on curvature from the polyhedral Gauss map

G. Echeverria & L. Alboul

Materials and Engineering Research Institute, Sheffield Hallam University, Sheffield, UK

ABSTRACT: This paper presents an improvement on methods to simplify, de–noise and identify feature regions of a triangular mesh, based on curvature estimations. Computations of curvature are obtained from the polyhedral Gauss Maps of the individual vertices. This allows identification of positive and negative curvature components, thus determining the Total Absolute Curvature (TAC) of a vertex. Using this new measure the curvature computed is more reliable and provides more information on the features of the polyhedral surface. The TAC of a region is obtained by summing the TAC's of the vertices in the region, which also provides identification of regions of similar curvature on a mesh model. In order to use this information for mesh simplification we introduce the *weighted total absolute curvature* measure, abbreviated as WTAC, which takes into consideration not only the curvature but also the area of the region, normalised in a specific way. We apply then the triangle decimation of a mesh, by removing vertices with the smallest WTAC. The decimation algorithm automatically performs also a smoothing (de–noising) operation by deleting outliers. By setting thresholds on the WTAC one can obtain various degrees of decimation (smoothness). All operations are linear with respect to the number of vertices in the mesh.

Keywords: Mesh decimation, Curvature, Polyhedral Gauss Map, Spherical geometry.

1 INTRODUCTION

The most common representation for an object in computer aided applications, specially in 3D graphics, is still the polygonal mesh, where a collection of points in space are joined together by edges and faces to form a shape. This provides great flexibility in representing a variety of shapes, from simple to very complex ones, and in varying degrees of detail. Modern methods for generating 3D objects involve high definition scanning of data, providing very detailed meshes of the target objects. Often this produces an enormous amount of data, which is difficult to handle by current computer systems for application in real time. Meanwhile, the data acquired with these methods may be redundant or excessive to portray the shape of the object.

For these reasons, it is desirable to simplify the polygonal geometry in a model. Polygon simplification is the process of decreasing the number of components from a mesh, so that it will be less demanding of computational resources, while at the same time still providing a good visual representation of the original object. In this paper we will address two types of polygonal simplification: decimation and de–noising.

The main objective of a decimation algorithm is to remove the elements of the mesh with the least impact on the shape of the object being portrayed. Several methods exist to accomplish this (Cignoni et al. 1998), (Luebke 2001), with emphasis on removing edges or vertices. De–noising is similar to decimation, but it removes outliers, i.e. features that are possibly prominent, but do not reflect the real shape of the object.

The curvature of an object is a good measure of the behaviour of the shape, and thus an important characteristic to consider when modifying a mesh model. Curvature can be calculated for the whole object, specific regions, or individual vertices or edges. The curvature of a vertex is a measure of how far the vertex is 'pulled away' from a plane. It is a good estimate of how relevant is the vertex for the general shape of the object.

The method of polygonal simplification, proposed here, concerns only vertex removing, while keeping remaining vertices in place, and is based on the computation of the curvature of a vertex, called the Total Absolute Curvature, abbreviated as TAC.

The Total Absolute Curvature is capable of identifying even hidden features of a vertex, not recognised by other methods. Thus it is ideal for correctly evaluating the importance of a vertex for the shape of the mesh.

The Total Absolute Curvature of a mesh is obtained by summing the TAC's of all vertices and gives a complete characterisation of the shape of a polygonal surface, formed by the mesh. The algorithm is fast and works for all kinds of vertices, including self-intersecting ones. It allows to describe the shape of polyhedral surfaces of various complexity.

The *angle deficit*, or *discrete Gaussian curvature*, a common estimation of the curvature of polygonal meshes, does not fully reflect the local shapes and therefore is not an ideal candidate for decimation based on curvature.

The paper (Alboul and Echeverria 2005) provides a detailed theoretical background of both curvature measures and provides more references. The method to compute and visualise the Total Absolute Curvature for complex polyhedral surfaces is also presented in more detail there.

In order to perform the simplification of the mesh, we introduce a new measure, called the *weighted total absolute curvature* and denoted as WTAC, as is takes into consideration not only the TAC of a vertex but also the area of the surrounding region, normalised in a sound way. This measure weighs the complexity of the vertex with the visual impact it has on the overall surface of the object. It also filters out some degenerate cases that may occur.

The WTAC measure can be applied to existing polygon simplification algorithms, based on decimation or edge collapse, to improve their selection of the vertices to simplify. Here it is applied to a simple decimation program to show its advantages. Having WTAC measures for each vertex, it is straightforward to simplify a mesh by removing the vertices with the smallest WTAC measure.

The measure proposed to drive the decimation will also automatically remove noise from the model. Because of the choice made for the normalisation of the areas, very outstanding vertices, like those present in a model due to noise, will produce a very small WTAC, and make these vertices be selected for removal during the decimation process.

After applying WTAC measure to the mesh and removing a corresponding vertex with all adjacent edges, the obtained 'hole' is then re-triangulated. As our aim is not to increase the overall TAC of the simplified mesh with respect to the initial one, among all possible triangulations of the hole the one with the smallest TAC is chosen. This is performed with the aid of the optimisation procedure based on edge swapping (see, for example, (Alboul 2003)).

Besides the theoretical considerations, we show also experimentally the advantage of the use of TAC for decimation in comparison with the methods based on the angle deficit. We also discuss properties of the decimation method based on the WTAC measure in comparison with other decimation methods.

2 PREVIOUS RESEARCH ON POLYGON SIMPLIFICATION

Extensive research has already been done in the field. There are several different approaches to simplifying a polygon mesh. (Cignoni et al. 1998) and (Luebke 2001) have written papers reviewing the most important of these methods, and comparing their advantages and shortcomings.

Cignoni et al. do a comparison of the error difference, from the original models to the simplified ones, and the time taken to obtain the results, using several different algorithms. To do this they created the 'Metro' tool (Cignoni et al. 1998), which has since been used by others to compare their algorithms with existing ones.

As explained in these papers, simplification techniques vary in: *optimisation goal*, which can be size minimisation given an error bound, or error minimisation given a target object size; *local* or *global optimisation*; *preservation of the original object's topology*; *maintaining the original vertices* or *re–meshing the model*.

Some simplification methods also perform 'view dependant' decimation. That is, decimate a model more in areas far away or hidden from the current perspective. This permits a larger reduction of the mesh, but needs to be recalculated if the view direction changes.

The algorithms tested include Mesh Decimation, Simplification Envelopes, Multiresolution Decimation, Mesh Optimization, Progressive Meshes and Quadric Error Metric Simplification. In the tests, Mesh Decimation produces the largest error with respect to the original model, although it is by far the fastest algorithm. Other techniques perform better in most respects, specifically Quadric Error Matrix (QEM) (Garland and Heckbert 1997), which is generally considered as a very effective simplification method and used as a parameter of comparison for newer techniques.

The first method for decimation was proposed by (Schroeder et al. 1992), and is based on progressive removal of specific vertices from a mesh. All vertices are classified according to their local topology, and are handled accordingly. Vertices are labelled as 'simple', 'complex' or 'boundary'. Complex ones are generally non-manifold vertices, and are left untouched. Simple and boundary vertices are treated differently. Simple vertices are further classified according to the number of feature edges connected to the vertex. Vertices are selected for decimation according to a distance metric. For simple vertices the parameter is the distance from an average plane; for border vertices, it is the distance to the line connecting the neighbours along the border. The vertex with the shortest distance is selected for removal, along with the triangles surrounding it. The hole produced in the mesh is filled using a recursive splitting of the remaining region into triangles. This method generates a subset of the original vertices, not adding any new ones. It also preserves the topology of the object.

In (Kim et al. 2002) a measure based on curvature is used to assign a cost to the edges, and to select the ones to be collapsed. The authors make use of both the gaussian and mean curvatures on the mesh. After deciding on an edge to collapse, a new vertex is generated in place of the two edge vertices. The location of this new vertex is found using a butterfly subdivision mask. Using curvature to decide on the geometry to eliminate, they prove that important features of the object are preserved after heavy simplification.

(Hussain et al. 2004) propose a simplification driven by half-edge collapses that keep at least one of the vertices of the edge removed. They use a metric based on the angle difference from the original faces to the ones that will be created after the collapse; effectively an approximation of the curvature of the region, although not very accurate, but useful again in preserving important geometry. This implementation competes in performance with QEM but claims to require less memory to store data.

Our method can be considered, to a certain extent, as a generalisation of the one proposed by Schroeder et al. It is also a novel contribution to the selection of vertices, based on curvature measures, used to drive decimation methods.

3 DESCRIPTION OF THE POLYHEDRAL GAUSS MAP

The method employed to compute the curvature of a polyhedral surface is based on the area of the polyhedral Gauss Map. If V is the set of all the vertices in a mesh, for any individual vertex $v \in V$, we define the *star of the vertex* as the set of faces immediately surrounding v, ordered in counter clockwise direction, and denote it as *star(v)*.

The Polyhedral Gauss Map is computed individually for each vertex v in the triangular mesh using the normal vectors of the faces in *star(v)*. These vectors are placed in the centre of a sphere and joined in order by geodesic arcs, creating what is called the *spherical indicatrix* (see Figure 1).

In many cases the arcs in the indicatrix will intersect each other. It is necessary to identify where these intersections occur and separate them into individual non-intersecting spherical polygons, which might be overlapping. For instance, Figure 2 shows a vertex whose star is a non-convex cone. In this case, the concavity causes the normals of the faces to go in clockwise direction (opposite to the direction of the

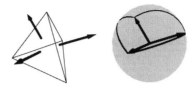

Figure 1. Using the normal vectors of faces in *star(v)* to create a Gauss Map.

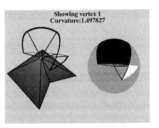

Figure 2. A vertex with non-convex neighbourhood (showing the indicatrix), and its corresponding Gauss Map.

faces), and this in turn produces a self intersection of the arcs. Separating this polygons produces two areas of opposing orientation.

Each individual polygon splits the surface of the sphere into two areas, one of positive orientation and one negative. Only one of these two areas corresponds to the vertex curvature, and it is chosen based on the characteristics of the vertex and its star. Positive polygons result from normal vectors that go in the same direction as their corresponding faces. If they go in the opposite direction, the area represents a negative curvature. Once the correct orientation has been determined for each of the spherical polygons, the curvature is computed as the sum of their areas.

Adding together the absolute values of the positive and negative curvatures, the result is what is called the Total Absolute Curvature.

The Polyhedral Gauss Map obtained gives a complete characterisation of the vertex; it identifies all folds in its star, thus emphasising 'hidden' curvature regions. It works also for self-intersecting vertices, and determines all types of self-intersections. Therefore, all irregularities of the mesh can be easily determined based on the Gauss Map only, without computationally expensive checking on the mesh itself.

A detailed description of the method to compute the TAC and visualise the Polyhedral Gauss Map can be found in (Alboul and Echeverria 2005).

4 ADVANTAGES OF USING TAC

In most cases, curvature of a vertex is measured simply as the angle deficit. This is also known as the (discrete)

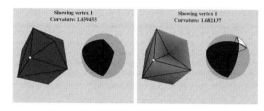

Figure 3. Gauss Map of a vertex in two different objects. Left: convex vertex (only positive curvature). Right: non-convex vertex (positive and negative curvature components).

Gaussian curvature. It normally identifies only two kinds of vertices: convex or saddle. Vertices which do not fit in these categories will present problems to the algorithm, and their true curvature can not be found. This is the case, for example, when a non-saddle vertex has concavities (folds) in its star.

The TAC provides a better understanding of the geometry of an object than the Gaussian curvature. It can identify more detailed features of the shape such as hidden regions of negative curvature that exist in the concavities of a cone and appear only when analysing both the positive and negative components of the curvature of the vertices.

The TAC algorithm can also handle non-manifold vertices, correctly identifying the curvature of vertices with self intersecting faces. These vertices will produce larger than normal areas in the Gauss Map.

The algorithm is fast as all computations are linear with respect to the number of vertices in a given mesh. All the required information about each individual vertex, such as the angle deficit, curvature, triangles in $star(v)$, etc., is obtained on the first pass, and stored in data structures.

While doing the computation of the Curvature of each vertex, other information is gathered along the way, such as the projection of the vertex star, computation of mean curvature, which is concentrated on the edges, and the type and number of self-intersections if any exist.

The Gauss Maps for two different vertices are shown in Figure 3. While angle deficit methods would conclude that the two vertices are equal, using our method it is immediately evident that both vertices have distinct curvature.

5 CURVATURE AS DECIMATION MEASURE

Below we introduce a new decimation measure based on curvature, which we call the *weighted total absolute curvature*, denoted as WTAC, and give some justifications for its choice.

For any vertex v, as well as the star $star(v)$, we also define $neighbours(v)$ as the set of vertices in $star(v)$ other than v.

N_v is an artificial vector, computed as the average of the sum of normals and areas of the incident triangles. It can be defined to be the normal to the vertex.

Then, the *projected star* of v is the polygon P_v obtained by perpendicularly projecting the vertices in $neighbours(v)$ on a plane that passes through v and is specified by the vector N_v. The projected star can also be defined for saddle-like vertices; in this case the vertices will be projected from both half-spaces determined by the plane. The area of the projected polygon is calculated, and then divided by the area of the faces in $star(v)$ to normalise it. This value is then multiplied by the absolute curvature of the vertex, to become the WTAC measure of the decimation.

Next we define T_v as the list of triangles in $star(v)$ of length n, so that $t_i \in T_v$, then the normalised area of v, N–area$_v$, is given by the expression:

$$\text{N–area}_v = \frac{area(P_v)}{\sum_{i=1}^{n} area(t_i)}$$

The WTAC of the vertex v is defined then as:

$$W(v) = TAC_v * \text{N–area}_v \qquad (1)$$

The advantage of using this parameter to drive the decimation is that it takes into account the complexity of the vertex and its visual importance in terms of the area.

The use of the N–area also ensures that vertices with similar stars at different scales will be treated in the same way. Otherwise using un–normalised area would make larger structures more important, reducing the relevance of the curvature.

Using both the curvature and the area also automatically takes care of degenerate cases, such as spikes, that normally appear on polygon meshes due to noise during the capturing process. For this kind of vertex the sum of the areas of the triangles t_i will be much larger than the area of the P_v, producing a very small N–area$_v$, and hence WTAC. This also eliminates non-manifold vertices if they appear during the polygon simplification. For example, degenerate vertices with glued triangles, as they will have a projected area equal to zero.

6 IMPLEMENTATION

For the purpose of testing the curvature metric on simplification, a simple triangle decimation algorithm is used, similar to the one used by Schroeder et al.

The algorithm makes several passes over the whole dataset, removing each time the vertex with the smallest weight. It does local minimisation of the curvature at each vertex. This preserves topology but imposes a limit on the amount of simplification that can be

Figure 4. Comparison of the triangulations obtained after removal of a saddle vertex. Left: Original vertex (TAC = 13.817166); Middle: Non-convex triangulation (TAC = 12.566371); Right: Convex triangulation (TAC = 14.177234).

applied to a model. Our implementation makes no extra attempt to correct degenerated cases outside of those inherently avoided by the use of WTAC.

The process of decimation is as follows:

The model to decimate is read, and lists are created for the vertices and faces. The lists are then converted into arrays that make reference to each other by their indices. This simplifies the search for specific vertices or triangles.

Initially, the TAC is computed for all vertices in the mesh, obtaining along the way all other required information for each vertex: angle deficit, list of triangles in the star, list of neighbours, artificial vector normal to the vertex, projection of the neighbours and feature edges. All of these are stored in an array of data structures.

A vertex v is selected for removal if it has the smallest value for the WTAC measure. For the special case of vertices on the border of an open object, the parameter used for selection also considers the angle deficit, since the Gauss Map area in these vertices can be equal to zero.

The vertex v and all the triangles $t_i \in star(v)$ are marked as deleted. Then, for each $v_j \in neighbours(v)$, the triangles t_i are removed from the lists of their corresponding stars.

The hole left in the place of $star(v)$ has to be triangulated. The initial triangulation is done on the projection of the star using an algorithm for triangulation of planar polygons. The result is then translated to the coordinates of the corresponding neighbours, to create new triangles on the mesh. The curvature for all v_j is recomputed using their new stars.

A list of edges is created for the new triangulation, keeping only the edges not on the border of the polygon. Each edge structure contains the vertices that define it, the two neighbour triangles and the two opposite vertices. Edge flipping is then used to improve on the initial triangulation, applying as a parameter the TAC of the 4 vertices directly affected by an edge. An edge will be flipped if the resulting configuration will minimise the sum of the TAC of

Figure 5. Venus model: shaded (top), wireframe (middle), curvature coded (bottom). Left: original (1418 triangles). Right: after 400 steps of decimation (618 triangles).

the neighbourhood. An example of vertex removal is given in Figure 4.

Once a final triangulation has been obtained that does not create degeneracies, for every vertex v_j the curvature and areas are updated according to the new triangulation, and the new faces are added to the appropriate lists.

7 EXPERIMENTAL RESULTS

Figure 5 shows the results of using curvature to decimate a mesh. Here regions of similar curvature are shown black for positive curvature, white for negative, and grey for mixed (both positive and negative) curvature. Using the same model, we also do a comparison of the overall TAC and the total surface areas, after various levels of decimation. Curvature and area both decrease with the loss of geometry, but curvature is more affected, because it is more sensitive to the change in the shape of the object.

Decimation (%)	Triangles	Curvature	Area
0	1418	80.120426	82096.70
28	1018	78.099026	82102.32
55	618	71.517657	81704.49
85	218	55.130329	81170.04

179

To evaluate the benefit of using TAC as the curvature measure instead of the angle deficit, the same decimation algorithm was applied to some models, using both curvature measures. In both cases the formula to compute the weight for the vertices is the same, but substituting angle deficit for TAC in Equation 1. The resulting simplified models were compared using the 'Metro' tool, using the default parameters. In the table we refer to the distances as going from the original to the decimated model ($O \rightarrow D$), or from the decimated to the original ($D \rightarrow O$). The parameters compared are the mean and maximum distances, as well as the Hausdorff distance.

	TAC		Anle deficit	
	28%	55%	28%	55%
Faces	1018	638	1018	638
Mean $O \rightarrow D$	0.13440	0.43985	0.18260	0.55382
Max $O \rightarrow D$	2.34320	8.80321	8.07986	8.80321
Mean $D \rightarrow O$	0.11785	0.39895	0.21156	0.53477
Max $D \rightarrow O$	2.32480	5.45336	14.83785	14.83785
Hausdorff	2.34320	8.80321	14.83785	14.83785

As can be seen from the distances measured, the use of TAC instead of angle deficit as an estimation of curvature produces a smaller error, and the difference in performance increases as the decimation goes further. We have not done tests to compare the time difference to do similar levels of decimation using both measures. Obtaining the TAC is more complex than angle deficit, but still has linear complexity, once all the information required has been extracted from the model and stored in appropriate data structures.

8 FURTHER WORK

At the moment the algorithm works progressively with vertices. The next step is to move from vertices to regions, by grouping together regions of similar curvature and removing all vertices within that region

if it has minimal WTAC. The Gauss Map allows us identification of such regions.

We also need to evaluate alternative approaches to optimise the re-triangulation of holes.

For any vertex, it is possible that the projection of the star can have self intersections, mainly in the case of saddle vertices. This can result in new triangles being inserted inside of the object, thus creating non-manifold vertices. Another effect of self intersections is that the computation of the polygon area will be incorrect. A possible solution to this problem is to do the projection of the vertex star on a sphere, and use the same techniques employed to obtain the Gauss Map to triangulate the polygon, and calculate its area.

REFERENCES

Alboul, L. (2003). Optimising triangulated polyhedral surfaces with self-intersections. In M. J. Wilson and R. R. Martin (Eds.), *IMA Conference on the Mathematics of Surfaces*, Volume 2768 of *Lecture Notes in Computer Science*, pp. 48–72. Springer.

Alboul, L. and G. Echeverria (2005). Polyhedral gauss maps and curvature characterisation of triangle meshes. In R. R. Martin, H. E. Bez, and M. A. Sabin (Eds.), *IMA Conference on the Mathematics of Surfaces*, Volume 3604 of *Lecture Notes in Computer Science*, pp. 14–33. Springer.

Cignoni, P., C. Montani, and R. Scopigno (1998, February). A comparison of mesh simplification algorithms. *Computers & Graphics 22*(1), 37–54. ISSN 0097–8493.

Cignoni, P., C. Rocchini, and R. Scopigno (1998). Metro: Measuring error on simplified surfaces. *Computer Graphics Forum 17*(2), 167–174. ISSN 1067–7055.

Garland, M. and P. S. Heckbert (1997, August). Surface simplification using quadric error metrics. *Computer Graphics 31*(Annual Conference Series), 209–216.

Hussain, M., Y. Okada, and K. Niijima (2004, February). Efficient and feature-preserving triangular mesh decimation. In V. Skala (Ed.), *Journal of WSCG*, Volume 12, Plzen, Czech Republic. UNION Agency – Science Press.

Kim, S.-J., C.-H. Kim, and D. Levin (2002, October). Surface simplification using a discrete curvature norm. *Computers and Graphics 26*(5), 657–663.

Luebke, D. P. (2001, May/June). A developer's survey of polygonal simplification algorithms. *IEEE Computer Graphics and Applications 21*(3), 24–35.

Schroeder, W. J., J. A. Zarge, and W. E. Lorensen (1992, July). Decimation of triangle meshes. In E. E. Catmull (Ed.), *Computer Graphics (SIGGRAPH '92 Proceedings)*, Volume 26, pp. 65–70.

Computational Modelling of Objects Represented in Images –
João Manuel R.S. Tavares & R.M. Natal Jorge (eds)
© 2007 Taylor & Francis Group, London, ISBN 978-0-415-43349-5

3D reconstruction of the middle ear for FEM simulation

Fátima Alexandre, Fernanda Gentil, Casimiro Milheiro & P. Martins
IDMEC-polo FEUP, Faculty of Engineering of University of Porto, Portugal

A.A. Fernandes, R. Natal Jorge, A.J.M. Ferreira & M.P.L. Parente
Faculty of Engineering of University of Porto, Portugal

Teresa Mascarenhas
Faculty of Medicine of University of Porto, Portugal

ABSTRACT: For a better knowledge of the human middle ear functions we conducted dynamic simulations studies to obtain its real mechanical behavior. The aim of this work is to create a finite element (FE) model for the human middle ear. The analysis and evaluation of the bone structures of the middle ear is a hard task when based on images obtained from the Computed Tomography (CT). All images obtained from CT were saved in DICOM (Digital Imaging and Communications in Medicine) format.

In this study we tried to use the ANALYZE software, but as the DICOM files used in this study contained information of an entire human head, they didn't have the necessary resolution of the structures of the human middle ear, and therefore, the results obtained for the reconstruction, using this software were not satisfactory. The human middle ear contains the smallest human bones, for example the malleus is only 9 mm length. CAD software, was used to solve this problem and the images were segmented to construct a 3D geometric model. FE models of each part were generated separately, based on the respective geometrical models obtained. For this purpose we used the software FEMAP, a commercial pre and post processing finite element program.

Keywords: Middle ear, 3D images, CT, FE model, ossicles, tympanic membrane.

1 INTRODUCTION

Nowadays, the recent images definition, resolution and technology evolution lead us to the Computed Tomography (CT) and Magnetic Resonance Imaging (MRI). This new availability of technology helps the radiologists and the otolaryngologists to diagnostic diseases and pathologies [1]. However, the complex anatomy of the middle ear and adjacent structures (Fig. 1) makes difficult the task of looking at cross-sectional 2D CT images [2] (Fig. 2).

The bones of the middle ear – the malleus, incus and stapes are the smallest bones in the body and are among the ones with most complicated functionally [3].The sound is collected in the ear canal and when it hits the tympanic membrane, the membrane starts vibrating, making the three ossicles in the middle ear move. Understanding the function of the auditory ossicles is essential in order to design protheses to replace them [4].

The development of a 3D geometric model of the middle ear will provide surgeons, radiologists and

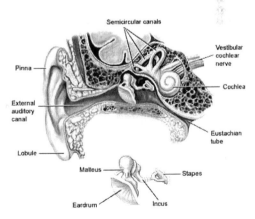

Figure 1. Anatomy of the ear and the ossicles (Malleus, Stapes, Incus) and the Tympanic Membrane (Eardrum) [3].

audiologists a self guided tour of detailed and specific ear anatomy.

For three-dimensional (3D) reconstruction of human middle ear, and its geometric model, we

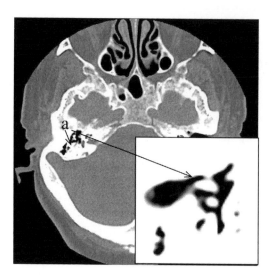

Figure 2. Axial 2D source image of the Head with the middle ear and the ossicles (a).

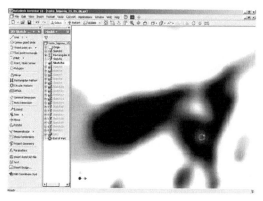

Figure 3. Segmentation of the malleus and incus by the tools "Image Edit" in ANALYZE (Zoom × 4).

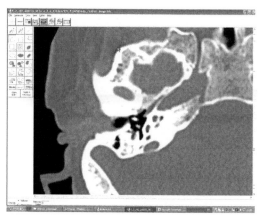

Figure 4. Incus Segmentation first slice.

have taken advantage of the existing technology and some commercial software available (Analyze – Software develop for the Mayo Clinic; CAD software – Autodesk Inventory Series 10; 3D Finite Elements Software – FEMAP 8.0).

2 MATERIALS AND METHODS

CT scan images of the middle ear, obtained from a 65 years old woman were used for this work. The image volume contained a number of bad slices with noise, so the volume was filtered to eliminate the noise prior to further image analysis. That task, i.e. noise reduction, was obtained from the Analyze software tool image edit with manual segmentation process and rendering. The manual options was used because these images and structures were not sufficient distinct in the original data. This methodology was applied often to obtain more detail, but, unfortunately, the resolution of the DICOM (Digital Imaging and Communications in Medicine) images was so poor (Fig. 3) that was quite impossible to have a clear shape of the ossicles and tympanic membrane. A direct copy of many slices, one by one, was performed with Analyze software tool. Each image (slice) was saved in BMP (bitmap image) format, in the same sequence that was used in Analyze. Those images were imported directly into a CAD software (Autodesk) in same order that it was saved.

Another methodology was used to obtain 3D images of the ossicles and tympanic membrane. Generally, the main different methods used in 3D images gen the main different methods used in 3D images generations are, first one, based on surface (CAD software) and the other one is based on volume (ANALYZE software). The new reconstruction of 3D objects process that was adopted is based upon their transversal section [6].

1.1 The following steps were adopted within the methodology established:pre-processing of the bi-dimensional images,

1.2 surface reconstruction between outlines and STL (Stereo Lithography) data generations files,

1.3 visualization of the generated images

2.1 Pre-processing

The scope of pre-processing is to extract, for each slice, a collection of representative object boundary points (Fig. 4). A triangulation process is then made, based on that collection of points between consecutive slices [5].

2.2 Surface reconstruction

Surface representation is obtained by connecting the object boundaries through triangular elements. The

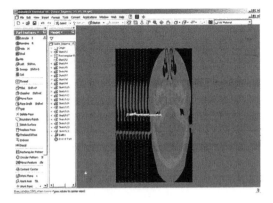

Figure 5.   Incus Triangulation process.

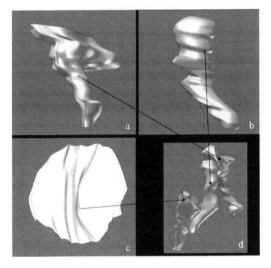

Figure 6.   Three-dimensional representation of the objects –
Incus (6a); Malleus (6b); Tympanic Membrane (6c); Partial
reconstruction of the Middle Ear.

triangularization process generates a collection of triangular patches between consecutive pairs of contours, forming a precise approximation to the original object surface. The triangulation algorithm (Fig. 5) is able to handle cases where several contours in each slice must exist, known as multiple branching problems.

Basically the reconstruction process consists of obtaining a three-dimensional representation of the object under investigation (malleus, incus and the tympanic membrane), allowing not only its visualization, but also a more detailed comprehension of its structure through the analysis of the geometric parameters of the object (Fig. 6a-d).

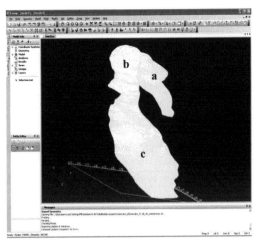

Figure 7.   Finite elements model of the Incus (a); Malleus
(b) and Tympanic Membrane (c).

2.3 Visualization of the generated images

After the reconstruction, the new 3D models must be converted to a standard format that can be read by a finite element software.

The STL format was used in this work because, first of all, it is an open-source format and an easy format to handle and manipulate. Once it was generated, the STL model was read by FEMAP (commercial pre and post processing finite element program) (Fig. 7). The FEMAP software has the ability to generate different layers and also to rebuilding the incomplete 3D elements [7]. Later on, that archive can be used directly into a fast prototype machine to obtain the prototype object, if that is the intention.

3   RESULTS

The simple analysis of the generated images is the basic validation of the methodology proposed in the present work. The precision of the reconstruction algorithm of the different available initial images is assessed by the 3D FE objects. The accuracy of the final 3D object is the result of the successful image generation process, and this accuracy is directly proportional to the higher CT images quality. Another related conclusion is that the time spent for each task i.e. time of pre-processing, surface reconstruction and visualization of the generated images is proportional to the CT images quality. CT images of reduced quality means a lot of work and more time lost for the same final result.

4 MAIN CONCLUSIONS

This work presented a methodology of surface three-dimensional reconstruction of both the ossicles (Malleus and Incus) and the tympanic membrane (Iadrum) based on transversal sections of CT images.

These 3D models can give a better comprehension of complex human anatomy, as the case for the middle ear. They also give us a kind of detail that is impossible to obtain with ordinary computer images.

Substantial improvements can be achieved in the prototype and prosthesis area, because the obtained models are now much more close to the real geometric dimensions of the bones of the middle ear.

ACKNOWLEDGEMENTS

The support of Ministério da Ciência e do Ensino superior (FCT and FSE) (Portugal) and the funding by FEDER under grant POCTI/ESP/46835/2002, are gratefully acknowledged.

REFERENCES

[1] Matheus M.de Carvalho, Ricardo M. Czekster, Isabel H. Manssour. Uma ferramenta interativa para visualização e extração de medidas de imagens médicas. *Faculdade de Informática-Pontifícia Universidade Católica do Rio Grande do Sul,* Brasil

[2] T.Rodt, P.Ratiu, H.Becker, S.Bartling, D.F.Kacher, M.Anderson, F.A.Jolesz, R.Kikinis (2002) 3D Visualization of the middle ear and adjacent structures using reconstructed multi-slice CT images datasets, correlating 3D images and virtual endoscopy to the 2D cross-sectional images. *Neuroradiology* 44: 783–790

[3] Frank H. Netter, M.D.(1997). Atlas of Human Anatomy.A. *CD-ROM Atlas Interativo de Anatomia Humana*

[4] M.D.Seemann, O.Seemann, H.Bonél, M.Suckfüll, K.H.Englmeier, A.Naumann, C.M.Allen, M.F.Reiser (1999) Evaluation of the middle and inner ear structures: comparison of hybrid rendering, virtual endoscopy and axial 2D source images. *Eur. Radiol.* 9. 1851–1858

[5] Fopefolu O. Folowosele, Jon J.Camp, Robert H.Brey, John I.Lane, Richard A.Robb, 3D imaging and modeling of the middle and inner ear. *Biomedical Imaging Resource, Audiology, Radiology,* Mayo Clinic, Rochester, MN.

[6] M.A. de Souza, T.M.Centeno, H. Pedrini. Integran (2003) Integrando reconstrução 3D de imagens tomográficas e prototipagem rápida para a fabricação de modelos médicos. *Revista Brasileira de Engenharia Biomédica, v.19, n.2, p.103–115*

[7] Q. Sun, R.Z. Gan, K.-H. Chang, K.J.Dormer (2002) Computer-integrated finite element modeling of human middle ear. *Biomechan Model Mechanobiol*: 109–122

Computational Modelling of Objects Represented in Images –
João Manuel R.S. Tavares & R.M. Natal Jorge (eds)
© 2007 Taylor & Francis Group, London, ISBN 978-0-415-43349-5

Acquisition of 3D regular prismatic models in urban environments from DSM and orthoimages

João Ferreira & Alexandre Bernardino

Instituto Superior Técnico, ISR Lisboa, Portugal

ABSTRACT: We present a novel method for fitting polygons to segmented image regions. The method is applied to the automatic reconstruction of 3D prismatic building models in urban areas, using regions segmented from the Digital Surface Map (DSM) and color orthoimages obtained from Google™-Earth. Polygon selection methods with complexity penalty terms are developed for the acquisition of simplified building outline models. Regular polygons such as rectangles, right trapezoids and general trapezoids are most common in urban areas, and special methods are developed for fitting these shapes to the segmented regions. Notwithstanding, irregular polygons are also considered since some ancient urban areas are plenty of buildings with non-standard shapes. Results are presented on the 3D reconstruction of the city of Lisboa, Portugal, with simple prismatic models.

Keywords: Urban modelling, regular polygons, model selection, 3D reconstruction, digital surface maps (DSM), orthoimages.

1 INTRODUCTION

In this paper we address the problem of automatic generation of 3D building models in urban areas. Several cities around the world have built virtual models of certain parts of their territory. These virtual models are important for many purposes, e.g urban modelling, telecommunications for transmitter placement, environmental planning, simulation etc. However, the large majority of the models are obtained by manual methods, requiring many man-hours of work for introducing data which makes the creation and maintenance of such models too costly and time consuming. Therefore there is a great deal of effort put in the research for automatic or semi-automatic tools for the acquisition, validation and update of cadastral data in urban areas. Classically the main input data for the production of 3D city models were aerial images, terrestrial images, map data, and surveying data. In the last years LIDAR (LIght Detection And Ranging) has become a very attractive alternative for the acquisition of 3D information since this technique directly provides a high density of 3D points. However, the obtained maps are usually smoothed and irregular, and should be complemented with other sources of information. Several works have proposed the use of aerial images to complement the LIDAR maps and present methods to combine their information toward 3D building reconstruction [3,4]. Our work fits in the same approach but the main problem addressed here is the acquisition of

regular shapes for buildings' outlines. This is important in order to obtain succinct representations of buildings, allowing light storage cost and fast manipulation in visualization systems.

Figure 1 shows a diagram with the overall processing pipeline for our reconstruction method.

We use LIDAR data in the form of a digital surface map (DSM), and color orthoimages obtained from Google™-Earth. In section 2 we describe how these two sources of information can be effectively used for the segmentation of buildings' regions, including the between-map registration problem, DSM normalization, removal of vegetation areas, and map binarization. A first step consists in identifying roads in the DSM by region growing processes. The Digital Terrain Map (DTM) is then obtained by interpolating a smooth surface supported in the identified roads. This allows a first segmentation of the buildings by comparison to the DSM. However, spurious segmentations are obtained due to vegetation.

A color segmentation processing on the orthoimages allows the detection of such cases and removing most of the segmentation errors. After this segmentation process we obtain a binary map with regions corresponding to buildings but these regions are still very irregular to be used for 3D reconstruction - they would render very unrealistic building models, as well as requiring a large amounts of information to represent them. Therefore, in section 3 we present methods to obtain simplified representation of the buildings'

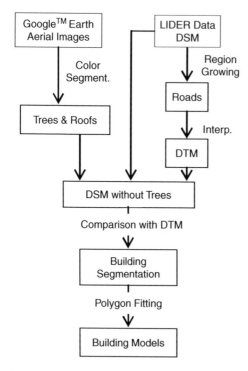

Figure 1. Architecture of the proposed building reconstruction method. In future work the information of building roofs will be used both for texture mapping and roof modeling and reconstruction.

outlines consisting regular polygons whenever adequate. This is a reasonable assumption since most buildings in urban areas follow simple geometric models. In [1], buildings are modeled by rectangles or combinations of rectangles. Instead we define a larger set of 2D regular geometric shape hypothesis with different complexities (from rectangular to trapezoidal shapes). These models are fit in the segmented data, using as criteria, both a goodness of fit term and a complexity term. Initial shapes for the regions are obtained with some heuristics and a final iterative minimization procedure is used to select the best hypothesis. The introduction of a complexity penalty term is important to reduce the amount of information required to represent the models, thus allowing faster visualization and lighter hardware requirements. If none of the predefined shapes fit well in the regions, the classical Douglas-Peucker line simplification algorithm [2] is used to fit a general polygon to the building's boundary. In section 4 we explain the model selection procedure, i.e. the choice of the best polygon, regular or non-regular, to represent a certain building. The obtained polygons are then used to create 3D virtual prismatic models of the buildings by extruding the 2D shape to a height provided by DSM information.

Along the paper we present the results obtained by applying the different phases of the processing pipeline to available DSM data from Lisbon city (Portugal) and orthoimages obtained from Google™-Earth. In section 6 we present the final results obtained illustrating the performance of the proposed polygon fitting and selection mechanisms. Finally, in section 7 we present the conclusion of this paper and ideas for future work.

2 INITIAL SEGMENTATION

To perform the acquisition of regular prismatic models for blocks of buildings[1], the data provided by DSM and aerial orthoimages is processed to create an initial binary image indicating the regions belonging to buildings with high likelihood. In the next lines we succinctly describe each of the modules of the processing pipeline shown in Figure 1.

2.1 Obtaining the digital terrain model

Digital Surface Model data is used to obtain the DTM. First, a region growing algorithm, based on height differences, is applied in the DSM, thus getting a representation of all connected streets and open spaces from initial seed points used in the region growing algorithm. The region growing algorithm is applied several times for different initial seed points and, by joining the several results, a good representation of all the streets and open spaces is obtained. Results of the region growing procedure applied to a part of the Lisbon city DSM are shown in Figure 2. After the ground level points have been determined, the DTM is obtained by interpolating a smooth surface supported on the ground level points (roads and open spaces with its heights obtained from the DSM).

2.2 Color segmentation

The color segmentation of the orthoimages is based on the Mahalanobis distance of a color pixel to a single Gaussian color model [5].

For the roof detection, the RGB color model is the one with the most Gaussian profile (for both light and dark roofs) while for the vegetation we use the UVL color model.

After the segmentation process, two masks are obtained, one for the roofs and other for the vegetation. Note that the roof mask is only used for future work. Results of the vegetation segmentation are presented in Figure 3.

---

[1] In this paper we do not consider single city buildings but connected blocks of buildings. Whenever we apply the word "building" it should be understood a "block of buildings".

Figure 2. Result of region growing on the DSM. Dark areas correspond to the roads and ground plane regions. Bright areas correspond to buildings, trees and other above ground structures.

Figure 3. Green color segmentation. Dark areas correspond to regions containing vegetation with high likelihood.

2.3 Building segmentation

Comparing the DSM and DTM, a first segmentation is obtained based on the heights of structures, which include spurious segmentations due to vegetation.

Figure 4. Building Segmentation. Dark areas correspond to blocks of buildings, with high likelihood.

With the application of the vegetation mask on the first building segmentation, followed by some morphological operations and elimination of small area objects, a good building segmentation is then obtained. Results of this process are shown in Figure 4.

3 SHAPE MODELING

For each one of the objects in the final building segmentation, four types of shapes are considered: regular polygons like rectangles, rectangular trapezoids, trapezoids and general non-regular polygons.

The first step for the modeling process is to obtain a contour representation of the object which is made by edge detection using the canny method, followed by the arrangement of the boundary pixels. Then, polygons of several types are fit in the contour, each one with a different number of edges. The end result of this procedure is to associate to each building a set of polygons with a number of sides ranging from 4 to a number determined by the Douglas-Peucker algorithm [2]. Then, in the shape selection step (section 4), we will choose the best model taking into account the fitness error and a complexity term proportional to the number of edges.

In both the shape modeling and the shape selection phases, the overlap error is used either for shape fitting and selection. Let us define the overlap error as the sum of pixels inside the estimated polygon but not on the object, plus the sum of pixels in the object but not

Figure 5. From right top to left bottom: initial contour; estimated contour using Douglas-Peucker algorithm; contour simplified to 10 corners; contour simplified to 4 corners.

inside the contour. This measure is then associated to each one of the estimated polygons.

3.1 Non-regular polygon estimation

The non-regular contour is modeled by the Douglas-Peucker (DP) algorithm [2]. This algorithm is modified to, iteratively, compute contours with lower number of points. The initial contour has the maximum number of points returned by the DP algorithm. Then, based on the cost and depth associated to each point (cost and depth are intermediate computations of the DP algorithm, see [2]), we iteratively remove one edge at a time from the polygon: the point with the lower cost, above a certain depth, is removed at each step. This allows to obtain simpler models of the general contour (Figure 5), ranging from the original, non-simplified contour, to a quadrilateral (four corner polygon). All polygons estimated in this sequence are stored and the *overlap error* is associated to each one for posterior use.

3.2 Regular polygon estimation

In this work we consider that most of the buildings' outlines have regular shapes and have developed specialized algorithms for fitting rectangles, right trapezoids and general trapezoids to the segmented contours.

The fitting algorithms consist in an initialization phase based in the region first and second order moments (centroid, orientation, major and minor axes length), and an iterative phase where the length of the

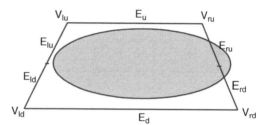

Figure 6. Error measurements computed in a general trapezoid.

polygon edges and its orientation are refined in an interleaved way.

For rectangles, an initial estimation is made based on the image blob's first and second order moments (centroid, orientation, major and minor axes length), which roughly indicate its position, orientation and dimensions. For the trapezoids, the initial estimation is made based on the first four corners returned from the Douglas-Peucker algorithm.

In the iterative phase, the position of the edges and the overall orientation of the polygon are updated step by step. The coordinates are first normalized so that two parallel sides are horizontal. Then, for each side of the polygon, it is associated one error measures at each edge (or half edge for the non-orthogonal sides in the trapezoids). The error measure is the average difference between the edge and the closest point in the blob orthogonal to the edge. Blob points to the inside of the current polygon have a negative contribution to the error measure while points to the outside have a positive contribution. Then each edge vertex is shifted orthogonally to the edge orientation an amount proportional to its error measure. If the error is positive the vertices are moved to the outside of the polygon and *vice-versa*. The trapezoid non-orthogonal sides are splint in two halves and each vertex has its own error measure, being shifted independently of the other. Let us analyze more closely the more complex case of a general trapezoid, as shown in Figure 6. Six error measurements are obtained: up edge (E_u), down edge (E_d), lower part of left edge (E_{ld}), upper part of left edge (E_{lu}), lower part of right edge (E_{ru}) and lower part of right edge (E_{rd}). Then, the horizontal (x) and vertical (y) coordinates of the trapezoid vertices are updated according to the following expressions:

$$\begin{cases} V_{lu}^{x} = V_{lu}^{x} + E_{lu}, & V_{lu}^{y} = V_{lu}^{y} + E_{u} \\ V_{ru}^{x} = V_{ru}^{x} + E_{ru}, & V_{ru}^{y} = V_{ru}^{y} + E_{u} \\ V_{ld}^{x} = V_{ld}^{x} + E_{ld}, & V_{ld}^{y} = V_{ld}^{y} + E_{d} \\ V_{rd}^{x} = V_{rd}^{x} + E_{rd}, & V_{rd}^{y} = V_{rd}^{y} + E_{d} \end{cases}$$

After the edges have been updated, the overall orientation of the polygon is refined by rotating the

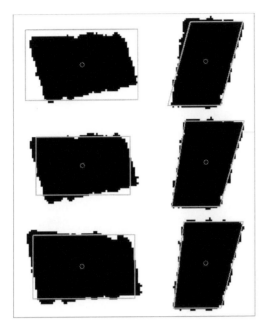

Figure 7. First column – rectangle; Second column – trapezoid. Rows top to bottom: initial estimation; edge size update; orientation update.

Figure 8. Fitted polygons to all building regions. Rectangular buildings are shown in black, trapezoidal buildings in dark grey and general polygons in light grey.

estimated contour left and right and evaluating the evolution of the *overlap error* to decide whether to update the orientation at this step or not. Examples of intermediate steps and final estimation are shown on Figure 7.

4 SHAPE SELECTION

After the four different contours are estimated for each one of the objects, a selection is made, based on the overlap error and its complexity so the best model is chosen to represent each object.

For the non-regular polygons one must first determine the best compromise between overlap error and complexity for each one of the different estimated models. First, the overlap error is normalized by the area of the object and then, the cost function is computed based on the normalized overlap error and complexity of the model (number of edges). The best non-regular polygon is the one with the minimum cost.

For regular polygons, the cost function is computed similarly where complexity is not directly associated with the number of edges but with the number of degrees of freedom (rectangle-4, right trapezoid-5, general trapezoid-6). However the proportionality factor affecting the complexity term in the cost function of regular models is lower than for non-regular models in order to favor the selection of regular polygons.

This parameter is manually tuned based on visual inspection of the results.

The cost functions of all model hypotheses for a given building are compared and the one with the minimum value is the model chosen to represent the object. Manual corrections are possible using a simple graphical user interface (GUI).

5 RESULTS

The whole processing pipeline is applied to available DSM data from Lisbon city (Portugal) and orthoimages obtained from Google™-Earth. Results are presented in Figure 8 illustrating the output of the algorithm. In general, buildings are well estimated and classified. The only exceptions are some buildings with small non rectangular corners, small buildings and trapezoidal buildings with near rectangularity which are classified as rectangular (the last ones can be manually adjusted).

We have applied the fitted polygons' mask of Figure 8 to the original orthoimage in order to measure the number of false positives (where ground or trees are inside the building area), which is very low aside the patios inside the buildings. In the other hand, applying the inverted mask, the number of true negatives (where buildings are not classified as buildings) can be measured. This number is higher, although some of them are due to the fact that when the DSM was

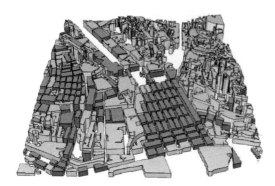

Figure 9. Reconstructed 3D model of Lisboa downtown with simple prismatic models.

obtained the buildings still didn't exist. The main sources of true negatives are the tree removal process and the DSM/DTM comparison. As can be seen in Figure 4, after the application of the vegetation mask, sometimes important information is removed because some trees are hiding important structures. This is not a simple problem to be solved automatically and will be included in future work. Also some trees aren't well removed in the tree removal process mainly because the DSM data is older than the orthoimages.

The automation of the process depends on some user defined constants: threshold on region growing, DSM/DTM comparison threshold, color segmentation threshold, control constants in the estimation algorithm and finally the complexity weight constants. While the first tree constants have great impact in the detection of buildings, the last ones have their impact in the classification of buildings.

In general, the computation time per building estimation is relative quick although larger objects take more time to estimate.

Finally, the 3D reconstruction is made by extruding the 2D model to a height equal to the mean of the building height without patios, which is provided by the DSM and the building segmentation, as shown in Figure 9.

6 CONCLUSIONS AND FUTURE WORK

In this paper we have presented a semi-automated method for the acquisition of simplified 3D models of urban areas from DSM and aerial image data. The whole processing pipeline is presented and illustrated with intermediate results of the several processing steps and applied to the 3D reconstruction of a significant part of the Lisbon city. We present the methods used for road detection, building segmentation, vegetation detection and 3D reconstruction, but special emphasis has been put on the modeling of the

buildings with regular prismatic models, allowing succinct representations of the data and reducing the complexity of managing its storage and visualization. The overall process has good results and the level of automation is very high (just needs some supervision of the results and adjustments of the type of polygon).

In future work several issues should be addressed. Since many buildings can be represented by the composition of regular shapes, we will study the possibility of such models. For instance an L shaped building could be modeled by two connected rectangles.

Also many buildings on the city have inner patios and the inner building contour should also be considered. Finally, since the results presented in this work reconstruct blocks of buildings instead of individual buildings, in future work we will evaluate the possibility of segmenting blocks into buildings looking for the characteristics of roofs both from aerial images, inspecting the edges, and from DSM data, adjusting planes to the roofs.

ACKNOWLEDGMENTS

This work was partially funded by the Portuguese Foundation for Science and Technology (FCT) through project POCTI/AUR/48123/2002 – Spatial validation of complex urban grids in virtual immersive environments.

REFERENCES

[1] S. Vinson, L. Cohen and F. Perlant. Extraction of Rectangular Buildings in Aerial Images. In Proc. Scandinavian Conference on Image Analysis (SCIA'01), June 2001, Bergen, Norway.

[2] D.H. Douglas and T.K. Peucker. Algorithms for the reduction of the number of points-required to represent a line or its caricature, The Canadian Cartographer, 10, (2), 112–122(1973).

[3] Martin Huber, Wolfgang Schickler, Stefan Hinz, Albert Baumgartner. Fusion of LIDAR Data and Aerial Imagery for Automatic Reconstruction of Building Surfaces. 2nd GRSS/ISPRS Joint Workshop on Remote Sensing and Data Fusion over Urban Areas, 2003.

[4] F. Rottensteiner, J. Trinder, S. Clode, K. Kubik and B. Lovell. Building Detection by Dempster-Shafer Fusion of LIDAR Data and Multispectral Aerial Imagery. In Proc. ICPR'04.

[5] Jean-Christophe Terrillon, Mahdad N. Shirazi, Hideo Fukamachi and Shigeru Akamatsu. Comparative Performance of Different Skin Chrominance Models and Chrominance Spaces for the Automatic Detection of Human Faces in Color Images. Proceedings of 4th IEEE International Conference on Automatic Face and Gesture Recognition, pp. 54–61, 2000.

Computational Modelling of Objects Represented in Images –
João Manuel R.S. Tavares & R.M. Natal Jorge (eds)
© 2007 Taylor & Francis Group, London, ISBN 978-0-415-43349-5

Automated generation of 3D bone models from planar X-ray images for patient-specific finite element modeling applications

P.E. Galibarov, A.B. Lennon & P.J. Prendergast

Trinity Centre for Bioengineering, School of Engineering, Trinity College Dublin, Dublin 2, Ireland

ABSTRACT: Replacement of arthritic human joint with a prosthesis is a complicated surgical procedure with variable outcome. Even very experienced surgeons cannot guarantee successful results. Behavior of the prosthesis within the bone can be predicted by building an appropriate 3D model of prosthetic joint and running realistic computer simulations. However, rapid creation of 3D models prevents application of patient-specific computer simulation on a pre-operative basis. The aim of this project is to develop an automated system for generating patient-specific 3D models for pre-operative planning, i.e. the system should be able to create a robust finite element model from 2D medical images for future mechanical analysis. The major parts of such a system are developed in this work. For creation of a 3D model from a planar X-ray image the following method was used: (i) an accurate contour of the patient's femur was extracted from the X-ray image using Image Processing techniques, such as "filtering" and "edge detection"; (ii) three generic models of different sizes were created using a standardized femur model; (iii) and the contour received was compared with different-angle profiles of generic models, the best matching generic model was chosen and it was warped to fit the patient's actual dimensions.

1 INTRODUCTION

Information and communication technologies are widely applied in all areas of medical science; however computer simulation tools are not as widely used as they could be because of the difficulties inherent in patient-specific responses. In the orthopaedic domain, surgeons perform numerous kinds of joint replacement operations with variable outcomes (Prendergast, 2001). In 2000 2.9% of patients had to undergo revision operations within three months following the operation, 2.5% reported a dislocation according to the UK National Total Hip Replacement Outcome study. Appropriately designed mechanical models and accurate simulations can help extend lifetime of such replacements, thereby preventing undesired consequences of operation. A barrier to computer simulation, in particular finite element analysis, is generating a suitable 3D model of the patient anatomy. In some cases 3D imaging techniques, e.g. CT or MRI, can be used to generate computer models but such technologies are not routine for many procedures, e.g. hip replacement. A method generating 3D models from a 2D image would therefore aid computer simulation of those surgical procedures.

This work consists of three parts: (i) femur contour extraction from X-ray image (ii) generation of database of generic models, and (iii) choosing the best fit generic model for the extracted patient contour.

2 ALGORITHMS DESCRIPTION

2.1 *Femur contour extraction from X-ray image*

Nowadays the most popular radiographic data source due to the low cost and simple realization is an X-ray machine. Permanently evolving world and improvement of X-ray devices, in particular, led to transition from hard copies of X-ray images to their digital analogue. Digital standards, like DICOM, were developed for convenient work with those images. Subsequently a large number of communication systems (PACS).

In our work we consider planar X-ray images of the pelvis and proximal femur. The femur of our interest can be recognized by an artificial mark on the X-ray image: "R" letter for right femur and "L" for the left femur. The image is split into two parts. We consider only the left femur in this study, right femur image can be symmetrically reflected to give the same starting point for the algorithm. A regular digital X-ray image is a matrix of point values in the range from 0 to 255. Vertical and horizontal Gaussian filters are applied to the image for edge extraction purposes, in other words,

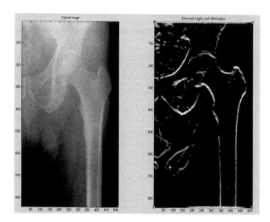

Figure 1. Original image and result of edge detection.

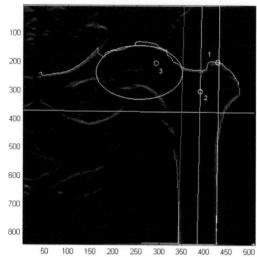

Figure 2. Detection of femur-specific anatomical landmarks.

original image is recalculated to form a matrix where homogeneous areas turn to be empty or zero areas and inconstant regions produce non-zero sections. Norm of the image gradient is calculated (Fig. 1).

Next, a threshold procedure is applied for ambiguous and fuzzy edge removal and removal of noise. This helps to get the most useful data for building the femur's contour. The threshold value was found experimentally.

Once edges have been found they should be classified. For these purposes we run a ray-tracing algorithm horizontally from both sides in the bottom part of the image to get an array of point-pairs. This array is investigated according to the femur dimension distributions found by Noble et al., 2003. Pairs of points which lie in an acceptable range are taken (between minimum and maximum diameter of femur at lesser trochanter +20 millimeters area). This assumption gives the distal border points of the femur at mediolateral region, which will be used as a starting point for building the required contour. Points of this array form two vertically oriented lines (parts of distal borders in mediolateral region) and they can be calculated using Least Squares algorithm, (Fig. 2). A half of the slopes sum and a point in the middle of any pair will give us a central line for the femur. A contour following algorithm is applied to locate further profile points. On the right side the anticlockwise length-increasing segment is rotated until it meets a new point; the same procedure, but with the clockwise scanning, is performed on the left side. If the length of segment exceeds some established value then the algorithm stops. The next step is to find an approximate location of the lesser trochanter – difference between left approximated line and a result of previous step is analyzed, and the maximum value is taken. From the femur dimension distribution we can say that the entire length of the femur along the central line doesn't exceed maximum femoral head position length plus femoral head radius with respect

to probable error in the calculation of lesser trochanter position. Following the central line and right border towards the bottom of the image we will intersect a contour of the saddle part, which is horizontal and will be extracted by vertical Gaussian very well (Fig. 2, point 1).

Using this point we start contour following again in the medial and lateral directions. Points found on the left will intersect a part of the femoral head contour. A line perpendicular to the central line is located within a distance equal to average femoral head position minus 20 millimeters. A point lying between the saddle point and perpendicular is taken and rough centre of femoral head is taken according to average neckshaft angle and average femoral neck length (Fig. 3). All the points from the contour following algorithm within circle with radius equal to maximum femoral head radius are taken. A circle is fitted to these points - a centre of this circle will accord to a real centre of the femoral head.

Next we take all the points received by contour following algorithm, newly received femoral head circle with slightly extended radius and build a convex hull using built-in Matlab function, Figure 3.

This convex hull contour will be an initial step for the active contours algorithm which is used to calculate accurate contour.

A traditional active contour or "snake" is a parameterized curve $x(s) = [x(s), y(s)]$, $s \in [0,1]$ that deforms within the spatial domain of an image until the energy functional:

$$E = \int_0^1 \frac{1}{2}(\alpha \mid x'(s) \mid + \beta \mid x''(s) \mid) + E_{ext}(x(s))ds \quad (1)$$

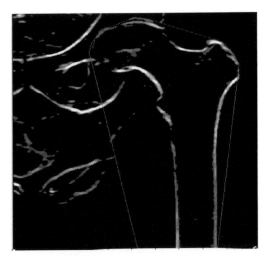

Figure 3. Convex hull, initial contour for the active contours, "snake" algorithm.

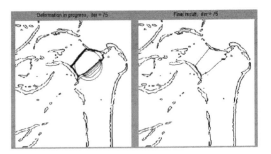

Figure 4. An example of GVF snake work: "snake's" iterations (left); the result contour (right).

is minimized; α and β are weighting parameters that control "snake's" tension and rigidity respectively, and E_{ext} is an external energy function derived from the image so that it takes on its smaller values at the features of interest, such as boundaries (Xu & Prince, 1997). A disadvantage of this approach is that the snake cannot move into a concave boundary region, which is common for the femur, but a modification of this algorithm, based on Vector Gradient Flow (GVF) field proposed by Xu & Prince, 1997, is able to solve this problem and was implemented for the contour calculation (Fig. 4). GVF field is calculated for external energy function in Equation 1.

Once the contour of the femur has been found we should calculate a secondary axis, which is similar to the midline of the femoral neck. Points of received contour in the vicinity of femoral head are taken and a Least Squares algorithm is applied to find a secondary axis. The point of central and secondary axes intersection is taken as a femoral origin and the entire

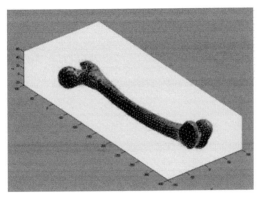

Figure 5. Standardized femur model from BEL repository.

contour is rotated according to central line slope and shifted to merge the femoral origin with the origin of the Cartesian coordinate system for convenience of further calculations.

2.2 Generic model design

For simplicity of calculation and testing purposes, different sizes of a single model are used to create a small database of generic models. The standardized femur model from the Biomechanics European Laboratory (BEL) was used (Viceconti et al., 1996). The model is stored in Facets file format, containing 3749 vertices and 69268 facets in total (Fig. 5):

Three different sized models were created according to femur dimensions distributions and deviations mentioned by Noble et al., 2003. Femoral length was used as a reference dimension for model scaling due to the high correlations with the most important dimensions, femoral head diameter, in particular. Each model is oriented such that the central axis of femur is merged with the Z axis, secondary axis belongs to the ZX plane and the point of central and secondary axes intersection is in the origin of coordinate system. This space orientation of the model gives us a way to rotate it about the central axis imitating various angles of the patient's femur during the X-ray taking procedure.

2.3 Choosing the best matching model

The contour received from the X-ray image is scaled to the original size and placed into ZX plane, femoral origin lies in the coordinate system origin. After that each model is rotated from -10 to 10 degrees about Z axis with the step of 1 degree; cuts by ZX plane are taken and compared to the contour. These cuts and contours are sets of connected segments. The comparison algorithm is described below. The contour and a cut are taken. They both are located and similarly oriented

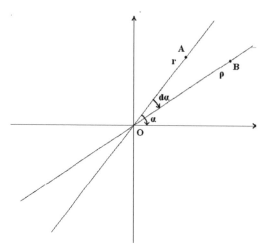

Figure 6. Schematic rotations of the femoral neck and head.

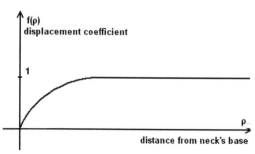

Figure 7. Desirable function for multiplying displacements.

in the XZ plane. A boundary rectangle for these two contours is calculated. The lower horizontal border is calculated according the contour received from the X-ray image. This rectangle is split into N by M smaller rectangles such that N is a number of horizontal rectangles and M is a vertical number correspondingly. This division makes a net which covers the rectangle. Each node of the net is checked for hitting the figures bounded by the cut and the contour. Two Boolean matrices of N by M size are the results of this checking. A correlation coefficient is calculated between these matrices and the result is a measure of similarity for our contours. The best matching cut, with the largest positive correlation, gives us the closest model and the best 3D space orientation which can be used for further creation of model according to patient's geometry.

2.4 Warping generic model

Warping the model involves two complicated steps: deformation of femoral neck with the head to accord patient's one and scaling the model or its parts to fit patient's geometry.

For rotation we turn all points of femoral head and neck by angle difference between the secondary axes of model and the contour (Fig. 6).

Point A, centre of femoral head of generic model, should be moved to new required location, point B. The rotation angle is $d\alpha$, see Figure 6. Except rotating points by angle difference we should scale the length of femoral head. For this purpose overall distance from the origin to the point is multiplied by ratio ρ/r. But moving all these points will lead to decreased smoothness of the model. To avoid unnecessary lack of smoothness we take points to rotate from the base of femoral neck towards the head. Desirable displacements at the neck's base are minimal ones, for

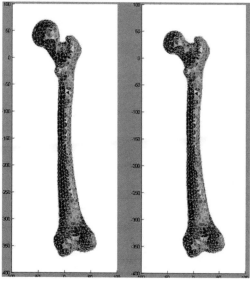

Figure 8. An example of transforming femoral head and neck.

this reason displacements are multiplied by function smoothly growing from 0 to 1 (Fig. 7):

A result of such transformation can be seen on Fig. 8:

Further transformations include scaling femur parts from mediolateral region up to lesser trochanter and greater trochanter. The idea of scaling is as follows: a slice of model is taken, two planes bounding this slice will define segments on the contour of the model and the one received from X-ray image, the figure that cut by one of these planes can be proportionally extended or shrunk, according to deformations performed on boundary figures points lying between these planes can be processed using linear approximation of scale value.

Scale value consists of left and right values, they make scaling half-ellipses – with the bigger radius equal to segments ratio: A_1O_1/A_2O_1, C_1O_u/C_2O_u, etc, see Fig. 9. The lower radius will be equal to average

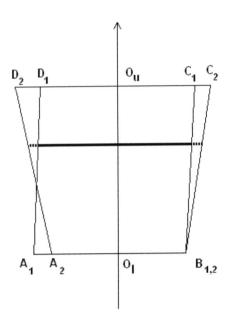

Figure 9. Scheme of slice to be scaled.

value of left and right radii. For each angle on the plane perpendicular to Z axis we are able to calculate displacement for every point on the surfaces and inside the slice building simple linear approximation of scaling half-ellipses. The same function for displacements multiplier can be used for internal points to keep internal smoothness.

Once we have scaled all possible slices of the model along central and secondary axes - we can run smoothing procedure importing obtained model into automated CAD systems.

3 CONCLUSIONS

In the present paper we described algorithms used to create, in an automated way, a 3D model reconstruction of a complex bone shape from planar X-ray image. This system may be essentially improved by adding a library of generic models: models of different sizes, with a wide specter of neck-shaft angles, etc. A very important and easy improvement is to formulate strict requirements for taking X-ray images, i.e. fix brightness, length of mediolateral parts, pelvic positions and so on. Nevertheless this system covers a very large part of patient cases. This process can be integrated with the computer simulations developed to predict implant performance pre-operatively, see Lennon et al. (in press).

4 FUTURE WORK

The automated system created for generation of patient-specific anatomical geometry from planar X-ray image will be used as a part of more complex one, which will deal with designing finite-element models from different types of radiographic sources – MRI, CT scans, X-ray images. Another part of this system will do further steps with the finite element models received. These models will be used for performing virtual implantation, investigating internal stress field during the action of various external forces, e.g. walking, running, cycling, stairs-climbing, etc., and as a result giving some recommendations to the surgeon. Such system would prolong a life-time of the embedded prostheses and reduce a probability of operation failure.

ACKNOWLEDGMENTS

This work was supported by Science Foundation Ireland, Research Frontiers Programme.

Authors would like to express their gratitude to Ruairi Mac Niocaill, Cappagh National Orthopaedic Hospital, for providing samples of planar X-ray images and consultations.

REFERENCES

Chen Y., Ee X., Leow W.K. et al., 2005, Automatic Extraction of Femur Contours from Hip X-ray Images, Proceeding of Computer Vision For Biomedical Applications, 200–209

Gregg P., Revees B., 2000, National Total Hip Replacement Outcome Study – Final report to the Department of Health, England

Lennon A.B., Britton J.R., MacNiocaill R.F. et al., 2006, Predicting revision risk for aseptic loosening of femoral components in total hip arthoplasty in individual patients – a finite element study, J. Orthop. Res. (in press)

Noble P.C., Alexander J.W., Lindahl L.J. et al., 1988, The Anatomic Basis of Femoral Component Design, Clin Orthop Relat Res. (235):148–65

Prendergast P.J., 2001, Bones and Prostheses, Bone mechanics handbook

Viceconti M., Casali M., Massari B. et al., 1996, The "Standardized Femur Program". Proposal for a reference geometry to be used for the creation of finite element models of the femur, J. Biomech., Vol. 29, 1241

Xu C., Prince J.L., 1997, Gradient Vector Flow: A New External Force for Snakes, IEEE Proc. Conf. on Comp. Vis. Patt. Recog. (CVPR'97)

Computational Modelling of Objects Represented in Images –
João Manuel R.S. Tavares & R.M. Natal Jorge (eds)
© 2007 Taylor & Francis Group, London, ISBN 978-0-415-43349-5

Physical simulation using FEM, modal analysis and the dynamic equilibrium equation

Patrícia C.T. Gonçalves, Raquel R. Pinho & João Manuel R.S. Tavares

Optics and Experimental Mechanics Laboratory – LOME, Portugal
Mechanical Engineering and Industrial Management Institute – INEGI, Porto, Portugal
Faculty of Engineering of the University of Porto – FEUP, Porto, Portugal

ABSTRACT: This paper presents a physical approach to simulate objects deformation in images. To physically model the given objects the finite element method is used, and to match the objects' nodes modal analysis is considered. The desired displacement field is estimated through the dynamic equilibrium equation. To solve this differential equation different integration methods can be used. In this paper we present and discuss the results obtained using four numerical integration methods: central difference, Newmark's and mode superposition allied with the former two. Some improvements are introduced in this work to allow the physical simulation even when not all of the objects nodes are successfully matched.

Keywords: Deformable objects, deformation simulation, dynamic equilibrium equation, finite element method, modal analysis.

1 INTRODUCTION

All real objects are deformable; that is why most of them cannot be accurately modelled if they are considered rigid bodies. Deformable objects simulation may be achieved using deformable models, which can be a challenging task because different application areas have different requirements: some require accuracy, like medical image analysis; others require real time interactivity, like virtual environments. Although many methods have been proposed to accurately simulate deformable objects at interactive rates, few are currently able to do so.

In this paper, we use the approach proposed by Terzopoulos et al. (1987, 1988) to do realistic deformation simulations considering an elastic model based on the resolution of the dynamic equilibrium equation. Therefore, we consider: Sclaroff's (1995) and Sclaroff & Pentland's (1995) isoparametric finite element to physically model each object; Shapiro and Brady's (1992) modal shape description to match the nodes (data points) of the objects; and Pentland and Horowitz' (1991) decomposition of object deformation into rigid and non-rigid modes.

In this paper, we propose a solution to apply this physical approach to objects that do not have all of their nodes successfully matched. Furthermore, we also verify that the integration method used to solve the dynamic equilibrium equation may influence the computation speed and the obtained simulation.

2 PHYSICAL MODELLING

The approach considered in this work for physical modelling can be applied to 2D/3D objects represented in images. To build each objects' physical model we employ the finite element method (FEM), a standard engineering technique whose interpolation functions are developed to allow continuous material properties along the model (Pentland & Horowitz 1991). Namely, we employ Sclaroff's (1995) isoparametric finite element that uses a set of radial basis functions that allows an easy insertion of the data points; thus, Gaussian interpolants are used and the model nodes do not need to be previously ordered. In consequence, with this finite element, when a 2D or 3D object is modelled it is as if each of its feature points are covered by an elastic membrane or a blob of rubbery material, respectively (Tavares 2000, Sclaroff & Pentland 1995).

Starting with a collection of m sample points $X_i(x_i, y_i, z_i)$ of the object to be modelled, the interpolation matrix H (which relates the distances between object nodes) of Sclaroff's isoparametric finite element (Sclaroff 1995, Sclaroff & Pentland 1995) is built using:

$$g_i(X) = e^{-\|X - X_i\|^2 / 2\sigma^2}, \tag{1}$$

where σ is the standard deviation (that controls the nodes interaction. Then, the interpolation functions,

h_i, are given by:

$$h_i(X) = \sum_{k=1}^{m} a_{ik} g_k(X),\qquad(2)$$

where a_{ik} are coefficients that satisfy $h_i = 1$ at node i and $h_i = 0$ at the other $m-1$ nodes. These interpolation coefficients compose matrix A and can be determined by inverting matrix G defined as:

$$G = \begin{bmatrix} g_1(X_1) & \cdots & g_1(X_m) \\ \vdots & \ddots & \vdots \\ g_m(X_1) & \cdots & g_m(X_m) \end{bmatrix}.\qquad(3)$$

Thus, for a 2D object (for the 3D approach see Tavares 2000, Sclaroff 1995), matrix H will be:

$$H(x) = \begin{bmatrix} h_1 & \cdots & h_m & 0 & \cdots & 0 \\ 0 & \cdots & 0 & h_1 & \cdots & h_m \end{bmatrix},\qquad(4)$$

and the mass matrix of the 2D Sclaroff's isoparametric element is defined as (Tavares 2000, Sclaroff 1995, Sclaroff & Pentland 1995):

$$M = \begin{bmatrix} \mathrm{M} & 0 \\ 0 & \mathrm{M} \end{bmatrix},\qquad(5)$$

where M is a sub-matrix $m \times m$ defined as $\mathrm{M} = \rho\pi\sigma^2 A^\mathrm{T} \Gamma A = \rho\pi\sigma^2 G^{-1}\Gamma G^{-1}$ (because A is symmetric, $A^T = A$), ρ is the mass density, and the elements of matrix Γ are the square roots of the elements of matrix G.

On the other hand, the 2D stiffness matrix is given by:

$$K = \begin{bmatrix} K_{11} & K_{12} \\ K_{21} & K_{22} \end{bmatrix},\qquad(6)$$

where K_{ij} are symmetric $m \times m$ sub-matrices depending on constants that are functions of the virtual material adopted for the object (Tavares 2000, Tavares et al. 2000, Sclaroff 1995, Sclaroff & Pentland 1995).

In this work we use Rayleigh's damping matrix, C, which combines the mass and stiffness matrices with constraints based upon the chosen critical damping (Bathe 1996, Cook et al. 1989).

3 MATCHING OBJECTS NODES

To match the nodes of the initial and target objects, each generalized eigenvalue/eigenvector problem is solved using:

$$K\Phi = M\Phi\Omega,\qquad(7)$$

where Φ is the modal matrix of the shape vectors ϕ_i (which describe the modal displacement (u, v) of each 2D node due to vibration mode i), and Ω is the diagonal matrix whose entries are the squared eigenvalues increasingly ordered.

After building the modal matrix for each object, the two objects nodes can be matched comparing the displacement of each node in the respective modal eigenspace (Shapiro & Brady 1992). The main idea is that low order modes of two similar shapes will be very close even in the presence of affine transformation, non-rigid deformations, local shape variations or noise. Thus, to match the nodes of the initial object, I, with the ones of the target object, T, an affinity matrix, Z, is built whose entries are defined as:

$$Z_{ij} = \left\| u_{I,i} - u_{T,j} \right\|^2 + \left\| v_{I,i} - v_{T,j} \right\|^2,\qquad(8)$$

where the affinity between nodes i and j will be 0 (zero) if the match is perfect, and increases as the match worsens.

In this work, to find the matches two search methods are considered: a local method and a global one. The local method was proposed by Sclaroff (1995), Sclaroff & Pentland (1995) and Shapiro & Brady (1992), and consists in searching each row and each column of the affinity matrix for their lowest values. This method has the main disadvantage of disregarding the object structure as it searches for the best match for each node. On the other hand, the global method proposed by Bastos & Tavares (2006) and Tavares & Bastos (2005) consists in describing the matching as an assignment problem, and solving it using an appropriate optimization algorithm.

4 DYNAMIC EQUILIBRIUM EQUATION

To estimate the objects deformation, and thus to obtain their transitional shapes, attending to physical properties, we solve the second order ordinary differential equation commonly known as Lagrange's dynamic equilibrium equation:

$$M\ddot{U}^t + C\dot{U}^t + KU^t = R^t,\qquad(9)$$

for each time step t, where U, $\dot{U}$ and $\ddot{U}$ are respectively the nodal displacement, velocity and acceleration vectors, and R is the vector of the loads.

In this paper, to solve the dynamic equilibrium equation we used four integration methods: central difference, Newmark's and mode superposition in conjunction with each of the former two.

4.1 Central difference method

The central difference method is a direct and explicit integration method, with second order precision if the mass matrix is diagonal (Bathe 1996, Cook et al. 1989). However, using Sclaroff's isoparametric finite element the mass matrix is not usually diagonal. The velocity and acceleration vectors are approximated by:

$$\dot{U}^{t-\Delta t/2} = \frac{1}{\Delta t}\left(U^t + U^{t-\Delta t}\right),$$ (10)

$$\ddot{U}^t = \frac{1}{\Delta t}\left(\dot{U}^{t+\Delta t/2} + \dot{U}^{t-\Delta t/2}\right),$$ (11)

making the dynamic equilibrium equation to suffer a half time step delay in velocity, transforming $C\dot{U}^t$ in $C\dot{U}^{t-\Delta t/2}$. Thus, Equation 9 may be approached by:

$$\frac{1}{\Delta t^2}MU^{t+\Delta t} = R^t + \left(\frac{1}{\Delta t^2}M - K\right)U^t +$$

$$\left(\frac{1}{\Delta t}M - C\right)\dot{U}^{t-\Delta t/2}.$$ (12)

A disadvantage of this algorithm is that it has only first order precision, because the viscous forces, $C\dot{U}^{t-\Delta t/2}$, are delayed half a time step. Another disadvantage of the central difference method is that it is conditionally stable, thus the time step used must be small or the system will diverge (Bathe 1996).

4.2 Newmark's method

In this work we also considered Newmark's method (originally proposed by Newmark 1959) that has been modified and improved by many researchers. Newmark's method is a direct integration process that considers:

$$U^{t+\Delta t} = U^t + \dot{U}^t \Delta t + \left[\left(\frac{1}{2} - \chi\right)\ddot{U}^t + \alpha U^{t+\Delta t}\right]\Delta t^2,$$ (13)

$$\dot{U}^{t+\Delta t} = \dot{U}^t + \left[(1-\delta)\ddot{U}^t + \delta\ddot{U}^{t+\Delta t}\right]\Delta t,$$ (14)

where χ and δ are chosen in order to control stability and accuracy (Bathe 1996).

Thus, the dynamic equilibrium equation (for time $t + \Delta t$) may be rewritten as:

$$\left(\frac{1}{\chi\Delta t^2}M + \frac{\delta}{\chi\Delta t}C + K\right)U^{t+\Delta t} = R^t +$$

$$\left(\frac{1}{\Delta t}M + \delta C\right)\frac{1}{\chi\Delta t}U^t + \left[\frac{1}{\Delta t}M - (\chi-\delta)C\right]\frac{1}{\chi}\dot{U}^t +$$

$$\left[\left(\frac{1}{2\chi}-1\right)M - \left[(1-\delta)\Delta t - \left(\frac{1}{2\chi}-1\right)\Delta t\delta\right]C\right]\ddot{U}^t.$$ (15)

This numerical method is unconditionally stable if $2\chi \geq \delta \geq 0.5$. In this work we used $\delta = 0.5$ and $\chi = 0.25$, which means that no numerical damping was introduced and Newmark's method was employed as a second order implicit scheme.

4.3 Mode superposition method

The mode superposition method proposes the transformation of the modal displacements, V, into the nodal displacements, U, using the eigenvectors matrix Φ: $U = \Phi V$. Thus, using the generalized coordinates the equilibrium equation can be rewritten as:

$$\ddot{V}^t + \Phi^T C\Phi \dot{V}^t + \Omega V^t = \Phi^T R^t,$$ (16)

because $\Phi^T K \Phi = \Omega$ and $\Phi^T M \Phi = I$, where I is the identity matrix. $\dot{V}$ and $\ddot{V}$ are, respectively, the first and second order derivatives of the modal displacement vector.

This integration method obtains new stiffness, mass and damping matrices with smaller bandwidth (Bathe 1996), allowing the resolution of the dynamic equilibrium equation with just a part of the models vibration modes. This initiative reduces the involved computational cost by ignoring the local components of the transformation between the shapes, essentially associated with noise. However, in this paper, we used the mode superposition method with all the vibration modes.

As an indirect integration method, to proceed with the resolution of the dynamic equilibrium equation, we have to apply a direct integration method to solve the transformed uncoupled equations set. Thus, we considered the two methods previously described: central difference and Newmark's.

4.4 Implicit loads and initialization

For the implicit loads applied on each matched node i, we consider that each load is proportional to the associated displacement (Pinho & Tavares 2004):

$$R(i) = k\left(X_{F,i} - X_{j,i}\right),$$ (17)

where $R(i)$ is the ith component of the load vector, $X_{F,i}$ is the coordinates of node i in the target object, $X_{j,i}$ represents the coordinates in object j (i.e. the object obtained in the jth iteration) and k is a global stiffness constant.

However, some nodes of the initial object may not be successfully matched with any of the nodes in the target object. Thus, suppose that B is an unmatched node between nodes A and C, matched with nodes A' and C' of the target object, respectively (Fig. 1). If B is the ith node in the jth shape, then the ith component of the load vector can be given by:

$$R(i) = k\left(\sum_{\substack{p \,(\text{nodes between} \\ A' \text{ and } C')}}\left[W_p\left(X_{F,p} - X_{j,B}\right)\right]\right),$$ (18)

where W_p is the weight of node p, according to its matching affinity with node B provided by

199

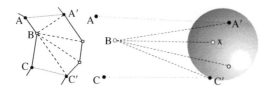 

Figure 1. Estimation of the implicit loads for unmatched nodes in 2D cases (left) and 3D cases (right).

Figure 2. From left to right: the two original images (the selected isobar contours are pointed out); matches found.

Equation 8 – thus, the higher the matching affinity, the lesser the weight.

For the 3D case, A and C are the closest matched nodes to B matched to A' and C', respectively (Fig. 1). A' and C' lie inside a sphere centred at point x, the rigid geometric transformation of point B (estimated in this work using Horn's (1987) method), and with the distance from x to C' as the radius. Hence, in Equation 18, node p is each one of the unmatched nodes lying inside that sphere.

The solution we adopted to estimate the initial displacement is to consider it in terms of the implicit loads (Pinho & Tavares 2004):

$$\begin{cases} U^0(i) = \dfrac{c_u}{k} R(i) & if \ k \neq 0 \\ U^0(i) = 0 & if \ k = 0 \end{cases}, \qquad (19)$$

where $U^0(i)$ represents the ith component of the initial nodal displacements vector, and c_u is a constant to be defined by the user.

The initial modal velocity vector is considered in terms of the initial modal displacement vector (Pinho & Tavares 2004):

$$\dot{U}^0(i) = c_v U^0(i), \qquad (20)$$

where c_v is another user defined constant.

5 EXPERIMENTAL RESULTS

Consider the two contours obtained from real images of weather isobars with 27 and 39 nodes respectively, in Figure 2. Using the described matching approach with global search and rubber as the virtual material, all the 27 nodes of the initial object are successfully matched. To simulate the given objects deformation, we considered critical damping between 1% and 2% and chose to stop the computation process when the Euclidean norm of the nodal displacements vector is lesser than 5×10^{-4} pixels. Other relevant constants were chosen as follows: $\Delta t = 0.05$ s, $k = 1000$ N/m, $c_u = 0$, $c_v = 0$. Under these circumstances, the steps and the time needed to complete the computation process by each one of the integration methods are indicated in Table 1,

Table 1. Results obtained to achieve a nodal displacement vector norm lesser than 5×10^{-4} pixels.

Method	Number of steps	Distance to target object (% of initial distance)	Time (seconds)
Central difference	48136	2.90	146
Newmark's	48134	2.90	141
Mode superposition and central difference	48137	2.90	144
Mode superposition and Newmark's	48134	2.90	149

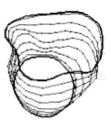

Figure 3. Results obtained with Newmark's method. In black are the initial and target objects, and in grey the intermediate ones obtained after 500, 1300, 2600, 5600, 11000, 20000 and all steps (from bottom to top).

as well as the distance between the last estimated object and the target one. (In this work, we used a computer with an Intel Pentium D at 3 GHz and 2 GB of RAM.)

In Figure 3 are the results obtained with the Newmark method in the estimation of the involved deformation, which are visually similar to the other ones but slightly faster (Table 1).

If, instead of stopping the computation process when the nodal displacements norm is lesser than 5×10^{-4} pixels, we stop it when the distance between an estimated shape and the target object is, for example, 4% of the distance between the original objects, we have a decrease in number of steps as well as in time (Table 2) and visually there are little differences between the final results.

The fact that we need more 12% of the steps and 20 seconds more to reach a decrease of only 1.1% in

Table 2. Results obtained to achieve a distance lesser than 4% of the initial distance.

Method	Number of steps	Time (seconds)
Central difference	42560	127
Newmark's	42558	125
Mode superposition and central difference	42561	127
Mode superposition and Newmark's	42558	131

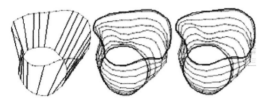

Figure 4. From left to right: the original contours and the 21 matches found; results obtained by the mode superposition with Newmark's integration method when only 21 nodes of the initial object are matched; and when all nodes are matched.

Table 3. Results obtained to achieve a nodal displacement norm lesser than 5×10^{-4} pixels considering only 21 nodes matched.

Method	Number of steps	Distance to target object (% of initial distance)	Time (seconds)
Central difference	48045	2.72	145
Newmark's	48044	2.72	141
Mode superposition and central difference	48046	2.72	142
Mode Superposition and Newmark's	48044	2.72	148

distance is a consequence of the fact that the nodal displacements norm decreases as the number of steps increases. This means that the biggest part of the initial image displacements happens during the first steps of the computation process.

If we use the local search method for the modal matching, only 21 of the 27 nodes of the initial object are successfully matched with nodes of the target one (Fig. 4). This means that Equation 18 will be used instead of Equation 17 to determine the implicit loads of the 6 unmatched nodes. Again, to achieve a nodal displacements norm lesser than 5×10^{-4} pixels, the results obtained by the different integration methods are very similar, being Newmark's method the fastest one (Table 3).

Table 4. Results obtained to achieve different distances using Newmark's method or the mode superposition with Newmark's method when all or only 21 nodes are successfully matched.

Distance to target object (% of initial distance)	All nodes matched		21 nodes matched	
	Number of steps	Time (seconds)	Number of steps	Time (seconds)
1	698	2	652	2
0.5	899	3	786	2
0.1	1625	5	1264	3

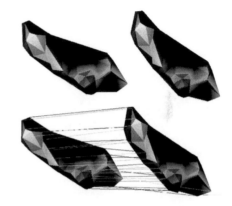

Figure 5. From left to right and top to bottom: pedobarography 3D object (intensity surface); target object obtained by applying a rigid geometric transformation to the first one; the 37 matches found between the objects' nodes.

In Figure 4 we can compare the results obtained by the mode superposition with Newmark's integration method when only 21 nodes of the initial object are matched; and when all nodes are successfully matched.

As mentioned in the previous section, the central difference method needs a small time step to preserve stability; but Newmark's method used as a second order implicit scheme is unconditionally stable and we can use any time step. Thus, changing the time step to 1s and the stiffness constant to 10^5 N/m, we can not use the central difference method, but we can achieve excellent results using Newmark's method alone or combined with the mode superposition method, Table 4.

For a 3D example, consider the first object displayed in Figure 5 obtained from a real pedobarography image (Tavares et al. 2000), and the second one obtained by a rigid transformation applied to the first (15° rotation along the zz axis). In this example it is easy to evaluate the accuracy of the obtained results: the intermediate objects should have the same shape but be in different positions.

Table 5. Results obtained to achieve a nodal displacements norm lesser than 5×10^{-4} pixels.

Method	Number of steps	Distance to target object (% of initial distance)	Time (seconds)
Central difference	7774	7.35	173
Newmark's	7681	7.38	170
Mode superposition and central difference	7812	7.33	172
Mode superposition and Newmark's	7681	7.38	170

Figure 6. Results obtained with Newmark's method after 80, 200, 460 and 7681 steps (from left to right and top to bottom). In the back is the target object for comparison.

Using the modal matching with local search described earlier and rubber as the virtual material for the objects, 37 of the 47 nodes of the initial object are successfully matched (Fig. 5). To simulate the given objects deformation, we used critical damping between 1% and 2% and chose iterations to stop when the nodal displacements norm is lesser than 5×10^{-4} pixels. Other relevant constants used are: $\Delta t = 0.3$s, $k = 10^6$ N/m, $c_u = 0$, $c_v = 0$. Under these conditions, the steps and the time each one of the integration methods needed to complete the computation process are displayed in Table 5, as well as the distance between the last estimated shape and the target object. In Figure 6, we can see some intermediate shapes estimated with Newmark's method, one of the fastest ones.

Changing the time step to 100 s and the stiffness constant to 10^7 N/m, we can not use the central difference method because the time step is too large, but we achieve a distance of 5% of the initial one in only 39 seconds using Newmark's method or the mode superposition with Newmark's method. In Figure 7, we can

Figure 7. Final estimated object obtained with Newmark's method (or mode superposition with the Newmark method) to achieve a distance between the final estimated object and the target one lesser than 5% of the original distance.

see the final results obtained with those methods compared with the target object – the fact that we can not see it means the results are good.

6 CONCLUSIONS

In this paper we described a physical approach to simulate objects deformation in images. In this approach we used two stopping criteria to end the resolution of the dynamic equilibrium equation: one based in the nodal displacements norm, the other based in the distance between objects. This last stopping criterion allows a significant decrease on computational cost without a great loss in accuracy.

We proposed a solution that enables the application of the used approach on objects that do not have all of their nodes successfully matched. That solution consists on applying to those nodes implicit loads that depend on their matching affinities.

The experimental results obtained in the matching process and in the estimation of the involved deformation, some presented in this paper, are coherent with the physically expected behaviour of the objects modelled, validating the used approach.

As the central difference method has first order precision and its stability is conditioned by the chosen time step; when the mode superposition method is employed with it the results are of second order but the accuracy still depends on a small time step. However, if the mode superposition method is used with Newmark's method, then the results are equal to the ones obtained by Newmark's method but can spend a few more seconds in the computation process.

Hence, when comparing the results obtained by the used integration methods, we would recommend Newmark's method to solve the dynamic equilibration equation. This suggestion is based on the accurate results that can be obtained and on the speed of the associated computation process.

Although our results are quite satisfactory, the computation process is not very fast when it comes to 3D objects. So, in the future, we could try parallel implementations and different integration methods to solve the dynamic equilibrium equation faster. Also, we can try other approaches to determine the implicit loads in the unmatched nodes.

ACKNOWLEDGMENTS

The presented work was partially done in the scope of the project "Segmentation, Tracking and Motion Analysis of Deformable (2D/3D) Objects Using Physical Principles", with reference POSC/EEA-SRI/55386/2004, financially supported by *FCT – Fundação para a Ciência e a Tecnologia*.

The second author would like to thank the support of the grant SFRH/BD/12834/2003 of the *FCT*.

REFERENCES

Bastos, L. & Tavares, J. 2006. Matching of objects nodal points improvement using optimization. *Inverse Problems in Science and Engineering* 14 (5): 529–541.

Bathe, K.-J. 1996. *Finite Element Procedures*. New Jersey: Prentice-Hall.

Cook, R., Malkus, D. & Plesha, M. 1989. *Concepts and Applications of Finite Element Analysis*. New York: John Wiley and Sons.

Horn, B. 1987. Closed-Form Solution of Absolute Orientation using Unit Quaternions. *Journal of the Optical Society of America* 4 (4): 629–642.

Newmark, N. 1959. A Method of Computation for Structural Dynamics. *ASCE Journal of the Engineering Mechanics Division* 85 (3): 67–94.

Pentland, A. & Horowitz, B. 1991. Recovery of Nonrigid Motion and Structure. *IEEE Transactions on Pattern Analysis and Machine Intelligence* 13 (7): 730–742.

Pinho, R. & Tavares, J. 2004. *Morphing of Image Represented Objects Using a Physical Methodology*. 19th ACM Symposium on Applied Computing, Nicosia, 14–17 March 2004.

Sclaroff, S. 1995. *Modal Matching: a Method for Describing, Comparing, and Manipulating Digital Signals*. PhD Thesis. Massachusetts Institute of Technology.

Sclaroff, S. & Pentland, A. 1995. Modal Matching for Correspondence and Recognition. *IEEE Transactions on Pattern Analysis and Machine Intelligence* 17 (6): 545–561.

Shapiro, L. S. & Brady, J. M. 1992. Feature-based correspondence: an eigenvector approach. *Image and Vision Computing* 10 (5): 283–288.

Tavares, J. M. 2000. *Análise de Movimento de Corpos Deformáveis usando Visão Computacional*. PhD Thesis. Faculdade de Engenharia da Universidade do Porto.

Tavares, J. M., Barbosa, J. & Padilha, A. 2000. *Matching Image Objects in Dynamic Pedobarography*. RecPad'2000 – 11th Portuguese Conference on Pattern Recognition, Porto, 11–12 May 2000.

Tavares, J. M. & Bastos, L. 2005. Improvement of Modal Matching Image Objects in Dynamic Pedobarography using Optimization Techniques. *Electronic Letters on Computer Vision and Image Analysis* 5 (3): 1–20.

Terzopoulos, D., Platt, J., Barr, A. & Fleischer, K. 1987. Elastically deformable models. In M. C. Stone (ed.) *Proceedings of the 14th Annual Conference on Computer Graphics and interactive Techniques*. SIGGRAPH '87, Anaheim, 27–31 July 1987. New York: ACM Press.

Terzopoulos, D., Witkin, A. & Kass, M. 1988. Constraints on deformable models: recovering 3D shape and nongrid motion. *Artificial Intelligence* 36 (1): 91–123.

Computational Modelling of Objects Represented in Images –
João Manuel R.S. Tavares & R.M. Natal Jorge (eds)
© 2007 Taylor & Francis Group, London, ISBN 978-0-415-43349-5

An evolution model of parametric surface deformation using finite elements based on B-splines

Manuel González-Hidalgo
Computer Graphics and Vision Group, Maths and Computer Science Department,
University of the Balearic Islands, Spain

Arnau Mir & Gabriel Nicolau
Mathematics and Computer Science Department, University of the Balearic Islands, Spain

ABSTRACT: In this paper we introduce a dynamic evolution model in order to deform parametric surfaces. In order to do it, we present the associate variational formulation to the problem of minimize an energy functional. Its numerical resolution is developed using a finite element method based on B-splines. We compute the spatial discretitation where the finite elements are defined and we show that a reduced number of control points are deformed instead of all the surface points.

1 INTRODUCTION

The deformation models include a large number of applications, and they have been used in fields as the edge detection, computer animation, geometric modelling, and so on. In this work, a deformation model will be introduced that uses B-splines as finite elements. This theory was introduced by Höllig in (Höllig 2003). In fact, we have used a variational formulation similar to the used one in (Cohen 1992) changing, among other things, the selected finite elements.

The most used finite elements for surfaces are the triangles, squares, among others; with them, it is easy to make a mosaic that it fills all space, but this technique needs long computation time since we must use some big data structures to solve our problem. This data structure are bound by the quantity of surface points that we must to take to obtain a good approximation of the surface.

In this work, we have applied the finite elements method based on B-splines to solve numerically a partial differential equation problem. The advantage to use B-splines finite elements is that it combines the computational advantage of B-splines and standard mesh-based elements. Thus, we will obtain a data structure smaller than the obtained one using the usual finite elements.

This work is organized as follows. In section 2, we define the uniform B-splines used to introduce finite elements. First, we define one dimensional B-splines and we display a recurrence relation to evaluate the numerical value in a point of the B-spline. Moreover we present a recurrence formula in order to compute the B-splines derivatives. Next, we define the multivariate B-splines. To finish this section, we define a B-spline parametric surface. Section 3 is devoted to introduce the model of surface deformation to which we will apply the finite elements methods using B-splines. The numerical resolution is focused in section 4. First, we raised the variational formulation to follow with its numerical resolution, introducing the finite element B-splines. Next, we compute the spatial and temporal discretization where the finite elements are defined. We will show that a subset of the control points are deformed instead of all the surface points. The model evolution will be also introduced. In the next section, several numerical computer deformations are displayed using the evolution model with different forces. Finally, some conclusions are exposed.

2 B-SPLINES

The B-splines are piecewise polynomial functions. It has been verified, with others approximations functions technics (Piegl 1997) that the polynomials provide a good local approximation for smooth functions.

However, if we use large intervals, accuracy of the approximation can be very low, the exactitude of the approach could be very low and local changes have global influence. Therefore, it is natural to use piecewise polynomials, defined on a fine partition of the function domain. For this reason, we will use piecewise polynomials approximation, and of between all of them, we choose the B-splines.

2.1 B-splines functions

Uniform B-splines can be defined in several ways (de Boor 1978) (Farin 1997) (Piegl 1997) (Höllig 2003). In this work we have taken the definition given by Höllig in (Höllig 2003), which we describe next.

Definition 2.1 *(Höllig 2003) An uniform B-spline of degree n, b^n, is defined by*

$$b^n(x) = \int_{x-1}^{x} b^{n-1}(t)dt$$

starting with $b^0(x) = \begin{cases} 1, & x \in [0,1), \\ 0, & \text{otherwise}. \end{cases}$

The previous definition is not adapted for numerical evaluations. In order to be able to evaluate the B-splines in a simple form and fast computationally, we can use the recurrence equation. This equation was given by De Boor (de Boor 1978) and Cox (Cox 1972), and it is a linear combination of smaller degree B-splines.

$$b^n(x) = \frac{x}{n}b^{n-1}(x) + \frac{(n+1-x)}{n}b^{n-1}(x-1) \quad (1)$$

In order to construct the finite elements bases, we will use a scaled and translated uniform B-spline. They are defined by transforming the standart uniform B-spline, b^n, to the grid $h\mathbb{Z} = \{\ldots, -2h, h, 0, h, 2h, \ldots\}$, where h is the scaled step.

Definition 2.2 *The transformation for $h > 0$ and $k \in \mathbb{Z}$ is $b_{k,h}^n(x) = b^n(\frac{x}{h} - k)$. The support of this function is $[k, k+n+1)h$*

In order to make a variational formulation of a differential equation problem, we will need the derivatives of the finite elements. From the definition 2.1 we obtain that the first order derivative of degree n B-spline is given by

$$\frac{d}{dx}b^n = b^{n-1}(x) - b^{n-1}(x-1)$$

with $b^n(0) = 0$ (Höllig 2003). If we apply the transformation given in definition 2.2, the first order derivative of the transformed B-spline is given by

$$\frac{d}{dx}b_{k,h}^n = h^{-1}(b_{k,h}^{n-1}(x) - b_{k+1,h}^{n-1}(x)) \quad (2)$$

(Höllig 2003) We also need the derivatives of any order. These ones are given by a linear combination of lower degree B-splines. The differentiation formula can be expressed in a compact form as follows.

Theorem 2.1 *The m-th derivative of a degree n transformed B-spline following the definition 2.2 is given by the recurrence relation*

$$\frac{d^m}{dx^m}b_{k,h}^n(x) = h^{-m}\sum_{i=0}^{m}(-1)^i \begin{pmatrix} m \\ i \end{pmatrix} B_{k+i,h}^{n-m}(x) \quad (3)$$

Obviously, this equation has sense if $m \le n$ since in others cases the derivative is 0.

2.2 Multivariate B-splines

There is no unique generalization of one dimensional B-splines. These generalizations differs in the underlying partition for the polynomial segments (de Boor, Höllig, and Riemenschneider 1993) (Piegl 1997), (Höllig 2003),. A possibility is to form products of one dimensional B-splines, as described in the following construction. The N-variate B-spline of degree $\mathbf{n} = (n_1, \ldots, n_N)$, of index $\mathbf{k} = (k_1, \ldots, k_N)$ and the space discretization $\mathbf{h} = (h_1, \ldots, h_N)$ is defined as

$$B_{\mathbf{k},\mathbf{h}}^{\mathbf{n}}(\mathbf{x}) = \prod_{i=1}^{N} b_{k_i,h_i}^{n_i}(x_i). \quad (4)$$

The support of this function is $\prod_{i=1}^{N}[k_i, k_i + n_i + 1)h_i$.

Applying basic properties of differential calculus and applying theorem 2.1 we can obtain a compact

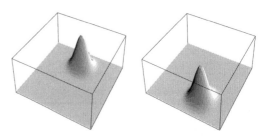

Figure 1. Bicubic B-splines with scaled step $h = 1/6$. Left: using a translation $k = (4, 9)$. Right: using a translation $k = (9, 0)$.

expression for any partial derivative of multivariate B-spline. We note that using theorem 2.1, we can evaluate the derivatives and they are obtained with a smaller computational cost, since less recurrences are applied.

In the next figure we show the graph of several bidimensional B-splines.

2.3 Parametric surfaces with B-splines

A parametric surface is defined as $S : \Omega \subset IR^2 \to IR^3$, $(u, v) \mapsto S(u, v) = (x(u, v), y(u, v), z(u, v))$ with the necessary degree of differentiability (do Carmo 1976), where Ω is a bounded bidimensional subset. This surface will be B-spline if we can put it as a linear combination of bidimensional B-splines. That is,

$$S(\mathbf{x}) = \sum_{\mathbf{k} \in \mathbb{Z}^2} P_{\mathbf{k}} B_{\mathbf{k},\mathbf{h}}^{\mathbf{n}}(\mathbf{x}) \qquad (5)$$

where $\mathbf{n} \in IN^2$, and $\mathbf{h} \in IR^2$ with positive coordinates. The coefficients $P_{\mathbf{k}} \in IR^3$ are called *control points* and they are the elements that determine the B-spline surface.

In order to be able to work with finite elements, we will need bases with a finite number of elements. The parametric surfaces we will use must have a bounded domain. Consequently, all $P_{\mathbf{k}}$ will be zero except a finite number of them. In order to find this set of control points, we must find the relevant B-splines. These ones fulfill $Sup(B_{\mathbf{k},\mathbf{h}}^{\mathbf{n}}) \bigcap \Omega \neq \varnothing$.

The relevant B-splines of our surface are determined by the spatial discretization, since the B-splines support depends on them. This problem will be addressed in the section of numerical resolution (section 4).

3 DEFORMATION MODEL MINIMAL SURFACES

The deformation model is based on an associated energy to one surface, that it checks the shape of it.

The energy function is a non convex function with a local minimum. The goal is to achieve this minimum using an evolution model. This minimum depends on the initial surface and the used evolution model.

The associated energy functional, $E : \Phi(S) \to IR$, $S \mapsto E(S)$, is defined as

$$E(S) = \int_{\Omega} \omega_{10} \left| \frac{\partial S}{\partial u} \right|^2 + \omega_{01} \left| \frac{\partial S}{\partial v} \right|^2 + \omega_{11} \left| \frac{\partial S}{\partial u \partial v} \right|^2$$

$$+ \omega_{20} \left| \frac{\partial^2 S}{\partial u^2} \right|^2 + \omega_{02} \left| \frac{\partial^2 S}{\partial v^2} \right|^2 + \mathcal{P}(S(u, v)) du dv$$

(Terzopoulos 1986), (Cohen 1992), (Montagnat, Delingette, and Ayache 2001), where $\mathcal{P}$ is a potential of the forces that works on the surface. Using the equations of Euler-Lagrange, it can be proved (Cohen 1992) that an energy local minimum must be satisfied:

$$-\omega_{10} \frac{\partial^2 S}{\partial u^2} - \omega_{01} \frac{\partial^2 S}{\partial v^2} + 2\omega_{11} \frac{\partial^4 S}{\partial u^2 \partial v^2} + \omega_{20} \frac{\partial^4 S}{\partial u^4} \qquad (6)$$

$$+ \omega_{02} \frac{\partial^4 S}{\partial v^4} = -\nabla \mathcal{P}(S(u, v)) + \text{boundary conditions}$$

The surface domain is $\Omega = [0, 1]^2$ and we take as boundary conditions $S(u, 0) = (u, 0, 0)$, $S(u, 1) = (u, 1, 0)$, $S(0, v) = (0, v, 0)$, $S(1, v) = (1, v, 0)$.

4 NUMERICAL RESOLUTION

4.1 Variational formulation

With the purpose of establishing the variational formulation of the boundary value problem done by (6), we recall the definition of a Sobolev Space of order two $H^2(\Omega)$,

$$H^2(\Omega) = \{ S \in L^2(\Omega) : \frac{\partial^{(\alpha_1 + \alpha_2)} S}{\partial x_1^{\alpha_1} \partial x_2^{\alpha_2}} \in L^2(\Omega),$$

$$0 \leq \alpha_1 + \alpha_2 \leq 2, \alpha_1, \alpha_2 \in IN \}$$

We will consider the set of functions $(H^2(\Omega))^3$ satisfying the previous boundary conditions. We will denote this set by $\mathcal{H}$.

The weak formulation of the equation (6) is:

$$\int_{\Omega} \left(\omega_{10} \frac{\partial S}{\partial u} \frac{\partial T}{\partial u} + \omega_{01} \frac{\partial S}{\partial v} \frac{\partial T}{\partial v} + 2\omega_{11} \frac{\partial^2 S}{\partial u \partial v} \frac{\partial^2 T}{\partial u \partial v} \right.$$

$$\left. + \omega_{20} \frac{\partial^2 S}{\partial u^2} \frac{\partial^2 T}{\partial u^2} + \omega_{02} \frac{\partial^2 S}{\partial v^2} \frac{\partial^2 T}{\partial v^2} \right) du dv$$

$$= - \int_{\Omega} \nabla \mathcal{P}(S) T du dv \qquad (7)$$

where the functions S, T belongs to $\mathcal{H}$ and u, v are the spatial variables.

It can be proved (Cohen 1992) (Raviart 1992) that to solve equation (6) is equivalent to find an element $S \in \mathcal{H}$, such that $a(S, T) = L(T)$ for all $T \in \mathcal{H}$. Where $a(\cdot, \cdot)$ is a bilinear form defined as

$$a(S, T) = \int_{\Omega} \left(\omega_{10} \frac{\partial S}{\partial u} \frac{\partial T}{\partial u} + \omega_{01} \frac{\partial S}{\partial v} \frac{\partial T}{\partial v} + \right.$$

$$\left. 2\omega_{11} \frac{\partial^2 S}{\partial u \partial v} \frac{\partial^2 T}{\partial u \partial v} + \omega_{20} \frac{\partial^2 S}{\partial u^2} \frac{\partial^2 T}{\partial u^2} + \omega_{02} \frac{\partial^2 S}{\partial v^2} \frac{\partial^2 T}{\partial v^2} \right) du dv$$

(8)

and $L(\cdot)$ is the following linear form

$$L(T) = -\int_\Omega \nabla P(S)T\,du\,dv$$

4.2 Discretization

We want to find a function $S \in \mathcal{H}$ such that

$$a(S,T) = L(T) \ \forall T \in \mathcal{H} \qquad (9)$$

In order to do this, the surface domain will be discretized. But, first of all, we have to find a set of functions of finite dimension. The B-splines defined in section 2 will be the finite elements that we will use as the base of our function set. The problem is to find the relevant B-splines. That is, the B-splines satisfying $Sup(B^n_{k,h}) \bigcap \Omega \neq \varnothing$, and from this, the set of index $\mathbf{k}$ of the B-splines that satisfies the boundary conditions. Therefore, we want to evolve $N_1 \times N_2$ control points of the B-spline surface S. To do it, we need $N_1 \times N_2$ bidimensional B-splines such that its support will be in Ω.

The solution $S \in \mathcal{H}$ is a B-spline surface of degree $n = (n_x, n_y)$. The surface domain is discretized by $h_1\mathbb{Z} \times h_2\mathbb{Z}$ where $h_1 = \frac{1}{N_1+n_x-1}$ and $h_2 = \frac{1}{N_2+n_y-1}$. This spatial discretization will fix the control points that are not zero. The index $\mathbf{k} = (k_1, k_2)$ belongs to the set $\{-n_x, \ldots, N_{1+n_x-1}\} \times \{-n_y, \ldots, N_{2+n_y-1}\}$. So, the B-spline surface will come determined by the relevant B-splines, and they are specified in the following equation

$$S(u,v) = \sum_{k_1=-n_x}^{N_1+n_x-1} \sum_{k_2=-n_y}^{N_2+n_y-1} P_{(k_1,k_2)} B^{\mathbf{n}}_{(k_1,k_2)\mathbf{h}}(u,v) \quad (10)$$

In addition, deforming only the corresponding $N_1 \times N_2$ control points of the B-spline surface (10), we made sure that the boundary conditions are satisfied.

The set of finite elements of finite dimension that determines our B-spline bases is given by $V^{\mathbf{n}}_{\mathbf{h}} = <\{(B^{\mathbf{n}}_{\mathbf{k,h}}(u,v),0,0) : \mathbf{k} \in \mathcal{I}\} \cup \{(0, B^{\mathbf{n}}_{\mathbf{k,h}}(u,v),0) : \mathbf{k} \in \mathcal{I}\} \cup \{(0,0, B^{\mathbf{n}}_{\mathbf{k,h}}(u,v)) : \mathbf{k} \in \mathcal{I}\} >$ where $\mathcal{I} = \{0, \ldots, N_1-1\} \times \{0, \ldots, N_2-1\}$.

Thus, taking into account the boundary conditions, the control points $P_{\mathbf{k}}$ associated to B-splines belonging to the set $V^{\mathbf{n}}_{\mathbf{h}}$ are the unique ones that are computed using the equations (11) and (12) (see below).

Using equations (9) and (10) we can obtain three linear systems, one for each coordinate:

$$AP_i = L_i, \quad i = 1,2,3, \qquad (11)$$

where A is a square matrix and their elements are:

$$a((B^n_{k,h}, 0, 0), (B^n_{j,h}, 0, 0))_{(k,j) \in \mathcal{I} \times \mathcal{I}},$$

P_i is a vector of component i of each control point and L_i is a vector with components $L_1 = L((B^n_{k,h}, 0, 0))_{k \in \mathcal{I}}$, $L_2 = L((0, B^n_{k,h}, 0))_{k \in \mathcal{I}}$, $L_3 = L((0,0, B^n_{k,h}))_{k \in \mathcal{I}}$.

The static problem has been introduced. Next, we will construct the evolution model. The classical dynamical model of evolution has been applied (Cohen 1992), (Qin 1997), (Montagnat, Delingette, and Ayache 2001), (González, Macaró, Mir, Palmer, and Perales 2001), (Mascaró 2002) (Mascaro, Mir, and Perales 2002):

$$M\frac{d^2 P_i}{dt^2} + C\frac{dP_i}{dt} + AP_i = L_i, \ i = 1,2,3. \qquad (12)$$

where M and C are the mass and damping matrices respectively and are diagonal matrices. The dynamic system (12) has been discretized in time by the finite difference scheme using central differences, obtaining an explicit numerical scheme.

When we applied the dynamic model, the surface depends on time. So, we have $S(u,v,t)$. Nevertheless, this dependency only affects to the control points which is an advantage since in each iteration we do not have to calculate all the surface. Therefore, we only must calculate the new control points. The used numerical scheme in the dynamic model depends on two previous iterations $P^{t-\Delta t}$ and P^t, but as it is not a physical method, we have taken the same P^0 and P^1, where $P^0 = S(u,v,0)$.

5 EXAMPLES

This section shows several examples of deformations obtained applying our dynamical model. All the experiments displayed in this section has been made using $\omega_{10} = \omega_{01} = 0.1$ and $\omega_{11} = \omega_{20} = \omega_{02} = 0.01$ and taking a temporal step $t = 0.1$.

In Figure 2, we show several iterations obtained using the dynamic model with bicubic B-splines, a positive force in the direction $(0,0,1)$ and $N_1 \times N_2 = 25$, The force is applied in an unique point of the surface.

Figure 3 shows the deformation obtained using bicubic B-splines, two forces simultaneously in opposite sense in the direction $(0,0,1)$ and $N_1 \times N_2 = 25$. The force is applied in two nearly points of the surface.

Not only we can apply forces in vertical directions. Also, we can apply forces in other directions as we can see in Figure 4, where we have applied over all the surface domain an oblique force in the direction $(1,1,4)$, with module 106, taking bicubic B-splines and $N_1 \times N_2 = 49$.

As we can observe in equation 8, the bilinear form has five parameters. They represent the resistance to the deformation. In Figure 5, we can see the same

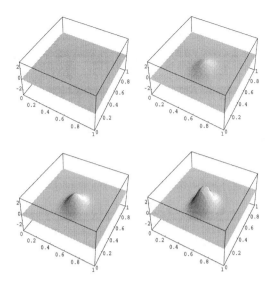

Figure 2.   Dynamic simulation of a plane deformation using a positive constant force with direction $(0, 0, 1)$.

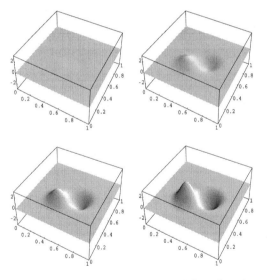

Figure 3.   Dynamic simulation of a plane deformation using two constant forces simultaneously in opposite sense.

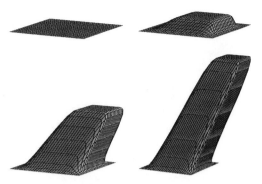

Figure 4.   Dynamic simulation of a plane deformation using a force in the direction $(1, 1, 4)$.

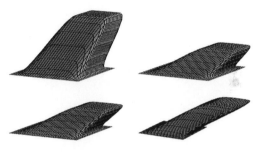

Figure 5.   Several iterations of a plane deformation using the force $(1, 1, 4)$ and different values of the bilinear form parameters. For details see the text.

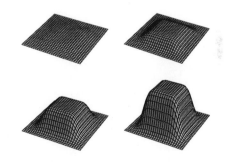

Figure 6.   Several iterations of a plane deformation using the vertical force $(0, 0, 200)$ over all the surface.

iteration (iteration 6) of the experiment displayed in Figure 4 changing the bilinear form parameters. From top to buttom and from left to right these paremeters are: $\omega_{10} = \omega_{01} = 0.1$ and $\omega_{11} = \omega_{20} = \omega_{02} = 0.01$ (see Figure 4); $\omega_{10} = \omega_{01} = 0.1$, $\omega_{11} = 1$ and $\omega_{20} = \omega_{02} = 0.01$; $\omega_{10} = \omega_{01} = 10$, $\omega_{11} = 1$ and $\omega_{20} = \omega_{02} = 0.01$ and $\omega_{10} = \omega_{01} = 100$, $\omega_{11} = 1$ and $\omega_{20} = \omega_{02} = 0.01$, respectively.

Figure 6 shows several iterations obtained using the vertical force $(0, 0, 200)$ over all the surface, using

bicubic B-splines and $N_1 \times N_2 = 36$. We can compare this deformation with the obtained one in Figure 2.

The next figures, Figure 7 and Figure 8, display experiments using sinusoidal forces. In Figure 6 we use the force given by $\nabla \mathcal{P}(u, v) = (50 \sin v\pi, 50 \sin u\pi, 200 \cos u\pi \sin v\pi)$, B-splines of degree $\mathbf{n} = (4, 4)$ and $N_1 \times N_2 = 36$. In Figure 7 we take the force defined by $\nabla \mathcal{P}(u, v) = (-50 \sin u\pi, 0, 200 \cos v\pi \sin u\pi)$, taking bicubic B-splines and $N_1 \times N_2 = 36$.

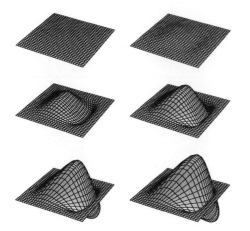

Figure 7. Dynamic simulation of a plane deformation using a sinusoidal force. Details in the text.

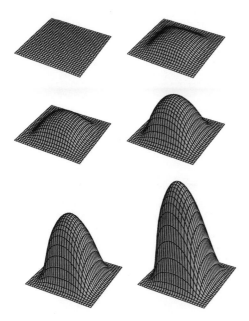

Figure 8. Deformations obtained using other sinusoidal force different from the applied one in the figure 7. Details in the text.

6 CONCLUSIONS AND FUTURE WORK

As we can see with the examples, we have shown the viability to make deformations of parametric surfaces using the variational formulation of the model, and solving it numerically using finite elements based on B-splines. In addition, as a certain number of control points only evolves, a very efficient scheme is obtained. In the context of future work we would like to highlight the following: a study of the stability

and complexity of the model and a complete study of the influence of each parameter in the deformation obtained by the model. That is, the influence of them in the resistance to length deformation, to shear deformation and to bend deformation, as well as the use of other energy functional, among others. At the present, we are working on the implementation of this model using C++ and Coin3D, a 3D modeling toolkit which simplifies visualization and scene composition tasks.

ACKNOWLEDGEMENTS

This work is supported by the project TIC-2004-07926-E, INEVAI3D, of the Spanish Government. The authors would like to thank to the department of Mathematics and Computer Science of University of the Balearic Islands.

REFERENCES

Cohen, I. (1992). Modèles Déformables 2-D et 3-D: Application à la Segmentation d'Images Médicales. Ph. D. thesis, Université Paris IX, Dauphine.

Cox, M. G. (1972). The numerical evaluation of b-splines. *IMA Journal of Applied Mathematic 10(2)*, 134–149.

de Boor, C. (1978). *A practical guide to splines*. New York: Springer Verlag.

de Boor, C., K. Höllig, and S. Riemenschneider (1993). *Box Splines*. New York: Springer Verlag.

do Carmo, M. (1976). *Differential Geometry of Curves and Surfaces*. Prentice-Hall.

Farin, G. (1997). *Curves and Surfaces for Computer- Aided Geometric Desing: A Practical Guide*. Academic Press.

González, M., M. Mascaró, A. Mir, P. Palmer, and F. Perales (2001). Modeling and animating deformable objects. In *Proceedings of IX Spanish Symposium on Pattern Recognition and Image Analysis,* Benicasim, Castellón, Spain, pp. 279–290. AERFAI Society.

Höllig, K. (2003). *Finite element methods with B-Splines*. Frontiers in Applied Mathematics. Philadelphya: SIAM.

Mascaró, M. (2002). *Modelo de Simulación de Deformaciones de Objetos Basado en la Teoría de la Elasticidad*. Ph. D. thesis, Universitat de les Illes Balears.

Mascaro, M., A. Mir, and F. Perales (2002). P^3DMA : A physical 3D deformable modelling and animations system. *LNCS 2492*, 68–79.

Montagnat, J., H. Delingette, and N. Ayache (2001). A review of deformable surfaces: topology, geometry and deformation. *Image and Vision Computing 19*(14), 1023–1040.

Piegl, L. & Tiller, W. (1997). *The NURBS book*. Berlin: Springer Verlag.

Qin, H & Terzopoulos, D. (1997). Triangular nurbs and their dynamic generalizations. *Computer Aided Geometric Design 14*, 325–347.

Raviart, P. A. & Thomas, J. M. (1992). *Introduction à l'analyse numérique des équations aux derivées partielles*. Paris: Masson.

Terzopoulos, D. (1986). Regularization of inverse visual problems involving discontinuities. *IEEE PAMI 8*(4), 413–424.

Computational Modelling of Objects Represented in Images –
João Manuel R.S. Tavares & R.M. Natal Jorge (eds)
© 2007 Taylor & Francis Group, London, ISBN 978-0-415-43349-5

Influence of sphericity parameter on the detection of singularities in synthetic images

Céline Gouttière[1], Ghislain Lemaur[2] & Joël De Coninck[3]

University of Mons-Hainaut, CRMM Parc Initialis, Avenue Copernic 1, Bât. Materia Nova, Mons, Belgium

ABSTRACT: It has been demonstrated previously that isotropy of wavelets is advantageous for the detection of pointwise singularities. The influence of the isotropy is studied here for a broader class of operators. We compare Gabor filters, which provide a relatively high directionality, the spherical wavelet and the Mexican Hat function, which is isotropic. From experiments on different backgrounds and with defects of varying size, we observe a clear influence of sphericity on the detection of small singularities.

1 INTRODUCTION

One of the key aspects in the computer analysis of images is the extraction of the information from the image. Usually, most of the useful information is provided by the singular part of the signal. Wavelet analysis is one of the preferred tools of image analysis for extracting this part (Antonini et al. 1992; Hsieh et al. 1997). The choice of the best wavelet basis function has been addressed previously particularly for singularity detection (Mallat and Hwang 1992; Lemaur 2003; Lemaur et al. 2003). A new property of separable filters was introduced: the sphericity. This new criterion proved useful for creating new wavelet filters. These filters make it possible to detect singularities which have different intensities and sizes, in synthetic and real images (like mammograms) with great efficiency. The influence of the isotropy is studied here for a broader class of operators: the nearly isotropic wavelets, the Mexican Hat filters and the Gabor filters.

2 SYNTHETIC IMAGES

Synthetic images are chosen to have a tunable roughness. Only one parameter, denoted h, is used to vary the roughness of the background texture. Three values of h are tested: 0.4, 0.6 and 0.8. An example of these images is shown in figure 1.

Defects are added in these images. These defects are pointwise singularities and are represented by a cone.

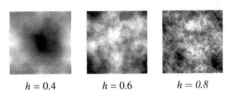

$h = 0.4$ $\qquad$ $h = 0.6$ $\qquad$ $h = 0.8$

Figure 1. Synthetic images with various roughnesses.

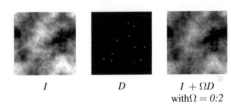

I $\qquad$ D $\qquad$ $I + \Omega D$
with $\Omega = 0.2$

Figure 2. Synthetic image with singularities.

Their radius and height can be adjusted. The studied radius of the defects to detect lie in the range 1–10 pixels. The height can be modified by a parameter Ω. A defect is less visible if Ω is low, see example in figure 2. The image to be filtered is the synthetic image with the attenuated defects.

3 EXAMINED FILTERS

Three families of wavelets and filters are examined. These are the nearly isotropic wavelets, the Mexican Hat filters, and the Gabor filters.

3.1 *Nearly isotropic wavelets*

A new family of wavelet filters with a specific property of rotational invariance has been created (Lemaur

[1] celine.gouttiere@crmm.umh.ac.be

[2] ghislain.lemaur@crmm.umh.ac.be

[3] joel.de.coninck@crmm.umh.ac.be

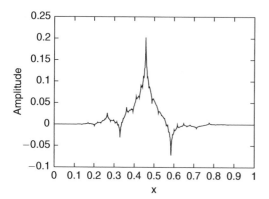

Figure 3. ψ.

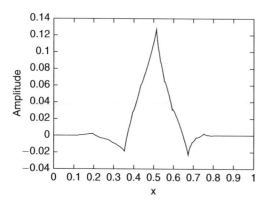

Figure 4. ϕ.

2003). The nearly isotropic wavelets are characterized by a wavelet function and a scale function with an optimized sphericity, denoted ψ and ϕ (see figures 3 and 4). Their efficiency to detect singularities was previously demonstrated in the case of synthetic images, and also in mammograms (Lemaur et al. 2003). Details about the construction of these wavelets can be found elsewhere (Lemaur and De Coninck 2003).

3.2 Mexican Hat Filters

The Mexican Hat filters are completely isotropic, but are not separable. These filters have already been used for the defects detection (Chen and Qiu 2001). The Mexican Hat function is

$$\psi(x,y) = \left(1 - \frac{x^2+y^2}{2\sigma^2}\right) e^{-\frac{x^2+y^2}{2\sigma^2}}. \qquad (1)$$

Table 1. Sphericity values of the examined filters.

	Sphericity values
Gabor $\lambda = \mathbf{2}$	0.180
Gabor $\lambda = \mathbf{1.5}$	0.239
Gabor $\lambda = \mathbf{1.25}$	0.286
Gabor $\lambda = \mathbf{1}$	0.354
ϕ	0.802
ψ	0.840
Mexican Hat	0.919

3.3 Gabor filters

The Gabor filters are used in various applications of the image processing like defect detection (Tsai Lin, and Huang 2005) or texture analysis (Manthalkar, Biswas, and Chatterji 2003; Tanaka, Yoshida, Fukami, and Nakano 2004). These filters present the advantage to be tunable in scale and orientation. Indeed, the Gabor function consists of an sinusoid modulated by a gaussian envelope in the space domain. This 2D function has the following general form:

$$\psi(x,y) = \frac{1}{2\pi\sigma_x\sigma_y} e^{-\frac{1}{2}\left(\frac{x^2}{\sigma_x^2}+\frac{y^2}{\sigma_y^2}\right)} \cos(2\pi W x), \qquad (2)$$

where W is the modulation frequency of the filter, and σ_x, σ_y define the gaussian envelope size.

The frequencies are multiplied by a parameter λ to obtain several shapes of the Gabor function, without changing the envelope, and so to have different values of sphericity. The tested values of λ are 1, 1.25, 1.5, and 2.

4 EXPERIMENTAL RESULTS

4.1 Sphericity of examined filters

The first step of this study is the computation of the sphericity measure of the examined filters. This measure is obtained from a comparison between the filter with an isotropic separable function, which is the gaussian function. The sphericity measure S_f and the sphericity values of the studied filters are presented in the equation 3 and in the table 1.

$$S_f = \max_{\sigma \in R_0^+} \left| \frac{\langle f, g_\sigma \rangle}{\| f \| \| g_\sigma \|} \right|, \qquad (3)$$

where $g_\sigma = e^{-\frac{x^2}{\sigma}}$.

4.2 Singularities detection

The comparison between the filters is based on the mean and standard deviation of $\Delta\Omega$ which is the

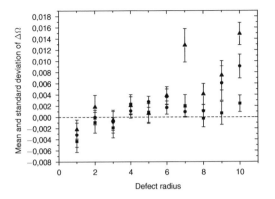

Figure 5. Comparison of the families of Gabor filters with $\lambda = 1$: (▲) $\lambda = 2$; (•) $\lambda = 1.5$; (■) $\lambda = 1.25$.

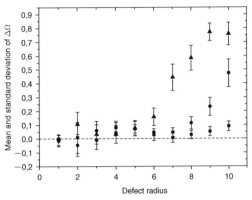

Figure 7. Comparison of the families of Gabor filters with $\lambda = 1$: (▲) $\lambda = 2$; (•) $\lambda = 1.5$; (■) $\lambda = 1.25$.

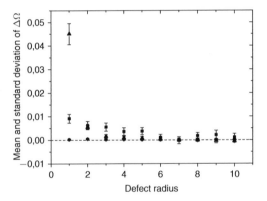

Figure 6. Comparison of all families of filters with ϕ: (▲) Mexican Hat; (•) ψ; (■) Gabor $\lambda = 1$.

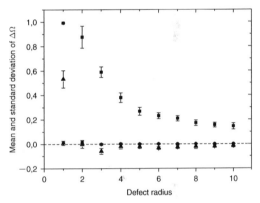

Figure 8. Comparison of all families of filters with ϕ: (▲) Mexican Hat; (•) ψ; (■) Gabor $\lambda = 1$.

difference between the minimal Ω obtained for the detection and a filter of reference. The first comparison concerns the families of Gabor filters with $\lambda = 1$ being the reference. For the second comparison, all three families of filters are compared with ϕ chosen as reference.

4.2.1 Synthetic images with a low roughness background

The graphs of the figures 5 and 6 presents a summary of the results obtained for the detection on synthetic images with a low roughness background. In the first graph with the comparison of the different families of Gabor filters, we observe that a smaller λ leads to a better detection for defect radius higher than 5 pixels. From the second graph, we notice the same order in the families of filters for the detection of high defects than those found with the sphericity values.

4.2.2 Synthetic images with a high roughness background

The tendencies observed for the detection of singularities in synthetic images with a low roughness background are even more visible when the synthetic images have a high roughness background. The graphs of the results are presented in the figures 7 and 8.

5 CONCLUSION

A link exists between the sphericity of the filters and the efficiency of the detection: filters which have a greater sphericity are also more efficient at detecting singularities. Two important parameters influence the efficiency of the detection. These are the level of the roughness of the synthetic images and the radius of the defects. The order of the filters is the same with regards to values of both sphericity and efficiency of the detection, particularly for synthetic images with a high roughness background.

ACKNOWLEDGMENTS

The authors would like to acknowledge the support provided by the *Fonds pour la Formation à la Recherche dans l'Industrie et dans l'Agriculture* in Belgium and the *Belgian Federal Science Policy Office*.

REFERENCES

Antonini, M., M. Barlaud, P. Mathieu, and I. Daubechies (1992). Image coding using wavelet transform. *IEEE Transactions on Image Processing 1*(2), 205–220.

Chen, C. and G. Qiu (2001). Detection algorithm of particle contamination in reticle images with continuous wavelet transform. In *Proceedings of the British Machine Vision Conference 2001*, Manchester, UK.

Hsieh, J.-W., M.-T. Ko, H.-Y. M. Liao, and K.-C. Fan (1997). A new wavelet-based edge detector via constrained optimization. *Image and Vision Computing 15*, 511–527.

Lemaur, G. (2003). *On the choice of the wavelet basis function for image processing*. Ph. D. thesis, University of Mons-Hainaut, Belgium.

Lemaur, G. and J. De Coninck (2003). Sphericity of wavelets may improve the detection of singularities in images. In *Proceedings of Computing Engineering in Systems Applications*, Lille, France.

Lemaur, G., K. Drouiche, and J. De Coninck (2003). Highly regular wavelets for the detection of clustered microcalcifications in mammograms. *IEEE Transactions on Medical Imaging 22*, 393–401.

Mallat, S. G. and W. L. Hwang (1992). Singularity detection and processing with wavelets. *IEEE Transactions on Information Theory 38*(2), 617–643.

Manthalkar, R., P.-K. Biswas, and B.-N. Chatterji (2003). Rotation invariant texture classification using even symmetric gabor filters. *Pattern Recognition Letters 24*(12), 2061–2068.

Tanaka, H., Y. Yoshida, K. Fukami, and H. Nakano (2004). Texture segmentation using amplitude and phase information of gabor filters. *Electronics and Communications in Japan part IIIFundamental Electronic Science 87*(4), 66–79.

Tsai, D.-M., C.-P. Lin, and K.-T. Huang (2005). Defect detection in coloured texture surfaces using gabor filters. *The Imaging Science Journal 53*(1), 27–37.

Computational Modelling of Objects Represented in Images –
João Manuel R.S. Tavares & R.M. Natal Jorge (eds)
© 2007 Taylor & Francis Group, London, ISBN 978-0-415-43349-5

3D morpho-topological analysis of asymmetric neuronal morphogenesis in developing zebrafish

S. Härtel, J. Jara, C.G. Lemus & M.L. Concha

Anatomy and Developmental Biology Program, Faculty of Medicine, Universidad de Chile, Santiago, Chile

ABSTRACT: We applied *in vivo* confocal microscopy of GFP-transgenic zebrafish in combination with 3D image analyses to study the asymmetric morphogenesis of the diencephalic parapineal organ on a supra-cellular, cellular, and sub-cellular level. Following a rough manual segmentation of the respective regions of interest (ROIs), the morphology of generated surface meshes was refined by an active surface model which iteratively adjusts the mesh towards the morphology of the cellular structures. This procedure is essential for a precise morpho-topological analysis, mostly because of the adversarial diffraction limited resolution in the z-dimension of confocal image stacks. 3D Morphology and topology of the reconstructed cellular and supra-cellular structures during morphogenesis was quantified by principal axis transformations and 3D moment invariants. Our data indicates that migration of the parapineal organ is accompanied by a rapid transition between predominantly parallel cell orientations towards predominantly perpendicular orientations, a phenomenon which requires a precise control of cell shape and polarity. The orientational transition is followed by a phase of polarized cell motility in which membrane protrusions in the form of blebs and filopodia become oriented in the direction of the asymmetric migration. The morpho-topological descriptors unveil information that is not perceptible for a direct visual analysis of the microscopical data sets. This approach becomes essential to access morphogenetic mechanisms which control asymmetry and migration.

1 INTRODUCTION

1.1 *Left-right asymmetry in the zebrafish brain*

Despite our increasing understanding of the mechanisms that control left-right asymmetry in the heart, little is known of the morphogenetic mechanisms that establish lateralized circuitry in the vertebrate brain. In recent years we have studied the mechanisms that control the asymmetric development of neuronal nuclei in the zebrafish brain (Concha et al. 2000/2003; Concha 2004). Development of brain asymmetry is preceded by left-sided expression of a number of genes, and is characterized by the allocation of a single midline nucleus on the left side (the parapineal organ), and by left-right differences in the pattern of neuronal differentiation within a paired nucleus (the habenulae). Asymmetric morphogenesis of the parapineal organ is a key event as it induces further asymmetric morphogenesis, and involves the migration of parapineal precursors from their place of origin at the midline towards the left side of the brain (Concha et al. 2003; Gamse et al. 2005). To understand the morphogenetic mechanisms responsible of this phenomenon, we have recently initiated a 3D morphological and topological description of the cell behaviours underlying the asymmetric positioning of the parapineal organ.

1.2 *Morpho-topological analysis of the parapineal organ*

Image processing routines have become indispensable for the reliable detection and quantification of light-microscopic data. Sophisticated algorithms not only restore optical information on the level of data acquisition (see 2.3 below), they become increasingly important for the characterization and quantification of morpho-topological information in biological structures. In the past, we have developed diverse approaches to access biochemical and structure related information on a 2D level (e.g. Carrer et al. 2003; Härtel et al. 2003/2005a/b; Alvarez et al. 2005). In this contribution we continue our approach to reveal biologically relevant information on a 3D level. As shown recently, regulation of cellular volume and morphology can be monitored with great precision in cultured HeLa cells (Castro et al. 2006). In order to derive morpho-topological information from the parapineal cells and the parapineal organ during the asymmetric morphogenesis in zebrafish embryos, we present an approach based on principle axis transformation and the calculation of rotation, translation, and scale invariant moments. This method has recently been suggested for morphometry of cortical sulci by Mangin et al. (2004). For 3D presentation of cell surfaces and calculation of surface related parameters, we adopt an active surface model introduced by Ahlberg (1996) based on the formulation from Kass et al. (1988). Analytical characterizations are performed on different scales of cellular organization: a supra-cellular scale (pineal and parapineal organ), a cellular scale (parapineal cells), and a sub-cellular scale (blebs and filopodia).

2 MATERIAL & METHODS

2.1 Manipulation of GFP-transgenic zebrafish

Transgenic zebrafish expressing green fluorescent protein (GFP) in the dorsal diencephalon were generated previously (Concha et al. 2003), and kept under standard laboratory conditions in the Fish Facility of the Faculty of Medicine, Universidad de Chile. Embryos between 24 and 38 hours post-fertilisation were manually dechorionated, anesthetized with Tricaine (3-amino benzoic acid ethy lester, Sigma), and mounted in a custom-made chamber in 1% agarose dissolved in embryo medium (Concha et al. 2003). After the end of the experiment, embryos were removed from the chamber and sacrificed with a Tricaine overdose.

2.2 Acquisition of confocal microscopic images

The morphogenesis of the parapineal organ in living GFP-transgenic zebrafish embryos was followed by confocal epifluorescence microscopy (Leica TCS SP) under controlled temperature conditions (28°C). 3D Image stacks, intensity $I \in [0, 255]$, were captured with a 63x (NA 0.9) water-immersion objective, excitation/emission at 488/505–560 nm, with typical xyz stack $[512 \times 512 \times 70$ voxels] and voxel $[0.116 \times 0.116 \times 0.5 \, \mu m/voxel]$ dimensions. The selected vertical sampling distances guaranteed reliable volume determinations and shape analysis.

2.3 Deconvolution and image analysis

Huygens Scripting (Scientific Volume Imaging BV, www.svi.vl, Hilversum, Netherlands), using a Maximum Likelihood Estimator (MLE) algorithm was used to deconvolve 3D confocal data in order to reveal hidden details of the biological structures and to improve the signal to noise ratio significantly. All further image processing routines for visualization and morpho-topological analysis were written in our laboratory, using Interactive Data Language (IDL, ITT, www.ittvis.com/idl/, CO, USA).

2.4 Cell segmentation and surface reconstruction

A first approximation of ROIs including sub-cellular structures (blebs or filopodia) and parapineal cells was obtained manually by drawing object contours into the xy-frames of the deconvolved image stacks, using an Interactive Pen Display (Cintiq-15X, Wacom). Binary masks of the ROI were generated with a custom-made macro written for the public domain image analysis software Image-SXM (http://www.liv.ac.uk/~sdb/imageSXM). The definition of cellular contours in 3D was improved significantly by the application of a modified version of an active surface model that was originally presented by Ahlberg (1996) expanding the 2D active contour model developed by Kass et al. (1988). The active surface model parameterizes internal forces like elasticity (α) or rigidity (β), which counteract line tension or curvature and mimic intrinsic physical properties of a deformable surface.

Following Xu & Prince (1998), surface tension and curvature are induced by external force fields, which are derived from the intensity gradients and laplacians of the image data by an iterative algorithm. The external forces are parameterized by so-called Generalized Gradient Vector Flows (GGVF) and consist of 3D vector fields, which attract surface points towards the object borders. Force balance between internal and external forces is solved by the Euler-Lagrange condition for the minimization of an energy functional E for a parametric surface $\underline{C}(s) = [x(s), y(s), z(s)], s \in [0,1]$.

$$E = \int_0^1 0.5 \cdot [\alpha |\underline{C}'(s)|^2 + \beta |\underline{C}''(s)|^2] + E_{ext}(\underline{C}(s)) ds \quad [1]$$

For the cellular structures presented in this work, an initial surface mesh is derived from the manually approximated ROI contours.

Precise assimilation of the active surface mesh towards the morphology of the cellular structures was supervised interactively by setting the appropriate parameter combination to the following coefficients: α, β, viscosity (γ), external force (k) and iterations (t).

2.5 Calculation of surface curvature

Surface curvature (κ) was calculated for each mesh node in respect to its adjacent neighbours. For this purpose spheres with radius ($r = \kappa^{-1}$) were fitted to each node and 3 of its neighbours. For nodes with more than 3 neighbours, we calculate κ for all possible 3 neighbour combinations and derived a mean value for κ.

2.6 Invariant moments

Eigen vectors ($\underline{e}_{1,2,3}$) and Eigen values ($\lambda_{1,2,3}$) of segmented objects were calculated on three levels of spatial organization: (*i*) on the sub-cellular level (blebs and filopodia, $\varnothing = 1$–$3\,\mu$m), (*ii*) on the cellular level (individual parapineal cells, $\varnothing = 5$–$15\,\mu$m), and (*iii*) on the supra-cellular level (parapineal organ, $\varnothing = 20$–$100\,\mu$m). Eigen vectors and values were derived from the inertia tensor T (or covariance matrix) by Householder reduction and the QL method (based on the routine tqli described by Press et al. 1992). T is required in order to determine the rotation of a rigid body around its centre of mass. In analogy to mechanical physics, the segmented structures were considered to have a uniform mass distribution $\rho(x,y,z) = 1$ in each segmented voxel. The Eigen values and vectors were ordered in respect to the size of λ_i, which directly represents the rotational inertia in respect to each axis $\underline{e}_i$. The Eigen systems $\underline{E} = [\lambda_1 \cdot \underline{e}_1, \lambda_2 \cdot \underline{e}_2, \lambda_3 \cdot \underline{e}_3]$ were used to access object morphologies (morphometry), spatial orientation between objects of the same level of organization (Fig. 1), and spatial organization between the different levels of organization (Fig. 2). In addition, segmented parapineal cells were transformed into their corresponding Eigen systems $\rho(x,y,z) \rightarrow \rho(x',y',z')$ and aligned in box pattern in order to improve the visual perception of more subtle geometric features.

Object morphometry was parameterized by translation-, rotation-, and scale-invariant descriptors (μ'_{ijk} invariant moments of order *ijk*), which were calculated according to Castleman (1996):

$$\mu_{ijk} = \int\limits_{-\infty}^{\infty}\int\limits_{-\infty}^{\infty}\int\limits_{-\infty}^{\infty}(x'-\bar{x})^i\,(y'-\bar{y})^j\,(z'-\bar{z})^k\,\rho(x',y',z')dx'dy'dz' \quad [2]$$

$$\mu'_{ijk} = \mu_{ijk}\,/\,\mu_{000}^{((i+j+k)/3)+1} \quad [3]$$

3 RESULTS

Fig. 1 shows a temporal series of 3D reconstructions in which the topological reorganization of the parapineal organ is depicted at supra-cellular and cellular levels. The initial phase of asymmetric migration is revealed as a gradual movement of the parapineal organ to the left in respect to the mayor symmetry axis of the entire pineal complex. This phenomenon is accompanied by oscillations in the orientation of the 1st principal axis of the parapineal organ and by changes in the orientation of the 1st principal axes of individual parapineal cells: at early stages principal axes show a predominantly parallel alignment, which is gradually lost during morphogenesis.

Fig. 2 shows a quantitative analysis of the temporal reorganization of the alignments of the 1st principal axes of the parapineal cells in respect to each other (red circles) and in respect to the 1st principal axes of the entire parapineal organ (white squares), which accompanies the asymmetric cell migration described in Fig. 1. Both parameters indicate a reorganization of a

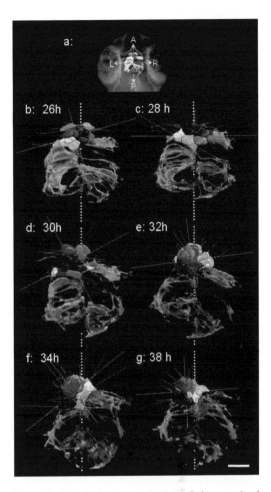

Figure 1. Topological reorganization of the parapineal organ during asymmetric morphogenesis. [a] Scheme shows the transparent head of a developing fish embryo and the fluorescence of the GFP-transgenic pineal complex (green). The parapineal organ is located at the anterior-most end of this complex. [b–g] Pineal complex (green) and colour-coded active surface models of individual cells of the parapineal organ at 26, 28, 30, 32, 34, and 38 hours post-fertilisation. Left-right symmetry axes of the pineal complex are shown by dotted, vertical lines. Thick and thin grey lines depict principal axes of the entire parapineal organ and of individual reconstructed cells, respectively. Abbreviations: anterior (A), posterior (P), left (L), right (R). Scale bar = $20\,\mu$m.

predominantly parallel axes alignment (26–28 h post-fertilization) towards a predominantly perpendicular alignment (32–38 h), undergoing an intermediate phase of random distribution (30 h). The transition between the parallel alignment toward the predominantly perpendicular orientation between the principal axes of individual cells and the axis of the entire parapineal organ occurs within a time span of 2–4 h.

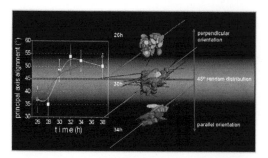

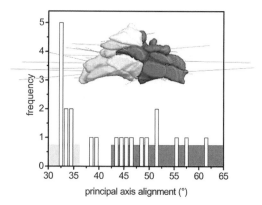

Figure 2. Principal axes alignment during parapineal morphogenesis. [Left] Alignment between the 1st principal axis of individual parapineal cells (red circles) is compared to the alignment between the 1st principal axis of the individual cells with the 1st principal axis of the entire parapineal organ (white squares). The parameter 'principal axis alignment' represents the mean values calculated from the respective angular distributions in three independent experiments. Mean values were connected by β-spline curves. Error bars represent standard deviations. [Centre-Right] Representative surface reconstructions of the parapineal organ based on active contours in combination with the respective 1st principal axis of individual parapineal cells (red lines) and of the entire parapineal organ (grey lines) at 26, 30, and 34 hours post-fertilisation. The reconstructions open a direct visual access to the data presented in the plot. The data reveals a transition of the organization of parapineal cells from a predominantly parallel orientation towards predominantly perpendicular orientation.

Figure 3. Asymmetric distribution of the axial alignment of parapineal cells during morphogenesis. The histogram plots the frequency distribution of the parameter 'inter cellular alignment of the 1st principal axis (e_1)' for a representative parapineal cell population at 28 h post fertilisation. Three intervals of the angular alignment were colour coded according to the predominant orientation: parallel (green), random (yellow), and perpendicular (red) (see Fig. 2). Colours were projected to the surface of the corresponding parapineal cells.

Further information about the process of reorganization during the orientational transition presented in Fig. 2 was obtained by a more detailed analysis of the distribution of the angles between the 1st principal axes of individual parapineal cells. Fig. 3 connects the spatial allocation within the 3D reconstructions with the parameter distribution of a sample population that enters the phase of orientational transition (28 h).

As can be observed, cells with a predominantly parallel axes alignment form the left wing of the parapineal organ (green), while cells that diverge from the parallel alignment constitute the right wing of the organ (red). In addition, cells with an intermediate degree of alignment are located close to the centre (yellow).

On a final level of 3D topological analysis, we analysed the distribution of sub-cellular structures (membrane protrusions) within cells of the parapineal organ. Fig. 4 visualizes the sub-cellular structures for a wild type and a mutant embryo and shows the spatial distribution of membrane protrusions in the form of blebs (red) and filopodia (yellow). In the wild type, parapineal membrane protrusions show a polarised behaviour during the asymmetric morphogenesis as they concentrate on the left side to which the parapineal organ migrates. In contrast to the wild type,

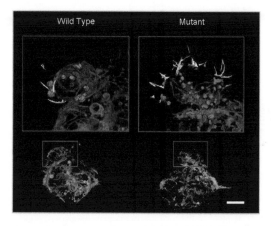

Figure 4. Spatial distribution of sub-cellular structures within the parapineal organ of wild type (left) and mutant (right) zebrafish embryos. As can be observed in the upper 3D reconstructions, membrane protrusions in the form of blebs (red), and filopodia (yellow) are oriented in the direction of the asymmetric migration toward the left side of the wild type pineal complex. In contrast, blebs and filopodia are symmetrically oriented in the mutant. Scale bar = 20 μm.

protrusion polarisation is lost in a mutant embryo that is characterized by an impaired parapineal migration.

4 DISCUSSION

Principal axes transformation and invariant moments have been introduced a few decades ago for the characterisation and recognition of pattern in 2D images by Hu (1962), and for the analysis of 3D data sets by Lo & Don (1989). The method presents a standard mathematical tool that is applied in a wide range of disciplines including statistics, quantum mechanics, or classical mechanics. In biological and medical science however, the application of this attractive tool is rather sparse.

Recently, 3D invariant moments have been reported to provide good discriminatory power to identify handedness- and sex-correlated shapes of the cortical sulci in brain morphometry (Mangin et al. 2004). The general low acceptance of invariant moments in biological and medical science might be caused by the slightly abstract formalism (see [2,3]). Nevertheless, the invariants up to the power of 2 have a direct physical relevance: they define the principal rotational axis of a solid body of homogeneous or heterogeneous mass distribution ($\rho(x,y,z) = 1$ or $\rho(x,y,z) = f(x,y,z)$).

The calculated Eigen system $\underline{E} = [\lambda_1 \cdot \underline{e}_1, \lambda_2 \cdot \underline{e}_2, \lambda_3 \cdot \underline{e}_3]$ directly presents the rotational symmetry axes ($\underline{e}_{123}$, axes where the angular momentum is unchanged, unless an external torque is applied) in combination with its respective rotational inertia (λ_{123}). Length and orientation of projected axes ($\lambda_i \cdot \underline{e}_i$) therefore represent and quantify an essential visual characteristic of a solid body or cellular surface (see Fig. 1–3): the longest principal axis ($\lambda_1 \cdot \underline{e}_1$) generally aligns with the elongated axis of a cellular body.

Principal axes become explicitly powerful when it comes to characterize the orientation of individual cells which form a complex supra-cellular structure. In case of the developing parapineal organ, an agglomeration of 15–20 cells already makes it difficult to visually explore the orientation of each individual cell (see Fig. 1–3). Therefore the morpho-topological descriptors unveil information that is not perceptible for a direct visual analysis of the microscopical data sets. For example, the transition of the orientational organisation within the parapineal organ could not be perceived visually without the projection of the principal axes (Fig. 1).

The rapid access to statistical properties of the axes components on a cellular and a supra-cellular level permitted to determine the temporal interval for the orthogonal transition in parapineal cell organisation (Fig. 2). Finally, the back projection of colour coded parameter intervals revealed a supra-cellular pattern inside the parapineal organ (Fig. 3). Altogether, the analysis allowed us to detect a morphogenetic asymmetry prior to the lateral migration of the organ (compare 28 h and 30 h in Fig. 1 with Fig. 3). The detected internal cellular pattern directly leads to the question of the underlying biological mechanisms.

In conclusion, a relatively simple mathematical analysis unveils new biological questions in respect to the first phenomena of asymmetry during parapineal morphogenesis. We are presently expanding the analysis in all levels of organization and include mutant zebrafish embryos (Fig. 4) to address the biological implication of the observed phenomena.

ACKNOWLEDGEMENTS

This work was supported by FONDECYT 1060890, PBCT ACT 47, and ICM P04-068-F.

REFERENCES

Ahlberg, J. 1996. Active Contours in Three Dimensions. Thesis project done at Computer Vision Laboratory, Linköping University, Sweden.

Alvarez, M.; Godoy, R.; Heyser, W. & Härtel, S. 2005. Anatomical-physiological determination of surface bound phosphatase activity in ectomycorrhiza of Nothofagus obliqua based on image processed confocal fluorescence microscopy. *Soil Biology and Biochemistry* 37(1): 125–132.

Castleman, K.R. 1996. *Digital Image Processing*. Prentice Hall, Englewood Cliffs, NJ, USA.

Castro, J.; Ruminot, I.; Porras, O.; Flores, C.; Hermosilla, T.; Verdugo, E.; Härtel, S. & Barros, L.F. 2006. ATP steal: a novel mechanism linking Na+ with the onset of necrotic Ca2+ overload. In press: *Cell Death & Differentiation*.

Carrer, D.; Härtel, S. & Maggio, B. 2003. Ceramide Modulates the Lipid Membrane Organization at Molecular and Supramolecular Levels. *Chemistry and Physics of Lipids* 122: 147–152.

Concha, M.L.; Burdine, R.D.; Russel, C.; Schier, A.F. & Wilson, S.W. 2000. A Nodal signalling pathway regulates the laterality of neuroanatomical asymmetries in the zebrafish forebrain. *Neuron* 28: 399–409.

Concha, M.L.; Russell, C.; Regan, J.C.; Tawk, M.; Sidi, S.; Gilmour, D.; Kapsimali, M.; Sumoy, L.; Goldstone, K.; Amaya, E.; Kimelman, D.; Nicolson, T.; Grunder, S.; Gomperts, M.; Clarke, J.D. & Wilson, S.W. 2003. Local tissue interactions across the dorsal midline of the forebrain establish CNS laterality. *Neuron* 39: 423–438.

Concha, M.L. 2004. The dorsal diencephalic conduction system of zebrafish as a model of vertebrate brain lateralisation. *Neuroreport* 15: 1843–1846.

Gamse, J.T.; Kuan, Y.S.; Macurak, M.; Brosamle, C.; Thisse, B.; Thisse, C. & Halpern, M.E. 2005. Directional asymmetry of the zebrafish epithalamus guides dorsoventral innervation of the midbrain target. *Development* 132: 4869–4881.

Härtel, S.; Zorn-Kruppa, M.; Tykhonova, S.; Alajuuma, P.; Engelke, M. & Diehl, H. 2003. Staurosporine-Induced Apoptosis in Human Cornea Epithelial Cells In Vitro. *Cytometry* 08: 15–23.

Härtel, S.; Fanani, M.L. & Maggio, B. 2005a. Shape transitions and lattice structuring of ceramide-enriched domains generated by sphingomyelinase in lipid monolayers. *Biophysical Journal* 88: 287–304.

Härtel, S.; Rojas, R.; Räth, C.; Guarda, M.I. & Goicoechea, O. 2005b. Identification and Classification of Di- and

Triploid Erythrocytes by Multi-parameter Image Analysis: A New Method for the Quantification of Triploidization Rates in Rainbow Trout (Oncorhynchus mykiss). *Archivos de Medicina Veterinaria* 37(2): 147–154.

Hu, M.K. 1962. Visual pattern recognition by moment invariants. *IRE Transactions on Information Theory* 8 (February): 179–187.

Kass, M.; Witkin, A. & Terzopoulos, D. 1988. Snakes: Active Contour Models. *International Journal of Computer Vision* 1: 321–331.

Lo, C.-H. & Don, H.S. 1989. *3D moment forms: their construction and application to object identification and positioning. IEEE PAMI* 11 (October): 1053–1064.

Mangin, J.F.; Poupon, F.; Duchesnay, E.; Riviere, D.; Cachia, A.; Collins, D.L.; Evans, A.C. & Regis, J. 2004. *Brain morphometry using 3D moment invariants.* Medical Image Analysis 8: 187–196.

Press, W.H.; Teukolsky, S.A.; Vetterling, W. & Flannery, B.P. 1992. *Numerical Recipes in C: The Art of Scientific Computing* (2nd edition), section 11.3, Cambridge University Press.

Xu, C. & Prince, J.L. 1998. Generalized gradient vector flow external forces for active contours. *Signal Processing* 71: 131–139.

Computational Modelling of Objects Represented in Images –
João Manuel R.S. Tavares & R.M. Natal Jorge (eds)
© 2007 Taylor & Francis Group, London, ISBN 978-0-415-43349-5

Applied methods for transparent materials inspection

K. Horák & I. Kalová

Brno University of Technology, Department of Control and Instrumentation, Brno, Czech Republic

ABSTRACT: Main topic of the article is application methods for defect detection on transparent materials. Paper describes entire evaluation process from image scan procedure right to object classification. In article are describe techniques for image scanning, techniques for elimination of different optic attributes of transparent materials, furthermore techniques for image processing and finally classification techniques. In part describing image attribute classification procedure is shown standard rule classifier and besides that also classifier based on neural network. In this paper individual techniques are introduced at the same sequence as a real process is executed on the each inspected object.

1 INTRODUCTION

Processional automation is more and more required on whole world. Branch monitoring quality of industrial processes sets for academic workplaces the complicated task: development of autonomous inspection systems. Visual inspection systems which use programmable cameras are most technical sophisticated systems today. Their disadvantages are ordinarily high price and other cost and necessary higher sophisticated approach. Vice versa advantages are high reliability, speed, accuracy and last but not least also academic research and development support. One of them is visual system for glass bottle inspection. Many defects with various characteristics features can occur on one concrete bottle. Broken neck, dirt on bottle side, strange thing thrown into a bottle, bad bottle type, all of this can be considered to be a defect. Size of enumerated and some others faults is ordinarily only a few millimetres. Because of this tiny defects and high performance of manufacturing plant can inspection process effectively realizes only with the assistance of latest cameras systems. Image processing methods are very heavy on computational performance due to necessary use special hardware devices as control cards, industrial computers etc. Individual methods for acquiring image, image processing and classification which can be applied to industrial process are subject of interest on almost all technical college.

2 STATE-OF-ART

2.1 Image acquiring methods

Each image scan procedure requires three fundamental components: sensor (camera), light source and inspected object. There are two basic categories for acquiring image of transparent material in machine vision domain. First of them is based on reflection technique. Camera is scanning image created by light reflection on the object. If camera with CCD sensor is used we can to take advantage of spectral sensitivity in infrared zone. Spectral sensitivity for ordinary silicon CCD chips is from 400 up to 1100 nm. Ergo light sources are realized as IR sources (for example as LED matrices with dominant wavelength go beyond 800 nm). Coupling of CCD camera and IR light source is convenient also from perspective of scanned material (glass bottle). Metal-plate surfaces of non-eroded bottle's necks these wavelengths perfectly reflect, whereas eroded bottle's necks not (metal-plate layer missing). Generalized spectral sensitivity of CCD chip is shown on Figure 1.

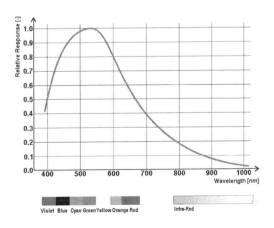

Figure 1. Spectral sensitivity of CCD sensor.

Figure 2. (a) Bottle's image without polarising filter in front of camera, (b) Bottle's image with polarising filter.

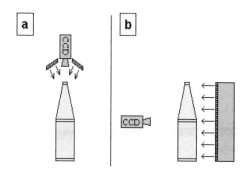

Figure 3. (a) Reflection, (b) Trans-illumination technique.

In this scanning method can take advantage of using circular polarising filter placed in front of camera lens. This one filters reflection of foreign sources of visible-radiation. Result of polarisation filter usage is shown on Figure 2. Bottle's illuminated by external source of laser radiation. Source such as this can occur very easily in industry plant (for example sensor or detector lights).

Second category takes advantage of transparent material basic property and it's called as trans-illumination scanning technique. Object (bottle) is brightened by light source towards camera sensor in this case. Both basic techniques are shown on following Figure 3.

For acquiring image of neck can use reflection method, vice versa trans-illumination method can be used for bottom and bottle side image acquiring. Place of image scanning must be always protected from surroundings sources of lights (ambient light, foreign emitter, etc.).

2.2 Elimination of optic attributes

Images must be very homogenous for correct and robust image processing. Transparent materials often have different optic attributes, which have to be

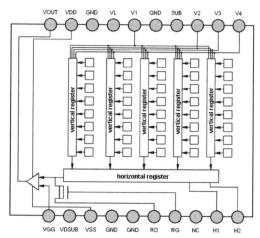

Figure 4. Control signals of CCD sensor.

eliminated already on analogue level of image scan. Different optic attributes of glass bottles are caused by glass colour (apart from other things).

Compensation can be generally non-linear and is always based on exposure control (differential bottle's optic attributes can't be changed evidently). Be in existence three primary techniques. First is control of camera shutter, second is control of flash time (flash of light source) and last one is control of intensity of flash of light source. Usually camera shutter is driven with the assistance of analogue circuits and optic probes. Exposure time is short for light materials and on the contrary long for dark materials. This simple principle secures homogenous input data to image processing methods. Principle of compensation can be simply described by following relation:

Where E(t) is intensity of chip illumination and integral represents total chip illumination at exposure. Compensation is solved if total chip illumination is invariable for all scanned objects (bottles).

CCD camera exposure is technically solved as periodically suction of electric charge from photo-sensitive layer. This is realized by SUBST signal (substrate clocks). Control signals and matrix architecture of CCD sensor is shown on following Figure 4.

$$\int_{start\_exp}^{stop\_exp} E(t) \cdot dt = konst.$$

Exposure is in progress if electric charge isn't suction from silicon structure. In moment of charge suction is exposure stopped. Exposure time and moment of latent image transfer to vertical registers are shown on Figure 5.

As is obvious from picture above, exposure can be started and stopped in unvarying steps 64 μs (in case

222

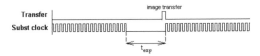

Figure 5. Exposure control signals.

Figure 6. (a) Image of dark and light bottles without analogue elimination of optic attributes, (b) Image of the same bottles with analogue elimination of optic attributes.

of industrial camera Modicam with Lattice signal box and sensor Sony ICX DLA). Transfer of latent image from matrix structure to vertical registers begin tightly before end of exposure time (before begin of charge suction from silicon substrate). Then matrix structure is again ready to accumulate electric charge to photosensitive cells.

3 SYSTEM DESCRIPTION

3.1 Image acquiring sequence

As described above, for acquiring image is used matrix camera with CCD chip, which provide ability to change control signals. Maximum of shutter time is limited due to high linear speed of moving bottles on industrial belt. Maximal exposure time value depends on physical size of detected faults. Bottles are moving about 36000 units per hour usually. It means moving of bottle over one millimetre in one millisecond. Limitary exposure time is one millisecond if in detection process is needed to find faults sized about one millimetre. Exceeding this border could cause slurred image what means worse processing. Shutter time is control according to amount of light pass through the bottle. This light quantum is measured by optic probe and subsequently analogue processed. Shutter time is control by feedback. There are shown examples of images in relation to exposure time on the Figure 6.

Light source is fastening electronically. Its active state begins before exposure start and always takes the same time (constant period of flash). Final exposure

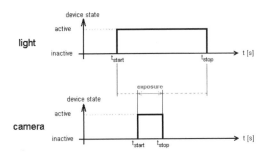

Figure 7. Final exposure time.

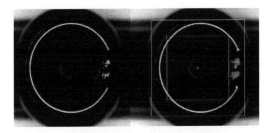

Figure 8. (a) Original image, (b) Founded clusters.

is given by intersection of activity of light source and camera (Figure 7.).

Camera exposure always finishes before end of flash of light source. In this way is final exposure given by camera exposure time. On the contrary intensity of ambient light has to be matchlessly lower than intensity of special light source.

3.2 Image processing

Finding relevant indications at the picture is necessary for correct classification process. In first step goes image through noise filters and thereafter non-homogenous areas are founded by edge detection and cluster analysis methods. Structure of indication properties (such as absolute or relative image level, size, gradient difference etc.) is assigned to each individual non-homogenous area. Clusters are subsequently classified in accordance with this value.

Clusters finding is realized as suitable and functional sequence arrangement of typical image processing methods (for example simple or adaptive thresh-holding, edge detection, noise filtration, correlation, clusters analysis, etc.). A priori properties are also used – for instance knowledge about existence of light stripe on bottle's necks. Neck location on the image is exactly determined basely this stripe. Clusters on damaged bottle's neck are in graphical form shown on the right half of Figure 8.

Figure 9. Quantitatively and quantitative identical clusters (a) of false faults, (b) of factual faults.

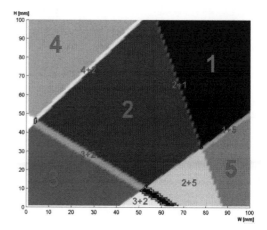

Figure 11. (a) Original image of bottle's bottom, (b) Classification with graphically marked fault area.

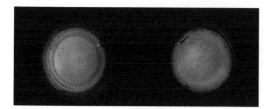

Figure 10. (a) Original image of bottle's side, (b) Classification result with graphically marked fault areas.

Figure 12. Two-dimensional space – width x height.

3.3 Classification

Two basic techniques are used for classification procedure. First of them is rule method. General clusters space can be divided to several linear sub-space by limited sequence of rules. Object is marked as suitable for next processes only if all clusters meet range of parameters. This method is typical example of using brute force, however for some simple cases is convenient and elegant.

Second classification method is based on neural network and in this time is in research sphere. Clear necessity of using neural network is given by very complicated clusters space. Clusters of some image objects are very similar and there is no way to simply separate them by rule classifier. As illustrate example is fault on the bottle's bottom, which is very easily interchangeable with bottle's bottom border feature (see on Figure 9.).

Such clusters are correct classified only by neural network system. Neural network inputs represent relevant indications. In reverse neural network outputs represent classification classes (big defect, small defect, noise etc.). Illustration of faulty bottle classification is shown on following Figure 10 and Figure 11.

Clusters space has quite high dimension. Meanly variant of algorithm is cluster space dimension from four up to eight or nine. This is depending on number of detected clusters. For example simple two-dimensional space determined by neural network is shown on Figure 12. Individual dimensions are represented by width and height of detected fault. Faults are classified to five predetermined classes – small fault (2), big fault (1), thin vertical fault (4), thin horizontal fault (5) and noise (3).

Crossing areas between individual classes are also shown on the graph. In these areas is necessarily determine winning class by some special procedure (for example by max method or weighted max method). Surfaces such on Figure 12 are totally 256 for next sign – brightness level e.g. (image is saved in 8 bit colour depth). It causes formation of three-dimension space. Next signs permanently increase dimension of classifier (clusters space). Example above of classification to five classes was simulated by neural network with these properties:

- Three layers (input, hidden and output layer)
- Two neurons in input layer (representation of fault width and height)

- Five neurons in output layer (output classes – Figure 12)
- Variable number of neurons in hidden layer
- Variable learning rate
- Transfer function: sigmoid
- Topology: multilayer perceptron
- Learning: LM (Levenberg-Marquardt) with back-propagation algorithm BP.

4 CONCLUSION

Designed and described methods were verified in laboratory and real conditions. Visual system using just introduced methods and techniques (see above) was successfully applied into industrial manufacturing plant with cooperation with developmental company CAMEA Ltd. Czech Republic. For example system is placed in Czech Republic (Samson, Hlinsko), Russia (Kirov), Poland (Jurajska), Latvia (Liepaja) etc. Visual system can generalize for inspect of random transparent materials due to programmable hardware and modifiable software. In this time is theoretically and practically too solved using neural network as classifier in general case of defect detection.

REFERENCES

Russ, J.C. 1995. *The Image Processing Handbook*, Florida: CRC Press.
Vernon, D. 1991. *Machine Visions.* New York: Prentice Hall.
Bajcsy, R., Lieberman, L. 1976. Computer Graphics and Image Processing.
Bates, R.H.T., MCDONNEL, M.J. 1986. *Image Restoration and Reconstruction*. Oxford: Clarendon Press.

Computational Modelling of Objects Represented in Images –
João Manuel R.S. Tavares & R.M. Natal Jorge (eds)
© 2007 Taylor & Francis Group, London, ISBN 978-0-415-43349-5

Region growing segmentation approach for image indexing and retrieval

Suhendro Irianto[1,2], JianMin Jiang[2] & Stan S. Ipson[2]
[1] *SPIE Student Member,* [2] *School of Informatics, University of Bradford, Bradford, United Kingdom*

ABSTRACT: We use region growing technique to segment the images. Based on the segmented region, we then select the size of the region to construct indexing keys. By using region growing technique on DCT image we reduce the number of region which is on the segmented regions only. Based on these regions, we then construct the indexing keys to match the images. Our technique will reduce the process time of constructing indexing keys. The indexing keys will then be constructed by calculating the regions distance. Our proposed of recursive region growing is not a new technique but its application on DCT images to build indexing keys is quite new and has not been presented by many other authors.

Keywords: Region growing, segmentation, image indexing, image retrieval.

1 INTRODUCTION

In the field of digital imaging, image segmentation plays a vital role as a preliminary step for high level image processing. To understand an image, one needs to isolate the objects in it and find relation among them. The process of image portioning is referred as image segmentation (Jain, 1998). In other words, segmentation is used to extract out the significant objects from the image.

Deng and Manjunath (Deng, 2001) proposed a JSEG algorithm to segment the image based on multiscale 'J-images'. The images which correspond to the measurements of local homogeneities at different scales are called 'J-images'. The system has the ability to segment colour-textured images without supervision. First the colour inside the image is quantized into several classes. The pixels are then replaced by their corresponding colour class label which forms the class map of the image. A region growing method is then used to segment the image based on multiscale 'J-images'.

Histogram thresholding is one of the common techniques for monochrome image segmentation (Pal, 1988; Weszka, 1978). This technique considers that an image consisting of different regions corresponds to the grey level ranges. The histogram of an image can be separated using peaks (modes) corresponding to different regions. A threshold value corresponds to the valley between two adjacent peaks that can be used to separate these objects (Sahoo,1988). One of the weaknesses of this method is that, it ignores the spatial relationship information of the pixels.

Boskovitz & Guterman (2002) proposed a neural network based on adaptive thresholding segmentation algorithm for monochrome image. The main advantage of his method is that, it does not require a priori knowledge about number of objects in the image.

To humans, an image is not just a random collection of pixels or DCT coefficients; it is a meaningful arrangement of regions and objects. There is also exist a variety of images such as natural scenes, and paintings. Despite the large variations of these images, humans have no problem to interpret them. Image segmentation is the first step in image analysis and pattern recognition. It is a critical and essential component of image analysis system, is one of the most difficult tasks in image processing, and determines the quality of the final result of analysis. Image segmentation is the process of dividing an image into different regions where each region is homogeneous.

Many content-based image retrieval (CBIR) systems have been developed since the early nineties. Most of the CBIR projects aimed at general-purpose image indexing and retrieval systems focus on searching images visually similar to the query image or a query sketch. They do not have the capability of assigning comprehensive textual description automatically to the images because of the great difficulty in recognizing a large number of objects.

Many researchers have attempted to use machine-learning techniques for image indexing and retrieval (Minka & Picard, 1997; Wiederhold & Wang, 1998). A system developed by Minka and Picard included a learning component. The system internally generated much segmentation or groupings of each image's regions based on different combination of features, then learned which combination best represented the semantic categories given as examples by the user. The system requires the supervised training of various parts of the image.

The remainder of the paper is organized as follows. Section 2 describes some of the basic JPEG compression. Section 3 discusses the method of indexing key based on region growing segmentation. Section 4 describes the experiment result. Section 5 as the last section, present conclusions and some remark for the future work.

2 JPEG IMAGE BASICS

To this point, we have defined functions to compute the DCT of a list of length n = 8 and the 2D DCT of an 8 × 8 array. We have restricted our attention to this case partly for simplicity of exposition, and partly because when it is used for image compression, the DCT is typically restricted to this size. Rather than taking the transformation of the image as a whole, the DCT is applied separately to 8 × 8 blocks of the image.

To compute a blocked DCT, we do not actually have to divide the image into blocks. Since the 2D DCT is separable, we can partition each row into lists of length 8, apply the DCT to them, rejoin the resulting lists, and finally transpose the whole image and repeat the process.

DCT-based image compression relies on two techniques to reduce the data required. Firstly, a quantization which is the process of reducing the number of possible values of a quantity, thereby reducing the number of bits needed to represent it. Secondly, entropy coding that is a technique for representing the quantized data as compact as possible. A function is then developed to quantize images and to calculate the level of compression provided by different degrees of quantization.

JPEG (Wallace, 1991) is a joint CCITT and ISO standard for compressing images developed by the Joint Photographic Experts Group. JPEG uses a combination of spatial-domain and frequency-domain coding. The image is divided into 8 × 8 blocks, each of which is transformed into the frequency domain using the discrete cosine transform (DCT). Each block of the image is thus represented by 64 frequency components. The signal tends to concentrate in the lower spatial frequencies, enabling high-frequency components, many of which are usually zero, to be discarded without substantially affecting the appearance of the image.

The main source of loss of information in JPEG is in a quantization of the DCT coefficients. A table of quantization coefficients is used, one per coefficient. It is usually related to human perception of different frequencies. The quantized coefficients are ordered in a "zig-zag" sequence, starting at the upper left (the DC component), where the most of the energy lies in the first few coefficients. The last step is entropy coding of the coefficients.

3 REGION GROWING BASED IMAGE INDEXING

Image segmentation is the first key process in numerous applications of computer vision. It partitions the image into different meaningful regions with homogeneous characteristics using discontinuities or similarities of image components, the subsequent processes depend on its performance. In most cases, the segmentation of colour image demonstrates to be more useful than the segmentation of monochrome image, because colour image expresses much more image features than monochrome image. In fact, each coefficient is characterized by a great number of combinations of R, G, B chromatic components. However, more complicated segmentation techniques are required to deal with rich chromatic information in the segmentation of colour images.

A variety of segmentation techniques have been proposed, recently. However, most techniques are kind of "dimensional extension" directly inherited from the segmentation of monochrome image (Bhanu,1994). The spatial compactness and colour homogeneity are two desirable properties in unsupervised segmentation, which leads to image-domain and feature-space based segmentation techniques. According to the strategy of spatial grouping, image-domain techniques include split-and-merge, region growing and edge detection techniques.

The segmentation of images has always been a key problem in computer vision. Up to the early nineties bottom-up techniques like edge detection and split-and-merge algorithms were the primary focus of research. However, by that time people realized that "perfect" segmentation would not be possible without incorporation of higher level knowledge. Thus the focus shifted towards model based on techniques like snakes (Kass & Witkin, 1996) and methods based on geometric models (Pope, 2004).

Region growing algorithms start from an initial, incomplete segmentation and try to aggregate the unlabelled DCT coefficient to one of the given regions. The initial regions are usually called seed regions or seeds. The decision whether a pixel should join a region or not is based on some fitness function which reflects the similarity between the region and the candidate pixel. As proposed in (Adams & Bischof, 1994) the order in which the pixel or DCT coefficients in this experiment is determined by a global priority queue which sorts all candidate pixels by their fitness values. This approach elegantly mixes local (fitness) and global (pixel order) information.

There is an abundance of researches on segmentation, and a number of review articles highlighting them (Koller & Sahami, 1996; Wang & Li, 2001). Some methods also have been defined for post processing the low-level segmentation to further regularize the segmentation output, such as Markov Random Fields (Luo & Guo, 2003).

Automatic image segmentation is one of the primary problems of early computer vision, it has been intensively studied in the past (Sonka & Hlavac, 1999). The existing automatic image segmentation techniques can be classified into four approaches, namely: thresholding techniques, boundary-based methods, region-based methods, and hybrid techniques.

Region-based techniques rely on the assumption that adjacent DCT coefficients in the same region have similar visual features such as grey level, colours value, or texture. A well-known technique of this approach is split and merge (Haralick & Shapiro, 1985; Hijjatoleslami & Kittler, 1998). Obviously, the performance of this approach basically depends on the selected homogeneity criterion.

Instead of tuning homogeneity parameters, the seeded region growing (SRG) technique is controlled by a number of initial seeds (Chang & Li, 1994; Jain, 1998). Given the seeds, SRG tries to find an accurate segmentation of images into regions with each connected component of a region meeting exactly one of the seeds. Moreover, high-level knowledge of the image components can be exploited through the choice of the seeds. This property is very attractive for semantic object extraction toward content-based image database applications. However, SRG suffers from another problem, how to select the initial seeds automatically for providing more accurate segmentation of the images.

Markov chain Monte Carlo algorithm for image segmentation has drawn considerable attention due to its ability to integrate texture, colour, and edge information in an optimal manner to devise a robust labeling of the image into homogeneous regions (Lee & Cohen, 2004; Tu & Zhu, 2002). These methods still depend on the assumption that the pixels belonging to the object of interest share a common set of low-level image attributes, thereby allowing the object to be extracted as a single entity. If an object is composed of multiple regions of differing texture or colour then the object is divided into regions corresponding to each of these, and these sub regions must then be re-assembled through some contextual-based post processing to segment the complete object from the image. By employing an additional constraint upon the segmentation that encourages a human to find, it would be possible to only extract the regions corresponding to the human in the image. This additional constraint can be provided through information regarding the desired shape of the final retained region.

The query process has been established as follows: Firstly, a user queries an image in the system, the image then is converted into a grayscale image. Secondly, by utilizing region growing algorithm this image will be segmented into regions. Finally based on these regions, the minimum distance between them will be calculated and compared to the image regions in the database. The examples segmented images retrieved with RGB image query can be illustrated in Fig. 3.

Once a query is specified, the system scores each segmented image based on how closely it satisfies the query. The score i for each atomic query (segmented image) is calculated by this using the equation.

$$d(H_q - H_k) = \frac{1}{64} \sum_{i=0}^{63} |h_{qi} - h_{ki}| \qquad (1)$$

where H_q and H_k are the query indexing keys and image indexing keys, respectively. The distance is equal to 0, if the image is identical in all the regions. We then rank the images according to the overall score and return to the twenty best matches.

4 EXPERIMENTAL RESULTS

In our experiment we use 10,000 jpeg images as our database ground truth which consists of 10 classes including "bear", "bike", "building", "car", "cat", "flower", "model", "mountain", "sky", and "texture".

We evaluate only the top twenty images ranked in terms of the similarity measures by using precision and recall parameters. Precision is the ratio of the number of relevant images retrieved to the total number of irrelevant and relevant images retrieved. Whilst, Recall is the ratio of the number of irrelevant images retrieved to the total number of relevant images in the database.

$$\text{Precision} = \frac{\text{number of relevant images retrieved}}{\text{number of images retrieved}} \qquad (2)$$

$$\text{Recall} = \frac{\text{number of irrelevant images retrieved}}{\text{number of relevant images in the class}} \qquad (3)$$

Fig. 1 illustrates the precision and the recall of image retrieval of ten image classes with two methods. Image retrieval on grayscale images and image retrieval based on segmented images using region growing technique.

Fig. 2 statistically shows the precision and the recall of two methods. The image retrieval based on segmented images shows higher in average precision which is 0.75 compare to the image retrieval on grayscale images which is 0.65.

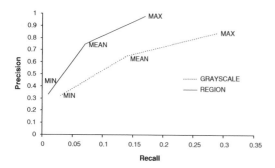

Figure 1. Mean, maximum, and minimum of precision and recall on grayscale and segmented images.

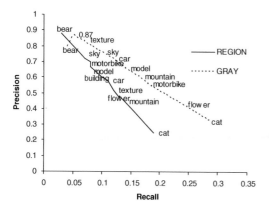

Figure 2. The effectiveness of image retrieval on grayscale and segmented images (using region growing segmentation).

Table 1. Precision and recall with grayscale and segmented methods.

Class	Gray scale images		Segmented images	
	Precision	Recall	Precision	Recall
Bear	0.88	0.03	0.98	0.01
Bike	0.70	0.08	0.92	0.02
Build	0.60	0.10	0.88	0.03
Cars	0.58	0.11	0.93	0.04
Cat	0.25	0.19	0.78	0.06
Flower	0.47	0.13	0.72	0.07
Model	0.67	0.08	0.70	0.08
Mount	0.48	0.13	0.68	0.09
Sky	0.72	0.07	0.53	0.12
Texture	0.52	0.12	0.33	0.17

The highest precision of 0.98 has been achieved by utilizing region growing technique for bear images, and the lowest precision of 0.25 for texture images. Interesting result has been found that the greyscale images show the highest precision of 0.88 also for

Table 2. Statistically precision and recall with grayscale and segmented methods.

Statistics	Segmented		Grayscale	
	Precision	Recall	Precision	Recall
Mean	0.75	0.07	0.65	0.14
Maximum	0.98	0.17	0.88	0.29
Minimum	0.25	0.01	0.32	0.03

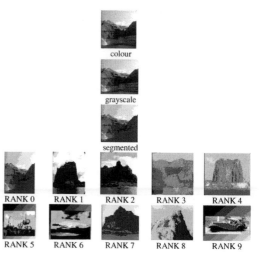

Figure 3. An examples of images retrieved with RGB image query, RGB image converted into grayscale, the grayscale image then partioned by region growing technique.

bear, and the lowest precision of 0.32 for cat images as shown in Table 2.

The experimental result shows that our propose method image retrieval on segmented images present good precisions which are higher than 0.50 on all classes excluding texture class. For further result detail, it is shown in Table 1.

As Table 2 illustrates the highest (maximum), the lowest (minimum), and the average (mean) of the precision and recall of every query, we found that the application of region growing technique (segmented images) gives better maximum and mean precision for all ten image classes in the database.

5 CONCLUDING REMARK

New approach has been proposed for image retrieval system based on region growing segmentation technique on DCT compress domain. It is presented as an alternative way to develop image indexing by using of DCT feature descriptors. The method has been carried out for compressed images database to verify its performance in JPEG standard stream line.

From the experiments, it could be concluded that segmentation, while imperfect, is a useful approach in building indexing keys. In summary, this indexing key method is a promising method for image retrieval on compress domain. This new approach could be used for image indexing by using other segmentation methods.

For the near future, we are applying this method on one of the DCT coefficients which is a DC coefficient only as representation of the whole image in order to simplify algorithm and improve speed of image indexing.

REFERENCES

Adams, R. & Bischof, L.1994. Seeded Region Growing. *PAMI* 16(6): 641–647.

Bhanu, B. 1994. Genetic Learning for Adaptive Image Segmentation. Norwell. MA: Kluwer.

Boskovitz, V. & Hugo Guterman. 2002. An adaptive neuro fuzzy system for automatic image segmentation and edge detection. *IEEE Transaction fuzzy systems*, 10(2): 247–262.

Chang, Y.L. & Li, X. 1994. Adaptive image region-growing. *IEEE Trans. Image Processing*, Vol. 3: 868–872.

Haralick, R.M. & Shapiro, L.G. 1985. Survey: Image segmentation techniques. Computer *Vision Graph. Image Processing*, Vol. 29:100–132.

Hijjatoleslami, S.A. & Kittler, J. 1998. Region growing: A new approach. *IEEE Trans. Image Processing*, Vol. 7:1079–1084.

Jain, A.K. 1998. *Fundamentals of Digital Image processing*. Upper Saddle River, NJ: Prentice Hall.

Kass, M. & Witkin, A. 1996. Snakes: Active Contour Models. *Proc. 1st Intl. Conf. on Computer Vision*: 259–269.

Koller, D. & Sahami, M. 1996. Toward optimal feature selection. *Proceeding 13th International Conference Machine Learning*: 197–243.

Lee, M.W. & Cohen, I. 2004. Proposal maps driven MCMC for estimating human body pose in static images. *Proceeding IEEE Conference Computer Vision and Pattern Recognition*: 334–341.

Luo, J. & Guo, C. 2003. Perceptual grouping of segmented regions in colour images. *Pattern Recogniion*, Vol. 36: 2781–2792.

Minka, T.P. & Picard, R.W. 1997. Interactive Learning Using a Society of Models.*Pattern recognition*, Vol. 30, No. 3: 565.

Pal, S.K. & Rosenfeld, A. 1988. Image enhancement and thresholding by optimization of fuzzy compactness. *Pattern Recognition Letter*, 7: 77–86.

Pope, A.R. 2004. Model-Based Object Recognition: A Survey of Recent Research. Univ. of British Columbia, Dept. of Computer Science, Techn. *Report CS-TR 94-04*.

Sahoo, P.K:A. 1988. Survey on Thresholding Techniques. *Computer Vision Graphics Image Processing*, 41: 233–260.

Sonka. M. & Hlavac V. 1999. Image Processing Analysis and Machine Vision. London, U.K.: Chapman & Hall.

Tu, Z & S.-C. Zhu, C.S. 2002. Image Segmentation by Data-driven Markov chain Monte Carlo. *IEEE Trans. Pattern Anal. Mach. Intell.*, Vol. 24, No. 5: 657–673

Wallace, G.K. 1991. The JPEG Still Picture Compression Standard. *Comm. ACM*, Vol. 34: 30–44.

Wang, J.Z & Li, J. 2001. SIMPLICITY: Semantics Sensitive Integrated Matching for Picture Libraries. *IEEE Transaction on Pattern Analysis and Machine Intelligence*, Vol. 23, No. 9: 947–963.

Wiederhold, G. & Wang, J.Z. 1998. Content-Based Image Indexing and Searching Using Daubechies' Wavelets. *International Journal Digital Libraries (IJODL)*, Vol. 1, No. 4: 311–328.

Weszka, J.S. 1978. A Survey of Threshold Selection Techniques. *Computer Graphics Image process*, 7: 259–265.

Deng, Y. & B.S. Manjunath:. 2001. Unsupervised Segmentation of Colour Texture Region in Images and Video, *Final Manuscript to PAMI: 1–25* (http://citeseer.ist.psu.edu/deng01unsupervised.html)

Computational Modelling of Objects Represented in Images –
João Manuel R.S. Tavares & R.M. Natal Jorge (eds)
© *2007 Taylor & Francis Group, London, ISBN 978-0-415-43349-5*

Vehicles measuring and classification

I. Kalová & K. Horák

Brno University of Technology, Department of Control and Instrumentation, Brno, Czech Republic

ABSTRACT: The paper is focused on a domain of collecting and subsequent analysis of traffic data. The article describes the system that is able to measure some vehicles parameters (length, height profile, actual velocity) without constraint of vehicle speed (free-flow system) and without necessity to put some device into the vehicle. According to these measured parameters is vehicle classified into the predetermined classes. The principle of this system is based on an active one-dimensional triangulation technique.

1 INTRODUCTION

With permanent accumulation of strain on the road net (number of vehicles was increased by 30% in the last ten years), which brings first of all the augmentation of traffic accidents, communication failures by local overexertion and air pollution, the pressure on useful management and systematization of transport is increased. This makes bigger demands on systems designated for traffic data collecting (relevant to traffic density or basic vehicles parameters – height, length, weight, category), automated analysis of traffic state and recognition of important events in traffic (accidents, traffic jams) or interception of traffic rules breaking (the red-light crossing, speeding). There are a lot of different systems and methods (tried in practice or progressive) in this domain. New solutions are designed and tested in several national and international developmental workplaces.

2 PRESENT SYSTEMS FOR TRAFFIC DATA COLLECTION

The systems that are used today are usually aimed just at measuring of one parameter of vehicle (weight or speed most often). Several individual solutions have to be combined for the complete measuring. But it brings the increasing in price and strenuousness of whole solution. A lot of systems involve disruption of the roadway. The system which can solve the connection of measured parameters with license plate automatically, without any device inside of the vehicle, must obtain the camera system. These systems are necessary for potential documentation of given situation too. The systems that are used at present (Mencl 2003):

– inductive loops, piezoelectric sensors, weighs
– radars, microwave sensors
– optical gates
– laser sensors (Time-of-Flight)
– camera systems, . . .

A very useful data (e.g. for traffic statistics creation or for electronic fee collection) is a category into which given vehicle falls. The results from different present-day measuring methods can be applied to classification. The difference is which data are used to vehicle classification, what is the reliability and into how many categories can be vehicles separated.

2.1 Loop sensors

Data about number of axles, total weight or weight of each axle are used for classification by techniques based on infrastructure direct in roadway (inductive loops, piezoelectric sensors). The vehicles are distinguished according to weight (less than 3.5 t, 3.5–12 t, more than 12 t) very often but it is not possible to class the lorry to the lower category just for the fact that it doesn't carry the freight now (its actual weight is low). Furthermore these methods use the intervention into the roadway that brings damages and mostly complicated installation. But these methods are simple and cheap solution.

2.2 Radars, microwave sensors, optical gates

Radar and microwave sensors can provide data just about vehicles length. But length is unfit to cope with reliable classification. For example towing vehicle itself without trailer corresponds to category – passenger cars with its length even if it can fall into the category – more than 12 t. Similarly unsuitable is mere information about vehicle height which can be measured by optical gates.

2.3 Camera systems

The camera systems can be used for classification too. Their advantage is that they use the same technology utilized for licence plate recognition. These systems can detect vehicles even with wheels in taken images and they can determine the length and height (respectively the whole height profile) based on projection. Nevertheless full scene is very complicated (reflections, other vehicles, surrounding scene – buildings, kerbs, road infrastructure etc.) and that is the reason why classification is very difficult. Beside this, the camera has to be placed on the road side (for number of wheels detection), which makes these systems not usable for measuring of communications with more than two traffic lanes (vehicle goes in outside lane can block the view on vehicle goes in middle lanes).

2.4 Laser sensors (Time-of-flight)

To this intent a technique based on time-of-flight (TOF) of modulated light measuring (Haußecker & Geißler 1999) appears the most suitable. These systems can measure the whole vehicle profile. It is certain that the vehicle shape can effectively characterize the category. The advantage of work with profile is, that individual vehicles can be separated much easier due to the measuring of course of vehicle height and a priori knowledge of possible vehicles shapes. The errors like disconnecting of the towing vehicle and trailer or cab and semitrailer can be eliminated by careful signal analysis.

Unfortunately the TOF method has some drawbacks. The first disadvantage is that the method is relatively slow. For example the time response of system offered by company SICK (SICK AG. 2003, consortium Kapsch that solves the control systems for electronic fee collection in Germany, Austria and Switzerland) is 13.3 ms for scanning of 100° by 1° (101 values).

Another drawback of this method is that for one profile obtaining it is necessary the sender, accurate optics, receiver and electronics for signal processing and data transmission to the central system. Several of these expensive systems are required for providing of sufficient amount of data for measurement on the whole traffic lane width. That is the reason why it is used a consecutive scanning of space instead of more sensors. The sender with receiver and optics are situated on rotary joint (or mirror is used) and this system measures distances e.g. by 1°. Providing of rotary movement brings partly bigger possibility of failures and partly aggradation by uncertainty. The rotation prolongs the whole measuring too. The samples are not acquired just in the line of vehicle movement but "zigzag".

All above mentioned methods have some advantages and disadvantages. Which method will be used depends on experiences (with separate technology), on claims on classification quality (reliability, number of classes) and of course on finances. Applying of more methods in combination is more appropriate but it is also more expensive.

3 MEASURING PRINCIPLE

Our Group of Computer Vision on Department of Control and Instrumentation (FEEC, Brno University of Technology) with company Camea (CAMEA s.r.o. 2004) have been engaged in design of camera systems and their using in industry, even in traffic industry, for many years. They devised several traffic systems for example for detection of vehicles which cross the junction on red lights, for average speed measuring in road segment or detection of critical states (tunnels, junctions) etc. We have also experiences with applications of triangulation technique (Kalová & Lisztwan 2006) e.g. for weld quality inspection of passenger car wheels (Kalová & Richter 2004) for Hayes Lemmerz Autokola in Ostrava; for volume measuring of viscose glue droplet in production of tantalic capacitors (Honec et al. 2001) for AVX Lanškroun or for acquisition of 3-D model of hat form (study for Tonak in Nový Jičín). That is the reason why our working group made a resolution about using the acquired experiences and putting together two problems in a design of system for detection and classification of vehicles.

3.1 Triangulation technique

An active triangulation method is most often applied in practice for its simplicity and robustness. This method is based on photogrammetric reconstruction of measured object by illumination its surface and contemporaneous scanning by CCD sensor.

The light source, the detector and the illuminated part of the measured object form a triangulation triangle (Fig. 1). The light source ray angle α is fixed whereas the angle on the detector side β changes. Based on the knowledge of two angles and one side of the triangle (base-line b), the distance can be determined (Haußecker & Geißler 1999, Kalová & Lisztwan 2005).

According to the light source we distinguish these three variants:

- 1-D triangulation – light point
- 2-D triangulation – light stripe
- 3-D triangulation – light volume.

3.2 Mathematical description

Proposed system is based on 1-D variant. Principle of measuring is shown on Figure 1. If the reflected ray of the light is projected to the nth pixel from the total number pxl of pixels, than the size of corresponding projection a [mm] can be calculated by formula (1) (Kalová and Lisztwan 2005), where c is the chip size [mm].

$$a = \frac{c \cdot n}{pxl} \text{ [mm]}, \tag{1}$$

The projection a is used to angle β computation (f ... focal length [mm]).

$$\beta = actg\left(\frac{c/2 - a}{f}\right) + 90° \text{ [°]}, \tag{2}$$

The distance l can be determined due to angles α, β and the size of base b:

$$l = \frac{b \cdot \sin \alpha \cdot \sin \beta}{\sin(180° - (\alpha + \beta))} \text{ [mm]} \tag{3}$$

If the investigated object is situated nearer to the camera and light source, the projection a is bigger. This means that reflected light ray is projected to the pixel more right (according to Fig. 1). In this way, the distance of the object can be obtained just from the position of the light point in the image from the camera. The measuring accuracy and also the usability of this method depend on several parameters. The better resolution (smaller difference of measured distances of two neighbouring pixels – lesser discretisation error) can be obtained by increasing size of base b, camera resolution pxl (number of photosensitive elements) and objective focal length f and on the other hand by lessening chip size c. Fine accuracy is achieved in shorter measuring distance l too. Important is angle α too.

3.3 Calibration

These formulas are very sensitive to accuracy of inserted data. Small shortcoming on input side brings big inexactitude on output. That's the reason why a transformation formula obtained by calibration is usually used. The transform matrix (dimension 2×2) is used. Each constituent member represents mutual relation between pixel position n and measured distance l. This relation is characterized by following equation (4).

$$\begin{pmatrix} hl \\ h \end{pmatrix} = \begin{pmatrix} t_{11} & t_{12} \\ t_{21} & 1 \end{pmatrix} \begin{pmatrix} n \\ 1 \end{pmatrix} \tag{4}$$

For matrix members (t_{11}, t_{12}, t_{21}) determination is necessary to make measuring in needed range. Distance and pixel position is determined for each laser and each distance. With these data the transformation matrix is established by minimization of differences between set and computed values.

Based on defined mathematical model (get from Fig. 1) or transformation (acquired by calibration of the system) is able to determine the distance between the system and point of reflection (respectively the actual vehicle height) just from position of bright point in the image.

4 CONFIGURATION DESIGN

Principle of measuring demands two systems (will be explained later) each with one linear camera and four lasers (sufficient number for providing of the whole traffic lane width). All these components have to be selected and positioned carefully for obtaining the best accuracy together with the smallest dimensions. Designed configuration is on Figure 2 (camera is placed on the left, lasers on the right).

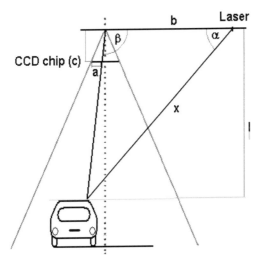

Figure 1. Principle of measuring by triangulation technique.

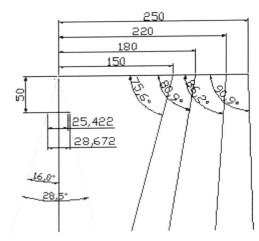

Figure 2. Design of base-line lengths.

Figure 3.   Photo of the platform with camera and four lasers.

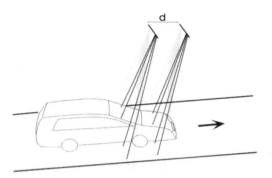

Figure 4.   Two measuring units placement.

The supposed base-lines are in sequence 150, 180, 220, 250 mm. Selected camera uses chip SONY ILX751 (effective pixels 2048, pixel size $14 \times 14\,\mu\text{m}$, clock frequency 5 MHz).

Lens system Pentacon MC 1.8/50 is used. Lasers consist of simple laser diode and focusing lens. Laser wavelength is 650 nm. Thin band filter is used for influence reducing of other light sources and sunlight thus it is set according to laser wavelength. The filter is placed between camera and lens. All these components are fixed on a duralumin platform (Fig. 3).

It is imperative that second system (camera and light sources) situated behind the first in a defined distance (Fig. 4, $d = 0.25$ m) must be used for determining the vehicle velocity and thus the distance between individual samples. The vehicle velocity is calculated from the time shift between the two corresponding profiles (Fig. 5).

This system (two measuring units – camera and lasers) with control electronics is inserted in waterproof resilient case (Fig. 6) and this is placed above the roadway (perpendicular to vehicle movement direction) and monitors the area below.

The beam of light is reflected on the road respectively on the vehicle surface and the position n of the illuminated point in camera row changes according to the distance l (see Fig. 1).

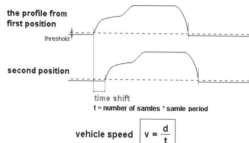

Figure 5.   Principle of velocity definition.

Figure 6.   Photo of open measuring device.

5   DATA PROCESSING

5.1   Acquired data

Individual camera rows (2048 pixels in 256 grey shade) are composed consecutively to operating blocks (images). These come in algorithm.

Figure 7 shows example of real unprocessed image data from camera. Four laser tracks can be seen on the image. Vehicle beginning is pictured on the image – lasers (except for first) start to fluctuate right.

5.2   Laser track detection

Each laser can change just in determined area that corresponds to height 0–5 m above the road. Laser tracks can't cross each other. So lasers can be processed separately. Standard algorithms (filtering, thresholding etc.) of image processing are used for this purpose.

On the each camera row segment is one sample. With the row frequency of used camera are obtained other samples.

5.3   Profile processing

Acquired data sequences are filtrated (median, average filters) for short errors elimination (reflection,

Figure 7. Example of image data from linear camera.

Figure 8. a) Photo of measured vehicle, b) Image acquired from linear camera with detected laser track (image is rotated for better clearness and presentation).

vehicles spotlights, snow etc.). Smooth samples are compared with threshold defined road position continually. If current laser beam position is bigger (pixel more right), it is vehicle profile beginning. Other samples are included in vehicle profile until the whole vehicle is measured (Fig. 8b – one laser). A photo of passing vehicle is made by matrix camera placed in the corner of the portal (Fig. 8 a).

The corresponding profile from both measuring systems are correlated and time shift between them is determined (Fig. 9). The shift is used for vehicles velocity calculation. Real samples distances and so

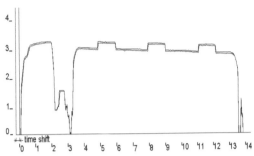

Figure 9. Matched height profiles from two measuring positions.

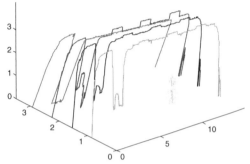

Figure 10. 3D model of measured cistern.

vehicle length can be computed due to velocity knowledge.

The best profile can be chosen from four separated vehicle profiles then. This is used as a final result of measuring and for classification. All profiles can be used for 3D vehicles model set-up (Fig. 10).

6 VEHICLE CLASSIFICATION

The triangulation technique usage is very suitable in this domain. The technique allows to measure the vehicle height profile in movement without constraint of its speed. Height profile is probably the best input for classification. It represents the vehicle well.

Chosen registered vehicles (1268 vehicles) were used for composition of testing series for design of classification procedure (attributes choice, generation of training set, classification method choice and classifier setting). The testing vehicles were classified into eleven categories (passenger car, van, bus, camion, lorry etc., see Fig. 11).

From every single acquired profile a set of eight attributes is computed (length, average height, position of centre of gravity, standard deviation etc.). The attributes were selected prudently according to their

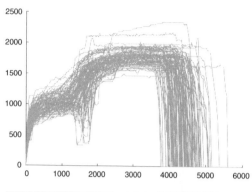

Figure 11. Example of category – 2. big passenger cars: graph of all vehicles in this category, the smallest and biggest member.

Table 1. Composition of testing and training set (vehicles number).

Category	Testing	Training
1	291	21
2	110	23
3	293	46
4	52	26
5	49	25
6	83	42
7	245	33
8	18	9
9	132	18
10	12	6
11	37	11
Σ	1268	260

dissimilarity between categories (overlay of histograms as small as possible).

Then the training set of characterful members from each category is created (260 vehicles in all, see Table 1).

Several methods of classification were tested (classification trees, fuzzy logic, algorithms based on instances recording, several types of neural network etc.). Best results were achieved with multiple feedforward layers backpropagation neural network.

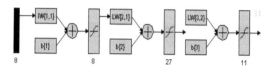

Figure 12. Neural network structure.

Table 2. Successfulness of classification procedure.

Category	Testing	Success
1	291	99.7
2	110	97.3
3	293	97.9
4	52	98.1
5	49	100
6	83	98.8
7	245	99.2
8	18	100
9	132	100
10	12	100
11	37	100
Σ	1268	98.9

Number of neurons in the input layer is set resolutely according to the number of attributes, similarly number of neurons in output layer is given by number of output categories. Optional is just number of inner layers and neurons in them. Sufficient net is with three layers and 27 neurons in middle layer (Fig. 12).

Just vehicles from training set were used for network learning. The whole group of vehicles was put to the resultant net test. Each vehicles is classified to the category which gives the highest output value (output of correspond neuron). Successfulness of classification procedure is summarized in Table 2.

7 CONCLUSION

The main aim of this project is a development of system based on active 1D triangulation technique which is able to measure significant vehicle parameters (length, height profile, actual speed) without constraint of its speed and threating driver safety. According to these parameters is vehicle classified (e.g. passenger car, van, lorry, bus, camion, …).

The developed system is possible to integrate to the whole visual system for monitoring of traffic parameters (connecting with system for licence plate recognition) and in this way the complex system for a plenty of applications could be created. An example of using can be: parking and reception systems, systems for entry inspection (according to the height or vehicle type), control system for electronic fee collection, systems for creation of statistics and prognosis etc.

ACKNOWLEDGEMENTS

This paper has been prepared with the support of CAK (Research Centre of Application Cybernetic), concretely project No. 1M0567.

REFERENCES

CAMEA s.r.o. 2004. UNICAM Camera-based traffic monitoring system [online]. last rev. March 2005. www.unicam.cz/en/.

Haußecker, H. & Geißler, P. 1999. *Handbook of Computer Vision and Applications*. San Diego: Academic press.

Honec, P. et al. 2001. 3D Object Surface Reconstruction. In: *13th International Conference of Process Control*, Bratislava: Slovak University of Technology.

Kalová, I., Richter, M. 2004. Inspection of welding seams. In *15th International DAAAM Symposium*. Vienna: DAAAM International, Vienna 2004.

Kalová, I. & Lisztwan, M. 2005. Active Triangulation Technique In *5th International Conference of PhD Students*. Miskolc: University of Miskolc.

Kalová, I. & Lisztwan, M. 2006. Industrial Applications of Triangulation Technique. In *Proceedings of IFAC Workshop on Programmable Devices and Embedded Systems – PdeS*. Brno: Brno University of Technology.

Mencl, J. 2003. Systémy a metody pro zjišťování údajů o dopravě a pro automatické rozpoznávání událostí: co nabízí trh? [online]. last rev. 2003 [2005-04-11]. www.datis.cdrail.cz/EDICE/Telema/IDS2004/ids4_04.pdf.

SICK AG. 2003. Laser Measurement systems – technical description [online]. last rev. June 2003 [2005-04-20]. www.sick.com/saqqara/IM0012759.PDF.

Computational Modelling of Objects Represented in Images –
João Manuel R.S. Tavares & R.M. Natal Jorge (eds)
© 2007 Taylor & Francis Group, London, ISBN 978-0-415-43349-5

Textual description of images

S. Larabi

Computer Science Department, USTHB University, BP 32 EL Alia, Algiers, Algeria

ABSTRACT: We present in this paper a new method for image description based on the description of regions that contains. Image is firstly segmented in meaningful regions and in the next step these regions are represented semantically in a tree structure. Regions are described by means the geometry of their outline contours and the description of encompassed internal regions. An XML language XLWDI (XML Language for Writing Descriptors of Images) is proposed to write these descriptions. This method may be used for image coding in multimedia applications or for image description in computer vision applications. Experiments conducted over real images are presented.

1 INTRODUCTION

Few methods have been proposed for the description of images, the most known is the method used in the MPEG-7 standard that allowing the description of image as a still region, which is described by creation, usage and media information, color histogram and texture descriptor. This initial region is decomposed into individual regions for which are associated the feature type, color histogram and textual annotation. A tree structure is associated to the image.

Various methods have been also developed in order to represent silhouettes and shapes of image in an abstract and efficient way while still preserving important shape feature. The most interesting methods may be classified as follows:

– Part-based methods (Kim et al. 2005) (Pitas & Venetsanopoulos 1990) (Rosin 1999) (Siddiqi & Kimia 1995) (Yu & Wang 2004),
– Aspect-graph methods (Cyr & Kimia 2004) (Koenderink & Doorn 1976),
– Methods that use the medial axis of silhouettes (Geiger et al. 2003) (Ruberto 2004) (Zhu & Yuille 1995),
– Methods based on the shock graph (Siddiqi & Kimia 1996) (Sebastian et al. 2004),
– Methods using graph for shape representation (Badawy & Kamel 2002) (Lourens & Wurtz 1998),
– Approximation of outline shape by 2D features (Cronin 2003) (Grosky & Mehrota 1990) (Nelson & Selinger 1998) (Petrakis et al. 2002),
– Methods based on the reference points of outline shape (Berretti & Bimbo 2000) (Mokhtarian & Mackworth 1992) (Mokhtarian 1995) (Shao & Kittler 1999),
– Methods based on the attributes of outline shape (Arica & Vural 2003) (Bernier & Landry 2003) (Orrite & Herreo 2004) (Sethi et al. 2004),
– Method based on the shape context (Belongie et al. 2002),
– Appearance-based methods (Murase & Nayer 1995),

A review of shape representation methods can be found in (Campbell & Flynn 2001) (Zhang & Lu 2004).

In order to homogeneous all databases of image descriptions and to offer accessibility to all users, a common and compact format of representation is recommended. In addition, this format must be easy to index, easy to compare and efficient for computation and storage. Our aim in this paper is to propose a new image descriptor verifying the above recommendations.

We assume that image is segmented into a set of regions corresponding either to shape of object or to area in the viewed world. These regions are represented semantically in a tree structure and described by means the geometry of their outline contours and the description of encompassed internal regions. Our approach for image description is based on the part-based methods for silhouette description (Larabi et al. 2003), shape description (Larabi 2005) and on the basic principle of image representation proposed by Z. Tu et al (Tu et al. 2005).

The proposed scheme can provide the following properties:

– Coding of Images with text following XML language
– Reduced size of computed description

- Easiness for index extraction
- Preservation of perceptual structure with loss of information
- Complete storage of all object information
- Coding and visualization of images in multimedia applications.

This paper is organized as follow: Section 2 describes the basic principle of our method for representing images. In section 3, we present the XML language XLWDI for writing image descriptors. In section 4 we discuss results of experiments conducted over real images demonstrating the useful of this description for images coding.

2 DESCRIPTION OF IMAGES

2.1 Basic principle of the method

We assume that image is segmented onto regions using the one of known methods. (Deng & Majunath 2001) (Felzenszwalb & Huttenlocher 2005) are two examples of segmentation methods. This task is not discussed in this paper.

Image is considered as an initial region encompassing all other regions (see figure 1) and each region may encompass other regions or may be an elementary region.

To describe images we must indicate: List of all regions, the inclusion relation between regions, the position of each region in the image and the description of each region (the geometry of its outline contour and its colour).

To describe the relation of inclusion, we use a tree structure where nodes are regions and children nodes of any node are the set of regions that it contains.

We associate to each region:

- The level zero (0) if it not contains internal regions.
- The level one (1) if it contains regions of level 0.
- The level (n) if it contains regions of level less or equal to (n − 1) (see figure 1).

We will note $R_{i,j}$ the jth region of level i. Image of figure 1 contains two regions of level 0 ($R_{0,0}$, $R_{0,1}$), one region of level 1 ($R_{1,0}$) containing $R_{0,0}$, $R_{0,1}$ and one region $R_{2,0}$ of level 2 containing $R_{1,0}$.

The position of regions is expressed by the x and y-coordinates of its left highest contour point (see figure 1).

To describe silhouettes and regions we use the methods published in (Larabi et al. 2003) and (Larabi 2005). We give an overview of these methods in section 3.

2.2 Language for writing descriptors of images

We present in this subsection the grammar of the language XLWDI (**XML Language for Writing**

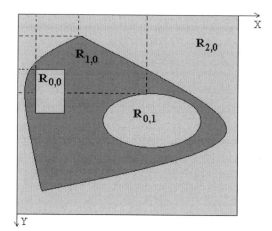

Figure 1.   Different regions of image.

Descriptors of **I**mages) that allow writing the description of images following XML syntax.

Description of Image →
   <XDLWDI>
      <Name = Name-of-image/**>**
      ¿Description of all regions!
   </XDLWDI>

¿Description of all Regions! →
   ¿Description of Region!
   ¿Description of all Regions! /
   ¿Description of Region!

¿Description of Region! →
   <Shape>
   ¿Position of Outline Region!
   ¿Description of Outline Region!
   ¿Description of Internal Regions!
   </Shape>

¿Description of Internal Regions! →
   <Internal Shape>
   ¿Position of Outline Internal Region!
   ¿Description of Outline Internal Region!
   ¿Description of Internal Regions!
   </Internal Shape>/ ε

¿Description of Outline Region! →
   ¿Description of Silhouette!

¿Position of Outline Region! →
   ¿X-Coordinate of Beginning Point!
   ¿Y-Coordinate of Beginning Point!

¿Position of Outline Internal Region! →
   ¿X-Coordinate of Beginning Point!
   ¿Y-Coordinate of Beginning Point!

¿Description of Outline Internal Region! →
   ¿Description of Silhouette!

242

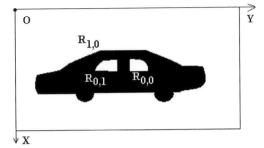

Figure 2. Example of decomposition of image into regions.

¿X-Coordinate of Begin Point! →
  INTEGER VALUE
¿Y-Coordinate of Begin Point! →
  INTEGER VALUE

For example, the image of figure 2, contains two regions of level 0: $R_{0,0}$ and $R_{0,1}$ and one region of level 1: $R_{1,0}$.

The descriptor of this image is written as follow:

<DXLWDI><Name = Image of figure 3/>
<Shape>111 127 **Description of $R_{1,0}$**
<Internal Shape>123 174 **Description of $R_{0,0}$**
</Internal Shape>
<Internal Shape>124 143 **Description of $R_{0,1}$**
</Internal Shape> </Shape>
</DXLWDI>

In bold character are written the description of silhouettes that will be done using the method published in (Larabi et al. 2003) and explained in section 3.

3 DESCRIPTION OF REGIONS

3.1 Overview of the method for describing outline of regions (Larabi et al. 2003)

To describe outline of regions we use the part-based method published in (Larabi et al. 2003). Its basic principle is the following:

The sweep up of silhouette from the top to bottom allows locating concave points for which the direction of the outer contour changes following top-bottom-top or bottom-top-bottom (see figure 3). These points are used to partition silhouette onto parts, junction and disjunction lines: either, two parts or more are joined with a third part through a junction line, or a part is joined with two parts or more through a disjunction line. This process applied to silhouette of figure 3 gives five parts, one junction and one disjunction line.

Description of silhouette is the grouping of descriptions of its elements.

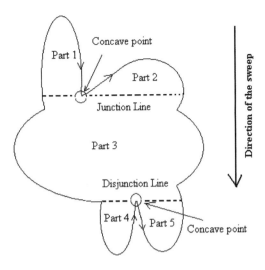

Figure 3. Partition of region onto parts, junction and disjunction lines.

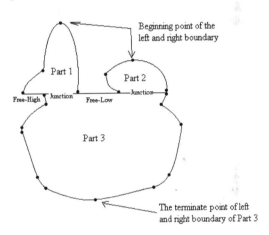

Figure 4. Left and right boundary of parts.

Part is defined by its two boundaries (left and right) which begin at the highest left point and terminate at their lowest points. Using the curvature points (Chetverikov, 2003), these boundaries are segmented into a set of primitives (line, convex and concave contours) and described by the parameters: type (line, convex or concave curve), degree of concavity or convexity, angle of inclination and length (see figure 4).

Junction and disjunction lines are decomposed onto segments. Each segment is described with three parameters: type, the reference numbers of parts where it appertains and its length. Types of segment are: **Junction** if it is common for two parts, **Free-High** if it belongs only to the high part or **Free-Low** if it belongs to the low part (see figure 4).

From this description of the outline region, it is easy to draw the correspondent region without ambiguity. The use of absolute lengths produces an absolute description of regions. Such description is recommended for coding images in multimedia applications. For computer vision applications as recognition of objects, relative lengths may be used to guaranty the scalability property.

3.2 Overview of the XML language for writing descriptors of outline regions (Larabi 2005)

The proposed XML language for writing descriptors of outline regions is presented here. Firstly we give the set of rules allowing the writing of description of parts, junction and disjunction lines.

¿Part! → <**P num = x**>
    ¿Left boundary! ¿Right boundary!
    </**P num = x**>
¿LeftBoundary! →
    <**L**> ¿Contour 1!...¿Contour n! </**L**>
¿RightBoundary!→
    <**R**> ¿Contour 1!...¿Contour n! </**R**>
¿Contour i ! →
    <**CV degree** = x **Inclin** = x **Length** = x/> /
    <**CC degree** = x **Inclin** = x **Length** = x/> /
    <**LN Inclin** = x **angle** = x **Length** = x/>
where **CV, CC, LN** indicate respectively convex, concave, and line contour.

¿JunctionLine!→
    <**J**> ¿Segment 1! .... ¿Segment n! </**J**>
¿DisjunctionLine!→
    <**D**>¿Segment1! ...¿Segment n! </**D**>
¿Segment i! →
    <**JN Numpart1** = x **Numpart2** = x **Length** = x/>/
    <**W Numpart1** = x **Length** = x />/
    <**H Numpart1** = x **Length** = x />

Where **JN**, **W** and **H** denote respectively the attributes: **Junction**, **FreeLow** and **FreeHigh**.

The composition of different elements of the silhouette produces the **composed part** defined as the set of two (or more) parts joined to another part using junction or disjunction line, and written as follow:

Composed Part→
<**CP**> ¿SetParts! ¿JunctionLine! ¿Part! </**CP**>/
<**CP**>¿Part! ¿DisjunctionLine! ¿SetParts! </**CP**>
¿SetParts! → ¿Part! ¿SetParts!/ ¿Part!

Recursively, a composed part is considered as a part and can constitutes with other elements another composed part.

The write the descriptor of silhouettes we use:

Silhouette→ <**DLWDOS**> <**Name** = Object name/> Composed Part </**DLWDOS**>

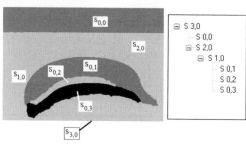

Figure 5. Segmentation and representation by a tree structure of banana image.

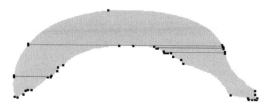

Figure 6. Example of computation of silhouette descriptor.

4 EXPERIMENTS

Images used for the validation of our method are images of industrial logos and real images. Each image is segmented using JSEG method (Deng & Majunath 2001). The processing of the segmented image produces its representation by a tree structure of regions.

Figure 5 illustrates an example of real image, the segmented image and the tree structure associated where $S_{3,0}$ corresponding to the entire image is composed by the region $S_{0,0}$ and the $S_{2,0}$. $S_{2,0}$ is contains the region $S_{1,0}$ composed by three elementary regions: $S_{0,1}$, $S_{0,2}$ and $S_{0,3}$.

A textual descriptor is computed for each image as described in section 2.2. The descriptor of each region (elementary or composed) of the image is computed firstly using the Chetverikov's algorithm (Chetverikov 2003) in order to locate the curvature points using. Two parameters are necessary: the angle and distance. Figure 6 illustrates the result of the $S_{0,1}$ partitioning into parts, junction and disjunction lines.

```
<DXLWDOS>
<Name>S01</Name>
<CP>
 <P1>
 <L>CV 9 23 46</L>
 <R>CV 12 164 46</R>
 </P1>
 <D1>W P2 1 S P1 P2 123 H P1 18 S P1 P3 122 W P3 1</D1>
 <CP>
  <P2>
  <L>CV 7 65 41</L>
  <R>CC 11 19 27 CC 13 23 4 R 79 6 R 41 7</R>
  </P2>
  <D4>S P2 P9 47 H P2 3 S P2 P10 1 H P2 2</D4>
  <P9>
  <L>CV 1 92 27 CV 24 166 2</L>
  <R>CV 5 38 18 R 0 6 R 18 2 R 59 6 CC 25 27 5</R>
  </P9>
  <P10>
  <L>R 90 1</L>
  <R>R 90 1</R>
  </P10>
 </CP>
 <CP>
  <P3>
  <L>CC 5 177 2</L>
  <R>R 90 2</R>
  </P3>
  <D2>H P3 9 S P3 P4 1 H P3 1 S P3 P5 90 W P5 1</D2>
  <P4>
  <L>R 90 1</L>
  <R>R 90 1</R>
  </P4>
  <P5>
  <L>CC 12 150 5</L>
  <R>R 117 5</R>
  </P5>
 </CP>
</CP>
</DXLWDOS>
```

Figure 7. The computed descriptor for $S_{0,1}$ silhouette.

Figure 8. BP logo and its decomposition into regions.

Table 1. Sizes (KB) of used images for different formats.

Image	BMP	GIF	JPG	Scalar	XLWDI
Logo of Esso	762	10.6	25.1	21.4	3.03
Logo of Peugeot	644	40.9	24.7	15.1	9.06
Logo of BP	541	12.5	55.8	21.1	1.94
Banana	322	9.8	9.2	–	3.09

The computed descriptor for image includes all descriptors of internal regions and their (x,y) positions.

Figure 8 illustrates another image (BP logo) and the result of representation by a tree structure of regions.

For each image we give in table 1 the size of the computed XLWDI descriptor and the size of the different formats. We can see that in addition of the semantic contained in the XLWDI description, the correspondent size is also interesting.

5 CONCLUSION

In this paper we have presented a contribution for image representation and description. We presented also the XML language noted XLWDI that permit to write textually these descriptors. The proposed method can be also used to describe image textually for computer vision application. Regions must be described relatively to the minimum rectangle encompassing it in order to guaranty the invariance to rotation of shape of this description.

Another interesting application of our representation method is the coding of images using the XLWDI language. The XML text generated can be used for image visualisation.

REFERENCES

Arica N. & Vural F. T. Y. 2003. BAS: a perceptual shape descriptor based on the beam angle statistics, Pattern Recognition Letters, vol. 24, issues 9–10, June 2003, 1627–1639.

Badawy O. E. & Kamel M. 2002. Shape representation using concavity graphs, Proceeding of the 16th International Conference on Pattern Recognition.

Belongie S., Malik J., Puzicha J. 2002. Shape Matching and Object Recognition Using Shape Contexts, IEEE Transactions on Pattern Analysis and Machine Intelligence, vol. 24.

Bernier T. & Landry J. A. 2003. A new method for representing and matching shapes of natural objects, Pattern Recognition 36, 1711–1723.

Berretti S. & Bimbo A. 2000. Retrieval by shape similarity with perceptual distance and effective indexing, Transactions on Multimedia, vol. 2, N° 4.

Larabi S., Bouagar S., Trespaderne F. M., Fuente E. L. 2003. LWDOS: Language for Writing Descriptors of Outline Shapes, Lecture Notes in Computer Science, Proceeding of Scandinavian Conference on Image Analysis, June 29–July 02, Gotborg, Sweden, Vol. 2749, pp. 1014–1021.

Larabi S. 2005. Textual description of shapes, paper submitted to Image and Vision Computing, 2005.

Campbell R. J. & Flynn P. J. 2001. A Survey of Free-Form Object Representation and Recognition Techniques, Computer Vision and Image Understanding, 81, 166–210.

Chetverikov D. 2003. A Simple and Efficient Algorithm for Detection of High Curvature Points in Planar Curves. 10th International Conference, CAIP 2003, Groningen, The Netherlands, August 25–27.

Cronin T. M. 2003. Visualizing concave and convex partitioning of 2D contours, Pattern Recognition Letters, 24, 429–443.

Cyr C. M. & Kimia B. B. 2004. A similarity-based aspect-graph approach to 3D object recognition, International Journal of Computer Vision, 57(1), 5–22.

Deng Y. & Majunath B. S. 2001. Unsupervised segmentation of color-texture regions in images and video, IEEE Transactions on Pattern Analysis Machine intelligence, 23(8): 800–810.

Felzenszwalb P. F. & Huttenlocher D. P. 2005. Pictorial structures for object recognition. International Journal of Computer Vision, 61(1), 55–79.

Geiger D., Liu T., Kohn R. V. 2003. Representation and Self-Similarity of shapes, IEEE Transactions on Pattern Analysis and Machine Intelligence, Vol. 25, N° 1.

Grosky W. I. & Mehrota R. 1990. Index-based recognition in pictorial data management. Computer Vision Graphics and Image Processing, vol. 52, pp. 416–436.

Kim D. H., Yun I. D., Lee S. U. 2005, A new shape decomposition scheme for graph-based representation, Pattern Recognition, 38, 673–689.

Koenderink J.J. & Doorn V. 1976. The internal representation of solid shape with respect to vision, Biol. Cyber., 32.

Leibe B. & Schiele B. 2003. Analyzing appearance and contour based methods for object categorization, Proceedings of the IEEE Computer Society Conference on Computer Vision and Pattern Recognition.

Lourens T. & Wurtz R. P. 1998. Object recognition by matching symbolic edge graphs, In the proceeding of ACCV'98, pp. 193–200.

Maragos P. A. & Schafer R. W. 1986. Morphological skeleton representation and coding of binary images, IEEE transactions on Acoustic, Speech Signal Processing, 34(5).

Mokhtarian F. & Mackworth A. K. 1992. A Theory of Multiscale, Curvature-Based Shape Representation for Planar curves, IEEE Transactions on Pattern Analysis and Machine Intelligence, 14(8), 789:805.

Mokhtarian F. 1995. Silhouette-Based isolated object recognition through curvature scale space, IEEE Transactions on Pattern Analysis and Machine Intelligence, 17(5): 539–544.

Murase H. & Nayer S. K. 1995. Visual learning and recognition of 3-D objects from appearance, International Journal of Computer Vision, vol. 14, Issue 1.

Nelson R. C. & Selinger A. 1998. Large-scale tests of a keyed, appearance-based 3-D object recognition system, Vision Research, Vol. 38, N° 15–16.

Orrite C. & Herreo J. E. 2004. Shape matching of partially occluded curves invariant under projective transformation, Computer Vision and Image Understanding, vol. 93(1).

Petrakis E. G. M., Diplaros A., Milios E. 2002. Matching and Retrieval of Distorted and Occluded Shapes Using Dynamic Programming, IEEE Transactions on Pattern Analysis and Machine Intelligence, vol. 24, N° 11.

Pitas I. & Venetsanopoulos A. N. 1990. Morphological shape decomposition, IEEE Transactions on Pattern Analysis and Machine Intelligence, vol. 12(1), pp. 38–45.

Rosin P. L. 1999. Shape partitioning by convexity, British Machine Vision Conference.

Ruberto C. 2004. Recognition of shapes by attributed skeletal graphs, Pattern Recognition, 37, pp. 21–31.

Sebastian T. B., Klein P. N., Kimia B. B. 2004. Recognition of Shapes by Editing Their Shock Graphs, IEEE Transactions on Pattern Analysis and Machine Intelligence, Vol. 26, N° 5.

Sethi A., Renaudie D., Kriegman D., Ponce J. 2004. Curve and surface duals and the recognition of curved 3D objects from their silhouette, International Journal of Computer Vision, 58(1), 73–86.

Shao Z. & Kittler J. 1999. Shape representation and recognition based on invariant unary and binary relations, Image and Vision Computing, 17, 429–444.

Siddiqi K. & Kimia B. B. 1995. Parts of visual form computational aspects, IEEE Transactions on Pattern Analysis and Machine Intelligence, vol. 17(3), pp. 239–251.

Siddiqi K. & Kimia B. B. 1996. A shock grammar for recognition, Conference of Computer Vision and Pattern Recognition.

Tu Z., Chen X., Yuille A. L., Zhu S. 2005. Image parsing: Unifying Segmentation, Detection, and Recognition, International Journal of Computer Vision, 63(2), 113–140.

Yu M. H., Lim C. C., Jin J. S. 2003. Shape similarity search using XML and portal technology, Workshop on visual information processing, VIP 2003, Sydney.

Yu L. & Wang R. 2004. Shape representation based on mathematical morphology, Pattern Recognition Letters.

Zhang D. & Lu G. 2004. Review of shape representation and description techniques, Pattern Recognition, 37, 1–19

Zhu S. C. & Yuille A. L. 1995. FORMS: A flexible object recognition and modeling system, Fifth International Conference on Computer Vision, June 20–23, M.I.T. Cambridge.

Computational Modelling of Objects Represented in Images –
João Manuel R.S. Tavares & R.M. Natal Jorge (eds)
© 2007 Taylor & Francis Group, London, ISBN 978-0-415-43349-5

Optimal image registration via efficient local stochastic search

Q. Li, I. Sato & Y. Murakami

National Institute of Advanced Industrial Science and Technology (AIST), Higashi, Tsukuba, Japan

ABSTRACT: This paper focuses mainly on an efficient local stochastic search approach for the optimal registration transformation. A simultaneous perturbation stochastic approximation technique is successfully implemented on optimizing mutual information based similarity measures. The hill climbing search and the Nelder-Mead simplex direct search are also considered for the comparative purpose. Our registration experiments are associated with the pairs of optical sensor images, synthetic aperture radar images and medical multimodality images, which are misaligned by the rigid or affine transformations. The experimental results show that in general the local stochastic search effectively yields significant improvements on the optimal solution over the conventional search technique in terms of accuracy and robustness. The main contribution of this work is the first accomplishment of an efficient local stochastic search strategy on the mutual information based affine image registration scheme.

1 INTRODUCTION

Image registration is one of the basic image pre-processing. With the digital image era coming, many image applications in the field of the computer vision, image aided medical operation and earth observation mission (EOS) need to firstly register multiple images of the same scene acquired by different sensors, or images taken by the same sensor at different times. In short, image registration is defined as a mathematical modeling which determines a mapping that is the best match of two or more images (Li et al. 2006a).

The core process of image registration is to optimize the similarity measure. The mutual information based similarity measure, which is proposed independently by Voila & Wells (1995) and Collignon et al. (1995), is used in our study. The choice to use this similarity measure is strongly based on the outstanding merits of the mutual information that it does not depend on any assumption on the image set and does not assume specific relationship between image intensities in different modalities. To apply the mutual information criterion for a realistic image registration, an important module is needed. That is how to find the global maximum of the mutual information based registration function efficiently and robustly. This is in general a complicated optimization problem. Because most of imaging applications assume that the initial misregistration is relatively small enough, a local optimization algorithm is almost the first choice. In the past research, many local optimization algorithms have been proposed for image registration (Press et al. 2002). In this paper, a simultaneous perturbation stochastic approximation algorithm is proposed for optimizing the mutual information based similarity measure. This local stochastic search algorithm is firstly proposed by J.C. Spall (2003), and it has been widely and successfully used to solve many kinds of optimization problems including the image registration. The main difference between our work and the other researchers' (Cole-Rhodes et al. 2003, Yoo 2004, Mueller et al. 2005) is that we firstly conclude three aspects of the effectiveness of Spall's local stochastic search algorithm over other two algorithms: hill climbing and Nelder-Mead simplex direct search. Especially, we firstly and successfully applied it to optimize a mutual information based pseudo-metric for simultaneously registering the six-parameter affine transformation for the multimodality images.

The paper is organized as follows: Section 1 introduces the research background. Section 2 presents a unified registration framework by using the mutual information based similarity measure associated with the histograms of images. Section 3 describes the implementation process of three local search algorithms. Section 4 presents some comparative experimental results and discussions of the efficient local stochastic search technique. Section 5 gives some concluding remarks.

2 UNIFIED FRAMEWORK OF OPTIMAL IMAGE REGISTRATION

2.1 Unified registration framework

A unified framework for an automatic image registration system is shown in Figure 1. It can be seen that the optimization module is the driver for the whole registration process. It is also known from the system view that the image registration is an iteration process.

In our study, the rigid (translation and/or rotation only) and affine transformations are considered to register the pairs of two dimensional gray-scale images. The rigid transformation matrix can be written as:

$$T_p^{rigid} = \begin{bmatrix} \cos(\theta) & -\sin(\theta) & 0 \\ \sin(\theta) & \cos(\theta) & 0 \\ tx & ty & 1 \end{bmatrix} \tag{1}$$

where (tx, ty) and θ denote the translation and rotation respectively.

The general affine transformation matrix can be written as:

$$T_p^{affine} = \begin{bmatrix} a & b & 0 \\ c & d & 0 \\ e & f & 1 \end{bmatrix} \tag{2}$$

In the affine registration experiment, we decompose T_p^{affine} into a product of scale, shear and rotations plus translations. The decomposition is listed as follows:

$$T_p^{affine} = \begin{bmatrix} a & b & 0 \\ c & d & 0 \\ 0 & 0 & 0 \end{bmatrix} + \begin{bmatrix} 0 & 0 & 0 \\ 0 & 0 & 0 \\ e & f & 1 \end{bmatrix} = T_p^{affine,I} + T_p^{affine,II} \tag{3}$$

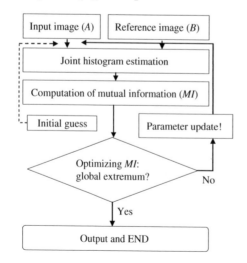

Figure 1. A unified framework for image registration.

$$T_p^{affine,I} = \begin{bmatrix} 2^s & 0 & 0 \\ 0 & 2^s & 0 \\ 0 & 0 & 0 \end{bmatrix} R(\theta) \begin{bmatrix} 2^t & 0 & 0 \\ 0 & 2^{-t} & 0 \\ 0 & 0 & 0 \end{bmatrix} R(\phi) \tag{4}$$

$$R(\theta) = \begin{bmatrix} \cos(\theta) & -\sin(\theta) & 0 \\ \sin(\theta) & \cos(\theta) & 0 \\ 0 & 0 & 0 \end{bmatrix} \tag{5}$$

$$R(\phi) = \begin{bmatrix} \cos(\phi) & -\sin(\phi) & 0 \\ \sin(\phi) & \cos(\phi) & 0 \\ 0 & 0 & 0 \end{bmatrix} \tag{6}$$

where s and t are scale and shear parameters, θ and ϕ are two rotation parameters, and e and f are two translation parameters.

Our goal is to find the optimal parameters discussed above which define the relative position and orientation of the two sensed images (the input image and the reference image) in an efficient and robust way.

2.2 Mutual information

Based on the information theory (Cover & Thomas 1991), the mutual information, $MI(A, B)$, of two random variables A and B can be calculated as follows:

$$MI(A,B) = H(A) + H(B) - H(A,B) \tag{7}$$

where $H(A)$ and $H(B)$ are the marginal entropies of A and B, and $H(A, B)$ is their joint entropy. Considering A and B as two images, the mutual information is maximal when the two images are totally geometrically aligned by a certain transformation matrix. In our rigid registration, the standard mutual information criterion is used. But in practice, many pseudo metrics based on mutual information are educed for the different purposes. In our affine registration experiment, a normalized pseudo metric (Zhang & Rangarajan 2004) is used and defined as follows:

$$pMI(A,B) = \frac{2H(A,B)}{H(A) + H(B)} - 1 \tag{8}$$

In the computation of mutual information, the marginal entropies and joint entropy can be calculated as follows:

$$H(A) = \sum_a - p_A(a) \log p_A(a) \tag{9}$$

$$H(B) = \sum_b - p_B(b) \log p_B(b) \tag{10}$$

$$H(A,B) = \sum_a \sum_b - p_{A,B}(a,b) \log p_{A,B}(a,b) \tag{11}$$

where $P_A(a)$ and $P_B(b)$ are the marginal probability density functions (PDF), and $P_{A,B}(a,b)$ is the joint probability density function. These PDFs can be computed from the histogram of images as follows:

$$p_{A,B}(a,b) = \frac{h(a,b)}{\sum_a \sum_b h(a,b)} \qquad (12)$$

$$p_A(a) = \sum_b p_{A,B}(a,b) \qquad (13)$$

$$p_B(b) = \sum_a p_{A,B}(a,b) \qquad (14)$$

where h is the joint histogram of the image pair. The value of $h(a,b)$ is the statistic numbers of corresponding pairs having intensity value a in the input image A and intensity value b in the reference image B (Chen et al. 2003). In the whole experiments, the partial volume interpolation was used for the joint histogram estimation.

3 LOCAL SEARCH ALGORITHM

3.1 Hill climbing search

The hill climbing is a graph search algorithm where the current path is extended with a successor node which is closer to the solution than the end of the current path. In simple hill climbing, the first closer node is chosen whereas in steepest ascent hill climbing all successors are compared and the closest to the solution is chosen. It fails if there is no closer node. This may happen if there are local maxima in the search space which are not solutions (http://foldoc.org/).

3.2 Nelder-Mead simplex direct search

The Nelder-Mead simplex direct search is a commonly used nonlinear optimization algorithm for minimizing an objective function in a many-dimensional space (Nelder & Mead 1965). The method uses the concept of a simplex, which is a polytope of $N + 1$ vertices in N dimensions: a line segment on a line, a triangle on a plane, a tetrahedron in three-dimensional space and so forth. It approximately finds a locally optimal solution to a problem with N variables when the objective function varies smoothly.

The search procedure terminates when the standard deviation of each transformation parameter of the final simplex is less than the residual settings.

3.3 Spall's stochastic search

The stochastic search implemented in our study is the simultaneous perturbation stochastic approximation (SPSA) algorithm, which is firstly put forward by

J.C. Spall. It becomes seriously popular for solving some challenging optimization problems. The prominent merit of this algorithm is that it does not require an explicit knowledge of the gradient of the object function, or measurements of this gradient. At each iteration, it only needs an approximation to the gradient via simultaneous perturbations. The gradient approximation is based on only two function measurements regardless of the dimensions of the search parameter space. The SPSA algorithm is a very strong optimizer, which can get through some local maxima of the objective function to successfully find the global maximum because of the stochastic nature of the gradient approximation (Spall 2003).

The implementation of the standard SPSA algorithm is used in our current study. The idiographic implementation on the image registration can be found in detail (Li et al. 2006a, b).

4 EXPERIMENTAL RESULTS

4.1 Time-effective demonstration

The first experiment is designed to register one parameter transformation. The simplest rigid transformation with the rotation only is considered. Figure 2 (i) and (ii) show a pair of the Phased Array type L-band Synthetic Aperture Radar (PALSAR) images to be registered. The Figure 2 (ii) is generated from (i) by rotating 5 degrees clockwise with the cubic interpolation. Figure 3 shows the mutual information as a function of the rotation angle after the registration with plotting the search trajectory to the ground truth. The initial guess starts from −5 degrees for each search algorithm.

By observing Figure 3 (i), (ii) and (iii), it can be found that the Spall's stochastic search algorithm shows the best performance/cost to the rotation registration. The hill climbing search difficultly gets across the local maxima. The Nelder-Mead simplex direct search can jump over the local maxima, but the computation of simplex costs too much time.

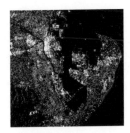 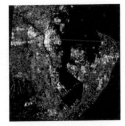

(i) PALSAR image   (ii) Transformed PALSAR image

Figure 2. A pair of PALSAR images of shizuoka city, Japan. The rotation angle is shown in degrees (size: 256 × 256).

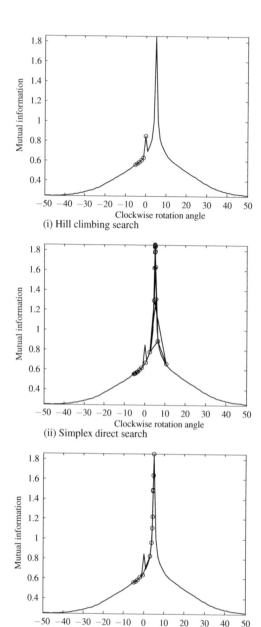

(i) Hill climbing search

(ii) Simplex direct search

(iii) Stochastic search

Figure 3. The mutual information registration function along the rotation axis. The search trajectory is shown in circle and dash line.

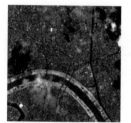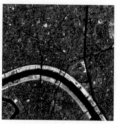

(i) With cloud contamination   (ii) Without cloud contamination

Figure 4. ASTER VNIR image pair (size: 256×256).

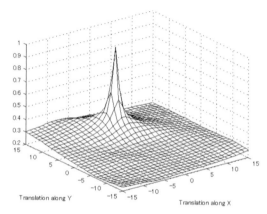

Figure 5. Mutual information registration function along two direction translations.

4.2 Accuracy-effective demonstration

The second experiment is to register two ASTER (Advanced Spaceborne Thermal Emission and Reflection radiometer) VNIR (visual near-infrared) images in a two-translation transformation by using a standard mutual information criterion. The image pair is shown in Figure 4. The mutual information registration function is plotted in Figure 5 for the image pair along two-direction translations. The true solution of the image pair is (0, 7) if the image without cloud contamination is regarded as the reference image and the bottom left corner as the coordinate origin.

In order to demonstrate the performance of the local stochastic search algorithm, ten initial guesses near the global maximum are randomly generated in the square region from $(-5, 5)$ to $(5, 15)$. These initial guesses are listed in Table 1 and plotted in Figure 6. The experimental results are shown in Figures 7 and 8. Figure 7 shows a typical stochastic search trace starting from the No. 1 initial guess. The absolute registration errors for translation in X and Y directions are plotted in Figure 8 (i) and (ii) respectively. The simplex direct search succeeds only for experiment No. 3. Both the hill climbing and local stochastic search achieves good results, but the mean registration error of local stochastic search is less than 0.03 pixel, obviously better than the 0.34 pixel of the hill climbing search. The 500 iterations were fixed in the local stochastic search.

Table 1. Random initial guess for the translation registration.

Experiment number	X	Y
1	4.2948	10.0030
2	2.8396	5.7396
3	0.4030	9.8188
4	2.1581	10.4194
5	4.0608	11.4797
6	4.5350	11.6998
7	−3.7055	13.6334
8	−4.5914	7.8322
9	−2.6475	13.5747
10	−0.9674	7.5768

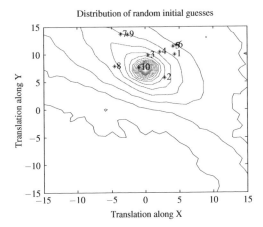

Figure 6. Distribution of initial guesses denoted by asterisk.

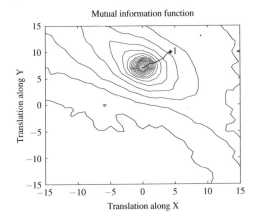

Figure 7. A typical local stochastic search trace starting from No.1 initial guess.

4.3 *Affine-effective demonstration*

In the third experiment, the stochastic search algorithm is used to register the multimodality images with the affine transformations. In the experiments,

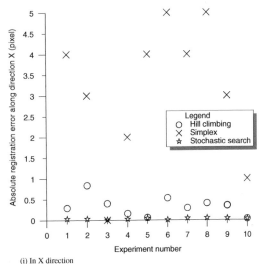

(i) In X direction

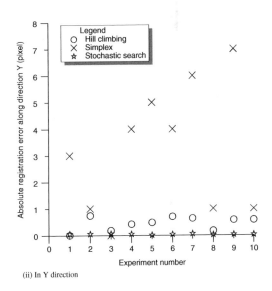

(ii) In Y direction

Figure 8. Absolute registration error for translation in X and Y directions.

the powerful Brainweb simulated MRI (magnetic resonance imaging) volumes for a normal brain are used. The main advantage to use the simulated MRI database is that the ground truth is known (http://www.bic.mni.mcgill.ca/brainweb/).

In the experiments, one triplet of 2D slice of 3D PD (proton density), T2 and T1 MR volume images along sagittal direction for a normal brain is used. The T2 MR imaged is transformed in an affine transformation $\hat{T}_2^{affine}$ with $s_1 = t_1 = 0.1$, $\theta_1 = \phi_1 = 5$ degrees, $e_1 = f_1 = 3$. The T1 MR image is transformed

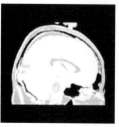
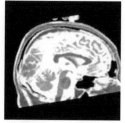
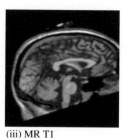

| (i) PD | (ii) MR T2 | (iii) MR T1 |

Figure 9.   2D slice of 3D PD and MRI volumes along sagittal direction for a normal brain (size: 256×256).

Table 2.   Registration results of 2D slice PD and MR T2.

Affine parameters	Ground truth	Stochastic search
s	0.1	0.1
t	0.1	0.1
θ	5	4.9
ϕ	5	5
e	3	3
f	3	3

Table 3.   Registration results of 2D slice PD and MR T1.

Affine parameters	Ground truth	Stochastic search
s	0.2	0.2
t	0.2	0.2
θ	10	9.9
ϕ	10	10
e	5	5
f	5	4.7

in an affine transformation $\hat{T}_1^{affine}$ with $s_2 = t_2 = 0.2$, $\theta_2 = \phi_2 = 10$ degrees, $e_2 = f_2 = 5$. During the generation process of the transformed T2 and T1 MR images, the linear interpolation was used. The PD, transformed T2 and transformed T1 MR images are shown in Figure 9.

The measure pMI in Equation 8 was used to register the 2D PD and transformed T2 MR images, and the 2D PD and transformed T1 MR images in sequence. The initial guess for both affine registration experiments is (0,0,0,0,0,0). Table 2 shows the registration results of 2D PD and transformed T2 MR slices by using the local stochastic search. Table 3 shows 2D PD and transformed T1 MR slices by using local stochastic search. The 10000 iterations were fixed in the local stochastic search. The experimental results demonstrate a proof of an efficient local stochastic search

approach to multimodality image registration with an affine transformation.

5 CONCLUSIONS

The experimental results show that in general the local stochastic search yields significant improvements on the optimal solution over the conventional search techniques in terms of accuracy and robustness. An efficient local stochastic search strategy on optimizing the mutual information based similarity measure was firstly and successfully accomplished to simultaneously register the six-parameter affine transformation for the multimodality images.

REFERENCES

Chen, H., Varshney, P.K. & Arora, M.K. 2003. Mutual information based image registration for remote sensing data. *International Journal of Remote Sensing* 24(18): 3701–3706.

Cole-Rhodes, A.A., Johnson, K.L., LeMoigne, J. & Zavorin, I. 2003. Multiresolution registration of remote sensing imagery by optimization of mutual information using a stochastic gradient. *IEEE Transactions on Image Processing* 12(12): 1495–1511.

Collignon, A., Maes, F., Vandermeulen, D., Suetens, P. & Marchal. G. 1995. Automated multimodality image registration using information theory. In Y. Bizais, C. Barillot, & R. Di Paola (eds), *Information Processing in Medical Imaging*: 263–274. Dordrecht: Kluwer.

Cover, T.M. & Thomas, J.A. 1991. *Elements of information theory*. New York: John Wiley & Sons.

Li, Q., Sato, I. & Murakami, Y. 2006a. Automated image registration using stochastic optimization strategy of mutual information. *Dynamics of Continuous Discrete and Impulsive Systems, Series B: Application and Algorithms* An Added Volume: 2872–2877.

Li, Q., Sato, I. & Murakami, Y. 2006b. Simultaneous perturbation stochastic approximation algorithm for automated image registration optimization. *Proceeding of 2006 IEEE international geoscience and remote sensing symposium & 27th Canadian symposium on remote sensing (IGARSS 2006), Denver, Colorado, 31 July–4 August 2006*. USA: IEEE (in press).

Mueller, U., Hesser, J. & Maenner, R. 2005. Optimal parameter choice for automatic fast rigid multimodal registration. In J.M. Fitzpatrick & J.M. Reinhardt (eds), *Medical Imaging 2005: Image Processing*. V5747: 163–169. USA: SPIE.

Nelder, J.A. & Mead, R. 1965. A simplex method for function minimization. *The Computer Journal* 7(4): 308–313.

Press, W.H., Teukolsky, S.A., Vetterling, W.T. & Flannery, B.P. 2002. *Numerical recipes in C++: the art of scientific computing (2nd edition)*. New York: Cambridge University Press.

Spall, J.C. 2003. *Introduction to stochastic search and optimization: estimation, simulation and control*. New Jersey: John Wiley & Sons.

Viola, P.A. & Wells III, W.M. 1995. Alignment by maximization of mutual information. *Proceedings of the fifth international conference on computer vision, Boston, Massachusetts, 20–23 June 1995*. Cambridge: IEEE.

Yoo, T.S. 2004. *Insight into images, principles and practice for segmentation, registration, and image analysis*. Massachusetts: A K Peters.

Zhang, J. & Rangarajan, A. 2004. Affine image registration using a new information metric. *Proceeding of IEEE computer society conference on computer vision and pattern recognition (CVPR'04), Washington, DC, 27 June–02 July 2004*. USA: IEEE.

Computational Modelling of Objects Represented in Images –
João Manuel R.S. Tavares & R.M. Natal Jorge (eds)
© *2007 Taylor & Francis Group, London, ISBN 978-0-415-43349-5*

Rapid prototyping technology in medical applications: A critical review

P. Lopes
Biomedical Engineering MSc student, Mechanical Engineering Department, University of Minho, Portugal

P. Flores & E. Seabra
Mechanical Engineering Department, University of Minho, Portugal

ABSTRACT: This paper aims to present and discuss the main features concerning with the application of the rapid prototyping technology in medicine. A brief introduction to the genesis of rapid prototyping technology, in general, is made throughout this work. In the sequel of this process, the most relevant rapid prototyping technologies and their implications on the medical applications are presented and discussed, namely in what concerns their main advantages and limitations, as well as to understand the future possibilities related to the development of this technology in Portugal. Furthermore, a simple and fundamental case study is used as an application example, in order to demonstrate the procedures and methodologies adopted in the present work. Finally, the main conclusions from this work are drawn and the perspectives for future research are outlined.

1 INTRODUCTION

This work deals with the Rapid Prototyping (RP) technology applied to medicine. In the field of design and development of medical models for implants, prostheses, design of auxiliary devices for medical training it is observable that the Portuguese industry presents accentuated deficiencies, being a specific market that has a high potential of growing. Actually, in Portugal there are very few companies dedicated to design and development of medical equipment. Thus, it is expected that the RP in medicine will call attention to industrial investors which might create high tech jobs, and might even enlarge its application to other medical products, namely auxiliary devices for physical handicap people.

Medical Rapid Prototyping (MRP) is defined as the manufacture of dimensionally accurate physical models of human anatomy derived from medical image data using a variety of RP technologies. Thus, physical model of human tissues can easily be obtained, with all crucial time and cost savings. MRP is a multi-discipline area, which applies the biomedical modeling and RP to develop medical applications. It involves human resources from the fields of Reverse Engineering (RE), design and manufacturing, biomaterials and medicine (Hieu et al. 2005). MRP processes are used in three main areas of application: (i) design and manufacturing of biomodels; (ii) development of surgical training models, surgical aid tools and implants; (iii) design and manufacturing of scaffolds for tissue engineering (Gebhart 2003).

The most significant effort applying RP to medical modeling, both in number of programs and scope of effort is in Europe. The major focus is the BRITE EuRAM PHIDIAS (acronym for Basic Research of Industrial Technologies for Europe, European Research on Advanced Materials Program) project administered by Materialise (Belgium), in which an RP service bureau set up in the technology park on the campus of the Catholic University of Leuven. Other centers focus their work on surgical applications of RP, namely: the Institute of Medical Physics at the University of Erlangen-Nürnberg in Erlangen, Germany; the University of Zurich (CH), and the University of Leeds (UK). In addition, there is a significant number of locally funded application evaluations underway throughout Europe, including France, Austria, Denmark, and Germany (Lightman 1997).

Despite, all the progresses of MRP technology worldwide, in Portugal, the application of RP to the medical field is almost inexistent. The great majority of medical models produced as an aid to surgical procedures are imported from Spain, Belgium and Germany, contributing to an increased cost and time. Companies such as SOLIDtech and MedMat, as well as some institutes, including, University of Minho, among others, have already cooperate with some particular clinical cases; however, the lack of knowledge regarding these technologies is almost unanimous between the Portuguese medical community. Thus, it will be necessary a great effort to divulge the capabilities of MRP technology and reduce the existing gaps.

2 RAPID PROTOTYPING DESCRIPTION

2.1 Genesis

Rapid Prototyping technology has its root at the end of 80's of the last century, and has been coming to be used almost exclusively to produce prototypes and, in some cases, to directly or indirectly produce tools. RP technology has changed the way products are being designed and manufactured. Being first introduced in 1987, this technology has made a tremendous impulse in all aspects over the last two decades (Alves et al. 2001). The building speed has been drastically increased, part's dimensional accuracy has been significantly improved and a wide variety of new building materials have also been identified. With these improvements, new applications for RP have been explored. In recent years, medicine is, indubitably, on of the most relevant and mentioned areas that gained directly from RP technology. Figure 1 shows the percentage of RP applications according to activity sector (Lee et al. 2002).

2.2 Concept

All processes by which 3D models and components are produced additively, that is, by fitting or mounting volume elements together are called generative production processes. However, this generic expression is seldom used in practice. More common are other expressions, namely, solid freeform manufacturing, sometimes called solid freeform fabrication, desktop manufacturing and rapid prototyping (Gebhart 2003). The expression rapid prototyping process is probably not the best; however, it has an indomitable practical advantage. It is engraved in everyone's memory and is undoubtedly the most used expression (Gebhart 2003).

Onuh & Yusuf (1999) established five criteria to help deciding if a process can be classified as a RP process, namely, (i) the process should take in raw material in some shapeless form such as blocks, sheets or a fluid, and produce solid objects with a definite shape; (ii) the process must do this without a significant amount of human interaction; (iii) the process must produce shapes with some degree of three dimensional geometrical complexity; (iv) the process must not involve the manufacture of new tools for each different shape to be generated; (v) each item produced must be a single object not an assembly of component parts thus eliminating joining operations such as gluing, welding and riveting.

3 RAPID PROTOTYPING PROCESSES

3.1 Stereolithography

Patented in 1986, Stereolithography (SL) started the rapid prototyping revolution. Stereolithography apparatus (SLA) machines have been made since 1988, by 3D Systems of Valencia, and it is the most well-known and used RP technique for all applications, including medical model building (Popat et al. 1998). An SL RP system consists of a bath of photosensitive resin, a model-building platform and an ultraviolet laser for curing the resin. A 3D CAD model of the desired part is sliced, in software, into a series of adjacent 2D slices. This data is used to control a laser beam that induces a chemical reaction, bonding large number of small molecules and forming a highly cross-linked polymer. The photosensitive resin becomes cured or solidified when exposed to the ultra-violet (UV) radiation. The model is built onto a platform situated just below the surface in a vat of liquid epoxy or acrylate resin. The first layer of the model, the base, is drawn by the laser to form a solid layer, typically 0.1 mm thick. A low-power highly focused UV laser traces out the first layer, solidifying the model's cross section while leaving excess areas liquid.

The platform then descends in the bath to allow new liquid resin to cover the cured layer and the next model slice is constructed above it. In this way the whole of the model is built from the base up. This process is repeated until the prototype is completed. Afterwards, the solid part is removed from the vat and rinsed clean of excess liquid. Support structures are

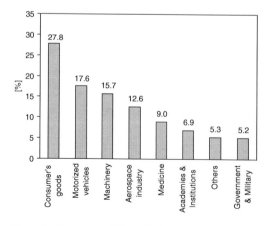

Figure 1. Percentage of RP applications according to activity sector {adapted from Lee et al. 2002}.

removed by hand after the model is fully cured. It is worth noting that this process is a labor-intensive and time-consuming process (Bartolo & Mitchell 2003, Winder & Bibb 2005).

Overall, the distinguishing factors between this technology and other RP technologies for medical models include: (i) ideal compromise between model size needed and overall accuracy for medical models; (ii) very reliable technique; (iii) using acrylate materials, the technique allows high transparency, fast building and possible medical grade models with selective coloration of regions of interest (Popat et al. 1998).

3.2 Fused deposition modeling

Fused deposition modeling (FDM) is one of the few commercially available rapid prototyping technologies offering real possibilities of producing solid objects in a range of different materials, including metals and composites (Masood 1996). FDM uses a similar principle as SL. It builds models on a layer-by-layer basis. In this technique, filaments of heated thermoplastic are extruded from a tip that moves in the x-y plane. The controlled extrusion head deposits very thin beads of material onto the build platform to form the first layer. Once a layer has been deposited, the nozzle is raised between 0.18 to 0.36 mm and the next layer is deposited on the top of the previous layer. This process is repeated until the model is completed. The platform is maintained at a temperature lightly above (approximately 0.5°C) the build material's melting point, so that it solidifies within a very short time (approximately 0.1 s) after extrusion and cold-welds to the previous layer (Lee et al. 2002, Winder & Bibb 2005).

As with SL, support structures are required for FDM models as it takes time for the thermoplastic to harden and the layers to bond together. The supports are added to the model at the design stage and built using a different thermoplastic material, extruded through a second nozzle. The support material is a different color than the building material and does not adhere to it. This enables easy identification and subsequent removal of the supports by hand after the model is completed. A more recent development is the availability of a soluble support material, which enables support structures to be dissolved from the model in an agitated water bath (Winder & Bibb 2005).

In the FDM process, the accuracy, surface finish quality, cost and time of production of the part greatly depends on several factors requiring the designer's knowledge, skills and judgment. This technique is more suited for placement in a hospital environment than the SL, since a FDM machine is a compact machine with dry materials, which are easy to handle in a hospital environment. It's possible to choose materials with different colors, and medical grade ABS that can be gamma sterilized.

3.3 Laminated object manufacturing

Laminated object manufacturing (LOM) technique, developed by Helisys of Torrance, California (USA), creates objects by bonding and cutting out sheet materials and is very interesting for foundry applications and for large volumes (e.g. Orthopedic) since the complete volume does not have to be scanned (Popat et al. 1998).

The original material consists of paper laminated with heat-activated glue and rolled up on spools. A feeder/collector mechanism advances the sheet over the build platform, where a base has been constructed from paper and double-sided foam tape. Next, a heated roller applies pressure to bond the paper to the base. The sliced CAD data is used to control a focused laser which cuts the outline of the first layer into the paper and then cross-hatches the excess area. Cross-hatching breaks up the extra material, making it easier to remove during post-processing. During the process, the excess material provides excellent support for overhangs and thin-walled sections. After the first layer is cut, the platform lowers out of the way and fresh material is advanced. The platform rises to slightly below the previous height, the roller bonds the second layer to the first and the laser cuts the second layer. This process is repeated as needed to build the part, which will have a wood-like texture. Since the models are made of paper, they must be sealed and finished with paint or varnish to prevent moisture damage.

3.4 Selective laser sintering

Developed by Carl Deckard for his master's thesis at the University of Texas, Selective laser sintering (SLS) was patented in 1989. This technique uses a laser beam to selectively fuse powdered materials. Parts are built upon a platform which sits just below the surface in a bin of the heat-fusible powder. A laser traces the pattern of the first layer, sintering it together. The platform is lowered by the height of the next layer and powder is reapplied. This process continues until the part is completed (Chatterjee et al. 2003).

SLS is emerging as a popular technique among solid freeform fabrication processes for RP because of its capability to form solid parts of arbitrary geometry without part-specific tooling. Furthermore, SLS can be carried out on metal powders so that functional prototypes can be produced in considerably shorter cycle times. This process also has the advantage of inherent supports and will be comparable in speed with SLA. Surface quality and model details will be somewhat less than with SLA models but the range of different materials (polystyrene, nylon, wax, etc.) which can be used on the machine is a great advantage. Also FDA approved medical grade Nylon which can be sterilized is available for use in making medical models with this technology.

This RP technique is believed to be more suitable for orthopedic type applications rather than transparent models of skull, etc. (Popat et al. 1998).

3.5 Three dimensional ink-jet printing

3D printing (3DP), developed and patented by Massachusetts Institute of Technology (MIT) and licensed to Z Corporation in 1994, was the first ink-jet based technology, being compact, accurate and suitable for small parts that can be built at low cost. In this RP process, the completed CAD solid model is transferred to a Stereolithography Text File (STL) file and parts are built upon a platform situated in a bin full of powder material. An ink-jet printing head selectively deposits or prints a binder fluid to fuse the powder together in the desired areas. Unbound powder remains to support the part. The platform is lowered, more powder added and leveled, and the process repeated. When finished, the green part is then removed from the unbound powder, and excess unbound powder is blown off (Dimitrov et al. 2006).

The appearance of the final model is similar to that produced with RP technologies such as SL, LOM and SLS, namely, a stair stepped appearance. This is an undesirable effect which is created by trying to duplicate freeform shapes with discrete layers. By constructing models of thinner layers the stair stepping effect is reduced, but thicker layers may still be acceptable within concept modeling.

Models constructed using the 3D printer method are weak and can easily be damaged and distorted. In this case the models can be strengthened by infiltrating with wax, and ink can be added to the initially transparent wax to produce parts that have a variety of colors. Temporary support structures are required to hold the part before being completed. The main reason for supports is that unconnected islands and thin overhangs can move and sag, respectively, before they cure. Supporting structures further elevate the part from the build platform, making it easier to remove the part once completed.

4 RAPID PROTOTYPING IN MEDICINE

4.1 Design and manufacturing biomodels

Biomodels can be defined as physical models of anatomical structures fabricated by RP from 3D digital anatomical data. Instead of using 2D images and 3D virtual computer models, medical doctors nowadays can use the accurate replica of parts or anatomical structures or region of interest of the human body for preoperative planning diagnosis and treatment (Hieu et al. 2005). Given the visualization capabilities of sophisticated software packages, the production of physical models may, firstly, seem superfluous; however, this is not the case, since the depiction of a 3D volume on a 2D screen does not provide surgeons with a complete understanding of patient's anatomy.

There are several visualization issues that are addressed but not resolved by virtual computer models, providing the motivation for the construction of physical models of bones: (i) 2D screen displays do not provide an intuitive representation of 3D geometry; (ii) unusual or deformed bone geometry may be hard to comprehend on-screen; (iii) the integration of multiple bone fragments is difficult to visualize on-screen; (iv) planning complex 3D manipulations on the basis of 2D images is difficult.

A physical model manufactured from X-ray Computed Tomography (CT) or Magnetic Resonance Imaging (MRI) data can be held and felt, offering surgeons a direct, intuitive understanding of complex anatomical details which cannot be obtained from imaging on-screen. A precise physical model can offer an accurate prediction of the shape, orientation, relative location and size of internal anatomical structure, providing hands-on surgical planning and rehearsal. Proposed cuts or bone fragment realignments can be marked and/or tried on the model to establish an optimal surgical plan. The model can then serve as a custom, patient-specific intra-operative reference guide.

Medical models have a number of characteristics that are secondary or without importance for product development in other sectors of application, namely, large amount of data, large models, transparency, biocompatibility, unconnected model parts, just to mention few. The use of medical models has been applied to a wide range of medical specialties, including craniomaxillofacial surgery, dental implantology, tumor removal, neurosurgery and orthopedics.

4.2 Development of surgical training models

Medical RP and biomedical modeling technologies evolutionally contribute to the development of new surgical training models, which are playing an important role in improvement of undergraduate and postgraduate training skills in medical education. One example is the Primacorps, a surgical training model which replaces the human form for the practice of surgery. Surgical aid tools are very important for surgeons during the operation. They are used for enhancing surgeon's skills, reducing the operation time, and increasing the accuracy and safety of the operation. The most commonly used surgical aid tools are the drilling guides for dental and spine surgery (Hieu et al. 2005).

The areas of Medical Imaging, Computer Aided Design, Finite Element Analysis and Rapid Prototyping, can, as an integrated approach, provide a powerful tool for the biomechanical analysis of anatomical structures and the design of medical implants. Although implants can be prepared preoperatively based on biomodels, the implant design process is required for the ones that are used for complex defects. From 3D anatomical data, custom-made implants, as well as surgical aid tools can be exactly designed to avoid the mismatch of using the standard ones for the patient. This is very useful for the defect areas that need cosmetic surgery. The typical examples are cranioplasty implants and titanium membranes for bone reconstruction.

4.3 Design and manufacturing of scaffols

Tissue Engineering (TE) is an emerging inter-disciplinary field that involves the application of engineering and life science principles to develop biological substitutes that restore, maintain or improve tissue functions, from native or synthetic sources. Most of the TE approaches towards creating functional replacement tissues or organs employ temporary porous scaffolds to guide the proliferation and differentiation of seeded cells in vitro and in vivo as well as to provide for the mechanical stability of the biological substitutes in a three dimensional organization.

A successful scaffold should possess the following characteristics: (i) a suitable macrostructure to promote cell proliferation and cell-specific matrix production; (ii) an open-pore geometry with a highly porous surface and microstructure that enables cell ingrowth; (iii) optimal pore size employed to encourage tissue regeneration and to avoid pore occlusion; (iv) suitable surface morphology and physiochemical properties to encourage intracellular signaling and recruitment of cells; (v) being made from a material with a predictable rate of degradation, with a nontoxic degraded material (Tan et al. 2003).

Currently, a variety of conventional or advanced processing methods are available for the fabrication of tissue scaffolds. Conventional fabrication techniques include fiber bonding and fiber meshes, gas foaming, phase separation, solution casting, emulsion freeze drying, melt molding, membrane lamination, solvent casting and particulate leaching. However, much of these techniques have serious shortcomings that restrict their scope of applications. Inconsistency in pore size reproduction, irregularity in pore distribution, indispensable usage of toxic organic solvents, labor-intensive and deficiency in scaffolds' mechanical strength are some of the problems commonly associated with these fabrication techniques (Tan et al. 2003, Yeong et al. 2004).

Researchers try to modify the conventional techniques to overcome these inherent process limitations. Due to such limitations, the potential in using RP is tremendous as most RP techniques enables the tissue engineer to have full control over the design, fabrication and modeling of the scaffold being constructed, providing a systematic learning channel for investigating cell-matrix interactions. Additionally, indirect RP methods, coupled with conventional pore-forming techniques, further expand the range of materials that can be used in tissue engineering. RP has the distinct advantage of being able to build objects with predefined microstructure and macrostructure. This distinct advantage gives RP the potential for making scaffolds or orthopedic implants with controlled hierarchical structures.

The most used RP techniques within this field are 3DP and FDM. 3DP was the first RP technique to be proposed for biomedical and tissue engineering purposes. Scaffolds using different materials, as well as with different morphologies have been produced by means of using this technique, showing good cell-material interactions. The main disadvantages of this technique include the fact that the pore size of the fabricated scaffolds is dependent on the powder size of the stock material, closure of the pores by the stock material, and the use of organic solvents as binders for the traditional poly (α-hydroxy acids). Furthermore, and because the final structure is a combination of several stack up powdered layers, the mechanical properties can also be a problem.

5 EXAMPLE APPLICATION

This section presents an example application with the intent to demonstrate how to construct medical models based on 2D medical images, creating 3D virtual models that can, consequently and eventually be used to build a real prototype. Further, since the 3D virtual model is available, a posterior structural analysis based on FEM can be performed. During this work stage, enterprises, research groups and medical doctors have been contacted. Some CT/MRI data have been acquired, but none of them contained any pathological problem. The CT/MRI data have been acquired from a female child's skull.

At the same time, available evaluation versions of different commercial image processing programs were used and compared, namely, Mimics, AnatomicsPro, 3DDoctor, In Vesalius, among others.

Figure 2 shows a 3D virtual model obtained from 2D medical images and converted to 3D by using the commercial program Mimics. This virtual model was then imported to Solidworks program that allowed the structural analysis of the model.

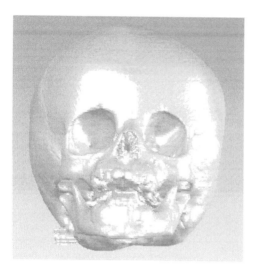

Figure 2. 3D model of a female child's skull obtained with evaluation version of Mimics.

6 CONCLUSIONS AND FUTURE DIRECTIONS

In this paper, an attempt to demonstrate the application of Rapid Prototyping technology in medicine was presented and discussed. Further, the main RP processes were also briefly offered throughout this work. In addition, Medical Rapid Prototyping (MRP) processes used in three main areas of application were presented: (i) design and manufacturing of biomodels; (ii) development of surgical training models, surgical aid tools and implants; (iii) design and manufacturing of scaffolds for tissue engineering.

An example application was used to demonstrate how to construct 3D virtual medical models based on 2D medical images, which can be the basis to build a real prototype. This academic example is quite elucidative how easy the construction of virtual and real models can be obtained from 2D medical images. However, the RP processes beside their availability, they are quite expensive. It should be mentioned that a simple physical model of a full skull can cost about two thousand euros.

This work reports on an initial phase that corresponds to creation of a research work at the University of Minho, entitled ModMed (acronym for Models Medical), which will develop 3D virtual medical models and physical prototypes applying RP techniques.

The ModMed team will work in cooperation with companies and institutions, and can be a stimulant initiative to the creation on new enterprises and the development of the MRP technology in Portugal.

It should also be highlighted that the first author is currently developing research work at the Katholieke Universiteit Leuven (Belgium) under the topic "Selective Laser Sintering of a mono-bloc implant for hip reconstruction", which main objective is to establish of a bridge between virtual implant design for hip surgery and the SLS production technique.

REFERENCES

Alves, F.J.L. et al. 2001. *Protoclick – Prototipagem Rápida*. Ed. Protoclick, Porto.

Bartolo, P. & Mitchell, G. 2003. Stereo-Thermal-Lithography: A New Principle for Rapid Prototyping. *Rapid Prototyping Journal* 9(3), 150–156.

Chatterjee, A. et al. 2003. An Experimental Design Approach to Selective Laser Sintering of Low Carbon Steel. *Journal of Materials Processing Technology* 136, 151–157.

Dimitrov, D. et al. 2006. Advances in Three Dimensional Printing – State of the Art and Future Perspectives. *Rapid Prototyping Journal* 12(3), 136–147.

Gebhart, A. 2003. *Rapid Prototyping*. Hanser Gardner Publications, Inc. Munich. 23–218, 267–275, 315–364.

Hieu, L. et al. 2005. Medical Rapid Prototyping Applications and Methods. *Assembly Automation* 25(4), 284–292.

Lee, P. et al. 2002. Prosthetic Sockets Fabrication Using Rapid Prototyping Technology. *Rapid Prototyping Journal* 8(1): 53–59.

Lightman, A. 1997. Medical Applications. Rapid Prototyping in Europe and Japan. *Rapid Prototyping Association of the Society of Manufacturing Engineers*, Vol. I. Analytical Chapters. 129–135.

Masood, S. 1996. Intelligent Rapid Prototyping with Fused Deposition Modelling. *Rapid Prototyping Journal* 2(1), 24–33.

Onuh, S. & Yusuf, Y. 1999. Rapid Prototyping Technology: Applications and Benefits for Rapid Product Development. *Journal of Intelligent Manufacturing* 10, 301–311.

Popat, A. et al. 1998. Rapid Prototyping and Medical Modelling. *Phidias, Rapid Prototyping in Medicine* 1, 10.

Tan, K. et al. 2003. Scaffold Development Using Selective Laser Sintering of Polyetheretherketone Hidroxyapatite Biocomposite Blends. *Biomaterials* 24, 3115–3123.

Winder, J. & Bibb, R. 2005. Medical Rapid Prototyping Technologies: State of the Art and Current Limitations for Application in Oral and Maxillofacial Surgery. *Journal of Oral and Maxillofacial Surgery* 63, 1006–1015.

Yeong, W. et al. 2004. Rapid Prototyping in Tissue Engineering: Challenges and Potential. *Trends in Biotechnology* 22(12), 643–652.

Computational Modelling of Objects Represented in Images –
João Manuel R.S. Tavares & R.M. Natal Jorge (eds)
© 2007 Taylor & Francis Group, London, ISBN 978-0-415-43349-5

Road extraction from high-resolution satellite images using level set methods

Zhen Ma, Ji-Tao Wu & Zhong-Hua Luo
LMIB, Key Lab of Education Ministry, Department of Mathematics, Beihang University, Beijing, China

ABSTRACT: Through establishing proper models, our paper presents a road extraction method based on the level set methods. An initial contour is first extracted by the combination of the threshold of intensity gradient and the fast marching method. Then the contour is moved with several speed models to remove the leakages, eliminate the noises and improve the smoothness until the curve finally reaches the accurate road edge. Our method makes full use of the characteristics of the roads in the high-resolution satellite images. The initialization process is simple and no priori topology information is needed. The level set methods allow us to get a better and more complete road network and handle the topological changes during the curve evolution. Numerical experiments have further demonstrated the effectiveness and advantages of our method.

1 INTRODUCTION

Road information contained in the satellite images is very important for the update of GIS (Geographic Information System). Many efficient and effective methods have been proposed to extract roads in the low-resolution satellite images, for example, Geman & Jedynak (1996) and Steger (1998). While as the advances of remote sensing technologies, more and more high-resolution satellite images become publicly available. While some assumptions made by the former extraction methods, such as the roads have linear structure and negligible widths, do not hold in the high-resolution images, so most of those methods are not applicable any more. Information retrievals from high-resolution images gradually become the research focus.

Compared with the low-resolution satellite images, roads in the high-resolution ones form a much more complex network. As the roads have appreciable width, extraction process is a little similar to the shape detection. Among the methods introduced in recent years, for example, Lee et al (2000), Péteri & Ranchin (2003) and Zhu et al (2004), the most frequently used techniques are the morphologic method and the snake method. But some characteristics of the two methods decrease the feasibilities of their applications in this field. The main drawback of the morphologic method is that we can hardly exclude the objects with similar intensities to roads even when they are not connected. Besides, using structuring elements in the morphological operations may break the smoothness of the contour; the snake method has the difficulty to get the prior topology information of the roads, and its strict requirement that initial road network must be put in a position near the roads greatly constrained its application. While these mentioned problems can be easily handled through using the level set methods.

The level set method was first introduced in Osher & Sethian (1988) to track the interface of a closed curve moving in a direction normal to itself. It has many excellent characteristics such as the ability of handling topological changes and the high computing efficiency through using the narrow band method and the fast marching method. Due to these excellent characteristics, the level set methods have been applied in many areas especially in image processing during the last ten years. Our paper makes the attempt to apply this method to the field of extracting roads from the high-resolution satellite images. Finally, we note that a different study also applying the level set methods has been performed by Keaton & Brokish (2002).

2 EXTRACTION PRINCIPLE

2.1 *Standard level set method*

The central idea of the level set method is to represent the evolving curve as the level set $\{\phi(x, y, t) = 0\}$ of a function ϕ, which means to embed a 2-D curve into a 3-D space. Thus when we differentiate the two sides of the level set, we get the level set equation:

$$\phi_t + F \mid \nabla \phi \mid = 0 \qquad (1)$$

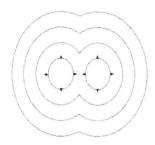

Figure 1. Evolving process of two circles.

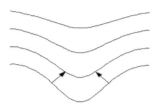

Figure 2. Evolving processes of the concave part.

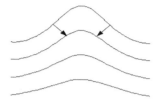

Figure 3. Evolving process of the convex part.

$$\phi(x, y, t) = 0 \qquad (2)$$

where the F is the velocity of the curve. To propagate the level surfaces, we initialize the level set function using the distance function, and the plus (minus) sign is chosen if the point is inside (outside) the initial curve. The intrinsic geometric properties of the front can be easily calculated through the level set function. For example, the normal vector and the curvature can be expressed as:

$$\bar{n} = \nabla \phi / |\nabla \phi|, \quad \kappa = \nabla \cdot \bar{n} = \nabla \cdot (\nabla \phi / |\nabla \phi|).$$

As the discussion in Osher & Sethian (1988), schemes used to solve the hyperbolic conservation law can be borrowed to solve the level set equation. The implicit calculations of the intrinsic properties of the curve greatly enhance the stabilities of the schemes. This method can handle the topological changes easily and give the correct solutions that satisfy the entropy condition. For details of the standard level set method, see Osher & Sethian (1988). Figure 1 gives us an example of the computing results of the evolution of two circles through using the level set method. We can see that the ability of handling topological changes and the correctness of the level set method.

2.2 Narrow band method

Because the standard level set method suffers a draw back of low computing efficiency, Chopp proposed the narrow band method to handle the problem. The main idea of his approach is to use only the points close to the curve at every time step so that the calculating is constrain to a narrow band. If the front has $O(N^2)$ points in three dimensions, the narrow band method can decrease the operation count from $O(N^3)$ to $O(kN^2)$, where k is the number of cells in the narrow band. Although the re-initialization step of the narrow band method is still time-consuming, this method does save the computing time especially when the curve moves in a large region. For details and expository reviews of the narrow band method, please refer to Chopp (1993).

2.3 Fast marching method

When the speed F depends only on position and does not change its sign. The level set equation reduces to the Eikonal equation:

$$F |\nabla T| = 1 \qquad (3)$$

$$\Gamma(t) = \{(x, y) \mid T(x, y) = t\}, F > 0 . \qquad (4)$$

where $T(x, y)$ is the arrival time as the curve crosses the point (x, y). One scheme used to solve the equation is:

$$[\max(D_{ij}^{-x}T, 0)^2 + \min(D_{ij}^{+x}T, 0)^2 + \max(D_{ij}^{-y}T, 0)^2$$
$$+ \min(D_{ij}^{+y}T, 0)^2] = \frac{1}{F_{ij}^2} \qquad (5)$$

where D is the difference operator. Having observed the upwind difference structure of the scheme (5), Adalsteinsson and Sethian introduced the fast marching method to improve the computing efficiency. We can avoid the replicate calculating through using their method. For detailed procedure, please see Adalsteinsson & Sethian (1995) and Sethian (1996).

2.4 Curve movements used in the extraction

Our method mainly uses two movements of the curve: the closed curve moving outward and inward in a direction normal to itself. Figure 2 and Figure 3 show the evolving processes of the concave part and the convex part when the curve moves outward and inward respectively. We notice that both the irregular parts become smooth during the evolution. This is due to the existence of the entropy condition: once the overlaps are eliminated, they will not appear in the later evolution,

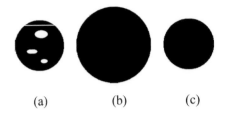

(a)   (b)   (c)

Figure 4. Evolving process of the multi-connective curves: (a) original curves; (b) curves after expansion; (c) curves after contraction.

even if the curve is moved opposite to the original direction. The irregularity of these parts can be greatly decreased and some parts with large curvatures will be removed.

Figure 4 explains the principle we use to remove the holes and discontinues inside the curves. We color the insides of the curves with black. The initial curves are multi-connected. As the curves move outward, the two curves merge and the holes inside the below curves diminish. When the merged curves move inward, we get a single curve with no discontinues and no holds inside it.

3 EXTRACTION ALGORITHM

The backgrounds of the high-resolution satellite images are very complex. To better process the road information, we need a method that can extract both the road contour and the road region. The level set method offers us a good way to do so. The contour curve represents the edge of the roads with its inside representing the road region. Given the connectivity among roads, we establish four velocity models to handle the retrieval problem. Our method includes three phases. We first extract an initial contour of the roads through the fast marching method and the threshold of the intensity gradient, then remove the leakages caused by the blurred parts of images, and finally eliminate the noises and discontinuities caused by the objects on roads.

3.1 Extract the initial contour of the roads

We define the speed function as follows:

$$F_1(x,y) = \begin{cases} 1 & \text{if } |\nabla(G_\sigma * I(x,y))| < \nabla I_{threshold} \\ \varepsilon & \text{if } |\nabla(G_\sigma * I(x,y))| \geq \nabla I_{threshold} \end{cases}, \quad (6)$$

where I is the image intensity, $\nabla I_{threshold}$ is the threshold of the gradient intensity, G_σ is a Gauss kernel with width σ, ε is a small positive real number. Use the image pixels as the mesh points and put the initial

curves to an area on roads without highly affected by the noises. Calculate the average intensity of the points on the initial curve and denote the value as I_0. Let the initial curve move with speed F_1. Use the fast marching method to solve the equation (3) and (4) to get the arrival time. Stop the calculation when all speeds of the points on the curve equal to ε.

This phase is to extract an initial shape of the roads. Since the roads have certain widths, the extraction process is similar to the shape detection. Malladi et al (1995) first use the level set method to perform the edge detection in the medical images. They put an initial curve inside the object and let it move with speed $F_0(x,y) = 1/(1 + |\nabla(G_\sigma * I(x,y))|)$. But this model is not applicable there, for the high resolution images usually contain huge amount of pixels. Improper placement of the initial curves might lead to the result that while one side of the curve does not reach the edge of roads, other sides might have massively leaked, which means different placements lead to different results.

In fact, since we only concern the final position of the evolving curves, we can combine the velocity with the threshold of intensity gradient and define it as above. The number of the initial curves used for the fast marching method can be one or more. Theoretically, because of the definition of velocity and the ability of handling topological changes of the fast marching method, our method assures that as long as the initial curves are placed in the areas of roads that are not too much affected by the noises, the final results will be the same. In the ideal condition, one initial curve will be enough no matter what shape the curve has (in order to facilitate the process, one can use a single point as the choice). While in practice, we usually use several initial curves and distribute them in different roads, because some roads might be so seriously affected by noises that the evolving curves might not be able to move into these regions with the speed we define. More initial curves mean higher possibility of getting complete road network. If some roads extracted in this phase are not connected with others due to the noises, we can reconnect them through the later processes. So the dependency to the initial positions of the curves is greatly reduced, and the fast marching method allows a high computing efficiency.

3.2 Remove the leakages of the contour

Define the speed function as:

$$F_2(x,y) = \begin{cases} -1 & \text{if } Ave(x,y,h) \notin [I_0 - \delta, I_0 + \delta] \\ F_0(x,y) & \text{if } Ave(x,y,h) \in [I_0 - \delta, I_0 + \delta] \end{cases} \quad (7)$$

where $Ave(x,y,h)$ is the average of intensity in a disk of the radius $R = kh$ centered around the point (x,y). Initialize the level set function using the contour curve and let the curve move with speed F_2. Use the narrow band method to solve the equation (1) and (2) to get the values of the level set function at the time t. Stop the calculation after getting the steady state of the curve.

The blurred parts of image caused by the weather or the objects on roads might decrease the intensity changes. While in order to get the complete road network, the $\nabla I_{threshold}$ defined in 3.1 should not be too small. For propagating contour curve, the decreased values of intensity gradient may lead the curve to leak into other objects. To remove the leakages, we notice that the leakages are caused by the points near the edges. Although the intensity gradients of these points are not large, the gradual changes of intensity still make the appreciable difference between the intensity of the roads and other objects. So we can use the difference to remove the leakages. Since the intensities of road region are quite uniform, the I_0 defined in the 3.1 can be used as the average intensity of road region and we use $[I_0 - \delta, I_0 + \delta]$ as the range of road intensity. The leakage parts with their intensity do no belong to the given range should contract, while the other parts should move with the velocity F_0 to reach the accurate position of the road edge. The inward and outward movements will achieve a steady-state after the leakages contract to the reasonable positions. Further calculation will not get appreciable changes of the curve.

When the curve achieves the steady state, the leakage parts will be diminished and most parts of the contour curve will approach to the edge of roads. The narrow band method guarantees the topological changes and improves the computing efficiency in this phase. The state of stability can be achieved quickly because the curves have arrived at the neighborhood of the edges after the movement of phase 3.1.

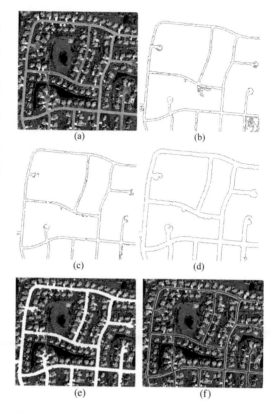

Figure 5. Extraction process: (a) original image; (b) initial contour; (c) contour without leakages; (d) contour after the outward movement; (e) extracted road region; (f) linear structure network.

3.3 Eliminate the noises

Move the contour outward with speed $F_4 = 1$ until the time t_0, then move the expanded curve inward with speed $F_5 = -1$ until the time $2t_0$.

After the movements of 3.1 and 3.2, most parts of the curve have reached the edge of roads. But the noises, such as the holes and discontinues, still exist in the contour. In the paper of Keaton & Brokish (2002), the authors once proposed to use the speed $F_6 = \min(k, 0)$ to eliminate the noises. While under this speed model, different parts will move with different speeds. Some parts with large curvatures might move so drastically that the whole state of the curves might become unstable. Besides, the holes and discontinuities inside the contour do not always have negative curvatures, for example, the curvature of the breakages caused by the crosswalk is zero. So noises sometimes may not be completely removed and the parts with positive curvatures can not be improved under this model. What we use to handle this problem is based on the discussions in the 2.4 and the fact that although the noises are outside the curves, they are surrounded by them. So we can let the curves first move outward and then move inward to eliminate these noises. The entropy condition ensures that once those noises enter the inside of the curves during the outward movement, they will not appear when the curves move inward.

After this phase, curves with dynamic changes will be smoothed, and the break among roads can be reconnected (For a case in point, please refers to Fig. 5). The procedure of this phase is a little similar to the morphological method, but it needs no structuring elements, and the smoothness of the edges will be improved rather than be disturbed. The value of t_0 mainly depends on the size of objects in the satellite images. The small value of t_0 may cause the movements unable to completely eliminate the noises; while the large value of t_0 might lead to the wrong result that unconnected roads are combined through the merges of curves and the entropy condition. According to our practices, the range from 5 to 15 is a preferable choice.

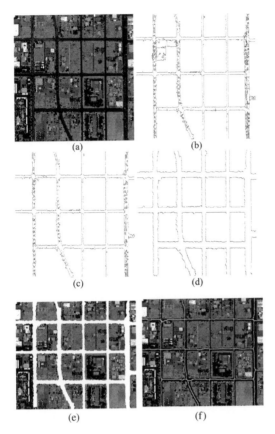

Figure 6. Extraction process: (a) original image; (b) initial contour; (c) contour without leakages; (d) contour after the outward movement; (e) extracted road region; (f) linear structure network.

We get a more complete and smoother curve in this phase. The curve represents the final contour with its inside representing the road region.

4 NUMERICAL EXPERIMENTS

We have tested our methods on several Landsat 7 satellite images and aerial images. Figure 5 and Figure 6 show two representative ones (urban and suburban). The original size of the images are 667×631 and 1024×743 respectively. The evolution processes of the curve are shown in Figure 5.(b)–(e) and Figure 6.(b)–(e). For both the experiments, we use one curve for the fast marching method. To better exhibit the results, we place the final result together with the original images and color the road regions white.

From the results, we can see our method handles the topological changes naturally through using the level set methods. When the road has branches, the curves split; when the roads are crossed, the curves merge. Through the curve movements, the leakage parts are removed and the noises inside the curves are eliminated. The smoothness of the contour is greatly improved and the parts with large curvatures diminished. After getting the road regions, we use the method introduced in Pratt (1991) to extract the linear structure network of roads. Figure 5.(f) and Figure 6.(f) show the linear structure network with the original images.

5 CONCLUSION AND DISCUSSION

In this paper, we present a road extraction method based on the level set methods. Compared with other methods, our method has a simple procedure of initialization, and needs no priori knowledge about the topology of roads. Both the contour of the roads and road regions can be extracted through three phases of the movements. Our method offers several speed models to handle the extraction, remove the leakages, eliminate the noises and improve the smoothness of the contour. We use the level set methods to handle the topological changes during the contour evolves. Numerical experiments have further proved the efficiency and effectiveness of our method.

However, as can be seen from the procedure, our method is a semi-automation method. Although the fast marching method and the narrow band method have greatly improved the computing efficiency, sometimes selecting a series of prefect parameters is still a time-consuming thing. Besides, our method may encounter some difficulties to deal with the objects that are connected with roads and have similar intensities to the roads. Our next aims are to improve the automation of the method and enhance its general applicability to different satellite images with different conditions.

REFERENCES

Adalsteinsson, D. & Sethian, J.A. 1995. A fast level set method for propagating interfaces, *J. Comp. Phys* 118(2): 269–277.

Chopp, D.L. 1993. Computing minimal surfaces via level set curvature flow, *Jour. of Comp. Phys.* 106: 77–91.

Geman, D. & Jedynak, B. 1996, An Active Testing Model for Tracking Roads in Satellite Images, *IEEE Trans. on Pattern Analysis and Machine Intelligence* 18(1): 1–14.

Keaton, T. & Brokish, J. 2002. A level set method for the extraction of roads from multispectral imagery. *Proceedings of the 31st applied imagery pattern recognition workshop*: 141–147.

Lee, H.Y. et al. 2000. Towards knowledge-based extraction of roads from 1m-resolution satellite images. *Proc. of the IEEE southwest symposium on image analysis and interpretation. Austin, Texas*: 171–176.

Malladi, R. et al. 1995. Shape modeling with front propagation: A level set approach, *IEEE Trans. On Pattern Analysis and Machine Intelligence* 17(2): 158–175.

Osher, S.J. & Sethian J.A. 1988. Fronts propagating with curvature-dependent speed: Algorithms based on Hamilton-Jacobi formulations. *J. Comp. Phys.* 79: 12–49.

Péteri, R. & Ranchin, T., 2003. Multiresolution snakes for urban road extraction from Ikonos and Quickbird. *23nd earsel annual symposium remote sensing in transition. Ghent, Belgium, 2–5 June 2003.* Rotterdam: Millpress.

Pratt, W.K. 1991. *Digital image processing.* Hoboken: John Wiley & Sons Inc.

Sethian, J.A. 1996. *Level set methods and fast marching methods: Evolving interfaces in computational geometry, fluid mechanics, computer vision and materials sciences.* London: Cambridge University Press.

Shi, WZ. & Zhu, CQ. 2002. The line segment match for extracting road network from high-resolution satellite images. *IEEE Trans Geos & RS* 40(2): 511–514.

Steger, C. 1998. An unbiased detector of curvilinear structures, *IEEE Trans. on Pattern Analysis and Machine Intelligence* 20(2): 113–125.

Zhu, CQ. et al. 2004. Road extraction from high-resolution remotely sensed image based on morphological segmentation. *Acta Geodaetica et Cartographic Sinica* 33(4): 347–351.

Computational Modelling of Objects Represented in Images –
João Manuel R.S. Tavares & R.M. Natal Jorge (eds)
© 2007 Taylor & Francis Group, London, ISBN 978-0-415-43349-5

Multimedia interface with an intelligent wheelchair

Bruno Martins, Eduardo Valgôde, Pedro Faria & Luís Paulo Reis
FEUP – Faculty of Engineering of the University of Porto, Portugal
LIACC – Artificial Intelligence and Computer Science Laboratory of the University of Porto, Portugal

ABSTRACT: Many of the physically injured use electric wheelchairs as an aid to locomotion. Usually, for commanding this type of wheelchair, it is required the use of one's hands and this poses a problem to those who, besides being unable to use their legs, are also unable to properly use their hands. The aim of the work described here, is to create a prototype of a wheelchair command interface that do not require hand usage. Facial expressions were chosen instead, to provide the necessary visual information for the interface to recognize user commands. The facial expressions are captured by means of a digital camera and interpreted by an application running on a laptop computer on the wheelchair. The software includes digital image processing algorithms for feature detection, such as colour segmentation and edge detection, followed by the application of a neural network that uses these features to detect the desired facial expressions. A specific command language for controlling the intelligent wheelchair was also developed in the scope of this work, as well as a simple wheelchair simulator for validating the final results.

1 INTRODUCTION

1.1 *Motivation*

Several physical disabilities or diseases result in severe impairments to mobility. Spinal chord damage, cerebral palsy and several other conditions can result in the loss of movement of some or all four limbs. Usually people in these situations are either completely dependant on others, or have suitable technological help for moving from place to place. The most common aid for this kind of disability is the electric wheelchair, which is quite good in what concerns motion, but frequently inadequate in the user interface. The standard command method available is the joystick that can only be used by people with relatively good hand dexterity, and impossible to be used by people in the situations mentioned above.

The main objective of this work was to create an interface that allows a person to drive a wheelchair, using only facial expression in an easy, practical and robust fashion.

The hardware used to achieve these goals, a digital camera and a PC, and the software developed in C++, were the main tools used to put this idea into practice.

1.2 *Facial expressions*

Facial expressions may be interpreted basically as a communication language, used voluntarily but also,

quite often, involuntarily, by people, to show their emotions. It is considered that they are inborn and cultural independent. This can be observed in blind people since birth, as they show the same relationship between facial expressions and emotions, as normal people.

Although there is a strong relationship between emotion and facial expression, there is a wide range of face movements not connected with emotions, such as eye blinking or raising only one eyebrow.

Ekman and Friesen developed the Facial Action Coding System, which describes thoroughly all the visually perceptible face movements. This system is particularly useful because of its objectivity. It defines all individual face movements, and facial expressions as their possible combinations. The FACS system was used to select some basic expressions to be used in the command language.

1.3 *Constraints*

The prototype developed has to meet certain conditions in order to achieve good results.

– *Time*: the software development process bared in mind that it should be as fast as possible. There is no need to have exact timing, but it is mandatory that the time interval between the command and the action is as small as possible.

- *Lighting conditions*: image processing, with varying illumination, demands more complex algorithms in order to achieve good results. Since there is a time constraint, for this work, the restriction that the system should only operate indoors in good light conditions was imposed.
- *Regular background*: the face detection is done through colour detection and therefore the background must be of a different colour.
- *Face Type*: in order to use colours as much as possible in feature detection, the face must have dark hair, dark eyes and somewhat lighter skin colour.

2 RELATED WORK

Most of the work done in this area (facial processing) concerns facial recognition or facial expression recognition for facial modelling.

These two main fields differ mainly in the fact that one aims to identify a face regardless of the expression, while facial expression recognition's target is to identify facial expressions regardless of identity.

Human-Computer interaction and psychological tools are the areas where most of the facial expression recognition work was performed.

FACS ("Facial Action Coding System") was used as base of work in many projects [6]. The first step is, in most cases, to separate background from areas of interest or to detect the face in the image. Two approaches to this problem are common. One is to recognize the face as a whole, using models or templates [1, 3, 6, 20] and the other is to detect relevant details (eyes, mouth, skin) and from that information obtaining the whole face or simply gathering information about those details [1, 4]. On other situations, some authors have chosen to build face models [2]. Once gathered the image information, there is a wide variety of methods that can be used. One of the most frequent is to build a database or eigenspace where images or characteristic data, typical of each expression to detect, can be stored followed by the computing of the expression to be identified, using comparison [1, 10]. Other method is to process the information in a network (Neural, Bayesian, etc) whose outputs, after training, are used to determine the expression [7, 9]. Another approach uses local or global classifiers (K *nearest neighbour, kernel, Gabor*) [3, 4, 20]. Finally, there are also authors that base their work in Markov Models (*Hidden Markov Models, Pseudo 3D Hidden Markov Models*) [2].

With the evolution of the psychological studies, muscle movement based systems began to be developed. One of the options, in this area, is to build a representation of the muscle activation time from facial models. This representation is characteristic from each expression; this method is called "*Motion Energy*".

3 DEVELOPED WORK

3.1 *System architecture*

The architecture that was chosen for implementing the multimedia interface is resumed in the following logical blocks:

- *Data acquisition*: the process of capturing the image with a digital camera and pre-processing at the camera's embedded software level.
- *Pre-Processing*: after acquiring the digital picture it is necessary to extract useful information from the picture.
- *Identification*: as soon as the feature information is gathered, an intelligent algorithm does the facial expression deduction.
- *Interface*: the command language is implemented on this level.

3.2 *Data acquisition*

The system is ready to operate with any webcam connected to an USB port. The camera used in the experimental setup was the *Logitech Sphere* webcam, using a resolution of 320×240 pixels and 24 bits of colour. *Logitech Sphere* has motion capability and zoom. The driver provides automatic histogram and white-balance correction.

3.3 *Pre-processing*

There are basically two types of information in an image, which are relevant for describing the facial expression: low spatial frequency components and high frequency components.

Low spatial frequency information is used to determine where important sets are located, mainly the face, while the high frequency components provide information about contours and shapes.

The face detections and segmentation is achieved through colour classification and the contours are extracted using the Canny algorithm [16].

3.3.1 *Noise filter*
The captured image is invariably polluted by noise, no matter how good the sensor is. The main types of noise that affect the image are: thermal noise, sensibility variations in each pixel and random noise. Usually these kinds of noise have a relatively high spatial frequency, so in order to attenuate it, a Gaussian filter was implemented in the space domain.

3.3.2 *Face tracking and segmentation*
In order to identify interest zones in an image it was decided proceed with colour segmentation [11, 17]. This method allows, for instance, identifying all face pixels as being of the same category. In this way, it is possible to extract from the image all the relevant zones in a fast and efficient way (Figure 1).

Figure 1. Segmentation results.

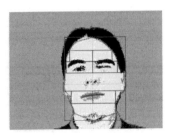

Figure 2. Relevant zones.

To improve the efficiency of this technique, a "Look Up Table" that associates each group of colours (usually 8 bits leap for memory and time consuming efficiency), to a given category with a given trust level, was created.

Once the table is built, segmenting an image is done by analysing it and, for each pixel, matching the category with the colour and then replacing the colour value for the category code. To enhance the system's performance in a brilliance varying environment, an intensity correction method was developed. In the HSI colour system all it takes to correct brilliance is adjusting the I component.

Once the image is segmented, the next step is to track the face. This is done in a progressive way. First the centre of the face is computed simply as being the average of the points of the category associated to the face.

$$Centre(x, y) = \left(\sum \frac{x_i}{i}, \sum \frac{y_i}{i}, \right)$$

Equation 1: Face Centre

At this point we have the centre of the face, so what is done next is expanding in four directions (axis oriented) to detect face limits (upper, lower, right and left).

Now relevant zones of the face have to be defined, aiming to extract information to input the neural network. This is achieved dividing the height of the face in order to obtain the zones where eyebrows, eyes and mouth are expected to be found (Figure 2).

On the next step the face, eyebrows, mouth and no category points are counted in each zone. The relationship between those values and the total points of each

Table 1. Face zones (values are fractions of the face height).

	Eyebrows	Eyes	Mouth
Upper	7/8	3/4	3/8
Lower	3/4	5/8	1/4

Figure 3. Canny results.

zone is a characteristic of a given facial expression. Note that eyes, eyebrows and mouth are not tracked. Once tracked the face and established the 6 interest zones, a simple count of the points of all categories (eyebrow, mouth, face . . .) is performed in each zone.

These zones are obtained by dividing the height of the face in equal parts in a successive way, and next, chose the upper and lower limits of the zones, as presented in Table 1.

If no points of the expected category are found in a given zone, eyebrow (black points for example) points are not found in the eyebrow zone, for instance, this is also a characteristic input of an expression for the neural network.

This happens if the head is turned up, down or left, right.

3.3.3 Edge detection

Edge detection is made in order to capture the shapes inside the face. This process consists in extracting the high frequency components and for this purpose, the Canny operator [16] is used. The canny operator is an optimal algorithm for edge detection according to the criteria of detecting every edge, getting an accurate edge location and detecting each edge only once. This criteria favours precision and accuracy to processing speed. Using a Gaussian mask of 5 by 5 pixels of dimension, a Sobel operator of 3 by 3 and 8 connectedness for thresholding, gives about 135 operations per pixel. In this implementation it took about 200 ms per image of 320 by 240 pixels to produce the final result: a binary image with only the most important edges. Since the real time constraint here is not very strict, a better relationship between the extracted information and the facial expression was preferred to a faster but no so accurate method (Figure 3).

The following process aims to resume the obtained information, as much as possible, without losing much accuracy.

In order to better describe the image, several patterns for each expressions are taken during the configuration phase (Figure 4).

These patterns are superimposed in order to obtain a binary image that contains the main traits that define the expression. During the execution cycle the contour image is compared with each saved pattern. The comparison consists in counting the number of coincident points with each saved pattern. To make this process faster, each contour image is sampled using a tight grid (1% of the height and width of the rectangle containing the face).

3.4 *Identification*

Here the measures taken in the pre-processing are translated into facial expressions. The data deployed to this block is relatively large, 23 inputs, and there is no guarantee of a linear relationship with the outputs. Establishing a relationship manually, even using fuzzy logic, would be very inefficient.

The identification process with this type of inputs can be posed as a pattern classification problem. Hence that the outputs of pre-processing are subject to noise due to several factors such as little variations in the rectangle that contains the face or even the image noise. A mulitlayer perceptron with a continuous activation function can be used to compute the probability of an expression, for a given input pattern, after learning automatically with a training set. An Artificial Neural Network takes some advantage over other methods with automatic learning facing this problem.

The MLP type of network was chosen because it is possible to have good results, even if not using the best parameters, in an almost real time environment. At this stage only a good solution was sought and optimizations to the identification process are left for future development.

The network uses 22 inputs, 11 outputs (expressions) 1 hidden layer, 10 hidden neurons all with sigmoid activation function, trained with simple backpropagation and 110 training patterns are used.

Concerning the number of hidden layers, one is enough for universal approximation, and it is also less prone for local minima to occur than using 2 hidden layers.

There is no exact method for calculating the necessary number of hidden neurons, but the following heuristic helps: $h \geq p - 1/n + 2$, in which p is the number of training examples and n the number of input neurons. In this case this tells that the number of hidden neurons should be at least 4. Ten neurons are used, in order to improve precision in the outputs without overfitting.

The number of training patterns, 10 per expression to be identified, intends to be large enough data to minimize the influence of noise and to allow slightly different possibilities to be identified as the same expression. The Vapnic Chervonenkis dimension [19], a quantitative measure of the set of functions that a neural network can compute, was also taken in consideration. The VP dimension is less than 4533.5 given by $2|W| \log_2(eN)$ – with W being the number of weights and N the number of neurons – and greater than 271 given by $n_1 n_2 + n_2(n_3 - 1)/2 + 1$, in which n_i is the number of neurons in the respective layer [19]. The number of training examples used is 110 which is within range.

The network was trained using simple backpropagation. Training was done so that the final training error would be less than 6E−5 (experimentally it was found to be a good value), and it takes about 5 minutes (approx. 100000 iterations) to achieve that result. The network must be trained for each user. Seldom did a network trained for one person, perform well on another.

3.5 *Interface*

After identifying the facial expression, the corresponding command is checked in the command language definitions and sent to the (simulated) control system. In fact, for a command to be validated the corresponding facial expression has to be almost constant for about 0.5 s. This prevents interferences and casual facial expressions to be misunderstood for a command.

For orders to be validated, the instruction mode must first be entered. The instruction mode is entered and left with relatively complicated facial expression, such as raising only one eyebrow. Only in this mode the facial expressions with commands associated, are validated. Once out of the instruction mode, all commands other than the "instruction mode" command are ignored. In this way, the necessary level of concentration, while driving, can be reduced.

Figure 4.   Pattern adding.

The driver can watch the captured image, with and without processing. Finally, it is possible to create a new training set for the NN.

3.6 *Simulator*

A simple graphic simulator was created in order to evaluate the final results. The simulator consists in a wheelchair representation moving with an open field. It was taken into consideration the dynamic behaviour of the wheelchair as well. At the end of the whole process, all the wheelchair commands were inserted directly in the simulator.

4 RESULTS

The image processing and identification takes about 500 ms (using a 3.0 GHz processor). The facial expression identification results of one good test are presented in the following confusion matrix as well as the analysis of the matrix. A satisfactory average accuracy was obtained in the results presented in Table 2.

It is clear that the frowned expression had a weak accuracy and precision. This is not unexpected since two very similar expressions were chosen on purpose to determine how well the system can discriminate between slightly different expressions. "Frowned" and "Frowned and wrinkled nose" are quite similar and judging from the results the system can tell the difference between them but not in a very reliable way.

Table 3. Confusion Matrix analysis. acc – accuracy, tpr – true positive rate, fpr – false positive rate, tnr – true negative rate, fnr – false negative rate, p – precision.

	ACC	TPR	FPR	TNR	FNR	P
Opened Mouth	94,8	100,0	1,0	99,0	0,0	90,0
Frowned	54,7	100,0	6,5	93,4	0,0	30,0
Frowned and Wrinkled nose	89,4	100,0	1,9	98,0	0,0	80,0
Leaned Right	100,0	100,0	0,0	100,0	0,0	100,0
Leaned left	100,0	100,0	0,0	100,0	0,0	100,0
Normal	95,3	90,9	0,0	100,0	9,1	100,0
Raised Right Eyebrow	54,8	100,0	6,5	93,6	0,0	30,0
Raised Left Eyebrow	94,9	100,0	1,0	99,0	0,0	90,0
Raised Eyebrows	95,5	90,9	0,0	100,0	9,1	100,0
Turned Right	100,0	100,0	0,0	100,0	0,0	100,0
Turned left	89,4	100,0	2,0	98,0	0,0	80,0
AVERAGE	88,1	98,3	1,7	98,3	1,7	81,9

Table 2. Confusion Matrix for 10 tests per expression.

	Opened mouth	Frowned	Frowned and wrinkled nose	Leaned right	Leaned left	Normal	Raised right eyebrow	Raised left eyebrow	Raised eyebrows	Turned right	Turned left
Opened Mouth	9	0	0	0	0	0	0	0	0	0	0
Frowned	0	3	0	0	0	0	0	0	0	0	0
Frowned and Wrinkled nose	0	0	8	0	0	0	0	0	0	0	0
Leaned r	0	0	0	10	0	0	0	0	0	0	0
Leaned l	0	0	0	0	10	0	0	0	0	0	0
Normal	0	1	0	0	0	10	0	0	0	0	0
Raised Right Eyebrow	0	0	0	0	0	0	3	0	0	0	0
Raised left Eyebrow	0	0	0	0	0	0	0	9	0	0	0
Raised Eyebrows	0	0	0	0	0	0	1	0	10	0	0
Turned r	0	0	0	0	0	0	0	0	0	10	0
Turned l	0	0	0	0	0	0	0	0	0	0	8
Unident	1	6	2	0	0	0	6	1	0	0	2

Table 4. Identification results using only colour segmentation or contours.

	ACC	TPR	FPR	TNR	FNR	P
Colour seg avg	82,6	90,9	2,2	97,8	9,1	77,3
Contour avg	79,6	85,7	2,2	97,8	5,2	76,4

Table 5. Training and testing results.

	Segmentation only	Contours only	Both
Training error	4,65E−03	5,40E−03	5,80E−05
Test error	1,50E−01	1,65E−01	8,57E−02

As for the "Raised Right Eyebrow", its below average results are probably due to some casual error during the extraction of training patterns, which can happen naturally.

In order to have an estimate of the performance of each type of features, colour segmentation and edge comparison, the identification process using the same network architecture (apart from the number of input neurons) was done using them separately. The average results are shown in Table 4.

Using only colour segmentation, the identification results are quite good, and it is less time consuming (less than 150 ms). The contour measures alone are not as good as the colour segmentation ones (Table 5) though they are more robust to lighting variations. The combination of both types is better than only one set.

5 CONCLUSIONS

The main conclusion of this work is that it is possible to drive a wheelchair, by using facial expressions, in a very comfortable way, using the implemented system. This conclusion is based on the very good results achieved by our facial expression detection process, together with the general empirical feel attained by the project members and a volunteer when driving our simulated wheelchair. Videos of the tests can be observed at the project's webpage: http://paginas.fe.up.pt/~ee99041/IMCRI/EN.

The main limitations to the good performance of our system are presently located in the pre-processing stage. The colour segmentation is much too sensitive to large light variations and slight colour shifts, and the shape extraction should have better precision without increasing the processing time. These problems affect the robustness of the system and its response time.

All the other parts of the system have not posed any practical limitation to the system's performance.

In order to overcome the problems mentioned, future developments will be first focused on the pre-processing aiming to allow the system to better perform in outdoor environments or rough lightning conditions.

Other future developments will include improvements in our wheelchair simulator. It is necessary to obtain objective indicators of the global performance, and this can be achieved by including some simple skill tests, such as obstacle avoiding, in the simulator.

This work will also be continued in the context of "INTELWHEELS – An Intelligent and Configurable Wheelchair for helping Quadriplegia and Cerebral Palsy Handicapped" project. In the context of this project the system will be improved and implemented in a real wheelchair driven by quadriplegia and cerebral palsy handicapped people.

REFERENCES

1. Aas, Kjersti 1996. *Audio-Visual Person Recognition: A Survey*, Report no. 911, Norwegian Computing Center
2. Alekic, Peter et al 2005, *Automatic Facial expression recognition using facial animation parameters and multi-stream HMMs*; Available at:http://ivpl.ece.northwestern.edu/Publications/Conferences/2005/WIAMIS-NEW-1.pdf
3. Bartlett, M. et al 2003. *Real Time Face detection and facial expression recognition: Development and Applications to Human Computer Interaction*; Available at: http://mplab.ucsd.edu/~movellan/mypapers/BartlettCVPR2003.pdf
4. Beurel, Fabrice et al. *Recognition of Facial Expressions in the presence of Occlusion*; Available at: http://www.bmva.ac.uk/bmvc/2001/papers/48/accepted_48.pdf
5. Bruce, J. et al 2000. *Real-time machine Vision Perception and Prediction*, Available at: http://citeseer.ist.psu.edu/cache/papers/cs/20670/http:zSzzSzwww.cs.cmu.eduzSz~jbrucezSzcmvisionzSzpaperszSzJBThesis00.pdf/bruce00realtime.pdf
6. Cohen, Ira et al 2002. *Facial expression recognition from video sequences*; Available at: www.ifp.uiuc.edu/~iracohen/publications/ICME02Cohen.pdf
7. Cohen, Ira et al. *Semi-Supervised Learning for Facial Expression Recognition*; Available at: staff.science.uva.nl/~nicu/publications/MIR03.pdf
8. Ekman, Friesen 1978, *FACS – Facial Action Coding System*, Available at: http://www.cs.cmu.edu/afs/cs/project/face/www/facs.htm
9. Franco, Leonardo 2001. *A Neural Network Facial Expression Recognition System using Unsupervised Local Processing*, Available at: www.lcc.uma.es/~lfranco/B4-Franco+Treves01.pdf
10. Frank, Carmen & Nöth, E. 2003. *Automatic Pixel Selection for Optimizing Facial Expression Recognition using Eigenfaces*. Available at: http://www.smartkom.org/reports/Report-NR-41.pdf

11. Gonzalez, Rafael, C. & Woods, Richard, E. 2002. *Digital Image Processing*, Prentice-Hall

12. Green, Bill 2002. *Canny Edge Detection Tutorial*, Available at: http://www.pages.drexel.edu/~weg22/can_tut.html

13. Hager, Joseph, C. & Ekman, P. 1983. *The Inner and Outer Meanings of Facial Expressions*, chapter 10, in *Social Psychophysiology: A Sourcebook*, The Guilford Press

14. Ham, Fredric, M. & Kostanic, I. 2001. *Principles of Neurocomputing for Science & Engineering*, McGraw-Hill

15. Haman, M. et al 2004. *Real-time shape estimation for continuum robots using vision*; Available at: www.ces.clemson.edu/~ianw/robotica.pdf

16. Haykin, Simon 1999. *Neural Networks – A Comprehensive Foundation*, Prentice-Hall

17. Jain, Anil, K. 1989. *Fundamentals of Digital Image Processing*, Prentice-Hall

18. Jones, M. Tim 2005. *AI Application Programming*, Charles River Media

19. Kasabov, Nikola, K. 1998. *Foundations of Neural Networks, Fuzzy Systems, and Knowledge Engineering*, The MIT Press

20. Lyons, M. et al 1998. Coding Facial Expressions with Gabor Wavelets; Available at: http://dx.doi.org/10.1109/AFGR.1998.670949

21. Morse, Bryan, S. 2000. *Lecture 7: Shape Description (Contours)*, Available at: http://homepages.inf.ed.ac.uk/rbf/CVonline/LOCAL_COPIES/MORSE/boundary-rep-desc.pdf

22. Michel, P. et al 2005. *Facial Expression Recognition using Support Vector Machines*; Available at: https:// bscw.ercim.org/pub/bscw.cgi/S4498a329/d34261/VCIP2005_Kotsia.pdf

Computational Modelling of Objects Represented in Images –
João Manuel R.S. Tavares & R.M. Natal Jorge (eds)
© 2007 Taylor & Francis Group, London, ISBN 978-0-415-43349-5

Tracking the 3D deformation of hexahedric samples, using video information from two perpendicular faces. Application to uniaxial tension tests of hyperelastic materials

P. Martins & R. Natal Jorge
IDMEC-Polo FEUP, Faculdade de Engenharia da Universidade do Porto, Porto, Portugal

A. Ferreira
DEMEGI, Faculdade de Engenharia da Universidade do Porto, Porto, Portugal

ABSTRACT: The authors propose a method to study the $3D$ evolution of hexahedric samples of hyperelastic materials. The methodology starts with video acquisition from two orthogonal directions. After that, frame synchronization is performed to the two video signals. The synchronized frames suffer digital image processing techniques and the information from both (orthogonal) frames is combined in order to build a $3D$ mesh. The video acquisition and processing software is developed using MATLAB®. The material used to validate the procedure is silicone-rubber, which presents hyperelastic behavior. Application to soft biological tissues will be an extension of the present work.

1 INTRODUCTION

Digital image correlation techniques (Hild and Roux 2006), are a useful tool to analyze the dynamical behavior of materials subjected to external loads. The type of materials whose mechanical behavior is treated with this kind of techniques, goes from the biomaterials like human skin (Marcellier et al. 2001) and rabbit ligaments (Rasanen and Messner 2000), to wood (Farruggia and Perré 2000).

Deformation and strain fields are often acquired via digital image correlation for uniaxial and biaxial tension tests (Kabir et al. 2005; Chevalier et al. 2001). However, if $3D$ information is required, techniques like photogrammetry (Zhou et al. 1996; Tyson et al. 2002) are needed.

With this paper, the authors aim to provide an efficient, yet affordable, solution to evaluate the spatial (volumetric) behavior of hyperelastic material samples, subjected to uniaxial tension tests. The process is based on video information taken in two orthogonal planes.

2 METHODOLOGY

On the basis of the process lays the video recording of the mechanical tests (uniaxial tension tests). For that purpose, a custom made application was built

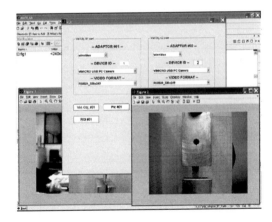

Figure 1.   Video capturing application.

(Fig. 1) using MATLAB's *Image Processing Toolbox* (IPT) functionalities.

2.1   Video acquisition

The application enables the capture of two video signals at the same time. It has automatic configuration capabilities in order to ease the process of recording setup assembly. Video is recorded in two directions orthogonal to each other, and coincident with sample's frontal and lateral faces (Fig. 2).

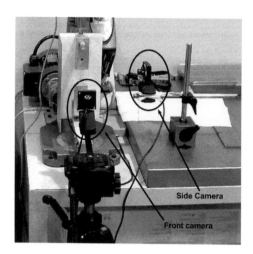

Figure 2. Video capturing setup.

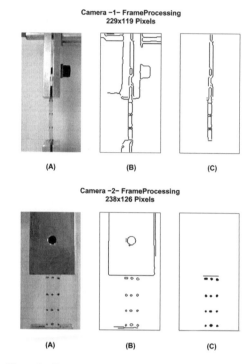

Camera –1– FrameProcessing
229x119 Pixels

(A)   (B)   (C)

Camera –2– FrameProcessing
238x126 Pixels

(A)   (B)   (C)

Figure 3. Frame processing: Side View - Camera 1; Front View – Camera 2.

In order to preserve the time-stamp information of the acquired frames (for each camera c), the video acquisition application records the individual frame (f) time stamp data (eq. 1) in a MATLAB's data file (.mat).

The video information (frames from both cameras) was stored in the same fashion (eq. 2), like a sequence of individual frames (RGB images).

$$\{t_f^c\} \quad (s) \tag{1}$$

$$\{I_f^c\} \quad c = 1, 2 \; ; f = 1, \ldots, nr.\, frames \tag{2}$$

This approach allowed better synchronization of the signals from the two cameras, because the acquisition is performed taking one frame (from each camera) per software trigger.

2.2 Frame synchronization

Since there's no hardware triggering mechanism involved in the video acquisition process, a time shift appears between the two video signals. The time shift (ΔT_f) value is expected to be 'small', since the trigger command acts 'simultaneously' on both cameras, however, this makes frame synchronization a requirement.

For the fth pair of frames (corresponding to the fth trigger), the time shift (eq. 3) is calculated with

$$\Delta T_f = \left| t_f^1 - t_f^2 \right| \; (s) \tag{3}$$

A statistical criteria is taken for selecting the valid (I_f^1, I_f^2) frame pairs. The maximum ΔT_f allowed is the sum of the mean time shift $\overline{\Delta T_f}$ with the standard deviation σ (eq. 4).

$$\Delta T_f \; \leq \; \overline{\Delta T_f} + \sigma \tag{4}$$

$$\overline{\Delta T_f} \; = \; \frac{\sum\limits_{f=1}^{n} \Delta T_f}{n}$$

$$\sigma \; = \; \sqrt{\frac{1}{n-1} \sum_{f=1}^{n} (\Delta T_f - \overline{\Delta T_f})^2}$$

2.3 Geometry extraction and mesh construction

The pairs of synchronized frames (I_f^1, I_f^2) are stored in 8 bit ($2^8 = 256$ colors) format which correspond to a *uint8* MATLAB variable. In order to reduce the frame size, the capturing application developed by the authors allows the definition of rectangular ROI's (regions of interest). Frame size reduction is an important issue since the USB bandwidth is a limitation to the volume of video data processed per second (480 Mb). This also explains why the frame dimensions differs for the two cameras (Fig. 3).

The RGB images represented by $3D$ matrices (Gonzalez et al. 2004), are treated according to the procedure shown in Figure 3.

The sequence of procedures applied to both frame subsets, $\{I_f^1\}$ (camera 1) and $\{I_f^2\}$ (camera 2) is the following:

- (A) Conversion from RGB images to grayscale images
- (B) Use of image segmentation, via *edge* detection filter
- (C) Cleaning of image areas without relevant information (markings)

With the images processed by the method $(A) \Leftrightarrow (C)$ is now possible to use an object identification algorithm that extracts the coordinate values for the center of each mark (X_m, Y_m). The procedure is repeated for every frame (f) from each camera (c) (eq. 5).

$$(X_m, Y_m)_f^c \qquad (5)$$

$$c = 1, 2$$

$$f = 1, \ldots, nr.\,frames$$

$$m = 1, \ldots, nr.\,marks$$

At this point, it is possible to combine the two orthogonal face geometries in order to build a $3D$ mesh.

The assembly of the tridimensional mesh, requires that solid planes parallel to the measured surfaces have the same geometry. In particular, directly opposed faces are assumed to have equal geometries, which, conditioned to the care in sample preparation is a valid assumption for homogeneous materials (metals, composites and even hyperelastic materials like rubbers).

3 EXPERIMENTAL PROCEDURE

The uniaxial tension tests were performed using a custom made mechanical testing machine. The main application of this solution (fig. 4) is the study of hyperelastic materials (Martins et al. 2006b), particularly biological soft tissues (Martins et al. 2005; Martins et al. 2006a).

The sensor array allows the capture of load and displacement information, with the aid of a LABVIEW software interface. The displacement rate is controled independently with a control unit (upper left corner of Fig. 4). For the present work it was used a slow displacement rate, ≈ 5 mm.s^{-1}.

Video feed from the 2 USB webcams is processed using the application presented on section 2.1.

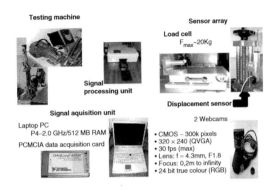

Figure 4. Acquisition sytem.

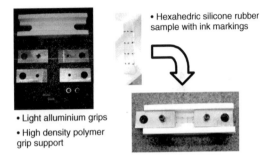

Figure 5. Sample preparation.

3.1 Sample preparation

The hexahedric silicone rubber samples (Fig. 5) with dimensions $9.0 \times 3.0 \times 50.0$ mm, were marked with a 'regular' grid of dots. For the markings the authors used a STAEDTLER Lumocolor waterproof black marker, with a thin tip (**F**).

In order to obtain a better alignment of the sample (on the testing machine), the grip-sample system was assembled using a grip support (see Fig. 5).

4 RESULTS

The total number of acquired frame pairs during the experiment was 3739, which according to equation 2 is also the dimension of the set $\{(I_f^1, I_f^2)\}$. This means that the total number of acquired frames was in fact $2 \times 3739 = 7478$.

The video acquisition time domain (equation 1) was determined by ($\forall f \in \{1, \ldots, 3739\}$):

$$min(t_f^1, t_f^2) \leq t_f \leq max(t_f^1, t_f^2)$$

$$\Rightarrow t_f \in [0.038, 261.570]\,(s)$$

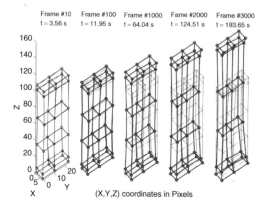

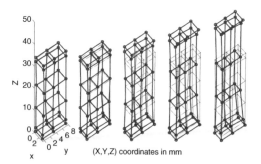

Figure 6.   Mesh evolution in time using first frame mesh as reference.

The mean *fps* value (frames per second) was 14, however, the authors verified that this number suffered fluctuations during the acquisition ($9 \leq fps \leq 25$).

According to equation 4, the maximum time shift allowed between the components of frame pair is $\Delta T_f = 0.0219\,(s)$. From the original frame pair set of dimension 3739, only 3226 pairs meet the time shift criteria, to form the set of synchronous frames.

From these frame pairs only those that pass the geometric quality evaluation (after the processing sequence of section 2.3) will be used for mesh construction. Finally, the set of mesh generating frame pairs was obtained with a population of 2823 elements.

4.1   Mesh construction

In order to illustrate the method's application, 6 meshes were built based on the following frame pairs, $\{(I_i^1, I_i^2)\}: i = 1, 10, 100, 1000, 2000, 3000$.

The first mesh, from (I_1^1, I_1^2), is used as a comparing term for the deformation. The results can be seen in figure 6 both in pixels and in mm. All the calculation where performed in pixels since those are the natural coordinates of digital image matrix representation.

The results were converted to mm afterward using the adequate conversion factor for each coordinate.

The X and Z coordinate values agree with observation, however, Y coordinate values are systematically ≈ 1 mm inferior to the actual ones. The authors view on the matter is given on the next section.

In order to draw the $3D$ meshes, the authors used a MATLAB script that requires the assembly of the connectivities matrix and element construction, on the present case for elements of type $B8$. This functionality is built in on the geometry processing code, and will certainly ease the usage of the meshes in finite element codes.

5   CONCLUSIONS

The process works with samples that possess a well defined geometry. The example of the paper, a silicone rubber sample, shows that this approach is a suitable and effective mean of studying the volumetric deformation on hyperelastic materials, subjected to uniaxial tension tests.

However, on the case of inhomogeneous materials like soft tissues, the lack of geometry definition and material homogeneity are issues difficult to overcome, leading necessarily to a quality loss of the generates $3D$ meshes.

One issue that has to be carefully addressed in future developments is the dot marking technique. The authors believe that the over sized markings on the frontal view (camera 2) were responsible for the systematic 1 mm variation on Y coordinate values. This effect, although easily correctable has to be avoided.

This project also presents limitations concerning the mechanical testing's velocity. We used a constant displacement velocity of ≈ 5 mm.s^{-1} with all the samples. With displacement values of this magnitude, it is possible to use low video frame rates, which present obvious advantages since the video data volume stays under the USB bandwidth limitation. The latter does not occur, however, if an higher frame rate must be used in order to accommodate higher mechanical testing velocities.

6   FUTURE WORK

The application of this methodology on the study of biological soft tissues, with hyperelastic behavior, presents itself as a natural evolution of the present work.

The method also enables the direct calculation of mechanical quantities such as the Poisson ratio v. It is also possible to obtain strain fields on two perpendicular planes.

Another application, is the improvement of finite element or meshless simulations (Guo and Qin 2005). Better geometry accuracy of the $3D$ meshes should provide an increase on simulations' fidelity to the observed experimental phenomena.

ACKNOWLEDGEMENTS

The support of Ministério da Ciência, Tecnologia e Ensino superior (FCT and FSE) (Portugal) and the funding by FEDER under grant POCTI/ESP/46835/2002, are gratefully acknowledged.

REFERENCES

Chevalier, L., S. Calloch, F. Hild, and Y. Marco (2001). Digital image correlation used to analyze the multiaxial behavior of rubber-like materials. *European Journal of Mechanics, A/Solids 20*(2), 169–187.

Farruggia, F. and P. Perré (2000, June). Microscopic tensile tests in the transverse plane of earlywood and latewood parts of spruce. *Wood Science and Technology 34*(2), 65–82.

Gonzalez, R. C., R. E. Woods, and S. L. Eddins (2004). *Digital Image processing using MATLAB*. Pearson Prentice Hall.

Guo, X. and H. Qin (2005). Real-time meshless deformation. *Computer Animation and Virtual Worlds 16*(3–4), 189–200.

Hild, F. and S. Roux (2006). Digital image correlation: from displacement measurement to identification of elastic properties – a review. *Strain 42*(2), 69–80.

Kabir, S., G. Attenburrow, P. Picton, and M. Wilson (2005). The application of image analysis to determine strain distribution in leather. In *Proceedings of the Seventh IASTED International Conference on Signal and Image Processing, SIP 2005*, University College Northampton, Boughton Park Road, Northampton, United Kingdom, pp. 505–510.

Marcellier, H., P. Vescovo, D. Varchon, P. Vacher, and P. Humbert (2001). Optical analysis of displacement and strain fields on human skin. *Skin Research and Technology 7*(4), 246–253.

Martins, P., R. Jorge, and A. Ferreira (2006a, June). Experimental study of the middle ear biological support structures. In *ECCM2006*.

Martins, P. A., R. M. Natal Jorge, A. Ferreira, A. Fernandes, M. Figueiredo, and R. Silva (2005, September). Modeling the mechanical behavior of soft tissues using hyperelastic constitutive models. In *II International Conference on Computational Bioengineering – ICCB*, Volume 1, Lisbon, pp. 403–410.

Martins, P. A. L. S., R. M. N. Jorge, and A. J. M. Ferreira (2006b). A comparative study of several material models for prediction of hyperelastic properties: Application to silicone-rubber and soft tissues. *Strain 42*(3), 135–147.

Rasanen, T. and K. Messner (2000). Estrogendependent tensile properties of the rabbit knee medial collateral ligament. *Scandinavian Journal of Medicine and Science in Sports 10*(1), 20–27.

Tyson, J., T. Schmidt, and K. Galanulis (2002). Biomechanics deformation and strain measurements with 3d image correlation photogrammetry. *Experimental Techniques 26*(5), 39–42.

Zhou, W., R. Brock, and P. Hopkins (1996). A digital system for surface reconstruction. *Photogrammetric Engineering and Remote Sensing 62*(6), 719–726.

Computational Modelling of Objects Represented in Images –
João Manuel R.S. Tavares & R.M. Natal Jorge (eds)
© *2007 Taylor & Francis Group, London, ISBN 978-0-415-43349-5*

Automatic analysis of dermoscopy images – a review

T. Mendonça, A.R.S. Marçal, A. Vieira & L. Lacerda
Faculdade de Ciências da Universidade do Porto, Portugal

C. Caridade
Instituto Superior de Engenharia de Coimbra, Portugal

J. Rozeira
Hospital de Pedro Hispano, Matosinhos, Portugal

ABSTRACT: Dermoscopy is a non-invasive diagnostic technique for the in vivo observation of pigmented skin lesions used in dermatology. There is currently a great interest in the prospects of automatic image analysis methods for dermoscopy, both to provide quantitative information about a lesion, which can be of relevance for the clinician, and as a stand alone early warning tool. The effective implementation of such a tool could lead to a reduction in the number of cases selected for exeresis, with obvious benefits both to the patients and to the health care system. The standard approach in automatic dermoscopic image analysis has usually three stages: (i) image segmentation, (ii) feature extraction and feature selection, (iii) lesion classification. This paper presents a review of the dermoscopic image analysis systems currently available, and an evaluation of the performance of one such systems, the Tuebinger Mole Analyser, with 83 dermoscopic images of melanocitic nevus.

1 INTRODUCTION

Dermoscopy (dermatoscopy or skin surface microscopy) is a non-invasive diagnostic technique for the in vivo observation of pigmented skin lesions used for dermatology. This diagnostic tool allows for a better visualization of surface and subsurface structures and permits the recognition of morphologic structures not visible by the naked eye, thus opening a new dimension of the clinical morphologic features of pigmented skin lesions (Argenziano et al. 2000). In the last few years there have been significant developments in both dermoscopy and telemedicine, allowing for improved clinical diagnosis of cutaneous lesions. At present there is great interest in the prospects of automatic image analysis systems for dermoscopy. The benefits of such systems are two fold: (1) to provide quantitative information about a lesion that can be relevant for the clinician; (2) to be used as a stand alone early warning tool.

The dermoscopic diagnosis of pigmented skin lesions is based on various analytic approaches and algorithms that have been set forth in the last few years, namely, pattern analysis, ABCD rule and seven-point checklist (Argenziano et al. 2000). The common denominator of all these diagnostic methods is particular dermoscopic criteria that represent the backbone for the morphologic diagnosis of pigmented skin lesions (Argenziano et al. 2000). However, most of these criteria are based on qualitative features and thus difficult to implement on a computer based algorithm. Although there is still considerable work to be done in order to establish a link between the human based criteria and image extracted features, advances along this path can lead to the implementation of an effective fully automatic system for early warning diagnosis of skin lesions.

The purpose of this paper is to present an overview of the existing automatic dermoscopic image analysis systems and methods, reported in the scientific literature, and to present an evaluation of the performance of one such systems – the Tuebinger Mole Analyser software (Mole Analyzer, 2006).

2 AUTOMATIC DERMOSCOPIC IMAGE ANALYSIS SYSTEMS

The standard approach in automatic dermoscopic image analysis is generally composed of three stages: (i) image segmentation, (ii) feature extraction and feature selection, (iii) lesion classification. This is illustrated schematically in Figure 1.

The segmentation stage is not a straightforward task due to the great variety of lesions, skin types, presence of hair and so forth. Various image segmentation

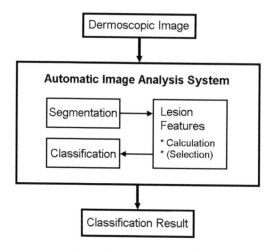

Figure 1. Schematic representation of the standard approach for automatic dermoscopic image analysis systems.

Table 1. List of papers reporting automatic analysis systems.

Paper	Authors	Year	Image Analysis System
#1 –	Elbaum et al.	2001	(proposed by Grutkowicz 1997)
#2 –	Ganster et al.	2001	global/dynamic thresholding + 3D Colour clustering
#3 –	Rubegni et al.	2002	Laplacian filter, zero crossing
#4 –	Gerger et al.	2003	Tissue counter analysis
#5 –	Hoffmann et al.	2003	Statistical clustering + region-growing
#6 –	Sboner et al.	2003	Leica software
#7 –	Blum et al.	2004	(proposed by Ritter 1996)
#8 –	Burroni et al.	2004	Laplacian filter, zero crossing
#9 –	Burroni et al.	2005	Laplacian filter, zero crossing
#10 –	Seidenari et al.	2005	(proposed by Grana 2003)
#11 –	Tomatis et al.	2005	Thresholding, region-oriented
#12 –	Celebi et al.	2006	B-spline

Table 2. Summary of main results presented in the literature.

Paper #	No. of features	Classification methods	No. of lesions Total	No. of lesions (Mal)	ROC SE %	ROC SP %
#1 –	822	LDA	246	(63)	100.0	85.0
		noLDA			100.0	73.0
#2 –	122	K-NN	5363	(96)	87.0	92.0
#3 –	48	SLP-ANN	588	(217)	94.3	93.0
		ANN	147	(57)	93.8	92.8
#4 –	70	CART, K-NN	136	(NS)	100.0	99.0
#5 –	NS	ANN	968	(187)	80.0	88.0
#6 –	38	LDA, K-NN	152	(42)	75.0	64.0
		Decision tree			86.0	89.0
#7 –	64	Logistic regres.	837	(84)	87.5	85.7
#8 –	49	LDA	840	(391)	95.0	78.0
		K-NN			98.0	79.0
#9 –	48	Logistic regres.	174	(38)	55.3	68.4
		(with different features)			71.0	72.1
#10 –	NS	Pigmented distr.	459	(95)	87.5	85.7
#11 –	7	ANN	348	(53)	80.4	85.0
		(with only the large lesions)			100.0	95.0
#12 –	28	Decision tree	224	(108)	NS	NS

ROC – Receiver Operating Characteristic
SE – Sensibility (fraction of correctly classified melanomas)
SP – Specificity (fraction of correctly classified benign lesions)
NS – Not specified in the paper
ANN – Artificial Neural Network
CART – Classification and Regression Trees
K-NN – K-Nearest Neighbour
SLP – Single-Layer Perceptron
LDA – Linear Discriminant Analysis

methods have been proposed, such as thresholding and gradient vector flow snakes (Erkol et al. 2005), Laplacian filters with zero-crossing (Rubegni et al. 2002a, Burroni et al. 2004), hybrid algorithms (Hoffman et al. 2003, Tomatis et al. 2005).

Some segmentation methods are semi-automatic (Celebi et al. 2005), requiring an interaction between the user and the software in order to achieve the proper segmentation. Once there is an identification of the skin lesion, a great number of features can be extracted from the image. The features are generally grouped in three main categories: (i) geometric (such as area, perimeter, polar measures and bounding box), (ii) morphologic (such as symmetry, roundness and shape) and (iii) colour related features. Some authors do feature selection, using a subset of all available features for classification, while other use all the features produced. For example, Blum used the 6 (out of 64) more important features (Blum 2004), and Elbaum used small subsets (less than 15 features) from the total number of candidates (822) for lesion classification (Elbaum 2001).

Several papers in the literature report the results of lesion classification using automatic algorithms. These studies investigate automatic systems, based in EpiLuminescence Microscopy (ELM) images processing techniques, using different statistical approaches like: Linear Discriminant Analysis (LDA) (Elbaum et al. 2001, Sboner et al. 2003, Burroni et al. 2004), decision trees (Dreiseitl et al. 2001, Sboner et al. 2003, Sboner et al. 2004) and neural networks (Ganster et al. 2001, Rubegni et al. 2002b, Hoffman et al. 2003, Burroni et al. 2004). A summary of the most relevant aspects of these studies is presented in Tables 1 and 2. Table 1 presents the authors, year, image analysis system, and a number that relates to Table 2. Table 2 presents the number of features and the classifiers used, the number of cases (total and malignant) and the Receiver Operating Characteristic (ROC): the Sensibility (SE), which correspond to the fraction of correctly classified melanomas (true positives), and the Specificity (SP), which is the fraction of correctly classified benign lesions (true negatives).

In the reported studies, the number of cases available for training and testing the algorithms very considerably (from 147 to 5363) as does the number of features (7 to 822). The sensibilities reported for the classifiers are often over 90%, and reach 100% for some cases. However, these figures are somehow misleading as in most cases only perfect quality images are used for classification, with those images contaminated by noise, such as hair or air bubble, rejected (Hoffman 2003, Tomatis 2005). The classifiers are generally tuned to prevent false negatives, which results in an increase of the sensibility. However, this will also increase the number of false positives, thus reducing the specificity.

Overall, the results reported in the literature are somehow disappointing. The high sensibility values reported, and sometimes also the high specificity, refer to favourable conditions, usually where only a subset of the initial dataset is used for training and testing. Amongst the various reasons that can justify the poor results, the most obvious one is the great variety of lesions and types of images tested. The difficulty in establishing the relevant features to properly characterise a lesion seems also to be a likely reason for the inability of the methods to properly discriminate the different types of lesions. In summary, it can be said that at present the automatic methods are still in the experience stage for clinical diagnosis and are an expert second opinion for clinical use (Menzies 2006).

3 EVALUATION OF THE MOLE ANALYZER

The Tuebinger Mole Analyzer software was developed at the University of Tuebingen in Germany, based on the software program FotoFinder from TeachScreen Company, Griesbach, Germany. Its purpose is to perform an analysis of pigmented moles surrounded by pale skin (Mole Analyzer, 2006). There are two main aims of the Tuebinger Mole Analyzer image analysis program: (1) to detect changes in pigmented lesions during the follow-up of patients; (2) to judge the dignity of the pigmented lesions. However, the second objective was not yet achieved, as there is not yet a clear cutoff-point for determining malignancy (Mole Analyzer, 2006).

The software works almost as a black box, with very little options for user intervention. Once a dermoscopic image is selected, the system provides

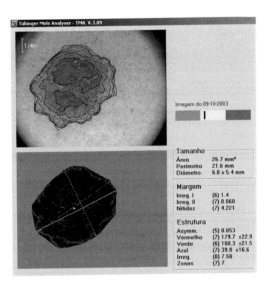

Figure 2. Example display from the Mole Analyser software.

an automatic identification (or segmentation) of the lesion, which aims at identifying the lesion and separates it from the background. The border of the lesion is determined by approximation from both sides (Mole Analyzer, 2006). Subsequently, the area and other geometric parameters are measured and the border and its regularity and sharpness are determined. The features used in this program (metric and others) are used to distinguish between different pigmented moles and are examined as criteria to judge the malignancy of pigmented moles. The features are rated between 1 and 8, and higher scores than 6 may indicate developing malignancy (Mole Analyzer, 2006). An example of the display produced by the Mole Analyser software is presented in Figure 2. The top left part of this figure shows the dermoscopic image of the skin lesion with the segmented contours overlaid. The bottom left part shows an image of the extracted lesion and symmetry axes. The feature values are shown on the bottom right part of Figure 2, and on the top right there is a label indicating the degree of malignancy of the lesion on a colour bar.

3.1 Experimental setup

The clinical database of the Hospital Pedro Hispano (HPH) has over 4000 cases with dermoscopic images of various types of lesions. A total of 83 melanocitic nevus were selected randomly from the database. Each case was analysed using the Mole Analyser software in order to evaluate the performance of the system. The cases were also diagnosed by a dermatologist with over 8 years of experience in dermoscopic image analysis (Dr. J. Rozeira).

Table 3. Mole Analyser performance for lesion segmentation.

Segmentation result	No. cases	(%)
Correct (without user intervention)	65	(78.3%)
Correct after 1 user interaction	7	(8.4%)
Correct after 2 user interactions	4	(4.8%)
Correct after 3 user interactions	1	(1.2%)
Failure	6	(7.2%)

Table 4. Mole analyser feature summary.

Feature	Average	Min.–Max.	St. Dev.
#1 – Area (mm^2)	25.1	4.9–68.3	16.6
#2 – Perimeter (mm)	20.6	8.6–42.6	7.8
#3 – Irregularity I	1.5	1.1–7.0	0.8
#4 – Irregularity II	0.06	0.008–0.48	0.08
#5 – Sharpness	4.1	0.7–22.8	3.8
#6 – Asymmetry	0.06	0.02–0.24	0.04
#7 – Red	167.4	92.8–245.3	33.6
#8 – Green	109.0	59.4–178.4	26.8
#9 – Blue	79.1	37.4–158.3	28.2
#10 – Structure irreg.	7.6	6.6–8.7	0.4
#11 – Zones	4.9	1.0–12.0	5.0

3.2 Lesion segmentation

The initial processing step of the Mole Analyser is the segmentation of the image. The system provides an initial estimate, allowing for the user to confirm it or to suggest an increase or decrease in the lesion area. This often allows for a correct identification of the lesion after one or more interactions from the user. However, in some cases, the software is unable to properly identify the lesion even after several indications from the user. The performance of the Mole Analyser software for the 83 test cases is presented in Table 3. The table presents the number of user interventions required to achieve the adequate lesion identification (segmentation).

The majority of cases (78.3%) were identified correctly by the software at the first attempt, and most of the remaining cases were identified after a direct intervention from the user (14.4%). However, the system was unable to properly segment the skin lesion in 6 images (7.2%), even after various interactions from the user. These cases were thus removed from the subsequent analysis, which was carried out on the remaining 77 images.

3.3 Feature extraction

The feature values provided by the Mole Analyzer software were registered for all 77 lesions. The range of values, the average and the standard deviation are presented in Table 4 for each feature. The range of values covered by the different features is very different, from as little as 0.20 for Asymmetry (#6) to over 100 for the 3 colour features (#7, #8 and #9).

The covariance and correlation matrices were computed for the dataset. The correlation matrix is presented in Table 5, where the high values (above 0.80) are displayed in bold. The two irregularity features (#3 and #4) are highly correlated (0.95), so one of these features is redundant. These two features are also strongly correlated with Assymetry (feature #6). It is interesting to note that Red (#7) is highly correlated with Green (#8), and Green is highly correlated with Blue (#9), but the correlation between Red and Blue is low (0.53). This is not surprising, as it is well known that neighbouring spectral bands tend to be more correlated than bands further apart in the electromagnetic spectrum.

The information displayed in Table 5 indicates that the number of features is too high for the data. A principal component analysis was carried out in order to reduce the data dimensionality. The variance and the cumulated variance of the first 5 Principal Components (PC) is presented in Table 6. A total of 98.48% of the data variance is retained by the first 3 PCs. The classification of the data using only the first few PCs should provide better results than using the original 11 features, due to the following reasons: (i) the original features have very different numeric ranges, (ii) some features are clearly redundant, and (iii) the original data dimensionality (11) is rather high.

3.4 Lesion classification

The Mole Analyser performs a lesion classification, according to the estimated risk of malignancy. Each feature is rated with a score, an integer between 1 and 8, with scores higher than 6 indicating a higher risk of developing malignancy (Mole Analyzer, 2006). An inspection of the relation between the feature values and the corresponding score was carried out for each feature. As an illustration of the typical relationship between the features and the assigned scores, Figure 3 shows a plot of the Asymetry feature (#6). The relation between the two is not clear, as there are lesions with the same feature value assigned to different scores, as it can be seen in the figure. Apart from this odd behaviour, the relation is roughly linear, with higher Asymmetry values corresponding to greater risk of malignancy. The liner relation is saturated on level 8 for asymmetry values over 0.09. The relationships between the other features and the corresponding scores are generally of the same type as that displayed by the Asymmetry feature.

In addition to the malignancy score assigned to each feature, the Mole Analyser software also provides an overall score, presented as a mark in a colour bar with

Table 5. Correlation matrix between features.

	#1	#2	#3	#4	#5	#6	#7	#8	#9	#10
#2	**0.86**									
#3	0.00	0.48								
#4	−0.01	0.46	**0.95**							
#5	0.18	0.53	0.64	0.62						
#6	−0.08	0.37	**0.89**	**0.93**	0.55					
#7	0.24	0.35	0.33	0.40	0.19	0.39				
#8	0.24	0.39	0.39	0.45	0.32	0.44	**0.89**			
#9	0.18	0.34	0.37	0.39	0.35	0.41	0.53	**0.82**		
#10	0.02	−0.06	−0.14	−0.09	−0.32	−0.01	0.41	0.34	0.23	
#11	**0.81**	0.72	0.04	0.02	0.16	−0.03	0.10	0.06	−0.02	−0.07

Table 6. Variance of the first 5 Principal Components.

Component	Variance (%)	Cumulated variance (%)
PC 1	73.89	73.89
PC 2	14.63	88.52
PC 3	9.95	98.48
PC 4	0.72	99.20
PC 5	0.62	99.81

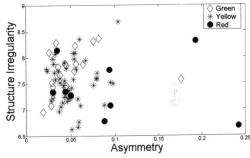

Figure 4. Location of the lesions on the 2D feature space for Asymmetry (#6) and Structure irregularity (#10).

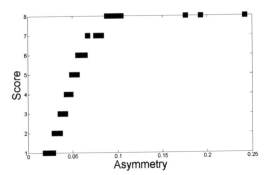

Figure 3. Plot of feature #6 (Asymmetry) values vs. 1–8 labels.

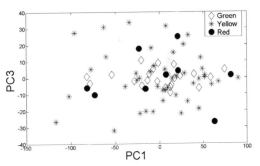

Figure 5. Location of the lesions on the 2D feature space for PC1 and PC3.

3 levels (green, yellow, and red). An example of such score can be seen in Figure 2 (in black & white). In this case the lesion was classified as yellow, although the label is in the section of the yellow bar closer to the green bar (left).

Several plots of pairs of features were produced, to analyse the distribution of the green, yellow and red labelled lesions on the various 2D feature spaces. An example of such plots is presented in Figure 4, for Asymmetry (#6) and Structure irregularity (#10). The correlation between these features is very low (−0.01), which can be confirmed by the plot of Figure 4. However, the 2D feature space is shared by all 3 classes (red, yellow and green), which indicates that it is not possible to discriminate between these classes using

only these two features. The same conclusion could be drawn from the various pairs of features attempted. A similar analysis was carried out using the first few PCs. As an illustration, the plot of the PC1 versus PC3 is presented in Figure 5. Although in this plot the 2D feature space is more evenly occupied, there is again no clear separation of the classes. The same type of behaviour occurs for PC1 vs PC2 and PC2 vs PC3.

The classification provided by the Mole Analyser for the 77 nevus tested was registered, with 9 levels (each colour section was sub-divided in three regions,

285

Table 7. Mole Analyser performance for lesion classification.

Label	Benign	Doubtful	Malignant
Green−	**8**	0	0
Green	**5**	0	0
Green+	**8**	0	0
Yellow−	30	**8**	0
Yellow	1	**0**	0
Yellow+	4	**1**	0
Red−	1	0	**0**
Red	0	0	**0**
Red+	7	1	**0**

e.g. green−, green, green+). For example, the lesion classification illustrated in Figure 2 is labelled as yellow-. The results provided by the Mole Analyser and the clinical evaluation of the lesions, based exclusively on the dermoscopic images, are presented in Table 7. The results overall are rather poor. The doubtful cases were adequately labelled in yellow by the Mole Analyser, except for 1 case out of 10 (10.0%). However, less than one third of the cases diagnosed as benign were labelled as green by the Mole Analyser (21 out of the 64 or 32.8%) and 8 cases (12.5%) were even labelled as red. This is not much better than a random labelling of the cases. The k coefficient has a value of 1 when there is a perfect classification, and a value of 0 when the result is totally random. The classification results produced by the Mole Analyser software for our dataset correspond to a k of 0.326, a very low value.

In order to compare the performance of the Mole Analyser software with other systems (section 2) in the task of lesion classification, the labelling was grouped in positives (yellow and red labels) and negatives (green labels). This is a reasonable criterion, as an automated system should select all the lesions that might be malignant so that they can be inspected by a dermatologist. In this context, the Mole Analyser produced 10 true positives, 21 true negatives, 43 false positives and 0 false negatives. The resulting sensibility is 100.0% and the specificity is 32.8%. The classifier parameters of the Mole Analyser seem to be tuned to prevent false negatives, which results in an excellent sensitivity but a poor specificity. It is also worth mentioning that all the lesions tested have the same pathology (melanocitic nevus), which is not the most challenging of lesions for an automatic diagnosis system.

4 CONCLUSIONS

The computer based automatic image analysis systems have great potential for dermoscopy. The systems can provide meaningful quantitative information to assist the clinical evaluation, and, at a further level, to perform an automatic early screening of skin lesions. This later task is not yet feasible by the existing systems, as the classification accuracy is still not satisfactory. A number of systems are reported in the literature that are able to classify skin lesions with very high sensibilities (few false negatives). However, in most cases only perfect quality images were used for classification, with those images contaminated by noise, such as hair or air bubble, rejected.

The evaluation of the Tuebinger Mole Analyzer software was done with 83 dermoscopic images of melanocitic nevus. The system performance was reasonably good in terms of lesion segmentation, but less satisfactory for lesion classification. An accurate fully automatic identification of the lesion was achieved by the Mole Analyzer in 78.3% of the cases, with a further 14.4% after one or more corrections by the user. Only in 7.2% of the cases the system was unable to select the lesion from the background. However, the lesion classification results – sensibility of 100.0%, specificity of 32.8%, k coefficient of 0.326 – are not satisfactory, although the positive fact that no false negatives cases were registered. The Mole-Analyser software is nonetheless a useful tool for the clinical follow-up of lesions (change in perimeter, axes dimensions and colour).

As a final remark, it can be said that the desired wide use of dermatoscopy, tele-dermatoscopy and tele-medicine requires effective computer based early warning diagnosis systems. However, further developments are still required in order to have a robust and reliable computer based diagnosis tool.

REFERENCES

Argenziano, G., Soyer, H.P., De Giorgi, V., et al. 2000. Dermoscopy, an interactive atlas. Milan, Italy: EDRA Medical Publishing. (http://www.dermoscopy.org)

Blum, A., Luedtke, H., Ellwanger, U., Schwabe, R., Rassner, G. & Garbe, C. 2004. Digital image analysis for diagnostic of cutaneous melanoma. Development of a highly effective computer algorithm based on analysis of 837 melanocytic lesions. British Journal of Dermatology, 151:1029–1038.

Burroni, M., Corona, R., Dell'Eva, G., Sera, F., Bono, R., Puddu, P., Perotti, R., Nobile, F., Andreassi, L. & Rubegni, P. 2004. Melanoma computer-aided diagnosis: reliability and feasibility study. Clinical Cancer Research, 10 (6):1881–1886.

Celebi, M.E., Kingravi, H.A., Aslandogan, Y.A. & Stoecker, W.V. 2006. Detection of Blue-White Veil Areas in Dermoscopy Images Using Machine Learning Techniques. To appear in SPIE Medical Imaging 2006.

Dreiseitl, S., Ohno-Machado, L., Kittler H., Vinterbo, S., Billhardt, H. & Binder, M. 2001. A comparison of machine learning methods for the diagnosis of pigmented skin lesions, Journal of Biomedical Informatics, 34: 28–36.

Elbaum, M., Kopf, A.W., Rabinovitz, H.S., Langley, R.G.B., Kamino, H., Mihm, M.C., Sober, A. J., Peck, G.L., Bogdan, A., Gutkowicz-Krusin, D., Greenebaum, M., Keem, S., Oliviero, M. & Wang, S. 2001. Automatic differentiation of melanoma from melanocytic nevi with multispetral digital dermoscopy: a feasibility study. *Journal of the American Academy of Dermatology*, 44 (2): 207–218.

Erkol, B., Moss, R.H., Stanley, R.J., Stoecker, W.V. & Hvatum, E. 2005. Automatic lesion boundary detection in dermoscopy images using gradient vector flow snakes. *Skin Research & Technology*, 11: 17–26.

Ganster, H., Pinz, A., Röhrer, R., Wildling, E., Binder, M. & Kittler, H. 2001. Automated melanoma recognition. *IEEE Transactions on Medical Imaging*, 20 (3): 233–239.

Gerger, A., Stolz, W., Pompl, R. & Smolle, J. 2003. Automated epiluminescence microscopy-tissue counter analysis using CART and 1-NN in the diagnosis of melanoma. *Skin Research & Technology*, 9 (2): 105–110.

Grana, C., Pellacani, G., Cucchiara, R., Seidenari, S. 2003. A new algorithm for border description of polarized light surfacemicroscopic images of pigmented skin lesions. *IEEE Transactions on Medical Imaging*; 22: 959–964.

Gutkowicz-Krusin, D., Elbaum, M., Szwajkowski, P., Kopf, A.W. 1997. Can early malignant melanoma be differentiated from atypical melanocytic nevi by in-vivo techniques? Part II: Automatic machine vision classification. *Skin Research & Technology*, 3: 15–22.

Hoffmann, K., Gambichler, T., Rick, A., Kreutz, M., Anschuetz, M., Grunendick, T., Orlikov, A., Gehlen, S., Perotti, R., Andreassi, L., Bishop, J. N., Césarini, et al. 2003. Diagnostic and neural analysis of skin cancer (DANAOS). A multicentre study for collection and computer-aided analysis of data from pigmented skin lesions using digital dermoscopy. *British Journal of Dermatology*, 149: 801–809.

Menzies, S.W. 2006. Cutaneous melanoma: making a clinical diagnosis, present and future. *Dermatologic Therapy*, 19: 32–39.

Mole Analyzer, 2006. Computer dermatoscopy – Tubinger Mole Analyzer. (www.moleanalyzer.com)

Ritter, G.X., Wilson, J.N. 1996. Handbook of Computer Vision Algorithms in Image Algebra. Boca Raton: CRC Press.

Rubegni, P., Cevenini, G., Burroni, M., Perotti, R., Dell'Eva, G., Sbano, P., Miracco, C., Luzi, P., Tosi, P., Barbini, P. & Andreassi, L. 2002a. Automated diagnosis of pigmented skin lesions. *International Journal of Cancer*, 101 (6): 576–580.

Rubegni, P., Burroni, M., Cevenini, G., Perotti, R., Dell'Eva, G., Barbini, P., Fimiani, M. & Andreassi, L. 2002b. Digital dermoscopy analysis and artificial neural network for the differentiation of clinically atypical pigmented skin lesions: a retrospective study. *The Journal of Investigative Dermatology*, 119 (2): 471–474.

Sboner, A., Eccher, C., Blanzieri, E., Bauer, P., Cristofolini, M., Zumiani, G. & Forti, S. 2003. A multiple classifier system for early melanoma diagnosis. *Artificial Intelligence in Medicine*, 27: 29–44.

Sboner, A., Bauer, P., Zumiani, G., Eccher, C., Blanzieri, E., Forti, S. & Cristofolini, M. 2004. Clinical validation of an automated system for supporting the early diagnosis of melanoma. *Skin Research & Technology*, 10 (3): 184–192.

Seidenari, S., Pellacani, G. & Grana, C. 2005. Colors in atypical nevi: a computer description reproducing clinical assessment. *Skin Research & Technology*, 11 (1): 36–41.

Tomatis, S., Carrara, M., Bono, A., Bartoli, C., Lualdi, M., Tragni, G., Colombo, A. & Marchesini, R. 2005. Automated melanoma detection with a novel multispectral imaging system: results of a prospective study. *Physics in Medicine and Biology*, 50: 1675–1687.

Computational Modelling of Objects Represented in Images –
João Manuel R.S. Tavares & R.M. Natal Jorge (eds)
© 2007 Taylor & Francis Group, London, ISBN 978-0-415-43349-5

Fast estimation of skeleton points on 3D deformable meshes

Julien Mille, Romuald Boné, Pascal Makris & Hubert Cardot
Université François-Rabelais de Tours, Laboratoire Informatique, France

ABSTRACT: Segmentation of volumetric images by means of 3D deformable meshes is a widespread technique. Extracting the skeleton of such meshes can present great interest for modeling the shape of the segmented area. In this paper, we propose a new method performing fast approximation of skeleton points of a triangular mesh, driven by greedy energy-minimization technique. The skeletonization method uses available data relative to the mesh, such as geometrical and topological structures as well as motion information.

1 INTRODUCTION

Deformable models are powerful tools used in image segmentation. Among these, active surfaces, which are 3D extensions of the well known active contour model (Kass, Witkin, and Terzopoulos 1987), are often employed for segmentation in 3D volumetric images. From an initial location, the surface deforms iteratively according to an evolution algorithm, in order to fit the object boundary. Several evolution methods were used in 3D, like the variational approach (McInerney and Terzopoulos 1995), the greedy algorithm (Williams and Shah 1992) (Bulpitt and Efford 1996) and the physics-based method (Montagnat and Delingette 2005) (Park, McInerney, and Terzopoulos 2001). For surveys of deformable models representation and evolution methods, see (Montagnat, Delingette, and Ayache 2001) and (Mille, Boné, Makris, and Cardot 2005). A widespread discrete explicit implementation of the active surface is the mesh, unlike implicit surfaces usually implemented through level-set techniques (Malladi, Sethian, and Vemuri 1995). Among different types of meshes, in this paper, we deal with triangular meshes (Lachaud and Montanvert 1999), which are polyhedrons made up of vertices and adjacent triangles.

Given the mesh data, we consider the skeleton, a compact representation of geometry and topology, which makes it suitable for structural pattern recognition (Loncaric 1998). Skeletonization methods are divided into two groups, whether they are based on pixels or not. Pixel-based methods use the whole set of pixels inside the shape in the skeletonization process, starting from the assumption that the shape is fully discretized. These approaches generally use thinning algorithms, based for example on distance transforms or mathematical morphology (Lam, Lee, and Suen 1992). Significant work in this area

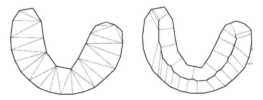

Figure 1. Constrained delaunay triangulation (left) and inner voronoi diagram (right).

include veinerization (Deseilligny, Stamon, and Suen 1998) and distance maps combined with level-sets (Kimmel, Shaked, Kiryati, and Bruckstein 1995). Conversely, non-pixel-based methods only consider the shape boundary and are more suitable to polygonal representations. In this context, the Voronoi diagram is the usual basis for the computation of a geometric skeleton (Brandt and Algazi 1992). The Voronoi graph (see figure 1) is the topological dual equivalent of the Delaunay triangulation, which can be computed on any set of unorganized points. The polygon-specific Delaunay triangulation, in which polygon edges are fixed, is known as Constrained Delaunay Triangulation (CDT) (Chew 1989). An efficient implementation of the CDT in *n*-dimension was presented in (Shewchuk 2002). Given the Voronoi graph, the geometric skeleton is computed as the set of Voronoi edges which are totally inside the shape. In (Amenta, Sunghee, and Kolluri 2001), one may find an algorithm for computing 3D Voronoi graphs.

Let us consider the case where the skeleton information is needed at each step of the mesh deformation. Having such data may be useful for integrating prior knowledge about the shape, by making the skeleton intervene in the motion of the deformable mesh. For example, when segmenting branching objects (Ferley, Gascuel, and Attali 1997), a distance function between the surface points and the growing skeleton could be minimized, thus making the surface tubular-shaped. In this case, the surface and its skeleton should grow simultaneously, which implies to update the skeleton after each deformation step. We believe that having skeleton data available during the whole deformation can be useful for further investigation (in this paper, we focus on computing a dynamic skeleton, without making it influence the deformation yet). To do this, the above-described skeletonization techniques may suffer from several drawbacks. Pixel-based methods need the surface to be discretized with voxelization techniques (Kaufman 1987), which would be prohibitively time-consuming if performed several times. Boundary-based methods using Voronoi diagram could be used, however computing the Voronoi points after each iteration may be tedious. Moreover, like in many deformable mesh models, remeshing is performed, i.e. vertices are added or deleted to keep constant vertex distribution along the surface. Integrating this feature in the Voronoi-based approach without recomputing the whole diagram is difficult.

Hence, we propose a new method for estimating skeleton points on the deformable mesh. This method takes advantage of the geometrical and topological data included in the mesh, as well as knowledge about motion and remeshing, allowing the skeleton to be approximated during deformation. In this paper, we first describe the 3D deformable model, in terms of geometry and topology. The evolution based on energy minimization with greedy algorithm and the remeshing technique are presented. Then, we describe how skeleton points can be approximated in this context. Finally, we present experimental results obtained on tubular shapes extracted from both artificial and real medical images.

2 THE 3D DEFORMABLE MESH

The 3D deformable model, also described in (Mille, Boné, Makris, and Cardot 2005), is a triangular mesh made up vertices and adjacent triangles. In what follows, we will denote $\{i, j\}$ the vertex indices and $\{u, v\}$ the triangle indices. Each vertex i has 3D coordinates $\mathbf{p}(i) = (p_x(i), p_y(i), p_z(i))^T$, a unit vector $\vec{\mathbf{n}}(i)$ (the inner normal to the surface), a set of neighboring vertices $N(i)$ and a set of neighboring triangles $U(i)$. As stated in (Park, McInerney, and Terzopoulos 2001), the normal of a vertex is the normalized sum of the normals of the neighboring triangles. Moreover, the normal of a given triangle u, denoted $\vec{\mathbf{m}}(u)$, is obtained with the normalized cross product between two of its edges:

$$
\begin{aligned}
\vec{\mathbf{m}}(u) &= \frac{\vec{\mathbf{v}}(1) \times \vec{\mathbf{v}}(2)}{\|\vec{\mathbf{v}}(1) \times \vec{\mathbf{v}}(2)\|} \\
\vec{\mathbf{v}}(1) &= \mathbf{p}(\tau(u, 2)) - \mathbf{p}(\tau(u, 1)) \\
\vec{\mathbf{v}}(2) &= \mathbf{p}(\tau(u, 3)) - \mathbf{p}(\tau(u, 1))
\end{aligned}
\tag{1}
$$

where $\tau(u, k), k = 1, 2, 3$ denotes the three vertices defining triangle u. The initial mesh is obtained from successive subdivisions of an icosahedron (McInerney and Terzopoulos 1995) (Park, McInerney, and Terzopoulos 2001). To perform segmentation, the mesh is endowed with an energy functional, being the sum of the energies of each vertex. The greedy approach is an energy-minimizing algorithm, originally introduced for 2D contours in (Williams and Shah 1992), consisting in successive local optimizations. A 3D extension of the greedy algorithm applied on a triangular mesh was proposed in (Bulpitt and Efford 1996). At each iteration, a cubic window of width w around each vertex is considered. The energy is computed at each voxel belonging to the window and the vertex is moved to the location leading to the lowest energy. In contrast with $\mathbf{p}(i)$ (the initial position), we denote $\mathbf{p}'(i)$ the current tested position, whose energy is a weighted sum of internal and external energies, normalized on the whole window.

$$
\begin{aligned}
E(\mathbf{p}'(i)) &= \alpha E_{cont}(\mathbf{p}'(i)) + \beta E_{curv}(\mathbf{p}'(i)) \\
&+ \gamma E_{grad}(\mathbf{p}'(i)) + \delta E_{bal}(\mathbf{p}'(i))
\end{aligned}
\tag{2}
$$

These are mesh-suitable extensions of the energies commonly used in 2D active contour models. Coefficients $(\alpha, \beta, \gamma, \delta)$ are user-defined weights. The continuity E_{cont} and curvature E_{curv} are internal energies maintaining the geometrical regularity of the contour, whereas the gradient energy E_{grad} and balloon E_{bal} are external energies driving the mesh towards the object boundary. We define the continuity energy as follows:

$$
E_{cont}(\mathbf{p}'(i)) = \sum_{j \in N(i)} \left| \overline{d^2} - \|\mathbf{p}'(i) - \mathbf{p}(j)\|^2 \right|
\tag{3}
$$

where $\overline{d^2}$ is the mean squared distance over all neighbors. Minimizing it reduces the distance deviation, so that the vertices remain evenly spaced along the surface. In 2D active contours, the curvature is approximated as the squared distance between the vertex and the middle of its two neighbors (minimizing it has a smoothing effect). Extending this principle to the mesh, the curvature of the tested point $\mathbf{p}'(i)$ is the squared distance between $\mathbf{p}'(i)$ and the centroid of the neighbors of $\mathbf{p}(i)$.

$$
E_{curv}(\mathbf{p}'(i)) = \left\| \mathbf{p}'(i) - \frac{1}{|N(i)|} \sum_{j \in N(i)} \mathbf{p}(j) \right\|^2
\tag{4}
$$

The gradient energy E_{grad} is a function of the normalized gradient magnitude g of image function I, making the vertices locate on salient edge voxels. In presence of noisy data, the image is smoothed with a gaussian filter prior to gradient operation. In the following equations, G_σ is a gaussian mask with standard deviation σ. Real 3D edge detection is obtained by convolving I with the Zucker-Hummel operator (Zucker and Hummel 1981).

$$g(\mathbf{p}) = \|\nabla I(\mathbf{p}) * G_\sigma\| / g_{max}$$

$$E_{grad}(\mathbf{p}'(i)) = -g(\mathbf{p}'(i)) \qquad (5)$$

To increase the capture range, allowing the mesh to be initialized far from the object boundaries, we add a balloon energy E_{bal}. It is derived from the inflation force proposed in (Cohen 1991) and makes the mesh propagate along its normal direction. The sign of weight δ (see equation 2) controls the orientation of the motion and should be set according to the initial location of the contour, whether it is inside or outside the object. The window width w is used to make the extremity of the resulting vector point outside the window.

$$E_{bal}(\mathbf{p}'(i)) = \|\mathbf{p}'(i) - (\mathbf{p}(i) + w\vec{\mathbf{n}}(i))\|^2 \qquad (6)$$

3 ADAPTIVE REMESHING

To maintain a stable vertex distribution along the surface, adaptive remeshing is performed (Lachaud and Montanvert 1999) (Slabaugh and Unal 2005). The surface is allowed to add or merge vertices to keep the distance between neighboring vertices homogeneous. It ensures that every couple of neighbors $(\mathbf{p}(i), \mathbf{p}(j))$ satisfies the following constraint:

$$d_{min} \leq \|\mathbf{p}(i) - \mathbf{p}(j)\| \leq d_{max} \qquad (7)$$

where d_{min} and d_{max} are two user-defined thresholds. Since adding or merging vertices modifies local topology, topological constraints should be verified. Let us consider the couple of neighbors $(\mathbf{p}(i), \mathbf{p}(j))$. To perform vertex adding or merging, $\mathbf{p}(i)$ and $\mathbf{p}(j)$ should share exactly two common neighbors:

$$N(i) \cap N(j) = \{a, b\} \qquad (8)$$

When $\|\mathbf{p}(i) - \mathbf{p}(j)\| > d_{max}$, a new vertex is created at the middle of line segment $\mathbf{p}(i)\mathbf{p}(j)$ and connected to $\mathbf{p}(n+1)$ and $\mathbf{p}(b)$ (see middle part of figure 2).

The size of neighborhood is not modified for $\mathbf{p}(i)$ and $\mathbf{p}(j)$, whereas $\mathbf{p}(a)$ and $\mathbf{p}(b)$ gain one neighbor. When $\|\mathbf{p}(i) - \mathbf{p}(j)\| < d_{min}$, $\mathbf{p}(j)$ is deleted and $\mathbf{p}(i)$ is translated to the middle location (see right part of figure 2). The neighbors of $\mathbf{p}(j)$ become neighbors of $\mathbf{p}(i)$. Vertex merging prevents neighboring vertices from getting too close, which might result in vertex overlapping and intersections between triangles. Using the greedy algorithm, two neighbors $\mathbf{p}(i)$ and $\mathbf{p}(j)$ might overlap at iteration t only if their respective windows overlapped at iteration $t-1$. However, two square windows of width w do not overlap if their centers are w pixels away in at least one dimension. Using the infinite norm $\|\mathbf{p}(i) - \mathbf{p}(j)\|_\infty$, we redefine the remeshing criterion given in equation 7.

$$w < \|\mathbf{p}(i) - \mathbf{p}(j)\|_\infty < 2w \qquad (9)$$

Hence, the parameter w does not only control the search space width of the greedy algorithm, but also the mesh fineness, since small w values will lead to dense vertex distribution along the surface.

4 ESTIMATION OF SKELETON POINTS

The method described in this section provides fast skeleton estimation of the deformable mesh. Our approach generates a set of points, approximating the skeleton, which evolves simulaneously as the mesh does, taking advantage of available surface geometrical and topological data, such as normal vectors and vertices neighborhoods. As a basis, we consider the continuous definition of the medial axis: a smoothed closed curve in the plane has a symmetry set, which is the closure of the locus of the circle centers having tangency at more than one place. The medial axis is the part of the symmetry set for which the corresponding maximal circles are entirely inside the shape. Thus, given a point $\mathbf{p}$ on the curve and a maximal circle having tangency at point $\mathbf{p}$, it is obvious that the circle center $\mathbf{q}$ (a local symmetry center and thus, a skeleton point) is on the half-line starting from $\mathbf{p}$ along its inward normal $\vec{\mathbf{n}}$ (see left part of figure 3).

Extending this principle to our 3D discrete implementation, the skeleton is approximated by finding local symmetry centers: considering each vertex, the part of the mesh facing $\mathbf{p}(i)$ is the triangle intersected by the half-line starting from $\mathbf{p}(i)$ along vector $\vec{\mathbf{n}}(i)$.

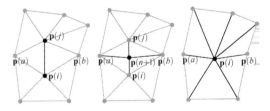

Figure 2.   Remeshing operations between vertices $\mathbf{p}(i)$ and $\mathbf{p}(j)$.

Figure 3.   Skeleton point on the inward half-line normal, for a continuous 2D closed curve (left) and the triangular mesh (right).

Then, the corresponding estimated skeleton point $\mathbf{q}(i)$ is the middle of line segment defined by $\mathbf{p}(i)$ and the intersection point (see right part of figure 3). In this way, we generate a point cloud skeleton in which each skeleton point is associated with a mesh vertex. Moreover, this guarentees that all estimated skeleton points are inside the mesh. This is an advantage over Voronoi diagram computation, generating Voronoi vertices outside the shape, which have to be eliminated afterwards.

Initially, the facing triangle u of each vertex i should be determined. We introduce the boolean function $f(i, u)$, true if triangle u is facing vertex i, false otherwise. On condition that $\vec{\mathbf{m}}(u) \cdot \vec{\mathbf{n}}(i) \neq 0$, a projection point is computed (on the contrary, $\vec{\mathbf{m}}(u)$ is parallel to the plane which u belongs to, no further work on u is needed). Point $\mathbf{r}(i)$ is the projection of $\mathbf{p}(i)$ along its normal, on the plane defined by triangle u:

$$\mathbf{r}(i) \leftarrow \mathbf{p}(i) + \frac{\vec{\mathbf{m}}(u) \cdot (\mathbf{p}(\tau(u,1)) - \mathbf{p}(i))}{\vec{\mathbf{m}}(u) \cdot \vec{\mathbf{n}}(i)} \vec{\mathbf{n}}(i) \quad (10)$$

To check if $\mathbf{r}(i)$ is inside triangle u, projection is made into 2D plane by discarding the coordinate component that has the greatest absolute value in the triangle normal. We denote $\text{proj}_u \mathbf{p}$ the projection of $\mathbf{p}$ on the plane defined by triangle u. A 2D point $\mathbf{d}$ is inside the triangle defined by three vertices $\mathbf{a}, \mathbf{b}, \mathbf{c}$ if $\mathbf{d}$ is on the left side of edges $\mathbf{ab}, \mathbf{bc}$ and $\mathbf{ca}$ considered clockwise (this relation is expressed with 2D cross product $\mathbf{a} \times \mathbf{b} = a_x b_y - a_y b_x$). Thus, $f(i, u)$ is evaluated as follows:

$$\begin{aligned} f(i, u) &= (\mathbf{d}-\mathbf{a}) \times (\mathbf{b}-\mathbf{a}) \geq 0 \\ &\wedge (\mathbf{d}-\mathbf{b}) \times (\mathbf{c}-\mathbf{b}) \geq 0 \\ &\wedge (\mathbf{d}-\mathbf{c}) \times (\mathbf{a}-\mathbf{c}) \geq 0 \end{aligned} \quad (11)$$

where $\mathbf{a}$, $\mathbf{b}$, $\mathbf{c}$ are respectively $\text{proj}_u \mathbf{p}(\tau(u,k))$, $k = 1, \ldots, 3$ and $\mathbf{d} = \text{proj}_u \mathbf{r}(i)$. At initialization, exhaustive search is performed for each couple (i, u). We denote $e(i)$ the facing triangle assigned to i (i.e. $e(i) = u$ if $f(i, u) = \text{TRUE}$). When a complete evolution step has been performed (i.e. vertex motion with greedy algorithm, remeshing and normal vector correction), every $e(i)$ (and consequently $\mathbf{q}(i)$) should be updated. Prior to updating, skeleton calculation is integrated into the remeshing process, based on the assumption that neighboring vertices are likely to have close facing triangles: when a new vertex (of index $n+1$) is created between vertices i and j, its facing triangle is set temporarily equal to the facing triangle of one of its new neighbors (i, j, a or b on figure 2). When a vertex j is deleted, triangle formed by $u = (i, j, a)$ and $v = (i, j, b)$ are also deleted. This requires to check if u and v are assigned as facing triangles. Since it is time-consuming to look at each vertex k to check the condition $e(k) = u$ or $e(k) = v$, a set of facing vertices is assigned to the triangles, denoted

$F(u) = \{i | e(i) = u\}$. $F(u)$ is never computed "as is", but instead should be added with index i when $e(i)$ is set to u (this makes redundant data, however we found it necessary to achieve good performance). When a triangle u is deleted, all vertices in $F(u)$ have their e value assigned to a neighbor of u. We denote $M(u)$ the "wide" neighborhood of u, which is the set of triangles different from u sharing at least one vertex with u.

$$M(u) = \bigcup_{i=1}^{3} U(\tau(u,i)) - \{u\} \quad (12)$$

The following algorithm updates the previously-assigned temporary facing triangles. The knowledge that the distance between neighboring vertices is small (as regards the mesh total area), and vertex motion is small between two successive iterations of the greedy algorithm allows us to perform local search in order to update facing triangles without excessive computations of equations 10 and 11. To do so, each vertex has its e value updated through a search among the local neighborhood of $e(i)$. Local search is done according to a layer approach, the layers L being successive neighbor sets: $L_1(u)$ contains the neighbors of u, $L_2(u)$ contains the neighbors of neighbors of u (excluding u itself), ... The search over the layers is bounded with a user-defined threshold l_{max}. The intersection condition (equation 11) is checked consecutively on the triangles until it is verified or l reaches l_{max}. The set V stores the visited triangles, to avoid multiple checkings on the same triangle.

Algorithm 1 Updating facing triangles and skeleton points

1: **if** $f(i, e(i)) = \text{FALSE}$ **then**
2:   triangle_found $\leftarrow$ FALSE
3:   $l \leftarrow 0$, $L_l(e(i)) \leftarrow \{e(i)\}$, $V \leftarrow \{e(i)\}$
4:   **while** triangle_found $= \text{FALSE} \wedge l < l_{max} \wedge L_l(e(i)) \neq \emptyset$ **do**
5:     $L_{l+1}(e(i)) \leftarrow \emptyset$
6:     **for all** $u \in L_l(e(i))$ **do**
7:       **for all** $v \in M(u), v \notin V$ **do**
8:         **if** $f(i, v) = \text{TRUE}$ **then**
9:           triangle_found $\leftarrow$ TRUE
10:           $\mathbf{q}(i) \leftarrow \dfrac{\mathbf{p}(i) + \mathbf{r}(i)}{2}$, $e(i) \leftarrow v$
11:         **else**
12:           $L_{l+1}(e(i)) \leftarrow L_{l+1}(e(i)) \cup \{v\}$
13:           $V \leftarrow V \cup \{v\}$
14:         **end if**
15:       **endfor**
16:     **endfor**
17:     $l \leftarrow l+1$
18:   **end while**
19: **end if**

292

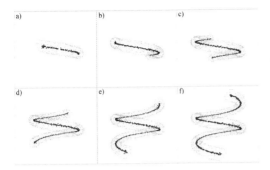

Figure 4. Building skeleton points on a 3D helicoidal shape.

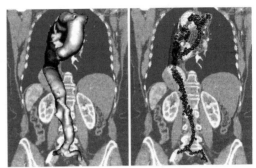

Figure 5. Segmentation results of the aorta (ascending, arch and descending). Left: Smooth surface. Right: wires with skeleton point cloud.

5 EXPERIMENTAL RESULTS

Segmentation and skeletonization results on an artificial grey scale image are shown in figure 4. The image has a $400 \times 400 \times 400$ voxel resolution and represents a helicoidal shape in a uniform background, corrupted with gaussian noise. The mesh was initialized as a sphere with 12 vertices inside the helicoidal shape. Weights values were $\alpha = 0$, $\beta = 0.3$, $\gamma = 2$ and $\delta = -1$. Several values of w were used, generating different mesh densities. For comparison purpose, we generated the skeleton with our algorithm and a Voronoi diagram-based approach[1]. To validate our method, the skeleton was updated after each deformation step (according to algorithm 1). In order to limit the search in the triangle updating algorithm, l_{max} was set to 3. The following table lists computational times (in seconds) obtained with a C++ implementation running on a Pentium IV 2.8 Ghz.

	#Iterations	/wo skel.	/w skel.	Voronoi
$w = 3$	350	3.23	13.41	552.33
$w = 5$	195	1.23	5.67	128.18
$w = 7$	140	1.09	3.58	57.09

Note that when w increases, the deformable mesh needs fewer iterations to converge to the object boundaries (the search window in the greedy algorithm is larger). It also generates a mesh with fewer vertices, decreasing skeletonization computational cost. Our method is significantly faster than Voronoi-based skeletonization, especially on the dense mesh obtained with $w = 3$. The generated point cloud is a good approximation of the skeleton.

---

[1] We used the PowerCrust method developed by Amenta *et al* (Amenta, Sunghee, and Kolluri 2001). Source code is available at http://www.cs.utexas.edu/~amenta/powercrust/welcome.html

Figure 5 represents the results on a medical volumetric grey scale image of the abdomen, generated by Magnetical Resonance Imaging. The image resolution is $512 \times 512 \times 810$ voxels. The 3D mesh was used to segment the inside of the whole aorta (ascending, arch and descending), starting from the heart boundary to the iliac bifurcation (the Y-shaped blood vessel). Such segmentation is made in the framework of abdominal aortic aneurysm diagnosis. As for the previous image, the mesh was initialized as a sphere totally inside the vessel and inflated afterwards. Computational times in seconds are listed in the table below. Our skeletonization method yields faster results than the Voronoi-based approach. In figure 5, the estimated skeleton represents well the overall shape of the blood vessel, pointing out the tubular structure of the area of interest.

	#Iterations	/wo skel.	/w skel.	Voronoi
$w = 3$	580	25.30	290.84	12932.15
$w = 5$	330	9.65	134.72	3021.50
$w = 7$	240	8.69	84.45	1457.74

6 CONCLUSION

In this paper, we proposed a fast skeletonization method applicable on dynamic triangular meshes used for 3D image segmentation. The skeletonization method uses available data relative to the mesh, such as geometrical and topological structures as well as motion information. Future work will include improvements on localization and thinness of the approximated skeleton, while maintaining speed. Skeleton data will be included in the motion equations, to model a priori knowledge about the shape of the target object. This aims at maintaining tubular geometry on the mesh, in order to segment blood vessels properly, especially in noisy MRI where the contrast between the vessel and nearby anatomical structures is insufficient.

REFERENCES

Amenta, N., C. Sunghee, and R. Kolluri (2001). The power crust, unions of balls, and the medial axis transform. *Computational Geometry: Theory and Applications 19*(2–3), 127–153.

Brandt, J. and V. Algazi (1992). Continuous skeleton computation by Voronoi diagram. *Computer Vision, Graphics, and Image Processing: Image Understanding 55*(3), 329–337.

Bulpitt, A. and N. Efford (1996). An efficient 3D deformable model with a self-optimising mesh. *Image and Vision Computing 14*(8), 573–580.

Chew, L. (1989). Constrained Delaunay triangulations. *Algorithmica 4*(1), 97–108.

Cohen, L. (1991). On active contour models and balloons. *Computer Vision, Graphics, and Image Processing: Image Understanding 53*(2), 211–218.

Deseilligny, M.-P., G. Stamon, and C. Suen (1998, May). Veinerization: a new shape description for flexible skeletonization. *IEEE Transactions on Pattern Analysis and Machine Intelligence 20*(5), 505–521.

Ferley, E., M.-P. Gascuel, and D. Attali (1997). Skeletal reconstruction of branching shapes. *Computer Graphics Forum 16*(5), 283–293.

Kass, M., A. Witkin, and D. Terzopoulos (1987). Snakes: active contour models. *International Journal of Computer Vision 1*(4), 321–331.

Kaufman, A. (1987). Efficient algorithms for 3D scanconversion of parametric curves, surfaces, and volumes. *Computer Graphics 21*(3), 171–179.

Kimmel, R., D. Shaked, N. Kiryati, and A. Bruckstein (1995). Skeletonization via distance maps and level sets. *Computer Vision and Image Understanding 62*(3), 382–391.

Lachaud, J. and A. Montanvert (1999). Deformable meshes with automated topology changes for coarse-to-fine three-dimensional surface extraction. *Medical Image Analysis 3*(2), 187–207.

Lam, L., S. Lee, and C. Suen (1992). Thinning methodologies – a comprehensive survey. *IEEE Transactions on Pattern Analysis and Machine Intelligence 14*, 869–885.

Loncaric, S. (1998). A survey of shape analysis techniques. *Pattern Recognition 31*(8), 983–1001.

Malladi, R., J. Sethian, and B. Vemuri (1995). Shape modeling with front propagation: a level set approach. *IEEE Transactions on Pattern Analysis and Machine Intelligence 17*(2), 158–175.

McInerney, T. and D. Terzopoulos (1995). A dynamic finite element surface model for segmentation and tracking in multidimensional medical images with application to cardiac 4D image analysis. *Computerized Medical Imaging and Graphics 19*(1), 69–83.

Mille, J., R. Boné, P. Makris, and H. Cardot (2005). 3D segmentation using active surface: a survey and a new model. In *5th Int. Conf. on Visualization, Imaging & Image Processing (VIIP)*, Benidorm, Spain, pp. 610–615.

Montagnat, J. and H. Delingette (2005). 4D deformable models with temporal constraints: application to 4D cardiac image segmentation. *Medical Image Analysis 9*(1), 87–100.

Montagnat, J., H. Delingette, and N. Ayache (2001). A review of deformable surfaces: topology, geometry, and deformation. *Image and Vision Computing 19*(14), 1023–1040.

Park, J.-Y., T. McInerney, and D. Terzopoulos (2001). A non-self-intersecting adaptive deformable surface for complex boundary extraction from volumetric images. *Computer & Graphics 25*(3), 421–440.

Shewchuk, J. (2002). Delaunay refinement algorithms for triangular mesh generation. *Computational Geometry 22*(1–3), 21–74.

Slabaugh, G. and G. Unal (2005). Active polyhedron: surface evolution theory applied to deformable meshes. In *IEEE Computer Vision and Pattern Recognition (CVPR)*, Volume 2, San Diego, USA, pp. 84–91.

Williams, D. and M. Shah (1992). A fast algorithm for active contours and curvature estimation. *Computer Vision, Graphics, and Image Processing: Image Understanding 55*(1), 14–26.

Zucker, S. and R. Hummel (1981). A three-dimensional edge operator. *IEEE Transactions on Pattern Analysis and Machine Intelligence 3*(3), 324–331.

Computational Modelling of Objects Represented in Images –
João Manuel R.S. Tavares & R.M. Natal Jorge (eds)
© 2007 Taylor & Francis Group, London, ISBN 978-0-415-43349-5

AD3RI a tool for computer – automatic Drusen detection

Fernando Moitinho[1], André Mora[1,2], Pedro Vieira[2] & José Fonseca[1,2]

[1]*Intelligent Robotics Center, Uninova, Portugal*
[2]*Faculty of Sciences and Technologies, New University of Lisbon, Portugal*

ABSTRACT: Drusens are indicators of macular degeneration, a disease characterized by accumulations of extra cellular materials under retina. The automatic study of the quantitative evolution of Drusen spots throughout a medical treatment constitutes a useful tool for ophthalmologists. Until now, drusen evaluation was done manually, based only on qualitative aspects. Also the analyses depended on the ophthalmologist and were not ease to reproduce.

Another important issue is that retina is not a plane surface and therefore light doesn't have a uniform distribution, producing images with non-uniform illumination and consequently with different contrast areas.

Computer aided tools that can help doctors in repetitive analysis are becoming more reliable and faster, contributing for its increasing acceptance. In this paper it is presented an algorithm based on smoothing splines for non-uniform illumination correction, and a software application based on Levenberg-Marquardt optimization algorithm for automatic modelling of drusen deposits in retina images. For improving the modelling process a customized algorithm for detecting image intensity maximums based on labelling gradient paths is presented. The results of applying this methodology to retina images containing medium sized Drusen will also be presented.

Keywords: Drusen detection, medical image processing, image modelling

1 INTRODUCTION

A very frequent disease appearing mainly on older people is called age-related macular degeneration (ARMD). It is considered the leading cause of irreversible blindness in developed countries and therefore is object of attention of ophthalmologists during a patient examination. The retina images are taken with imaging techniques like fundus photography or fluorescein angiography. However, in the future is expected that true color scanning laser ophthalmoscopes (Vieira, Manivannan et al. 2002) will be capable of providing better quality images. ARMD examination consists on detecting and quantifying yellowish spots in or around the macula. Those yellowish spots are called Drusens. According with Hageman et al. which gives in his paper (Hagemana, Luthertb et al. 2001) a good overview of Drusen biogenesis, those are accumulation of extra-cellular material beneath the basal lamina of the retinal pigment epithelium that builds small sized bubbles.

To improve the manually analysis made so far, a software application is proposed in this paper. This software application is able to automatically detect and model Drusens spots on retina images. The benefits of using this application are a more accurate analysis, a reduction of image non-uniform illumination

problems, independence of the image model to noise and the achievement of a reproducible process. The reliable evaluation of the disease in a sequence of images taken during a long-term treatment will certainly help clinicians to evaluate therapy effectiveness.

The main terminology to distinguish Drusen phenotypes used in grading systems such as the Wisconsin age-related maculopathy grading system (Klein, Davis et al. 1991) or the Alabama Age-Related Maculopathy Grading System (Curcio, Medeiros et al. 1998) is *hard* and *soft*. *Hard* Drusen are mainly characterized by small sized spots with less then $50\mu m$ diameter and sharp edges. *Soft* Drusens are medium sized ($\approx 250\,\mu m$) spots with smooth edges. So far, the software is parameterized to the detection of soft Drusens phenotype mainly because hard Drusen spots have reduced significance for the disease evaluation.

This paper is organized in seven sections, starting with a brief presentation of selected related works on Drusen detection. The following section will be dedicated to the Drusen Location Algorithm which is based on image gradient information (Mora, Vieira et al. 2004) to detect Drusen location and its area of influence. Sections four and five describe the modelling methodology, including non-uniform illumination correction. Section six presents the software application Automatic Drusen Deposits Detection in

Retina Images (AD3RI) that implements this methodology. The final section shows the results of applying these algorithms to retina images.

2 RELATED WORK

The problem of Drusen automatic unsupervised identification is not recent. There are papers dating from 1986 reporting to this subject. With the advances during the last decade in computing facilities and image processing techniques the interest in this subject has grown even though there are not many ongoing works on the subject. An overview of several previous works on Drusen detection is presented on the following paragraphs of this section.

The first important work was published by Peli and Lahav (Peli and Lahav 1986) in 1986 and further developed be Sebag et al. (Sebag, Peli et al. 1991) from New England Medical Center and Tufts University School of Medicine. It consisted in dividing the image in 8×8 pixel windows, computing all local windows threshold value, interpolate these values using a two-dimensional linear interpolation and then with the computed values applying a local window threshold. The results can be considered acceptable, but it has detected several small image irregularities as Drusen (false positives).

A fuzzy logic approach to the problem of Drusen detection was presented in 2000 by Thdibaoui et al. (Thdibaoui, Rajn et al. 2000) from the Université de Paris. The initial step of the algorithm is to divide the pixel into three classes: background, Drusens and ambiguous, according with their grey levels. The final step is to classify the ambiguous pixels into one of the other two classes using fuzzy logic approach.

A geodesic reconstruction for Drusen segmentation was proposed in 2001 by Sbeh et al. from University Paris-Dauphine (Sbeh, Cohen et al. 1997; Sbeh, Cohen et al. 2001). The algorithm, after applying a preprocessing step for contour enhancement detects regional maximums and from these maximums and neighbour pixels (geodesic neighbours) extracts the background. This is done by subtracting a pre-determined value to the regional geodesic maximum (maximum and neighbour pixels). A strategy for automatic determining the regional background value is presented. The overall results obtained can be classified as an adaptive threshold technique and therefore have the same problems of false positives as previous works presented.

With a study on the light reflectance of the macula, Smith et al. (Smith, Nagasaki et al. 2003) from Colombia University propose in 2003 an algorithm for levelling the background based on the elliptically concentric geometry of the reflectance on a normal macula. The Drusen segmentation and area measurements were performed by global threshold. The algorithm for background levelling produced good results and is well supported since it is based on the physical geometry of the macula. The results produced by the Drusen segmentation method are similar to the previous works that used image thresholds.

This small survey on previous works on Drusen detection shows that the use of other techniques rather then adaptive or global thresholds has not been studied yet. It is also important to notice that all these threshold methods have the same tendency for producing false positives, especially when the image has small irregularities. Computing Drusen total area (number of pixels) is the only proposed method for evaluating Drusen evolution along time using threshold methods.

3 DRUSEN LOCATION ALGORITHM

To reduce the processing time and achieve a more accurate model it was decided to give Drusen location points to the optimization algorithm. In order to achieve an accurate Drusen spots location, we propose a location algorithm based on labelling the maximum gradient path. In medium resolution images, there are more then just one pixel pointing to each intensity maximum. Therefore, it is possible to follow one or more ascending paths that inevitably reach an intensity maximum. The proposed algorithm consists in a labelling procedure similar to connect components labelling or to watershed segmentation (described in (Gonzalez and Woods 1992)), but optimized to maximums search.

The algorithm begins by determining the image gradient using a 3×3 Sobel operator, which evaluates intensity changes in the horizontal and vertical axis. The pixel gradient is a vector oriented to the ascending pixel (see Figure 1.b).

The first stage of the labelling procedure consists in examining every pixel in a top-left to bottom-right direction and assigning a new label to all pixels that are not marked yet. Whenever a label is assigned to a pixel it is propagated to the neighbour pixel that is in the gradient azimuth direction. This labelling continues on the propagated pixels until the next pixel is already marked. Figure 1.c presents the result of the labelling first stage.

Every time the label propagation finishes on the same label that is being propagated, that pixel is marked as an intensity maximum. In the same situation but when there is a different label, these are defined as being compatibles. This means that they belong to the same intensity maximum, i.e., the same Drusen spot.

The second stage of the labelling procedure is to apply the label compatibilities resulting on an image with as many labels as possible Drusens. The result is a segmented image where all pixels that contribute to a spot have the same label. Each group of pixels with the same label constitutes the spots area of influence.

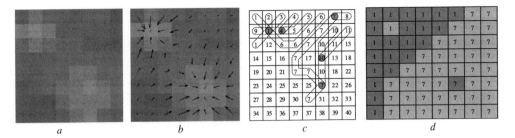

a   b   c   d

Figure 1. Example of Drusens center detection algorithm. (*a*) Original image (detail); (*b*) Image gradient; (*c*) Labelling 1st stage – initial label propagation – (with label propagation on the two upper rows); (*d*) Labelling 2nd stage – apply compatibilities – (two Drusens centered on the highlighted pixels).

It is important to notice that the algorithm is so far a non-parameterized algorithm. But a problem arises when flat valleys or flat hills occur. In these cases not all gradient paths end on the same intensity maximum pixel since there might exist more than one pixel for that intensity maximum. Consequently, the algorithm will generate more intensity maximums then it really should. For solving this problem, a merging method was adopted.

The merging method is more efficient but more time consuming. For every detected maximum, using a connected components approach, it is explored if there is another intensity maximum that can be connected by a path that doesn't goes lower than a predefined amplitude.

At this stage spots are detected and their area of influence defined. The AD3RI will use this information for the optimization algorithm in order to decrease processing time.

4 IMAGE MODELING

4.1 *Modeling function*

To model the images it is important to create a model of each detected Drusen spot. Thus, a mathematical function has to be chosen in order to model the Drusens spots. The genesis of Drusen deposits is an extra-cellular material accumulation between the basal lamina of the retinal pigment epithelium and the inner collagenous layer of Brush's membrane, causing the retinal pigment epithelium to raise with a smooth slop. Using the image two-dimension coordinates and the pixels intensity value it is possible to estimate a tri-dimensional surface where spots elevation can be perceived (Figure 2.a). It should be notice that it is an estimation and not a real tri-dimensional view since the depth coordinate is very dependent on illumination. Figure 2.a and 2.b shows the similarity between an isolated Drusen spot and a gaussian function, respectively.

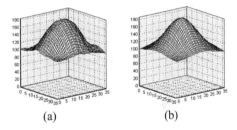

(a)   (b)

Figure 2. Drusen modeling (a) 3D original image view; (b) 3D Gaussian function view.

These observations motivated the use of a tri-dimensional model of the Gaussian function for modelling spots. The tri-dimensional Gaussian function is given by:

$$G(x,y) = a \times e^{-\left(\frac{X^2}{S_x} + \frac{Y^2}{S_y}\right)^{\frac{2}{d}}} + z_0$$

$$X = (x - x_0).\cos(\theta) + (y - y_0).\sin(\theta)$$

$$Y = -(x - x_0).\sin(\theta) + (y - y_0).\cos(\theta)$$

$$S_y = S_F.S_x$$

With parameters:

a – amplitude	θ – rotation
(x_0, y_0) – center coordinates	D – shape factor
Z_0 – background value	S_F – scale factor (S_x/S_y)
S_x – scale factor in x	S_y – scale factor in y

This is a modified version of the Gaussian function since it uses a shape factor d to allow another degree of freedom for the function to be adjusted to the image. Figure 3 shows the function's shape for different d values. The original Gaussian distribution can be obtained using $d = 2$.

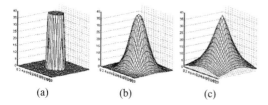

(a)   (b)   (c)

Figure 3. Gaussian function for different d values; (a) $d = 0, 1$; (b) $d = 2$; (c) $d = 3$.

4.2 Opitimization algorithm

For the drusen modeling Levenberg-Marquardt (Marquardt 1963) optimization algorithm was selected.

The chosen algorithm is a combination of steepest descent and the Gauss-Newton Method. When the current solution is far from the correct one the algorithm behaves like a steepest descent method. In the other case where the current solution is close to the correct solution, it becomes a Gauss-Newton method.

The Levenberg-Marquardt has become a standard of nonlinear least-squares routines (William H. Press 1999). For a given mathematical function depending on several parameters the algorithm adjusts those parameters in order to minimize the chi-square error function (χ^2).

$$\chi^2(a) = \sum_{i=1}^{N} \left[\frac{z_i - z(x_i; y_i; a)}{\sigma_i} \right]^2$$

This equation represents the distance between current solution ($z(x_i, y_i, a)$) and the correct one (z_i), with σ representing the standard deviation.

The Levenberg-Marquardt method is iterative: initiated at a starting point a_0, the method produces a series of vectors $a_1, a_2, \ldots$, that converge towards a local minimum. The difference between the first chi-square value (χ^2) and the new one at each iteration gives an idea of the optimization success. Another variable that shows the success of the procedure is λ. Because this value is directly implicated in the success of each iteration, λ decreases by a factor of ten if the new chi-square value is smaller than the previous, and grows by a factor of ten if not.

Because of the convergence problem that small variations make difference between actual chi-square and its previous we decided to impose two stop conditions. The first condition is if the chi-square value doesn't improve for four consecutive iterations and the second condition is if the chi-square value reaches a predefined minimum value.

4.3 Background correction

Because retina is not a plane surface, illumination has not an uniform distribution producing images with a

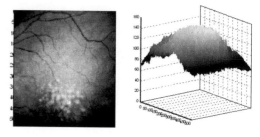

Figure 4. Image with non-uniform background.

non-uniform background. In figure 4 an example of an image with non-uniform background is shown. Since this problem can reduce the effectiveness of the Drusen modeling process. In the proposed methodology non-uniform background is estimated and removed.

To remove background we use a smoothing spline algorithm. The purpose of this algorithm is to smooth the image, so that just background become relevant. This algorithm has two input values: the grid interval and the level of smoothness desired. The grid interval is the length of the space between control points.

A removed background image is achieved from dividing the original image by the background image and then multiplying the resultant image by a factor of 100, so that the final image is centered on the intensity value 100.

4.4 Drusens modeling

The next step is to model Drusen spots. To accomplish a more accurate and quicker model, initial parameters values are given to the optimization algorithm. Those values are: center coordinates and maximum amplitude, determined by the Drusen Location algorithm presented before.

Another important issue is concerned with noise resulting from using image compression methods. These problems may affect the detection and the quantification process. Thus, the first step is to apply a noise reduction filter to reduce these effects. After testing other filters (median filter, k nearest-neighbour), the one that gave better results in reducing noise and especially *jpg* artefacts was the mean filter.

5 DEVELOPED APPLICATION

The AD3RI (Automatic Drusen Deposits Detection on Retina Images) is a C++ application developed for the automatic modelling of Drusen spots on retina images.

The AD3RI basic functionalities include: Image modelling; Background correction; Manual drusens location; Gaussian parameters I/O from text file; Save

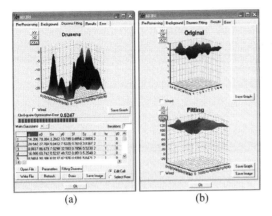

(a)   (b)

Figure 5.   (a) Drusens fitting panel; (b) Results panel.

modelled image; Save 3D graphs; Noise reduction filter.

The application is divided into five basic panels:

− Preprocessing – where noise reduction tool, Drusen location algorithm and image 3D view are located
− background – where background correction can be made
− Drusen Fitting – where Drusens spots are modelled
− Results – to visualize the modelling results
− Error – to mark the Drusen contour.

The gaussian function parameters are disposed in a table and can be entered automatically (using the values determined by the Drusen Location algorithm), manually or loaded from a text file. Figure 5.a shows Drusens fitting panel where a table can be seen with gaussian function parameters and some functions associated with the modelling procedure like fitting, refresh, draw and save image. The user has the possibility of saving the 3D graph view or the flatten modelled image.

Other important panels are, result and error tabs, where are summarized the results of the modelling process (figure 5).

6   EXPERIMENTAL RESULTS

For demonstrating results of the proposed Drusen detection and modelling algorithm, it will be applied to three different images. The images were taken using analogue Fundus photography and then digitalized using a scanner. The working images had approximately 1600×1100 pixels, but, in order to reduce the computation complexity, it has only been considered for the analysis a square region of interest with 540×540 pixels in the macula region.

The first step in the optimization process is background correction, executed by the algorithm

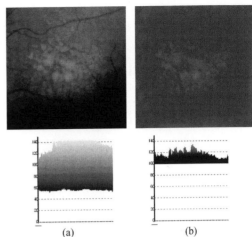

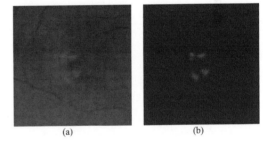

(a)   (b)

Figure 6.   (a) Image with non-uniform background; (b) Original image after background removal.

(a)   (b)

Figure 7.   (a) Original image after background correction; (b) Image modeling result.

explained in section 4.3 (background correction). The result can be seen in figure 6.

As it can be seen in figure 6.a background level is near 140 value, background correction levelled image and lower background value to 100, as can be seen in figure 6.b. Similar results were achieved for the other images.

Another feature of this correction method is the fact that it eliminates veins as well, resulting on an image with just the relevant elements that should be considered for drusen modeling.

The next procedure is to model the corrected image, with the initial values given by the drusens location algorithm. In figures 7 and 8 are presented the optimization result for three different images. The number of iterations taken by the Levenberg-Marquardt algorithm differs from an image to another, because the number of relevant drusen spots.

The results accomplished can be considered a good approximation of the real drusen spots. The process can be reproduced, independent of the clinician.

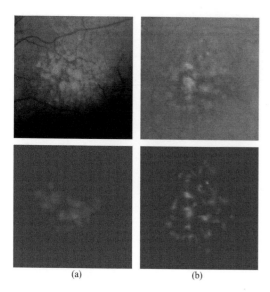

| (a) | (b) |

Figure 8. (a) Second image modelling process; (b) Third image modeling process;

7 CONCLUSIONS

The use of image processing techniques for Drusen automatic detection and quantification on retina images can significantly improve treatment effectiveness evaluation. In this paper, a three-step methodology for detecting and modelling of Drusen was proposed.

The first step is dedicated to the location of Drusen spots in the retina image used a segmentation technique based on the image gradient. This technique shown good results even when applied to images with non-uniform illumination. Iet detects Drusen spots in low or high contrast areas. And additionally, it provides Drusen area of influence and preliminary characterization that can be useful for future spots classification.

The second step consists in reducing the effects on non-uniform illumination by modelling the image background through a smoothing splines method and subtract it to the original image. This algorithm shown good results for the majority of images that were available.

The third step consists in modelling each Drusen spot, using the same optimization method with a modified Gaussian function. This can be a significant advance in Drusen detection algorithms since it can provide, not only shape consistent segmentation and more noise independence, but also a good spot characterization. Future work will be concentrated in tuning the algorithm parameters in order to improve the modelling quality and speed. The evaluation of other modelling functions for the modelling step is also expected to be undertaken. An automatic spot classifier for detecting non-Drusen spots and to classify Drusen spots in soft and hard classes is expected to be developed.

REFERENCES

Curcio, C. A., N. E. Medeiros, et al. (1998). "The Alabama Age-Related Macular Degeneration Grading System for donor eyes." *Invest Ophthalmol Vis Sci* **39**(7): 1085–96.

Gonzalez, R. and R. Woods (1992). *Digital Image Processing*, Addison-Wesley.

Hagemana, G. S., P. J. Luthertb, et al. (2001). An Integrated Hypothesis That Considers Drusen as Biomarkers of Immune-Mediated Processes at the RPE-Bruch's Membrane Interface in Aging and Age-Related Macular Degeneration. *Elsevier Science Ltd. – Progress in Retinal and Eye Research.* **20**: 705–732.

Klein, R., M. D. Davis, et al. (1991). "The Wisconsin age-related maculopathy grading system." *Ophthalmology* **98**(7): 1128–34.

Marquardt, D. W. (1963). "An algorithm for least-squares estimation of non-linear parameters." *Journal of the Society for Industrial and Applied Mathematics* **11**(2): 431–441.

Mora, A., P. Vieira, et al. (2004). *Drusen Deposits on Retina Images: Detection and Modeling.* MEDSIP-2004, Malta.

Pauleikhoff, D., M. J. Barondes, et al. (1990). "Drusen as risk factors in age-related macular disease." *Am J Ophthalmol* **109**(1): 38–43.

Peli, E. and M. Lahav (1986). Drusen Measurement from Fundus Photographs Using Computer Image Analysis. *Ophtalmology* **93**: 1575–1580.

Sbeh, Z. B., L. D. Cohen, et al. (2001). A New Approach of Geodesic Reconstruction for Drusen Segmentation in Eye Fundus Images. *IEEE TRANSACTIONS ON MEDICAL IMAGING.* **20**: 1321–1333.

Sbeh, Z. B., L. D. Cohen, et al. (1997). *An adaptive contrast method for segmentation of drusen.* International Conference on Image Processing.

Sebag, M., E. Peli, et al. (1991). Image analysis of changes in drusen area. *Acta Ophtalmologica.* **69**: 603–610.

Smith, R. T., T. Nagasaki, et al. (2003). "A Method of Drusen Measurement Based on the Geometry of Fundus Reflectance." *BioMedical Engineering OnLine* **2**(10).

Thdibaoui, A., A. Rajn, et al. (2000). *A fuzzy logic approach to drusen detection in retinal angiographic images.* 15th International Conference on Pattern Recognition, Barcelona, Spain.

Vieira, P., A. Manivannan, et al. (2002). "True colour imaging of the fundus using a scanning laser ophthalmoscope." *Physiological Measurement* **23**: 1–10.

William H. Press, S. A. T., William T. Vetterling, Brian P. Flannery (1999). *Numerical Recipes in C*, Cambridge University Press.

Computational Modelling of Objects Represented in Images –
João Manuel R.S. Tavares & R.M. Natal Jorge (eds)
© 2007 Taylor & Francis Group, London, ISBN 978-0-415-43349-5

Best multiple-view selection: Application to the visualization of urban rescue simulations

Pedro Miguel Moreira[1,4*], Luís Paulo Reis[2,4†] & A. Augusto Sousa[3,4‡]

[1] *ESTG-IPVC – Escola Superior de Tecnologia e Gestão de Viana do Castelo, Portugal*
[2] *LIACC – Laboratório de Inteligência Artificial e Ciência de Computadores da Univ. Porto, Portugal*
[3] *INESC-Porto – Instituto de Engenharia de Sistemas e Computadores do Porto, Portugal*
[4] *FEUP – Faculdade de Engenharia da Universidade do Porto, Portugal*

ABSTRACT: In this paper we address the problem of automatically computing a set of views over a simulated three dimensional environment. The viewing system aims at, for each moment, supplying the user with the most pertinent information in order to allow a good understanding of the evolving environment. Our approach relies on an innovative optimization architecture that enables intelligent optimization techniques based on simulated annealing and genetic algorithms. Reported experiments were performed in urban rescue scenarios from the RoboCup Rescue Domain. We outline the possible extension of the proposed architecture to other visualization problems and argue on how several problems within the fields of Visualization and Rendering can benefit from it.

1 INTRODUCTION

Automatic computation of best views is an important problem and has applications in several fields, such as: automatic object (Vázquez and Sbert 2003) and scene (Andújar et al. 2004) exploration, virtual cinematography (Drucker and Zelter 1995), image based modeling and rendering (Vázquez et al. 2003) and also in the improvement of rendering algorithms such as ray-tracing and radiosity. There are several computer graphics applications that can take benefit from the automatic computation of best-views. In medical applications, best views can be usefully used as starting points for 3D model inspection. Best-views can assist camera control in games by weighting objects with importance values. Using best-views as informative image descriptions for 3D model repositories can lead to very significant resource savings (e.g time and bandwidth) in the search and retrieval process. We refer the interested reader to a recent review of applications by Sbert et. al (Sbert et al. 2005).

In this paper we address the problem of automatically selecting a fixed set of views (best multi-view) over a three dimensional dynamical and evolving simulated environment. This multi-view should provide the user with useful information leading to an improved understanding of the on-going simulation. We report the application to the RoboCup Rescue Domain where we aim at obtaining, at each moment, the set of views which better describe the emergency situations and rescue operations.

We developed a model for estimating the multi-view quality. The problem is stated as an optimization problem. Finding optimal solutions is not usually feasible within the available time budget, specially if the solutions must be provided at interactive rates. Thus, our approach makes use of search heuristics to compute a best (sub-optimal) solution.

We describe a methodology that relies on the use of intelligent optimization techniques (Pham and Karaboga 2000) as tabu search, genetic algorithms and simulated annealing. Our approach is also based on an architecture comprising an optimization agent which operates autonomously from the main application. The underlying goal of is to make this architecture as general and modular as possible, allowing its use in several other distinct problems.

The rest of the paper is organized as follows. In Section 2 we briefly survey some relevant related work. Next, in Section 3, the RoboCup Rescue domain is described. Our approach, based on an optimization architecture is detailed in Section 4. Section 5 states the problem of automatically find a best multi-view over the application scenario. In Section 6 we give detail on relevant implementation issues and present the results. Finally, in Section 7, we conclude and point out future work.

---

* pmoreira@estg.ipvc.pt

† lpreis@fe.up.pt

‡ augusto.sousa@fe.up.pt

2 RELATED WORK

A fundamental concept in best view computation is *viewpoint quality*. Intuitively, viewpoint quality, can be defined as a measure of how pertinent is the obtained view to the scene understanding. Although intuitive, viewpoint quality does not have a precise definition. Sbert et al. (Sbert et al. 2005) state that a measure of viewpoint quality must consider: the number of visible surfaces, their area, orientation and distance in respect to the view parameters (e.g. position, direction, orientation and field-of-view). Despite their usefulness, purely geometric criteria may be insufficient to accurately estimate the goodness of a view. Taking into account non geometric information should lead to better results (Sokolov et al. 2006). Examples of such information are: object's importance to the user/task, perceptual factors (e.g. relative position in the obtained image) and lightning conditions.

In (Kamada and Kawai 1988) the authors describe a method which aims at minimizing the amount of degenerated faces in respect to orthogonal projection. Degenerated faces are those where edges project over the same straight line. For the purpose they suggest the minimization, over all the faces, of the maximum angle deviation between the normal vectors and the line of sight. Although the method performs interestingly for wire frame display it has a major drawback for more realistic rendering. As it is not aware of visibility, it does not take into account the amount of detail visible in the solution. Barral and Plemenos (Barral et al. 2000) present an extension of Kamada's method, formerly developed by Plemenos, which mainly consists in taking into account the amount of visible detail.

Colin (Colin 1988) presents a method, for scenes modeled as octrees, consisting in the maximization of the number of visible voxels.

Vázquez et al. (Vázquez et al. 2001) introduced *viewpoint entropy* as a metric for viewpoint quality. The viewpoint entropy concept is borrowed from Information Theory and the method aims at selecting the more informative view, i.e. the one with maximum entropy. In their formulation, information probabilities have correspondence on the ratio between the face's projected area and the area of the enclosing sphere surface where all viewpoints lie. Inherently the method takes into account the amount (number and size) of visible detail present in the solution. As brute force strategies are very time consuming, they developed an adaptive algorithm based on a heuristic search supported by a conservative viewpoint entropy predictor (Vázquez and Sbert 2003). Later (Vázquez and Sbert 2004) the use of OpenGL graphics hardware acceleration facilities has been explored leading to *on the fly* best view detection.

Lee et al. (Lee et al. 2005) introduced *mesh saliency* as perceptually-driven measure for viewpoint quality.

Figure 1. RoboCup Rescue Simulated Environment.

They argue that their measure allows for more visually pleasing results compared to those obtained using purely geometric criteria.

Object and scene exploration can be classified as *static* or *dynamic*. Static explorations compute a single view of the object or scene. For more complex scenes a single view is generally not enough. Thus, a minimal set of best views is computed and the camera is animated for the in-between positions. For static scenes, best views and paths can be computed off-line, whereas for dynamic scenes best view computation must occur in real-time. We refer to a recent survey on methods and data-structures for virtual world exploration (Sokolov et al. 2006).

Our problem differs from those we are aware of, as it aims at obtaining the best set of k simultaneous views which allow a good understanding of the ongoing simulation. Beyond the dynamics of the objects in the scene, their importance to scene understanding is subject to change during the simulation.

3 URBAN RESCUE SIMULATIONS

RoboCup was created as an international research and education initiative, aiming to foster artificial intelligence and robotics research, by providing a standard problem, where a wide range of technologies can be examined and integrated (Anonymous 2006). The huge success of the RoboCupSoccer international research and education initiative, led the RoboCup Federation to create the RoboCupRescue project focussing on Urban Search and Rescue (USAR) operations (Anonymous 2006).

The RoboCupRescue Simulation League consists of a simulated city (see Figure 1) in which heterogeneous simulated robots, acting in a dynamic environment, coordinate efforts to save people and property. Heterogeneous robots in a multi-robot system share a common goal, but have different abilities and specializations, adding further complexity and strategic options.

These systems can manifest self-organization and complex behaviors even when the individual strategies of all the robots are simple. The team-programmed robots are of three different types: Fire Brigades, Police Forces and Ambulance Teams. Fire Brigades are responsible for extinguishing fires; Police Forces open up blocked routes; and Ambulance Teams unbury Civilians trapped under debris. Each of these types of robots is coordinated by an intelligent centre responsible for communication and strategies. In order to obtain a good score, all these robots work together to explore the city, extinguish fires, and unbury Civilians.

There are several tools for visualizing RoboCup Rescue Simulations. Log viewers are used to track the evolution and result of rescue simulations. Some viewers have been written so far, enabling different viewing perspectives of the simulation, but all of them lack the functionality of a good debugging viewer (Arian Team 2005). Some teams have performed work in this area, but the need for a more comprehensive tool was only made more visible (Arian Team 2005). The Freiburg team, amongst others, has developed its own viewer, releasing it to the rescue community. Freiburg's 3D viewer (Kleiner and Göbelbecker 2005) is one of the most used by the community, second only to Morimoto's 2D viewer (Morimoto 2002), which is included in the official simulator package.

Our purpose was to develop a visualization tool to the Robocup Rescue Domain, that features a multi-view over the simulated environment. Camera positions are restricted to existing rescue agents or entities (such as buildings, police, fire brigades, etc.). Aerial views are also available. Users monitoring the rescue simulation should benefit from such tool since they are provided with a fixed (and small, e.g. four) number of views selected based on criteria that tries to optimize the relevance of the virtually captured imagery to the understanding of the evolving simulation. Our viewer is partially built on top of (Kleiner and Göbelbecker 2005).

4 OPTIMIZATION METHODOLOGY

We propose an optimization architecture relying on an optimization agent that operates autonomously from the main application. As showed in Figure 2, communication is achieved by means of a simple protocol (in the figure a simplified version is depicted). The optimization agent (OA) informs the visualization application (VA) that it is available with a connect message. The VA acknowledges the connection and sends a problem description. The OA requests data relevant to

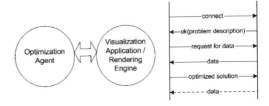

Figure 2. Optimization architecture.

the optimization process. As optimized solutions are computed, they are communicated to the VA which sets up and operates appropriately.

Further communication between the two processes should be well established by means of a proper language. Some desirable features of such a language are:

- Independence of implementation details;
- Small but easily extensible;
- Generality of its format and possible incorporation of different abstraction levels;
- Robustness, with simple data validation and override of usual errors.

The main advantages of this architecture are:

Generality The architecture is designed to be easily applied to a wide variety of problems.

Flexibility Its behavior should adapt to the demands of the visualization application.

Modularity The design of the supporting architecture enables its extension by modular pieces.

Portability Operating at a higher level and being independent from the implementation details of the visualization application and the rendering engine it is expected to be easily portable or used by many different systems.

As a main drawback there is some possible loss of performance comparatively to approaches were optimization is hardly embedded within the visualization processes or the rendering pipeline. This loss of performance is fundamentally due to the necessary communication overhead.

For time critical visualizations tasks an asynchronous behavior is desired as synchronism tends to degrade the performance of such systems. Producing solutions incrementally is also a desired behavior when synchronism is demanded.

5 BEST MULTI-VIEW SELECTION

The problem can be informally stated as:

In an urban rescue environment there are m visualization agents that can obtain views over the scene. The problem is to find the set of k views that better describe the simulation at each moment. The visualization agents are controllable in the sense that one can affect their viewing parameters.

We developed a model for estimating the quality of a multi-view. Our quality is a function of the *visibility*, *relevance*, *redundancy* and *eccentricity* of the entities represented in the set of selected views.

The optimization problem can be formalized as follows:

$$E = \{e_1, \dots e_n\}$$

$$V = \{v_1, \dots v_m\}$$

$$v_i = f(\vec{Pos}_i, \vec{VD}_i, \vec{VUP}_i, FoV_i)\ i \in \{1, \dots, m\}$$

$$MV = \{mv_1, \dots, mv_k\}\ \text{where}\ MV \subset V\ \text{and}\ mv_i \neq mv_j\ \forall i \neq j$$

MAXIMIZE: $Q(MV) =$
$$\sum_{j=1}^{k} \sum_{i=1}^{n} Vis(e_i^j).Red(e_i|MV)(W_1.Rel(e_i) + W_2.Ecc(e_i^j))$$

In the above formulation, E denotes the set of n entities that have relevance in the scene (buildings, agents, etc) and V is the set of different views (equals the number of agents/entities with viewing capabilities). Each view is characterized by usual camera parameters, as the position $\vec{Pos}_i$, view direction $\vec{VD}_i$, relative camera orientation $\vec{VUP}_i$ and field of view FoV_i. Aspect ratio is not being considered as it remains unchaged. A multi-view, MV, comprises a set of k distinct views from V.

The problem is to find the optimal MV set, with appropriate view parameters, that describes the rescue scenario with better quality. For quality Q estimation we have used the following criteria (note that e_i^j denotes the visual properties of an entity e_i in a image obtained by the view j)

Visibility: $Vis(e_i^j)$ This feature relates to the visibility of the relevant entities (i.e. the visible projected area). Several factors contribute to an entity's visibility such as the distance to the viewpoint, size, relative orientation, and also by how much partial occlusion it suffers from other objects.

Relevance: $Rel(e_i)$ A measure of how relevant is the entity for the purpose of the visualization. For example, if tracking emergency situations, a building on fire has a greater relevance than an unaffected building. The intrinsic importance of an object is also considered, e.g. hospitals, fireman headquarters, schools have more relevance than ordinary buildings.

Redundancy: $Red(e_i|MV)$ It is expected that the multiple set of views describe as much as possible distinct situations occurring during the simulation. Thus, redundant views over the same entities are penalized.

Eccentricity: $Ecc(e_j^i)$ A measure on how distant to the center of the image an object is displayed. This criterium has a perceptual foundation based on the observation that an user will pay more attention to image centered entities rather than to those projected in more peripherical regions.

6 IMPLEMENTATION & RESULTS

In order to evaluate usefulness of our approach, we are implementing the proposed architecture and conducting our first experiments. A scenario with 1035 relevant entities and 50 agents with viewing capabilities was used.

As it is expected that the quality function has several local optima, simulated annealing (Kirkpatrick et al. 1983) (SA) was chosen as the meta heuristic as it has the ability to continue the search even if a local optimum is found. Another contributing reason is its computational efficiency compared to other meta heuristics (e.g. genetic algorithms). At each iteration SA considers a neighbour of the current state, and probabilistically decides on moving to it or staying in the current state. The probabilities are chosen so that the system ultimately tends to move to states with better quality.

The exact visible set is necessary to determine whether an object is actually visible. We are using pre-processed from-region visibility and then using an item-buffer technique for exact visibility. Image analysis reveals the visible objects as well as coverage statistics. Histogram utilities from OpenGL can be used to improve performance. Further acceleration can be achieved if considered that objects with small visible areas are not relevant. In such case the above process can be done at smaller image resolutions. We are also investigating the use of OpenGL hardware occlusion queries. Eccentricity is estimated by projecting the center of the objects bounding boxes.

Our neighbour states are obtained, by exchanging one view from the multiview set or by changing one of the view parameters. In this experiment we are not changing the view position, but using the agent's position during the rescue simulation. We are also exploring the concept of adaptive neighborhood by adaptively defining the range of change in view parameters as a function of the evolution of quality. Note that for problems where the conditions vary with some kind of continuity, the optimization step (as well as visibility determination) can benefit from the exploration of spatial and temporal coherence.

In Figure 3 the evolution of quality, using simulated annealing, can be observed for an optimization of a multiview consisting in four views. The solution converged to a stable maximum on approximately 500 iterations.

A typical multi-view is shown in Figure 4.

7 CONCLUSIONS & FUTURE WORK

We have presented a method to automatically and dynamically select a best multi-view over a three

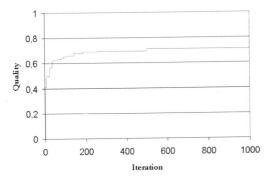

Figure 3. Evolution of the quality of the views along the optimization process.

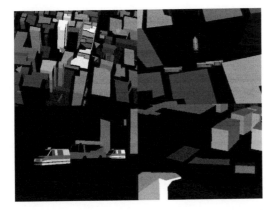

Figure 4. Typical 4-view obtained during the simulation.

dimensional scene. Best-view computation has several and interesting applications.

Our approach relies on an innovative optimization architecture which we have described and pointed out several advantages. The results are promising to demonstrate the adequateness and usefulness of our approach.

As future work we plan to define a framework based on the proposed optimization architecture suitable to be applied to several visualization and rendering problems.

A guideline of our on-going research concerns to automatic path advice leading to assisted exploration of scenes. Due to high model complexity or scene extension, unguided exploration can result on poor explorations, namely by missing relevant parts of the scene. The expected result will be a system that interactively generates and suggests to the user optimized paths according to the users interests, advertisements, etc.

Another important problem that could benefit from our approach is the remote rendering pipeline optimization (Teler and Lischinski 2001), where adaptive and optimized policies on when and what data transmit to the client, depending on the available bandwidth, storage and rendering capabilities, shall have significant impact on the system's performance.

ACKNOWLEDGMENTS

The first author acknowledges the support by European Social Fund program, public contest 1/5.3/PRODEP/2003, financing request no. 1012.012, medida 5/acção 5.3 – Formação Avançada de Docentes do Ensino Superior, submitted by Escola Superior de Tecnologia e Gestão do Instituto Politécnico de Viana do Castelo. This work was partially supported by the project Rescue: Coordination of Heterogeneous Teams in Search and Rescue Scenarios – FCT/POSI/EIA/63240/2004.

REFERENCES

Andújar, C., P. Vázquez, and M. Fairén (2004). Way-finder: Guided tours through complex walkthrough models. *Comput. Graph. Forum 23*(3), 499–508.

Anonymous (2006). Robocup international project homepage. available at: http://www.robocup.org/.

Arian Team (2005). Official arian team home page. Sharif University of Technology, http://ce.sharif.edu/~arian/, Visited in Nov, 10, 2005.

Barral, P., G. Dorme, and D. Plemenos (2000). Scene understanding techniques using a virtual camera. In A. de Sousa and J. C. Torres (Eds.), *Proceedings of Eurographics'00 (Short Presentations)*.

Colin, C. (1988). A system for exploring the universe of polyhedral shapes. In *Proceedings of Eurographics'88*, Nice, France, pp. 209–220. Elsevier Science Publishers.

Drucker, S. M. and D. Zeltzer (1995). Camdroid: a system for implementing intelligent camera control. In *SI3D '95: Proceedings of the 1995 symposium on Interactive 3D graphics*, New York, NY, USA, pp. 139–144. ACM Press.

Kamada, T. and S. Kawai (1988). A simple method for computing general position in displaying three-dimensional objects. *Comput. Vision Graph. Image Process. 41*(1), 43–56.

Kirkpatrick, S., C. D. Gelatt, and M. P. Vecchi (1983). Optimization by simulated annealing. *Science 220*(4598), 671–680.

Kleiner, A. and M. Göbelbecker (2005). Rescue3d: Making rescue simulation attractive to the public. Technical report, Institut für Informatik, Universität Freiburg. Available at http://kaspar.informatik.unifreiburg.de/~rescue3D/3dview.pdf.

Lee, C. H., A. Varshney, and D. W. Jacobs (2005). Mesh saliency. *ACM Transactions on Graphics (Proceedings of SIGGRAPH'05) 24*(3), 659–666.

Morimoto, T. (2002, Nov.). *How to Develop a RoboCupRescue Agent for RoboCupRescue Simulation System version 0*. The RoboCup Rescue Technical Committee.

Pham, D. and D. Karaboga (2000). *Intelligent Optimization Techniques*. London: Springer Verlag.

Sbert, M., D. Plemenos, M. Feixas, and F. Gonzalez (2005, May). Viewpoint quality: Measures and applications. In *Computational Aesthetics in Graphics, Visualization and Imaging*, Girona, Spain, pp. 185–192.

Sokolov, D., D. Plemenos, and K. Tamine (2006). Methods and data structures for virtual world exploration. *The Visual Computer Journal 22*(7), 506–516.

Teler, E. and D. Lischinski (2001). Streaming of complex 3D scenes for remote walkthroughs. *Computer Graphics Forum (Proceeding of Eurographics '01) 20*(3), 17–25. ISSN 1067-7055.

Vázquez, P.-P., M. Feixas, M. Sbert, and W. Heidrich (2001). Viewpoint selection using viewpoint entropy. In *VMV*, pp. 273–280.

Vázquez, P.-P., M. Feixas, M. Sbert, and W. Heidrich (2003). Automatic view selection using viewpoint entropy and its applications to image-based modelling. *Comput. Graph. Forum 22*(4), 689–700.

Vázquez, P.-P. and M. Sbert (2003, Jan). *Lecture Notes in Computer Science*, Chapter Fast Adaptive Selection of Best Views, pp. 295–305. Number 2669. Springer Verlag. (Proc. of ICCSA'2003).

Vázquez, P.-P. and M. Sbert (2004). On the fly selection of best views using graphics hardware. In *4th IASTED International Conference on Visualization, Imaging and Image Processing (VIIP 2004)*.

Computational Modelling of Objects Represented in Images –
João Manuel R.S. Tavares & R.M. Natal Jorge (eds)
© 2007 Taylor & Francis Group, London, ISBN 978-0-415-43349-5

Automatic vertebra detection in X-ray images

Daniel C. Moura
INEB – Instituto de Engenharia Biomédica, Laboratório de Sinal e Imagem, Porto, Portugal
Instituto Politécnico de Viana do Castelo, Escola Superior de Tecnologia e Gestão, Viana do Castelo, Portugal

Miguel V. Correia & Jorge G. Barbosa
INEB – Instituto de Engenharia Biomédica, Laboratório de Sinal e Imagem, Porto, Portugal
Universidade do Porto, Faculdade de Engenharia, Dep. Eng. Electrotécnica e de Computadores, Porto, Portugal

Ana M. Reis, Manuel Laranjeira & Eusébio Gomes
Instituto de Neurociências, Porto, Portugal

ABSTRACT: In this paper we will describe our experiments with x-ray image analysis for vertebra detection in juvenile/adolescent patients with idiopathic scoliotic spines. We will focus on detecting vertebrae location in a anterior-posterior x-ray image in a fully automatic way. For accomplishing this, we propose a set of techniques for (i) isolating the spine by removing other bone structures (e.g. ribs), (ii) detecting vertebrae location along the spine using an hierarchical and progressive threshold analysis, and (iii) detecting vertebrae lateral boundaries.

1 INTRODUCTION

In this paper we will describe our experiments with x-ray image analysis for vertebra detection. The input of the image analysis process is a pair of high-resolution grayscale images obtained by x-ray examinations. Both images are obtained at the same time and capture the spine of a given person, although from different perspectives: anterior-posterior (Figure 1 and lateral (Figure 2).

Our goal is to detect the 3D geometric location of each vertebra for constructing a 3D model of the spine. We intend to accomplish this by using a common and affordable diagnosis examination such as x-ray images. The main objective is to enable physicians to visualise in a 3D perspective a model that approximates the spine of their patients, obtained after processing of standard diagnosis examinations already available. Additionally, the vertebrae detection process should be able to deal with examinations of juvenile/adolescent patients with idiopathic scoliosis (condition that involves an abnormal side-to-side curvature of the spine). However, the methods presented here are not yet prepared for handling severe conditions of scoliosis.

Some work has been developed in the area of vertebra detection. A considerable part of the methods developed for vertebra segmentation in x-ray images need a set of images manually labeled by experts.

These images are used as training sets for the program to build a model of the vertebrae. This model is then used to determine vertebrae location and form in other images (de Bruijne and Nielsen 2004b; de Bruijne and Nielsen 2004a; Zamora, Sari-Sarrafa, and Long 2003; smyth, Taylor, and Adams 1999; Scott, Cootes, and Taylor 2003). Additionally, research is being developed in different types of examinations, such as, DXA scans (Smyth, Taylor, and Adams 1999; Scott, Cootes, and Taylor 2003) or CT scans (Ghebreab and Smeulders 2004). Apart from that, the authors usually choose to segment a small set of vertebrae, like the lumbar or the cervical. Our work differs from the previous, since we use x-ray images only, we intended to detect the location of the maximum number of vertebrae possible, and we do not need to have access to images labelled by experts. Using these same principles, Benameur and Pomero already were able to achieve interesting results. Benameur was able to reconstruct the lumbar and part of the thoracic spine using frontal and lateral x-rays with little user intervention (Benameur, Mignotte, Labelle, and Guise 2005). The user has to mark two landmarks in one of the vertebrae in both x-rays and the rest of the process is automatic. Pomero accomplished to reconstruct the vertebral body from T1 to L5 with a semi-automated method (Pomero, Mitton, Laporte, de Guise b, and Skalli 2004). The user has to identify the corners of every vertebra in both lateral and front projections.

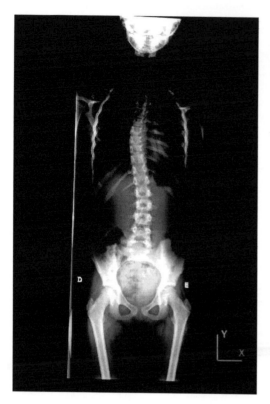

Figure 1. X-Ray anterior-posterior (AP) projection.

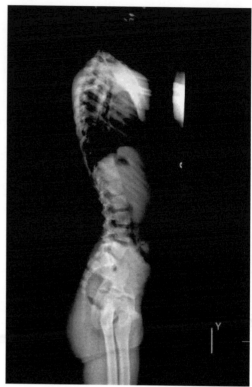

Figure 2. X-Ray lateral projection.

However, this process may be speeded-up using an algorithm that automatically calculates part of the landmarks based on the landmarks already marked by the user.

The method we propose here, tries to minimise user interaction. We are aiming for detecting the most of the vertebral body without the user having to insert any landmark in order to eliminate possible user related errors. For accomplishing this, we will start by analysing the anterior-posterior (AP) image. As we can see in Figures 1 and 2, the AP projection is much richer in information than the lateral projection, in it one is able to see and identify almost every vertebra. On the other hand, in the lateral projection the rib cage and the arms make it difficult to identify vertebrae, even for a human expert. Therefore, our strategy is to start by analysing the AP image to obtain the X and Y coordinates. We will then try to obtain the vertebrae depth (Z coordinate) using information about the curvature of the exterior boundary in the lateral projection.

In the next sections we will focus on the analysis of the front perspective image and we will demonstrate how we were able to determine vertebrae positions in the XY plane.

2 DETECTING VERTEBRA POSITIONS IN THE FRONT PERSPECTIVE

The x-ray images that feed the image analysis process have much more information besides the spine. There are a lot of bones structures present in the image that we do not need and that may difficult detecting vertebrae. Therefore, the first step consists in isolating the spine in order to remove undesired information, such as, the ribs, the head and legs. After isolating the spine we determine where each vertebra begins and ends along the spine (along the Y axis). For doing this, we "walk" through the spine searching for discontinuities that may indicate vertebrae limits. Finally, having identified the Y limits of every vertebra, we are able to determine their lateral limits using local information. In the next subsections, we will describe the process of vertebra detection in detail by addressing the issues of spine isolation and limits detection separately.

2.1 Isolating the spine

Isolating the spine is a crucial step in the process of vertebra detection. The main goal is to remove the major bone structures that are not vertebrae and whose presence may cause errors in the following steps.

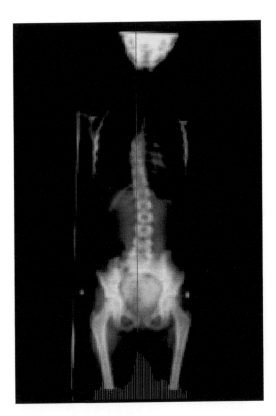

Figure 3.   Body center detection.

Next, we try to isolate the head, spine and legs, removing all other bone structures (e.g. ribs, clavicles, arms). For achieving this, we start by thresholding the image to remove the objects with low intensity. Usually, vertebrae have the pixels with highest intensity in their neighbourhood, although this brightness varies a lot among them. Therefore, the threshold is done locally, line by line. This operation is able to isolate part of the spine, but fails in areas where other bone structures have more intensity than the vertebrae. To solve this problem we analyse the image from the bottom to the top (legs to head) and we monitor the width of the body. From the moment that the width starts to decrease (more or less at the hips) we constrain its increase. This will allow to ignore ribs and other structures that are considerably wider. We also control the centre evolution to prevent deviations. Of course, with this constrain the head "becomes" very narrow, which difficults differentiating it from the spine. To avoid this, we repeat the same process, but now from the top to the bottom (head to legs) and we merge the two results. The result is presented in Figure 4a (after using morphing operations for closing some holes). Finally, we obtain the image illustrated in Figure 4b by upsampling the mask and performing a bitwise AND operation with the original image.

Removing the head, hip and legs is then accomplished by detecting large concentration of bright pixels in the top and bottom of the image respectively.

Currently, we are working with high resolution images (2448 × 3264) with very high detail but considerable noise. For detecting a big object, such as the spine, we built a multiresolution Gaussinan pyramid. This pyramid allow us to analyse the image at lower resolutions and therefore with less detail. Small objects disappear and the image is more regular, which facilitates detecting the spine boundaries. In Figure 3 we present the result of down sampling the original image 5 times.

In order to isolate the spine, we first have to know its location. As we can see in the previous figures, the image columns with more bright pixels are the columns where the spine is located. This happens because bright and large bone structures like the spine, head and hips are vertically aligned. Therefore, for determining the spine location in the X axis, we start by counting the number of bright pixels in every column and then we select the column with the highest counting. In spite of the simplicity of this technique, it turned out to be very robust in the available sample of images. Figure 3 shows a graphic of the column count (at the bottom) and a vertical line representing the selected centre in the X axis.

2.2 Detecting vertebrae limits in the Y axis

Having isolated the spine, the next task is to divide it in vertebrae. If we look closely, we can see that vertebrae are usually bright and the disks that separate them have lower intensity. Based on this observation, we built an algorithm that detects discontinuities along the spine and tries to figure out if they may indicate the presence of a disk separating vertebrae. However, vertebrae intensity vary a lot: cervical vertebrae usually have low intensity and lumbar vertebrae usually present very high intensity. This makes it difficult to classify regions as vertebrae because we cannot define pattern levels of intensity. The intensity of a vertebra depends of its position and of the image acquisition equipment. For tackling this problem our algorithm uses a progressive thresholding approach. The algorithm starts by counting the number of pixels per row at a very low threshold. Then, the threshold value is incremented at a slow rate and the counting process is repeated. Figure 5 illustrates in the right side the result of applying this technique, although we only included some threshold values for demonstration proposes. As we may observe, with low threshold levels we are able to isolate vertebrae with low intensity (typically at the cervical) and with higher threshold levels we accomplish to detect vertebrae with higher intensity.

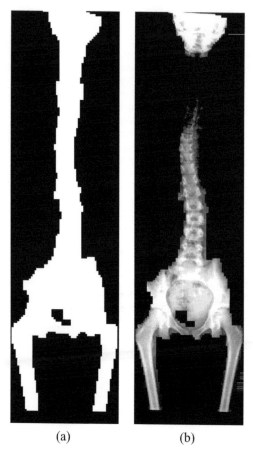

(a)   (b)

Figure 4.   Spine isolation. Isolation mask (a) and the result
of applying the mask to the original image (b).

Nevertheless, this algorithm has two issues that
must be solved in order to correctly divide the spine
into vertebrae: (i) vertebrae may become shorter and
shorter while incrementing the threshold, and (ii) vertebrae may be divided in several smaller regions due
to intensity variations along vertebrae. For handling
these problems we decided to use a tree data structure
to store the regions that the algorithm detects. Every
time a new region is found, it is added to the tree as a
child of the smallest region that entirely encloses the
new region. In order to control the tree size and to overcome the problem (i), before increasing the threshold
to search for new regions, we prune the tree by removing the leafs which have no siblings. Leafs with no
siblings are not interesting because they do not divide
the parent region. At most, they reduce the size of the
parent region which is not interesting because vertebrae should have maximum height in order to stay close
to each other. By the end of the algorithm, the tree is
fully constructed and its leafs should represent vertebrae, unless some vertebrae were over-divided. For

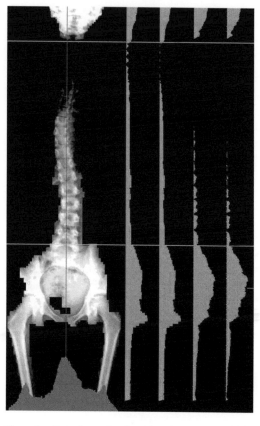

Figure 5.   Detecting nodes along the spine (thresholds from
left to right: 32, 48, 176, and 192).

detecting over-divided vertebrae we do two tests: (i) we
check if the gaps between vertebrae are not too large,
and (ii) we determine if the vertebra size is consistent
with its adjacent vertebrae (e.g. if it is not too small
compared to its largest adjacent vertebra). Whenever
one of the previous situations is detected, we test if the
leaf's parent is a better candidate for that vertebra. If
that is the case, we remove all the leaf's parent childes,
transforming the parent into a leaf and therefore in a
vertebra.

2.3   Detecting vertebrae limits in the X axis

After detecting where the vertebrae are located along
the spine, we must detect where they start and end
along the X axis. This operation may be more difficult
than what it seems because part of the ribs may still
be attached to vertebrae in the processed image. This
happens when ribs also show high intensity levels and
the spine isolation method is not precise enough to get
rid of them.

For detecting vertebrae X limits we divide them in
several clusters along its width (currently we divide it

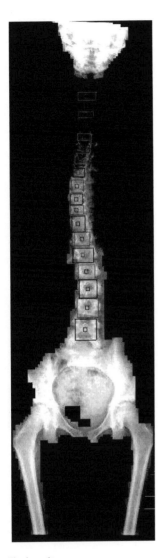

Figure 6. Final result.

nearest 4 vertebrae. If we detect a considerable deviation of the vertebra centre or an unexpected change in width, we try different combinations of the three candidate clusters and we select the ones that best fit the conditions.

Finally, we optimise the results by finding inside the elected clusters the largest concentration of bright pixels. Only then the process of detecting the X limits is completed. In Figure 6 we may see the nodes fully identified with the centre marked with red small squares.

The next step would be detecting the Z coordinate of each vertebra. For accomplishing this, we intend to use the body curvature that is observable in the lateral perspective (Fig. 2) and the already calculated Y coordinate, which tell us where to find each vertebra along the spine.

3 CONCLUSIONS

In this paper, we have proposed a set of techniques for detecting the vertebrae location in x-ray images in a fully automatic way. We started by isolating the spine for removing other bones structures. We then used a progressive thresholding algorithm for detecting vertebrae along the spine, which uses a tree data structure to store regions that may correspond to vertebrae. After pruning the tree, its leafs have the vertebrae location in the Y axis. Finally, the X boundaries of each vertebra is determined by performing an intensity analysis along the vertebra width.

So far, we have obtained promising results for detecting vertebrae in the anterior-posterior projection. Our next step is to improve the present process using domain specific information, such as, a spine model. We will then try to detect vertebrae location in the lateral projection, and use all captured features to produce a 3D model of the spine.

in 15 clusters). Then, we rank the clusters according to their intensity levels. Intensity levels are calculated using an exponential scale to give more preponderance to very high levels. This allow us to distinguish between clusters with high intensity structures surrounded by low intensity pixels, and more homogeneous clusters with average intensity levels. We then select the first three clusters with more intensity and we elect them as candidates for being the X limits. One of these candidates will represent the start of the vertebra and the other will represent the end. Initially, the two more intense clusters are selected. Then, for each vertebra, we compare its width and X centre with its

REFERENCES

Benameur, S., M. Mignotte, H. Labelle, and J. A. D. Guise (2005, December). A hierarchical statistical modeling approach for the unsupervised 3-d biplanar reconstruction of the scoliotic spine. *IEEE Trans Biomed Eng. 52*(12), 2041–2057.

de Bruijne, M. and M. Nielsen (2004a). *Image segmentation by shape particle filtering*, Chapter III, pp. 722–725. International Conference on Pattern Recognition. IEEE Computer Society Press.

de Bruijne, M. and M. Nielsen (2004b). *Shape particle filtering for image segmentation*, Volume 3216 of *Lecture Notes in Computer Science*, Chapter I, pp. 168–175. Springer.

Ghebreab, S. and A. Smeulders (2004, October). Combining strings and necklaces for interactive three-dimensional segmentation of spinal images using an integral deformable spine model. *IEEE Trans Biomed Eng. 51*(10), 1821–9.

Pomero, V., D. Mitton, S. Laporte, J. A. de Guise b, and W. Skalli (2004, March). Fast accurate stereoradiographic 3d-reconstruction of the spine using a combined geometric and statistic model. *Clinical Biomechanics 19*(3), 240–247.

Scott, I., T. Cootes, and C. Taylor (2003). *Shape particle filtering for image segmentation*, Volume 2732 of *Lecture Notes in Computer Science*, pp. 258–269. Springer.

Smyth, P., C. Taylor, and J. Adams (1999). Vertebral shape: Automatic measurement with active shape models. *Radiology 211*(2), 571–578.

Zamora, G., H. Sari-Sarrafa, and R. Long (2003). Hierarchical segmentation of vertebrae from x-ray images. In *Med Imaging: Image Process*, Volume 5032, pp. 631–642. SPIE Press.

Computational Modelling of Objects Represented in Images –
João Manuel R.S. Tavares & R.M. Natal Jorge (eds)
© 2007 Taylor & Francis Group, London, ISBN 978-0-415-43349-5

Line-based pose estimation through rotational motion analysis

A. Navarro & J. Aranda

Technical University of Catalonia, ESAII Department, Barcelona, Spain

ABSTRACT: Studying the relation between changes in image features derived by 3D transformations of a rigid object is a useful tool to deal with the 2D-3D pose estimation problem. Some applications in which the relative position and orientation of a moving object with respect to a fixed camera is necessary, may take advantage of this relation. Our purpose in this paper is to describe some properties of movement analysis over projected features of objects represented by lines. Then, having knowledge of the transformations applied to the object, we show that it is possible to estimate its orientation with only two different rotations, and also its position in the case length information is provided.

1 INTRODUCTION

Knowledge of 3D scene information obtained from a video camera is of relevant importance in many applications. The case where the position and orientation of the camera with respect to other objects in the scene are supplied allows to perform a diversity of 3D tasks in which 2D data are the only inputs. Some action-perception applications as robot navigation or manipulation rely in this kind of sensory input. The 2D-3D pose estimation problem serves to map this relation estimating the transformation between coordinate frames of objects and the camera.

There are several methods proposed to estimate the pose of a rigid object. The first step of their algorithms consists on the identification and location of some kind of features that represent an object in the image plane. Most of them rely on feature points and apply closed-form or numerical solutions depending on the number of correspondences between objects and image features. A solution for three non-coplanar and non-collinear points was proposed by Fischler & Bolles (1981) and Hung et al. (1985) proved that four coplanar and non-collinear points also give a unique solution using closed-form methods. On the other hand, Lowe (1987) and Haralick (1989) applied numerical optimization techniques and showed its usability for real-time algorithms. These point-based methods are commonly used and have been extensively studied.

Lines, however, are the features of interest in this work. As higher-order geometric primitives, describe objects where part of its geometry is previously known. These kinds of features have been incorporated to take advantage of its inherent stability to solve pose estimation problems. Approaches as the one by Navab & Faugeras (1993) proposed a solution using line correspondences representing them as Plücker lines, Homer (1991) discussed the problem for line-plane correspondences and Kumar & Hanson (1994) and Ansar & Daniilidis (2003) estimated the pose from a set of lines and studied its sensitivity. Other approaches as the one proposed by Liu et al. (1990) combine line and point features.

In the case of image sequences motion and structure parameters of the scene can be determined. Correspondences between selected features in successive images must be established. This provides motion information and can be used, as is our case, to estimate the pose of a moving object. The uniqueness of the structure and motion was discussed by Holt & Netravali (1996) for combinations of lines and points correspondences, and their result was that three views with a set of correspondent features, two lines and one point, or two points and one line give a unique solution.

It is of our interest the analysis of changes in the image plane determined through selected feature correspondences induced by 3D transformations applied to objects. Some properties of these motions are useful to estimate the pose of an object addressing questions as the number of movements or motion patterns required which give a unique solution. Straight lines are the features to be used in this paper and we show how knowing the transformations applied, it is possible to estimate their relative orientation with only two different rotations. The relative position can also be estimated if the length information of the line segment is provided.

The remainder of this paper introduces an analysis of motion from line correspondences, its advantages as a robust method for pose estimation and applications. Then a mathematical description of the proposed pose estimation algorithm using only rotations is presented, followed by a study of motion patterns that provide direct solutions, to finalize with concluding remarks.

2 MOTION ANALYSIS OF LINE FEATURES

In this work straight lines were the features extracted from a sequence of images. They provide motion information through its correspondences. There are several approaches that use this kind of features to estimate motion parameters. Some early works solve a set of nonlinear equations, as the one by Yen & Huang (1983), or use iterated extended Kalman filters, as showed by Faugeras et al. (1987), through three perspective views. Ansar & Daniilidis (2003) combined sets of lines and points for a linear estimation, and Weng et al. (1992) discussed the estimation of motion and structure parameters studying the inherent stability of lines and explained why two views are not sufficient.

An important property in using lines, as was reported by Mitiche et al. (1986), is their angular invariance between them. Then, our purpose was to study this property to provide a robust method to solve pose estimation problems. It is possible to compute the position and orientation of an object through the analysis of angular variations in the image plane induced by its 3D rotations with respect to the camera. It can be seen as an exterior orientation problem where objects in the scene are moved to calculate their pose.

Some action-perception applications can be seen as a fixed camera visualizing objects to be manipulated. In our case, these objects were represented by lines. Three views after two different rotations generate three lines in the image plane, each of them define a 3D plane called the projection plane of the line. These planes pass through the projection center and their respective lines. Their intersection is a 3D line that passes through the origin of the perspective camera frame and the centroid of the rotated object, as seen in Figure 1.

The motion analysis of angular variations between lines permitted us to estimate the pose of a given object. Therefore, we propose a robust method to compute the orientation through rotations. These rotations must be known. They could be sensed or given and controlled, as is the case of robotic applications. This information also gives rise to address the uniqueness issue and study motion patterns that permit to solve the problem in an easier way. Chen (1991) addressed questions that are useful for our analysis such as the number of movements needed, and the motion patterns required to obtain a unique solution for the relative

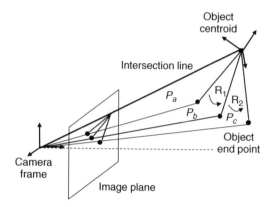

Figure 1. A 3D line through the origins is the intersection of the projection planes after two rotations of the object.

orientation and consequently to the relative position. These questions were applied in hand-eye calibration problems. In the case where the camera is not fixed to a robot gripper, these questions are similar to ours.

3 POSE ESTIMATION ALGORITHM

Motion analysis of feature lines was the base of our pose estimation algorithm. In this case known 3D rotations of a line and its subsequent projections in the image plane were related to compute its relative position and orientation with respect to a perspective camera. Vision problems as feature extraction and line correspondences are not discussed and we suppose the focal distance f as known. Our goal is, having this image and motion information, estimate the pose of an object represented by feature lines with the minimum number of movements and identify patterns that permits to compute a unique solution without defined initial conditions.

3.1 Mathematical analysis

A 3D plane is the result of the projection of a line in the image plane. It is called the projection plane and passes through the projection center of the camera and the 3D line. This 3D line, in this case, is the representation of an object. With three views after two different rotations of the object, three lines are projected in the image plane. Thus three projection planes can be calculated. These planes are P_a, P_b and P_c, and their intersection is a 3D line that passes through the projection center and the centroid of the rotated object. Across this line a unit director vector v_d can be determined easily knowing f and the intersection point of the projected lines in the image plane. Our intention is to use this 2D information to formulate angle relations with the 3D motion data.

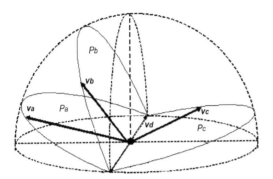

Figure 2. Unit vectors Va, Vb and Vc are constrained to lie on planes Pa, Pb and Pc respectively. Their estimation can be seen as a semi sphere where their combination must satisfy the angle variations condition.

Working in the 3D space permits to take advantage of the motion data. In this case where the object is represented by a 3D line, the problem could be seen as a unit vector across its direction that is rotated two times. In each position of the three views this unit vector lies in one of the projection planes as seen in Figure 2. It first is located in P_a, then it rotates an angle α_1 to lie in P_b and ends in P_c after the second rotation by an angle α_2. To estimate the relative orientation of the object we first must obtain the location of three unit vectors, v_a, v_b and v_c, that coincide with the 3D motion data and lie on their respective planes. To do this we know that the scalar product of:

$$v_a . v_b = \cos \alpha_1 \tag{1}$$

$$v_b . v_c = \cos \alpha_2 \tag{2}$$

Calculating the angle γ between the planes formed by $v_a v_b$ and $v_b v_c$ from the motion information, we have

$$(v_a \times v_b)(v_b \times v_c) = \cos \gamma \tag{3}$$

And applying vector identities

$$v_a . v_c = \cos \alpha_1 \cos \alpha_2 - \cos \gamma \tag{4}$$

With the set of equations conformed by (1), (2) and (4) we have the three unit vectors to calculate. However, there is not a unique solution, thus constraints must be applied.

3.2 Projection planes constraint

There are many possible locations where the three unit vectors can satisfy the equations in the 3D space. To obtain a unique solution unit vectors v_a, v_b and v_c must be constrained to lie in their respective planes. Unit vector v_a could be seen as any unit vector in the

plane P_a rotated through an axis and an angle. Using unit quaternions to express v_a we have

$$v_a = q_a v q_a^* \tag{5}$$

where q_a is the unit quaternion applied to v_a, q_a^* is its conjugate and v is any vector in the plane. For every rotation about an axis n, of unit length, and angle Ω, a corresponding unit quaternion $q = (\cos \Omega/2, \sin \Omega/2\, n)$ exists. Thus v_a is expressed as a rotation of v, about an axis and an angle by unit quaternions multiplications. In this case n must be normal to the plane P_a if both unit vectors v_a and v are restricted to be in the plane.

Applying the plane constraints and expressing v_a, v_b and v_c as mapped vectors through unit quaternions, equations (1), (2) and (4) can be expressed as a set of three nonlinear equations with three unknowns

$$q_a v_d q_a^* q_b v_d q_b^* = \cos \alpha_1 \tag{6}$$

$$q_b v_d q_b^* q_c v_d q_c^* = \cos \alpha_2 \tag{7}$$

$$q_a v_d q_a^* q_c v_d q_c^* = \cos \alpha_1 \cos \alpha_2 - \cos \gamma \tag{8}$$

The vector to be rotated is v_d, which is common to the three planes, and their respective normal vectors are the axes of rotation. Extending the equations (6), (7) and (8), multiplying vectors and quaternions, permits to see that there are only three unknowns which are the angles of rotation Ω_a, Ω_b and Ω_c.

3.3 Relative position and orientation estimation

Applying iterative numerical methods to solve the set of nonlinear equations, the location of v_a, v_b and v_c with respect to the camera frame in the 3D space are calculated. Now we have a simple 3D orientation problem that can be solved easily by a variety of methods as least square based techniques. However, in the case where motions could be controlled and selected movements applied, this last step to estimate the relative orientation would be eliminated. Rotation information would be obtained directly from the numerical solution. If we assume one of the coordinate axes of the object frame coincide with the moving unit vector and apply selected motions, as one component rotations, a unique solution is provided faster and easier.

The estimation of the relative orientation serves now to compute the relative position. It is necessary to track the projection of the end point of the 3D line that represents the object. The relation between the 3D position of this end point, (X, Y, Z), and its projection in the image plane, (x, y), can be expressed as $x = fX/Z$ and $y = fY/Z$. In our case where we know the 3D transformations applied to the object and its image projections, the relative position can be easily calculated rotating them to camera frame coordinates and by relating point

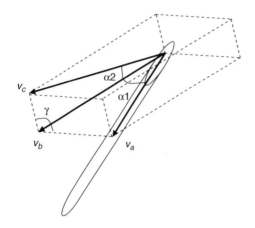

Figure 3. Two selected rotations through normal axes simplify computations.

position differences ΔX, ΔY and ΔZ and their projections Δx and Δy. Thus two views are sufficient to obtain the position of the centroid if the length to its end point is provided.

3.4 Uniqueness of solution and motion patterns analysis

The pose estimation algorithm explained above guarantee a unique solution. With two views, there would be many combinations of unit vectors constrained to their planes that satisfy the angular variations condition. The third view constrains the set of combinations, thus the solution is unique. It is estimated from two rotations and is considered as a robust orientation method due to the use of angle between lines. A pattern of motions analysis is useful in the case movements are controlled and can be selected to simplify calculations. This is the case of robotic surgery.

In this case where small rotations in different axes provide a unique solution and the object is represented by a 3D line, some selected rotations provide faster computations. If it is assumed that one of the axes of the coordinate frame of the object is the unit vector that describes the 3D line direction, with two different one component rotations, the relative orientation result comes directly from the numerical solution. With normal axes of rotations the angle between the planes formed by $v_a v_b$ and $v_b v_c$ is $\pi/2$, as seen in Figure 3. It also serves to reduce calculations. This is relevant when applications require real time performance, providing fast results which in many cases should be updated frequently.

4 SIMULATIONS

The angular variations analysis in the image plane due to 3D transformations permits to study motions

and patterns that can provide a direct solution. Rotations applied to the object must be large enough to be appreciated through image changes. This improves the feature extraction process and its correspondences establishment. However, in some applications is only possible to perform little movements. The possibility to observe the changes induced by different motions of the object and relate them is an important issue to tackle this problem. At the same time, proving our proposed algorithm in diverse cases permits us to validate it in different applications. The simulations performed served to carry out this study.

Simulations were performed through a graphics interface application where 3D transformations could be handled. The object to be transformed was a line segment where one of its end points was the centroid. The position in the 3D space of the view point was known and represented the camera. Input parameters were the respective angles for each rotation of the object, α_1 and α_2, and the angle between the planes generated by these rotations, γ. Image changes were identified through straight line detection algorithms. The parameters to measure from the 2D data were the angular variation, β_1 and β_2, corresponding to the two 3D rotations, and the unit vectors that describe the projection planes of each line to compute the relative orientation.

In Figure 4 can be observed the angular variations projected in the image plane, β_1 and β_2, through different 3D rotations of α_1 and α_2. There is an increment of the angular variation with greater changes of motion. With only variations of γ and static rotation angles, the angular difference β_2 also varies. And there are values of γ where the angular variation is higher with small variations of α. Thus γ can be employed to determine the minimum number of movements to estimate the orientation easier and decrease errors. In the case of 3D transformations with high angular variations the 2D difference decreases constantly. This case is not useful, motions with such a magnitude are not always applicable. However, observing the diversity of variations in the 3D space projected in the plane give us an idea of the information acquired to solve the orientation problem. In Figure 5 is shown the performance of the orientation estimation through the rotational motion analysis. It can be observed the relative error and the computation time with variations of α_1 and α_2. The relative error decreases with increments of α and provide good results for small and middle angles. On the other hand, the computation time increases with α increments, due to the greater number of iterations required to solve the nonlinear set of equations.

5 CONCLUSIONS

A robust method to estimate the relative orientation of an object with respect to a camera has been proposed.

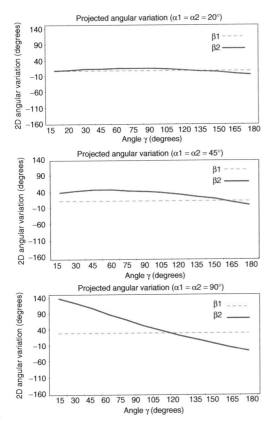

Figure 4. 2D angular variations induced by 3D rotations. With γ variable, different values of α_1 and α_2 produce changes in the 2D angular variation.

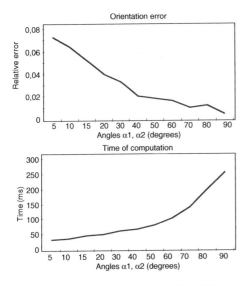

Figure 5. Relative error and computation time of the orientation using the rotational motion analysis algorithm.

The object assumed was represented by feature lines. 2D correspondences of a line due to known 3D transformations of the object were the information used to calculate its pose. We showed that with only two rotations the angular variation between lines provides sufficient information to estimate the relative orientation. The relative position can also be calculated by tracking the end point of the object after the orientation has been determined if its length from the centroid is given. This motion analysis led to address questions as the uniqueness of solution for the minimum number of movements and possible motion patterns to solve it directly. In the case of controlled motions, one component rotations through normal axes simplify calculations to provide a robust technique to estimate the relative orientation with no initial conditions defined.

REFERENCES

Ansar A. & Daniilidis K. 2003. Linear pose estimation from points and lines. In IEEE Trans. Pattern Analysis and Machine Intelligence. 25:578–589.

Chen H. 1991. A screw-motion approach to uniqueness analysis of head-eye geometry. In IEEE Proc. Computer Vision and Pattern Recognition. IEEE:145–151.

Faugeras O., Lustran F. & Toscani G. 1987. Motion and structure from point and line matches. In Proc. First Int. Conf. Computer Vision.

Fischler M.A. & Bolles R.C. 1981. Random sample consensus: a paradigm for model fitting with applications to image analysis and automated cartography. In Communications of the ACM, vol. 4, no. 6, pp. 381–395.

Haralick R.M. 1989. Pose estimation from corresponding point data. In IEEE Trans. Systems, Man, and Cybernetics, vol. 19, no. 6, pp. 1426–1446.

Holt J.R. & Netravali A.N. 1996. Uniqueness of solution to structure and motion from combinations of point and line correspondences. In Journal of Visual Communication and Image Representation, vol. 7, no. 2, pp. 126–136.

Homer H.C. 1991. Pose determination from line-to-plane correspondences: Existence condition and closed-form solutions. In IEEE Trans. Pattern Analysis and Machine Intelligence, vol. 13, no. 6, pp. 530–541.

Hung Y., Yeh P.S. & Harwood D. 1985. Passive ranging to know planar point sets. In IEEE Int. Conf. on Robotics and Automation, pp. 80–85.

Kumar R. & Hanson A.R. 1994. Robust methods for estimating pose and sensitivity analysis. In Computer Vision and Image understanding, vol. 60, pp. 313–342.

Liu Y., Huang T.S. & Faugeras O. 1990. Determination of camera location from 2-D to 3-D line and point correspondences. In IEEE Trans. Pattern Analysis and machine Intelligence, vol. 12, pp. 28–37.

Lowe D.G. 1987. Three-dimensional object recognition from single two-dimensional images. In Artificial Intelligence, vol. 31, no. 3, pp. 355–395.

Mitiche A., Seida S. & Aggarwal J.K. 1986. Interpretation of structure and motion using straight line correspondences. In Proc. Int. Pattern recognition.

Navab N. & Faugeras O. 1993. Monocular pose determination from lines: Critical sets and maximum number solutions. In Proc. Computer Vision and Pattern Recognition, New York, pp. 254–260.

Weng J., Huang T.S. & Ahuja N. 1992. Motion and structure from line correspondences; closed-form solution, uniqueness, and optimization. In IEEE Trans. Pattern Analysis and Machine Intelligence. Vol. 14, 3:318–336.

Yen B.L. & Huang T.S. 1983. Determining 3-D motion and structure of a rigid body using straight line correspondences. In Image sequence Processing and Dynamic Scene Analysis, Heidelberg, Germany: Springer-Verlag.

Computational Modelling of Objects Represented in Images –
João Manuel R.S. Tavares & R.M. Natal Jorge (eds)
© 2007 Taylor & Francis Group, London, ISBN 978-0-415-43349-5

A region growing approach for pulmonary vessel tree segmentation using adaptive threshold

D.A.B. Oliveira[1], G.L.A. Mota[1] & R.Q. Feitosa[1,2]
[1]*Computing Engineering Department, Rio de Janeiro State University – Rio de Janeiro*
[2]*Electrical Engineering Department, Catholic University of Rio de Janeiro – Rio de Janeiro*

R.A. Nunes
Medical Sciences Faculty, Rio de Janeiro State University – Rio de Janeiro, RJ, Brazil

ABSTRACT: Current computed tomography (CT) technology allows for high resolution imaging of the complete chest. These scans have become indispensable in thoracic radiography, but have also increased substantially the amount of data to be analyzed by radiologists. Automating the analysis of such data is a growing necessity. In CT images, inside the lungs, main pulmonary structures (such as airways and vessels) can be clearly observed. This paper presents an algorithm to segment the pulmonary vessel tree in CT images. Our method starts separating the lungs from the rest of the image. Then, the segmentation of the vessel tree inside the lung's area is performed by a region growing technique combined with hysteresis thresholding. An experimental evaluation based on real CT images demonstrated the outcome of the proposed method is generally consistent with a visual segmentation.

1 INTRODUCTION

Current computed tomography scanners allow for isotropic acquisition of the complete chest with millimeter resolution in a single hold breath. A major challenge accompanying this improvement is dealing with the enormous amount of data generated in the form of image sequences.

By and large the CT data analysis is performed visually by a radiologist. This is a time consuming task, whose accuracy depends essentially on the experience of the analyst. (Kakinuma et al., 1999; Li et al., 2002). Digital Image Processing techniques can be used to develop methods that automatically perform many of the tasks involved in the CT analysis, improving productivity and the overall accuracy.

The segmentation is the key step in these processes. Particularly in automatic nodule detection the number of false positives is mainly determined by the quality of the segmentation of the pulmonary vessel tree.

While blood vessels and nodules may share similar local gray-scale characteristics in CTs, global 3D information, as the continuity of blood vessels, can be used to discriminate between them. The segmentation may also provide the volume of interest for abnormalities that occur in vessels (e.g. pulmonary embolisms), reducing the search volume and eliminating false positives located outside the vascular structures (Masutani et al., 2002).

Many techniques have been proposed in the literature for the analysis of lung's CT scans. Wu et al. (2004) uses a regulated morphology approach to fuzzy shape analysis in order to segment the pulmonary vessel tree. The voxels belonging to the vessel tree and to the nodules are isolated based on their intensity. Then, regulated morphology operations are applied and the result is refined so as to build a fuzzy spherical object representation of blood vessels and nodules. Finally a tracking algorithm is used to connect the spherical objects into trees representing networks of blood vessels. This method has the inconvenience of rounding the volume by the fuzzy spheres, what limits its accuracy.

Kiraly et al. (2004) developed a semi-automatic method for segmentation in which a seed point is initially selected in the image and a region growing procedure is started until the entire lungs are segmented. Lung vessels are then segmented by including all voxels above a pre-defined threshold value within the lung mask. Small volume structures are eliminated. As a consequence small nodules may be undetected.

Masutani et al. (2004) firstly segments the major vascular structures, including the heart and the main branches of the pulmonary vessels, using hysteresis thresholding and connected-component analysis. Then, peripheral vessels are segmented using region-growing. Finally, back-growing is performed to fill the volume from the peripheral branches to the main branches, terminating at a pre-defined distance before the growth fronts reach the heart.

Yang et al. (2004) extracts vessel information based on a pre-defined Bayes's rule to classify the volumetric data into three homogeneous regions (myocardium, blood, and lung tissue), and find the one of interest. This method is sensitive to the problem of noise/leak of information as it segments the vessels with a single threshold.

This paper proposes an algorithm for the segmentation of the pulmonary vessel tree using an automated region growing approach.

It uses a 3D hysteresis thresholding technique inspired in Canny's Edge Detector (Canny et al., 1986). A heuristic derived from a series of practical experiments, provides invariance against contrast and average brightness levels.

The following sections describe our approach in greater detail. Section 2 discusses the details of our segmentation methods, section 3 reports our experimental analysis, and the main conclusions are presented in section 4.

2 THE 3D SEGMENTATION METHOD

The method works on each CT image slice performing two sequential steps. First, the area corresponding to the lungs is segmented from the rest of the image. Next, an adaptive procedure works on the lung's area obtained in the previous step and separates the vessel tree from the lung's tissues. Each step is thoroughly described in the following sub-sections.

2.1 Lungs segmentation

Thoracic tomography images contain three main types of tissues: osseous tissues, soft tissues and lungs tissues (mainly filled by air). These tissues can be roughly recognized by their typical gray level ranges. Osseous tissues, which are very dense, are represented by the highest gray levels; the soft tissues correspond to intermediate gray levels; and the lungs have the lowest gray level characteristics.

The lungs are segmented by the following sequential steps:

(a) First the Otsu's algorithm (Otsu et al., 1979) selects a threshold value used to separate osseous tissues from the rest of the image; Figure 1b shows the *osseous tissue mask* obtained this way,

(b) Otsu's algorithm is applied only on the non-osseous region. With this second threshold the *soft tissue mask* is created (Fig. 1c). Pixel on this mask outside the patient body will be later discarded,

(c) Next, we derive the *thorax mask* (Fig. 1d) by filling the holes inside the soft tissue mask,

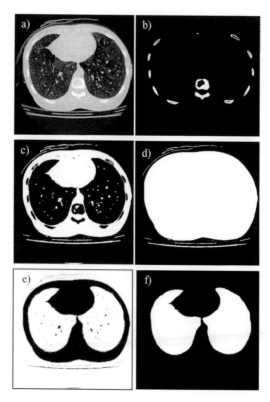

Figure 1. (a) Original image; (b) Osseous tissue mask; (c) Soft tissue mask; (d) Thorax mask; (e) Lungs tissue mask; (f) lungs mask.

(d) The *lung tissue mask* (Fig. 1e) is then computed by first "oring" the soft tissue mask (Fig 1c) with the osseous tissue mask (Fig. 1b), and then complementing the result.

(e) The *lungs mask* (Fig. 1f) is next calculated by "anding" the thorax mask (Fig. 1d) with the lung tissue mask (Fig. 1e). Note that the region external to the patient body contained in the lung tissue mask is eliminated in this step.

(f) We finally separate the lung's image from the original image multiplying the result by the corresponding TC slice, as shown in Figure 2.

2.2 Adaptive vessel tree segmentation

Tissues in the lung's area are composed by the pulmonary vessels, bronchi and lung parenchyma. Essentially, pulmonary vessels are cylinders with thin walls and filled by blood. Bronchi are cylinders with some thicker wall than vessels but filled by the same air that is around the body. Lung parenchyma is characterized by the air mixed with normally indistinguishable alveolus capillary tissues but which are enough to be responsible for an overall increase of radiological

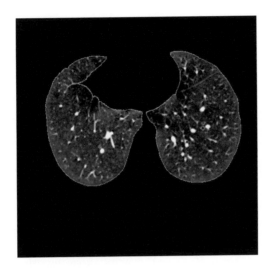

Figure 2. The lungs image.

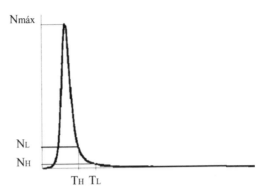

Figure 3. Threshold determination.

density in comparison with the air around the body. From all of three components only the pulmonary vessels are structures entirely without air. For this reason we use hereafter a single term, *pulmonary vessels*, to designate the soft tissues inside the lungs.

Lungs and pulmonary vessels can be distinguished based on their intensities. Lung's tissues are dark while pulmonary vessels are bright. A confusion arises on pixels having intermediary intensities which may belong to any of these tissues types. Therefore, a simple thresholding procedure is not able to accurately separate them.

This section proposes an adaptive method that uses hysteresis thresholding similar to the Canny Edge Detector (Canny et al., 1986) to segment the vessel tree from CT images.

Initially we select a threshold T_H, such that intensities above it identify unambiguously pulmonary vessel. A second threshold T_L ($T_L < T_H$) is further selected such that intensities below it clearly indicate lung tissue.

These two threshold values define three ranges of pixel intensities, namely:

- the *strong vessel range*, defined by intensities above T_H,
- the *weak vessel range*, comprising intensities between T_L and T_H, and
- the *lung tissue range*, covering intensities below T_L.

The construction of the *vessel tree* is performed by a region growing approach consisting of the following basic steps:

(a) Build the *weak vessel tree* defined by the pixels with values above T_L.

(b) Build the *strong vessel tree* defined by the pixels with values above T_H.
(c) Take the *strong vessel tree* computed in the preceding step as the initial *vessel tree* estimate.
(d) Add to the *vessel tree* estimate all *weak vessel* pixels adjacent to it in 3 dimensions.
(e) Repeat the previous step using the current *vessel tree* estimated until it stops growing.

Experiments using the above explained procedure have shown that the selection of the threshold values T_L and T_H is a key issue for an accurate segmentation. Appropriate values for these parameters change from image to image depending on the contrast and average brightness level.

We searched appropriate values for T_L and T_H manually through many experiments using different CT sequences. We observed that the histogram counts for the manually selected values stayed at a roughly constant ratio to the intensity corresponding to the maximum count.

Formally, if N_{max} is the maximum histogram count, and N_L and N_H are the counts corresponding respectively to T_L and T_H, the ratios $R_L = N_L/N_{max}$ and $R_H = N_H/N_{max}$ do not significantly change from CT to CT (see Fig. 3). In fact these ratios lied around $R_L = 0.05$ and $R_H = 0.01$ through all our experiments.

This regularity suggests the following procedure to select the lower and higher threshold values:

(a) Compute the histogram of the image region inside the lungs.
(b) Smooth the histogram using some 1D smoothing filter.
(c) Detect the maximum histogram count N_{max} and the corresponding intensity T_{max}.
(d) Multiply N_{max} by the constant factors R_L and R_H, and obtain the count values N_L and N_H.
(e) Search the smoothed histogram for the intensity values T_L and T_H corresponding to N_L and N_H, whereby both T_L and T_H are greater than T_{max}.

3 EXPERIMENTAL ANALYSIS

A software prototype implementing the proposed segmentation method was built for evaluation purpose.

Figure 4 shows an example of the *strong vessel tree* obtained by thresholding the lung image with T_H estimated by our method.

The *weak vessel* information obtained from the same image using T_L is shown in Figure 5.

Figure 6 shows the final segmentation.

The results obtained by our method on twelve CT sequences were evaluated visually by an experienced lung's CT analyst and were initially all considered satisfactory with recommendation to ameliorate the visualization of trunk vessels in the hilar area.

Our software prototype is also able to generate 3D models. It receives as input the structured points model correspondent to the segmentation of each image slice and the thickness of the CT slices provided in the DICOM (American College of Radiology, 1993) image summary.

Using the VTK library (Schroeder et al., 1998) our prototype implements the marching cubes algorithm contour filter (Lorensen et al., 1998) and a polygon data mapper the extracts the polygon data information from the structured points model. A 3D actor created from this polygon dataset is added to the openGL renderer provided by the library.

One 3D model generated by our prototype is shown in Figure 7. The model was build from four different pulmonary structures obtained by our segmentation method: the thoracic region, the osseous tissue, the lungs and the pulmonary vessel tree. A separate 3D model was build for each of these structures and later merged together as shown in Figure 7.

Figure 8 shows the 3D models: strong vessel tree (Fig. 8a), weak vessel information (Fig. 8a), and final segmented vessel tree (Fig. 8c).

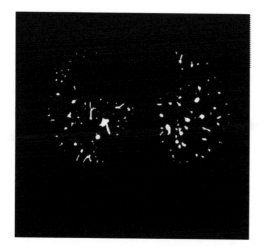

Figure 4.   Segmentation using the higher threshold.

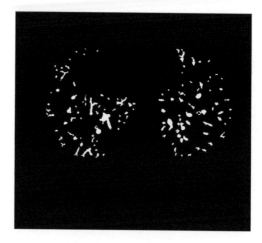

Figure 6.   Segmentation using the proposed method.

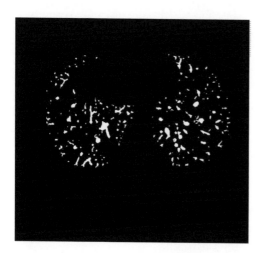

Figure 5.   Segmentation using the lower threshold.

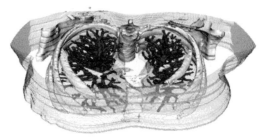

Figure 7.   3D pulmonary structures model.

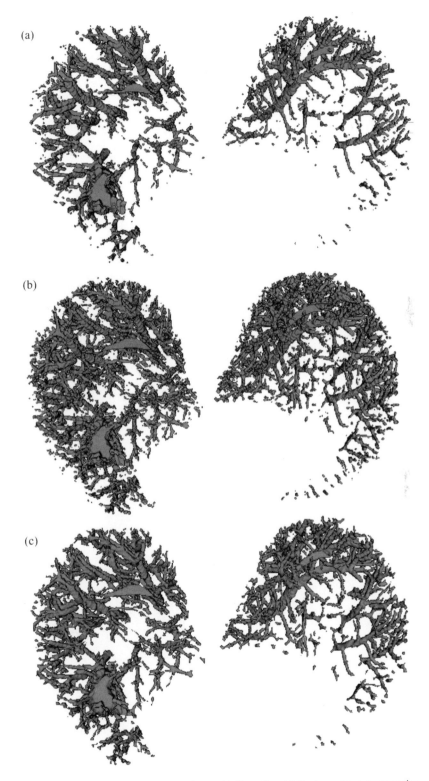

Figure 8.   3D models – (a) Strong vessel tree; (b) Weak vessel information; (c) Final vessel tree segmentation.

4 CONCLUSION

The present work proposed an algorithm to segment the pulmonary vessel tree in computer tomography (CT) images.

Preliminary results from an experimental evaluation confirm the efficiency of the proposal, albeit some improvement and adjustments must be still reached like better definition of the hilar vessels.

The method has the potential of becoming a useful tool in various applications. It can be used to generate 3D vessel tree representations to aid visual diagnostic. Volumes can be computed from the generated 3D models to monitor, for example, how pulmonary nodules grow along the time, providing important information for cancer diagnostic. Shape attributes other than volume may also be measured from the 3D model and explored in Computer Aided Diagnostic environments.

We are looking forward the development of a quantitative method for the evaluation of the segmentation quality. Another important topic is the implications of this research on nodule and embolism detection techniques. This is one of the next steps planned for the continuation of this research.

REFERENCES

R. Kakinuma, H. Ohmatsu, and M. Kaneko, "Detection failures in spiral CT screening for lung cancer: Analysis of CT findings." Radiology 212, pp. 61–66, 1999.

F. Li, S. Sone, H. Abe, H. MacMAhon, S. Armato III, and K. Doi, "Lung cancers missed at low-dose helical CT screening in a general population: Comparison of clinical, histopathologic, and imaging findings." Radiology 225, pp. 673–683, 2002.

Y. Masutani, H. MacMahon, and K. Doi, "Computerized detection of pulmonary embolism in spiral CT angiography based on volumetric image analysis," *IEEE Trans. Med. Imag.*, vol. 21, no. 12, pp. 517–1523, Dec. 2002.

C.Wu, G. Agam, A. S. Roy, and S. G. Armato, "Regulated morphology approach to fuzzy shape analysis with application to blood vessel extraction in thoracic CT scans," *Proc. SPIE*, vol. 5370, pp. 1262–1270, 2004.

A. P. Kiraly, E. Pichon, D. P. Naidich, and C. L. Novak, "Analysis of arterial subtrees affected by pulmonary emboli," *Proc. SPIE*, vol. 5370, pp. 1720–1729, 2004.

Y. Yang, A. Tannembaum & D. Giddens, "Knowledge-Based 3D Segmentation and Reconstruction of Coronary Arteries Using CT Images," *Proc. IEEE EMBS.*, pp. 1664–1666, Sep. 2004.

Otsu, N., "A Threshold Selection Method from Gray-Level Histograms," *IEEE Transactions on Systems, Man, and Cybernetics*, vol. 9, no. 1, 1979, pp. 62–66.

Canny, J., "A Computational Approach to Edge Detection, IEEE Trans. on Pattern Analysis and Machine Intelligence", vol. PAMI-8, no. 6, Nov. 1986, pp. 679–698.

American College of Radiology, National Electrical Manufacturers Association, "Digital Imaging and Communications in Medicine (DICOM): Version 3.0", Draft Standard, ACR-NEMA Committee, Working Group VI, Washington, DC, 1993.

W. Schroeder, K. Martin, and B. Lorensen, *"The Visualization Toolkit."* Prentice Hall, 2nd edition, 1998.

William E. Lorensen and Harvey E. Cline, "Marching Cubes: A High Resolution 3D Surface Construction Algorithm", Computer Graphics (Proceedings of SIGGRAPH '87), vol. 21, no. 4, pp. 163–169, 1998.

Computational Modelling of Objects Represented in Images –
João Manuel R.S. Tavares & R.M. Natal Jorge (eds)
© 2007 Taylor & Francis Group, London, ISBN 978-0-415-43349-5

Active surfaces acceleration methods

Julien Olivier, Julien Mille, Romuald Boné & J.-Jacques Rousselle

Université François Rabelais de Tours, Laboratoire Informatique, Tours, France

ABSTRACT: Active contours and surfaces are deformable models used for 2D and 3D image segmentation. In this paper, we propose two methods developed in order to accelerate 3D image segmentation process. These last ones are adaptations on active surfaces of two methods developed for 2D active contour. We use them on a discrete 3D surface model (mesh) evolving with the greedy algorithm. Those methods will be compared to the classical greedy algorithm and to a recent fast adaptation of the level set method.

1 INTRODUCTION

Active contours or snakes, initially developed by Kass *et al* in (Kass, Witkin, and Terzopoulos 1987), are powerful segmentation tools thanks to their noise robustness and ability to generate linked closed boundaries. Their 3D extensions, the active surfaces, were developed according to several implementations (see (Montagnat, Delingette, and Ayache 2001) (Mille, Boné, Makris, and Cardot 2005) for surveys on 3D deformable models). Among these implementations, meshes are explicit discrete representations (Zhang and Braun 1997), which models the surface as a set of interconnected vertices. The model is deformed by direct modifications of vertices coordinates. Several evolution algorithms have been developed to deform them. One of the most popular is the greedy algorithm (Williams and Shah 1992) because of its efficiency. An adaptation of the greedy algorithm on 3D surface was proposed by Bulpitt and Efford in (Bulpitt and Efford 1996).

Conversely, implicit implementations, based on the level set framework (Malladi, Sethian, and Vemuri 1995), handle the surface as the zero level of a hypersurface, defined on the same domain as the image (for 3D images, the hypersurface is a $\mathbb{R}^3 \longrightarrow \mathbb{R}$ application). Level sets are often chosen for their natural handling of topological changes and adaptiveness to any dimension. Their algorithmic complexity is a function of the image resolution, making them time-consuming. Despite the development of accelerating methods (like the narrow band technique (Malladi, Sethian, and Vemuri 1995), the fast marching method (Adalsteinsson and Sethian 1995) and the recent fast level set (Shi and Karl 2005)), their computational

cost remains high, preventing their use in time-critical applications. Moreover, mesh surfaces have several advantages over their implicit counterparts. Their representation is more intuitive, and thus allow easier modeling of *a priori* knowledge and user interaction. The main drawback is that meshes do not modify their topology naturally (techniques for detection of topological changes must be implemented beside the evolution algorithm).

In many applications, the topology of the area of interest is known in advance. When segmenting images in which prior knowledge about the object topology is available, we believe that mesh-based approaches should be privileged over implicit surfaces. In this paper, we deal with a 3D triangular mesh driven by greedy algorithm. The model is able to perform remeshing, thus providing geometrical versatility (in the same manner that 2D reparameterization techniques overcome the lack of geometrical flexibility of traditional snakes). Several methods have been developed in (Olivier, Boné, and Rousselle 2005) in order to accelerate 2D active contours. In this paper, we propose 3D adaptations of these methods in order to accelerate discrete active surfaces.

The outline of this paper is as follows: section 2 presents the 3D model and its energies, the greedy algorithm for active surfaces and the remeshing principle. Section 3 and 4 describe the shifted neighborhood method and the line search method. Section 5 shows our experimental results on 3D models, comparing the performances of our acceleration methods with the basic greedy algorithm. A comparison with a 3D implementation of the fast level set method is also provided. Section 6 concludes with our work expecting the future developments.

Figure 1. The basic icosahedron and its first two tesselations.

2 THE 3D ACTIVE SURFACE MODEL

To model the active surface, we use the discrete representation described in (Mille, Boné, Makris, and Cardot 2005), which is a triangular mesh made up of n vertices, denoted $\mathbf{p}_i = (x_i, y_i, z_i)^T \in IR^3$, and edges connecting the vertices (making a set of adjacent triangles). In order to represent the connectivity notion, each vertex $\mathbf{p}_i$ has a set of adjacent vertices, denoted A_i. The mesh is built from successive subdivisions of an icosahedron (McInerney and Terzopoulos 1995) (Park, McInerney, and Terzopoulos 2001), thus leading to a sphere-like surface with a homogeneous vertex distribution (see Figure 1).

The surface is endowed with a discrete energy functional, being the sum of the energies of each vertex (vertex energy will be described later). To perform image segmentation, the surface evolves in order to minimize this energy, attracting the vertices towards the object boundaries, while keeping geometrical smoothness. Initially developed for 2D active contours by Williams and Shah in (Williams and Shah 1992), the greedy algorithm is an energy minimizing method first proposed as an alternative to the variational method (Kass, Witkin, and Terzopoulos 1987) and the dynamic programming (Amini, Weymouth, and Rain 1991). It has been recently used for 2D segmentation in (Ji and Yan 2002) and (Lam and Yuen 1998). In (Bulpitt and Efford 1996) one may find an extension of this algorithm for 3D meshes.

Global energy minimization is performed via successive local optimizations: at each iteration, a cubic neighborhood of side length w around each vertex is considered. The energy is computed at each voxel belonging to the neighborhood and the vertex is moved to the location leading to the lowest energy.

Unlike in the classical greedy algorithm, the neighborhoods may not be centered around the vertices. To represent the location of a vertex in its neighborhood, we define a shift vector $\vec{\mathbf{s}}_i^{(t)} = (s_x, s_y, s_z)^T$ representing the coordinates of the ith vertex at iteration t relatively to an original voxel chosen in the neighborhood. At first, all vertices are neighborhood-centered, hence we have $\vec{\mathbf{s}}_i^{(0)} = (w/2, w/2, w/2)^T$.

We can now define the neighborhood of the ith vertex at iteration t:

$$\mathcal{N}_i^{(t)} = \left\{ \mathbf{p}_i^{(t)} + \mathbf{r} - \vec{\mathbf{s}}_i^{(t)} \mid \mathbf{r} \in [0, w-1]^3 \right\} \quad (1)$$

The initial position being $\mathbf{p}_i$, we denote $\mathbf{p}_i'$ a tested location in the neighborhood. Once all energies have been computed the new location of vertex $\mathbf{p}_i$ is chosen:

$$\mathbf{p}_i^{(t+1)} = \arg \min_{\mathbf{p}_k' \in \mathcal{N}_i^{(t)}} E(\mathbf{p}_k') \quad (2)$$

The energy of a vertex at location $\mathbf{p}_i'$ is a weighted sum of various internal and external energies, normalized on the whole neighborhood.

$$E(\mathbf{p}_i') = \alpha E_{cont}(\mathbf{p}_i') + \beta E_{curv}(\mathbf{p}_i') + \gamma E_{grad}(\mathbf{p}_i') + \delta E_{bal}(\mathbf{p}_i') \quad (3)$$

The coefficients $(\alpha, ..., \delta)$ are the weights defining the relative influence of the energies. The continuity E_{cont} and the curvature E_{curv} are internal energies helping to keep the geometrical smoothness of the surface. These two energies imply to take the euclidian distance between the adjacent vertices into account. In following equations, the squared distance is computed rather than the distance, in order to save computational time. We also distinguish $\|\cdot\|$ the magnitude of a vector from $|\cdot|$ the absolute value of a scalar. Thus $\|\mathbf{p}_i - \mathbf{p}_j\|$ represents the euclidian distance between vertices $\mathbf{p}_i$ and $\mathbf{p}_j$.

Let us describe the adaptation of the different energies to our 3D model. The energies are intuitive extensions of the 2D active contour ones, suitable to our mesh representation. The continuity E_{cont} is an internal energy maintaining the vertices evenly spaced along the surface. Minimizing it reduces the gap between the mean squared distance $\overline{d^2}$ and the distance between the considered vertex and its adjacent vertices.

$$E_{cont}(\mathbf{p}_i') = \sum_{j \in A_i} \left| \overline{d^2} - \|\mathbf{p}_i' - \mathbf{p}_j\|^2 \right|$$

$$\overline{d^2} = \frac{1}{n} \sum_{i=1}^{n} \frac{1}{|A_i|} \sum_{j \in A_i} \|\mathbf{p}_i - \mathbf{p}_j\|^2 \quad (4)$$

The second internal energy is the curvature E_{curv}, which minimization results in a local smoothing effect, by making the vertex get closer to the centroid of its adjacent vertices.

$$E_{curv}(\mathbf{p}_i') = \left\| \mathbf{p}_i' - \frac{1}{|A_i|} \sum_{j \in A_i} \mathbf{p}_j \right\|^2 \quad (5)$$

To attract vertices towards salient edges, the external energy E_{grad} is a function of normalized gradient magnitude g of image I. In presence of noisy data, the image is smoothed with a gaussian filter prior to gradient operation. In the following equations, G_σ is a gaussian kernel with standard deviation σ and $*$ is the convolution operator.

$$g(\mathbf{p}) = \|\nabla I(\mathbf{p}) * G_\sigma\| / g_{max}$$

$$E_{grad}(\mathbf{p}'_i) = -g(\mathbf{p}'_i) \qquad (6)$$

As regards gradient magnitude, real 3D edge detection is obtained by convolving the image with the Zucker-Hummel operator (Zucker and Hummel 1981), made up of three $3 \times 3 \times 3$ masks. For example, the following mask filters the image along the x-axis.

$$Z_x = \begin{bmatrix} -k_1 & 0 & k_1 \\ -k_2 & 0 & k_2 \\ -k_1 & 0 & k_1 \end{bmatrix} \begin{bmatrix} -k_2 & 0 & k_2 \\ -k_3 & 0 & k_3 \\ -k_2 & 0 & k_2 \end{bmatrix} \begin{bmatrix} -k_1 & 0 & k_1 \\ -k_2 & 0 & k_2 \\ -k_1 & 0 & k_1 \end{bmatrix} \qquad (7)$$

$$k_1 = \frac{\sqrt{3}}{3} \; ; \; k_2 = \frac{\sqrt{2}}{2} \; ; \; k_3 = 1 \qquad (8)$$

To increase the capture range, we introduce a balloon energy E_{bal} derived from the inflation force proposed in (Cohen 1991). It allows the mesh to be initialized far from the object boundaries.

$$E_{bal}(\mathbf{p}'_i) = \|\mathbf{p}'_i - (\mathbf{p}_i + k\vec{\mathbf{n}}_i)\|^2 \qquad (9)$$

where $\vec{\mathbf{n}}_i$ is the unit inward normal, defined at vertex $\mathbf{p}_i$. The normal of vertex $\mathbf{p}_i$ is the normalized sum of the normals of the neighboring triangles (Park, McInerney, and Terzopoulos 2001). Rigorously, the normal of a triangle t is the unit vector orthogonal to the plane defined by t. In the following expressions, T_i is the set of neighboring triangles of $\mathbf{p}_i$. The normal $\vec{\mathbf{n}}_t$ of a given triangle is the normalized cross product between two vectors belonging to the corresponding plane.

$$\vec{\mathbf{n}}_i = \frac{\displaystyle\sum_{t \in T_i} \vec{\mathbf{n}}_t}{\left\| \displaystyle\sum_{t \in T_i} \vec{\mathbf{n}}_t \right\|} \; ; \; \vec{\mathbf{n}}_t = s_t \frac{(\mathbf{p}_{t_2} - \mathbf{p}_{t_1}) \wedge (\mathbf{p}_{t_3} - \mathbf{p}_{t_1})}{\|(\mathbf{p}_{t_2} - \mathbf{p}_{t_1}) \wedge (\mathbf{p}_{t_3} - \mathbf{p}_{t_1})\|} \qquad (10)$$

where $\mathbf{p}_{t_j}, j = 1 \ldots 3$ are vertices of triangle t ($\mathbf{p}_i$ must be one of them). $s_t = \pm 1$ is the sign changing the orientation of $\vec{\mathbf{n}}_t$ insuring that it points towards the interior of the surface. Such a calculation of the normal vector is necessary to a correct balloon implementation. The motion resulting from the balloon energy

Figure 2. Cubic centered neighborhood.

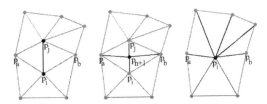

Figure 3. Remeshing operations: vertex adding and deleting.

minimization is either a dilatation or a retraction of the surface, depending on the sign of coefficient δ. This one must be chosen regarding the initial position of the surface with respect to the target object.

In order to adapt local topology, remeshing is performed after each iteration of the greedy algorithm. The mesh is allowed to add or delete vertices to keep the distance between adjacent vertices homogeneous, resulting in a stable vertex distribution (Lachaud and Montanvert 1999) (Park, McInerney, and Terzopoulos 2001) (Slabaugh and Unal 2005). It insures that every couple of adjacent vertices $(\mathbf{p}_i, \mathbf{p}_j)$ satisfies the constraint:

$$d_{min} \le \|\mathbf{p}_i - \mathbf{p}_j\| \le d_{max} \qquad (11)$$

where d_{min} and d_{max} are two user-defined thresholds. Adding or deleting vertices modifies local topology, thus topological constraints should be verified. To perform vertex adding or deleting, $\mathbf{p}_i$ and $\mathbf{p}_j$ should share exactly two common adjacent vertices: $|A_i \cap A_j| = 2$. When $\|\mathbf{p}_i - \mathbf{p}_j\| > d_{max}$, a new vertex is created at the middle of line segment $\mathbf{p}_i \mathbf{p}_j$ and connected to $\mathbf{p}_a$ and $\mathbf{p}_b$ (see middle part of Figure 2). When $\|\mathbf{p}_i - \mathbf{p}_j\| < d_{min}$, $\mathbf{p}_j$ is deleted and $\mathbf{p}_i$ is translated to the middle location (see right part of Figure 3).

3 THE SHIFTED NEIGHBORHOOD METHOD

In order to improve active surfaces completion time, we used the shifted neighborhood method developed for 2D snakes in (Olivier, Boné, and Rousselle 2005).

327

We adapted this greedy-based method on active surfaces. For each vertex and at each iteration, we are going to modify the neighborhood in order to direct the searching space of each vertex to the directions that seems the most interesting. To define where these directions are, we use the information of the direction followed by each vertex during the last iteration. So each neighborhood will be shifted from one voxel in the direction followed previously. At each iteration, we compute the next shift applied to the vertex with:

$$\vec{\mathbf{d}}_i^{(t+1)} = \mathcal{B}(-1, 1, \mathbf{p}_i^{(t+1)} - \mathbf{p}_i^{(t)}) \qquad (12)$$

The vector quantity $\vec{\mathbf{d}}_i^{(t)}$ represents the motion applied on the neighborhood of the ith vertex at iteration t. $\mathcal{B}$ is a shift limiting function, bounding the vector coordinates between two scalars:

$$\mathcal{B}(b_1, b_2, \vec{\mathbf{u}}) = \begin{pmatrix} \max(b_1, \min(b_2, u_x)) \\ \max(b_1, \min(b_2, u_y)) \\ \max(b_1, \min(b_2, u_z)) \end{pmatrix} \qquad (13)$$

The displacement $\vec{\mathbf{d}}_i^{(t+1)}$ given by equation (12) allows us to define the new shift vector $\vec{\mathbf{s}}_i^{(t+1)}$ for each vertex of the active surface. Thus we have:

$$\vec{\mathbf{s}}_i^{(t+1)} = \mathcal{B}(1, w-2, \vec{\mathbf{s}}_i^{(t)} - \vec{\mathbf{d}}_i^{(t+1)}) \qquad (14)$$

The next iteration of the greedy algorithm will be helded with these new neighborhoods. At this stage, we can define the algorithm for the shifted neighborhood method. This last consists in computing the new neighborhood $\mathcal{N}_i^{(t+1)}$ with equations (1), (12) and (14) at the end of each iteration, once all the vertices of the active surface have been modified. When included in the greedy algorithm for an active surface of n vertices and T iterations, the shifted neighborhood method is described in algorithm 1.

4 THE LINE SEARCH METHOD

The line search method (Olivier, Boné, and Rousselle 2005) originally developed for two-dimensional active contours allows to reduce completion time efficiently. We adapted this method on 3D active surfaces. The principle of this approach is to anticipate on the next iteration of the greedy algorithm using the information taken from the previous one. This method is launched at the end of each iteration of the greedy algorithm, once all the vertices have been translated.

Algorithm 1 Shifted Neighborhood Method: 3D Model

1: **for** $t \leftarrow 1$ to T **do**
2:    **for** $i \leftarrow 1$ to n **do**
3:       $\mathbf{p}_i^{(t+1)} = \arg\min_{\mathbf{p}_k' \in \mathcal{N}_i^{(t)}} E(\mathbf{p}_k')$
4:       $\vec{\mathbf{d}}_i^{(t+1)} = \mathcal{B}(-1, 1, \mathbf{p}_i^{(t+1)} - \mathbf{p}_i^{(t)})$
5:       $\vec{\mathbf{s}}_i^{(t+1)} = \mathcal{B}(1, w-2, \vec{\mathbf{s}}_i^{(t)} - \vec{\mathbf{d}}_i^{(t+1)})$
6:       $\mathcal{N}_i^{(t+1)} = \{\mathbf{p}_i^{(t+1)} + \mathbf{r} - \vec{\mathbf{s}}_i^{(t)} | \mathbf{r} \in [0, w-1]^3\}$
7:    **end for**
8: **end for**

The direction followed by each vertex $\mathbf{p}_i$ is memorized and we look toward it for a fixed number of voxels, which creates a linear neighborhood. These lasts are compared by computing their global energies in a similar way as it is done with the cubic neighborhood (see equation (4)). The voxel giving the lowest energy is then chosen as the new location of the current vertex. As a result, for each vertex, two neighborhoods are scanned consecutively: the cubic centered neighborhood and the linear neighborhood. The second algorithm describes the line search method integrated in the greedy algorithm for active surfaces. Let T be the number of iterations to be done by the greedy algorithm and l the number of voxels to be explored (length of the linear neighborhood).

Algorithm 2 Line Search Method: 3D Model

1: **for** $t \leftarrow 1$ to T **do**
2:    **for** $i \leftarrow 1$ to n **do**
3:       $\mathbf{p}_i^{(t+1)} = \arg\min_{\mathbf{p}_k' \in \mathcal{N}_i^{(t)}} E(\mathbf{p}_k')$
4:       Determine the direction $\vec{\mathbf{v}} = \frac{\mathbf{p}_i^{(t+1)} - \mathbf{p}_i^{(t)}}{\|\mathbf{p}_i^{(t+1)} - \mathbf{p}_i^{(t)}\|}$
5:       Line Search: $m = \arg\min_{k \in [0,l]} \{E(\mathbf{p}_i^{(t)} + k\vec{\mathbf{v}})\}$
6:       Update: $\mathbf{p}_i^{(t+1)} \leftarrow \mathbf{p}_i^{(t+1)} + m\vec{\mathbf{v}}$
7:    **end for**
8: **end for**

---

5 EXPERIMENTAL RESULTS

In this section, we present our experiments on 3D images. We compare the shifted neighborhood method and the line search method to the classical greedy algorithm. This comparative study also includes the results obtained with implicit surface modeling. To do so,

Figure 4. 2D slices of 3D noisy images (top) and surface results obtained with the line search method (bottom).

Image	Neighborhood	Method	# iterations	Time (s)
Spiral	w = 3	Greedy	400	0.30
(400 × 400 ×		LS	65	0.16
400 voxels)	w = 5	Greedy	195	0.52
		LS	63	0.30
		SN	138	0.31
	w = 7	Greedy	145	0.63
		LS	60	0.57
		SN	97	1.11
	Fast level set		1060	44.17
3 ellipsoids	w = 3	Greedy	47	0.63
(200 × 200 ×		LS	19	0.41
200 voxels)	w = 5	Greedy	25	1.05
		LS	12	0.69
		SN	16	0.98
	w = 7	Greedy	18	2.24
		LS	12	1.86
		SN	14	2.07
	Fast level set		120	30.82
Vase	w = 3	Greedy	155	0.98
(200 × 200 ×		LS	68	0.72
200 voxels)	w = 5	Greedy	120	1.27
		LS	50	0.85
		SN	92	1.08
	w = 7	Greedy	109	1.98
		LS	43	1.13
		SN	60	2.20
	Fast level set		210	47.04

Figure 5. Comparison of computational times between classical greedy algorithm, line search method, shifted neighborhood method and fast level set.

we use the fast level set method (Shi and Karl 2005), one of the recent breakthroughs in the level set field. Each tested image is made of several slices of gray objects embedded in white backgrounds, highly corrupted with gaussian noise. Different values of neighborhood width w are tested (obviously, the shifted neighborhood is not experimented with $w = 3$). For each image, the surface is initialized as a sphere with identical center and radius for all evolution methods, independently from its implementation (a mesh or a level set).

The first image represents a spiral and was chosen in order to test the methods when vertex adding is enabled. The active surface is initialized inside the 3D model with only 12 vertices. The final meshes for the three methods contain about one thousand vertices. We choose $\alpha = 0$, $\beta = 0.5$, $\gamma = 2$ and $\delta \in [-0.6, -1.1]$.

The second image is a 3D model of a vase and is interesting to test the infiltration of the model into the concavities. For this particular 3D dataset, remeshing of the model is disabled. We use $\alpha = 0$, $\beta = 0.5$ for the greedy algorithm, 0.4 for the shifted neighborhood method and 0.3 for the line search, $\gamma = 2$ and $\delta = 0.8$. We initialize the meshes with 2562 vertices.

The last image represents three ellipsoids and allows to have both salient and smooth angles on the same model. We prevent remeshing of the model and initialize it with 2562 vertices. The parameters are $\alpha = 0.5$, $\beta = 0.3$ for the greedy algorithm and 0.4 for the shifted neighborhood method and the line search, $\gamma = 2$ and $\delta \in [0.3, 0.7]$.

Figure 4 shows one slice of each image and the visual 3D results obtained with the line search method. Figure 5 lists computational times obtained for each 3D image, each method and each window width (tests were made on a Pentium IV 1.7 GHz with 512 Mb RAM).

6 CONCLUSION AND FUTURE WORK

As shown in Figure 5, these methods allow us to accelerate the greedy algorithm for active surfaces. Mesh-based methods (the greedy algorithm and our acceleration methods) leads to performances turning out to be far beyond the level set-based method ones. Accelerations methods tend to make completion time fall below 1 s (for the best configuration) whereas the fast level set approach exceeds 30 s.

The best acceleration method for 2D active contours was the shifted neighborhood but we detected the line search as the best to be applied on 3D active surfaces. The explication is that in three dimensions an exploration line stays a line whereas a square neighborhood becomes cubic, rising the completion times added by the shifted neighborhood method.

We also tested the Deformed Neighborhood method discribed in (Oliver, Boné, and Rousselle 2005) but it was not efficient on active surfaces for the same reasons.

We can also notice that the contribution of the shifted neighborhood method is better with a large neighborhood. Indeed, the shiftings are dependant of its size.

REFERENCES

Adalsteinsson, D. and J. Sethian (1995). A fast level set method for propagating interfaces. *Journal of Computational Physics 118*(2), 269–277.

Amini, A., T. Weymouth, and R. Rain (1991, September). Using dynamic programming for solving variational problem in vision. *IEEE Transactions on Pattern Analysis and Machine Intelligence 12*(9), 855–867.

Bulpitt, A. and N. Efford (1996). An efficient 3D deformable model with a self-optimising mesh. *Image and Vision Computing 14*(8), 573–580.

Cohen, L. (1991). On active contour models and balloons. *Computer Vision, Graphics, and Image Processing: Image Understanding 53*(2), 211–218.

Ji, L. and H. Yan (2002, April). Attractable snakes based on the greedy algorithm for contour extraction. *Pattern Recognition 35*(4), 791–806.

Kass, M., A. Witkin, and D. Terzopoulos (1987). Snakes: active contour models. *International Journal of Computer Vision 1*(4), 321–331.

Lachaud, J. and A. Montanvert (1999). Deformable meshes with automated topology changes for coarse-to-fine three-dimensional surface extraction. *Medical Image Analysis 3*(2), 187–207.

Lam, C. and S. Yuen (1998, April). An unbiased active contour algorithm for object tracking. *Pattern Recognition Letters 19*(5–6), 491–498.

Malladi, R., J. Sethian, and B. Vemuri (1995). Shape modeling with front propagation: a level set approach. *IEEE Transactions on Pattern Analysis and Machine Intelligence 17*(2), 158–175.

McInerney, T. and D. Terzopoulos (1995). A dynamic finite element surface model for segmentation and tracking in multidimensional medical images with application to cardiac 4D image analysis. *Computerized Medical Imaging and Graphics 19*(1), 69–83.

Mille, J., R. Boné, P. Makris, and H. Cardot (2005). 3D segmentation using active surface: a survey and a new model. In *5th IASTED Int. Conf. on Visualization, Imaging & Image Processing (VIIP)*, Benidorm, Spain, pp. 610–615.

Montagnat, J., H. Delingette, and N. Ayache (2001). A review of deformable surfaces: topology, geometry, and deformation. *Image and Vision Computing 19*(14), 1023–1040.

Olivier, J., R. Boné, and J.-J. Rousselle (2005, September). Comparison of active contour acceleration methods. In *5th IASTED Int. Conf. on Visualization, Imaging & Image Processing (VIIP)*, Benidorm, Spain.

Park, J.-Y., T. McInerney, and D. Terzopoulos (2001, June). A non-self-intersecting adaptive deformable surface for complex boundary extraction from volumetric images. *Computer & Graphics 25*(3), 421–440.

Shi, Y. and W. Karl (2005). A fast level set method without solving PDEs. In *IEEE Int. Conf. on Acoustics, Speech, and Signal Processing (ICASSP)*, Volume 2, Philadelphia, USA, pp. 97–100.

Slabaugh, G. and G. Unal (2005). Active polyhedron: surface evolution theory applied to deformable meshes. In *IEEE Computer Vision and Pattern Recognition (CVPR)*, Volume 2, San Diego, USA, pp. 84–91.

Williams, D. and M. Shah (1992). A fast algorithm for active contours and curvature estimation. *Computer Vision, Graphics, and Image Processing: Image Understanding 55*(1), 14–26.

Zhang, Z. and F. Braun (1997). Fully 3D active surface models with self-inflation and selfdeflation forces. In *IEEE Computer Vision and Pattern Recognition (CVPR)*, San Juan, Puerto Rico, pp. 85–90.

Zucker, S. and R. Hummel (1981). A threedimensional edge operator. *IEEE Transactions on Pattern Analysis and Machine Intelligence 3*(3), 324–331.

Computational Modelling of Objects Represented in Images –
João Manuel R.S. Tavares & R.M. Natal Jorge (eds)
© 2007 Taylor & Francis Group, London, ISBN 978-0-415-43349-5

Artery-vein separation of human vasculature from 3D thoracic CT angio scans

Sangmin Park & Chandrajit Bajaj
Computational Visualization Center, Department of Computer Sciences, Institute of Computational
Engineering and Sciences, University of Texas at Austin, Austin, Texas, USA

Gregory Gladish
Diagnostic Radiology, The University of Texas M.D. Anderson Cancer Center, Houston, Texas, USA

ABSTRACT: This paper presents vascular tree reconstruction and artery-vein separation methods from 3D thoracic CT-angiography (CTA) images. In the methods, the lungs, blood vessels and the heart are segmented by using intensity-based thresholds and morphological operations. After the distance transform for the regions of blood vessels and the heart, we compute seed points that are the maxima of distance values. At each seed point, spheres are inflated until hit a boundary. The spheres and the connections between overlapped spheres make a graph representation of the 3D image. Once the pulmonary trunk is detected by using directions in the graph, blood vessels are traversed toward the heart, while merging and verifying branches. If a branch is linked to the pulmonary trunk, then all subsegmental trees of the branch are classified into pulmonary arteries. Otherwise, it is a pulmonary vein.

1 INTRODUCTION

Human vasculature reconstruction from three-dimensional (3D) computed tomography (CT) images of the thorax is a critical step for computer-aided diagnosis (CAD) in disease domains such as lung nodules (Gady Agam 2005), coronary artery disease (Jamshid Dehmeshki and Qanadli 2004), and pulmonary embolism (PE) (Yoshitaka Masutani 2002). It remains a challenging problem even though there have been many published approaches (Kirbas and Quek 2004) (Katja Bühler and Cruz 2003).

Thoracic CT angiography (CTA) imaging is often performed for patients suspected of having pulmonary embolus (PE) that is defined as a thrombus (or a clot of blood) (Schoepf and Costello 2004). However, blood vessel extraction is often not enough for PE detection from CTA, since the position of thrombi is always in the pulmonary trunk and/or the subsegmental arteries (Hartmann and Prokop 2002). Differentiation of pulmonary arterial from venous trees significantly reduces false positive PE detection in pulmonary veins. While several authors have addressed the issue of artery-vein separation (AV separation) from magnetic-resonance angiography (MRA) (Cornelis M. van Bemmel and Niessen 2003) (Stefancik and Sonka 2001) (Tianhu Lei and Odhner 2001)

(Michael Bock 2000), Sluimer *et al.* mentions that AV separation from CTA is one of their primary future challenges (Ingrid Sluimer and van Ginneken 2006).

For AV separation from MRA, Bemmel *et al.* suggests a level-set framework (Cornelis M. van Bemmel and Niesen 2003). In their method, the central arterial and venous axes are determined by using a supervised learning (SL) procedure. Their SL method computes a minimum-cost path between two (or more) user-defined points in the arterial and venous parts of the vasculature. The high probability voxels which become part of the center of a vascular structure are enhanced using a multiscale filter. Stefancik and Sonka show that a graph search algorithm coupled to a knowledge-based approach could be applied to AV separation (Stefancik and Sonka 2001). Lei *et al.* utilized the principles and algorithms of fuzzy connected object delineation, for AV separation (Tianhu Lei and Odhner 2001). Bock *et al.* proposed correlation analysis between manually selected regions of interest and the acquired 3D MRA data sets, viewed as a time series (Michael Bock 2000). Since diverse medical imaging modalities require different approaches (Kirbas and Quek 2004), in this paper we focus on robust AV separation approach from 3D thoracic CTA images, knowing that the features of CTA images are quite different from MRA.

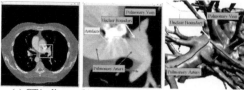

(a) CTA slice   (b) Enlgarged image   (c) Iso-surfaces

Figure 1.   Slice images of CT-angiography and Iso-surfaces: The yellow box of (a) is enlarged in (b) that shows the unclear boundary between the pulmonary artery and vein and the same region is visualized in (c) with iso-surfaces.

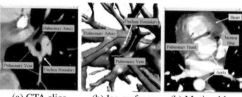

(a) CTA slice   (b) Iso-surfaces   (b) Motion blur

Figure 2.   There are many unclear boundaries in 3D CTA images. It makes it hard to separate pulmonary arteries and veins.

Fig. 1 (a,b) (image slices) and (c) (3D iso-surface) show some of the main problems that make it difficult to differentiate pulmonary arterial and venous trees from 3D thoracic CTA images. First, since the resolution of CTA images is limited and the artery and vein are very close to each other, there is an unclear boundary between them. The same problem is present in many other places even in a single data set. Fig. 2 shows another region that contains the similar problem. Second, the bright area in Fig. 1(b), which represents the vena cava, possesses many neighboring CTA artifacts. Since the artifacts are more distinct than the unclear boundary in Fig. 1(b), they are difficult to remove as unnecessary features, while identifying the unclear boundaries.

The unclear boundary problem is addressed by Bemmel et al. in (Cornelis M. van Bemmel and Niessen 2003). In their method, voxels are labeled arterial or venous based on the arrival time of the respective level-set fronts. The front evolution is depend on three external forces based on the information of intensity, gradient, and vesselness. However, the labeling approach does not work properly for the problem region of Fig. 1(b), since the pulmonary artery is thicker than the vein, and the evolution of level-set fronts can be hampered by high gradients at the artifacts. Furthermore, since there are many unclear boundaries between pulmonary arteries and veins in 3D thoracic CTA images, all the problem regions should be skeletonized before using the level-set framework, which requires heavy user interactions.

Kiraly et al. used branching angles to remove false branches for only arterial subtrees distal to a site of PE (Atilla P. Kiraly and Novak 2004). Finally, Bülow et al. proposed an algorithm for pulmonary arterial tree extraction. It is based on the fact that pulmonary arteries accompany the bronchial tree (Tomas Bülow 2005). The mislabeling of the algorithm increases as vessel radius decreases and it is not specifically evaluated.

Fig. 2(c) shows another problem in 3D CTA images. Since the heart keeps beating, when the images are taken, it causes motion blur and the heart boundaries are ambiguous. In this paper, we suggest AV separation methods that solve the problems of distinct artifacts, unclear boundaries between arteries and veins, motion blur effects.

This papers organized as follows. In Section 2, we present algorithms for the segmentation of lungs, blood vessels and the heart and address the pulmonary trunk detection from a graph. In Section 3, we describe the blood vessel tracking and AV separation methods. Finally, in Section 4, we show the results and evaluation of the AV separation and indicate the direction of future work.

2   PULMONARY TRUNK DETECTION

For AV separation, we use the fact that all pulmonary arteries are connected to the pulmonary trunk that is one of the main branches of the heart. The trunk comes out from the right ventricle of the heart and diverges to the left and right lungs. If there is a path from a branch to the pulmonary trunk within blood vessels, then we can say that the branch is a part of the pulmonary arterial trees. Once pulmonary arteries are extracted, it is assumed that all other blood vessels, which are linked with other parts of the heart, are pulmonary veins. Our AV separation method is composed of 5 steps.

First, we use multilevel thresholds to extract lungs and blood vessels. Automatic lung extraction has been addressed in several papers such as (Shiying Hu and Reinhardt 2001), (Binsheng Zhao and Schwartz 2003), and (Armato III and Sensakovi 2004). Our lung extraction method is similar to those algorithms in terms of using multilevel thresholds and component analysis, but we do not consider left-right lung distinction and the trachea and bronchi segmentation, since the segmented lungs are used to remove unnecessary bones that are the outside of the lungs and have similar intensity ranges to blood vessels. Fig. 3 shows the results of the lung extraction and the thresholding for blood vessels.

Second, to reduce the search space, the bones are removed by using smooth surfaces, which are defined by grid points. Each grid point moves along grid lines that are orthogonal to each initial surface. At the beginning, four flat surfaces are located at the left,

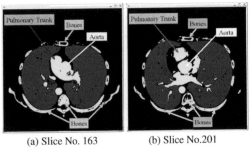

(a) Slice No. 163   (b) Slice No.201

Figure 3. Binary classification of lungs and blood vessels: Brown color represents the two lungs, while green color corresponds to many different objects such as arteries, veins, heart, and bones.

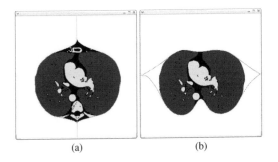

(a)   (b)

Figure 4. (a) Horizontally moving surfaces, (b) Vertically moving surfaces: To reduce the search space, bones are removed by using smooth surfaces. The planes start at left, right, top and bottom locations and move toward the center of the image until hit the lungs. After each surface is smoothed by a convolution, all voxels behind the surfaces are set to zero.

right, top and bottom data boundary and move toward the center of the data, until hit the extracted lungs (Brown color in Fig. 4). Each grid point location at $L(i,j)$ is adjusted by the smoothing convolution (Eq. 1) with the restriction that each surface patch composed by four grid points cannot penetrate the lungs. Once the locations of each grid point are computed, all voxels behind the surfaces are set to zero. The smoothing convolution is computed by

$$O(i,j) = \sum_{k=-m}^{m} \sum_{l=-n}^{n} L(i+k,j+l) \times K(k+m,l+n), \quad (1)$$

where $K()$ represents the kernel with the size of $(2m+1) \times (2n+1)$: The 5×5 and 7×7 kernels are used for the surfaces in Fig. 4(a) and Fig. 4(b) respectively.

Third, for the pulmonary trunk detection, we separate the heart area using dilation, which is

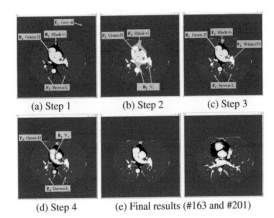

(a) Step 1   (b) Step 2   (c) Step 3

(d) Step 4   (e) Final results (#163 and #201)

Figure 5. Morphological Operations: In this operations, the heart area and blood vessels are separated and the unclear boundaries caused by the motion blur effects are redefined.

one of the morphological operations (Chen and Haralick 1995). We extend the dilation definition to 3D image space (Z^3) with foreground, $F \in Z^3$, and a structuring element, $S_r \in Z^3$, as following, $F \oplus S_r = \{c \in Z^3 | c = a + b \text{ for some } a \in F \text{ and } b \in S\}$, where r represents the radius of the sphere structuring element. In other words, after the dilation operation, a background voxel is set to foreground value, only when at least one voxel in the structuring element coincides with a foreground voxel.

In the first step of our morphological operations, two lungs (Brown-L) and the background voxels (Grey-B), which are set to zero in the bone removing step, work as the foreground (F_1) and the heart and blood vessels (Green-H) and gabs (Black-G) are the background (B_1) (See Fig. 5(a)). Unlike general dilation, the background colors are replaced with new colors that are not used in the image. The new voxels work as new groups in the following steps. Fig. 5(b) shows the new group voxels (V_1) computed by the first dilation step. V_1 is used as the background (B_2) and Grren-H and Black-G are the foreground (F_2) in the second step. After the second step, the heart area (green) and small blood vessels (white) are separated (See Fig. 5(c)). The third and forth steps are to remove the motion blur effects caused by the beating heart and redefine the heart boundaries. The new boundary may be different from the real heart boundary, but it is good enough for pulmonary trunk detection. The entire sequence of steps is as follows:

$F_1 =$ Brown-L $\cup$ Grey-B, $B_1 =$ Green-H $\cup$ Black-G, $V_1 = (F_1 \oplus S_{16}) \cap B_1$,

$F_2 =$ Green-H $\cup$ Black-G, $B_2 = V_1$, $V_2 = F_2 \oplus S_{16}$,

$F_3 =$ White-BV $\cup$ Black-G $\cup$ Brown-L, $B_3 =$ Green-H, $V_3 = F_3 \oplus S_5$, and

$F_4 =$ Green-H $\cup$ Brown-L, $B_4 = V_3$, $V_4 = F_4 \oplus S_4$.

Fourth, after the morphological operations, we apply the distance transform (Saito and Toriwaki 1994)

333

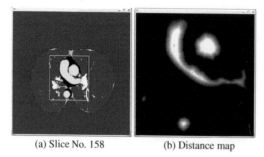

(a) Slice No. 158   (b) Distance map

Figure 6. Distance Transform: (b) shows the distance map of the left box of (a). The red voxels represent the local maxima of the distance values in the green region of (a). The blue local maxima are in the white area of (a).

to the green (the heart and its main branches) and white (blood vessels) regions in Fig. 5(e) to compute seed points that are local maxima of the distance values (See Fig. 6). At each seed point location, a sphere denoted by S_{id} is generated. Each sphere S_{id} is represented by the center C_{id} and the radius R_{id}. To compute R_{id}, spheres are inflated at C_{id} until any part of the sphere surface hits a boundary that is defined as the surface of the green area in Fig. 5(e). To reduce the number of spheres, some spheres with centers that are covered by a larger sphere are removed. When two spheres are overlapped, the spheres have a connection between them. All overlapped spheres with S_a are the neighbors of S_a. Let N_a be the neighbor set of S_a.

Some initial spheres generated by the above method can be separated from the main group that has the biggest number of connected spheres. To connect those spheres to the main group, the following two rules are applied, until all spheres have more than one neighbor.

Rule 1 If there exists a sphere, S_a, such that $|N_a| = 1$, then a new sphere, S_c, is generated. Let $N_a = \{S_b\}$ and $N_c = \emptyset$. To compute C_n and R_n, we define the set M_1 of voxel locations that satisfy the following conditions $R_a < D_{ao} \le R_a + 2$ and $V_{ao} \cdot V_{ba} > 0$, where D_{ao} is the Euclidean distance between C_a and a voxel location L_o and V_{ao} is the normalized vector from C_a to L_o. C_n is one of the elements of M_1 that maximizes R_c. If there are more than two voxel locations that have the same maximum radius, then we pick a voxel location that has the maximum value of $V_{ao} \cdot V_{ba}$. Once S_c is added, the neighbor sets of N_a and N_c change to $\{S_b, S_c\}$ and $\{S_a\}$ respectively.

Rule 2 If there exists an isolated sphere, S_d such that $|N_d| = 0$, then a new sphere, S_e, is generated at $C_d \in M_2$ that maximizes R_d, where M_2 is a set of voxel locations that have the distance between R_d and $R_d + 1$ from C_d.

Finally, after all the above steps, the 3D CTA image space (Fig. 7(a)) changes to a graph representation (Fig. 7(b)). To detect the pulmonary trunk in the graph,

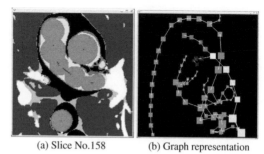

(a) Slice No.158   (b) Graph representation

Figure 7. Spheres and Medial Axis: The purple spheres are generated in the green heart area. The blue lines connecting the overlapped spheres are skeletons in (a). The graph representation of (b) shows the pulmonary trunk (green) and two branches that go to the left(red boxes) and right (blue boxes) lungs. The graph contains not only skeletons, but also blood vessel thickness at the sampled box locations.

we use direction vectors in the 3D space defined by three orthogonal axes X (horizontal), Y (vertical), and Z (perpendicular to slices). The origin is located at the top left corner of the first slice. The X and Y axes increase as go right and down in image slices respectively and the Z axis increases as the slice number increases. We search the sphere S_{ctr} (The yellow box in Fig. 7(b)) that has three branches and their direction vectors are (X_1, Y_1, Z_1), (X_2, Y_2, Z_2) and (X_3, Y_3, Z_3) that have 3–5 spheres per each branch and satisfy the following properties:

$$(\|Z_1\| > \|X_1\| \text{ or } \|Z_1\| > \|Y_1\|), \ Y_1 < 0, \ Z_1 > 0,$$
$$(\|X_2\| > \|Z_2\| \text{ or } \|Y_2\| > \|Z_2\|), \ X_2 < 0, \ Y_2 > 0,$$
$$\text{and}$$
$$(\|X_3\| > \|Z_3\| \text{ or } \|Y_3\| > \|Z_3\|), \ X_3 > 0, \ Y_3 > 0.$$

The green, red and blue boxes in Fig. 7(b) follow the above three criteria respectively. Once the pulmonary trunk and two branches are detected, all connections between the three branches are all other spheres are removed to separate the pulmonary arteries from veins.

3 BLOOD VESSEL EXTRACTION

To solve the ambiguous boundary problem (depicted in Fig. 1 and 2), we suggest a new vascular tree reconstruction algorithm as follows: First, spheres are generated at each blue seed point of Fig. 6(b) by using the same methods explained in the previous section (The fourth step) such as inflation, small radius sphere removals and neighbor connections. Second, we find all possible dead ends of blood vessels. In fact, blood vessels do not have dead ends in the human body, but as they branch and become more distant from the heart, the vascular diameter decreases and blood vessels disappear in 3D CTA images. Therefore, there are many

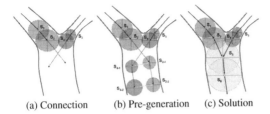

(a) Connection   (b) Pre-generation   (c) Solution

Figure 8. Branch Merging: To verify the connection between S_2 and S_4, spheres are pre-generated along the directions of V_{12} and V_{34}. If the lines connecting the sphere centers (C_{2-1}, C_{4-1}) and (C_{2-2}, C_{4-2}) are totally in the inside of blood vessel, then S_5 is generated and S_2 and S_4 are the neighbors of S_5.

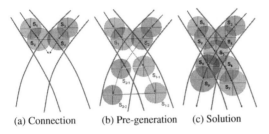

(a) Connection   (b) Pre-generation   (c) Solution

Figure 9. Branch Separation: If at least one line between the pre-generated sphere centers is in the outside of blood vessels (the red line in (b) between S_{2-2} and S_{4-2}), then there exist two different branches along V_{12} and V_{34}.

blood vessel dead ends in the image data. Third, at each dead end, spheres are added toward the opposite directions of the dead ends by Rule 1 defined in the previous section. When two spheres, S_2 and S_4, are overlapped (Fig. 8(a) and 9(a)), spheres are pre-generated along the blood vessel directions, V_{12} and V_{34}, (Fig. 8(b) and 9(b)) to verify the merging. If the lines between the pre-generated sphere centers are totally in the inside of blood vessels, then the two spheres, S_2 and S_4, should be merged and connected to S_5 (Fig. 8(c)). Otherwise, there exist two different branches along the two different directions (See Fig. 9(c)). Finally, each sphere, which is close to the heart, is connected to one of the heart spheres in Fig. 7 based on branch directions.

The above algorithm can solve the unclear boundary problem of Fig. 2, but Fig. 1, since the branch separation test of Fig. 9 is failed in the heart area. Therefore, we suggest a knowledge-based searching method for the problem of Fig. 1. In the method, we compute initial skeletons by connecting the local maxima of Fig. 6 along the ridges of the distance map, which is similar to Bitter *et al*'s algorithm (Ingmar Bitter and Sato 2001).

At the end sphere of the right pulmonary artery branch in Fig. 7(a), the direction of the thickest artery branch (X_r, Y_r, Z_r) has the following properties: $\|X_r\| > 0$, $\|Y_r\| > 0$ and $\|Z_r\| > 0$, while the direction

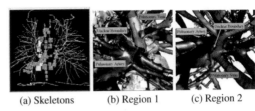

(a) Skeletons   (b) Region 1   (c) Region 2

Figure 10. AV separation results.

of the pulmonary vein (X_v, Y_v, Z_v) has the feature of $\|X_v\| < 0$, $\|Y_v\| > 0$ and $\|Z_v\| > 0$ (See Fig. 1(c)). Since the boundary between the pulmonary artery and vein is not clear, there exist a wrong skeleton that passes through the boundary. After searching the pulmonary artery and vein via the knowledge-based method, we can remove the wrong skeletons and re-define a rough boundary between them.

4 RESULTS

After all the above steps, all spheres that are connected to the pulmonary trunk are pulmonary artery. All others that have links to any part of the heart are pulmonary vein. Fig. 10(a) shows the spheres and skeletons. The blue color boxes represent pulmonary arteries that are connected to the pulmonary trunk. Fig. 10(b,c) contains the results of our AV separation algorithm. Fig. 10(b) is the same region as Fig. 1(c) and Fig. 10(c) shows the area of Fig. 2(b).

To verify our AV separation methods, we applied the methods to 6 datasets and counted wrong labeling and missing branches. The average number of incorrect labels is around 5 and the average number of missing branches is around 6. Our methods need to be evaluated on larger data sets. Automatic pulmonary embolus detection is the future work of this research.

ACKNOWLEDGMENTS

First and second authors are supported in part by NSF grants ITR-EIA-0325550, CNS-0540033 and NIH grants P20 RR020647, R01 GM074258-021 and R01-GM073087.

REFERENCES

Armato III, S. G. and W. F. Sensakovi (2004). Automated lung segmentation for thoracic ct: Impact on computer-aided diagnosis. *Academic Radiology 11*(9), 1011–1021.

Atilla P. Kiraly, Eric Pichon, D. P. N. and C. L. Novak (2004). Analysis of arterial sub-trees affected by pulmonary emboli. In *Proceedings of SPIE (Medical Imaging)*, Volume 5370, pp. 1720–1729. The International Society for Optical Engineering.

Binsheng Zhao, Gordon Gamsu, M. S. G. L. J. and L. H. Schwartz (2003). Automatic detection of small lung nodules on ct utilizing a local density maximum algorithm. *Journal of Applied Clinical Medical Physics 4*(3), 248–260.

Chen, S. and R.M. Haralick (1995). Recursive erosion, dilation, opening, and closing transforms. *IEEE Transactions on Image Processing 4*(3), 335–345.

Cornelis M. van Bemmel, Luuk J. Spreeuwers, M. A. V. and W. J. Niessen (2003). Level-set-based arter-vein separation in blood pool agent cemr angiograms. *IEEE Transactions on Medical Imaging 22*(10), 1224–1234.

Gady Agam, Samuel G. Armato, I. C. W. (2005). Vessel tree reconstruction in thoracic ct scans with application to nodule detection. *IEEE Transaction on Medical Imaging 24*(4), 486–499.

Hartmann, I. and M. Prokop (2002). Spiral ct in the diagnosis of acute pulmonary embolism. *Medica Mundi 46*(3), 2–11.

Ingmar Bitter, A. E. K. and M. Sato (2001). Penalized-distance volumetric skeleton algorithm. *IEEE Transactions on Visualization and Computer Graphics 7*(3), 195–206.

Ingrid Sluimer, Arnold Schilham, M. P. and B. van Ginneken (2006). Comptuer analysis of computed tomography scans of the lung: Survey. *IEEE Transactions on Medical Imaging 25*(4), 385–405.

Jamshid Dehmeshki, Xujiong Ye, F. W. X. Y. L. M. A. M. S. and S. Qanadli (2004, September). An accurate and reproducible scheme for quantification of coronary artery calcification in ct scans. In *Proceedings of the 26th Annual International Conference of IEEE EMBS*, pp. 1918–1921. IEEE: The International Society for Optical Engineering.

Katja Bühler, P. F. and A. L. Cruz (2003). Geometric methods for vessel visualization and quantification- a survey. In H. M. G. Brunnett, B. Hamann (Ed.), *Geometric Modelling for Scientific Visualization*, pp. 1–24. Springer.

Kirbas, C. and F. Quek (2004). A review of vessel extraction techniques and algorithms. *ACM Computing Surveys (CSUR) 36*(2), 81–121.

Michael Bock, Stefan O. Schoenberg, F. F. a. L. R. S. (2000). Separation of arteries and veins in 3d mr angiography using correlation analysis. *IEEE Transactions onMedical Imaging 43*, 481–487.

Saito, T. and J.-I. Toriwaki (1994). New algorithms for euclidean distance transformation of an *n*-dimensinal digitized picture with applications. *Pattern Recognition 27*(11), 1551–1565.

Schoepf, U. J. and P. Costello (2004). Ct angiography for diagnosis of pulmonary embolism: State of the art. *Radiology 230*(2), 329–337.

Shiying Hu, E. A. H. and J. M. Reinhardt (2001). Automatic lung segmentation for accurate quantitation of volumetric X-ray ct images. *IEEE Transactions on Medical Imaging 20*(6), 490–498.

Stefancik, R. M. and M. Sonka (2001). Highly automated segmentation of arterial and venous trees from three-dimensional magnetic resonance angiography (mra). *The International Journal of Cardiovascular Imaging 17*, 37–47.

Tianhu Lei, Jayaram K. Udupa, P. K. S. and D. Odhner (2001). Artery-vein separation via mra – an image processing approach. *IEEE Transactions on Medical Imaging 20*(8), 689–703.

Tomas Bülow, Fafael Wiemker, T. B. C. L. S. R. (2005). Automatic extraction of the pulmonary artery tree from multi-slice ct data. *Medical Imaging 2005: Physiology, Function, and Structure from Medical Images 6*(23), 730–740.

YoshitakaMasutani, H. M. J. D. (2002). Computerized detection of pulmonary embolism in spiral ct angiography based on volumetric image analysis. *IEEE Transaction on Medical Imaging 21*(12), 1517–1523.

Computational Modelling of Objects Represented in Images –
João Manuel R.S. Tavares & R.M. Natal Jorge (eds)
© *2007 Taylor & Francis Group, London, ISBN 978-0-415-43349-5*

LVQ acrosome integrity assessment of boar sperm cells

Nicolai Petkov[1], Enrique Alegre[2], Michael Biehl[1] & Lidia Sánchez[2]
[1] *Institute of Mathematics and Computing Science, University of Groningen, The Netherlands*
[2] *Department of Electrical and Electronics Engineering, University of León, Spain*

ABSTRACT: We consider images of boar spermatozoa obtained with an optical phase-contrast microscope. Our goal is to automatically classify single sperm cells as acrosome-intact (class 1) or acrosome-reacted (class 2). Such classification is important for the estimation of the fertilization potential of a sperm sample for artificial insemination. We segment the sperm heads and compute a feature vector for each head. As a feature vector we use the gradient magnitude along the contour of the sperm head. We apply learning vector quantization (LVQ) to the feature vectors obtained for 152 heads that were visually inspected and classified by a veterinary expert. A simple LVQ system with only three prototypes (two for class 1 and one for class 2) allows us to classify cells with equal training and test errors of 0.165. This is considered to be sufficient for semen quality control in an artificial insemination center.

1 INTRODUCTION

Image processing techniques have been applied to assess the quality of semen samples for therapeutic and fertilization purposes (Verstegen, Iguer-Ouada, and Onclin 2002; Linneberg, Salomon, Svarer, and Hansen 1994). Techniques that had been originally developed for human semen samples have meanwhile been adapted to other species. Most works on boar semen evaluation focus on measuring concentration and motility of spermatozoa or on detecting sperm head shape abnormalities, such as double heads, macro heads etc. Image processing techniques are deployed to automate such analysis in Computer Aided Sperm Analysis (CASA) systems (Suzuki, Shibahara, Tsunoda, Hirano, Taneichi, Obara, Takamizawa, and Sato 2002). The use of sperm motility (Quintero, Rigaub, and Rodríguez 2004) presents several disadvantages, such as sensitivity to temperature changes and unclear relation to fertility. In the case of sperm morphology, assessment is based on the state of the sperm cell structure: head, middle piece and tail, on computing morphometric measures to detect head shape abnormalities (Rijsselaere, Soom, Hoflack, Maes, and de Kruif 2004, Beletti, Costa, and Viana 2005; Ostermeier, Sargeant, Yandell, and Parrish 2001) and on detecting droplets in tails. Analysis of intracellular density distribution has been used in (Sánchez, Petkov, and Alegre 2005b; Sánchez, Petkov, and Alegre 2005a; Biehl, Pasma, Pijl, Sanchez, and Petkov 2006).

Veterinary experts believe that sperm fertility is related to the state of the acrosome, a cap-like structure that develops over the anterior half of the spermatozoon's head. It has its own membrane and contains enzymes. As the sperm approaches the oocyte, an acrosome reaction takes place during which the anterior head plasma membrane of the sperm fuses with the outer membrane of the acrosome, exposing the contents of the acrosome. The released enzymes are required for the penetration of sperm through a layer of follicular (cumulus) cells that encase the oocyte. The acrosome reaction also renders the sperm capable of penetrating through the zona pellucida (an extracellular coat surrounding the oocyte) and fusing with the egg. For these reasons, veterinary experts believe that a semen sample with a high fraction of acrosome-reacted sperm has low fertilizing capacity and cannot be used for artificial insemination.

The traditional techniques to assess acrosome integrity, such as visual inspection by veterinary experts or staining, are time consuming and have relatively high costs. Despite the broadly recognized importance of acrosome integrity evaluation in semen quality assessment, we are not aware of any image processing work on this topic. In this work we propose a new method to assess acrosome integrity based on the automatic analysis of grey level images acquired with a phase contrast microscope. Our approach is based on the observation that there are some characteristic differences in the gradient magnitude profiles along the contours of acrosome-intact vs. acrosome-reacted sperm heads. We extract a feature vector from the grey level image of a cell head and use this vector to classify the cell as acrosome-intact or acrosome-reacted by comparing it with prototype feature vectors. We determine these prototype vectors by applying Learning Vector Quantization (LVQ) to a training set of feature vectors obtained from the images of sperm heads that were marked by a veterinary expert by means of visual inspection as acrosome-intact or acrosome-reacted.

2 VECTORIZATION

2.1 *Pre-processing and segmentation*

Boar sample images were captured by means of a digital camera Nikon Coolpix 5000 mounted on an optical phase-contrast microscope Nikon Eclipse. The magnification used was $\times 100$ and the dimensions of each sample image were 2560×1920 pixels. A boar semen sample image contains a number of spermatozoa which can vary widely from one sample to the next and also present different orientations (Fig. 1).

Sperm head images were obtained manually by cropping from a boar semen sample image. These images were visually inspected by a veterinary expert and the spermatozoa were classified as acrosome-intact or acrosome-reacted. For each obtained image (Fig. 2a), we segment the sperm head by converting the image to a binary image using Otsu's method (Otsu 1979) and applying several morphological operations (dilations and erosions), Fig. 2b. We use the contour of the sperm head binary mask in the following. We also localize the point where the middle piece, from which the tail develops, connects to the head. This point is used as a reference point in the following.

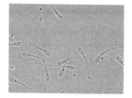

Figure 1. Boar semen sample images acquired using a phase-contrast microscope.

2.2 *Scale-dependent gradient computation and vectorization*

Let $f(x,y)$ and $g_\sigma(x,y)$ be a grey level distribution in an input image and a two-dimensional (2D) Gaussian function with standard deviation σ, respectively. The x- and y-components of the scale-dependent gradient of $f(x,y)$ are defined as convolutions of $f(x,y)$ with the x- and y-derivatives of $g_\sigma(x,y)$, respectively:

$$\nabla_{\sigma,x}f = f * \frac{\partial g_\sigma}{\partial x}, \quad \nabla_{\sigma,y}f(x,y) = f * \frac{\partial g_\sigma}{\partial y}. \quad (1)$$

This approach to gradient computation (Schwartz 1950) has been shown to reduce noise and discretisation effects (Canny 1986; Tagare and de Figueiredo 1990; Grigorescu, Petkov, and Westenberg 2004). In the following we use the magnitude $M_\sigma(x,y)$ of the scale-dependent gradient:

$$M_\sigma(x,y) = \sqrt{(\nabla_{\sigma,x}f(x,y))^2 + (\nabla_{\sigma,y}f(x,y))^2}. \quad (2)$$

Fig. 3 shows intracellular density distribution images typical for acrosome-intact and acrosome-reacted boar spermatozoa, respectively, and Fig. 4 shows the corresponding gradient magnitude images.

Next we determine the gradient magnitude along the cell head boundary as a 1D function of the boundary curve length from the reference point to a given contour point in a clock-wise direction. For a given contour point we take the local maximum of the gradient in a 5×5 neighborhood of that point. The resulting discrete function is a vector that we re-size by interpolation to a uniform length of 40 elements. We also normalize this vector by dividing it by its largest element, Fig. 5. The vectors obtained for different sperm heads are used for LVQ.

3 ANALYSIS BY LEARNING VECTOR QUANTIZATION

3.1 *LVQ training*

In the following we apply LVQ for the distance-based classification of the data. LVQ has been used in a

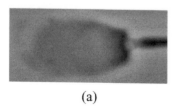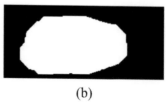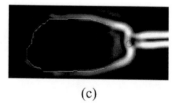

(a)	(b)	(c)

Figure 2. (a) Grey level image of a sperm head and a part of the middle piece protruding from the head. (b) Binary image of the sperm head obtained by thresholding and subsequent morphological processing. (c) Contour of the binary head mask (white line) superimposed on the gradient magnitude image. The reference point at the base of the middle piece is marked with a black circle.

variety of problems due to its flexibility and conceptual clarity (Kohonen 1995; Neural Networks Research Centre 2002).

We apply a heuristic training algorithm, so-called LVQ1 (Kohonen 1995), in order to determine typical representatives of each class from a (sub-) set of labeled training data $ID = \{\xi^\mu, S_T^\mu\}_{\mu=1}^P$ Here, the $\xi^\mu \in IR^N$ ($N = 40$) are the vectors of gradient magnitudes along the contour as described in the previous section. The corresponding class membership provided by the veterinary experts is denoted as

$$S_T^\mu = \begin{cases} +1 & \text{if } \xi^\mu \text{ was labeled as } \textit{acrosome-intact (class 1)} \\ -1 & \text{if } \xi^\mu \text{ was labeled as } \textit{acrosome-reacted (class 2)} \end{cases} \quad (3)$$

Fig. 5 displays three example profiles from each of the two classes.

In the set of prototypes $\{w^1, w^2, \ldots, w^M\}$ a vector $w^j \in IR^N$ is supposed to represent data with class membership $S^j \in \{+1, -1\}$. These assignments, as well as the number of prototypes are specified prior to training and remain unchanged.

At each time step t of the iterative training procedure, one example $\{\xi^\mu, S_T^\mu\}$ is selected randomly from ID. We evaluate the distances $d(j, \mu)$ of ξ^μ from all current prototype vectors $w^j(t)$. Here, we restrict ourselves to the standard (squared) Euclidean measure

$$d(j, \mu) = (\xi^\mu - w^j)^2 = \sum_i \left(\xi_i^\mu - w_i^j(t)\right)^2. \quad (4)$$

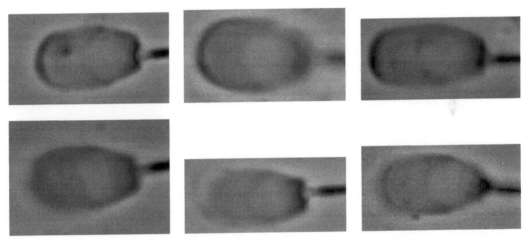

Figure 3. Grey level images characteristic of acrosome-intact (upper row) and acrosome-reacted (lower row) boar spermatozoa.

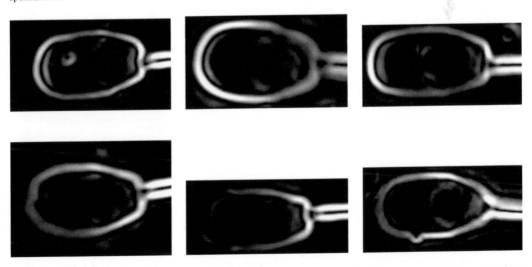

Figure 4. Gradient magnitude images characteristic of acrosome-intact (upper row) and acrosome-reacted (lower row) boar spermatozoa. The parameter σ used to compute the scale-dependent gradient is set to 0.03 of the cell head length.

Next we identify the minimal distance $d(J, \mu)$ among all prototypes and the corresponding *winnner*

$$\boldsymbol{w}^J(t) \text{ with } d(J, \mu) = \min_k \{d(k, \mu)\}. \quad (5)$$

In LVQ1, only this winner is updated according to

$$\boldsymbol{w}^J(t+1) = \boldsymbol{w}^J(t) + \eta(t) \left[S_T^\mu S^J \right] \left(\boldsymbol{\xi}^\mu - \boldsymbol{w}^J(t) \right). \quad (6)$$

Hence, the update is towards or away from the actual input $\boldsymbol{\xi}^\mu$, if the class labels of winner and example agree or disagree, respectively.

In the following studies, prototype vectors are initialized close to the mean of vectors $\boldsymbol{\xi}^\mu$ in the corresponding class. In order to avoid exactly coinciding $\boldsymbol{w}^j(0)$, small random displacements from the class conditional means are performed.

The learning rate $\eta(t)$ controls the step width of the iteration. It gradually decreases in the course of learning following a schedule of the form

$$\eta(t) = \eta_o / (1 + a\,t) \text{ with } a \text{ such that } \eta(t_f) = \eta_f. \quad (7)$$

Results presented in the next section were obtained with a schedule that decreases the learning rate from $\eta_o = 0.1$ to $\eta_f = 0.0001$ in 1000 sweeps through the training set, i.e. $t_f = 1000\,P$.

After training, the system parameterizes a distance-based classification scheme: any data $\boldsymbol{\xi}$ is assigned to the class S^J which is represented by the closest prototype.

3.2 Cross–validation

To obtain estimates of the performance after training we employ eight-fold cross-validation: We split the set of 152 available training data (105 from class 1 and 47 from class 2) randomly into disjoint subsets ID_i, $i = 1, 2 \ldots 8$, of equal size. For a given number of prototypes, each of eight identically designed LVQ systems, $n = 1, 2 \ldots 8$, is trained from the set $\cup_{i \neq n} ID_i$ containing $P = 133$ examples. Then, ID_n serves as a test set to evaluate the performance on novel data.

In the following, ϵ_{train} denotes the fraction of misclassified example data, obtained after training and on average over the eight systems. The test error ϵ_{test} quantifies the averaged performance with respect to the test set. Analogously we evaluate the class-specific test errors $\epsilon_{test}^{(1)}$ and $\epsilon_{test}^{(2)}$ as well as the training errors $\epsilon_{train}^{(1)}$ and $\epsilon_{train}^{(2)}$ with respect to only class 1 or class 2 data, respectively.

Although the training sets ID_i overlap, the corresponding standard deviations provide a rough measure of the expected variation of the classifier with different realizations of the training set. The main purpose of the cross-validation scheme is to compare the performance of different LVQ schemes, i.e. systems with different numbers of prototypes.

4 RESULTS

We have performed LVQ1 training following the above described scheme for systems with m and n prototypes

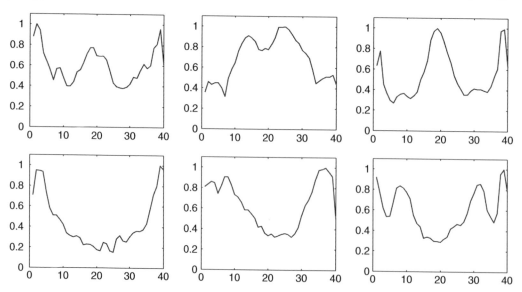

Figure 5. Example gradient magnitude profiles along the head boundary $\boldsymbol{\xi} \in I\!R^{40}$ from the class of acrosome-intact (upper row) and acrosome-reacted (lower row) spermatozoa. The displayed profiles correspond to the images shown in Figs. 3 and 4. These discrete 1D functions represent the vectors used for LVQ.

representing class 1 and class 2 data, respectively. The Table 1 summarizes the observed error measures in the simplest LVQ configurations. Numbers in parentheses give the corresponding standard deviations observed in the eight-fold cross-validation.

Table 1. Training and test errors observed using m and n prototypes of the class 1 and 2, respectively.

(m, n)	Training error	Test error
$(1,1)$	$\epsilon_{train} = 0.167\ (0.01)$	$\epsilon_{test} = 0.165\ (0.09)$
	$\epsilon_{train}^{(1)} = 0.061\ (0.01)$	$\epsilon_{test}^{(1)} = 0.054\ (0.09)$
	$\epsilon_{train}^{(2)} = 0.404\ (0.02)$	$\epsilon_{test}^{(2)} = 0.393\ (0.13)$
$(2,1)$	$\epsilon_{train} = 0.165\ (0.01)$	$\epsilon_{test} = 0.165\ (0.07)$
	$\epsilon_{train}^{(1)} = 0.057\ (0.01)$	$\epsilon_{test}^{(1)} = 0.055\ (0.06)$
	$\epsilon_{train}^{(2)} = 0.408\ (0.04)$	$\epsilon_{test}^{(2)} = 0.423\ (0.24)$
$(1,2)$	$\epsilon_{train} = 0.160\ (0.01)$	$\epsilon_{test} = 0.177\ (0.05)$
	$\epsilon_{train}^{(1)} = 0.070\ (0.01)$	$\epsilon_{test}^{(1)} = 0.080\ (0.07)$
	$\epsilon_{train}^{(2)} = 0.359\ (0.04)$	$\epsilon_{test}^{(2)} = 0.398\ (0.23)$
$(2,2)$	$\epsilon_{train} = 0.154\ (0.01)$	$\epsilon_{test} = 0.184\ (0.05)$
	$\epsilon_{train}^{(1)} = 0.064\ (0.01)$	$\epsilon_{test}^{(1)} = 0.092\ (0.08)$
	$\epsilon_{train}^{(2)} = 0.357\ (0.04)$	$\epsilon_{test}^{(2)} = 0.392\ (0.21)$
$(3,3)$	$\epsilon_{train} = 0.143\ (0.02)$	$\epsilon_{test} = 0.178\ (0.06)$
	$\epsilon_{train}^{(1)} = 0.038\ (0.01)$	$\epsilon_{test}^{(1)} = 0.064\ (0.08)$
	$\epsilon_{train}^{(2)} = 0.377\ (0.04)$	$\epsilon_{test}^{(2)} = 0.424\ (0.12)$

We would like to point out that even in the minimal setting ($m = n = 1$), the outcome of LVQ1 differs significantly from a naive representation of the two classes by their respective mean profiles. If we replace the two LVQ prototypes by class-conditional mean vectors, we obtain an average test and training error of $\epsilon_{test} \approx \epsilon_{train} \approx 0.184$. LVQ1 yields a better performance because the supervised training procedure detects and emphasizes the discriminative features in the data.

Fig. 6 (upper panel) displays two LVQ prototypes as obtained in one of the training runs. The lower panel shows an example outcome of LVQ1 with two class 1 and one class 2 prototype ($m = 2, n = 1$). Note how the class 1 prototypes have *specialized* to represent the two predominant types of acrosome-intact profiles which are also apparent in Figure 5 (low or high intensity around ξ_1 and ξ_{40}).

In general, test and training error are lower with respect to class 1, which reflects greater fluctuations in the class 2 data. Employing more prototypes for the second class yields a more balanced classifier, however, the overall test performance remains unchanged or even degrades. Note, for instance, that a system with one class 1 but two class 2 prototypes ($m = 1, n = 2$) clearly performs worse than the minimal configuration with $m = n = 1$.

When increasing the number of prototypes, i.e. the complexity of the LVQ system, we observe a

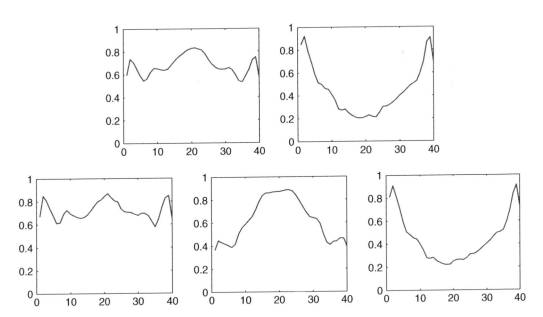

Figure 6. **Upper panel:** Two prototype profiles as obtained in LVQ1 from one set of 133 examples. The left and right prototype represent class 1 of acrosome-intact and class 2 of acrosome-reacted spermatozoa, respectively. **Lower panel:** Three prototype profiles as obtained in LVQ1 from one set of 133 examples. The leftmost and center prototypes represent class 1 (of acrosome-intact spermatozoa), whereas the rightmost profile corresponds to class 2 (of acrosome-reacted spermatozoa).

decrease of the overall training error ϵ_{train}. However, it is accompanied by a (moderate) increase of the test error which signals *over-fitting*: While the particular training set can be represented in greater detail, the generalization ability deteriorates. At the same time, the variance of the outcome tends to be larger in the overly complex systems.

Finally, we obtained essentially the same results with vectors of 20 elements that were obtained by taking every second element of the vectors of 40 elements used above.

5   SUMMARY AND OUTLOOK

We extract feature vectors from boar sperm head images and use these vectors for LVQ training and classification as acrosome-intact (class 1) or acrosome-reacted (class 2). As a feature vector we use the gradient magnitude along the contour of the sperm head. A simple LVQ system with only three prototypes (two for class 1 and one for class 2) allows us to classify cells with equal training and test errors of 0.165. This is considered to be sufficient for semen quality control in an artificial insemination center.

In future investigations we intend to increase the number of features by considering also the grey level distribution in the sperm head interior. We furthermore intend to apply more sophisticated cost function based schemes such as *Generalized Learning Vector Quantization* (GLVQ) as suggested in (Sato and Yamada 1995). Modified distance measures and *relevance learning*, see e.g. (Hammer and Villmann 2002), could be applied in order to obtain a better understanding of this classification task and to extract the most relevant features.

REFERENCES

Beletti, M., L. Costa, and M. Viana (2005). A comparison of morphometric characteristics of sperm from fertile Bos taurus and Bos indicus bulls in Brazil. *Animal Reproduction Science 85*, 105–116.

Biehl, M., P. Pasma, M. Pijl, L. Sanchez, and N. Petkov (2006). Classification of boar sperm head images using learning vector quantization. In M. Verleysen (Ed.), *Proc. European Symposium on Artificial Neural Networks (ESANN), Brugge, April 26–28, 2006*, pp. 545–550. d-side, Evere, Belgium.

Canny, J. F. (1986). A computational approach to edge detection. *IEEE Trans. Pattern Analysis and Machine Intelligence 8*(6), 679–698.

Grigorescu, C., N. Petkov, and M. A. Westenberg (2004). Contour and boundary detection improved by surround suppression of texture edges. *Image and Vision Computing 22*(8), 609–622.

Hammer, B. and T. Villmann (2002). Generalized relevance learning vector quantization. *Neural Networks 15*, 1059–1068.

Kohonen, T. (1995). *Self-organizing maps*. Springer, Berlin.

Linneberg, C., P. Salamon, C. Svarer, and L. Hansen (1994). Towards semen quality assessment using neural networks. In *Proc. IEEE Neural Networks for Signal Processing IV*, pp. 509–517.

Neural Networks Research Centre (2002). *Bibliography on the Self-Organizing Map (SOM) and Learning Vector Quantization (LVQ)*. Helsinki University of Technology.

Ostermeier, G., G. Sargeant, T. Yandell, and J. Parrish (2001). Measurement of bovine sperm nuclear shape using Fourier harmonic amplitudes. *J. Androl. 22*, 584–594.

Otsu, N. (1979). A threshold selection method from gray-level histograms. *IEEE Transactions on Systems, Man and Cybernetics 9*, 62–66.

Quintero, A., T. Rigaub, and J. Rodríguez (2004). Regression analyses and motile sperm subpopulation structure study as improving tools in boar semen quality analysis. *Theriogenology 61*, 673–690.

Rijsselaere, T., A. V. Soom, G. Hoflack, D. Maes, and A. de Kruif (2004). Automated sperm morphometry and morphology analysis of canine semen by the Hamilton-Thorne analyser. *Theriogenology 62*(7), 1292–1306.

Sánchez, L., N. Petkov, and E. Alegre (2005a). Classification of boar spermatozoid head images using a model intracellular density distribution. In M. Lazo and A. Sanfeliu (Eds.), *Progress in Pattern Recognition, Image Analysis and Applications: Proc. 10th Iberoamerican Congress on Pattern Recognition, CIARP 2005, Lecture Notes in Computer Science*, Volume 3773, pp. 154–160. Springer-Verlag Berlin Heidelberg.

Sánchez, L., N. Petkov, and E. Alegre (2005b). Statistical approach to boar semen head classification based on intracellular intensity distribution. In A. Gagalowicz and W. Philips (Eds.), *Proc. Int. Conf. on Computer Analysis of Images and Patterns, CAIP 2005, Lecture Notes in Computer Science*, Volume 3691, pp. 88–95. Springer-Verlag Berlin Heidelberg.

Sato, A. and K. Yamada (1995). Generalized learning vector quantization. In G. Tesauro, D. Touretzky, and T. Leen (Eds.), *Advances in Neural Information Processing Systems*, Volume 7, pp. 423–429. MIT Press.

Schwartz, L. (1950). *Théorie des Distributions*. Vol. I, II of Actualités scientifiques et industrielle. L'Institute de Mathématique de l'Université de Strasbourg.

Suzuki, T., H. Shibahara, H. Tsunoda, Y. Hirano, A. Taneichi, H. Obara, S. Takamizawa, and I. Sato (2002). Comparison of the sperm quality analyzer IIC variables with the computeraided sperm analysis estimates. *International Journal of Andrology 25*, 49–54.

Tagare, H. and R. de Figueiredo (1990). On the localization performance measure and optimal edge detection. *IEEE Trans. Pattern Analysis and Machine Intelligence 12*(12), 1186–1190.

Verstegen, J., M. Iguer-Ouada, and K. Onclin (2002). Computer assisted semen analyzers in andrology research and veterinary practice. *Theriogenology 57*, 149–179.

Computational Modelling of Objects Represented in Images –
João Manuel R.S. Tavares & R.M. Natal Jorge (eds)
© 2007 Taylor & Francis Group, London, ISBN 978-0-415-43349-5

Reconstruction of 3D models from medical images: Application to female pelvic organs

Soraia Pimenta, João Manuel R.S. Tavares & Renato Natal Jorge
Faculdade de Engenharia da Universidade do Porto, Porto, Portugal

Fátima Alexandre
IDMEC – Faculdade de Engenharia da Universidade do Porto, Porto, Portugal

Teresa Mascarenhas
Hospital S. João – Faculdade de Medicina da Universidade do Porto, Porto, Portugal

Rania F. El Sayed
Faculty of Medicine, University of Cairo, Cairo, Egypt

ABSTRACT: Physical anatomic models obtained from medical images can be a powerful aid in the treatment of several medical situations. In this work, a complete methodology, from the acquisition of medical images to materialisation of three-dimensional models, with a particular emphasis on image segmentation techniques, was studied, developed and applied to organs of the female pelvic cavity. This methodology is presented in this paper, including several experimental results of each involved step.

Keywords: models; segmentation; 3D objects reconstruction; prototyping; pelvic cavity; pelvic organs.

1 INTRODUCTION

Physical models of anatomic systems, organs and tissues can be extremely useful tools in many complex medical situations, referring to knowledge acquisition, diagnostic, simulation and even treatment of the related diseases.

In this work the reconstruction of three-dimensional models from medical images was applied to the female pelvic cavity. The main objective was to study, develop, employ, test and compare some available approaches for the three-dimensional reconstruction and building of physical models for medical purposes.

2 FEMALE PELVIC CAVITY

The female pelvic cavity is an interesting and challenging application for the reconstruction of virtual and physical models, both from medical and technical perspectives, [1–3].

Pelvic floor dysfunction is consistently associated to urinary incontinence and pelvic organ prolapse due to injuries and deterioration of the muscles, nerves and connective tissues that support and control normal pelvic organ function. The knowledge already achieved on the causers of pelvic floor disorders shows a great diversity of risk factors, as anatomic predisposition, childbirth, muscular and general injuries, surgeries, and the simple ageing.

Urinary Incontinence (UI) affects a large percentage of the adult world population, imposing significant adverse physical, psychological and financial effects on individuals, their families and caregivers (see Refs [1–3]). In the USA, studies show that more than 53% of the aged suffer from this disorder, and even among the people under 60 the incidence of the UI can reach 35%. Besides the decrease of the quality of life it imposes to the patients, the economic cost of the UI is very high, surpassing the 3500 € per patient and year.

In addition to these medical motivations, the female pelvic cavity also presents some characteristics that make it very interesting from technical point of view. In fact, pelvic cavity is a complex set of organs and tissues, presenting therefore some examples of both easily and hardly extracted (segmented) structures from medical images, which allows a progressive comprehension of the methodologies and procedures available to develop virtual and physical models as well as to identify their potentialities and limitations.

3 PROCEDURES

Usually, the methodologies used to obtain physical models of an anatomic structure represented in medical images follow a sequence of four fundamental procedures: 1) acquisition of the medical images of the structure to be reconstructed, namely sets of 2D slices; 2) segmentation of the interesting organs in each 2D slice; 3) construction of virtual 3D models from the segmentation results; and, finally, 4) materialization of the physical models, for instance, by rapid prototyping.

3.1 Acquisition of medical images

The acquisition of medical images, as the first stage of the process, has a preponderant role in the final result. In fact, the quality of the images of the structure to be reconstructed will dictate the success of the segmentation process, which is a critical stage of the complete procedure.

The medical images used in this work were acquired using systems of Magnetic Resonance Imaging (MRI). This radiology technique, characterized by the patient's exposition to magnetic fields and radio waves, is capable of recognizing accurately complex and thin objects, and is therefore widely used for the pelvic cavity, [3].

3.2 Segmentation

Image segmentation is usually the most complex procedure in the process of objects reconstruction from images, as it involves the identification of the shape to be extracted and its adequate partitioning in the image, in order to detach the same from its neighbourhood, [4–7]. This can be achieved using either manual or automatic methods, both presenting benefits and disadvantages.

Manual segmentation consists in the manual drawing of the shape to be extracted from each medical image, using the image as background. This process can make use of the tools provided by specific software for analysis of medical images or by a general CAD software, but the effort to identify the contours is completely done manually. Therefore, this technique is very versatile, but it demands a high manpower that might not be always available or justified.

Automatic segmentation avoids the greatest part of the manual process, as it appeals to the properties of the image to perform the identification of the shape to extract. The basic principles for automatic segmentation were established in [11], and are based on a dynamic contour that is attracted to the desired border of the object to be segmented by the minimization of its energy, under image forces (such as image gradients). In addition, several other techniques were studied in order to develop different technologies to increase the potential of the automatic segmentation, as using, for instance, statistical pattern recognition or 3D segmentation, [7]. Currently, the automatic segmentation is widely used in image processing, as it is very inexpensive both from time and manpower aspects.

However, automatic segmentation procedures cannot be faced as universal solutions for all anatomical structures: although able to present reliable results and be very efficiently when applied to organs with high contrast in images, it is ineffective if the organs to extract are thin, complex and/or hardly distinguished in the input images. Under these circumstances, semi-automatic or even manual segmentation is required, even though with a higher effort of the user and time consumption.

As it can be easily understandable, the major problem presented by the automatic methods for image segmentation is related to technical issues; it is sometimes very difficult to perform the segmentation with good quality without the knowledge of an experienced professional about the object to be segmented. Therefore, and regarding the greatest advantages presented by the automatic segmentation in comparison to the manual procedures, the study and development of methodologies for automatic segmentation procedures is a field currently in high progress.

As mentioned before, the female pelvic cavity presents a high variety of organs and characteristics that influence the segmentation process. Therefore, in this work both automatic and manual image segmentation techniques were applied.

3.3 Virtual 3D models

After the segmentation step, several 2D shapes representing the organ at different levels (slices) are available to be assembled in one 3D set; thus, the segmented contours are used to create a virtual three-dimensional body representing the extracted organ.

In this work, the technique to create a virtual organ cannot be detached from the one applied at the segmentation step. In fact, if the segmentation is done, whether manually or automatically, using specific software for analysis of medical images, the extraction of the organ from the set of images is completely automatic, but the quality of the 3D model is not very high, especially in terms of smoothness. On the other hand, if the segmentation is done manually using a CAD software, the 3D model might need also some manual work to align correctly the several slices and to produce the correct surface; however, this process has two great advantages: its high flexibility (allowing the model to be iteratively improved even respecting the segmentation process) and the high quality and smoothness of the 3D surface obtained.

Besides its application to the production of physical models, virtual 3D models can also be used as inputs for visualization, simulation and computational analyses, for instances, to be used in biomechanics analysis using finite elements packages. This last technique is a very interesting tool to study the mechanical behaviour of the muscles and tissues and damage resulting from several solicitations, providing a non-invasive methodology useful for prevention and treatment of diseases; in addition, it can also be used to simulate corrective surgical procedures.

3.4 *Physical 3D models*

Rapid prototyping is a modern technology especially indicated for production of prototype systems designed using CAD software. This technique uses the virtual 3D geometrical model to create a set of layers that will be materialized sequentially, creating a physical model almost identical to the virtual one.

This process is recognizable for being flexible, fast and producing prototypes with a very good quality that can be transparent or coloured according to some purpose, which are fundamental requirements for the production of physical models of anatomical structures.

4 APPLICATION TO FEMALE PELVIC ORGANS

The previously exposed methodologies were applied in this work to female pelvic organs, using medical images obtained from one volunteer and one patient.

When automatic image segmentation was possible, the commercial software for visualization and analysis of medical images Analyze 6.0, [8], was used for the whole process of the desired structures extraction from the input images. However, the pelvic floor proved to be impossible to segment automatically and, therefore, a different approach using the 3D CAD software Autodesk Inventor was applied, [9].

4.1 *Bladder*

Bladder is probably the easiest organ of the human's pelvic cavity to extract, because of its well defined contour and high contrast in the input medical images. Therefore, an automatic segmentation technique was successfully applied to this organ, using the tools provided by Analyze, Figure 1.

Although the results obtained with this process were considered adequate, for comparison purposes the manual technique (using the CAD software Autodesk Inventor) was also applied, Figure 2. In this image, it is

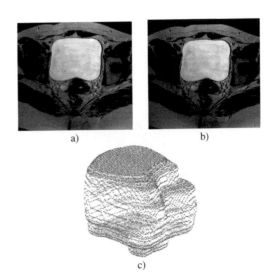

a)　　　　　　　　　b)

c)

Figure 1. Segmentation of the bladder: (a) bladder represented in a slice (white area); (b) automatic segmentation obtained; (c) virtual 3D model reconstructed.

Figure 2. 3D model obtained for the bladder using the manual segmentation technique.

easily verified that the quality of the model obtained using manual segmentation is considerably superior, but the higher effort required by this image segmentation procedure makes of course the first solution much more attractive.

4.2 *Pelvic floor*

The pelvic floor is the main responsible organ for supporting the intestines, the bladder and the uterus. It is composed by muscle fibres, and is thus a very thin, heterogeneous and badly defined structure, very difficult to distinguish in medical images, Figure 3. Consequently, all the attempts to segment automatically the pelvic floor did not produce acceptable results, so manual segmentation had to be applied.

Although Analyze has the required tools for manual image segmentation also, they are not as powerful,

345

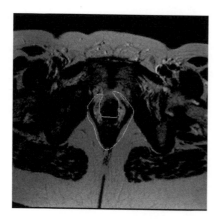

Figure 3. An example slice showing the pelvic floor (surrounded in green).

flexible and user-friendly as the tools offered by common 3D CAD software to construct and manipulate graphical entities (polylines, splines, bsplines, etc.), [9, 10]. Therefore, the segmentation of the pelvic floor in the input images was achieved using the Autodesk Inventor software.

The building of a 3D virtual model of this organ started with the adequate alignment of all medical images representing slices of the pelvic floor. After that, each slice was manually segmented using the Sketch tools (particularly *Splines*) of the Autodesk Inventor software, and then the corresponding 3D model was built using the *Loft* operation, that links all contours previously defined by an interpolate surface, Figure 4.

Besides the advantage of producing very smooth models, this technique evidenced an extremely useful advantage in the creation of virtual models by manual segmentation: its flexibility. With this procedure, it is possible to create a rough model at first and then refine and correct it, improving the segmentation of several slices accordingly to the needs evidenced by the 3D model obtained, by a fully iterative process. This flexibility is not possible using the usual medical software packages.

Two virtual three-dimensional models of the pelvic floor obtained using the Autodesk Inventor software from images of different women, can be seen in Figure 5.

4.3 Urethra

In terms of easiness of image segmentation and extraction, urethra is an intermediate case between the bladder and the pelvic floor: it has a regular and well defined shape, but its contrast in the input images, relatively to its neighbourhoods, is not very high. For that reason, and although it is not impossible

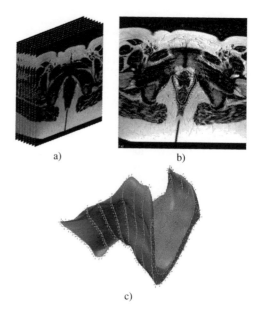

a)   b)

c)

Figure 4. Extraction of the pelvic floor: a) a set of slices; b) manual segmentation done in a slice; c) 3D virtual model obtained.

Figure 5. Two virtual 3D models of the pelvic floors obtained using manual segmentation.

to segment automatically the urethra, the manual procedure previously described was applied, Figure 6.

5 CONCLUSION

The work described in this paper is an introduction to medical image processing and bioengineering applied to the study of anatomic structures. It allowed the understanding of the basic concepts involved in the 3D reconstruction of organs and tissues from medical images, especially concerning the image segmentation process.

It was recognized that usual medical imaging software is a useful tool for those cases where automatic segmentation is possible, allowing a fast and effective creation of 3D virtual models. However, when manual segmentation is required, the medical imaging software (Analyze 6.0, in this work) has not the

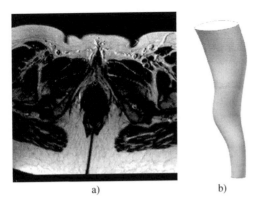

a)   b)

Figure 6.   Extraction of the urethra: a) manual segmentation; b) 3D virtual model built.

required flexibility and, for that reason, the use of a common 3D CAD software may present important advantages in the segmentation process, and thus in the quality of the obtained final 3D models.

Nevertheless, manual segmentation has a high cost, and therefore any effort to improve the automatic segmentation methodologies is a valuable help in this field.

ACKOWLEDGEMENTS

This work was partially done in the scope of the project "Segmentation, Tracking and Motion Analysis of Deformable (2D/3D) Objects using Physical Principles", with reference POSC/EEA-SRI/55386/2004, financially supported by FCT – Fundação para a Ciência e a Tecnologia from Portugal.

In addition, the support of Ministério da Ciência e do Ensino superior (FCT and FSE) (Portugal) and the funding by FEDER under grant POCTI/ESP/46835/2002, are gratefully acknowledged.

REFERENCES

T. Mascarenhas, R. Coelho, "Quality of life and psychological factors of urinary incontinence", Pelvic floor dysfunction- Investigations and conservative treatment, 423–427, Italy, 1999.
L. Viktrup, G. Lose, M. Rolf, K. T. Barfoed, "The frequency of urinary symptoms during pregnancy and puerperium in the primipara", Int. Urogynaecology, 4:27–30, 1993.
L. Hoyte, "MR-Based 3D Female Pelvic Foor Anatomy", USA, 2001.
Tim McInerney, D. Terzopoulos, "Deformable models in medical image analysis: a survey", Medical Image Analysis, 1996.
D. L. Pham, C. Xu, J. L. Prince, "Current Methods in Medical Image Segmentation", Annu. Rev. Biomed. Eng. 2000. 02:315–37.
A. Blake, Michael Isard, "Active Contours", Springer-Verlag, 1998.
T. S.Yoo, "Insight into Images – Principles and Practise for Segmentation, Registration, and Image Analysis", USA, 2004.
AnalyzeDirect, Inc – Mayo Clinic, "Analyze 6.0 - Users Manual", USA, 2004.
J. D. Bethune "Engineering Design and Graphics with Autodesk Inventor(R) 10", Prentice Hall, 2005.
G. Farin, "Curves and Surfaces for CAGD: A Practical Guide", Academic Press, 2001.
M. Kass, A. Witkin, D. Terzopoulos, "Snakes: Active Countour Models", Kluwer Academic Publishers, 1987.

Computational Modelling of Objects Represented in Images –
João Manuel R.S. Tavares & R.M. Natal Jorge (eds)
© 2007 Taylor & Francis Group, London, ISBN 978-0-415-43349-5

Efficient approximation of the mahalanobis distance for tracking with the Kalman filter

Raquel R. Pinho
FEUP – Faculdade de Engenharia da Universidade do Porto, Portugal
INEGI – Instituto de Engenharia Mecânica e Gestão Industrial, Portugal

João Manuel R.S. Tavares
FEUP (DEMEGI – Dep. de Engenharia Mecânica e Gestão Industrial), INEGI

Miguel V. Correia
FEUP (DEEC – Dep. de Engenharia Electrotécnica e de Computadores)
INEB – Instituto de Engenharia Biomédica, Portugal

ABSTRACT: We address the problem of tracking efficiently feature points along image sequences. To estimate the undergoing movement we use an approach based on Kalman filtering which performs the prediction and correction of the features movement in every image frame. In this paper measured data is incorporated by optimizing the global correspondence set based on efficient approximations of the Mahalanobis distances (MD). We analyze the difference between using the MD and its efficient approximation in the tracking results, and also examine the related computational costs. Experimental results which validate our approach are presented.

1 INTRODUCTION

Feature tracking is a complex problem whose automatic detection and execution evolved considerably in the past decade. Applications of movement tracking are numerous: surveillance, object deformation analysis, traffic monitorization, etc (see for example (Wren, Azarbayejani, Darell, and Pantland 1997; Cucchiara, Grana, Piccardi, and Prati 2000)). In the case of human movement analysis it can be used in medical diagnosis, physical therapy or sports, for example to study gait disorders related to knee or hip injury or pain, to control cycles of motion in rehabilitation or training processes.

Although the computational performance has improved significantly, the tracking systems which are able to capture and analyze the undergoing movement, for instance to track multiple features simultaneously or to obtain on-line results, often use some kind of simplification to speedup the process. So, a compromise must be achieved between the accuracy of the motion estimation and the related computational cost.

The Pfinder (Wren, Azarbayejani, Darell, and Pentland 1997) is a real-time system for tracking people and interpreting their behavior. It does not require accurate initialization and segments the person from the background in real-time with a standard computer workstation, but it expects only one person to be in the image scene, and the scene is supposed to be significantly less dynamic than the tracked person.

In (Jorge, Abrantes, and Marques 2004) Bayesian networks are used to perform tracking even in occlusions, group formation and splitting. The on-line object tracking is performed by gradually discarding the influence of past information on the current decisions avoiding a combinatorial explosion and keeping the network complexity within reasonable bounds.

On the other hand, many current works use a probabilistic representation of the uncertainty and stochastic filters to fuse and validate sensor information, and to estimate parameters describing the environment. In (Lin and Bar-Shalom 2004) the data association problem is combined with filtering to track ground targets using ground moving target indicator (GMTI) reports obtained from an airborne sensor.

The matching between the estimated features and the observations detected after a sensing operation is determined using data association techniques. Data association algorithms may include a hypothesis-validation step, which may be based on the Mahalanobis distance (MD) and its validation through the χ^2-distribution.

1.1 Related work

The use of the MD for associating data in tracking systems is usual as it gives good performance, but few approaches have tried to speedup this procedure. In (O'Malley, Nechyba, and Arroyo 2002) is presented a method for tracking and identifying moving persons from video images taken by a fixed field-of-view camera, and the MD is used to distinguish pixels either from foreground or background. To speed the calculation only the diagonals of the covariance matrix are used in the MD calculation. This corresponds to the same simplification we use, but applied in different tracking approaches.

On the other hand, the MD and its simplifications extensively are used for pattern recognition. As the computation time of the MD will reach $O(n^2)$ for n-dimensional feature vectors, to reduce the computational cost several approximations of the MD have been proposed as, for example, the quasi-Mahalanobis distance (Kurita, Tsuruoka, Yokoi, and Miyake 1983), the modified Mahalanobis distance (Kato, Abe, and Nemoto 1996), and the modified quadratic discriminant function (Kimura and Shridhar 1991). All these approaches were proposed to improve recognition accuracy but not to approximate the quadratic discriminant function.

In (Sun, Omachi, Kato, Aso, Kono, and Takagi 2000) an approximation of the MD is used in pattern recognition for example in Chinese and Japanese character recognition. Based on two kinds of approximations of the MD, the proposed algorithm consists of two stages which are feature vector division and dimensional reduction. The first stage of feature division is based on the characteristic of covariance matrix. The second stage is done by regarding the values of small eigenvalues as a constant. When compared to the well-known dimensional reduction method, the Karhunen-Loéve (K-L) expansion, experimental results showed that the proposed algorithm not only reduced the computational cost but also improved the recognition accuracy. But this approach requires computation of the involved eigenvalues.

1.2 Our approach

In our previous work we used the MD to evaluate the quality of correspondences, and the actual set of matching between predictions and measurements is obtained by optimizing the sum of all the involved MD (Pinho, Tavares, and Correia 2005).By doing so, the best global correspondence set is guaranteed. To track the captured movement we use a well-known statistical model: the Kalman filter (KF). The drawbacks of this stochastic process are due to its relatively high restrictive assumptions (Welch and Bishop 1995). But we combined the KF with optimization techniques for data association in order to increment the filters robustness to occlusion and non-linear movement. The correspondence between each features prediction and new measurement data is set upon the MD minimization. The MD ensures that the correspondence is done according to each features previously known behavior. Its approximation to the χ^2-distribution allows the choice of a significance level, from which features will be considered as unmatched. Therefore, even if the KF restrictions are not satisfied (a often situation in many tracking applications) the results obtained may be corrected with optimization techniques and the MD.

In this paper we propose an efficient approximation of the MD between features, in order to sort them and match corresponding features along image sequences for tracking. Experimental results which validate our approach are presented.

1.3 Paper overview

This paper is organized as follows. In the next section a brief introduction is made to the MD. In section 3, we describe the used efficient approximation to the MD and analyze the undergoing error. In section 4 we overview the tracking system which uses the efficient approximation of the MD for features correspondence along frames. In section 5 some experimental results are shown on synthetic and real image sequences. In the last section some conclusions will be held.

2 THE MAHALANOBIS DISTANCE

The MD, also known as statistical distance, is a distance that for each of its components (the variables) takes the variability of that variable into account when determining its distance from the center. So, components with high variability receive less weight than components with low variability. This is done obtained by rescaling the components. That is, for two points $X_i = (x_{1i}, x_{2i}, \ldots, x_{ni})$ and $Y_j = (y_{1j}, y_{2j}, \ldots, y_{nj})$ the MD is given by

$$d_M = \sqrt{(X_i - Y_j)^T C^{-1} (X_i - Y_j)} \qquad (1)$$

where $C_{(n \times n)}$ is a non-singular covariance matrix (therefore symmetric positive definite). Points with the same distance of X_i satisfy the equation of an ellipsoid centered about X_i, and those points of the ellipsoid defined by C have unitary MD.

The MD is a standard manner of associating data for tracking features in image sequences. However, this step is one of the most time-consuming operations of the matching process. After a sensing operation, M feature location estimates, and N measurements, are available. The problem is how to associate each measurement, $X_i (i = 1, \ldots, N)$, with a feature estimate, $Y_j (j = 1, \ldots, M)$.

This matching procedure is time-consuming because it involves a matrix inversion, and the computation of matrix C and vector $v = X_i - Y_j$ which is subject to linearizations. To save computational cost some data-association techniques perform a validation test for each pairing hypothesis, in order to work with only a reduced set of hypotheses. The process of the validation is performed using a statistical test based on the MD and its approximation by the χ^2-distribution:

$$v^T C^{-1} v \leq \chi^2 \tag{2}$$

where v is the vector between a predicted feature state and an acquired measurement. This test should theoretically be computed for $M \mathbf{x} N$ hypotheses.

3   AN EFFICIENT APPROXIMATION OF THE MAHALANOBIS DISTANCE

In our tracking system the captured measurements are composed by their position coordinates in the image plane, and each of the tracked features has its own KF. This means that the C and v are given by:

$$C = \begin{bmatrix} c_{11} & c_{12} \\ c_{12} & c_{22} \end{bmatrix} \tag{3}$$

and,

$$V = \begin{bmatrix} v_1 & v_2 \end{bmatrix}^T. \tag{4}$$

So the MD in this case is given by:

$$d_M = \frac{c_{22}v_1^2 - 2c_{12}v_1v_2 + c_{11}v_2^2}{c_{11}c_{22} - c_{12}^2}. \tag{5}$$

Rearranging the terms in the equation above we can obtain

$$d_M = \frac{v_1^2}{c_{11}} + \frac{v_2^2}{c_{22}} - \frac{2c_{12}v_1v_2}{c_{11}c_{22}} + \\ + \frac{c_{12}c_{22}v_1^2 + c_{12}c_{11}v_2^2 - 2c_{12}^3v_1v_2}{c_{11}c_{22}\left(c_{11}c_{22} - c_{12}^2\right)}. \tag{6}$$

So, the efficient approximation we propose to the MD consists on:

$$d_M \approx \hat{d_M} = \frac{v_1^2}{c_{11}} + \frac{v_2^2}{c_{22}}, \tag{7}$$

which is thereby affected of an error, $\Delta \hat{d_M}$, of

$$\Delta \hat{d_M} = -\frac{2c_{12}v_1v_2}{c_{11}c_{22}} + \frac{c_{12}c_{22}v_1^2 + c_{12}c_{11}v_2^2 - 2c_{12}^3v_1v_2}{c_{11}c_{22}\left(c_{11}c_{22} - c_{12}^2\right)}. \tag{8}$$

Although better approximations can be used for the MD their computational cost can be questioned, on the other hand, the approximation that we propose is quite efficient as it only involves 5 arithmetic operations for each pair of features; instead of the 18 operations involved in equation (5) with the complete calculation of the MD.

4   TRACKING WITH THE KALMAN FILTER AND ASSOCIATING DATA WITH THE MAHALANOBIS DISTANCE

The KF is an optimal recursive Bayesian stochastic method. It provides *optimal* estimates that minimize the mean of squared error of the modeled process. In a Bayesian stochastic viewpoint, the filter propagates conditional probability density of the system state conditioned on the knowledge of the actual data acquired by the measuring devices.

The equations for the KF fall into two steps: time update (or prediction) and measurement update (or correction). The time update equations are responsible for projecting forward (in time) the current state and error covariance estimates to obtain the *a priori* estimates for the next time step. The measurement update equations deal with the feedback, that is new measurements are incorporated into the *a priori* estimates to obtain improved *a posteriori* values (Welch and Bishop 1995).

The prediction step is based on the Chapman-Kolmogorov equation for a first order Markov process:

$$x_t^- = \Phi x_{t-1}^+ \tag{9}$$

where Φ relates the system state x_{t-1}^+ at the previous time step $t-1$ to the state x_t^- at the current step t. The superscripts $^+$ and $^-$ indicate if measurement data have been or not incorporated, respectively. The related uncertainty is given by:

$$P_t^- = \Phi P_{t-1}^+ \Phi^T + Q \tag{10}$$

where P is the prediction covariance matrix and Q models the process noise.

The correction equations that update the predicted estimates upon the incorporation of new u_t measurements are given by:

$$K_t = P_t^- H^T \left[H P_t^- H^T + R_t \right]^{-1} \tag{11}$$

$$x_t^+ = x_t^- + K_t \left[u_t - H x_t^- \right] \tag{12}$$

$$P_t^+ = \left[I - K_t H \right] P_t^- \tag{13}$$

where K is chosen to be the gain that minimizes the *a posteriori* error covariance equation, H processes

351

the coordinates transformation between the predicted and the measurement spaces, R is the measurement noise involved, and I is the identity matrix (Welch and Bishop 1995).

One of the drawbacks of the KF is the restrictive assumption of Gaussian posterior density functions at every time step, as many tracking problems involve non-linear movement. To minimize such ambiguities we evaluate all the $M \times N$ possible correspondences by estimating the MD, and optimize the global matching set. The efficient MD can sort correspondences according to their MD, and so the computational burden associated to the MD calculation in the matching process is overcome with a small error involved.

5 EXPERIMENTAL RESULTS

In each frame of the presented examples the predicted position is represented with a $+$, with uncertainty area circumscribed in a solid ellipse, each measurement is the center of the detected contour, and the corrected position is represented by a x. The association between each prediction/measurement is represented with a solid line.

For the first example, Figure 1, consider a synthetic sequence of 9 frames. In the beginning of the sequence only two blobs are visible. The circular blob will disappear definitively but the tracking approach keeps on trying to track it during the subsequent frames, although with gradually higher uncertainty (in frame (e) the uncertainty region surpasses the image border). In the second frame a triangular blob appears, and in the third frame the square blob disappears instantly. In the fourth frame the captured blobs overlap, and with the used image processing techniques only one measurement is captured and associated to a blob, but both features continue to be correctly tracked. From the seventh frame onwards 25 blobs are tracked. The results presented in Figure 1 were obtained by data association with the efficient MD but there exists no visual difference to those obtained with the usual MD calculation, but the computational cost associated to the efficient MD is obviously less, Figure 2. Indeed, as little features are tracked the computational load of the MD is not significant, but as the number of tracked features increases the advantages of the efficient MD are more notable.

In Tables 1 and 2 are represented the results obtained with the efficient approximation of the MD and those obtained with the usual MD, respectively, to associate the tracked features and the measurements in frame (d) of Figure 1. It can be noticed that the efficient approximation of the MD can be larger or smaller than the complete value, but differences are generally small, Figure 3.

Note that in frame (d) of Figure 1 three features are being tracked while only two measurements are

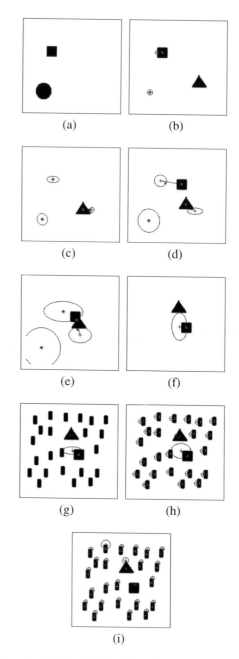

(a)

(b)

(c)

(d)

(e)

(f)

(g)

(h)

(i)

Figure 1. Tracking blobs in a 9 frame image sequence: (a) – original 1st frame; (b) to (i) – KF results: search area defined by solid ellipses, the predicted position for each marker is given by $+$, and the corrected position is represented with x.

captured, so to apply the optimization algorithm a fictitious variable is included, and the cost of correspondence with it is null (Pinho, Tavares, and Correia 2005).

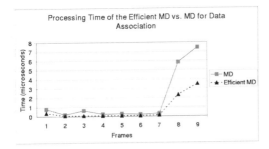

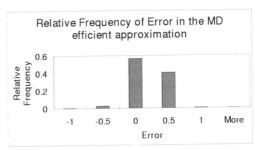

Figure 2. Data association time while using the efficient MD or the usual MD on a Mobile AMD Atlhon(tm) 4 at 1.20 GHz with 256 MB RAM.

Figure 3. Relative frequency of errors in the efficient approximation of MD.

Table 1. Associating tracked features and measurements in Figure 1 frame (d): Values of the efficient approximation of the MD.

Tracked features	Measurements		
	Xm_1	Xm_2	Xm_3
Xe_1	5.635	7.248	**0.0**
Xe_2	3.910	**1.678**	0.0
Xe_3	**0.406**	2.570	0.0

Table 2. Associating tracked features and measurements in Figure 1 frame (d): Values of the MD.

Tracked features	Measurements		
	Xm_1	Xm_2	Xm_3
Xe_1	5.930	7.816	**0.0**
Xe_2	3.652	**1.657**	0.0
Xe_3	**0.382**	2.416	0.0

Minimizing the overall correspondence costs evaluated in our tracking approach with the MD or its efficient approximation the same results are obtained: the third tracked feature, Xe_3 is associated to the first measurement captured; Xm_1; the second tracked feature and measurement are matched, Xe_2 and Xm_2 respectively; and the first tracked feature Xe_1 is associated to the fictitious variable Xm_3, which means that it is considered unmatched (Tables 1 and 2).

What is more advantageous to our tracking application is that the efficient approximation of the MD can efficient and adequately sort out the features to build the correct correspondences along the image sequences.

For the next example consider a sequence of real images presented in (Brown, Senior, Tian, Connell, Hampapur, Shur, Merkl, and Lu 2005) for the evaluation of performance of surveillance systems. In this example a person walks into the room and stops at the center, meanwhile two other persons enter, walk

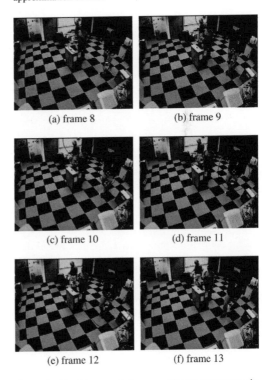

(a) frame 8 (b) frame 9

(c) frame 10 (d) frame 11

(e) frame 12 (f) frame 13

Figure 4. Tracking people in a surveillance system: results obtained with data association using the efficient approximation of the MD.

around the room and exit. Then, the remaining person walks out the room.

One usual difficulty related to tracking a person with most low level image processing techniques is the detection of 2 regions of movement, one of them corresponding to the peoples head and upper body, and the other to their legs and feet. In the images presented we did not try to remove those and other instantaneous noisy features in order to test the proposed data association algorithm. We tracked the moving features using both the MD as well as its efficient approximation and generally the same results are obtained, Figure 4. But

353

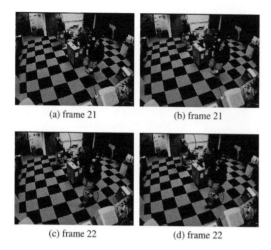

(a) frame 21   (b) frame 21

(c) frame 22   (d) frame 22

Figure 5. Results obtained with data association using the efficient approximation of the MD (left column) and the MD (right column).

in some rare cases the data association may differ as in Figure 5, but in the next frames the tracking process is correctly recovered.

6 CONCLUSIONS

To track features in image sequences we use a KF which performs the prediction and correction of the features movement in every image frame. With our approach the correspondence between each prediction made by the KF and its measurement in each frame are established by optimizing the global matching set based on the MD. To reduce the computational cost associated to the calculation of the MD we propose in this paper a simplification of the MD with an efficient approximation in order to reduce the computational cost of associating data in the KF for tracking. We have exemplified that this approach is especially advantageous when a large number of features are tracked, and have shown that generally the resulting correspondences are very similar to those obtained with the complete MD.

ACKNOWLEDGMENTS

This work was partially done in the scope of the project *Segmentation, Tracking and Motion Analysis of Deformable (2D/3D) Objects using Physical Principles*, reference POSC/EEA-SRI/55386/2004, financially supported by FCT – Fundação para a Ciência e a Tecnologia in Portugal.

The first author would like to thank the support of the PhD grant SFRH / BD / 12834 / 2003 of FCT.

REFERENCES

Brown, L., A. Senior, Y. Tian, J. Connell, A. Hampapur, C. Shu, H. Merkl, and M. Lu (2005). Performance evaluation of surveillance systems under varying conditions. In *IEEE International Workshop on Performance Evaluation of Tracking and Surveillance*, Colorado, USA.

Cucchiara, R., C. Grana, M. Piccardi, and A. Prati (2000). Statistic and knowledge-based moving object detection in traffic scenes. In *IEEE Conference on Intelligent Transportation Systems*, Dearborn, USA.

Jorge, P., A. Abrantes, and J. Marques (2004). On-line tracking groups of pedestrians with bayesian networks. In *Workshop PETS*, Prague, Czech Republic.

Kato, N., M. Abe, and Y. Nemoto (1996). A handwritten character recognition system using modified mahalanobis distance. *Transactions Institute of Electronics, Information and Communication Engineers (IEICE) J79-D*(1), 45–52.

Kimura, F. and M. Shridhar (1991). Handwritten numerical recognition based on multiple algorithms. *Pattern Recognition 24*, 969–983.

Kurita, M., S. Tsuruoka, S. Yokoi, and Y. Miyake (1983). Handprinted "kanji" and "hiragana" character recognition using weighting direction index histograms and quasi-mahalanobis distance. Technical report, IEICE Technical Report PRL82-79.

Lin, L. and Y. Bar-Shalom (2004). New assignment-based data association for tracking move-stop-move targets. *IEEE Transactions on Aerospace and Electronic Systems 40*(2), 714–725.

O'Malley, P., M. Nechyba, and A. Arroyo (2002). Human activity tracking for wide-area surveillance. In *2002 Florida Conference on Recent Advances in Robotics*, Miami, USA.

Pinho, R., J. Tavares, and M. Correia (2005). Human movement tracking and analysis with kalman filtering and global optimization techniques. In *II International Conference On Computational Bioengineering*, Lisbon, Portugal.

Sun, F., S. Omachi, N. Kato, H. Aso, S. Kono, and T. Takagi (2000). Two-stage computational cost reduction algorithm based on mahalanobis distance approximations. *International Journal of Pattern Recognition 2*, 2969–2973.

Welch, G. and G. Bishop (1995). *An Introduction to Kalman Filter*. Technical report, University of North Carolina at Chapel Hill.

Wren, C., A. Azarbayejani, T. Darell, and A. Pentland (1997). Pfinder: Real-time tracking of the human body. *IEEE Transactions on Pattern Analysis and Machine Intelligence 19*(7), 780–785.

Computational Modelling of Objects Represented in Images –
João Manuel R.S. Tavares & R.M. Natal Jorge (eds)
© 2007 Taylor & Francis Group, London, ISBN 978-0-415-43349-5

Defect detection in raw hide and wet blue leather

Hemerson Pistori, William A. Paraguassu, Priscila S. Martins & Mauro P. Conti
UCDB – Dom Bosco Catholic University, Campo Grande, Brazil
GPEC – Research Group in Engineering and Computing

Mariana A. Pereira & Manuel A. Jacinto
EMBRAPA – Brazilian Agricultural Research Corporation, Campo Grande, Brazil
EMBRAPA Beef Cattle Research Unit

ABSTRACT: This paper presents an important problem for the Brazilian economy, the classification of bovine raw hide and leather, and argues that this problem can be handled by computer vision and machine learning techniques. Some promising results, using standard techniques, like color based models and cooccurrence matrix based texture analysis, are reported. The paper also presents what seems to be the first major training and testing dataset of bovine raw hide and leather digital images.

1 INTRODUCTION

The importance of the bovine production chain for the Brazilian economy is a well acknowledged fact, as it has the largest cattle herd in the world (Matthey, Fabiosa, and Fuller 2004). In 2005, Brazil displaced the United States as the world's number one beef exporter (Valdes 2006). However, only recently government and private agents started planning proper policies to improve Brazilian role in the bovine leather markets.

The quality of Brazilian leather is far bellow the level of other important players. While in the United States, 85% of the leather are of highest quality, in Brazil, only 8.5% achieve this classification (da Costa 2002). Nonetheless, most of the problems that affects leather quality, like the use of barbed wire, brand marks made from hot iron in improper places, wrong transportation and flaying methods, have simple solutions.

Besides some lack of information about the leather importance for Brazilian economy, the main problem seems to be the absence of clear and fair policies and practices for raw hide (the untanned hide of a cattle) and leather commercialization. In most cases, bovine raw hides are treated as a by-product in the slaughtery, and for the raw hides, cattle ranchers are paid a fixed proportion directly related to the carcass weight. The creation of a remuneration system that rewards higher quality would stimulate more investments on leather care during cattle raising, when 60% of the problems that decrease leather value happens (da Costa 2002; Gomes 2002).

A fair remuneration system depends on standardized, non subjective and reliable classification and grading schemes. Brazilian government, by 2002, made a first attempt to create a national grading system for bovine raw hide[1]. A recent research conducted by the Brazilian Agricultural Research Corporation, EMBRAPA, suggests some improvements to the system and recommends the pursuing of automation to increase reliability. The work described in this paper is part of a scientific research and technological development project, DTCOURO[2], which envision the development of a completely automated system, based on computer vision, for bovine raw hide and leather classification and grading.

One of the DTCOURO project's goal is to propose and provide comparative studies of pre-processing, feature extraction, feature selection, segmentation and classification techniques. Among the feature extraction algorithms that will be experimented and compared are the ones based on Gabor Filters (Grigorescu, Petkov, and Kruizinga 2002), Windowed Fourier Transforms (Azencott, Wang, and Younes 1997), Wavelets (Sobral 2005), Cooccurrence Matrices (Jobanputra and Clausi 2004), Interaction Maps

---

[1] Normative instruction number 12, December 18th, 2002, Brazilian Ministry of Agriculture, Livestock and Food Supply.

[2] DTCOURO is an acronym, in Portuguese, for the project named Leather Defect Detection System, which has a Portuguese language website available at http://ommited for blind review.

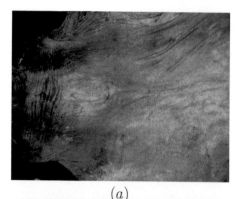

(a)

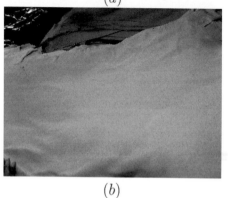

(b)

Figure 1. (a) Bovine raw hide and (b) Wet blue leather samples.

(Chetverikov 2000), Local Binary Patterns (Menp 2002) and Color Models. Figures 1.(a) and 1.(b) show images from raw hide and wet blue leather[3]. A large image dataset of defects in hides and leather is being prepared and used to train and test supervised machine learning algorithms. Classifiers based on Support Vector Machines, Artificial Neural Networks and Bayesian Inference, among others, will be implemented and experimented.

In this work, we present the first version of the leather defect dataset and the software that makes easier the creation of ground truth images and automatic comparison of different computer vision and machine learning techniques. We also show the results from the first defect inspection experiments, using cooccurrence matrices features and supervised learning. The experiments were based on a 2000 samples dataset, taken from 16 different raw hide and wet blue leather pieces and containing 4 defect types: brand marks made from hot iron, tick marks, cuts and scabies.

---

[3] Wet blue leather is a hide that has been tanned using chrominus sulphate. It is an intermediate stage between untanned and finished leather.

The results were very promising, with an overall correct classification rate above 94%. The experiments with bovine raw hide images, a much more complex problem than wet blue and finished leather defect inspection, are rarely cited in the specialized literature, which concentrates on tanned leather. Furthermore, some important raw hide defects for the Brazilian leather productive chain are not always of interest in different countries where this type of defects are not so common, and usually not considered.

The next section presents some work directly related to leather inspection based on computer vision. Section 3 details important aspects of raw hide and wet blue leather defects. A brief review of texture analysis based on cooccurrence matrices can be found in Section 4. The following sections explain the experiments, discuss the results and presents the conclusions and future developments.

2 RELATED WORK

Yeh and Perng propose and evaluate semi-automatic methods for wet blue raw hide defects extraction and classification. The system has been tested in a large tannery for six months with reliable and effective results, when comparing with human specialists. Their work also presents an interesting taxonomy for leather defects classification and grading, based on the shape and size of affected area. The defects are classified as (1) thin spots: hair root, pinhole, putrid spot, dermatitis; (2) circular spots: thorn scratch, nail mark, chrome stain, slat stain, cured stain, putrefied; (3) thin line: vein, wring felt mark; (4) strips: score knife, neck wrinkle; (4) holes: dig damage, grub hole, bullet mark; (5) patterns: brand mark; and (6) irregulars: wart, contamination, pipe grain, flay mark, putrefied, scratch, chafe mark, gear mark, parasitic speckled (tick, mange), dung stain (Yeh and Perng 2001). The main contribution of the work is a fully quantified grading system, called *demerit count reference standard for leather raw hides*, but the authors also point out that one of the drawbacks of their proposal is the need for human, specialized intervention, for counting the total number of demerits on a wet blue raw hide.

A leather inspection method, based on Haar's wavelets, is presented by Sobral. The smoothing component has been tuned for each kind of defect, using a manually classified defect sample. The system is reported to perform in real time, at the same level of an experienced operator (Sobral 2005) and to outperform previous methods, based on Gabor filters, like the one described in Kumar and Pang (Kumar and Pang 2002). Although not clearly stated in Sobral's paper, the system seems to have been experimented only on finished leather, a much simpler problem than raw hide or wet blue raw hide defect extraction.

A dissimilarity measure based on χ^2 criteria has been used to compare gray-level histograms from sliding windows (65x65 pixels) of a wet blue raw hide image to an averaged histogram of non-defective samples in (Georgieva, Krastev, and Angelov 2003). The results of the χ^2 test and an experimentally chosen threshold are used to segment defective regions of the raw hide. The approach has not been used to identify the defect type. The segmentation of defective regions from wet blue raw hide images, using histogram and cooccurrence based features, has been investigated in (Krastev, Georgieva, and Angelov 2004). This work also proposes the use of fuzzy logic to model leather defects, but do not give sufficient details or experimental results supporting the proposal.

3 RAW HIDE AND WET BLUE LEATHER DEFECTS

Bovine leather undergoes a long way from cattle raising to final industrial production of leather goods, like furniture, footwear, belts and so on. The problems that affect leather quality begin when the animal is still alive, and include, (1) cuts resulting from barbed wired, in-fighting among male members and thorn scratches and cuts; (2) brand marks made for ownership purposes, using hot iron; (3) holes and spots from infections and infestations, caused by ticks, horn flies, manges and bot flies, among others; (4) abscesses resulting from wrong vaccination techniques and natural growth marks or excess weight related problems, like furrows and wrinkles (Barlee, Lanning, and McLean 1999; Roberts and Etherington 1981).

During transportation, the animal skin may suffer deep injures from nails and wood splints in the truck. Before tanning, three important process, which can also cause leather damage, happen: bleeding, skinning and curing. Insufficient bleeding can cause vain marks, while wrong skinning techniques may result in flaying cuts that, in some cases, may turn unusable otherwise valuable parts of the leather. As the raw hide is subjected to putrefaction, as soon as the animal dies, the raw hide must suffer a curing process to protect it until the tanning process begins, which can take months. Improper curing may lead to rotting and putrefaction. Defects during tanning and post-processing are much less common, as they are controlled by the tanneries, which have in the leather quality their main business.

Even without defects, bovine raw hide has a very complex surface, presenting different textures, colors, shapes and thickness. Besides, in order to be useful, automatic classification system should function in very different environments, like farms, slaughteries and tanneries. Blood or water drops may turn the

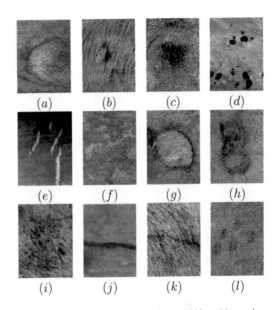

Figure 2. Examples of defects in raw hides: (a) vaccine abscess (b) bot fly open wounds (c) bot fly closed wounds (d) ticks marks (e) wrinkles (f) photo-sensibility (g) flay mark (h) brand marks made from hot iron (i) horn fly wounds (j) open cuts (k) closed cuts (l) scabies.

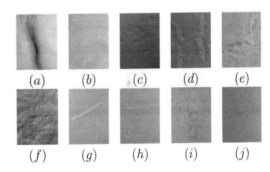

Figure 3. Examples of defects in wet blue leather: (a) vaccine abscess (b) bot fly wounds (c) ticks marks (d) wrinkles (e) brand marks made from hot iron (f) Haematobia irritans wounds (g) open cuts (h) closed cuts (i) scabies (j) veining.

task even more difficult if the raw hide is to be classified just after skinning or cleaning. Classification of living animals must deal with shadows from hair and the natural bovine anatomy.

Figures 2 and 3 illustrate the diversity of shapes, colors and texture that arise from some important defects that happens in Brazilian raw hides and leather, decreasing their market value. Images from raw hides, taken after skinning and before tanning, are shown in Figure 2. Figure 3 shows images from leather in the first stage of the tanning process, which are called wet blue leather.

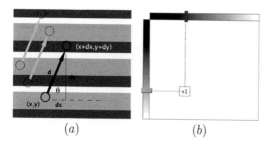

(a)   (b)

Figure 4. Construction of a GLCM: (a) Angle and displacement parameters in spatial domain (original image) (b) Gray-level coocurrence matrix.

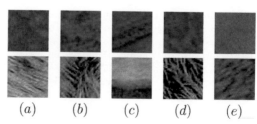

(a)   (b)   (c)   (d)   (e)

Figure 5. Examples of 40×40 image windows used in the wet blue leather (first line) and raw hide (second line) experiments: (a) ticks (b) brand marks (c) cuts (d) scabies (e) non defective.

4 GRAY SCALE COOCCURRENCE MATRICES

Image segmentation based on features extracted from gray-scale coocurrence matrices, GLCM, is a common and largely used technique in texture analysis (Singh and Singh 2002; Jobanputra and Clausi 2004; Latif-Amet, Ertuzun, and Ercil 2000). As wavelets, windowed Fourier or Gabor filter approaches, coocurrence matrices can represent information related to the frequential distribution content of the original, spatially represented, image. Several GLCMs must be constructed for each sliding window that scans the image during segmentation. Each GLCM has an associated angle and displacement, related to the direction and frequency that will be represented by this GLCM. The number of different angles and displacements, and consequently, the number of GLCMs to be constructed depends on the problem and computer power available.

Figure 4 illustrates the construction of a GLCM for a fixed angle, θ, and displacement, d. The GLCM, G, is an $m \times m$ accumulator, where m is the number of gray levels. For each pixel, (x, y), of the original image (or image window), the accumulator cell (i, j) is incremented, where $i = I(x, y)$, $j = I(x + dx, y + dy)$ and $I(.)$ is the pixel gray level. Interpolation must be used for certain angles and displacements, as $dx = d \times \cos(\theta)$ and $dy = d \times \sin(\theta)$ may be non-integers. Usually, the values of the GLCMs are not directly used as texture features, but some statistics and indices calculated from them, like entropy, contrast, angular second moment, inverse difference moment, energy and homogeneity.

5 LEATHER DATASET AND EXPERIMENTS

The DTCOURO's dataset includes, currently, images from 258 different pieces of raw hide and wet blue leather showing 17 different defect types. The images have been taken using a five megapixel digital camera during technical visits to slaughteries and tanneries located in the region of Mato Grosso do Sul, Brazil, by September 2005. A total of 66 pictures are from tanned leather in the wet blue stage, the other ones are from the raw hide stage.

For this first experiment, a set of four types of defect has been chosen: tick marks, brand marks made from hot iron, cuts and scabies. These defects have been chosen because they are very common in Brazil. Sixteen images, both from raw hide and wet blue leather, containing these defects (two image for each defect, both in raw hide and wet blue leather), were used. The defects were manually segmented with the help of a specialist and small samples from defective and non-defective areas have been extracted, using the software developed in DTCOURO's project. The samples used in these experiments are of windows of 10×10, 20×20, 30×30 and 40×40 pixels. A total of 2000 samples have been generated in this way, 400 for each defect and 400 for non-defective regions. Figure 5 illustrates some of the 40×40 size samples, from wet blue leather (first line) and raw hides (second line), that were used in the experiments.

From each sample, a set of 63 texture and 3 color features were extracted. The color features are the mean values of the histograms for the hue, saturation and brightness in HSB color space. The texture features are the entropy, inverse difference moment, dissimilarity, correlation, contrast, angular second moment and inverse difference entropy, calculated from the grey level coocurrence matrices, with angles 0, 45 and 90 (in degrees) and displacements 1, 5 and 10 (in pixels). A 10-fold cross-validation scheme was used to train and test three supervised learning approaches for this five class classification problem: (1) support vector machines (Keerthi, Shevade, Bhattacharyya, and Murthy 2001), (2) normalized Gaussian radial basis function network (Figueiredo 2000) and (3) k-nearest neighbours (Aha, Kibler, and Albert 1991). The experiment was repeated 5 times, resulting in a total of 50 runs, for each learning approach and each window size. The implementations for the machine-learning algorithms were taken from the Weka free-software, which were also used to produce performance statistics (Witten and Frank 2005).

The sequential minimal optimization algorithm (Keerthi, Shevade, Bhattacharyya, and Murthy 2001), implemented in Weka, has been used to train the support vector machines, with a third-order polynomial kernel. Pairwise classification were used to enable the application of this 2-class classification algorithm to our multi-class problem. Ridge estimators and logistic regression (Cessie and van Houwelingen 1992) were used to learn the normalized Gaussian radial basis functions, clustered by k-means, with $k = 5$ (4 defects and 1 clean leather). The k-nearest neighbour approach was tested with 5 neighbours, weighted by the inverse of their distance. All the parameters were experimentally chosen to enhance percentage correct classification rates for each learning scheme, using Weka's implementation.

6 RESULTS AND DISCUSSION

The correct classification rates, for each classifier, using defective (in four different types) and non-defective images from wet blue leather and raw hides are shown, respectively, in Tables 1 and 2. A two-tailed, t-student test, was used to infer improvement or degradation using support vector machines (SVM) performance as the null hyphotheses. It is clear, from the tables, that all three classifiers achieve excellent correct classifications rates when 20×20 or larger window size were used. Support vector machines achieved good results (above 94%) even with 10×10 windows.

These results must be considered with some care and as a first, initial, experiment that, nonetheless, should encourage further research. Each defect were taken from only two different pieces of leather and do not represent, mainly in the case of raw hides, all the possible configuration in which the defect could appear, as, for instance, different hair sizes, colors and directions. It is very important to note that processing time were not an issue in these initial experiments and the parameter extraction phase, applied to high resolution images of the full leather or hide piece, can take almost half an hour.

7 CONCLUSION

Some experiments have been conducted in order to verify the applicability of texture analysis and machine learning techniques to the problem of defect detection and classification in bovine hides and leather. Support vector machines trained with the sequential minimal optimization algorithm using attributes extracted from cooccurence matrices presented the highest correct classification rates. The detection of defects in bovine

Table 1. Percentual of correct classification in wet blue leather, for the three learning approaches experimented: Support Vector Machines (SVM), Radial Basis Functions Networks (RBF) and Nearest Neighbours (KNN).

Data Set	SVM	RBF	KNN
10×10	97.44	87.40•	90.42•
20×20	99.82	98.40•	99.36
30×30	100.00	99.58	99.66
40×40	100.00	99.74	100.00

o – improvement, • – degradation.

Table 2. Percentual of correct classification in raw hides.

Data Set	SVM	RBF	KNN
10×10	94.32	81.18•	91.56•
20×20	99.88	96.50•	99.20•
30×30	100.00	99.06	99.90
40×40	100.00	99.92	100.00

o – improvement, • – degradation.

raw hides was not, as far as the authors are concerned, handled in previous work, at least using computer vision approaches. One of the main contributions of this work is to present an important, but not extensively studied, problem.

The dataset will be enlarged, in the near future, with images from defects in live animals and skins with longer hair from the southeast regions of Brazil. Experiments with the complete dataset and other attributes extraction and learning techniques should also be conducted. It must be investigated if the same classification accuracy could be achieved with less computationally expensive approaches, other than cooccurence matrices.

ACKNOWLEDGMENTS

This work has been funded by the Dom Bosco Catholic University, UCDB, and the Brazilian Studies and Projects Funding Body, FINEP. The author of this paper and some of his advisees hold scholarships from the Brazilian National Counsel of Technological and Scientific Development, CNPQ.

REFERENCES

Aha, D. W., D. Kibler, and M. Albert (1991). Instance-based learning algorithms. *Machine Learning 6*, 37–66.

Azencott, R., J. Wang, and L. Younes (1997, February). Texture classification using windowed Fourier filters. *19*(2), 148–153.

Barlee, R., D. Lanning, and W. McLean (1999). The manufacture of leather. *Journal of Designer Bookbinders 19*, 48–59.

Cessie, S. L. and J. C. van Houwelingen (1992). Ridge estimators in logistic regression. *Applied Statistics 41*(1), 191–201.

Chetverikov, D. (2000). Structural defects: General approach and application to textile inspection. In *ICPR00*, Volume 1, pp. 521–524.

da Costa, A. B. (2002, Dezembro). Estudo da competitividade de cadeias integradas no brasil: Impactos das zonas de livre comrcio. Technical report, Instituto de Economia da Universidade Estadual de Campinas.

Figueiredo, M. A. (2000). On gaussian radial basis function approximations: Interpretation, extensions, and learning strategies. In *Proceedings of the 15th International Conference on Pattern Recognition*, Volume 2, pp. 2618–2622.

Georgieva, L., K. Krastev, and N. Angelov (2003). Identification of surface leather defects. In *CompSysTech '03: Proceedings of the 4th international conference conference on Computer systems and technologies*, New York, NY, USA, pp. 303–307. ACM Press.

Gomes, A. (2002). Aspectos da cadeia produtiva do couro bovino no Brasil e em Mato Grosso do Sul. In *Palestras e proposies: Reunies Tcnicas sobre Couros e Peles, 25 a 27 de setembro e 29 de outubro a 1 de novembro de 2001*, pp. 61–72. Embrapa Gado de Corte.

Grigorescu, S., N. Petkov, and P. Kruizinga (2002). Comparison of texture features based on Gabor filters. *IEEE Trans. on Image Processing 11*(10), 1160–1167.

Jobanputra, R. and D. Clausi (2004). Texture analysis using gaussian weighted grey level cooccurrence probabilities. In *Proceedings of the Canadian Conference on Computer and Robot Vision – CRV*, pp. 51–57.

Keerthi, S., S. Shevade, C. Bhattacharyya, and K. Murthy (2001). Improvements to platt's SMO algorithm for SVM classifier design. *Neural Computation 13*(3), 637–649.

Krastev, K., L. Georgieva, and N. Angelov (2004). Leather features selection for defects' recognition using fuzzy logic. In *CompSysTech '04: Proceedings of the 5th international conference on Computer systems and technologies*, New York, NY, USA, pp. 1–6. ACM Press.

Kumar, A. and G. Pang (2002, March). Defect detection in textured materials using gabor filters. *IEEE Transactions on Industry Applications 38*(2).

Latif-Amet, A., A. Ertuzun, and A. Ercil (2000, May). An efficient method for texture defect detection: Subband domain co-occurrence matrices. *Image and Vision Computing 18*(6), 543–553.

Matthey, H., J. F. Fabiosa, and F. H. Fuller (2004, May). Brazil: The future of modern agriculture. *MATRIC*.

Menp, M. P. T. O. T. (2002). Multiresolution gray scale and rotation invariant texture classification with local binary patterns. In *IEEE Transactions on Pattern Analysis and Machine Intelligence 24*, pp. 971–987.

Roberts, M. and D. Etherington (1981). *Bookbinding and the Conservation of Books: A Dictionary of Descriptive Terminology*. Library of Congress.

Singh, M. and S. Singh (2002). Spatial texture analysis: a comparative study. In *ICPR02*, pp. I: 676–679.

Sobral, J. L. (2005, September). Optimised filters for texture defect detection. In *Proc. of the IEEE International Conference on Image Processing*, Volume 3, pp. 565–573.

Valdes, C. (2006, April). Brazil emerges as major force in global meat markets. *Amber Waves – The Economics of Food, Farming, Natural Resources and Rural America*.

Witten, I. H. and E. Frank (2005). *Data Mining: Practical Machine Learning Tools and Techniques*. San Francisco: Morgan Kaufmann.

Yeh, C. and D. B. Perng (2001). Establishing a demerit count reference standard for the classification and grading of leather hides. *International Journal of Advanced Manufacturing 18*, 731–738.

Computational Modelling of Objects Represented in Images –
João Manuel R.S. Tavares & R.M. Natal Jorge (eds)
© 2007 Taylor & Francis Group, London, ISBN 978-0-415-43349-5

Biomodels reconstruction based on 2D medical images

P. Lopes
Biomedical Engineering MSc student, Mechanical Engineering Department, University of Minho, Portugal

P. Flores & E. Seabra
Mechanical Engineering Department, University of Minho, Portugal

ABSTRACT: Over the last few years, the combination of medical image processing and rapid prototyping technologies lead to the birth of Medical Rapid Prototyping (MRP), a technology that has been applied to construct physical models of a patient's anatomy from high resolution imaging data, such as Computed Tomography (CT) and Magnetic Resonance Imaging (MRI). These models, sometimes referred to as biomodels, are three dimensional (3D) sections of an individual patient such as the bone structure, soft tissue, vascular structures, foreign bodies and implants, and can be used in two major applications, namely, the visualization of invisible structures and surgical training. In this work, a general methodology for reconstruction of biomodels is presented and discussed. This approach is based on the 2D medical images such as CT and MRI. An elementary example is also presented in order to show how the process works, as well as discuss their main advantages and limitations.

1 INTRODUCTION

Biomodeling is a generic term closely associated to MRP (Medical Rapid Prototyping) that has been coined to describe the ability to replicate the morphology of a biological structure in a solid substance. Specifically, it is used to describe the process of using radiant energy to capture morphological data on biological structures and the processing of such data by a computer to generate the code required to manufacture the structure by a RP (Rapid Prototyping) apparatus.

A biomodel is the product of this process and real virtuality is the descriptive term coined for the visualization medium. The creation of medical models requires the performance of a number of steps, namely, data acquisition, data processing, geometrical modeling and design and production of the model (D'Urso et al. 1999; Hieu, et al. 2005).

Several processes exist to produce medical models. The selection of a particular technology will depend on the medical model application. The main factors in deciding which RP technology is the most appropriate for a clinical application include dimensional accuracy of the model, overall cost of the model, availability of the technology and model building material (Onuh & Yusuf 1999).

Reverse Engineering (RE) and medical image-based modeling technologies allow reconstruction of three-dimensional models of anatomical structures of the human body based on anatomical information from

scanning data such as Computed Tomography (CT) or Magnetic Resonance Imaging (MRI). With the use of RP and Rapid Tooling (RT) technologies, nowadays, it is possible to create complex physical objects in a wide range of materials and sizes, from plastics and metals to biocompatible ones and from big models to microstructures (Hieu et al. 2005). The creation of medical models requires the realization of four steps, namely: (i) acquisition of 3D image data of the anatomy to be modeled; (ii) image processing to extract the region of interest from the surrounding tissues; (iii) geometrical modeling and design; (iv) prototyping and production. The MRP (Medical Rapid Prototyping) process can be condensed in these four steps, which are object of study in the present throughout this work.

In this paper a general methodology for reconstruction of biomodels based on 2D medical images is presented and discussed. The remainder of this paper is organized as follows. In the section two, the 3D data image acquisition of the anatomy necessary to construct a model is presented. A short description of image processing to extract the region of interest from the surrounding tissues is given in section three. Section four deals with the geometrical modeling and design approaches. The prototyping and production is presented in section five. Some example applications are presented in section six. Finally, in the last section, the main conclusions from this work are drawn and the perspectives for future research work are outlined.

2 IMAGE DATA ACQUISITION

The input data for 3D geometrical modeling of anatomical structures are Reverse Engineering (RE) data (point clouds) and CT/MRI images. Other imaging modalities are possible, as long as they provide cross sectional images with a 3D correlation. To avoid the undercuts and to complete scanning of a whole object, different scanning angles and directions are used.

RE scanners are used to collect high-quality volumetric data for modeling of external shapes (face, foot, elbow, etc.), dental templates and dry bones. Transmissive scanning methods are used for modeling the internal structures, in which CT, for bone structures, and MRI, for soft tissues, are well known scanning technologies. The data are often presented in a slice format, showing cross-sections of the patient, with areas of similar brightness representing specific tissue types or regions of similar material.

Due to the fact that different CT and MRI systems from different vendors use their own native data formats, to read CT or MRI images, a common digital imaging standard – DICOM, acronym for Digital Imaging and Communications in Medicine, is required. To construct accurate 3D anatomical structures for medical applications, the slice scanning thickness, normally required, varies from 1 to 2 mm, using a spiral scanning mode (Gibson et al. 2006).

3 IMAGE PROCESSING

The volumetric or 3D image data required for MRP models have certain particular requirements. Specialized CT scanning protocols are required to generate a volume of data that are isotropic in nature. This means that the three physical dimensions of the voxels (image volume elements) are equal or approximately equal. This feature is achieved by introducing multi-slice CT scanners where in-plane pixel size is of the order of 0.5 mm and slice thickness is lower than 1.0 mm. Data interpolation is often required to convert the image data volume into an isotropic data set for mathematical modeling (Winder & Bibb 2005). RE data, obtained from different scanning angles and directions, are then registered to merge into one single point cloud. Noise and redundant data are filtered to optimize the construction of contours and triangle mesh models (Hieu et al. 2005).

Furthermore, image processing steps will be required to identify and separate the anatomy (segmentation) to be modeled from surrounding structures. Segmentation may be carried out by image thresholding, manual editing or autocontouring to extract volumes of interest. Based on the grayscale value of image pixels, segmentation by thresholding techniques

Table 1. Comparative table of some demo versions of medical image processing programs (−poor; +fair; ++good; +++very good; ++++excellent).

Program	Interface	Functionality	Easy to use	Tutorial
Mimics	+++	++++	++	++++
3DDoctor	++	++++	++	++++
AnatomicsPro	+++	++	++++	++++
Velocity^{2}Plus	+	++	+++	+
InVesaliues	−	++	++++	++

is used to define the region of interest (ROI) and the obtained segmentation object contains only those pixels of the image with a value higher than or equal to the threshold value. It is also possible to use two thresholds in which the segmentation volume is defined by all pixels with a gray value in between both threshold values. This technique is used for segmentation of soft tissue in CT images or for segmentation of several structures in MRI images. In many circumstances, the volume of the body scanned is much larger than that actually required for model making. In order too reduce the model size, and therefore the cost, 3D image editing procedures may be employed. Clearer and less complex models may be generated, making structures of interest more clearly visible. Some image processing functions such as smoothing, volume data mirroring, image addition, and subtraction should be available for the production of models (Hieu et al. 2005, Gibson et al. 2006).

Ideally, segmentation is performed by a radiologist. For some scans it may be necessary to correct distortion due to artifacts (such as metallic implants or prosthesis in the patient) or errors due to partial volume effects. When segmentation is completed, data are translated in machine instructions and, if necessary, structures to support a model during building are generated (Popat et al. 1998).

A number of commercial software packages is available for data conditioning and image processing for MRP, including Analyze, Mimics by Materialise (Leuven, Belgium), AnatomicsPro, by Anatomics (Brisbane, Australia), Velocity2Plus, by Image3, LLC (Salt Lake City, USA), 3DDoctor, by Able Software Corp. (Lexington, USA) and In Vesalius, by CenPRA (Campinas, Brazil), just to mention a few. Table 1 shows some comparative features of the different programs used in the present work.

4 GEOMETRICAL MODELING AND DESIGN

Outputs from the data processing step are slice contours or 3D triangle mesh models, which are normally stored, respectively, in the form of IGES (Initial Graphics Exchange Specification) and STL

(Sterolithography Text Language) format. The slice format shows cross-sections of the patient, with areas of similar brightness representing specific tissue types or regions of similar material. In order to generate the reconstructed virtual model, the patient image data must be processed to identify and extract the relevant materials corresponding to the regions of interest. The contours for each slice are interpolated and joined together to form a completed 3D object representation. In a similar way as the patient data, the models may also be processed using 3D CAD to incorporate other objects, like fixation devices and implants. The 3D CAD may also be used to produce patterns in order to make fixtures or templates. Once the virtual model has been completed, the data are translated into a triangular surface tessellated, STL, file (Popat et al. 1998, Winder & Bibb 2005).

With the intend to construct biomodels, resulted STL data of anatomical structures are directly used for the RP fabrication progress, since they can be directly constructed and optimized in MRP and RP software. The remaining medical RP applications normally require a geometrical modeling process to construct NURBS (NonUniform Rational B-Splines) surfaces or solids from slice contours and triangle meshes because of the two following reasons. Firstly, the design process is normally implemented in CAD packages and NURBS CAD modeling techniques are popular in CAD/CAM applications. Secondly, in CAD packages, many powerful modeling tools are available; in addition, CAD entities and design data are easily controlled and structured. Meanwhile, most of the RP and RE packages can only provide quite limited CAD and surface editing tools.

In order to design surgical aid tools and medical devices, first of all, triangle mesh models of anatomical structures or ROI are converted into the NURBS CAD surfaces models. This is done by using RE modeling techniques. The reference data such as the drilling position and direction, contact area, tumor boundary contour are stored in the form of point, line, contours and surfaces, which are imported into CAD modeling packages for the end-use modeling and design process. Tissue engineering scaffolds for bone replacement have to satisfy the biological, mechanical and geometrical constraints. Based on CT/MRI and Micro CT images, geometrical data of bone and anatomical microstructures are constructed and imported into CAD modeling packages for scaffold design. Library of unit cells of different internal architectures and structural properties, which are designed to meet intended biological purpose, are integrated with the shape of bone to form the bone tissue scaffold to match the actual replaced bone. It is also possible to superimpose 3D anatomical models with a designed porous architecture to build a complex scaffold that mimics anatomical structures (Hieu et al. 2005).

5 PROTOTYPING AND PRODUCTION

STL data of biomodels and final designs are transferred to the RP site and the building process starts. For biomodels in which the ROI is such as blood vessels and tumors, which need to be colored for better visualization, the color STL or VRML format is used as the input data. SLA and 3D printing techniques are normally used for making color biomodels. SLA gives better accuracy and allows visualization of the internal structures, but it is limited in the number of colors for presentation. Color RP models fabricated by 3D printing techniques can be used for presenting FEA results, especially in the design evaluation and biomechanics research (Lopes et al. 2006).

Surgical aid tools, such as drilling guides, can be directly fabricated by SLA techniques without requiring further tooling processes. The standard metal parts such as guiding tubes should be used together with the RP parts to strengthen the tools and stabilize the drilling operations. Tooling processes are used to make implants, surgical aid tools and medical devices from biocompatible materials. Selection of the right tooling techniques depends on the biomaterials to be used for the applications.

Tissue engineering scaffolds are directly fabricated by RP techniques or using the RP model to make a mould from which the scaffold is fabricated. Among the available RP techniques, 3D printing and fused deposition modeling are the most commonly used for fabricating tissue engineering scaffolds because of versatility of using scaffolding materials and they are able to overcome the limitations of conventional manual-based fabrication techniques (Winder & Bibb 2005, Hieu et al. 2005).

The final step in the process chain is the post-processing task. At this stage, generally some manual operations are needed therefore skilled operator is required. In cleaning, excess elements adhered with the part or support structures are removed. Sometimes the surface of the model is finished by sanding, polishing or painting for better surface finish or aesthetic appearance. Prototype is then tested and suggested engineering changes are once again incorporated during the solid modeling stage. Depending on the use of the model, it can be sterilized for assistance in an operating theatre (Popat et al. 1998).

The four steps summarized in sections two to five require significant expertise and knowledge in medical imaging, 3D medical image processing, computer-assisted design, and manufacturing software and engineering processes. The production of reliable, high-quality models requires a team of specialists that may include medical imaging specialists, biomedical and RP/RE/CAD/CAM engineers and surgeons (Winder & Bibb 2005). Surgeons are the ones who determine surgical procedures and clinical constraints

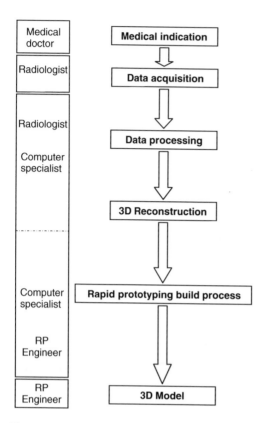

Figure 1. General procedures to create a medical model.

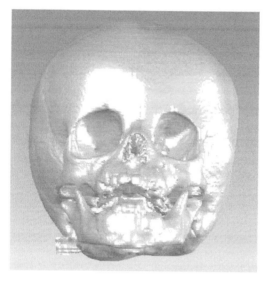

Figure 2. 3D model of a female child's skull obtained with evaluation version of Mimics.

for the design. The participation of surgeons and their suggestions about the design are very important. Biomodels and 3D computer simulations are helpful for preoperative planning, communication and discussions with surgeons. The design process is started based on clinical inputs given by surgeons via surgical procedures and clinical constraints.

The design must be carefully checked by both designers and surgeons before it is prototyped and produced. The best checking method is to evaluate the design based on CT/MRI data of the patient. For the design of implants and surgical aid tools, this is done by importing the design model into MIP packages and checking the design at every slice image. For complex and loading implants, surgical aid tools and medical devices, finite element analysis (FEA) and mechanical testing need to be implemented to evaluate the design (Popat et al 1998, Hieu et al. 2005).

Figure 1 shows the general procedures to create a medical model.

6 EXAMPLE APPLICATION

This section presents some example applications with the intention to demonstrate how to construct medical models based on 2D medical images, creating 3D virtual models that can, consequently and eventually be used to build a real prototype. Further, since the 3D virtual model is available, a posterior structural analysis based on FEM can be performed. During this work stage, enterprises, research groups and medical doctors have been contacted. Some CT/MRI data have been acquired, but none of them contained any pathological problem. The CT/MRI data have been acquired from a female child's skull, among others.

Figure 2 shows a 3D virtual model obtained from 2D medical images and converted to 3D by using the commercial program Mimics. This virtual model was then imported to Solidworks program that allowed the structural analysis of the model.

Figure 3 illustrates a 3D virtual model 3D model of a male adult's thorax obtained with the demo version of Anatomics.

Finally, figure 4 depicts a 3D model of a skull obtained with demo version of Velocity^{2}Plus.

7 CONCLUDING REMARKS

In this paper a general methodology for reconstruction of biomodels based on 2D medical images was presented and discussed. The necessary steps to obtain a medical model were also presented throughout this work, namely: (i) acquisition of 3D image data of the anatomy to be modeled; (ii) image processing to extract the region of interest from the surrounding tissues; (iii) geometrical modeling and design; (iv) prototyping and production. The MRP (Medical Rapid

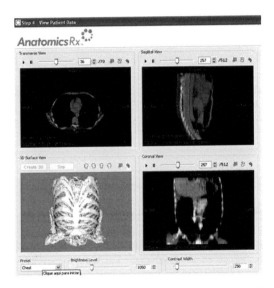

Figure 3. 3D model of a male adult's thorax obtained with the demo version of Anatomics.

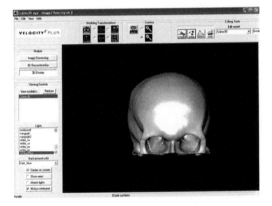

Figure 4. 3D model of a skull obtained with demo version of Velocity^{2}Plus.

Prototyping) process can be condensed in these four steps. Finally, some elementary example applications were presented and discussed.

This work reports on an initial phase that corresponds to creation of a research work at the University of Minho, entitled ModMed (acronym for Models Medical), which will develop 3D virtual medical models and physical prototypes applying RP techniques. The ModMed team will work in cooperation with companies and institutions, and can be a stimulant initiative to the creation on new enterprises and the development of the MRP technology in Portugal.

REFERENCES

D'Urso, P. et al. 1999. Biomodelling of Skull Base Tumours. *Journal of Clinical Neuroscience* 6(1), pp. 31–36.

Gibson, I. et al. 2006. The Use of Rapid Prototyping to Assist Medical Applications. *Rapid Prototyping Journal* 12(1) 53–58

Hieu, L. et al. 2005. Medical Rapid Prototyping Applications and Methods. *Assembly Automation* 25(4), 284–292.

Lopes, P., Flores, P. & Seabra, E. 2006. Rapid Prototyping Technology in Medical Applications: A Critical Review. *Proceedings of the CompIMAGE, Computational Modelling of Objects Represented in Images, Coimbra, Portugal, 20–21 October* (to be appeared).

Onuh, S. & Yusuf, Y. 1999. Rapid Prototyping Technology: Applications and Benefits for Rapid Product Development. *Journal of Intelligent Manufacturing* 10, 301–311.

Popat, A. et al. 1998. Rapid Prototyping and Medical Modelling. *Phidias, Rapid Prototyping in Medicine* 1, 10.

Winder, J. & Bibb, R. 2005. Medical Rapid Prototyping Technologies: State of the Art and Current Limitations for Application in Oral and Maxillofacial Surgery. *Journal of Oral and Maxillofacial Surgery* 63, 1006–1015.

Computational Modelling of Objects Represented in Images –
João Manuel R.S. Tavares & R.M. Natal Jorge (eds)
© 2007 Taylor & Francis Group, London, ISBN 978-0-415-43349-5

Robust vision algorithms for quadruped soccer robots

Luís Paulo Reis

FEUP – Faculty of Engineering of the University of Porto, Portugal
LIACC – Artificial Intelligence and Computer Science laboratory of the University of Porto, Portugal

ABSTRACT: This paper presents a robust vision system based on semi-automatic color calibration, color segmentation and image high-level analysis for quadruped soccer playing robots. The algorithms presented were used in the context of FC Portugal legged league team that participated since 2003 in RoboCup – Robotic Soccer World Championship. Controlled experiments with variable lightning conditions are analyzed in order to conclude the robustness of our vision system.

1 INTRODUCTION

RoboCup legged league uses teams of four AIBO (ERS210A or ERS7) robots (figure 1) that play a soccer match in a green carpeted field.

Figure 2 depict an ERS210A robot and its main sensors and actuators. The robots have a huge amount of sensors, including a CCD Color Camera (352×288 pixels), Stereo Microphone (16 kHz), Infrared Distance Sensors (10–90 cm), Touch Sensors spread throughout the body, Acceleration, Temperature and Vibration sensors.

The robot has 20 degrees of freedom, including three motors on each leg and three motors on the head (pan, tilt and roll). The robots also include a speaker and, depending on the model, several leds are also included, mainly with debugging purposes.

A legged league game has two halves of 10 minutes each, and is played in a $5,4 m \times 3,6 m$ field, as depicted in Figure 3. The robots must be completely autonomous and play without any human intervention. Most of the interesting objects in the field are colored in order to enable the robots to use as main sensorial source a real time vision system based on fast color segmentation.

FC Portugal legged league team research focus is mainly on coordination methodologies applied to the legged league and on developing a common approach to all RoboCup soccer leagues. In the context of our legged team we also perform research on high-level vision and automatic calibration, sensor fusion, multi-agent communication, navigation, localization and learning applied to teams of mobile robots.

The team started performing experiments with the German Team [21] simulator [5, 20, 21] and using our expertise in RoboCup simulation league and on

Figure 1. ERS210A (left) and ERS7 (right) AIBO Sony quadruped robotic platforms.

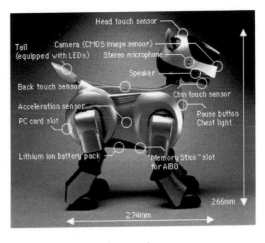

Figure 2. ERS210A sensors and actuators.

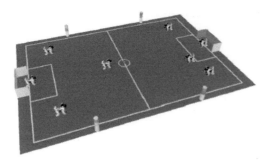

Figure 3. Legged league field showing colored goals and landmarks.

developing complex simulators [16, 17], we have built a very simple legged league simulator (with very simple models of the robots) enabling us to test different positioning strategies. Afterwards we have bought Sony ERS210A robots and moved our code from the simulator to the real robots. For that, we have used over the years, CMPack02 [23] and UNSW03 [7, 8, 22] codes as the base. We have applied over the base code, several previously researched methodologies developed and tested in our teams in other RoboCup leagues (Simulation, Small-Size, Middle-Size and Coach Leagues). From FC Portugal [10] (champion of RoboCup simulation league in 2000, European champion in 2000 and German Open Winner in 2001) we introduced simple versions of SBSP – Situation Based Strategic Positioning [17, 18], ADVCOM – Advanced Communications [11, 17] and DPRE – Dynamic Positioning and Role Exchange [17, 18]. From our FC Portugal Coach (Coach Competition champion in 2002), we have taken our tactical structure and coaching language [19]. From 5DPO teams [1] (small-size 3rd in RoboCup 1998, German Open Winners in 2001 and 2nd in 2002) we have taken the base vision system and most of our navigation algorithms [9, 15]. Our vision algorithms were then extended in order to enable robust color segmentation, automatic calibration and high-level image analysis. This paper describes briefly these extensions and the results achieved by our vision system in variable lightening conditions.

The paper is organized as follows. Section 2 presents our vision module. Section 3 describes the localization module and the world state information used for high-level decisions. Section 4 presents some results and section 5 the paper conclusions.

2   VISION MODULE

Based on the publicly available code from RoboCup 2002 legged league champions (CMPack) [23],

namely on its CMVision image processing library [3, 4] and on our experience and own code for small-size and middle-size leagues vision [9], we have developed a robust vision system including capabilities for color image segmentation, automatic calibration and object recognition.

The approach is based on the use of thresholds for performing color segmentation on a given color space. A color space is an abstract mathematical model describing the way colors can be represented as tuples of numbers, typically as three or four values or color components (e.g. RGB and CMYK are color models). Several color spaces are in wide use, including Red Green Blue (RGB), Hue Saturation Value (HSV) and YUV (YCbCr). The choice of color space for classification depends on several factors including the type of hardware used and the particular application.

RGB is one of the most used color spaces in image processing. It suffers from an important drawback regarding robotic vision applications based on color segmentation. In robotic soccer, for instance, features of the environment are identified by colors (e.g. the ball is orange, goals cyan and yellow). Color segmentation should be robust in the face of variations in the brightness of illumination. HSV and YUV systems have the advantage that chrominance is coded in two of the dimensions (H and S for HSV) or U and V for YUV) while intensity value is coded in the third. Thus a particular color can be identified as a column that uses all available intensities. These color spaces are therefore often used in simple robotic applications. However, using regions of arbitrary shape in RGB color space, enable us to identify exactly the same colors with a completely free region in the color same. Thus, that was our approach with only the special concern of achieving the possibility to perform color segmentation using arbitrary shapes in the RGB color space.

2.1   General description

This system is also capable of performing the generation of a high-level description of the image contents, including the identification of each object, its direction, distance, size, elevation and confidence. The steps performed by the vision module are the following:

- Construction of color calibration lookup tables by a semi-automatic process;
- Capturing an image and classifying pixels into the pre-defined color classes (basically by looking up into the previously defined table);
- Conversion of the image to RLE – Run Length Encoding (although in practice because of performance reasons this step is performed together with the previous step);

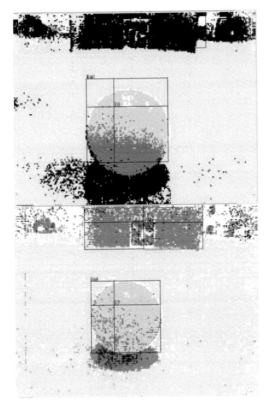

Figure 4. Color segmentation result for different lightning conditions.

- Image segmentation, finding blocks of the same color (blobs) and their characteristics (center, size and shape). This step is performed using hierarchical, multi-resolution algorithm;
- Object recognition and generation of an image high-level description: identifying objects based on color blobs and converting its own coordinates to world coordinates (relative distance and direction);
- Textual image description: changing the high level image description into a text description easily understandable by humans.

We use the eight typical colors of the legged league: pink (for the beacons), yellow and sky blue (for beacons and goals), dark red and dark blue (for the robots' uniforms), orange (for the ball), green (for the field carpet) and white (for the field lines and detection of ERS7 bodies). Our robots need to detect and discriminate these colors in order to recognize the appropriate objects. Figure 4 shows the result of the color segmentation process for completely different lightening conditions.

Image segmentation is performed based on the pre-processed image resulting from our color calibration module using the fast blob formation algorithm.

2.2 Automatic calibration

One of the main innovations of our team's vision system resides on the method developed for the construction of the color calibration tables. We have used our previous experience in designing vision systems for the small and middle-size leagues [1, 9] and designed a very simple automatic color calibration module. Using this module, we construct the color table, based on a set of significant images autonomously collected by the robot that walks around the field looking for colors similar to the ones available in its previous calibration. With the pictures autonomously gathered by the robot we assemble in a semi-automatic manner the color segmentation table. This system was demonstrated in RoboCup 2004 challenge (Lisbon) achieving 8th place in Challenge 1.

The system gathers in semi-automatic process 44 images for the final calibration process by using a moving robot on the field. The robot starts using a color table, previously constructed and that contains very large color areas. While collecting images, the robot also adjusts its initial color table in order to be able to cope with the lightening variations.

The robot starts on the middle of the field. Several balls are located on the field including a ball in each of the goals. Three robots of each team are located on the field, including one in each goal. The process steps are the following:

- Take pictures 1–4 of the field, one in each direction.
- Adjust global lightening of the color cube.
- Rotate to find the yellow/pink beacon and the pink/yellow beacon. Pink is the simplest color to identify on the field.
- Rotate to find the yellow goal.
- Adjust pink and yellow on the color table.
- Rotate to find the cyan/pink and the pink/cyan beacons.
- Rotate to find the cyan goal.
- Adjust pink and cyan on the color table.
- Take pictures (5–8, 9–12, 13–16, 17–20) from each beacon at different distances.
- Find the orange balls in both goals.
- Adjust orange on the color table.
- Find each of the four balls and take pictures at four different distances (21–24, 25–28, 29–32, 33–36).
- Find the red robot in the yellow goal.
- Adjust red.
- Find the blue robot on the cyan goal.
- Adjust blue.
- Take pictures (37–40, 41–44) from robots of each team at different distance.
- Use all images gathered to create the final color cube.

After this initial color calibration, steps are taken to verify that the calibration has been performed properly and, in addition, to further improving the accuracy of

369

the color calibration process. The following steps are performed:

- Re-Check Image Classification: This is a quick check that all the images have been classified correctly by the semi-automatic process. All the images are reviewed and checked to see that the ball has not been accidentally classified as a red dog for example. Also it is important that that the area outside the object has not been classified as this would result in the blob generated to expand into incorrect regions. Also the outside area of the field is checked for wrong positives that could mean objects detected outside of the field during the game.
- Examine the Colour Cube: The RGB colour cube is depicted graphically and checked analysing the eight color areas and comparing them with the "normal" areas for this application.
- Examine the Images Gathered in Real Movement: This check is performed to determine whether the robot has classified objects correctly. The robot analysis of the field in real movement is visualized on the screen and correct object detection is checked. Vital information is the capability to detect the ball, goals and localize using the beacons. Objects appear with a slightly different colour when the robot is moving fast (the images are more blurred and elongated as well). This is an important consideration. Although training images were collected with the robot also moving, since during the games the robots move in a faster way, this check enables to predict the real vision behaviour during real games.

If the calibration process is not sufficiently robust, new images may be gathered and classified manually to increase the robustness of the calibration process.

3 LOCALIZATION AND WORLD STATE

In the legged league the objects that must be tracked include: the four unique markers, the ball, the two goals and the other seven robots. Objects in the image are identified based on its color, shape and position. Color is the main feature used for object identification, an approach similar to that used by most other teams so far. Since our Image Segmentation module also gives some information about the shape and size of the blobs, we use this information to recognize objects and to estimate its distances, directions, elevations and headings relatively to the neck of the viewing robot.

Several localization algorithms have been proposed and tested in the context of RoboCup [12]. We started by using a localization method that works using high-level vision information with a Fuzzy Landmark-Based Localization algorithm similar to several RoboCup teams, inspired in [6] and moved to

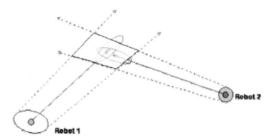

Figure 5. Limiting area for the presence of the ball, in conjunction with the extreme vectors.

a sensor resetting localization [13] whose source code was available on the web [14].

FC Portugal agent's internal representation of the world, results on the processing of various sensorial information sources fused with communicated information. The information contained in the world state includes:

- Self Position (Cartesian coordinates);
- Ball Position and Velocity (Cartesian coordinates);
- Teammates' Position and Velocities (Cartesian coordinates);
- Teammates' Behavior and State;
- Opponents' Position and Velocities (Cartesian coordinates);
- Opponents' Behavior and State.

All the above measures have an associated certainty factor, expressing the probability of the measure being accurate.

Our world state update method is somewhat different from other teams because since 2004 competition we use a trigonometric world state fusion method. Figure 5 shows a graphical description of this method. The area defined between the interceptions of two of the agents quantifies the uncertainty of the calculated ball position (the bigger the area the bigger the uncertainty). This area is the limiting area for the presence of the ball according to the sensorial information of the two involved robots (see details in [2]).

Our global World state update is similar to our simulation league team FC Portugal [10, 17]. It uses:

- High-level visual information: resulting from our vision module;
- Sensorial information: achieved using proximity and touch sensors' preprocessing;
- Communicated information: sent by other members of the team; and
- Action prediction: prediction of the effects of robot actions in the environment.

These four types of information are fused together to assemble our world-state used as an input in the high-level decision module.

Experience/Values	RES	ATT	BFP	OGFP
Auto vs CMU	8-0	90.3%	25	6
Auto vs UNSW	5-1	68.3%	38	3
Manual vs CMU	10-0	86.9%	24	2
Manual vs UNSW	6-1	67.2%	36	1

Figure 6. Results achieved in laboratory tests.

The communication between the robots complements the internal world state from all the robots, with additional information concerning team coordination, such as, position swap between team members or the existence of a ball pass to another team member.

4 RESULTS

To validate our approach six games, with 5 minutes duration, were performed against each of CMU 2003 and UNSW 2003 teams (configured with simple vision calibration). Our team used semi-automatic calibration in three games against each opponent and normal calibration (with the aid of CMVision) in the other three. The same localization, locomotion, decision and cooperation configuration was used in all experiences. The experiences were performed in a quite lightening variable environment in order to test the robustness of both vision systems.

Figure 6 show the results attained by our automatic calibration process compared with the results obtained using the completely manual calibration process. The results gathered, besides the global result (measured in the number of goals for each team), include, for each set of three experiences: the percentage of attack, ball false positives and opponent goal false positive detections.

From the results displayed it is clear that the performance of both teams is similar. With automatic calibration the performance is slightly worse, mainly in terms of the number of goals achieved, even with some improvement in the percentage of attack against both opponents. This performance loss in the number of goals scored seems to be explained by the number of false positives in the detection of the opponent goal, leading to shoots in the wrong direction. These false positives were mainly due to other objects around the field with color similar to yellow that were filtered in the manual calibration process but the automatic calibration process was not able to filter completely.

5 CONCLUSIONS

The results achieved showed that for normal play, the semi-automatic calibration process enables the robots to play without significant performance loss. Since

in these robotic applications, involving competitions, with limited time for setup and tuning, time is scarce resource, the semi-automatic calibration process is a major advantage. It enabled our team to participate in RoboCup 2003, 2004 and 2005 with only two partial-time team members and compete against teams with more than twenty full-time team members achieving very good results.

Our legged team in its three participations in RoboCup scored 40 goals and conceded also 40, achieving 5th place in Padua, Italy, and 9th place in Lisbon, Portugal. In all three competitions our vision system needed only one semi-automatic calibration in the beginning of the competition and was able to cope with different fields and lightning conditions without deteriorating significantly the team performance.

REFERENCES

1. 5DPO Home Page, online, available at: http://www.fe.up.pt/~robosoc
2. R. Afonso, T. Carvalho, L.P. Reis and E.Oliveira, Trignometric World State Fusion, Reis, L.P. et al. editors, Proc. Scientific Meeting of the Portuguese Robotics Open 2004, FEUP Edições, Colecção Colectâneas, Vol. 14, pp.157–162, ISBN 972-752-066-9, 23–24 April, 2004.
3. J. Bruce, T. Balch, M. Veloso, Fast and inexpensive color image segmentation for interactive robots, IROS-2000, Japan, October 2000.
4. J. Bruce, T. Balch, M. Veloso, CMVision, http://www.coral.cs.cmu.edu/vision
5. H.-D. Burkhard, U. Düffert, J. Hoffmann, M. Jüngel, M. Lötzsch, R. Brunn, M. Kallnik, N. Kuntze, M. Kunz, S. Petters, M.V. Risler, O. Stryk, N. Koschmieder, T. Laue, T. Röfer, Spiess, A. Cesarz, I. Dahm, Hebbel, M. Nowak, W and Ziegler, J. GermanTeam 2002.Technical report, 2002.
6. P. Buschka, A. Saffiotti and Z. Wasik, Fuzzy Landmark-Based Localization for a Legged Robot. Proc. of the IEEE/RSJ Int. Conf. Intelligent Robots and Systems (IROS) pp. 1205–1210. Takamatsu, Japan, 2000.
7. S. Chen, M. Siu, T. Vogelgesang, T. Yik, B. Hengst, Son B. Pham, C. Sammut: The UNSW RoboCup 2001 Sony Legged Robot League Team, in A. Birk, S. Coradeschi and S. Tadokoro, eds, RoboCup-2001: Robot Soccer World Cup V, Springer LNAI, pp. 39–48, Berlin, 2002.
8. Jin Chen et al. A Description for the rUNSWift 2003 Legged Robot Soccer Team, online, available at: http://www.cse.unsw.edu.au/~robocup/2003site/teamDesc2003.pdf, 2003.
9. P. Costa, P. Marques, A.P. Moreira, A. Sousa and P. Costa, Tracking and Identifying in Real Time the Robots of a F-180 Team in Manuela Veloso, Enrico Pagello and Hiroaki Kitano, editors. RoboCup-99: Robot Soccer World Cup III. Springer, LNAI, pp. 286–291, 2000.
10. FC Portugal Home Page, online, available at: http://www.ieeta.pt/robocup
11. R. Ferreira, L.P. Reis and N. Lau, Situation Based Communication for Coordination of Agents, Reis, L.P. et al. editors, Proc. of the Scientific Meeting of

the Portuguese Robotics Open 2004, FEUP Edições, Colecção Colectâneas, Vol. 14, pp.39–44, ISBN 972-752-066-9, 23–24 de April, 2004.

12. D. Fox, W. Burgard, F. Dellaert and S. Thrun. Monte Carlo localization: Efficient position estimation for mobile robots. In Proc. of the National Conf. on Artificial Intelligence (AAAI), 1999.

13. S. Lenser, M. Veloso, Sensor resetting localization for poorly modeled mobile robots, ICRA-2000, S. Francisco, 2000.

14. S. Lenser, Sensor Resetting Source Code, online, available at: http://www-2.cs.cmu.edu/~slenser/

15. A. Moreira, A. Sousa, P. Costa, Vision Based Real-Time Localization of Multiple Mobile Robots, 3rd Int. Conf. Field and Service Robotics, Helsinki, Finland, pp. 103–106, June 11–13, 2001.

16. A. Pereira, P. Duarte and L.P. Reis, Agent-Based Simulation of Ecological Models. In: H. Coelho & B.Espinasse (eds.) Proc. 5th Workshop on Agent-Based Simulation, pp. 135–140, Lisbon, May 3–5, 2004.

17. L.P. Reis, N. Lau, FC. Portugal Team Description: RoboCup 2000 Simulation League Champion, in [Stone, 2001], pp. 29–40, 2001.

18. L.P. Reis, N. Lau, E. Oliveira, SBSP for Coordinating a Team of Homogeneous Agents in M.Hannebauer et al. eds, Bal. React. Social Delib. in MAS, Sp. LNAI, Vol. 2103, pp. 175–197, 2001.

19. Luís Paulo Reis and Nuno Lau, COACH UNILANG – A Standard Language for Coaching a (Robo) Soccer Team, in A. Birk, S. Coradeschi and S. Tadokoro, eds, RoboCup-2001: Robot Soccer World Cup V, Springer LNAI, pp. 183–192, Berlin, 2002.

20. T. Röfer et al. An Architecture for a National RoboCup Team. In: RoboCup 2002. LNAI, RoboCup 2002: Robot Soccer World Cup VI Pre-Proceedings, 388–395, 2002.

21. T. Röfer, I.Dahm, U. Düffert, J. Hoffmann, M. Jüngel, M. Kallnik, M. Lötzsch, M. Risler, M. Stelzer, J. Ziegler. (2004). GermanTeam 2003. Proc. 7th Int. Workshop on RoboCup 2003 (Robot World Cup Soccer Games and Conferences), Springer LNAI, Padova, Italy, 2004.

22. C. Sammut, et al., UNSW 2003 Source Code, online, available at: http://www.cse.unsw.edu.au/~robocup/2003site/

23. M. Veloso, et al. CMPack02, CMU's Legged Robot Soccer Team, Carnegie Mellon University, October, 2002.

Computational Modelling of Objects Represented in Images –
João Manuel R.S. Tavares & R.M. Natal Jorge (eds)
© 2007 Taylor & Francis Group, London, ISBN 978-0-415-43349-5

Integrated graphical environment for support nonlinear dynamic software for the analysis of plane frames

H. Rodrigues, H. Varum & A. Costa
Department of Civil Engineering, University of Aveiro, Aveiro, Portugal

ABSTRACT: Nonlinear structural analyses allow reproducing in a more realistic sense the behaviour of structures subjected to several types of complex loading conditions, e.g. earthquakes. However, it is largely recognized that these analyses normally generate a considerable amount of results, being difficult its interpretation. Over the last years considerable progresses have been made in structural nonlinear behaviour modelling, associated to the fast growing development of numerical algorithms for structural analysis and computer capacities. However, a similar growth in the development of graphical results visualization tools has not been witnessed. To face this, a graphical processor called VISUALANL was developed for an existing nonlinear dynamic analysis program for plane frame structures, PORANL.

1 INTRODUCTION

Numerous programs developed over the past at the academic level for nonlinear structural analysis are not being used by engineers because, in most cases, these programs lack graphical interfaces, a downside creating strong barriers to their use. Traditionally, data input for these programs was carried out through extensive text files, with a pre-defined rigid sequence. Normally, the results are also presented in long text files, without any or poor graphical visualization.

To face this, a graphic tool called VISUALANL was developed to provide a user frontend for an existing nonlinear dynamic analysis program for plane frames, PORANL (Varum, 1996). By using VISUANL, introduction and modification of data are easier thanks to interactivity. Graphical representation at run-time is possible using dialogue windows. It should be noted that the philosophy followed in this graphical tool does not enforce a rigid sequence in the data input. For results interpretation, the program provides an easier visualization by supplying a series of graphical representations suited for different needs.

The development of VISUANL, which incorporates the necessary aspects for preparation, manipulation and visualization of a structural analysis problem, that may consider material nonlinearity and dynamic behaviour, was carried out in the programming language Visual Basic, version 6.0 (see for example Thayer, 1999) that enables an easy implementation of the necessary graphical concepts and tools.

Besides highlighting the potentialities of the graphical processor VISUANL at the levels of datageneration and results visualization, this article briefly describes the structural models available in the program PORANL.

2 AVAILABLE BEHAVIOUR MODELS

2.1 Structural behaviour models

Varum (1996) developed the initial version of the nonlinear analysis program PORANL in which a hysteretic behaviour model suited for seismic analysis. The non-linear model implemented is the Costa-Costa model (1987, 1996), a modified Takeda hysteretic model suited for seismic analysis of reinforced concrete (RC) frame elements predominantly subjected to bending.

The structural member modelling strategy associated to this hysteretic model considers that inelastic deformations are concentrated in the vicinity of the member's extremities, in which nonlinear behaviour is expected under earthquakes. Each structural element is a macro-element subdivided into three sub-elements (as represented in Figure 1-a), namely a central sub-element with linear elastic behaviour, connected at each end to a sub-element with nonlinear behaviour (plastic hinge). The behaviour of each plastic hinge under cyclic loading is represented by the referred hysteretic model. This model enables the numerical modelling of several known behaviour characteristics of members subjected to earthquake-type loading, namely: i) stiffness degradation with load reversals; ii) strength degradation due to cycling loading;

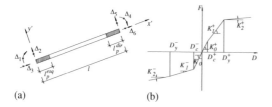

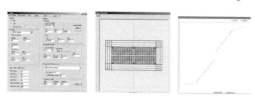

(a)   (b)

Figure 1. a) Macro-element of a structural member; b) Monotonic envelope for RC cross sections under bending.

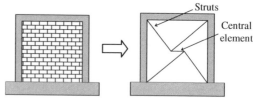

Figure 3. Proposed equivalent truss model.

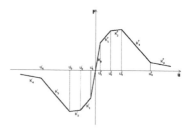

Figure 2. Graphical interface for the fibre model: data generation, section visualization, nonlinear monotonic envelope.

iii) pinching effects accounting for the importance of shear forces; iv) slipping effects accounting for reinforcement slippage; v) possibility to define different monotonic envelopes for positive and negative bending. The reader is referred to CEB (1996) for a more detailed description of these structural effects.

The behaviour of each plastic hinge under monotonic loading is represented by a trilinear moment-curvature envelope, Figure 1-b, for each bending direction, thus allowing the analysis of asymmetric RC cross sections. Trilinear moment-curvature envelopes of RC cross sections can be obtained using a fibre model approach as the one implemented in the program BIAX developed by Vaz (1996). The possibility to use the BIAX program was also included in the graphical framework of VISUALANL (see Figure 2).

In the fibre model approach, cross sections are divided into slices (in the case of uniaxial bending) or filaments (in the case of biaxial bending). In the fibre approach, the cross section constitutive behaviour is obtained from the uniaxial behaviour of the fibres but shear strains across the section are not modelled. The basic assumptions that cross sections remain plane after deformation and that only small deformations take place are also considered. Fibre modelling allows for the calculation of the field of axial extensions in the section by equilibrium as a function of the axial extension and curvatures. Based on the deformation of each fibre, stress at each fibre can be obtained from the stress-strain relation of each material. Then, integrating the stress field across the cross section yields the internal forces.

2.2 Infill masonry model

It is a common misconception that masonry infills in structural RC frames can only increase the overall

Figure 4. Monotonic envelope for the infill panel element.

lateral load capacity, and therefore, must always be beneficial to seismic performance. Even when relatively weak, masonry infills can drastically modify the global stiffness of the structure and consequently the structural response, attracting forces to parts of the structure that have not been designed to resist them (Paulay and Priestley, 1992).

A hysteretic model for the simulation of the cyclic behaviour of infill masonry panels based on the equivalent truss model, as presented in the Figure 3, was implemented in the program PORANL by Rodrigues (2005). In this model, each infill panel is represented by diagonal trusses with linear behaviour and a central element that simulates the nonlinear behaviour of the infill panel. The nonlinear behaviour of each panel defined by the central element is represented by an hysteretic model, contemplating important effects, such as: i) stiffness degradation with load reversals; ii) strength degradation due to cycling loading; iii) pinching and slipping effects accounting for the influence of shear forces.

Monotonic behaviour of the panels is characterised by a piecewise-linear envelope with five branches for each loading direction, as represented in Figure 4, thus allowing for the analysis of infill panels with non-symmetrical behaviour e.g., a panel with a non-centred opening.

3 PRE-PROCESSING DATA

The graphical framework VISUALANL integrates both pre and post-processing operations, necessary to carry out a structural analysis run. The basic steps of data input using this new interface, namely,

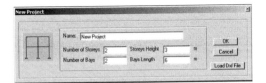

Figure 5.   Input of the geometry of the structure.

definition of geometry, constraints, loads, material characteristics (for members having linear or nonlinear behaviour) are described in the following. Description of the pre-processor steps will be carried out in a logical sequence for the definition of a structural problem. However, one of the major advantages of VISUALANL is the possibility of data definition without following to a rigid sequence.

3.1   Geometry of the structure

The geometry of the structure can be defined in two different ways (see Figure 5): i) defining an initial support mesh for the structure's geometry, establishing the number of storeys and bays, and corresponding dimensions; ii) importing a DXF file with the complete structural geometry. After either of the previously referred forms for geometry definition, the user is able to change the length of a bay or the height of a storey, add or delete frames or change the coordinates of the nodes.

3.2   Linear and nonlinear sections

As stated previously, each structural element having nonlinear behaviour is modelled with three sub-elements: a central element with linear behaviour and two sub-elements in the vicinity of the joints with nonlinear behaviour (see Figure 1-a). To define the characteristics of the sub-elements with linear behaviour (see Figure 6-a), the following parameters are necessary: width and depth of the member cross section, or its corresponding moment of inertia and area; volumetric mass of the material and its Young's modulus. For each sub-element with nonlinear behaviour are defined: the plastic hinges (Varum, 1996), parameters defining their monotonic behaviour (trilinear moment curvature envelope), the hysteretic behaviour rules and the damage index parameters. In addition, properties of the linear and nonlinear sub-elements can be imported from a text file to VISUALANL, and can be modified or deleted at any time.

3.3   Frame properties

The following properties are assigned to each frame or group of frames (Figure 7): i) the material defining the elastic properties of the member; ii) the length of the plastic hinges at the member ends; iii) the material defining the nonlinear behaviour of each plastic hinge.

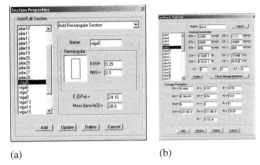

(a)   (b)

Figure 6.   a) Definition of elements with linear behaviour; b) Definition of elements with nonlinear behaviour.

Figure 7.   Frame element properties.

Figure 8.   Definition of nodal constraints.

3.4   Nodal constraints

Following the basic formulation of a 2D structural model, the user can restrict the degrees of freedom of a node, or group of nodes, namely, displacements in the x and y directions and the rotation (Figure 8). To facilitate the definition of constraints shortcut keys were defined for the different constraint types.

3.5   Infill panels

To add an infill masonry panel to the frame structure, the user has to select the 4 nodes surrounding

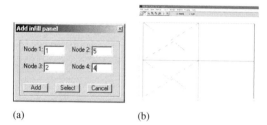

(a)   (b)

Figure 9. a) Infill panel definition; b) Infill panels representation.

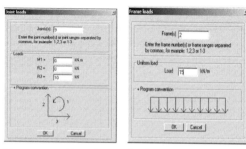

(a) concentrated in nodes   (b) distributed in frames

Figure 10. Static loads.

the panel. This operation can be performed in two different ways, namely, the user can write the node labels in a dialog box or directly selecting in the structure, with the mouse, the area where the infill panel must be considered (Figure 9-a). After adding the infill panel, VISUALANL represents it in the global structure (Figure 9-b), and the user is now able to define the masonry properties previously described.

3.6 Static loads

The static loads considered in the PORANL are concentrated loads in nodes and uniformly distributed loads in members. The concentrated loads applied directly to the nodes can be forces in x and y directions and also concentrated bending moments (see Figure 10-a). With respect to the distributed loads, only uniform loads applied on the total member length and normal to its axis are considered (as represented in the Figure 10-b). Any other different load case in members can be considered as a set of equivalent concentrated nodal forces. For a correct definition of the loading, the load sign conventions adopted in the PORANL program are displayed in the dialogue window.

4 PROBLEM CALCULATION

After completing the input of the structural problem using VISUALANL, the program prepares a series

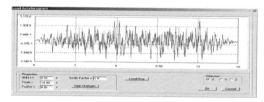

Figure 11. Accelerogram option window.

of text files that constitutes the text based input data for the analysis engine PORANL and runs it. Next, VISUALANL reads the results from the PORANL output text files, and, using a series of post-processing options, enables the fast graphical visualisation of the structural response results. As can be interpreted from the above, two different programs are used (VISUALANL and PORANL), although the user only works in the graphic environment VISUALANL without interacting directly with the analysis engine PORANL.

The program PORANL contemplates several analysis types, namely: i) linear elastic static analysis; ii) nonlinear static analysis, enabling to perform the pushover analyses; iii) nonlinear dynamic analysis; iv) nonlinear displacement controlled; and, v) calculation of the natural frequencies and vibration modes of the structure. Each one of these options was implemented in VISUALANL.

4.1 Static analysis

After the definition of the geometry, material properties and loading conditions, no additional data is necessary to perform a static analysis. Therefore, the user only has to choose whether the analysis is linear or nonlinear, though for the latter the user must supply the nonlinear material characteristics referred.

4.2 Dynamic analysis

To perform a nonlinear dynamic analysis the user has to define: i) the accelerogram (imported from a text file); ii) the dynamic equilibrium equation integration method to be used (Wilson-θ, Newmark or Central Differences Method); and, iii) the parameters for the definition of the Rayleigh-type damping matrix. These parameters can be calculated in VISUALANL as a function of an assumed damping value for two different natural frequencies.

After importing the selected accelerogram, it can be visualized (Figure 11) and the user can scale it to increase or reduce its intensity, alter the integration time-step or eliminate part of the accelerogram. In the dialogue window of the accelerogram definition, Figure 11, the user can also define the time interval for which the program PORANL will save results of the analysis in the output files.

376

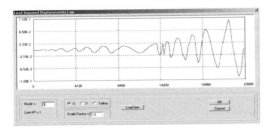

Figure 12. History of imposed displacements.

4.3 Displacement controlled nonlinear analysis

In this type of analysis, a set of nodal displacement evolution laws are imposed to one or more nodes in a specific direction. As for the accelerograms, the displacement evolution laws can be imported from text files, and scaling factors can also be applied to increase or reduce the intensity of each individual law (see Figure 12).

4.4 Natural frequencies and vibration modes

In this option natural frequencies and their corresponding vibration mode configurations are calculated for the structure. The frequencies are presented in a table and the vibration modes are represented graphically. After obtaining the natural frequencies, the Rayleigh-type damping matrix coefficients can be calculated.

5 POST-PROCESSING RESULTS

The post-processing operations consist in the visualization of the structural analysis results and imply that all pre-processing instructions have been carried out. The availability of the visualization options will depend on the type of analysis that was performed. For instance, for dynamic analyses it is possible to visualize the time-history evolutions of internal member forces, an option not available for results of a static analysis. Any of the graphics generated in the post-processing options can be exported to an image format or to a compatible MS Excel text file. Also, graphics can be visualized in detail within VISUALANL using common graphical tools like the zoom or pan operations.

5.1 Deformed shape plot

This option, Figure 13, represents both the original undeformed and the deformed shapes of the structure, for an automatically calculated scale factor.

However, the user can change it, in order to visualize the deformed structure in a more convenient scale. When the deformed shape is active, the user can select

Figure 13. Deformed shape of the structure.

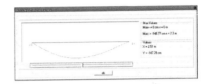

Figure 14. Deformed shape of an element.

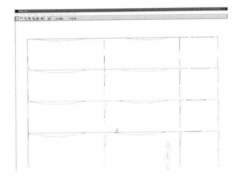

Figure 15. Distribution of bending moments across the structure.

a member with the mouse to additionally visualize the deformed shape of that member in the local reference axes, the maximum positive and negative values of the displacements and the value of the displacement at any point within the member (Figure 14).

5.2 Internal forces diagrams plot

Diagram visualization follows a procedure similar to that of the deformed shape. The user selects the time-instant, for dynamic analysis results, or the step number, for displacement controlled analysis results, for which the representation of the global internal forces diagrams is required (axial force N, shear force V or bending moment M). Figure 15 shows an example of a bending moment diagram.

377

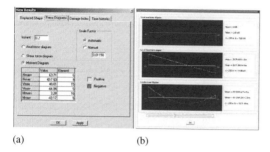

(a)   (b)

Figure 16. a) Options for diagrams of internal forces; b) Axial force, shear force and bending moment diagrams of a single frame element.

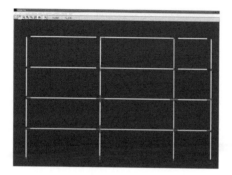

Figure 17. Local damage index distribution across the inelastic zones of the structure.

When the diagram plot is active, an options window enables the display of the maximum positive and negative internal force of the whole structure as well as the identification of the member where they occur (Figure 16-a). In addition, the user can select a member and visualize the three diagrams at once with indication of the maximum and minimum values (Figure 16-b). Internal force values at any point in the member can also be displayed.

5.3 Damage index plot

Following the previously referred damage index, two different types of results can be visualized: the evolution of the global damage index for the structure and the local damage index for each member for a certain time-instant, in the case of a dynamic analysis, or step number, in the case of a displacement controlled analysis (Figure 17). In addition, another measure of damage is also available for visualization. This measure of damage is the maximum inter-storey drift profile that represents the maximum relative displacement between consecutive floors normalized by the height of the storeys.

For the graphical representation of the different damage intensity values, VISUANL considers a predefined colour scale where each colour is assigned

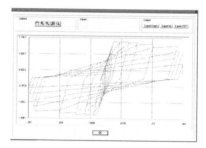

Figure 18. Moment-curvature diagram of a frame element under cyclic loading.

to a damage level. However, the damage limits for each damage level, as well as the colour scheme representing each damage level can be altered by the user.

5.4 Time-wise and step-wise evolutions

In this option, the user can visualize the time-wise evolution, in the case of a dynamic analysis, or the step-wise evolution, in the case of a displacement controlled analysis, of internal forces (N, V and M), of curvatures in a plastic hinge and of nodal displacements for a specific direction. In addition the user can also visualize the moment-curvature evolution plot at a plastic hinge (Figure 18).

6 FINAL COMMENTS

Structural analysis programs that include nonlinear models are valuable tools in the analysis and verification of structural safety, giving the engineer capacity to represent more precisely the real behaviour of the structures. For design of new structures or capacity assessment of existing ones, nonlinear analyses allow for a better representation of the structural response under any loading condition, and under earthquake loading in particular. Furthermore, nonlinear models can also be used in the calibration of more simplified numerical models.

Recent standards suggest nonlinear analyses as the reference analysis type to be used in capacity evaluation and verification of structural safety. For instance, Eurocode 8 (CEN, 2003) recommends the use of nonlinear time-history analysis for assessment of existent RC structures.

Nonlinear time-history analyses generate a considerable amount of information (nodal displacements, internal forces in frame elements, deformations, etc.), at each step of the analysis, thus making the treatment and interpretation of the results a rather difficult task. Most of the currently available tools for non-linear analysis are research programs lacking convenient user-interfaces, thus presenting several shortcomings from a practical point of view. Although a considerable number of structural programs include state-of-the-art nonlinear material models, few have adequate graphical tools for results visualization. Therefore, the development of a graphical processor like VISU-ALANL is considered to be a helpful tool for nonlinear analysis as a support for the design of new structures and, especially, in the capacity assessment of existing ones, under severe static and/or dynamic loads.

REFERENCES

H. Varum (1996), Modelo numérico para a análise sísmica de pórticos planos de betão armado, MSc Thesis, FEUP, Porto.

Thayer, R. (1999), Visual BASIC 6 Unleashed, Professional Reference Edition. Indianapolis, Indiana: SAMS Publishing.

A.G. Costa (1989), Análise sísmica de estruturas irregulares, PhD Thesis, FEUP.

C.T. Vaz (1996), Análise não-linear de pilares de betão armado sob cargas cíclicas, LNEC, Lisboa.

H. Varum and A.G. Costa (1997), Modelo não-linear para a análise estática e/ou dinâmica de pórticos de betão armado, 3° Encontro de Sismologia e Engenharia Sísmica, IST, Lisboa.

CEN (2003), Eurocode 8: Design of structures for earthquake resistance – Part 1-3: Strengthening and repair of buildings - European prEN 1998-1-3, European Committee for Standardization, Brussels, Belgium.

A.G. Costa and A.C. Costa (1987), Modelo histerético das forças-deslocamentos adequado à análise sísmica de estruturas, Relatório Técnico, Núcleo de Dinâmica Aplicada, L.N.E.C., Lisboa.

H. Rodrigues, H. Varum and A.G. Costa (2005), Modelo Numérico não-linear para painéis de Alvenaria de Enchimento em Pórticos de Betão Armado, Congreso Métodos Numéricos en Ingeniería, E.T.S. de Ingenieros de Caminos de la Universidad de Granada.

CEB (1996), RC frames under earthquake loading, Comité Euro-International du Béton, Bulletin n°231.

Computational Modelling of Objects Represented in Images –
João Manuel R.S. Tavares & R.M. Natal Jorge (eds)
© 2007 Taylor & Francis Group, London, ISBN 978-0-415-43349-5

Morphological dynamic study of human vocal tract

S.M. Rua
Escola Superior de Tecnologia da Saúde – IPP, Porto, Portugal

D. Freitas
Faculdade de Engenharia da Universidade do Porto, Portugal

ABSTRACT: Magnetic resonance imaging (MRI) has been successfully used in speech research, with the development of high B-value units and ultra-fast acquisition sequences, to obtain morphological and dynamic data of the vocal tract. This knowledge is useful for the understanding of speech mechanisms and can, in particular be employed for speech synthesis. The aim of our study was a combination of the morphological and dynamical characterizations of speech, resulting in the production of 3D articulatory models of some relevant Portuguese sounds and syllables. A set of static images in stacks and dynamic sequences were collected during the experiments from subjects that were instructed to vocalize during the acquisition, exploring the capabilities of the equipment. Analysis of the image stacks allowed the extraction of the vocal tract contours and a subsequent non-conventional 3D reconstruction by means of combination of orthogonal stacks allowed visualization and partial measurement of the sampled vocal tract shape.

Keywords: MRI of vocal tract, magnetic resonance imaging, speech production, morphological study, 3D vocal tract model.

1 INTRODUCTION

The human speech production process has long been a subject of interest, concerning morphological knowledge and speech acoustics, aiming to reach a useful understanding and modelling of all mechanisms involved. The main anatomic aspects and physiology of the vocal tract are common to all people, however speech production is a complex and highly individual mechanism. This recommends that modelling be done with enough accuracy for individual characterization.

Many approaches are available for measuring speech production (for example those based on muscular activity, articulator's movements and shapes, and signal analysis). Even so the available morphologic and dynamic information is scarce. A significant number of speech researchers have been using magnetic resonance imaging (MRI), that is a powerful tool for the study of the whole vocal tract with enough safety, through morphological measurements, to get multiplanar, high quality imaging of soft tissue, allowing the calculation of area functions which are recognized of interest for the understanding of the speech production mechanisms (Shadle et al. 1999, Kroger et al. 2000, Demolin et al. 2000, Engwall 2003,

Engwall 2004, Narayanan et al. 2004, Avila-García et al. 2004, Serrurier & Badin 2005).

More rarely, few known dynamic studies have been undertaken to date, mainly because of imaging device limitations. Presently with high B-value units, dynamic analysis becomes more feasible. Therefore, the aim of our study was a combination of the morphological and dynamical characterizations of speech, resulting in the production of 3D articulatory models of some relevant Portuguese sounds and syllables.

This research was carried out at the Faculdade de Engenharia da Universidade do Porto, with the collaboration of the Hospital de S. João and Faculdade de Medicina da Universidade do Porto.

2 METHODS

2.1 The equipment

Experiments for image acquisition were performed on a Siemens Magneton Symphony 1,5T system, with subjects lying in supine position. A head array coil was used. This study consisted firstly of a static phase, and secondly of a dynamic one.

2.2 The corpus and the subjects

The corpus consisted, initially, for the static study, of a set of images collected during sustained articulations of twenty five Portuguese sounds (oral and nasal vowels, and some consonants) uttered by two speech skilled subjects (one male and one female). For the dynamic study, four subjects produced several repetitions of sequences of three consonant-vowel syllables (/tu/, /ma/, /pa/) during the acquisition.

The corpus intends to cover the maximal possible range of the articulator's positions, namely, the tongue's, jaw's, lips and velum's, in order to explore a significant part of the articulatory space.

2.3 Static study

The sets of images obtained by MRI, called stacks, describe the object and represent the sampling of an anatomic volume by means of their regularly spaced slices.

Stacks of sagittal, coronal and axial MR, T1-weighted images, were recorded during 9 sec. and 9.9 sec. respectively, using Turbo Spin Echo sequences, and a 150 mm field of view. Three sagittal slices with 5 mm thickness, four coronal slices with 6 mm thickness, and 16 mm spaced and 4 axial slices were recorded. An acoustic recording of the produced speech just wasn't possible because of the high intensity of noise produced in the room during the MR acquisition.

Sagittal data obtained by MRI are particularly useful in the study of the whole vocal tract anatomy, shape and positions of some articulators, e.g. tongue, lips and velum (Figure 1).

This figure illustrates compared images from male to female subjects, showing vocal tract length differences, but the movements of articulator's are similar, namely the tongue's.

The static study was prepared to obtain morphologic data of the maximal range of articulator's positions for the characterization of Portuguese sounds.

Nasality is produced by the lowering of the velum, so part of the air stream is free to pass through the nasal cavity which functions as an anti-resonance chamber. Just adding this new feature alters the speech sounds significantly, making them more complex, which is very well known. In Portuguese there is a special interest in nasal sounds due to their frequent use in common speech.

The figure 2 depicts two situations with some image processing generated overlapped white contours and demonstrates the phenomenon of velum lowering (white arrow) during the sustained vocalization of two nasal vowels of Portuguese and its relationship with the tongue (blue arrows).

MRI coronal stacks bring specific lateral information, especially useful in the study of lips and tongue

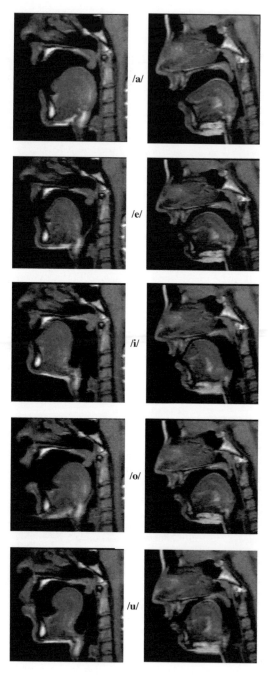

Figure 1. Midsagittal images of Portuguese oral vowels artificially sustained by a male (left) and a female (right).

shapes. The images of figure 3 show lips rounding (3a) by area filling in white, and tongue shape (3c) with bézier contour in yellow and the relationship between the palate and tongue (3b, 3d).

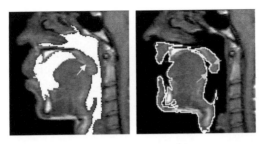

Figure 2.   Sagittal images of vocal tract during Portuguese nasal vowels: /~u/ (left), and /~i/ (right).

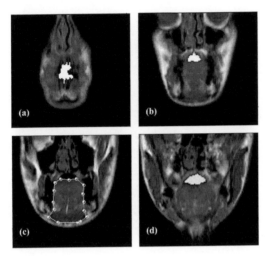

Figure 3.   Coronal stack of the Portuguese consonant /ch/.

2.4   Dynamic study

The dynamic study was performed using the same principle of MRI cardiac studies, with the modification of a FLASH 2D sequence using the patient's heart beat as trigger, and the acquisition parameters: TR = 60 ms, TE = 4.4 ms and a 300 mm field of view.

The subjects tried to synchronize the utterance of the syllables to their own cardiac rhythm by means of the simple ECG monitoring through a synchronous sound emission conveyed to the subject by an earphone. Each acquisition of a set of images from a single-slice (sagittal) of 6 mm thickness was done during 23 sec., resulting in a variable number of images that were achieved for each sequence, one acquired in each cardiac cycle with an increasing shift in synchronism from the start of the cardiac cycle.

This time undersampling method, necessary because of the important time overhead presented by the actual MR equipment in each single image acquisition was based on the assumption that the successive

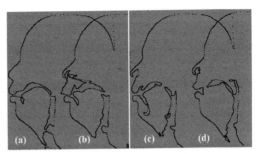

Figure 4.   Contours extracted from midsagittal images obtained in the dynamic study by the repetition of the sequence /tu/.

repetition of the utterances, one for each cardiac cycle was reasonably stationary and therefore a reasonable accuracy was expected and was judged the best way of achieving a good coverage of the whole articulatory gesture.

In figure 4 one of the obtained sequences obtained by the repetition of the sequence /tu/ is partially represented by some contours. The successive vocal tract shapes acquired present quite small differences that may pass undetected. In order to highlight the quantity and quality of the incremental modifications between successive images a set of contours were extracted and placed side by side. The differences are then quite well viewable. This illustration of the movement is therefore more observable as temporal modifications of the articulator's positions.

This sequence of images depicts the consonant /t/, which is characterized by a sequence starting from lips closure (4a) and then followed by the approximation of the tip of the tongue to the upper alveolar region (4b). After that, the tongue moves back with the opening of the mouth for the production of the vowel /u/ (4c), and starts the inverse process for the next repetition (4d).

When considering differences among the subjects, these dynamic studies demonstrate the variability in sounds production between subjects, not only due to anatomic differences, but also because each subject movement control is different.

2.5   Image analysis and 3D vocal tract models

Image analysis and 3D model construction was accomplished in two stages, namely (1) image segmentation using a 3D editing plugin for the Image J image processing software (developed by the National Institute of Health, http://rsb.info.nih.gov/ij/), and (2) a 3D reconstruction of each stack, using the Blender software for 3D graphics creation, version 2.41 (http://www.blender.org/cms/Home.2.0.html).

The image segmentation of the airway from the surrounding tissues for extraction of the contours of the

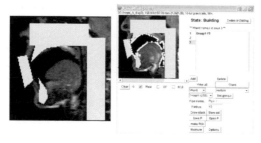

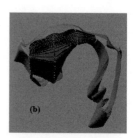

Figure 5.   Image segmentation of the airway.

vocal tract was performed by the following sequence procedures:

(a) Identification and closure of the vocal tract area of interest, mandatory closure of the mouth, larynx, vertebral column and velum;
(b) Manual pasting of teeth image (only on sagittal stacks), after extraction of the teeth contours from the previously acquired sagittal anatomic reference image;
(c) Extraction of the contours of the vocal tract, for each image of 2D slices using the Image J semi-automatic threshold technique.

Figure 5 depicts, at left, the result of the manual identification and closure of vocal tract of interest with solid boxes, and the pasting of the contours of the teeth. The extraction of contours of the vocal tract was then performed by outlining the area of interest, slice by slice. The result can be observed on the right part of the figure.

Each outline is defined by the minimum and maximum levels in the area, moving and controlling histogram values. The boundary is given after adjusting levels parameters to match the outline with the area of interest.

The outlines were subsequently used to generate a skin (a 3D object), after importing the contours in *shapes* format, into the Blender software.

For each articulatory position, the next phase is the combination of sagittal and coronal outlines (2D curves). To make this possible, it's required that the outlines be well aligned – this process is sometimes called image registration.

The image registration was performed using the data contained in the header of the image stacks containing the description of the DICOM (Digital Imaging and Communications in Medicine) protocol for the transmission of medical images. This protocol store all the metadata of each image and of each stack, including the specific attributes with positional and magnification interest: image orientation, image position, pixel spacing and slice thickness.

The alignment was obtained by the calculation of the reference pixel location, which is given by the

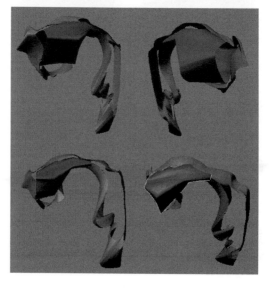

Figure 6.   Three-dimensional model of the vowel /u/.

product of the image position attribute values by the image resolution value.

2.6   *Some 3D models*

The following images (Figure 6) represent different perspectives of the 3D model obtained for the vowel /u/. The blue skin represents the union of the three outlines extracted from sagittal stacks (6a). The red skin represents the union of the four outlines extracted from coronal stacks (6b). It must be observed that these reconstructions are not yet closed as should be for a whole vocal tract reconstruction. This closing of the skin by unification of the different stacks is the next step in the processing.

Anyhow in this 3D model are demonstrated several essential features needed for the articulatory description of speech, namely:

– shape of the tongue;
– central position of the tongue in oral cavity, and upper position near the palate;

384

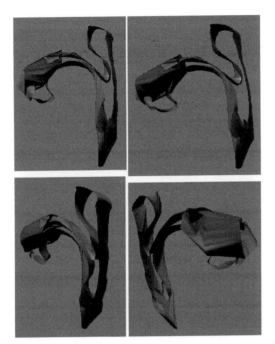

Figure 7.   Three-dimensional model of the vowel /~u/.

– mouth opening and rounding shape;
– lateral dimension of oral cavity;
– length of the vocal tract;
– velum position, which is up and closed.

Comparatively, we present the 3D model of the same vowel but for nasal sound production (Figure 7).The different view perspectives presented allow the identification of the velum lowering, and especially the partial closure of the oral cavity.

In these 3D models, some differences in vertical length between sagittal and coronal stacks, in some sounds were observed, resulting in some registration errors. This could reflect the specific variability of the speaker in sound production, due in this case, to the fact that acquisitions were separately taken (first sagittal images, and subsequently coronal images), to minimize subject effort in an extra long utterance. By another side, the segmentation process by the threshold technique has some implications in the determination of the vocal tract area contour as well.

3   DISCUSSION

From this study important problems have to be taken into account:

(1) Difficulties were encountered in the contours extraction, due to the non-visualization of the teeth by MRI, especially important in coronal images;

(2) Limitation on the number of images collected for a given acquisition time, due to the compromise between time and resolution and the machine's protocols, and to minimize subject effort;
(3) Dynamic acquisition time was limited by time factors namely by the trigger technique and functions of the equipment (e.g., time required to activate and deactivate gradients fields);
(4) Difficulties in acoustical recording because of the high acoustical noise level in the acquisition room.

The segmentation process is quite hard and time consuming, so the choice of segmentation tool is important for it. The threshold technique is the more useful in boundaries extraction, making models more realistic.

These collected images permit in theory the determination of areas useful for articulatory studies, including speech synthesis.

The morphologic study gives a better understanding about sounds production, about sound variability inter-speakers, and subject's comparisons.

The description of the movements of articulation involved in normal speech observed by the dynamic study, allow not only a better knowledge of all mechanisms involved, but also may contribute for understanding speech co-articulation phenomena.

4   CONCLUSION AND FUTURE WORK

The use of MRI can provide very useful and precise morphological information about the position and shape of the different articulators involved in speech production and also of their dynamics, although still, in the present work, with some speed restrictions due to the limited speed of present day available equipment and protocols.

The image material was analysed and processed and a set of 3D reconstruction data was produced and will be presented. These data can be used for several studies in articulatory phonetics, vocal tract modelling for speech synthesis, etc. The completion of the construction of the skins for the hybrid models made of sagittal and coronal stacks is the next step in the way to obtain a full 3D anatomical model of the vocal tract. The extension of the dynamic sequences obtained to other sequences is also important in terms of coverage of the study to be done in the near future.

Improvement and refinement of the protocols is also necessary to take still more advantage of the capabilities of the equipment.

ACKNOWLEDGMENT

Images were acquired at the Radiology Department of the Hospital S. João, Porto, with the collaboration of

Isabel Ramos (Professor of Faculdade de Medicina da Universidade do Porto and Department Director) and the technical staff, which is gratefully acknowledged.

REFERENCES

Avila-García, M.S., Carter, J.N., Damper, R.I. 2004. Extracting Tongue Shape Dynamics from Magnetic Resonance Image Sequences. *Transactions on Engineering, Computing and Technology V2, December*, ISSN, p. 288–291.

Demolin, D., Metens, T., Soquet, A. 2000. Real time MRI and articulatory coordinations in vowels. *In Proc. 5th Speech Production Seminar*. München, Germany.

Engwall, O. 2003. A revisit to the Application of MRI to the Analysis of Speech Production – Testing our assumptions. *Proceed. 6th Int Seminar on Speech Production*. Sydney.

Engwall, O. 2004. From real-time MRI to 3D tongue movements. *Proc of ICSLP 2004, vol. II: 1109–1112, October*. Jeju Island: Korea.

Kröger, B.J., Winkler, R., Mooshammer, C., Pompino-Marschall, B. 2000. Estimation of Vocal Tract Area Function from Magnetic Resonance Imaging: Preliminary Results. *Proceed. 5th Seminar On Speech Production: Models and Data, May 2000*. München: Germany.

Narayanan, S., Nayak, K., Lee, S., Sethy, A., Byrd, D. 2004. An Approach to Real-time Magnetic Resonance Imaging for Speech Production. *Journal Acoustical Society of America* 115(4): 1771–76.

Serrurier, A., Badin, P. 2005. A Three-dimensional Linear Articulatory Model of Velum based on MRI data. *InterSpeech, Lisboa*.

Shadle, C.H., Mohammad, M., Carter, J.N., Jackson, P.J.B. 1999. Multi-planar dynamic magnetic resonance imaging: new tools for speech research. *Proceed. 14th International Congress of Phonetic Sciences (ICPhS 99)*. S. Francisco: USA.

Computational Modelling of Objects Represented in Images –
João Manuel R.S. Tavares & R.M. Natal Jorge (eds)
© *2007 Taylor & Francis Group, London, ISBN 978-0-415-43349-5*

Implant trajectory estimation during femoral screw-plate osteosynthesis operation

T. Ben Saïd, F. Chaieb & F. Ghorbel
Ecole Nationale des Sciences de l'Informatique (ENSI), Manouba, Tunisia

V. Burdin
Ecole Nationale Supérieure des Télécommunications de Bretagne (ENST-B), Technopôle de Brest, France

H. Rajhi
CHU Charles Nicolle, Service Radiologie, Tunis, Tunisia

ABSTRACT: Femoral screw-plate Osteosynthesis is a surgical procedure that stabilizes and joins the ends of fractured femur bones (head or neck) by implants such as metal plates, pins or screws. In this paper, we deal with the screw trajectory estimation. In fact, we focus on position, direction and length estimation of the nail in order to avoid possible complications such as screw femoral neck effraction or screw articular penetration. These complications are due to measures calculated on antero posterior and lateral views. In our approach we propose to estimate implant trajectory from 3D femur patient shape rather than radiographies. This patient 3D shape was computed from the two orthogonal views and a statistical 3D model representing the population. At first, a 3D segmentation of the femur is carried out. We propose an interactive segmentation method which consists in positioning cut planes that separates the neck from the head of the femur. The choice of these cut planes requires a physician expert to perform a better extraction of the desired zones. Then, a safety zone for each segmented region is determined. The implant must be inside these zones to avoid complications. The safety zone of the femoral head is defined as the inscribed sphere in this head whereas the safety zone of the femoral neck is defined as the inscribed cylinder in this neck [2]. In order to control the position of the implant inside the femur, collision detection method is used. This method allows generating characteristics of the collision between the femur and the implant. These characteristics correspond to the suitable implant position, length and direction.

1 INTRODUCTION

Proximal femoral fractures include a broad group of common fractures of the femoral head and neck typically occurring in osteoporotic females. The Fixation of fractures of the femoral neck consists in inserting an implant inside the neck and the head of the femur. This operation generally requires antero-posterior and lateral views of the fractured femur. First, a pin-guide is inserted into the femur. Then, the two orthogonal views are acquired with pin-guide in place. While inserting this pin, we must make sure that the end of this pin is not intra-articular on the two views and we must materialize the thickness of the implant to be used in order to be sure that it will be located entirely in the femoral neck and that there will be no effraction of the posterior cortical. Once the position of the pin-guide is considered to be healthiest possible, the implant will be inserted in the same way as the pin-guide [1]. However, it proved that the rate of complications generated by this kind of operations is rather high as well for the intra-articular penetrations as for the effraction of the femoral neck [2]. So that many cases of femoral head necrosis has been observed due to screw misplacement. Figure 1 illustrates an example of imperfection of this method : Antero posterior view (a) and lateral view (b) let appear the implant as being inside the femur, whereas views (c) and (d) show the perforation of the femoral head and neck by the implant.

In order to reduce the rate of these complications, Hernigou proposes a study [2] which analyzes theoretically the position of an implant in the head and the femoral neck from the two views and to determine the risk of an intra-articular penetration or a cortical effraction by this implant. In this study, he defines safety zones which can be computed from the two orthogonal views. So, the pin-guide should be placed in these safety zones in order to avoid complications and guarantee more precision of the screw insertion. In spite of the reduction of complications rate, the

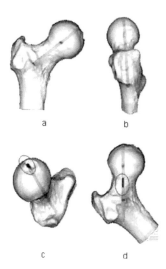

Figure 1.   Classic method imperfection.

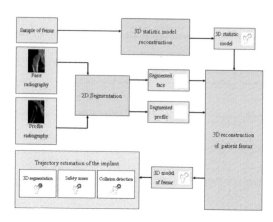

Figure 2.   General method diagram.

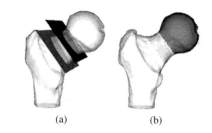

Figure 3.   3D segmentation of the femur.

Hernigou criteria is not able to eliminate completely these complications. Therefore, the use of a technique based only on two views seems to be insufficient. Thus, the use of a three dimensional femur shape could provide a better solution to articular penetration and head perforation problems. In section 2, an overview of the proposed method is presented. A description of the 3D segmentation of the femur is explained in section 3. In section 4, we describe femoral safety zones needed to avoid screw misplacement.We describe a collision detection process required to generate implant length and direction in section 5. Finally, the experimentation and the results of this work are presented and discussed in section 6.

2   METHOD OVERVIEW

In order to reduce the error rate raised in the femoral screw-plate osteosynthesis operation, we propose an approach which deals with three-dimensional (3D) shape of the femur. It consists on the reconstruction of a three-dimensional model of the femur from the two views obtained in intraoperative phase in order to eliminate the inaccuracies due to projection. The external surface of the femur is represented by a triangular mesh composed of a set of vertices defining the geometry of the mesh and of a set of faces determining its connectivity. The reconstruction of the femur 3D model from the postero-anterior and lateral views can be performed in two steps : At first, a statistical femur model is obtained by the fusion of 3D sample meshes representing a particular population. Then, a nonrigid 3D/2D registration of this model with contours extracted from the two radiographies is carried

out. Once the 3D femur model is computed an implant trajectory estimation procedure will be carried out in order to provide physician with the best implant length and direction that avoid articular penetration and head perforation problems. In this work we are concerned only with the implant trajectory estimation. We assume that the 3D femur model has been already computed. The diagram of the general method is illustrated by figure 2.

3   3D SEGMENTATION

The purpose of the 3D femur segmentation is to separate the head and the neck of the femur. The most of 3D segmentation methods [3,4,5,6,7] were studied and had generated interesting results for 3D meshes of synthetic forms. However, the results were not satisfying for natural forms and especially for the femur. Considering anatomical structure of the femur, we propose an interactive segmentation method which consists in positioning two cut planes separating the head and the neck of femur. This method requires physician expertise to put planes in the convenient emplacements. Figure 3 illustrates this segmentation method and shows the corresponding results.

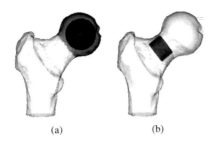

(a)   (b)

Figure 4. Approximation of the head and the neck.

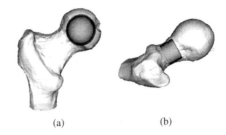

(a)   (b)

Figure 5. Safety zones of the femur.

4 FEMORAL SAFETY ZONES

In [2], Hernigou assumes that the femoral head is approximated by a perfect sphere of ray $R1$ and the femoral neck is considered as a perfect cylinder of ray $R2$. In postero-anterior and lateral views, the head is described by a circle and the neck by a rectangle. The femoral head safety zone is defined as the circle of radius $R1*sinus(45°)$ inscribed in the femoral head circle, whereas the femoral neck safety zone is assumed to be a rectangle of width $R2*sinus(45°)$ inscribed in the femoral neck rectangle. Hernigou suggests that the pin-guide should be placed in these safety zones in order to avoid complications and guarantee more precision of the screw insertion. Whereas Hernigou computes these zones from the two views, we propose to compute them from the 3D femur model. To perform this task, we have to approximate the femoral head by a sphere and the femoral neck by a cylinder.

4.1 *Approximation of the femur*

Due to its anatomical structure, the head of the femur can be approximated by a sphere. The center of this sphere can be chosen as the mean of the vertices of the femoral head. The ray of this sphere is determined as the minimum distance between the defined center and all head vertices. This sphere is entirely inscribed in the femoral head (see figure 4). Similarly, the femoral neck can be approximated by a perfect cylinder. To determine the axis of this cylinder, we proceed by cutting the femoral neck 3D mesh into parallel slices. We then, compute the center of each slice as the mean of contour vertices. The set of slices centers represents the cylinder axis. The cylinder ray is considered as the minimum distance between the axis and all vertices of the femoral neck (see figure 4).

4.2 *Safety zones*

The 3D femoral head safety zone can be defined as a sphere circumscribed in the femoral head sphere. The ray R_s of this sphere is given by : $R_s = \alpha R$ where α is a security factor chosen by the physician and R

the femoral head sphere ray. Similarly, the 3D neck safety zone is defined as the cylinder circumscribed in the femoral neck cylinder. The ray of the safety zone cylinder is equal to the femoral neck cylinder ray decreased by a security factor. The application of this method to the femur led to the following results (figure 5).

5 COLLISION DETECTION

The collision detection methods are used to detect the extremity of the implant once located outside safety zones. Several methods of collision detection were proposed in the literature. These methods depends on the representation model, the query type and the characteristics of the simulated environment. Among these methods, there are those which proceed by an incremental computation of the distance between convex polygonal objects [8,9]. Other methods are based on the exploration of the triangular mesh in order to determine the minimal distance [10]. BSPs [11] and the octrees [12] can also detect the collisions between the objects thanks to recursive subdivisions of space and the generation of the associated trees. The method proposed by A. Ehmann et al. can be also solution of this problem since it can manage several types of queries on general polyhedric models and don't require neither specialized data structures nor postprocessing [13]. In this work, we chose the proximities detection method by convex decomposition of surfaces [13]. In fact, this method deals with generic 3D models and provides a large varieties of queries. This method is able to carry out queries of intersection detection, distance computation and contact determination. It is composed of three steps : At first, the general polyhedron surface (convex or not) is divided into a set of convex patches. It is based on the exploration of the relations between faces and the use of a research incremental algorithm. Then, this decomposition is used to make a binary volumes hierarchy (BVH) with convex pieces as sheets. This step requires subdivision operations, division and face classification. Finally, a generic algorithm is used to carry out queries on a pair of polyhedrons by using their hierarchies. These

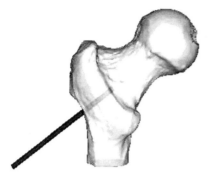

Contact: $(-3.4, 0.97, -1.2)$
Direction: $(-10, 6, -48)$
Distance: 0.28

Figure 6.   Example of collision detection.

proximities queries deals with intersection tests, tolerance verification and contact determination. However, this method does not compute depth penetration. So, we develop a module which computes interpenetration between the objects in collision.

This method can generate the characteristics of possible collision between the femur and the implant. These characteristics are especially:

Contact: The first point of contact between the femur and the implant following a collision.

Direction: The direction of the axis of the implant following a collision. This direction is given by the three angles of Euler which describes the rotations carried out according to various axis.

Distance: The length of the implant inside the femur.

Figure 6 shows an example of collision between the implant and the femur and the characteristics generated by this collision.

6   EXPERIMENT RESULTS

We have applied our method to the 3D model of the femur reconstructed from a set of CT scans by using marching cube technique. To implement collision detection method, we used the SWIFT++ Collision Detection/Proximity query package [14].

Our application loads 3D models of the femur and the implant. Then, the physician can move the implant inside the femur. A collision is detected once the implant arises from safety zones. So, collision characteristics such as the point of contact, the direction and the penetration depth of the implant are provided. This process could be repeated many times until the physician finds the best implant trajectory (see figure 7).

 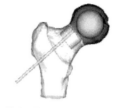

(a) Implant in safety zone of the neck.
Contact: $(-3.4, 0.97, -1.2)$
Direction: $(-10, 6, -48)$
Distance: 0.43

(b) Implant out of safety zone of the neck.
Contact: $(-3.4, 0.97, -1.2)$
Direction: $(-10, 6, -48)$
Distance: 0.58

 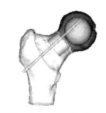

(c) Implant in safety zone of the head.
Contact: $(-3.4, 0.97, -1.2)$
Direction: $(-10, 6, -48)$
Distance: 0.67

(d) Implant out of safety zone of the head.
Contact: $(-3.4, 0.97, -1.2)$
Direction:$(-10, 6, -48)$
Distance: 0.94

Figure 7.   Experiment results.

7   CONCLUSION AND FUTURE WORK

In this paper, we were interested in femoral screw-plate osteosynthesis. We have focused mainly on the implant trajectory estimation problem by using a 3D model of the femur in order to avoid complications generated by the two views reparation techniques. The estimation of the direction and the length of the implant were then obtained by collision detection techniques. Moreover, we could visualize the safety zones as defined by Hernigou [2]. These zones are considered as guides for the physician to find better position of the implant without overflowing through concave cavities of the femur.

In a future work, we intend to perform the 3D reconstruction of patient femur from the postero-anterior and lateral views. Further more, we will focus on the best implant insertion position according to the femur anatomical structure and the bone density.

REFERENCES

[1]   F. Langlais and P. Burdin, *Ostéosynthèse par vis-plaque des fractures du fémur*, Revue de Chirurgie Orthopédique, 1991.

[2]   P. Hernigou, *Fractures du col du fémur: position de l'implant effraction ariculaire méconnue et ses*

conséquences, Revue de Chirurgie Orthopédique, vol. 80, pp. 503–519, 1994.

[3] L. Chevalier, F. Jaillet and A. Baskurt, *Segmentation and superquadric modeling of 3d objects*, WSCG, vol. 11, no. 1, 2003.

[4] S. Katz and A. Tal, *Hierachical mesh decomposition using fuzzy clustering and cuts*, ACM Transaction on Graphics, vol. 22, no. 3, pp. 954–961, 2003.

[5] G. Lavoue, F. Dupond and A. Baskurt, *Constant curvature region decomposition of 3d meshes by a mixed approach vertex-triangle*, WSCG, vol. 12, no. 2, pp. 245–252, 2004.

[6] R. Liu and H. Zhang, *Segmentation of 3d meshes through spectral clustering*, Proc. Pacific Conf. on Computer Graphics and Application, 2004.

[7] A. Ng, M. Jordan and Y. Weiss, On *spectral clustering: Analysis and an algorithm*, Advances in Neutral Information Processing Systems, no. 14, pp. 857–864, 2002.

[8] D. Dobkin and D. Kirkpatrick, *Determining the separation of preprocessed polyhedra*, Lecture Notes in Computer Science, vol. 443, no. 5, 1990.

[9] E. Gilbert, D. Johnson and S. Keerthi, *A fast procedure for computing the distance between objects in three-dimentional space*, IEEE International Conference on Robotics and Automation, 1988.

[10] C. Lin and F. Canny, *A fast algorithm for incremental distance calculation*, IEEE International Conference on Robotics and Automation, 1991.

[11] B. Naylor, J. Amantides and W. Thibault, *Merging bsp trees yields polyhedral set operations*, Proc. of ACM SIGGRAPH, 1990.

[12] K. Hamada and Y. Hori, *Octree based approach to real-time collision free path planning for robot manipulator*, IEEE-ACM, 1996.

[13] A. Ehmann and C. Lin, *Accurate and fast proximity queries between polyhedra using convex surface decomposition*, Department of Computer Science University of North Carolina at Chapel Hill, 2001.

[14] A. Ehmann and C. Lin, *SWIFT++, http://www.cs.unc.edu/ geom/SWIFT/*.

[15] G. Behiels, D. Vandermeulen and P. Suetens, *Statistical shape model-based segmentation of digital x-ray images*, IEEE-MMBIA, 2000.

[16] M. Fleute, *Shape reconstruction for computer assisted surgery based on non-rigid registration of statistical models with intra-operative point data and x-ray images*, 2001.

[17] W. E. Lorensen and W. E. Cline, *A high resolution 3d surface construction algorithm*, Proc. SIGGRAPH, vol. 21, no. 4, pp. 163–169, 1987.

[18] C. Oblonsek and N. Guid, *A fast surface-based procedure for object reconstruction from 3d scattered points*, Computer Vision and Image, vol. 69, no. 2, pp. 185–195, 1998.

Computational Modelling of Objects Represented in Images –
João Manuel R.S. Tavares & R.M. Natal Jorge (eds)
© 2007 Taylor & Francis Group, London, ISBN 978-0-415-43349-5

The use of virtual reality models in Civil Engineering training

A.Z. Sampaio & P.G. Henriques
Technical University of Lisbon, Dep. Civil Engineering and Architecture, Lisbon, Portugal

ABSTRACT: The use of virtual reality techniques in the development of educational applications brings new perspectives to the teaching of subjects related to the field of civil construction. In order to obtain models, which would be able to visually simulate the construction process of two types of construction work, the research turned to the techniques of geometric modeling and virtual reality. The applications developed for this purpose are concerned with the construction of a cavity wall and a bridge. These models make it possible to view the physical evolution of the work, to follow the planned construction sequence and to visualize details of the form of every component of the works. They also support the study of the type and method of operation of the equipment necessary for these construction procedures. These models have been used to distinct advantage as educational aids in first-degree courses in Civil Engineering.

1 INTRODUCTION

Normally, three-dimensional geometric models, which are used to present architectural and engineering works, show only their final form, not allowing the observation of their physical evolution. In the present study, two engineering construction work models were created, from which it was possible to obtain three-dimensional models corresponding to different states of their form, simulating distinct stages in their construction. The use of techniques of virtual reality in the development of these educational applications brings new perspectives to the teaching of subjects in the area of civil engineering.

The visual simulation of the construction process needs to be able to produce changes to the geometry of the project dynamically. It is then important to extend the usefulness of design information to the construction planning and construction phases. The integration of geometrical representations of the building together with scheduling data is the basis of 4D (3D + time) models in construction domain. 4D models combine 3D models with the project timeline (Leinonen, J. et al. 2003). 4D CAD models are being used more and more frequently to visualize the transformation of space over time. To date, these models are mostly purely visual models. On a construction project, a 4D CAD environment enabled the team, involved in the project, to visualize the relationships between time (construction activities) and space (3D model of the project).

In order to create models, which could visually simulate the construction process of two types of construction work, the authors turned to techniques of geometric modeling and virtual reality. The applications developed for this purpose refer to the construction of a masonry cavity wall and a bridge. These models make it possible to show the physical evolution of the work, the monitoring of the planned construction sequence, and the visualization of details of the form of every component of each construction. They also assist the study of the type and method of operation of the equipment necessary for these construction procedures. The virtual model can be manipulated interactively allowing the user to monitor the physical evolution of the work and the construction activities inherent in its progression.

One of the applications developed corresponds to the model of a masonry cavity wall, one of the basic components of a standard construction. To enable the visual simulation of the construction of the wall, the geometric model generated is composed of a set of elements, each representing one component of the construction. Using a system of virtual reality technologies, specific properties appropriate to the virtual environment are applied to the model of the wall. Through direct interaction with the model, it is possible both to monitor the progress of the construction process of the wall and to access information relating to each element, namely, its composition and the phase of execution or assembly of the actual work, and compare it with the planned schedule. This model had been used to distinct advantage as an educational aid in Civil Engineering degree course modules.

The second model created allows the visual simulation of the construction of a bridge using the cantilever method. The geometric model of the bridge deck was created through a bridge deck modeling system, implemented as part of a research project at the Institute of

Construction of the Higher Technical Institute/ Department of Civil Engineering and Architecture, Lisbon (ICIST/DECivil.) A system of virtual reality was used to program the visual simulation of the bridge construction activities. Students are able to interact with the model dictating the rhythm of the process, which allows them to observe details of the advanced equipment and of the elements of the bridge: (pillars, deck and abutments.) The sequence is defined according to the norms of planning in this type of work. The aim of the practical application of the virtual model of bridge construction is to provide support in those disciplines relating to bridges and construction process both in classroom-based education and in distance learning based on e-learning technology.

2 CONSTRUCTION OF A EXTERIOR WALL

2.1 Virtual 3D model

Described here are the processes both of the modeling of an exterior wall of a standard building and of the association of virtual properties with the created model, the intended outcome being the interactive exhibition of its construction (Sampaio et al. 2004). The model of the masonry cavity wall, including the structure of the surrounding reinforced concrete and the elements in the hollow area (bay elements), was created using a three dimensional graphic modeling system in widespread use in planning offices, namely, *AutoCAD*. The virtual environment was applied to the model through the computer program *EON Reality system* (EON 2003).

2.2 Geometric modeling of the construction elements

The representation of this model of an exterior wall of a conventional building comprises the structural elements (foundations, columns and beams), the vertical filler panels and two bay elements (door and window).

The structural elements of the model were created with parallelepipeds and were connected according to their usual placement in building works. Because this is an educational model, the steel reinforcements were also defined. In the model, the rods of each reinforcement are shown as tubular components with circular cross-section (Fig. 1).

The type of masonry selected corresponds to an external wall formed by a double panel of breeze-blocks, *11* cm, wide with an air cavity, *6* cm, wide (Fig. 2).

Complementary to this, the vertical panels were modeled, these comprising: the thermal isolation plate placed between the brick panels; the plaster applied to the external surface of the wall; the stucco applied on the internal surface; two coats of paint both inside and out and the stone slabs placed on the exterior surface.

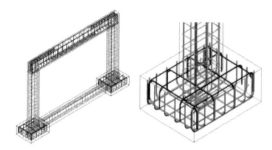

Figure 1. Te rods of each reinforcement are modeled as tubular components with circular cross-section.

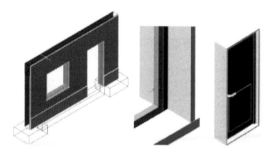

Figure 2. 3D models of the masonry wall components.

Finally, two usual bay elements (Fig. 2), a door and a window, were modeled. The completed model was then transferred to the virtual reality system *EON* (as a design file with *3ds* extension).

2.3 Programming the virtual construction

Figure 3 presents the main window of the *EON system*:

– The sub-window to the left contains a table of nodes or actions (of movement, sensors etc) available in this system;
– The center sub-window is designated a simulation tree; the objects making up a given scenario are presented in it, and the links between each object or group and the actions to be taken associated to each of them are also shown. It is therefore, through the use of this window that the simulation of the desired virtual action is programmed;
– In the sub-window on the right the links between the various nodes are established thus defining a network. In this network the nodes where the links originate and terminate are identified, as is the means of setting each action in motion.

In this system, the visual simulation of the building process of the wall, following a particular plan, was programmed. For this effect, 23 phases of construction were considered. The order in which components are

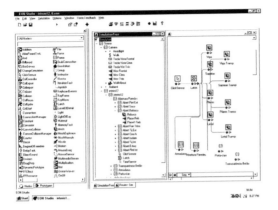

Figure 3.   The main window of the *EON studio* system.

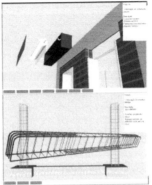

Phase 5
Cementing the lintel
and foundation

Description:
Cement B30

**Dimension of the
foundations:**
$100 \times 80 \times 40\,\text{cm}^3$

**Dimensions of
the lintel cross
section:**
$25 \times 40\,\text{cm}^3$

**Total volume of
cement:**
$0.98\,\text{cm}^3$

Figure 5.   Exhibition of phases in building evolution.

Figure 4.   Presentation in explosion of the vertical panels of
the wall model.

consecutively exhibited and incorporated into the virtual model, translates into the method of the physical evolution of the wall under construction.

The exhibition of the several vertical panels of the wall in explosion is a kind of presentation with a great didactic interest. Figure 4 includes two steps of this animation, the opened and closed situations. The translation displacement value attributed to each panel was distinct from each other in order to obtain an adequate explosion presentation. This type of presentation allows the student to understand the correct sequence of the panels in a wall and to observe the different thickness of the panels.

During the animation, the student can control the length of time that any phase is exhibited and observe the model using the most suitable camera and zoom positions for a correct perception of the details of construction elements. It is possible to highlight the component incorporated at each new phase and to examine it in detail, as shown in Figure 5.

Included, under the window in which the virtual scene is exhibited, is a bar, which shows the progress

of the construction. Throughout the animation, the bar is filled, progressively, with small rectangles symbolizing the percentage built at the time of the viewing of that particular phase, in relation to the completed wall construction (Fig. 5).

Simultaneously, with the visualization of each phase, a text is shown (in the upper right corner of the window, Fig. 5) giving data relating to the stage being shown, namely, its position within the construction sequence, the description of the activity and the characterization of the material of the component being incorporated.

3   CONSTRUCTION OF A BRIDGE DECK

3.1   *Virtual 3D model*

Throughout the bridge research project, a system of computer graphics was used, a system which enables the geometric modeling of a bridge deck of box girder typology (Sampaio et al. 1999). This system was used to generate, three-dimensional (3D) models of deck segments necessary for the visual simulation of the construction of the bridge (Sampaio et al. 2002 & Sampaio 2003). The attribution of virtual properties to the model of the bridge was implemented by using the virtual reality system *EON Studio* (EON 2003).

From amongst the examples of bridge decks modeled through the use of the graphic system for bridges, the North Viaduct of the Quinta Bridge (GRID 1995) in Madeira, Portugal, was the case selected for representation in the virtual environment. In cross-section, the deck of the viaduct shows a box girder solution and its height varies parabolically along its three spans (Fig. 6).

The most common construction technique for this typology is the cantilever method of deck construction. This method starts by applying concrete to a first segment on each pillar, the segment being long enough

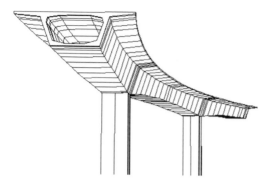

Figure 6. The North Viaduct of the Quinta Bridge in Madeira.

to install on it the work equipment. The construction of the deck proceeds with the symmetrical positioning of the segments starting from each pillar. The continuation of the deck, uniting the cantilever spans, is completed with the positioning of the closing segment.

In order to be able to manipulate the construction work of the bridge in virtual reality, in addition to the three-dimensional models of each segment, models of the pillars, form travelers, abutments and false work were made. The spans were created through the use of the representational system for bridges mentioned above and the remaining components through *AutoCAD*. All the components of the construction scenario were transposed to the virtual reality system, proceeding, then, to the definition of the desired animation.

3.2 *Geometric modeling of the construction scenario*

Geometric description can be entered directly into the deck-modeling program. To achieve this, the developed interface presents diagrams linked to parameters of the dimensions, so facilitating the description of the geometry established for each concrete case of the deck. The image included in Figure 8 shows the interface corresponding to the cross-section of the deck of the example.

The description of the longitudinal morphology of the deck and the geometry of the delineation of the service road, serving the zone where the bridge is to be built is carried out in the same way. The configuration and the spatial positioning of each are obtained with a high degree of accuracy.

Using the data relating to the generated sections, the system creates drawings and three-dimensional models of the deck. To obtain the definition of the deck segment models, consecutive sections corresponding to the construction joints are used. The configuration presented by the segment models is rigorously exact. Figure 8 shows one of the segments of the deck.

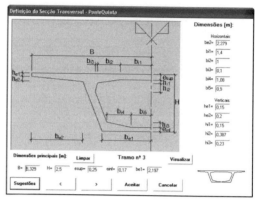

Figure 7. Interface of to describe cross-sections.

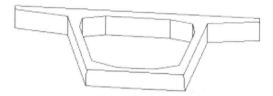

Figure 8. The 3D model of a deck segment.

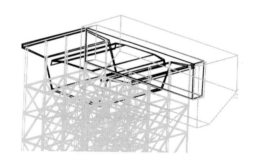

Figure 9. 3D models of the abutment and the scaffolding.

To complete the model of the bridge, the pillars and abutments were modeled using the *AutoCAD* system. Based on research in the literature concerning abutments for the typology of box-girder decks, a model was created as shown in Figure 9.

Then followed the modeling of the advanced equipment, which is composed not only of the form traveler, but also the formwork adaptable to the size of each segment, the work platforms for each formwork and the rails along which the carriages run (Fig. 10).

As, along with the abutments, the deck is concreted with the falsework on the ground, the scaffolding for placement at each end of the deck was also modeled (Fig. 9). Terrain suitable for the simulation of the positioning of the bridge on its foundations was also modeled.

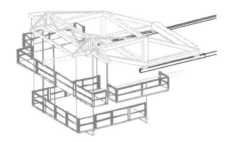

Figure 10.   3D models of the advanced equipment.

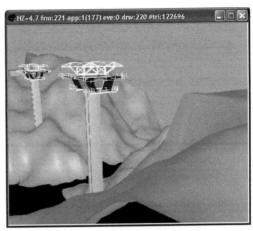

Figure 12.   Placing the four advanced equipments.

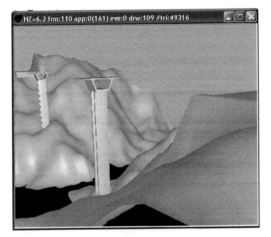

Figure 11.   Placing the pillars and the initial segments.

3.3   *Programming the construction sequence*

Once all the 3D models of the construction scenario had been generated, they were transposed, in *3ds* extension data file format, to the virtual reality system.

The definition of the construction sequence is based on a counter, which determines the next action when a mouse button is clicked. The first action consists of the insertion of the pillars in the first scenario, which is composed solely of the landscape. The next step is to place one of the segments on top of each of the pillars (Fig. 12).

After this, a form traveler is placed on each segment. For the simulation of the first cantilever segment (in each span), the four form travelers, the corresponding work platforms and the formwork components are included in the scenario. Once the first segments have been concreted, the construction of the cantilevered deck takes place. In each phase, two pairs of segments are defined. The construction of the deck is defined symmetrically in relation to each pillar and simultaneously.

For each new segment the following steps are established: raising the form traveler; moving the rails in the

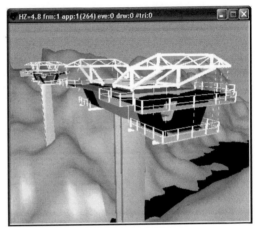

Figure 13.   Movement of the advanced equipment.

same direction as the construction (relocating them on the latest segment to have been concreted); moving the form traveler on the rails, positioning it in the zone of the next segment to be made; concrete the segment (Fig. 13).

Finally, the zone of the deck near the supports is constructed, the false work resting on the ground (Fig. 14).

Moving the camera closer to the model of the bridge and applying to it routes around the zone of interest, it is possible to visualize the details of the form of the components involved in the construction process. In this way, the student can interact with the virtual model, following the sequence specifications and observing the details of the configurations of the elements involved.

397

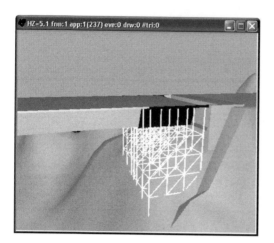

Figure 14. Concreting above the false work near the abutment.

4 LEARNING ASPECTS

The models are actually used in face-to-face classes of disciplines of Civil Engineering curriculum: Technical Drawing (1st year), Construction Process (4th year) and Bridges (5th year). The teacher interacts with the model showing the sequence construction and the constitution of the modeled building element.

As in Technical Drawing, students have to define and represent structural plants using architectural layouts, they better understand the relations between the architectural configurations and the structural elements in a building, following the exhibition of the wall's construction. In Construction Process and Bridges, in order to prepare students to visit real work places, the teacher shows the construction animation and explains some aspects of the construction process of the wall or the bridge that in the work place they are going to see.

Essentially, the models are used to introduce new subjects. The students reflected on their evaluation works a better understanding of subjects concerning structures, bridges and construction. For instance, in a structural plant the representation of columns and beams is now better aligned. Or, the students' reports concerning visits at work places include now some aspects shown in the virtual models.

The students can also interact with those models. For that, the models were posted on the Internet pages of undergraduate Civil Engineering. The student will be able to interact with the application *EonX*, which can be accessed at http://download.eonreality.com.

5 CONCLUSIONS

It has been demonstrated, through the examples presented here, how the technology of virtual reality can be used in the elaboration of teaching material of educational interest in the area of construction processes.

The models generated represent building in two standard situations. The student can interact with the virtual model in such a way that he can set in motion the construction sequence demanded by actual construction work, observe the methodology applied, analyze in detail every component of the work and the equipment needed to support the construction process and observe how the different pieces of a construction element mesh with each other and become incorporated into the model.

These models are used in disciplines involving construction in courses in Civil Engineering and Architecture administered by the Higher Technical Institute of the University of Lisbon. They can be used in classroom-based education and in distance learning supported by e-learning technology.

REFERENCES

EON. 2003. Introduction to working in EON Studio, *EON Reality, Inc.*

GRID. 1995. Planning office, Graphical documentation of the design of North Viaduct of the Quinta Bridge, 1st phase, Lisbon, Portugal.

Leinonen, J., Kähkönen, K. & Retik, A., 2003. New construction management practice based on the virtual reality technology, In Raja R.A. Issa, Ian Flood William & J. O'Brien (eds), *4D CAD and Visualization in Construction: Developments and Applications*: 75–100. Balkema.

Sampaio, A.Z., Braz, H., Silva, L., Lopes, B. & Gouveia, R., 2002. Graphic representation of bridges system: right bridges (in portuguese), *report DTC/ ICIST n° 06/02*, Lisbon, Portugal.

Sampaio, A.Z., Henriques, P. & Studer, P. 2006. Virtual Reality Models used in Civil Engineering Education, in A.C. Boucouvalas (ed.), *the 2nd IASTED Conference on Internet and Multimedia Systems and Applications, EuroIMSA 2006*, 119–124 (paper n° 516–028), Innsbruck (Austria), *13–15 February 2006*. ISBN: 0-88986-566-3.

Sampaio, A.Z. & Recuero, A., 2004. A geometric modelling of box girder deck for integrated bridge design. Inderscience Enterprises Ltd (ed.), In *International Journal of Computer Applications in Technology, IJCAT, Special issue on Interoperability for SME-based environments*, 20 (1–3): 54–61, ISSN: 0952-8091.

Sampaio, A.Z., Reis, A., Braz, H. & Silva, L., 1999. Project program report: Automatically generating model of the graphic representation of bridges, *ICIST/FCT*, Lisbon, Portugal.

Computational Modelling of Objects Represented in Images –
João Manuel R.S. Tavares & R.M. Natal Jorge (eds)
© 2007 Taylor & Francis Group, London, ISBN 978-0-415-43349-5

A system for the acquisition and analysis of the 3D mandibular movement to be used in dental medicine

Isa C.T. Santos, João Manuel R.S. Tavares & Joaquim Mendes
Faculty of Engineering of Porto University, Porto, Portugal

Manuel P.F. Paulo
Faculty of Dental Medicine of Porto University, Porto, Portugal

ABSTRACT: The peculiar construction of the temporomandibular joint allows the mandible to move according to six degrees of freedom. In Dental Medicine, is essential to know this movement to simulate the temporo-mandibular joints, to position tooth moulds in articulators and to reproduce the mandibular movements in order to insure a satisfactory occlusion. In this work a facial arc, commonly used in Dental Medicine, was adapted to use electromagnetic sensors with the purpose of acquire the 3D mandibular movement. Some parts of the chosen facial arc were redesigned, and it was developed a specific support for the sensors employed. To visualise and analyse the acquired movement it was developed a software application using *LabVIEW*.

1 INTRODUCTION

The human mandible and the temporomandibular joint form an interesting and complex biomechanical system that performs several functions, and have the capacity to make high forces with great precision. In Dental Medicine, is essential to know the mandibular cinematic to simulate the temporomandibular joints, to position teeth moulds in articulators, and to reproduce the mandibular movements in order to insure a satisfactory occlusion.

2 MANDIBULAR MOVEMENT

The movement of the mandible is usually described by the incisal point which is located between the two lower incisors. The maximal area including all the points the incisal point can reach is described by the Posselt's diagram, Fig. 1. The Posselt's diagram is the convex hull of all these points and displays the position of the incisal point for the extreme constellations of the joint. This is called the border movement of the incisal point.

The sagittal Posselt's diagram, Fig. 2, can be divided into 4 main segments. In the first segment (the maximal rear opening period), the jaw rotates about 10° around the hinge axis. If the mandible moves further downwards, a protrusion starts. The protrusion is a feed of the condyls on a circular path. Therefore, the final rear opening period can be considered a combinated movement of the hinge axis and a protrusion. After this period the maximal opening is reached. The

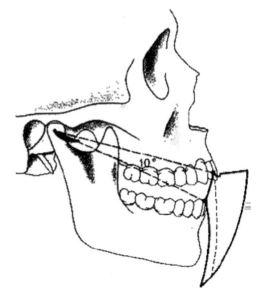

Figure 1. Posselt' diagram, sagittal view (image from (Weingärtner, 1997)).

maximal frontal path can be described as an rotation around the hinge axis by a maximal protrusion. At the upper bound, only a protrusion takes place. If the maximal frontal position is reached, the condyls are located under the tuberculum articulare. Finally, the jaw is gliding back into its resting position.

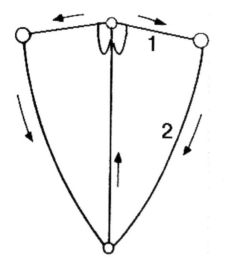

Figure 2. Posselt's diagram, frontal view (image from (Weingärtner, 1997)).

This is the border movement the incisal point can perform, projected on a sagittal plane. The path of the fourth segment is affected by the structure of the teeth, and the other paths by the muscles and ligaments.

Similar to the sagittal Posselt's diagram, each lateral movement can be divided into two segments: the lateral movement with occlusion and the maximal lateral position during an opening of the mandible, Fig. 2. Depending on position or malposition of the teeth, different pathologic movements are possible. The paths can differ significantly from patient to patient.

3 STATE OF THE ART

The *Jaw Motion Analyzer* described by Enciso (Enciso, 2002) and Hugger (Hugger, 1999) and the *ARCUS-digma* presented by Pröschel (Pröschel, 2002) are two commercial systems developed for the acquisition of the 3D mandibular movement that use ultrasonic sensors to acquire the trajectory of the incisive point. Leader (Leader, 2002), Mesnard (Mesnard, 2003) and Kinuta (Kinuta, 2003) use optical systems to record the same movement, and Garcia (Garcia, 2003) describes the *K7 Evaluation System* which is a commercial system that uses electromagnetic sensors for the same purpose.

Currently, both commercial and academic devices can be considered very expensive and difficult to use in common clinical situations. Considering these disadvantages, during this work it was developed a new prototype system for the acquisition, visualization and analysis of the 3D mandibular movement that is economical, easy to use and sufficiently precise (Santos, 2005).

The development of our prototype system was divided in three phases: (1) the choice of the technology to be used in the acquisition of the movement; (2) the conception of the support structure for the sensors selected; (3) the development of a computational application with an adequate graphical interface to visualize and analyze easily the 3D data obtained.

In the following section of this paper is described the prototype developed for the acquisition of the 3D mandibular movement. Next, is presented the computational application developed to visualize and analyze the movement acquired. Some conclusions and perspectives of future work are presented in the sixth and last section of the paper.

4 PROTOTYPE DEVELOPED

In order to obtain a product with higher quality, it was adopted a structured product development process in the development of our prototype system (Ulrich, 2000).

Discussing with medical dentists, the costumer needs were determined and, after a process of market research, the product specifications were defined; the concept and the product architecture were selected and finally, using rapid prototyping techniques, prototypes pieces were materialized and tested.

The first step in the creation of our prototype was the selection of the technology to be used in the movement acquisition, because the physical structure of the system would strongly depend on it. In our prototype, were employed electromagnetic sensors to evaluate the magnetic field created by a small magnet placed inside the patient's mouth (near to the incisive point). Using this type of sensors is possible to acquire the mandibular movement in an economical way and without any type of contact.

After choosing the technology to be used, the support structure for the acquisition system was defined. A common facial arc used in Dental Medicine was adapted as the main support structure of our system. With this approach, the development time, the development costs and the final price of the new system were considerably reduced. Among the commercial facial arcs, it was adapted the *Arcus* from *Kavo*, Fig. 3, (www.kavo.com). This arc was selected because is one of the most recent models and is compatible with several common articulators.

The CAD models of the facial arc was created using the parametric software *Autodesk Inventor Professional*, Fig. 4.

As the *Arcus* facial arc was originally designed to be used in static measurement procedures, the first step in the adaptation to its new function was the redesign of the parts that could difficult the dynamic measurement or even harm the patient during the medical exam. For that, it was necessary to adapt its auricular parts, Fig. 5.

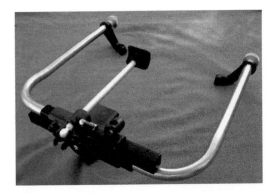

Figure 3. Facial arc *Arcus* from *Kavo* used as the main support structure of our prototype system.

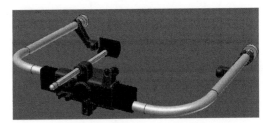

Figure 4. 3D CAD model of the facial arc *Arcus* from *Kavo*.

Figure 5. The original auricular part.

The general dimensions and the groove of the original auricular parts were maintained, but the original shape was replaced by a more organic one, to offer the patients a higher sensation of comfort. As a result, the end part of the original auricular was replaced by a sphere to avoid hurting the patients during the medical exams. The new auricular parts, that were materialized using a stereolitography process, have also the advantage to be easier to clean, Fig. 6.

The support of the electromagnetic sensors employed to acquire the 3D movement was designed

Figure 6. Prototype of the redesigned auricular part.

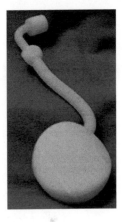

Figure 7. Support developed for the sensors used.

to meet the following requirements: to be adaptable to the facial arc used, to accommodate two circuit boards with the magnetic sensors mounted orthogonally and to be light. The most appropriate shape to satisfy these needs was also an organic one, Fig. 7.

To attach the support of the electromagnetic sensors designed to the facial arc adopted, it was used an already existing groove in the adopted facial arc; to prevent the translation and the rotation of the sensors support, it was developed a specific notch in the sensors support. The end of the sensors support was also enlarged to accommodate the connection between the wires coming from the sensors and the data acquisition's cable.

As the selected electromagnetic sensors were very small, to be easier to handle them it were developed the two circuit boards already referred. These two boards are mounted perpendicularly in the cavity of the sensors support by pressure, and the access to them is carried through a sliding cover, that presents also a hole for a led that is used to indicate if the acquisition system is on or not, Fig. 8.

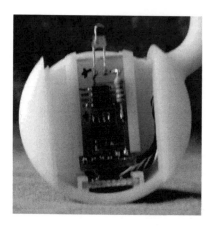

Figure 8. The two circuit boards developed mounted in the sensors support.

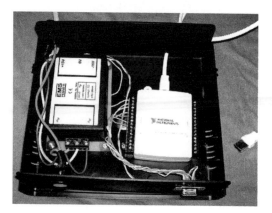

Figure 9. Box with the DAQ and the external power supply used for the magnetic sensors employed in our prototype system.

To acquire the signals from the electromagnetic sensors employed in the acquisition system was used a USB data acquisition device (DAQ). In spite of the DAQ could also be used as a power source for the three electromagnetic sensors employed, it was considered an external power supply instead in order to increase the signals' sensibility. The DAQ and the external power source were then mounted in a box, Fig. 9.

5 APPLICATION DEVELOPED

To visualize and analyze the 3D mandibular movement acquired by the electromagnetic sensors used, it was built a computational application with an adequate graphical interface, using the developing tool *LabVIEW* from *National Instruments*, Fig. 10. In the upper part of the interface designed, the user can write the patient's personal data: the name, the age, the sex and the missing teeth. In the lower part of the same

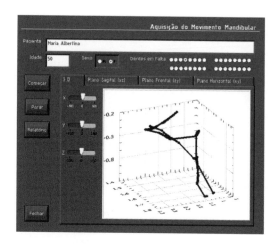

Figure 10. Graphical interface of the application built to visualize and analyse the 3D movement acquired.

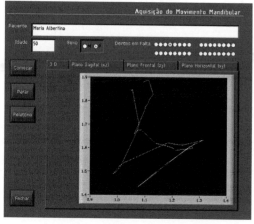

Figure 11. Computational application developed for the 3D mandibular movement acquisition and analysis: a 2D graphic in the sagital plane.

main window there are the control buttons of the acquisition process and the graphical area were the results are displayed.

The measurement results are displayed in four independent graphics: one in 3D that the user can rotate and zoom, and three in 2D corresponding to the projections of the movement acquired in the sagital, frontal and horizontal plans, Fig. 11. By pressing the button "Relatório" (Report) a report of the actual medical exam is created, Fig. 12, and the user can then print or save it.

To know the trajectory of the magnet used it was necessary to convert the electromagnetic sensors output voltages in 3D cartesian coordinates. For that conversion, were used neural networks (Demuth, 2005; Marques, 1999). Thus, the output voltages

402

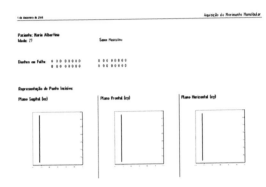

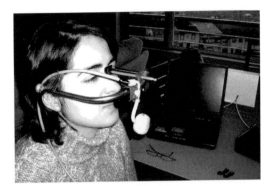

Figure 14.   Application of our developed system.

Figure 12.   A report example obtained using the application developed to visualize and analyse the 3D movement acquired.

Figure 13.   The developed prototype system for the acquisition and analysis of the 3D mandibular movement.

corresponding to several well defined 3D points were recorded and then this information was used to create and calibrate a radial neural network using the software *MatLab*.

The Fig. 13 and Fig. 14 show the prototype system developed in this work for the acquisition, visualization and analysis of the 3D mandibular and its application.

As can be seen in Fig. 14 our system is simple, light, small, with a pleasant design and easy to use.

6   CONCLUSIONS AND FUTURE WORK

In this work, to acquire the 3D trajectory of the incisive point were used three electromagnetic sensors; mainly, because they are economical and allow the development of a simple and comfortable support structure without any type of contact. The magnetic field to be evaluated by the sensors is created by a small magnet placed inside the patient's mouth.

To reduce the development costs and the final price of the new prototype system, a common facial arc was adapted as its main support structure. Some parts of the adopted facial arc were redesigned to be more comfortable for the patients during the medical exams, and it was also designed a specific support for the three electromagnetic sensors used.

To visualize and analyse the acquired signals associated to the movement of the magnet used it was built, using the developing tool *LabVIEW*, a application program with an adequate graphical interface were the operator can register the patient's personal data. With the same application is also possible to create a medical report with the patient's personal data and the results of the clinical exam done.

The new prototype system created is small, has a pleasant design and can be commercialized with an affordable price.

In the near future we are expecting to do several medical exams using our prototype system for validation purpose. Another task that we want to address is the building of a more precise neural network, to be used in the conversion between the voltages output of the three electromagnetic sensors used in the corresponding 3D cartesian coordinates of the magnet.

REFERENCES

(Demuth, 2005) Demuth, H., M. Beale and M. Hagan. Neural Network Toolbox For Use with MATLAB. The Mathworks.

(Enciso, 2002) Enciso, R., A. Menon, D. Fidaleo, U. Neumann and J. Mah. The virtual Craniofacial Patient 3D Jaw Modeling an Animation. Studies in Health Technology Informatics 94, 65–71.

(Garcia, 2003) Garcia R., V. Oliveira and A. Cury. Short term evaluation of interocclusal distance during speech after new removable prosthesis insertion. Journal of Applied Oral Science 11(3), 216–222.

(Hugger, 1999) Hugger A., E. Bölöni, U. Berntien and U. Stüttgen. Accuracy of an ultrasonic measurement system for jaw movement recording. 35th Annual Meeting of the Continental European Division of the International Association for Dental Research, Montpellier, France.

(Kinuta, 2003) Kinuta S., K. Wakabayashi, T. Sohmura, T. Kojima, M. Nagao, T. Nakamura and J. Takahashi. Simple system to record jaw movements by a home digital camcorder. The International Journal of Prosthodontics 16(5), 563–569.

(Leader, 2002) Leader J., J. Boston, R. Debski and T. Rudy. Mandibular kinematics represented by a non-orthogonal floating axis joint coordinate system. Elsevier.

(Marques, 1999) Marques J. Reconhecimento de padrões métodos estatísticos e neuronais. IST Press.

(Mesnard, 2003) Mesnard M., A. Ballu and M. Cid. Méthode et validation pour l'Articulation Temporo-Mandibulaire. 16ème Congrès Français de Mécanique, Nice, France.

(Pröschel, 2002) Pröshel P., T. Morneburg, A. Hugger, B. Kordaβ, P. Ottl, W. Niedermeier and M. Wichmann. Articulator-Related Registration – A simple concept for minimizing eccentric occlusal errors in the articulator. The International Journal of Prosthodontics 15(3), 289–94.

(Santos, 2005) Santos I. Desenvolvimento de um Sistema Protótipo para a Aquisição e Análise do Movimento Mandibular. MSc Thesis in Industrial Design of the University of Porto, Portugal.

(Ulrich, 2000) Ulrich K. Product Design and Development – 2nd edition. McGraw Hill.

(Weingärtner, 1997) Weingärtner T, Dillmann R. Simulation of jaw movements for the musculoskeletal diagnoses. Studies in Health Technology and Informatics, vol.39, p.p. 401-10-1997

Computational Modelling of Objects Represented in Images –
João Manuel R.S. Tavares & R.M. Natal Jorge (eds)
© 2007 Taylor & Francis Group, London, ISBN 978-0-415-43349-5

Hierarchical morphological analysis for generic detection of shapes in grayscale images

Alexandre G. Silva
Universidade do Estado de Santa Catarina, Joinville, Brazil

Roberto A. Lotufo
Universidade Estadual de Campinas, Campinas, Brazil

ABSTRACT: This work proposes a generic algorithm of shape detection in a grayscale image from its Max-tree. This structure organizes different connected components in all scales of the image in a hierarchical way offering more semantic information and less elements to process in relation to the traditional grid of pixels representation. Filtering methods and recognition of shapes, such as lines, straight lines, circles, and archs, are proposed through searching nodes of interest in the tree. The run time of this method is not necessarily increasing by the number of parameters of the geometric shape. Moreover other more complexer objects (non-parameterizable) can be detected by morphological analysis.

1 INTRODUCTION

An image is usually viewed as a set of pixels placed on a rectangular grid. However it is possible to understand an image as regions organized in a hierarchical way. Grouping pixels by similarities and understanding an image as a set of regions which have inherent relationships is an alternative. Some models of hierarchical structures are useful for the image processing based on region, allowing the representation of semantic aspects (Abdel-Mottaleb et al. 1999): Components Tree (Mattes et al. 1999), Tree of Critical Lakes (Carvalho 2004), and Max-tree (Min-tree) (Garrido 2002). This last one is chosen, because it presents three important characteristics: (i) it is not necessary to define any initial parameter in its construction process; (ii); all the connected components, in each level, are considered without redundancy; (iii) a considerable discarding of nodes to be analyzed in the detection of shapes is possible using the hierarchical organization of regions. This work proposes an generic algorithm of filtering and detection of objects of interest based on Max-tree information. The determining of shapes in images has great importance in several applications: roads and markings in GIS maps, references to guide a robot vehicle, fingerprint analysis, among other examples. The Hough transform (Hough 1962) is an interesting and computationally efficient (if the geometric shape has sufficient few parameters) procedure for detecting lines and curves (Duda and Hart 1972) and has been widely used to solve this problem mainly for binary images. However grayscale images can be considered, for instance, for detection of straight lines (Aggarwal and Karl 2006). Some important definitions, such as connectivity, morphological operators and the tree in studying, are briefly described in Section 2. In Section 3, we have some ideas of reduction of complexity of the image or filtering of nodes by any hierarchical criteria. Section 4 illustrates the proposed generic algorithm and the modeling of some shapes for searching and detection. In Section 5, experimental results are shown. Finally, the conclusions are commented in Section 6.

2 PRELIMINARY DEFINITIONS

A *grayscale image* is a rectangular matrix I of *pixels*. The intensity of a pixel p is denoted by $I(p)$. If there are only two intensities, $I_{bin}(p) = 0$ or $I_{bin}(p) = 1$, then it corresponds to a *binary image*. Let $\mathcal{N}_E(p)$ a *neighbourhood* of p, *path* from p_1 to p_n is defined as a sequence $\mathcal{P} = (p_1, p_2, \ldots, p_n)$, where $p_i \in \mathcal{N}_E(p_{i+1})$, $\forall i \in [1, n]$. A *componnected component* is a subset $\mathcal{C}$, where there is always a path $\mathcal{P}$, $\forall p_a, p_b \in \mathcal{C}$. A *flat zone* is a connected component $\mathcal{Z}$ of the image, such that, $\forall p_a, p_b \in \mathcal{Z}$, $I(p_a) = I(p_b)$. A *regional maximum* is a flat zone M such that $I(m) > I(n)$, $m \in M$, $n \in N, N \in \mathcal{F}_M$, where $\mathcal{F}_M$ is a set of all flat zones adjacent to M.

2.1 Binary morphological operators

Negation $v(\mathcal{C})$ consists of $1 - \mathcal{C}$. *Dilation* is a morphological operator $\delta(\mathcal{C})$ increases the connected component size, according to an intersection rule between the image and a neighborhood defined (Heijmans 1999). *Erosion* $\gamma(\mathcal{C})$, by other side, decreases this size, and it is defined by dilation dual $v(\delta(v(\mathcal{C})))$. *Morphological gradient* $\Psi(\mathcal{C})$ corresponds to a dilation followed by a erosion $\gamma(\delta(\mathcal{C}))$ and determines the connected component boundary (Dougherty and Lotufo 2003). *Reconstruction* consists of "infinite" (until stability) number of dilations of a marker S conditioned to a connected component $\mathcal{C}$. In other words, $\delta_S^\infty(\mathcal{C})$ is the reconstruction of an region $\mathcal{C}$ from a seed S (Dougherty and Lotufo 2003). *Distance transform* $\mathcal{D}(\mathcal{C})$ consists in the minimum distances from all pixels which are not zero until the boundary of the connected component $\Psi(\mathcal{C})$. It can be implemented by the sum of a sequence of erosions $\sum_{i=0}^{\infty} \gamma_{\mathcal{C}_i}$, where $\gamma_{\mathcal{C}_0}$ is the original binary image, until the erosion to result a null image. *Centroid* of a connected component $\mathcal{C}$ is only one pixel x_c, where $x_{c_{lin}}$ is the average line and $x_{c_{col}}$ is the average column of the pixels which are not zero in $\mathcal{C}$.

2.2 Max-tree representation

The Max-tree MT_I (Garrido 2002) consists of a hierarchical representation of regions based on relief of an image I. Figure 1 illustrates an example of construction of this structure for an image of height 4 and width 8 (a). Its relief is defined by a new axis as a graylevel function of the pixels (b). The tree root $NODE_{root}$ corresponds to region $\mathcal{C}_{NODE_{root}}$ formed by pixels whose intensities are equal or greater than the lowest intensity of I. The tree leaves corresponds to the regional maximums of I (supposing the pixel has 8 neighbours). The intermediary nodes follow the skeleton of the topographical surface (Fig. 1c). It is important to observe the sons of an node $NODE$ determine regions are subset of its region: $\forall i \in [1, n]$, $\mathcal{C}_{son_i(NODE)} \subset \mathcal{C}_{NODE}$, or in other words, the area of the connected component of $NODE$ is greater than the sum of areas of the connected components of its

n sons: $area(\mathcal{C}_{NODE}) > \sum_{i=1}^{n} [area_i(\mathcal{C}_{son_i(NODE)})]$. The idea of this work is to represent an image I in a Max-tree structure MT_I, where each node $NODE$ corresponds to a binary image $\mathcal{C}_{NODE}$ (same size of I) of a connected component.

3 PROPOSED FILTERING

The pre-selection of nodes of the Max-tree MT_I of an image I by any criteria is important therefore it reduces the number of nodes of the shape recognition will be presented in the next section. In other words, a simplified image can be treated and the time of processing can be improved.

3.1 Salient nodes

This filter consists of the remotion of all leaves which do not have brothers (except the root) recursively until there are only leaves which have at least one brother. The purpose is to remove all salient peaks or prominences of the image. Figure 2a illustrates the selection of nodes (coloured) of an image I_1 using this approach. Figures 5b,c show examples of filtering of salient peaks for a synthetic image I_2. Figures 2d,e are photographic original image I_3 and the filtering result by salient peaks remotion respectively. While I_3 has 7560 pixels, its Max-tree MT_{I_3} has 842 nodes or regions. After this filtering, 246 nodes were discarded in 0.02 second.

3.2 Nodes without brothers

This filter removes the intermediary increasing gray regions between two nodes which have at least one brother (including the previous remotion of salient nodes). The idea is to select only source regions of a tree branch. Figure 3a consists of the selection of nodes (coloured) of an image I_1 by this technique. Figures 3b,c refers to filtering of nodes without brothers of a synthetic image I_2. Figures 3d,e are photographic original image I_3 and the filtering result by this approach respectively. After this last filtering,

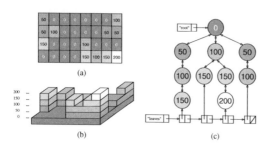

(a)

(b)

(c)

Figure 1. Max-tree construction. (a) Image I. (b) Topographical surface of I. (c) Max-tree of image or MT_I.

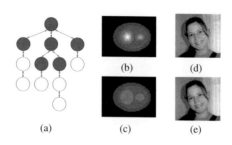

(a) (b) (d)

(c) (e)

Figure 2. Salient nodes filtering. (a) Selection of nodes of MT_{I_1} for an image I_1. (b) Original image I_2. (c) Remotion of salient peaks of I_2. (d) Original image I_3. (e) I_3 filtered by salient peaks.

542 nodes of I_3 were discarded (more double than the first one) in 0.02 second too.

3.3 Nodes in a degree interval

The degree of a node corresponds to number of its sons. This filtering implements the selection of nodes based on their degrees. Figure 4a consists of the selection of nodes (coloured) of an image I_1 by this issue. Figures 4b,c,d,e refers to original image I_2 and its filtering by selection of nodes in a degree variation. The box drill image I_2 in Figure 4b has 24975 pixels and its MT_{I_2}, 520 nodes. 513, 517 and 519 nodes were discarded in Figure 4c,d,e respectively in average 0.12 second. Observe in this last filtering only a node remain. The number of possible degrees is normally less than the maximum graylevel of an image. A thresholding based on this parameter can be useful in a segmentation algorithm.

3.4 k-max

All nodes distant of k_{up} steps from the leaves can be selected. Therefore, an ascendant searching in the tree is possible and allows finding regions near the regional

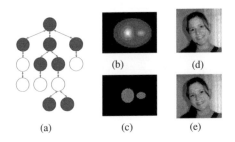

(a)   (c)   (e)

(b)   (d)

Figure 3. Nodes without brothers filtering. (a) Selection of nodes of MT_{I_1} for an image I_1. (b) Original image I_2. (c) Remotion of nodes without brothers of I_2. (d) Original image I_3. (e) I_3 filtered by nodes without brothers.

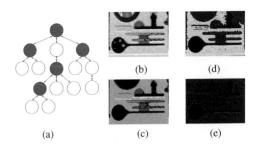

(b)   (d)

(a)   (c)   (e)

Figure 4. Filtering by degrees of nodes. (a) Selection on MT_{I_1} (for an image I_1) of nodes whose degrees are equal or greater than 2. (b) Original image I_2. (c) Filtering on I_2 of nodes whose degrees are 5. (d) Nodes of I_2 whose degrees are equal or greater than 15. (e) Nodes of I_2 whose degrees are between 15 and 20.

maximums. The upward processing must stop when the last visited node is not at less than k_{down} steps of the tree root. Figures 5a,b illustrates the selection of nodes (coloured) of an image I_1. Figures 5c,d,e,f show examples of k-max filtering for some different values of k_{up} (default value of k_{down} is 1). Figures 5g,h,i,j are original and filtered computerized tomograph images of 5041 pixels of loamy soil (Silva et al. 2006). In the first soil image, there were 113 nodes and 106 were discarded. The second had 108 nodes and 102 were discarded. Both filtering were done using $k_{up} = 4$ in 0.02 second.

3.5 Area

Sometimes there are excessive number of candidate nodes for a specific shape. Those nodes $NODE_i$ whose components C_{NODE_i} are in a certain interval of area, $area_{min}$ and $area_{max}$, can be selected.

4 PROPOSED SHAPE DETECTION

This section consists of the development of a generic algorithm for detection of shapes from regions of the Max-tree. The diagram of Figure 6 illustrates some elements related with this process. After the construction of the Max-tree MT_I of an image I, the idea is searching nodes that represent connected components of interest. The node $NODE$ associated to shape of search, considering a certain tolerance between d_{min} and d_{max} (for distances, diameters, rays, etc), is translated to its connected component C_{NODE} and it is added to binary image resulting of the segmentation I_{shape}. The searching in depth does not continue from

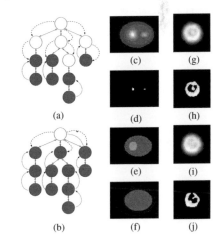

(c)   (g)

(a)   (d)   (h)

(e)   (i)

(b)   (f)   (j)

Figure 5. k-max filtering. (a) On MT_{I_1} (for an image I_1) with $k_{up} = 1$ and $k_{down} = 2$. (b) On MT_{I_1} with $k_{up} = 2$ and $k_{down} = 1$. (c) Original image I_2. (d) Regional maximums of I_2 or k-max with $k_{up} = 0$. (e) $k_{up} = 3$ on I_2. (f) $k_{up} = 5$ on I_2. (g) Image I_3. (h) $k_{up} = 4$ on I_3. (i) Image I_4. (j) $k_{up} = 4$ on I_4.

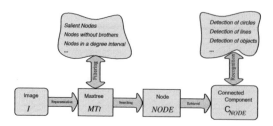

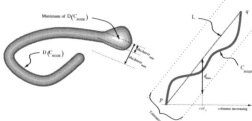

Figure 6. Stages of the detection of shapes.

a selected node because its branches are subsets of its connected component as seen in the Section 2.2. The process repeats until that there are not more nodes to analyze. Algorithm 1 implements this method.

Algorithm 1: Algorithm for detection of shapes.

INPUT: I, d_{min}, d_{max}, $Detect(\ )$
OUTPUT: I_{shape}

INITIALIZATION:
1    $I_{shape} \leftarrow \bigcirc_{heigth(I) \times width(I)}$
2    $C_{NODE_{root}} \leftarrow root(MT_I)$

DETECTION_OF_SHAPES
3    $stack.push\ (C_{NODE_{root}})$
4   **while** $stack \neq \emptyset$
5        $C \leftarrow stack.pop(\ )$
6   **if** $Detect\ (C_{NODE}, d_{min}, d_{max})$
7            $I_{shape} \leftarrow I_{shape} \cup C_{NODE}$
8   **else**
9            $stack.push(sons(NODE))$
10   **return** I_{shape}

The recognition stage of Figure 6 is implemented by the function $Detect(\ )$ of Algorithm 1 in two ways: **by shape analysis** from the algorithms of geometric figures with simple formulation or morphological criteria. Examples: lines, circles, straight lines, archs, specific curves, etc; **by template matching** to known shapes without algorithm simplified for their description. Examples: characters, ideograms, symbols, etc. Geometric shape analysis and segmentation in graylevel images are studied using the first idea. Functions $Detection(\ )$, returning *true* (shape found) or *false* (shape did not find), are presented in the remaining of this section. Each shape has a different meaning for the tolerance d_{min} and d_{max}. The detection of non-parameterizable lines (using morphology), and of parameterizable straight lines, circles and archs, is developed to follow.

4.1 Lines

A connected component of a node C_{NODE} is a line if all the pixels are minimum distant of 1 of its contour $\Psi(C_{NODE})$ or the maximum of its distance transform is 1. Considering certain tolerance: $thickness_{min} \leq \max(\mathcal{D}(C_{NODE})) \leq thickness_{max}$.

Figure 7. Line (left) and straight line (right) recognition approach.

Figure 7 schematizes this issue. The detection of lines usually is applied on a preprocessed image with enhanced contours by a high-pass filter or gradient operator and other refinements can be done (Silva and Lotufo 2004).

4.2 Straight lines

Considering two pixels which are not zeros of a connected component C_{NODE}, p and q, where the first one has the minimum column (or line) value – more on the left (or top) – and the other has the maximum column (or line) value – more on the right (or bottom), a digital straight line $\mathcal{L}$ between them, with angular coefficient $(p_{col} - q_{col})/(p_{lin} - q_{lin})$, is traced. If $|p_{col} - q_{col}| > |p_{lin} - q_{lin}|$ then the columns from p until q are increased, otherwise, the lines are increased. For each increment, the value of the pixel in $\mathcal{L}$ is compared with the value of the correspondent pixels in C_{NODE}. If this distance is between d_{min} and d_{max} for all the pixels of C_{NODE}, then C_{NODE} is a straight line. d_{min} is normally 0. d_{max} is equal for both sides of $\mathcal{L}$ and is related with the tolerance distance from a pixel of C_{NODE} to $\mathcal{L}$. Figure 7 exhibits this approximation.

4.3 Circles

A connected component is a circle if all the pixels x_i of its contour $\Psi(C_{NODE})$ have the same distance until the centroid pixel x_c. Considering certain tolerance: $ray_{min} \leq dist(x_i, x_c) \leq ray_{max}$, $\forall\, i \in [1, b]$, where $b = area(\Psi(C_{NODE}))$ or number of the boundary pixels. Figure 8 represents this consideration.

4.4 Archs

Finally, a connected component of a node C_{NODE} can be considered an arch. For this, the centroid x_c of C_{NODE} is calculated. The nearest pixel from C_{NODE} until x_c is selected. A disk (of proportional size in relation to ray_{max} is chosen) around this one is defined. A new centroid y_c of all pixels of C_{NODE} in this disk is calculated. After this, a straight line segment is defined from y_c passing through x_c. In some place, a point is equidistant of all boundary pixels $\Psi(C_{NODE})$ may exist.

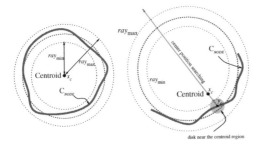

Figure 8.   Circle (left) and arch (right) recognition approach.

Therefore this searching is implemented considering a tolerance of rays between ray_{min} and ray_{max}. Figure 8 illustrates this purpose.

5   EXPERIMENTAL RESULTS

In this section, a set of tests on grayscale images is done to demonstrate the shape recognition purpose of this work. Notations are used in the bellow figures for each image i: p_i is the total number of pixels; N_i is the total number of nodes (or regions) of the Max-tree; n_i is the total number of selected nodes (or regions) represent a specific geometric shape; T_i is the total run time (including the Max-tree construction); t_i is only the Max-tree construction time. Figure 9 exemplifies the detection of lines. The thickness tolerance is: between 1 and 5 for fractures in soil detection (first image negated); between 1 and 5 for the second. Filtering by area greater than 50 also is done in rule image negated. Figure 10 illustrates the detection of straight lines. The maximum distance d_{max} from the pixels in $\mathcal{C}$ to the support straight line $\mathcal{L}$ is: 2 for the first image (plus filtering by area 50); 1 for the car image (plus filtering by area 30). Searching of circles is shown in Figure 11, where tolerances of rays are established: between 10 and 30 for the aspirin tablets detection (first image); between 3 and 7 for the tennis ball (second). Finally, Figure 12 consists of detection archs of rays: between 100 and 115 for the corridor image; and between 30 and 50 for the box drill.

Statistics about some processed grayscale images are presented in Table 1. The columns shows: the average relation of the total of pixels and the total of nodes (or regions) of the image; the average percentage of selected nodes in the filtering and shape detection; average relation of the Max-tree construction time and the number of nodes; and the average time after Max-tree construction reserved for recognition processing of each selected node. These results demonstrate there are less elements in relation to pixel based representation (until 58.8 times). Moreover a lot of nodes can be removed in filtering and segmentation processing (for instance, only 11.05% of all nodes

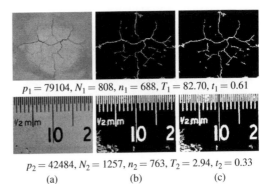

$p_1 = 79104, N_1 = 808, n_1 = 688, T_1 = 82.70, t_1 = 0.61$

$p_2 = 42484, N_2 = 1257, n_2 = 763, T_2 = 2.94, t_2 = 0.33$

(a)   (b)   (c)

Figure 9.   Segmentation of lines. (a) Original image I. (b) Image from nodes $NODE_i$ selected of the MT_I. (c) Binary image $\bigcup_{i=1}^{n} \mathcal{C}_{NODE_i}$.

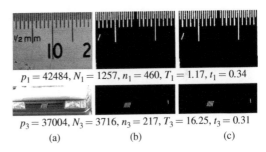

$p_1 = 42484, N_1 = 1257, n_1 = 460, T_1 = 1.17, t_1 = 0.34$

$p_3 = 37004, N_3 = 3716, n_3 = 217, T_3 = 16.25, t_3 = 0.31$

(a)   (b)   (c)

Figure 10.   Segmentation of straight lines. (a) Original image I. (b) Image from nodes $NODE_i$ selected of the MT_I. (c) Binary image $\bigcup_{i=1}^{n} \mathcal{C}_{NODE_i}$.

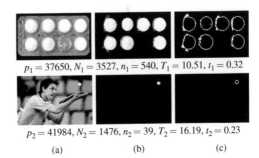

$p_1 = 37650, N_1 = 3527, n_1 = 540, T_1 = 10.51, t_1 = 0.32$

$p_2 = 41984, N_2 = 1476, n_2 = 39, T_2 = 16.19, t_2 = 0.23$

(a)   (b)   (c)

Figure 11.   Segmentation of circles. (a) Original image I. (b) Image from nodes $NODE_i$ selected of the MT_I. (c) Binary contour $\bigcup_{i=1}^{n} \Psi(\mathcal{C}_{NODE_i})$.

are used in the examples of circles detection). The Max-tree construction time depends strongly on the number of image regions (this average time per node is $2.86 \cdot 10^{-4}$s for the examples). Finally, the recognition average time per selected node indicates that straight lines and arches are 7.8 times faster than lines and circles in the cases seen. All source code was implemented in C++ at Pentium$^{©}$ 4, 3.2 GHz.

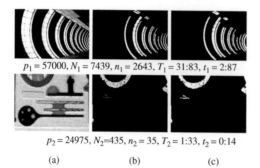

$p_1 = 57000, N_1 = 7439, n_1 = 2643, T_1 = 31:83, t_1 = 2:87$

$p_2 = 24975, N_2 = 435, n_2 = 35, T_2 = 1:33, t_2 = 0:14$

(a) (b) (c)

Figure 12. Segmentation of archs. (a) Original image I. (b) Image from nodes $NODE_i$ selected of the MT_I. (c) Binary image $\bigcup_{i=1}^{n} C_{NODE_i}$.

Table 1. Statistics of the grayscale images processed. Columns sequence: shape, pixel per node, selected nodes percentage, 10^4 times Max-tree time per node, recognition time per selected node.

Shape	$\frac{\sum p_i}{\sum N_i}$	$100 \frac{\sum n_i}{\sum N_i}$	$10^4 \frac{\sum t_i}{\sum N_i}$	$\frac{\sum T_i - \sum t_i}{\sum n_i}$
Lines	58.88	70.27%	4.55 s	0.0584 s
S. Lines	12.62	55.06%	1.51 s	0.0054 s
Circles	24.66	11.05%	1.56 s	0.0716 s
Archs	10.41	34.01%	3.82 s	0.0113 s

6 CONCLUSIONS

The searching by shapes of interest can be used in several applications: quality control in industrial processes, geological fractures segmentation, vehicle logos detection, sporting markings and objects identification, robot trajectory recognition, and diverse other image processing and computational vision problems. The proposed algorithm is based on Max-tree regions analysis, where the hierarchy is explored for better performance. Each filtering described in this work reduces the number of nodes, simplifies an image, and allows a lower processing time without compromising the extraction of the information sought. It normally takes little time for the Max-tree construction (considering images around 256×256), the filterings are frequently faster, and the recognition time depends on the shape analysis performed. Future work is required to improve the Max-tree construction probably using hashing table to find the correct insertion of a node faster. The recognition of ellipses or another parameterizable shape can be easily added. The function to detect the connected component shape can be adaptation's to allow better precision in the segmentation. Advantages of the proposed method on traditional

Hough-based techniques can be cited: there is not necessary a previous image binarization; the complexity does not grow with the number of parameters of the geometric figure; non-parameterizable shapes might be proposed by morphological analysis of the connected components or template matching. However, a more detailed comparison with other methods must be done. And, if an image is sufficient large and complex, in other words, if it has many regions (nodes in its Max-tree), then the computation required may be prohibitive. A parallel algorithm model can be proposed since the shape analysis on a node is completely independent of another node.

ACKNOWLEDGMENTS

The authors thank Viviane Baschirotto for revision of this work.

REFERENCES

Abdel-Mottaleb, M., N. Dimitrova, L. Agnihotri, S. Dagtas, S. Jeannin, S. Krishnamachari, T. McGee, and G. Vaithilingam (1999, August). MPEG 7: A Content Description Standard Beyond Compression. In *IEEE Midwest Symposium on Circuits and System*, USA.

Aggarwal, N. and W. C. Karl (2006). Line Detection in Images Through Regularized Hough Transform. *IEEE Transactions on Image Processing* 15(3), 582–591.

Carvalho, M. A. G. (2004, Janeiro). *Análise Hierárquica de Imagens Através da Árvore dos Lagos Críticos*. Dissertação de Doutorado, Universidade Estadual de Campinas.

Dougherty, E. R. and R. A. Lotufo (2003). *Handson Morphological Image Processing*. SPIE.

Duda, R. O. and P. E. Hart (1972). Use of the Hough Transformation to Detect Lines and Curves in Pictures. *Communications of the ACM* 15(1), 11–15.

Garrido, L. O. (2002, April). *Hierarchical region based processing of images and video sequences: application to filtering, segmentation and information retrieval*. Ph. D. thesis, Universitat Politcnica de Catalunya, Barcelona.

Heijmans, H. J. A. M. (1999). Introduction to Connected Operators. In E. R. Dougherty and J. T. Astola (Eds.), *Nonlinear Filters for Image Processing*, pp. 207–235. SPIE.

Hough, P. (1962). In *Method and Means for Recognizing Complex Patterns*. U.S. patent 3069654.

Mattes, J., R. Mathieu, and J. Demongeot (1999, March). Tree Representation for Image Matching and Object Recognition. In *8th International Conference on Discrete Geometry for Computer Imagery*, France, pp. 298–312.

Silva, A. G., S. C. Felipussi, R. A. Lotufo, and G. L. F. Cassol (2006, January). k-max – Segmentation based on selection of Max-tree deep nodes. In *Proc. of IS&T/SPIE Electronic Imaging*, Volume 6064, USA, pp. 195–202.

Silva, A. G. and R. A. Lotufo (2004, October). Detection of Lines Using Hierarchical Region Based Representation. In *Proceedings of XVII SIBGRAPI*, Curitiba, Brazil, pp. 58–64. IEEE.

Computational Modelling of Objects Represented in Images –
João Manuel R.S. Tavares & R.M. Natal Jorge (eds)
© *2007 Taylor & Francis Group, London, ISBN 978-0-415-43349-5*

LSAVISION a framework for real time vision mobile robotics

Hugo Silva, J.M. Almeida, Luís Lima, A. Martins, E.P. Silva & A. Patacho

Laboratório de Sistemas Autónomos, Instituto Superior de Engenharia do Porto, Porto, Portugal

ABSTRACT: This paper propose a real-time vision architecture for mobile robotics, and describes a current implementation that is characterised by: low computational cost, low latency, low power, high modularity, configuration, adaptability and scalability. A pipeline structure further reduces latency and allows a paralleled hardware implementation. A dedicated hardware vision sensor was developed in order to take advantage of the proposed architecture. A new method using run length encoding (RLE) colour transition allows real-time edge determination at low computational cost. The real-time characteristics and hardware partial implementation, coupled with low energy consumption address typical autonomous systems applications.

Keywords: Real-time, autonomous system, image processing, pipeline processing, vision architecture.

1 INTRODUCTION

Artificial vision systems are key elements in robotics navigation and localization systems. This is due, to the their great sensing capabilities and low cost. The following paper describes real-time vision system architecture with a mobile robotics scenario in mind.

The presented implemented architecture has low computational cost, low latency, low power, high modularity, configurability, adaptability, and scalability. The architecture allows pipelined processing, with modules in hardware embedded parallel implementation (Lima 2004) or software implemented.

Although, with the Robocup scenario has a benchmark scenario and the MSL ISePorto team (Almeida 2003) as a initial application, the system is designed to be used in all kinds of autonomous systems, land, air or sea. In order to achieve this goal a global strategy needs to be followed. This is of particular importance because most of the vision systems are developed for a unique application.

One of the advantages of an integrated solution is that the different modules can be replaced or adapted to different systems and to different environmental conditions (indoor or outdoor, structured or not).

A dedicated hardware vision system sensor was developed implementing part of the overall vision architecture. There has been a development of vision systems for mobile robots, since the use of active vision systems mounted in pan tilt heads (Hai 2003) & (Peig 2000) to the use of real-time human tracking methods for autonomous mobile robots (Doi 2001).

The vision systems dealing with a more and more complex and dynamic environments, which will lead to the expansion and refinement of the image processing algorithms and their hardware embedded implementation.

Also a new and innovative solution to edge identification based on the notion of run length encoding (RLE) transition will be presented. Topological image information is improved with edge detection in real-time and without total image processing, in our system it performs faster, with less latency and with minor energy and computational costs than any of the methods presented in (Heath 1996).

2 VISION SYSTEM ARCHITECTURE

The architecture used see figure 1 follows a image processing pipeline approach. Where some of the layers can be hardware or software implemented.

The bottom acquisition layer (three left blocks) takes care of all hardware related and image acquisition specifications, camera settings and so on. The system is programmed to acquire frames from different kinds of cameras.

It detects which type of camera is plugged in and automatically starts to acquire frames through that device. This layer also contains the colour interpolation module, whose function is to assigned colour information to image pixels, the system can also use IR and monochromatic image types.

The image processing layer is responsible for the refinement and abstraction of the image data. After the image pixels have been given texture info, they will be segmented into previously defined color clusters following a method proposed by (Bruce 2000).

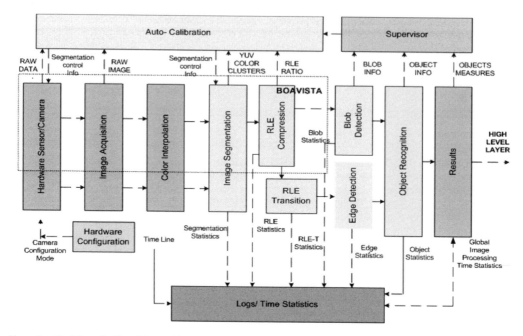

Figure 1. Real time pipeline vision architecture for mobile robotics.

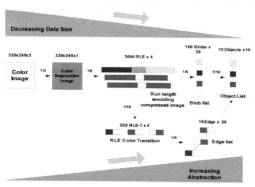

Figure 2. Data size in the system pipeline.

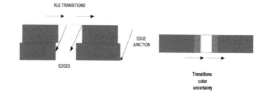

Figure 3. Example of a RLE colour transition with corresponding edge formation.

Furthermore, any other type of segmentation method is still valid and due to system modularity, can be implemented only changing the segmentation module. After that, the pixels with colour information will be compressed into run length encoding (RLE). This compression method is used due to is effectiveness in conserving the information and reducing data size.

One of the most positive and innovating aspects of our approach is extraction of structured edge information. After all data have been compressed into RLE format, a run is executed to search previously sanctioned colour transitions. When one of this transitions occurs, a transition RLE is created and stored. It will have the same information as a normal RLE: image position, colour and number of pixels. The method also deals

with color transactions uncertainty. These may occur due to interpolation issues, occlusion or illumination problem.

One of the key points of the method is that the number of RLE transitions are not directly attached to the number of image pixels, but are attached to the number of image RLE. So the computational cost of this method compared to other methods of edge finding in the same scenario (Hundelhansen 2003) is residual. In our system is about 1/8 of the computational cost of the RLE module and only stores 1/15 of the RLE data.

After all types of RLE have been processed, the RLE will be grouped into similar colour regions (BLOBS), and the transition RLE will be grouped into an edge list.

Once all the RLE are grouped into BLOBS further processing is done in the top architecture layer, the high level data layer (Object Recognition block).

The high level data layer is constituted by the modules that detect features for the robot localization

412

Figure 4.   Example of robot vision blob algorithm and edge finding algorithm.

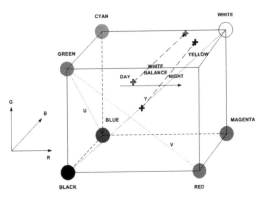

Figure 5.   RGB and YUV calibrations with white balance calibrations with illumination shifting.

and navigation sub-systems, thus closely related with the application. Lower level layers are relatively application independent and can be used in multiple autonomous scenarios. Image information at this stage already contains edge and blob identification, allowing particular object search.

In (Almeida 2003) the ball, robot markers, dark, blue and yellow objects are detected using a blob-based algorithm in conjunction with edge vectors. This provides information regarding ball centroid and bounding box for all possible clusters.

Besides the data processing modules, the system also has some auto-calibrations tools. The auto-calibration tools are used to help the vision system deal with environment changes. Some of the segmentation and color interpolation parameters are sent to a calibration module, that detects colour clusters shifting. In our system a white balance calibration is done to allow perception of the illumination changes, these changes are detected through the variations of the YUV and RGB cubes.

A statistical method based in the mean values of the image colour pixels is used. In case a variation, occurs information is sent to the acquisition module to change camera settings. An example of a tool can be found in (Browning 2005).

Furthermore the high level architecture modules are being continually improved. At this moment new features involving pattern recognition using graph vectors (Wilson 2005) are being implemented.

The connectivity graphs will allow structured image search, diminishing the computational cost of the object recognition modules. High level stereo is also currently under development, this module will not work at pixel level but will merge high level objects information, leading to a change in some of the higher architecture modules. One of its most relevant applications would be obstacle avoidance in Robocup scenario.

3   HARDWARE VISION SENSOR

The hardware vision sensor was developed to free system resources from processing the most heaviest data processing. In order to do so, a FPGA platform is used to process the image on the fly from the CMOS sensor.

This sensor allows the implementation of lower architecture levels at a fraction of power required in standard CPUs by taking advantage of the inherent parallel nature of image information and architecture pipeline structure.

The substantial power reduction constitutes a fundamental advantage to the use in autonomous systems. It is thus possible implement advanced sensing capabilities in low power systems and widen the range and scope of applications.

This image processing layer within the FPGA is divided into different modules, allowing access by the overall system to different kinds of data. As a result, the vision sensor can provide four types of data: raw data, RGB mode data, segmented data and RLE data.

In raw data mode, the sensor sends only raw image in Bayer pattern. This is used in a software complementary statistical module that provides the system self-calibration capabilities. In RGB mode, data is sent to allow image gain control. In segmented mode, the data is sent to search possible color clusters. In RLE mode, the image is sent in a compressed lossless form.

In figure 6 an information processing pipeline time diagram description is presented. A maximum latency of 500 μs is achieved from the initial pixel acquisition in the CMOS sensor to the processed data reception at the user level application.

This maximum latency includes processing and communication delays (USB bus). The interpolation, segmentation and RLE modules are similar to the software ones, for more information see (Lima 2004).

Moreover its also capable of having more than one CMOS sensor connected to the same platform. This capabilities can in future research be applied in stereo vision.

The communication link between the hardware sensor and the software application is done by USB connection. Due to the lack of resources within the FPGA device a communication protocol is implemented through a microcontroller device, whose protocol does not consume lots of resources.

4 OBJECT DETECTIONS FEATURES

The object recognition module is where the high level features are recognized and interpreted by the other robot sub-systems. In Robocup scenario these objects are the landmarks and the world targets, detected using a deterministic object descriptor.

These object descriptors uses a specific language for each of the objects. Taking the goals as an example. A probabilistic combination of edge and blob data is used to decrease the level of false detections. When this procedures are satisfied bearing and distance measures to the object are taken. A goal post bearing distribution measurements diagram is presented in figure 7, with the corresponding probability measures in figure 8. This results where taken at zero velocity.

The results show, that for measures to a distance up to 3 meters the bearing variations of the system is about 0.05 degrees.

Another application of the object recognition module is field line detection. The field line detection algorithm takes advantage of edge topology in the creation of edge vectors to allow field line recognition.

In our application in Robocup scenario, the lines are used to build the landmark figures necessary for the robot localization. Field lines are important because they provide reliable distances and bearing measures for the robot self-localization mechanism.

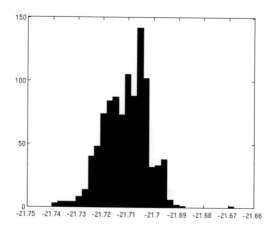

Figure 7.   Bearing measures between a robot an a goal post in degrees.

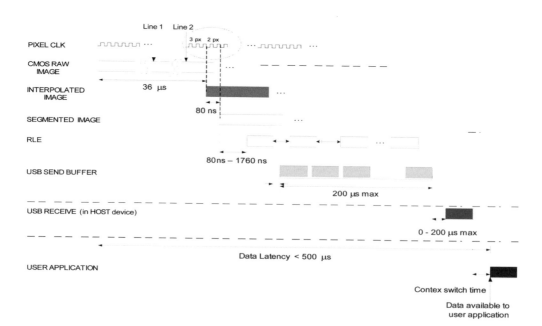

Figure 6.   Temporal diagram detailing hardware vision sensor modules.

5 RESULTS

In this chapter the overall system performance in terms of computational cost is detailed, and a comparison between two types of implementation is made.

One of the biggest advantages of the hardware implementation is the reduced latency when compared to the software implementation (see table 1). This is due to the software implementation, use of image compression for dispatching image data to the USB connection.

In Table 2. is shown the amount of average computational resources required to process a single frame. With the hardware vision system acquiring at higher resolution (640 × 480), the global image processing is still minor than with the software solution (320 × 240).

6 CONCLUSION

In this work we presented a real-time vision system for mobile robotics applications. The presented architecture allows latency reduction in sensor data reception. Very low power consumption solutions can be integrated. System organization allows hardware and software transparent solutions implementation.

A dedicated hardware vision sensor was developed to implement the more time consuming processing steps, taking advantage of image information parallelism. A high performance programmable logic device (FPGA) was used to process data from a CMOS sensor capable of VGA resolutions at 60 fps. This sensor can use different image sensors with a higher framerate, resolutions and High Dynamic Range Image capabilities for used in outdoors applications.

Power consumption reduction was significant, being possible now to segment and compress image for less than 1 W.

A new and innovative solution to edge identification based on the notion of RLE transition was implemented. This solution improves topological image information and allows edge detection in real-time and without total image processing and also proved to be a successful method in field line extraction.

Information coherence is maintained through different levels of abstraction in the architecture with pluggable module integration.

The vision architecture provided clear advantages to mobile robot navigation and advanced image perception systems, having also been applied to fire detections with Unmanned Autonomous Vehicles.

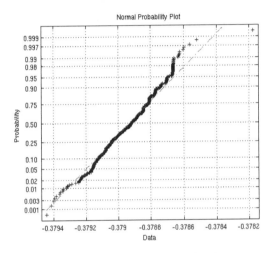

Figure 8. Normal probability distribution function.

Table 1. Latency comparison between the hardware and software implementations.

	Latency
Hardware vision system	<0,5 ms
Software (standard USB cam)	>50 ms

ACKNOWLEDGEMENTS

This research work has been partly supported by POSI and Fundação de Ciência e Tecnologia

Table 2. Comparison of the computational cost of the different system modules between a hardware sensor plus software implementation or only software solution.

	Image segmentation	RLE	RLE transition	Blobs	Edges	Objects	G. image processing
Hardware vision system (640 × 480)	–	–	0,75	3,5	3,0	1,2	8,45 ms
Software (320 × 240)	3,0	2,5	0,3	1,4	1,2	1,0	9,40 ms

under project BoaVista – "A Dedicated Vision System for Autonomous Mobile Robot Navigation", /POSI/ROBO/43914/2002.

REFERENCES

Almeida, J., Martins, A., Silva, E., Baptista, J., Patacho, A., Lima, L., Cerqueira, V., Almeida, C., Picas, R., 2003, "ISePorto Robotic Soccer Team for Robocup 2003", In: *RoboCup 2003 Int. Symposium*, Padua, Italy.

Bruce, J., Balch, T., Veloso, M., 2000: Fast and Inexpensive Color Image Segmentation for Interactive Robots. *IEEE/RSJ International Conf. On Intelligent Robots and Systems*, 3(2000) 2061–2066.

Browning, B., Veloso, M., 2005: Real Time Adaptive Color based Robot Vision. In: *Proceedings of IROS2005*.

Doi, M., Nakakita, M., Aoki, Y., Hashimoto, S., 2001: Real Time Vision System for autonomous mobile robotics. In: *IEEE International Workshop on Robot and Human Interaction Communication* 2001.

Hundelhansen, F., Rojas, R., 2003: Tracking regions and edges by shrinking and growing. In: *Computer Vision Winter Workshop* (CVWW03).

Heath, M., Sarkar, S., Sanocki, T., Bowyer, K., 1996: Comparison of edge detectors : In *IEEE Computer Vision and Pattern Recognition* 1996.

Hai, Z., Kui, Y., Jindong, L., 2003: A Fast and Robust Vision System for Autonomous Mobile Robots. In: *Proceedings of IEEE Intelligent Conference on Robotics, Intelligent Systems and Image Processing* 2003 China

Lima, L., Almeida, J., Martins, A., Silva, E., 2004: Development of a dedicated hardware vision system for mobile robot navigation. *robotica 2004* international conference.

Peiig, J., Skrikaew, A., Wilkes, M., Kawamura, K., Peters, A., 2000 : An active vision system for Mobile Robots In: *IEEE* (2000).

Rowe, A., Rosenberg, C., Nourbakhsh, I., 2002: A Low Cost Embedded Color Vision System. In: *Proceedings of IROS2002*.

Wilson, R., Hancock, R., Luo, B., 2005: Pattern Vectors from Algebraic Graph Theory. *IEEE transactions on Pattern Analysis and Machine Intelligence*, vol 27, no.7, July 2005.

Computational Modelling of Objects Represented in Images –
João Manuel R.S. Tavares & R.M. Natal Jorge (eds)
© 2007 Taylor & Francis Group, London, ISBN 978-0-415-43349-5

A gait analysis laboratory for rehabilitation of patients with musculoskeletal impairments

Daniela Sofia S. Sousa
INEGI – Instituto de Engenharia Mecânica e Gestão Industrial, Porto, Portugal, FEUP – Faculdade de Engenharia da Universidade do Porto, Porto, Portugal

João Manuel R.S. Tavares
INEGI, DEMEGI – Departamento de Engenharia Mecânica e Gestão Industrial, FEUP

Miguel Velhote Correia, Jorge G. Barbosa & Carolina Vila Chã
INEB – Instituto de Engenharia Biomédica, Porto, Portugal, DEEC – Departamento de Engenharia Electrotécnica e de Computadores, FEUP

Emília Mendes
CRPG – Centro de Reabilitação Profissional de Gaia, Arcozelo, Portugal

ABSTRACT: This paper describes the technology (software & hardware) used in the Gait Analysis Laboratory of CRPG – *Centro de Reabilitação Profissional de Gaia*. This laboratory is mainly intended to support CRPG with quantitative data for clinical rehabilitation of patients with musculoskeletal pathologies, namely in the lower limbs. Usually, for the gait analysis is considered data obtained from several acquisition sensors, such as: image cameras, electromyographic devices, pressure devices and force plates. This paper also presents briefly how quantitative motion analysis can be used for a clinical point of view and it presents the influence of image algorithms in gait rehabilitation.

1 INTRODUCTION

Clinical gait analysis is the systematic measurement and interpretation of the biomechanics parameters that characterize human locomotion and facilitates the identification of motion abnormalities, so that proper rehabilitation procedures can be applied (Baker 2006, Davis 1997, Davis 1988). It is also considered has a tool to predict the results of clinical interventions (Baker 2006).

According to several authors, the predominant methods for clinical gait analysis involve capturing the movement of human body segments with multiple image devices, monitoring the interaction with the ground using force plates and pressure devices, and recording muscle activity using electromyographic (EMG) devices during several gait cycles (Baker 2006, Chau 2001, Gavrila 1999, Davis 1997, Davis 1988). These data are used to analyse the gait pattern of a patient through the calculation of joint kinematics and kinetics in absolute 3D coordinates (global referential on laboratory environment) and relative

3D coordinates (local referential on patient's body), through the knowledge of the ground reaction forces components and distribution of pressure over the plantar surface and also through the evaluation of muscle activation patterns.

CRPG provides rehabilitation services (i.e. physiotherapy, evaluation of the patient's functional capability, adaptations on the job place, design, fit and test of prostheses and orthoses, etc.) for physical and psychological recovery (i.e. musculoskeletal deficiencies, neuromuscular disorders, hearing and vision impairments, etc.) and to promote a lifestyle of maximum independence.

The main goal of CRPG with the gait laboratory is to assist clinicians (prosthetists, physical therapists, etc.) in the medical diagnostic and treatment of patients with musculoskeletal problems, namely in lower limbs. Examples of the tasks are: the design and prescription of orthoses and prostheses, the monitoring of orthoses and prostheses effectiveness, the understanding of the abnormalities present, in order to formulate adequate treatment, etc.

Image processing and analysis algorithms can be found in software applications used for tracking and recovering 2D/3D trajectory of markers placed on the patient (Pinho et al. 2005a, b, Figueroa et al. 2003). Image sensors can also be used for capturing and interpreting the plantar pressure (optical pedobarograph, dynamic pedobarography). Adequate software applications for motion and deformation analysis are therefore needed (Tavares & Bastos 2005, Tavares et al. 2000, Urry 1999).

This paper will present the options made to mount the gait analysis laboratory of the CRPG in terms of devices and software considered. Special attention will be made on the tracking and analyse of features along image sequences. It will also be discussed the usefulness of quantitative gait analysis for CRPG rehabilitations procedures.

2 CRPG'S GAIT LABORATORY

In CRPG, gait analysis is accomplished by computerized measurements of kinematics, kinetics, muscular activity and foot pressure.

Kinematics is the measurement of movement without reference to the forces involved. Kinematics systems are used to calculate linear and angular displacements, velocities and accelerations (Winter 2005). Cadence, stride length and speed of the gait may also be provided by the kinematic system (Whittle 1996).

Four standard camcorders (PAL format, sampling frequency $= 25$ Hz, resolution $= 768 \times 576$ pixels) are used to track the 3D trajectories of reflective markers placed on patient skin. Simi Motion software used is able to provide an effective sampling rate of 50 Hz. The cameras use visible light close to the lens, so that reflective markers are easily identifiable. If at least two cameras capture a marker and the position and orientation of these cameras are known (by calibration procedures), then it is possible to calculate the 3D position of the marker (Cappozzo et al. 2005).

Data is collected in quiet standing and during the walk. The images are processed by Simi Motion to present graphs of the kinematics evolved.

Markers are tracked automatically or manually for each camera frame. 3D marker coordinates are computed from the markers positions in 2D images. The referential is defined during the calibration; x-axis has the movement direction, x-axis and y-axis define the plane of the movement and z-axis is points into the ceiling (right handed referential).

The locations of at least three non-collinear markers (Figure 1) on each body segment collected during quiet standing are used to calculate joints locations and center of mass for each body segment. Further from the markers positions a segmental coordinate system is

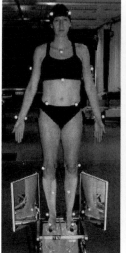

Figure 1. Simi marker set (adapted from Simi Motion Manual).

defined for each segment which origin is in the centre of mass of the respective segment. The location of the center of mass for each body segment is determined with respect to the joint centers based on regressions equations (NASA 1978). The x-axis points from dorsal to frontal, the y-axis is lateral to medial and medial to lateral for the right limbs and left limbs respectively and the z-axis direction points in distal-proximal direction. Joint angles are computed in Simi Motion software from the relative orientations of the segmental coordinates systems using Cardan angles in a xyz sequence (Shabana 2001). Finally, linear or angular velocities or accelerations require the differentiation of position data.

The output data curves can either be described relative to the body local coordinate system (segmental coordinates systems) or to the global coordinate system (defined during the calibration operation). In the case of local coordinate system the time curves are often easier to interpret.

Kinetics describes the forces (internal and external) acting on a moving body. Internal forces come from muscle activity, ligaments or from friction in the muscles and joints. External forces come from the ground or from external loads (e.g. wind resistance, gravitational attraction, etc.) (Winter 2005).

Several kinetic analyses can be done: the moments of force produced by muscles crossing a joint, the mechanical power flowing to or from those same muscles, and the energy changes of the body that result from this power flow (Winter 2005).

Internal forces and internal moments are calculated by Simi Motion using inverse dynamics based on kinematic and force platform data, mass and moment of inertia of each limb segment and the location of the center of mass of each segment. The moment of inertia of body segments are also estimated using NASA regressions equations (NASA 1978). Then, using the intersegmental load actions (joint moments and joint forces), the joint angular and translational velocities, it is computed the energy and power transmitted from one body segment to another.

The resultant ground reaction force is measured using one force plate (Kistler Instruments, Winterthur, Switzerland) at a sampling frequency of 1000 Hz. Force plate data is not acquired simultaneously with 3D marker coordinates, so manual synchronization is needed between these data.

Plantar pressure consists in distribution of the force over the sole of the foot (Rosenbaum & Becker 1997). Plantar pressure is essential in determining what is happening to individual parts of the foot (Roy 1988). For instance, plantar pressure data has been recognized an important element in the assessment of diabetes patients, it also can be used to alter footwear, foot orthoses, exercise programs, etc. (Orlin & McPoil 2000, Rosenbaum & Becker 1997).

A Footscan 2D (RSscan Internacional, Olen, Belgium) pressure platform with 1×0.4 m (length $\times$ width) was mounted on top of the force plate. This platform consists of two capacitive sensors in each cm^2, these sensors are calibrated by the manufacturer. The highest sampling frequency is 500 Hz, pressure range is $0.27-127$ N/cm^2. Footscan 2D is a stand-alone pressure system, since there is not synchronization between pressure measurements and the other motion data (EMG, kinetic/kinematic, force plate).

During testing, several precautions are taken in order to minimize the influence of several anthropometric and kinematic factors: patients are asked to walk at their normal rhythm, more than one contact from each foot are recorded and averaged in every experiment, shoes shape is controlled, etc. Footscan Gait Scientific (RSscan Internacional, Olen, Belgium) is used for analyses and visualization of collected data. Several types of information can be presented to the clinic: force and pressure versus time, force and pressure/time integrals, contact areas, displacement of center of pressure, etc. These parameters can be obtained for the whole plantar surface or certain anatomically defined regions, for one step or several steps and for separated phases of the gait cycle. The graphics can be in 2D and 3D representations. RSscan divides the foot in the ten anatomical areas: medial and lateral zone under the rear foot, mid foot, five metatarsals, hallux and the other toes. The value of the parameters obtained from pressure distribution depends on the clinical aims (Rosenbaum & Becker 1997).

The purpose of electromyography in clinical gait analysis is to define the muscle activity that controls joint motion (Merletti et al. 2004). Electromyography records the electrical signals resulting from the chemical activation of muscle fibers by the central nervous system.

The measurement of the dynamic EMG data is assured by a Biovision EMG system (Biovision, Wehrheim, Germany). Each analog channel have a single differential pre-amplifier module (bandwidth = 10–500 Hz, input impedance = 1200 GΩ, CMRR = 120 dB, input noise = 1 μV and an adjustable gain of 1000 or 2500 or 5000), which are connected to self-adhesive Ag/AgCl snap electrodes with a circular shape (diameter = 10 mm).

The skin is prepared and the surface electrodes are placed over muscles of interest for gait analysis, according with recommendations of SENIAM project (Hermens et al. 1999), movement of electrodes and cables should be avoided. The data is collected at a sampling frequency of 1000 Hz and stored on a computer for further analyses. The EMG data is collected synchronously with force plate data.

The raw EMG signal is presented by Simi Motion software package (Simi Reality Motion Systems, Unterschleissheim, Germany) and it can be processed through several Simi Motion options: integration, rectification, filtration, etc. There is a general agreement that processing the raw EMG signal is useful (e.g. for inter-subject and intra-subject comparison, to reduce noise effects, etc.) (Sutherland 2001). Usually, the raw data may be presented as such or filtered, smoothed, rectified, integrated, ensemble averaged and/or scaled to other references (e.g. maximum amplitude while walking) (Sutherland 2001, Davis 1997). After processing, this information can be observed in Simi Motion through EMG plots in time and in frequency. The EMG data typically used for clinical interpretation includes timing of muscle activity within the gait cycle (e.g. onset and cessation of each muscle's activity relative to the limb motion, time of peak effort) and general indication of the contraction intensity (e.g. muscle more or less active, muscle on or off) (Perry 1998, Whittle 1996).

3 IMAGE PROCESSING AND ANALYSIS ALGORITHMS IN CLINICAL GAIT ANALYSIS

Nowadays, typically the core of clinical gait analysis is image processing and analysis operations. Anatomical landmarks are placed on patients and motion is capture with image sensors. Image analysis techniques allow the calculation of kinematic parameters and support the description of the forces acting on the body.

According with (Baker 2006, Sutherland 2002) commercial motion tracking system (Simi Motion, Peak Motus, etc.) have accomplished a satisfying performance in markers identification and tracking. For instance, in a controlled environment such as CRPG laboratory, the main problems on automatic tracking are occlusion problems. Normally, automatic tracking has to be supervision and when a point is incorrectly defined, it has to be manually discarded and redefined. At the beginning, all points must be identified in the first frame. When the recognition of the point lies below the confidence area, the automatic tracking is aborted. It is possible to adjust the qualities of the points to a more successful tracking. This problem can be minimized increasing the number of cameras, adjusting cameras position, etc. However, it would be interesting eliminating the use of markers and decreasing the number of cameras in gait analysis (Sutherland 2002). This solution could be beneficial in terms of cost and in terms of accuracy and precision in markers placement. It would solve the difficulty of placing markers in anatomical regions close to each others (e.g. foot). It would also permit the center of mass calculation of each body segment in a repeatable way. The use of a markerless "marker set" would also eliminate soft tissue movement errors. The soft tissue movement errors are a consequence of skin movements, in other words, marker movements in relation to the skeleton. At least, with a markerless solution it would provide a reference points for the analyses of segment movements.

It has been done some work on markerless optical methods. For instance, by tracing the silhouette of subject it is possible to obtain the 3D contours of that subject and the next step is determination of the segmental coordinate systems (Baker 2006).

The 3D calibration process of currently motion tracking systems is fast and easy (Baker 2006). For example, in Simi Motion there is a need to record one frame with the calibration object for each camera, then the coordinates of each point must be entered (at least 8 points), after that the various points have to be manually identified in each frame and finally the software checks the calibration data.

In (Sisto 1998) is referred the problem of error magnification when velocity and accelerations are calculated from derivations of 3D markers coordinates. A possible solution would be not taking in account markers coordinates to obtain these data.

Optical pedobarographic platforms are other measuring devices which can be used in gait analysis. The optical pedobarograph consists on an elastic surface on top of a glass plate. The glass plate is illuminated from the sides and when the patients walks the critical angle of the glass plate increases, consequently, light passes through this surface. An image sensor captures the image obtained bellow the glass plate, light intensity is proportional to the applied pressures (Urry 1999). Image processing software converts the image information into foot pressures at various stages of the gait (Tavares & Bastos 2005, Tavares et al. 2000).

Although in gait analysis, this kind of technology does not replace pressure plates made of small sensor's matrix (e.g. Footscan platform) because technical specifications and price do not compensate the change.

4 QUANTITATIVE GAIT ANALYSIS IN REHABILITATION

Locomotion is a very complex task, for which contribute the coordinated efforts of sensorial, muscle and skeletal systems. It results form a complicated process involving the brain, spinal cord, peripheral nerves, muscles, bones and joints which makes its assessment a very difficult task (Whittle 2003). Locomotion or alternated bipedal walking is a basic, key essential function as it allows humans to perform several other tasks. When this key function is altered, it impacts negatively all aspects of the daily life.

Usually rehabilitation processes are very much focused on function improvement and training, with the aim of restoring full social and economic participation. In locomotion and mobility it is important that all aspects are very well known and documented, in terms of the relevant parameters that need to be addressed, to help the effective rehabilitation planning and monitoring.

CRPG is a multidisciplinary rehabilitation centre, offering both physical/functional and vocational rehabilitation services, technical aids production and adaptation. The focus on the client needs and final results leads to a client centered planning and monitoring of all the aspects involved in rehabilitation plans. When providing services, the locomotion and mobility functions are of extreme importance and therefore need to be addressed thoroughly.

Gait analysis can be thought as the systematic study of human walking (Whittle 2003). The Movement and Function Analysis Laboratory at CRPG was though as a support service, providing both in house and external rehabilitation teams, a tool helping on the diagnostic, planning and monitoring of rehabilitation plans. This centre attends clients with different clinical and functional situations, causing different functional abnormalities. The use of quantitative data as a complementary tool of a more holistic approach to the clients needs is expected to allow a more efficient service and finally optimize the rehabilitation results.

Gait analysis involving the use of the methods described, is being implemented for:

(a) The assessment of lower limb amputee gait, using a static alignment technique based on the visualization of the ground reaction force in relation to the center of the knee joint (Schmalz et al. 2002).
(b) The assessment of gait parameters on persons with cerebral palsy.

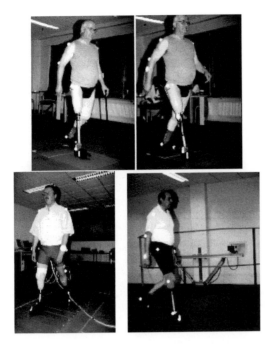

Figure 2. Clinical gait analysis in amputees (images from Otto Bock Health Care, Duderstadt, Germany).

(c) The assessment of foot pressure and insole production using a pedobarographic platform, and the evaluation of gait parameters of persons using lower limb orthosis (Figure 2).

5 CONCLUSIONS AND FURTHER WORK

In this paper was briefly described the current status of the Gait Analysis Laboratory for Rehabilitation of Patients with Musculoskeletal Impairments of CRPG.

All the software and hardware systems are already selected and fully operational; however, some work has to been done; mainly, on developing or adapting existing measurement protocols and guidelines for clinical interpretation, complete understand of quantitative data reliability and user-friendly reports with a complete description of patient gait data.

In terms of further challenges for gait analysis professionals it would be interesting a markerless motion tracking system or a marker active motion tracking system that avoids occlusion and reduces post-processing complexity. This last solution should not need cables to connect markers to a PC or a portable device carried by the patient, should avoid environmental noise, it should have reduced dimensions, sufficient sampling rate and the presence of active markers should not be uncomfortable to the patient.

Intramuscular pressure technology can also be used to obtain more muscle activity information.

Developments on models of the human motion/body could be used to predict the output of several interventions and help to evaluate the different clinical problems.

More standards of terminology, collection, analysis and visualization techniques and interpretations guidelines would improve clinical gait analysis efficiency. Databases with normal gait patterns and typically gait patterns of various motion abnormalities are needed for comparison purposes.

Accuracy of gait data could be improved with the elimination of soft tissue errors, joint centers calculations errors, joints angle calculation errors, errors introduced by the marker set, etc.

Clinical gait analysis is in an evolutionary stage, however in CRPG it is expected that with the present development state it would be an established tool in motion rehabilitation procedure.

ACKNOWLEDGMENTS

This work was done in the scope of the projects: "ACTIDEF – Avaliação Computacional e Tecnológica Integrada do Desempenho e Funcionalidade dos Cidadãos com Incapacidades Músculo-esqueléticas" (reference 242/4.2/C/REG, financially supported by POS-Conhecimento – *Programa Operacional Sociedade do Conhecimento*) and "Segmentation, Tracking and Motion Analysis of Deformable (2D/3D) Objects using Physical Principles" (reference POSC/EEA-SRI/55386/2004, financially supported by FCT – *Fundação para a Ciência e a Tecnologia*).

REFERENCES

Baker, R. 2006. Gait analysis methods in rehabilitation. *Journal of Neuro Engineering and Rehabilitation* 3(1): 4.

Cappozzo, A. & Della Croce, U. et al. 2005. Human movement analysis using stereophotogrammetry. Part I: theoretical background. Gait & Posture 21: 186–196.

Chau, T. 2001. A review of analytical techniques for gait data. Part 1: fuzzy, statistical and fractal methods. *Gait & Posture* 13(1): 49–66.

Davis, R. B. 1997. Reflections on clinical gait analysis. *Journal of Electromyography and Kinesiology* 7(4): 251–257.

Davis, R. B., III. 1988. Clinical gait analysis. *Engineering in Medicine and Biology Magazine, IEEE* 7(3): 35–40.

Figueroa, P. J. & Leite, N. J. et al. 2003. A flexible software for tracking of markers used in human motion analysis. *Computer Methods and Programs in Biomedicine* 72: 155–165.

Gavrila, D. 1999. The visual analysis of human movement: a survey. *Computer Vision and Image Understanding*, 73(1): 82–98.

Merletti, R. & Parker, P. 2004. *Electromyography: Physiology, engineering and noninvasive applications*. Wiley-IEEE Press.

NASA. 1978. *NASA Reference Publication*. Springfield: NASA.

Orlin, M. & McPoil, T. 2000. Plantar pressure assessment. *Physical therapy* 80(4): 399–409.

Perry, J. 1998. The contribution of dynamic electromyography to gait analysis. In J. DeLisa (ed.) *Monograph 002: Gait analysis in the science of rehabilitation*. 1998. Washington, DC.

Pinho, R. & Tavares, J. et al. 2005a. A Movement Tracking Management Model with Kalman Filtering, Global Optimization Techniques and Mahalanobis Distance. *ICCMSE 2005 – International Conference of Computational Methods in Sciences and Engineering; Loutraki, 21–26 October 2005*. Korinthos: Greece.

Pinho, R. & Tavares, J. et al. 2005b. Human Movement Tracking and Analysis with Kalman Filtering and Global Optimization Techniques. *ICCB2005 – II International Conference on Computational Bioengineering; Instituto Superior Técnico, Lisboa, 14–16 September 2005*. Lisboa: Portugal.

Rosenbaum, D. & Becker, H. 1997. Plantar pressure distribution measurements. Technical background and clinical applications. *Foot and Ankle Surgery* 3(1): 1–14.

Roy, K. 1988. Force, pressure and motion measurements in the foot: current concepts. *Clinics in Podiatric Medicine and Surgery* 5: 491–508.

Schmalz, T. & Blumentrit, S. et al. 2002. Energy expenditure and biomechanical characteristics of lower limb amputee gait: The influence of prosthetic alignment and different prosthetic components. *Gait & Posture* 16: 255–263.

Shabana, A. 2001. *Computational Dynamics*. NY: John Wiley.

Sisto, S. 1998. An overview of the value of information resulting from instrumented gait analysis for the physical therapist. In J. DeLisa (ed.) *Monograph 002: Gait analysis in the science of rehabilitation*. 1998. Washington, DC.

Sutherland, D. 2001. The evolution of clinical gait analysis part l: kinesiological EMG. *Gait & Posture* 14(1): 61–70.

Sutherland, D. 2002. The evolution of clinical gait analysis: Part II Kinematics. *Gait & Posture* 16(2): 159–179.

Tavares, J. & Bastos, L. 2005. Improvement of Modal Matching Image Objects in Dynamic Pedobarography using Optimization Techniques. *Electronic Letters on Computer Vision and Image Analysis, special issue on Articuled Motion & Deformable Objects* 5: 1–20.

Tavares, J. & Barbosa, J. et al. 2000. Matching Image Objects in Dynamic Pedobarography. *RecPad 2000 – 11th Portuguese Conference on Pattern Recognition; Porto, 11–12 Maio 2000*. Porto, Portugal.

Urry, S. 1999. Plantar pressure-measurement sensors. *Measurement Science and Technology* 10: 16–32.

Whittle, M. 1996. Clinical gait analysis: a review. *Human Movement Science* 15: 369–387.

Whittle, M. 2003. *Gait analysis an introduction*. Oxford, Boston: Butterworth-Heinemann.

Winter, D. 2005. *Biomechanics and motor control of human movement*. Hoboken: John Wiley & Sons.

Computational Modelling of Objects Represented in Images –
João Manuel R.S. Tavares & R.M. Natal Jorge (eds)
© 2007 Taylor & Francis Group, London, ISBN 978-0-415-43349-5

Visual assessment of image representations on mobile phones

W. Tang
School of Computing, University of Teesside, UK

A. Hurlbert
School of Biology and Psychology, University of Newcastle upon Tyne, UK

T.R. Wan
School of Informatics, University of Bradford, UK

ABSTRACT: Mobile technology industries are making personal mobile devices more environment aware and adaptive to individuals needs. This paper presents the results of assessing the quality of image representations on mobile phones and the problems associated with the small screen displays. Psychophysics experiments are designed to investigate the visual performance of people when they are reading images on mobile phone screens. A WAP (Wireless Application Protocol) application system is developed enabling the visual search experiments to be carried out on a recent Nokia 6230i mobile phone. A large number of participants aged between 17 to 83 years old are involved in the experiments, of whose test results are divided into the normal age group and the old age group for a comparative study. The results show the difference in performance with visual search tasks between the older adults and the young adults.

1 INTRODUCTION

1.1 Background

The estimated number of mobile phones in the world is three times higher than that of personal computers and today's most sophisticated phones have the processing power of a mid range personal computers in 1990's. World's leading telecom companies are undergoing extensive developments in new technologies for high resolution large displays (Microdisplay systems) and new types of networks for high speed multimedia data transmissions (GPRS on UMTS) [SIEMENS]. Many people believe that in the near future, with the increasing power and advances in network service technologies, mobile phones could be used as an alternative to desk top computers. Emerging digital technologies also enable PDAs and mobile phones to cater for a wide spectrum of users with various functionalities ranging from basic functions for making calls and text messaging to playing games and accessing the Internet. However, little research has addressed the visual communication aspect of different user groups, especially the special user groups such as disabled and older people.

A research team has conducted a set of experiments on visibility and characteristics of mobile phones for older people and concluded that screen displays in short vertical length of characters and low contrast ratio on mobile phones would affect the older people's reading performance [Omori].

Because visual communication is a prevailing part of every day life, new and effective ways of communicating with information visually would likely improve the quality of using mobile phones.

While young and trady users enjoy the advanced features on new types of mobile phones, many older people prefer their current phones to new models. As a natural progression, people's vision and motor skills start to decline when they get old. Assuming that two main problems associated with older people using mobile phones are the small screen displays and the complexity of functions of interface menus. Keys on mobile phones are too tiny and discrete with each key having various different functions to navigate and search for information. This paper presents our study on evaluations of visual qualities of the information displayed on the small mobile phone screens by assessing the visual performance of two groups of people with respect of reading the displays on mobile phones.

1.2 Methodology

Studies have shown that there are significant differences in vision characteristics of aged people compared with those aged below 50 years old. The major vision decline appears in most people who are over 50 years old due to changes in the optics of the eyes. The effects of physiological changes in aged eyes, such as lens become yellowier, could make the discrimination of blue colour more difficult and the changes in eye make less light entering the eyes and reaching the photoreceptors. Hence the lens and other optical media become opaque [HTTP:]. When viewing images, the same image seen by people over sixty years old are considerably darker than the visual effect perceived by people aged 20 years old. According to physiological studies on age-related changes in human eyes, we hypothesise that the vision decline may have some effects on people to communicate with digital information displayed on small mobile phone screens. However, researchers in the fields of neuropsychology and brain research suggest that, although vision functions start to decline as people getting older, brains become to adapt the changes cognitively in order to deal with the visual decline situation [Willis]. Therefore, it is not certain that wither or not such adaptation would assist older people to interpret the visual information on mobile phone screens hence as a result of such adaptive behaviour the visual performance of older people with mobile phone screens may be maintained at the same level as the young adults.

We take the vision decline and vision adaptation into account. In order to analysis and compare the visual performance of young adults and older people with the use of mobile phones, a set of experimental tests are carried out by developing WAP applications on a selected mobile phone using the latest mobile Java technologies. The design of visual search tasks in the experiments is based on the research work from scientific studies in vision research and each of these visual search tasks is intended to investigate correlations between vision characteristics of aged vision with the visual ability of older people in reference to perceiving the information displayed on mobile phone screens. The results of the experiments offer insights in visual performance of these two groups of people with the use of mobile phones, which in turn will form the basis for designing and developing new types of display functions on mobile devices that are adaptive and suitable for different people.

2 DESIGN OF VISUAL SEARCH EXPERIMENTS

Based on the review of the research work published in vision research [Chen, Tales], we devise a set of experiments to test the visual performance of two groups of people, i.e. young adult control group and older adult group.

The experiments are mainly concentrated on assessing the visual search task performance. Visual search is a common task used to determine attention-related

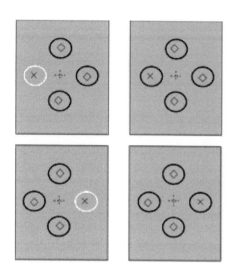

Figure 1.   Salient and non-salient visual search tasks.

performance in people in neuropsychology research. In such a task, participants are asked to detect or identify a target element, such as the white circle with a cross inside as shown in Figure 1, the target element is mixed with a set of non-target elements, such as the black circles with a diamond shape inside each of these circles. The level of difficulty of the visual search tasks varies depending on the design. Some tasks require more cognitive performance than others [(12)8]. Because we are interested in finding out the effects of visual declines in older adults on their visual ability of reading small mobile phone screens, the experiments are designed as such that some of the search tasks are of simple feature search with less interference from the non-target elements, whilst other tasks are more difficult with the target and non-target elements sharing similar features. In this way, the level of difficulty of the visual search task is associated with different level of visual abilities in the search performance.

2.1   Arrangements of visual search elements on screen

Taking into account of the display features seen on current mobile phones, we use the similar approach in the design of the visual search tasks as described in [Tales]. The performance on visual search tasks is not only related to the number of search elements, but also to the arrangement of the elements in the visual field i.e. the display. We designed eleven tasks and nine of the tasks are kept the same judgments as shown in Figure 1. We use the 'clock face' configuration and in each of these eleven tasks, there are four circles positioned around the centre cross that is the fixation at the centre of the mobile phone screen.

The elements arranged in a circle centred at a fixed point so that all the elements are equidistance from the centre gaze. This configuration allows the elements to have a constant distance from one another. Using four elements to construct the visual display is a general design representation of menus that is often seen on web design and mobile phone menus. Having four elements reduces the effects of the common confound between the size and density of the elements duo to the small size of the display screen on mobile phones. Each of these elements is positioned from the intersection of the vertical and horizontal lines forming the centre cross and equally spaced out from the centre. All the target, non-target elements and the centre cross are appeared on the mobile phone screen simultaneously. The image remains on screen until a response is entered on the keys on the mobile phone.

2.2 Design of stimuli

In each search task, the visual field configuration contains only one target circle with a cross drawn inside and three non-target circles with a diamond shape drawn inside each of these distractors. In each task participants are asked to identify whether the position of the target circle on the left or on the right of the centre cross by pressing the left or the right key on a mobile phone. The last two tasks are designed with one target element as a red dot embedded with various numbers of distractors as white dots. Participants are asked to locate the position of the target element by indicating whether the target is on the left side of right side of the centre cross. All the target and non-target elements are drawn as the same size with the same luminance value.

Figure 1 illustrates the design of a salient and non-salient search task. In the top and bottom left images in Figure 1, the target element is drawn in white and the non-target elements are drawn in black colour, whilst all the target and non-target elements in the top and bottom right images in Figure 1 are black. These two pairs of images refer to the salient feature search. In each trail, the reaction times (RTs) of the participant are recorded on these two conditions.

The purpose of the test is to assess the visual performances of the participant with the salient and non-salient conditions in order to provide evidence on whether or not the salient conduction would assist visual search. If it dose, we expect the RTs in both young and older age groups to decrease.

By alternating the salient settings between the target and non-target elements as shown Figure 2, we carry out further tests on the visual performance of the participants with different salient and non-salient conditions.

The colour salient search tasks are also designed to assess whether or not there exists difference in visual search performance between young and older adults

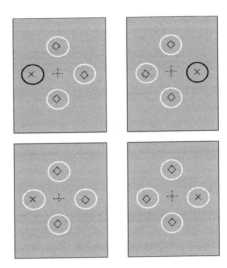

Figure 2.   Alternating salient and non-salient conditions.

in persevering colours on mobile screen. The visual search performance of the two groups is evaluated in terms of RTs in red colour salient visual search and in blue colour salient visual search in representing the long and short wavelength lights respectively.

2.3 Visual search performance as the function of the number of distractors

In this set of visual search tasks, on the mobile phone screen display, the target element (a red dot) is embedded within the non-target element served as distractors (random scatted white dots). The participants are required to track the location of the target red dot by indicating whether it is on the left or right side of the centre cross. The task has two levels of difficulty as shown in Figure 3, the top two images have many more distractors than those in the bottom two images that have a few of non-target elements. The red target element is randomly placed on either left or right of the centre fixation cross at both levels of the task.

By increasing the level of visual search difficulty, the visual search task requires an increased degree of cognitive activities. Therefore such visual search task may be more associated with the increased brain activation in older adults that reflects the acquisition of compensatory cognitive and behavioural strategies. If the factor of the acquisition of compensatory cognitive is presented in participants in the older group, we expect to see similar tendency in the increase of RTs in both groups as the level of the task difficulty increasing. By evaluating the visual search performance as the function of the number of distractors and the function of the level of difficulty, tt is possible to compare the performance of the participants of the both groups in these tasks.

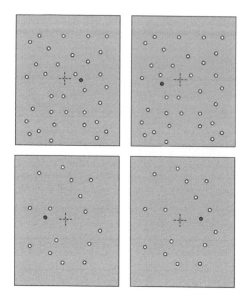

Figure 3. Visual search tasks as the function of number of distractors.

3 WAP SYSTEM DEVELOPMENT

3.1 *WAP communication system with mobile phones*

WAP (Wireless Application Protocol) is an independent operating system communication protocol optimized for handheld devices. WAP is developed and supported by the device industries including Nokia, Ericsson and Phone.com. Mobile phones as WAP enabled devices communicate with wireless network servers via a browser using WAP. In the similar way that PCs communicate with web servers, a WAP enabled handheld device such as a mobile phone can communicate with web server using Wireless Markup language (WML) that has similar structure as HTML used to communicate between web servers and desktop PC browsers. The wireless communication process is illustrated in Figure 4.

In our system, a HTTP server is setup to host our mobile applications. This web server supports for Java files that are required for our application and provides PHP and MySQL database support, which is essential for the project to manage testing data. A FTP client is used to send the Java application programs to the web server. The mobile phone used for our experiments can then type the URL where is the programs is located in the server to download and install the WAP application on to the phone. The mobile phone accesses the both GPRS and GSM data to setup a wireless mobile connection to the Internet. A MySQL database system is developed and a single MySQL database table is used. The PHP scripts are embedded in WML that

Figure 4. Wireless application process on handheld devices.

Figure 5. Java testing program running on PC emulator (left) and Two versions of MDIP applications on the Nokia 6230i testing mobile phone (middle: PNG images, right: vector graphics)

allow browsing or modifying the entries of the table to managing the testing data.

3.2 *Development of MIDP application on nokia 6230i mobile phone*

The implementation of our mobile application is based the latest mobile technology called MIDP (Mobile Information Device Profile). MIDP applications are small Java programs specifically designed for portable devices such as mobile phones and PDAs that operate on limited resource platforms. Java applications for mobile devices are compiled using a special development kit (SDK) called J2ME, which is designed for small and wireless devices. For our project, a Nokia 6230i mobile phone is chosen as the testing platform that is a very recent model allowing the J2SE SDK 1.3 and J2ME Wireless Toolkit 1.0.4 to work together.

During the development phase, the application programs can be tested on the emulator run on a desktop PC before it is uploaded to the web serve. The image in Figure 5 shows a test application program run on the emulator. A Nokia 6230i model mobile phone is selected for the tests in the project. Among various features of different mobile phones available in today's market, Nokia 6230i offers active TFT colour display which is noticeably brighter than the display models on other phones such as Motorola V3. It supports up to 65,536 colours with 208 × 206 screen pixels, which is ideal for ensuring the quality of the images displayed

426

in the tests. The display features is particular important for the purposes of the project. The large memory storage of the phone allows us to download the compiled different versions of MIDP Java applications from the server on to the mobile phone for testing and store the testing data. We have developed two versions of MDIP applications. One is the testing images drawn as vector graphics within the mobile phone and another version is the MDIP program loads PNG images and displays the images on the phone screen. In comparison, the quality of the displayed images with the vector graphics is higher than that of PNG images.

4 TESTING RESULTS AND EVALUATION

4.1 Participants and procedures

Fifty three participants who are aged between 17 to 83 years old are involved in the experiments. The results of the participants are divided into two age groups, the control age group and the old age group for a comparative study. There are 24 participants in the control group who are under 50 years old and the youngest in this group is 17 years old. Participants in the old group are aged above 51 years old, the oldest participant in this group is 83. All participants have normal or corrected to normal vision.

In each trail, eleven tasks as described in the chapter 2 were presented and for each task the participant was asked to identify the location of the target element by pressing the left or right key on key menu of the mobile phone. Before proceeding the trail, each participant was required to complete a short tick box questionnaire containing general questions about gender, age group, and whether or not the participant require glasses for correct vision. There are also questions on the general usage with mobile phones in his/her daily life. The participants were introduced to perform the task as quickly but as accurately as possible. Therefore, in this case, the reaction time (RT) of the participants for each task was captured the error occurred during the test is also captured into a PHP enabled WML page that is late transmitted to the web server after the trail. The date and time of each trial were also recorded into the database. At the beginning of each trail, the Java application program displays a brief instruction text, and participant is required to read through the introduction information and to get prepared to start the test. During the test, only the navigation menu and the centre select key are used to response to each task. At the end of each trail session, a message is displayed asking whether or not the results of the test should be transmitted to the serve to be saved into the database. A request for transmitting the test data can be issued using the top right key on the mobile phone. When an error data occurs, the test can be discarded and a new test can be started.

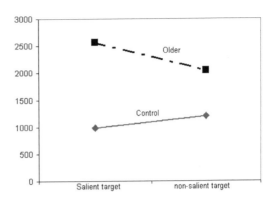

Figure 6. Performance of salient vs. non-salient visual search tasks.

4.2 Data analysis

The reaction time (RT) for each task performed by a participant was reordered in milliseconds and RTs for all the tasks were recorded for each participant. All the RTs for correct trials are grouped according to the visual search task conditions. Mean reaction times of the participants RTs are plotted for the two groups.

4.2.1 Performance of salient and non-salient search task

Figure 6 shows the mean RTs of the two groups of participants for the salient and non-salient search task. In this experiment the target element and non-target elements are switched on and off between the salient and non-salient conditions as shown in Figures 1 and 2.

The mean reaction times in the old group for both salient target and non-salient target conditions are increased by 161.6% and 70% respectively compared with the control group. The reaction time for non-salient target is increased for the control group compared with the RTs for salient target visual search task. It is interesting to see that the RTs for non-salient target search is decreased in the older group, contrary to the common assumption that the visual search performance of the older group would increase with salient target search. One possible explanation may be due to the sequence of the issued tasks. The image of the salient target was the first image in the sequence of the eleven images in each trail. The participants may require more time to get familiar with the test process. Further tests and research are required to make a conclusive conclusion, given that the results of young adults converge to the common assumptions. However, the performance of the older participants is considerably slower than that of the young adults.

In the colour salient visual search task, four tasks were issued in which the colour of the target element and non-target elements are switched between the red and black colours or between the blue and black

427

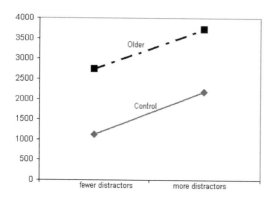

Figure 7. Performance of fewer vs. more distractors in visual search task.

colours. When the target was drawn in red colour combined with three black non-target elements or vice vas, the RTs in the control group suggest that the red salient target was slightly easier to be identified than the black target with three red non-target elements. On the contrary, the RTs for the same tasks in the old group show there is slightly reduced reaction time in the task with the black target and three red non-target elements. The overall performance difference in the two conditions for both groups is not significant. Again the task performance in the old group is slower than that in the control group.

4.2.2 RTs as the function of the number of distractors

This experiment is designed to test the visual search performance of the two groups of participants with respect to the number of distractors displayed simultaneously with the search target as shown in Figure 3. Participants are required to identify the location of the target element (the red dot) by indicating whether it is on the left side or on the right side of the centre fixation cross. In each trail, the locations of the target and non-target elements are randomly generated and the numbers of the distractors were also generated randomly within two different ranges, the low number range and the high number range, so that the image on the mobile phone screen is slightly varied in each trail.

The mean RT of each group for each task is plotted as the function of the number of distractors as shown in Figure 7. In both groups, there are linear increases in RTs as the number of distractors increasing. The values of RTs in the old group are considerably higher that of in the control group.

5 CONCLUSION

In this paper, we described our research work on the assessment of visual performance of different mobile

phone user groups. The experiments are focused on the image representations on the small mobile displays and the experimental data of the visual search performance tests are shown. The results of these tests have demonstrated good convergences with the results of the studies shown in the vision research field. Through the experiments, we have identified that the visual performances of older people are decreased across all the designed search tasks compared with the young people in the control groups.

In summary, the performance of older people with regard to reading mobile screens is decreased considerably compared with the young people. There is no solid evidence to define in which of the long wavelength light condition and the short wavelength condition older people would perform better. The visual search performance of older people is affected by the number of search elements displayed on the mobile screen. The difference in the performance is also evident in age. The performance of the participants over 70 years old is more affected than that of participants aged below 70.

There are many areas of research that we would like to extend our work into. One of these areas is the assessment of 2D vs. 3D graphics representations on visual performance evaluation. The current and future results will provide the basis for the development of adaptive image representation models on small screen mobile devices.

ACKNOWLEDGMENT

The work presented in this paper is funded by the One Northeast CODEWORKS Assistive Technology Lab.

REFERENCES

SIEMENS (Fall 2003): "Pictures of Future", *The Magazine for Research and Innovation*, pp.64-69

Omori, T., Watanabe, T., Takai, J., Takada, H. and Miyao, M. (2002) "Visibility and characteristics of the mobile phones for elderly people", *Behaviour & Information Technology*, Vol. 21, No. 5, pp. 313–316.

http://www.visualexpert.com/Resources/olderdrivers.html

Salthouse, T. A. (1985) "A theory of cognitive aging", Elsevier, Amsterdam

Willis, S. L. and Marsiske, M. (1990) "A life-span perspective on practical intelligence". In D. Tupper & K. Cicerone (Eds.), *The neuropsychology of everyday life* (pp.183–198). Boston: Kluwer Academic Publishers.

Chen, J., Myerson, J. and Hale, S. (2002) "Age-related dedifferentiation of visuospatial abilities", Neuropsychologia 40(2002) 2050–2056

Tales, A., Muir, J., Jones, R., Bayer, A. and Sonwden, R. J. (2004) "The effects of saliency and task difficulty on visual search performance in ageing and Alzheimer's disease", Neuropsychologia 42 (2004) 335–345.

Computational Modelling of Objects Represented in Images –
João Manuel R.S. Tavares & R.M. Natal Jorge (eds)
© 2007 Taylor & Francis Group, London, ISBN 978-0-415-43349-5

Mass lesion detection in mammographic images using Haralik textural features

S. Tangaro, F. De Carlo & G. Gargano
Istituto Nazionale di Fisica Nucleare (INFN) – Bari, Italy

R. Bellotti
Università and INFN di Bari, and Center of Innovative Technologies for Signal Detection and Processing, Italy

U. Bottigli
Dipartimento di Fisica, Università di Siena and INFN – Cagliari, Italy

G.L. Masala
Struttura Dipartimentale di Matematica e Fisica, Università di Sassari and INFN – Cagliari, Italy

P. Cerello
Istituto Nazionale di Fisica Nucleare – Torino, Italy

S. Cheran
Dipartimento di Informatica, Università di Torino and INFN – Torino, and ASP fellow, Italy

R. Cataldo
Dipartimento di Scienza dei Materiali, Università di Lecce and INFN – Lecce, Italy

ABSTRACT: In this article we present a classification system for an automatic detection of masses in digitized mammographic images. The system consists in three main processing levels: a) image segmentation for the localization of regions of interest (ROIs); b) ROI characterization by means of textural features computed from the Gray Tone Spatial Dependence Matrix (GTSDM), containing second order spatial statistics information on the pixel grey level intensity; c) ROI classification by means of a neural network, with supervision provided by the radiologist's diagnosis. The CAD system was developed and evaluated using a database of $N_I = 3369$ mammographic images: the breakdown of the cases was $N_{I_n} = 2307$ negative images, and $N_{I_p} = 1062$ pathological (or positive) images, containing at least one confirmed mass, as diagnosed by an expert radiologist. To examine the performance of the overall CAD system, receiver operating characteristic (ROC) and free-response ROC (FROC) analysis were employed. The area under the ROC curve was found to be $A_z = 0.78 \pm 0.008$ for ROI-based classification. When evaluating the accuracy of the CAD against the radiologist-drawn boundaries, 4.23 false positive per image (FPpI) are found at 80% mass sensitivity.

1   INTRODUCTION

The analysis of medical images is receiving, in the last years, a growing interest from the scientific community working at the crossover point among physics, engineering and medicine. The main purpose of this activity is to develop CAD systems for an automated

Address for correspondence: Dipartimento di Fisica, Università di Bari, Via Amendola 173, 70126 Bari, Italy. Email: sonia.tangaro@ba.infn.it

search for pathologies, which could be of great help for the physicians' diagnosis.

A typical example is the analysis of mammographic images which are widely recognized as the only imaging modality for an early detection of breast neoplasia (1; 2). Breast cancer is reported as the leading cause of woman cancer deaths in both United States and Europe. At present, screening programs are the best known method for an early diagnosis in asymptomatic women, thus allowing a reduction of the mortality (3; 4). Screening programs are based on a double visual inspection of the mammographic images, as double reading was shown to increase diagnostic accuracy (5). From this point of view, the use of a CAD system could provide valuable assistance to the radiologist.

In the present paper, a CAD system for mass detection will be described. The masses are often clear marks of a breast neoplasia. Masses are rather large ($d \simeq 1$ cm of diameter) objects with variable shapes, showing up with faint contrast. These textural characteristics can be exploited in both the definition of a ROI hunter procedure and the choice of the proper features to discriminate positive regions of the mammogram from negative ones.

This work fits in the more general framework of the MAGIC-5 Project (Medical Application on a Grid Infrastructure Connection) (7). The image collection in a screening program intrinsically creates a distributed database, as it involves many hospitals and/or screening centers in different locations. The amount of data generated by such periodical examinations would be so large that it would not be efficient to concentrate them in a single computing center. As an example, let us consider a mammographic screening program to be carried out in Italy: it should check a target sample of about 6.8 millions women in the 49–69 age range at least once every two years, thus implying a data flux of 3.4 millions mammographic exams/year. For an average data size of 50 MB/exam (4 images), the amount of raw data would be in the order of 160 TB/year. In addition, the amount of data linearly increases with time and a full transfer over the network from the collection centers to a central site would be large enough to saturate the available connections. On the other hand, making the whole database available to authorized users, regardless of the data distribution, would provide several advantages. This framework requires huge distributed computing efforts as for the case of the HEP (High Energy Physics) experiments, e.g. the CERN/LHC (Large Hadron Collider) collaborations. The best way to tackle these demands is to use the GRID technologies to manage distributed databases and to allow real time remote diagnosis. This approach would provide access to the full database from everywhere, thus making possible large-scale screening programs. The MAGIC-5 Project fits in this framework, as it aims at developing Computer Aided Detection (CAD) software for Medical Applications on distributed databases by means of a GRID Infrastructure Connection.

2 THE IMAGE DATABASE

The mammograms used in this study were collected in a network of hospitals belonging to the MAGIC-5 collaboration (9; 10). The images were acquired using different mammographic screen/film systems

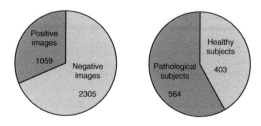

Figure 1. Database composition: images (left) and subjects (right).

and settings (all with molybdenum anode) and in the framework of different clinical applications, including both clinical routines carried out on symptomatic women, and screening programs addressed to asymptomatic women. All the images were digitized with a CCD scanner at a pixel size of 85 $\mu m \times 85\ \mu m$ with 12 bits gray level resolution (11). Each image is thus 2657×2067 pixels with $G = 2^{12} = 4096$ grey level tones. It should be stressed that no kind of normalization is applied to the images. The database consists of $N_I = 3369$ mammograms from $n_s = 967$ analyzed subjects. Some of the mammograms are different views (central, lateral, oblique) from the same subject and are treated as different case samples in the analysis.

We consider positive images the ones which contain at least one mass, as diagnosed by an expert radiologist and confirmed by biopsy; images with no mass at the first exam and after a follow up of at least three years are considered as negative, even if they contain some other pathology (e.g. microcalcifications). The breakdown of the cases is displayed in Figure 1 for both the images (positive/negative) and the analyzed subjects (pathological/healthy).

The classification of breast background is based only on the appearance of the breast parenchyma. No consideration was given to the skin, vascularity, presence or absence of masses, calcifications, lymph nodes, nor to parity, history of breast disease, age, family history.

3 METHODS

The CAD system consists of three main steps: segmentation, feature extraction and classification. The goal of the segmentation step is to locate, within the images, those suspicious regions, or ROIs, which are more likely to contain a mass. All the detected ROIs will be characterized by a proper set of features providing a second-order spatial statistics information on the pixel intensity. It should be stressed that, in general, a number of ROIs can be detected with different degrees of superimposition on the same mass, though being not overlapped among them. In few words, ROI-to-mass is not a one-to-one mapping.

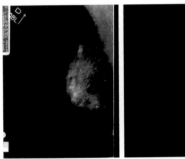

Figure 2. Left: the original image; right: the segmented image.

A tagging criterion, relying on the superimposition with the radiologist-drawn boundary, is adopted to define the true positive (TP) ROIs. This procedure is necessary to train the neural network with a ground truth based on the radiologist's diagnosis. Results are provided in terms of both ROC curve, displaying the neural network pattern (i.e. ROIs) classification and FROC curve, describing the overall CAD system (segmentation + feature extraction + neural network) mass detection performance.

In the following sections each of the above mentioned processing steps are reviewed.

3.1 Segmentation methods

Prior to be processed, the images are made anonymous and the borders of the breast are extracted by means of a threshold algorithm whose accuracy has been visually verified case by case. In this way, non-interesting portions of the mammogram reporting information about right/left breast, examination date and so on has been cut off.

A segmentation algorithm have been implemented to select the suspicious regions of the mammogram. The algorithm has been described in (6).

Figure 2 shows an example of the ROIs selected with the segmentation algorithm (right), together with the original image (left).

The number of ROIs detected from each image is not set a priori, rather it is related also to the texture properties of the mammogram. All the ROIs extracted from negative images are tagged as negatives, while the ROIs from positive images can be labeled as true positive (TP) or false positive (FP), depending if they meet or not of the radiologist-drawn boundary.

We point out that for the following classification step only TP and negative ROIs are used for both training and testing the neural network, according to the cross-validation technique (see section 3.3), while the FP ones are used for validation purpose only.

Table 1. Breakdown of the selected ROIs.

TP	FP	Negative	Total
1207	7642	13473	22322

The efficiency of the ROI hunter, computed as the percentage of masses correctly detected among those found by the radiologist, is 83.1%, corresponding to 1027 detected masses, with respect to 1236 radiologist-drawn boundaries. Moreover, 6.27 FPpI are obtained at this level, and the average area of the selected ROIs is 15% of the total area of the image. Table 1 reports the breakdown of the selected ROIs.

3.2 Feature extraction

Texture analysis can be used either to segment the image into areas indicating the mass, or to measure textural features to classify possible pathological sites. This approach has been used in (12) to experiment lung cancer nodule detection by means of textural features and neural network classification. This led us to follow a similar approach for the detection of the masses in mammography. The focus of the analysis is the computation of the Grey Level Co-occurrence Matrix (GLCM) (13), also known as Spatial Grey Level Dependence (SGLD) (14). To this purpose, we consider the minimal rectangular portion of the image which fully includes the ROI. As the name suggests, the GLCM is constructed from the image by estimating the pairwise statistics of pixel intensity, thus relying on the assumption that the texture content information of an image is contained in overall or average spatial relationship between pairs of pixel intensities (13). A co-occurrence matrix $\mathcal{M}$ is a $G \times G$ matrix, whose rows and columns are indexed by the image grey levels $i = 1, \ldots, G$, where $G = 2^n$ for an n-bit image. Each element p_{ij} represents an estimate of the probability that two pixels with a specified polar separation (d, θ) have grey levels i and j. Coordinates d and θ are, respectively, the distance and the angle between the two pixels i and j. In their seminal paper (13), Haralick et al. considered only displacements $d = 1$ at quantized angles $\theta = k\pi/4$, with $k = 0, 1, 2, 3$, thus having $\mathcal{M}_{d,\theta}(j,i) = \mathcal{M}_{d,\theta+\pi}(i,j)$. Symmetry is achieved by averaging the GLCM with its transpose, thus leading to invariance under π-rotations too. Textural features can be derived from the GLCM and used in texture classification in place of the single GLCM elements. In (13), 14 features are introduced, related to a textural property of the image such as homogeneity, contrast, presence of organized structure, complexity and nature of grey tone transitions. The values of these features are sensitive to the choice of the direction θ, given that the parameter d is fixed to 1 (greater values are rarely used). Invariance under rotation should be restored in order to avoid describing two images, one obtained by rotating the other, with different feature sets. This is achieved by considering mean and range of each feature values over the θ angles, thus obtaining a number of 28 textural variables, even if only few of them are used as inputs to a classifier (14; 15; 16).

As the texture is grey tone independent, either the image must be normalized or one should choose features which are invariant under monotonic grey level transformation. We select, among all GLCM features, the following features which are grey tone independent:

1. angular second moment:

$$f_1 = \sum_{ij} p_{ij}^2; \tag{1}$$

2. entropy:

$$f_2 = -\sum_{ij} p_{ij} \ln(p_{ij}); \tag{2}$$

3. information measures of correlation:

$$f_3 = \frac{f_2 - H_1}{\max\{H_x, H_y\}}; \tag{3}$$

$$f_4 = (1 - \exp\{-2(H_2 - f_2)\})^{1/2}, \tag{4}$$

where

$$P_x(i) = \sum_j p_{ij} \tag{5}$$

$$P_y(j) = \sum_i p_{ij} \tag{6}$$

$$H_1 = -\sum_{ij} p_{ij} \ln\{P_x(i)P_y(j)\} \tag{7}$$

$$H_2 = -\sum_{ij} P_x(i)P_y(j) \ln\{P_x(i)P_y(j)\} \tag{8}$$

$$H_x = -\sum_i P_x(i) \ln\{P_x(i)\} \tag{9}$$

$$H_y = -\sum_j P_y(j) \ln\{P_y(j)\} \tag{10}$$

For each of the above mentioned features $\{f_i\}$, $(i = 1, \ldots, 4)$, mean and range are computed over $\theta = k\pi/4$ angles, with $k = 0, 1, 2, 3$, thus obtaining a number of eight textural features.

3.3 Classification

We used a supervised two-layered feed-forward neural network, trained with gradient descent learning rule (17) for the ROI pattern classification.

Table 2. Breakdown of the patterns for the cross validation: first set A is used for training and set B for testing, then viceversa.

	set A	set B	validation
TP ROI	603	604	/
Negative ROI	604	603	13473
FP ROI	/	/	7642
Total	1207	1207	21115

The network architecture consisted of $n_i = 8$ input neurons and one output neuron. The size of the hidden layer was tuned in the range $[n_i - 1, 2n_i + 1]$ to optimize the classification performance.

All the TP ROIs ($N_{TP} = 1207$) and as many negative ones were used to train the neural network. To make sure that the negative training patterns were representative, they were selected with a probability given by the distribution of the whole negative ROI set, in the eight-dimensional feature space. With a random procedure we build up two sets (A and B), each one made of 1207 patterns, which are used, in turn, for both training and test, according to the cross validation technique (19): first, the network is trained with set A and tested with set B, then the two sets are reversed. All the other patterns (negative ROIs not selected for the training stage and FP) are used for validation only. The results presented in the following section (see ROC curves in section 4) refer to the classification of all the patterns at our hand. The breakdown of the patterns for the cross validation is reported in Table 2.

In all run set A and set B contains a balanced number of both the considered type of masses and the different kinds of tissue.

4 RESULTS AND DISCUSSION

The performance of the neural stage is provided in terms of ROC curve (20) analysis. The ROC curve is particularly suitable when testing a binary hypothesis: it is obtained by plotting the sensitivity (s, positive cases correctly recognized) against the false positive rate (FPR: fraction of misclassified negative patterns), at different values of the decision threshold.

In each case the classifications parameter (hidden neurons number, α) were changed so as the optimum performance was achieved.

Figure 3 displays a typical ROC curve obtained for the pattern classification. The area under the curve (AUC) is $A_z = 0.783 \pm 0.008$, where the error is computed as reported in Hanley et al. (21). The results are quite insensitive to the number N_h of the hidden neurons and α parameter.

As said in section 3, a number of ROIs can be superimposed to the same mass though not overlapping

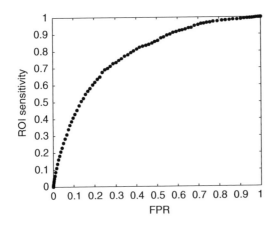

Figure 3. ROC curve for ROI-based classification. The area under the curve (AUC) is $A_z = 0.783 \pm 0.008$.

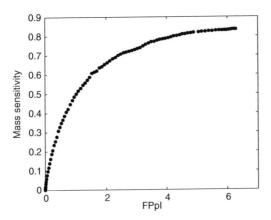

Figure 4. CAD FROC curve for mass-based classification.

5 CONCLUSIONS

A three-level classifier was developed and tested on the above described database. The first level is a segmentation-based procedure to extract ROIs. This routine consists in a dynamical threshold algorithm which allows to select iso-intensity contours around grey level maxima of the mammogram. The second level performs the feature extraction, based on the use of eight grey tone independent textural features measured from the GTSD matrix: these features carry second order spatial statistics information on the pixel intensity of the suspicious regions. The third level does the final classification by means of a two-layered feed-forward supervised neural network.

The results are provided in terms of ROC curve for the ROI-based neural classification, and FROC curve for the mass-based CAD classification.

Our scheme was developed and evaluated on a large database of mammographic images collected in the hospitals belonging to the MAGIC-5 Collaboration. The breast masses contained in the database span a wide range of shapes, sizes, and contrasts.

ACKNOWLEDGMENTS

We acknowledge the help of the staff and the institutions involved in the study. In particular, we thank prof. V. Lattanzio (Policlinico di Bari, Italy), prof. Bazzocchi (Dipartimento di Ricerce Mediche e Morfologiche, Universitá di Udine, Italy), prof. E. Zanon (Ospedale Valdese di Torino and INFN, sezione di Torino, Italy), prof. A. Sodano(Dipartimento di Scienze Biomorfologiche e Funzionali, Università Federico II, Napoli, Italy) and Prof. R. Lagalla, Dr. R. Ienzi (Istituto di Radiologia P. Cignolini, Università di Palermo) for the medical support in the data acquisition.

among them. For this reason, it should be more useful to provide the results in terms of the mass sensitivity, defined as the fraction of masses correctly detected by the CAD with respect to the total number of radiologist-drawn boundaries. In this way, the accuracy of the overall CAD system can be assessed against the radiologist's diagnosis. To this purpose, the following prescription is adopted: a mass is correctly detected by the CAD system if at least one ROI, among the ones superimposed to that mass, is classified as positive by the neural network.

A free-response ROC (FROC) curve can be drawn (see Figure 4), which reports the mass sensitivity of the overall system against the number of false positive per image (FPpI): 80% of mass sensitivity is achieved with 4.23 FPpI.

REFERENCES

[1] L.W. Bassett, et al., "History of the technical development of mammography, Syllabus: A Categorical course in Physics", RSNA (1994).

[2] L. Tabar, et al., "Reduction in mortality from breast cancer after mass screening with mammography randomised trial from the Breast Cancer Screening Working Group of the Swedish National Board of Health and Welfare", Lancet 829, 8433 (1985).

[3] S. Feig, "Increased benefit from shorter screening mammography intervals for women ages 40–49 years", Cancer 80, 2035–2039 (1997).

[4] R. Bird, et al., "Analysis of cancer missed at screening mammography", Radiology 184, 613–617 (1992).

[5] N. Karssemeijer, et al., "Computer-aided detection versus independent double reading of masses in mammograms", Radiology 227, 192–200 (2003).

[6] F. Fauci, et al., "Mammogram Segmentation by Contour Searching and Massive Lesion Classification with Neural Network", Proc. IEEE – Transaction on Nuclear Science, Vol 53, No. 4 (August 2006).

[7] R. Bellotti, et al., "The MAGIC-5 Project: Medical Applications on a Grid Infrastructure Connection", Proc. IEEE Nuclear Science Symposium, October 16–22 (2004), Rome, Italy.

[8] C.S. Cheran, et al., "Detection and Classification of Microcalcification Clusters in Digital Mammograms", Proc. IEEE Medical Imaging Conference, October 16–22 (2004), Rome, Italy.

[9] M. Bazzocchi, et al., "Application of a computer-aided detection (CAD) system to digitalized mammograms for identifying microcalcifications", Radiol. Med. 101, 334–340 (2001).

[10] M.E. Fantacci, et al., "Search of microcalcifications clusters with the CALMA CAD Station", Physics of the Medical Imaging, San Diego, USA, February 24–26, (2002).

[11] S.R. Amendolia et al., "Comparison of imaging properties of several digital radiographic system", Nuclear Instruments and Methods in Physics Research A 466, 95–98 (2001).

[12] G. S. Cox, F. J. Hoare, G. de Jager, "Experiments in lung cancer nodule detection using texture analysis and neural network classifiers", in Third South African Workshop on Pattern Recognition, (1992).

[13] R. M. Haralik, K. Shanmugam, I. Dinstein, "Textural Features for Image Classification", IEEE Transactions on Systems, Man and Cybernetics, Vol. SMC-3, 610–621 (1973).

[14] R.W. Conners and C.A. Harlow, "A Theoretical Comparison of Texture Algorithm", IEEE Transactions on Pattern Analysis and Machine Intelligence 2, 204–222 (1980).

[15] R.W. Conners, et al., "Segmentation of a High-Resolution Urban Scene using Texture Operators", Computer Vision, Graphics and Image Processing 25, 273–310 (1984).

[16] J.S. Weszka, C.R. Dyer, and A. Rosenfeld, "A Comparative Study of Texture measures for Terrain Classification", IEEE Transactions on Systems, Man and Cybernetics, 6, 269–285 (1976).

[17] J. Hertz, A. Krogh and R.G. Palmer, "Introduction to the theory of neural computation", Addison-Wesley (1991).

[18] D.E. Rumelhart and J.L. McClelland, "Parallel Distributed Processing", Vol.I MIT Press, Cambridge, MA (1986).

[19] M. Stone, "Cross-validatory choice and assessment of statistical predictions", Journal of the Royal Statistical Society B 36 (1), 111–147 (1974).

[20] J.A. Swets, "Measuring the accuracy of diagnostic systems", Science 240, 1285–1293 (1988).

[21] J.A. Hanley and B.J. McNeil, "The Meaning and Use of the Area under a Receiver Operating Characteristic (ROC) Curve", Radiology 143, 29–36 (1982).

Computational Modelling of Objects Represented in Images –
João Manuel R.S. Tavares & R.M. Natal Jorge (eds)
© *2007 Taylor & Francis Group, London, ISBN 978-0-415-43349-5*

Methodologies to build automatic point distribution models for faces represented in images

Maria João M. Vasconcelos & João Manuel R.S. Tavares

Faculdade de Engenharia da Universidade do Porto
Instituto de Engenharia Mecânica e Gestão Industrial
Laboratório de Óptica e Mecânica Experimental
Rua Drº Roberto Frias s/n, Porto, Portugal

ABSTRACT: This paper presents new methods to automatically build Point Distribution Models for faces represented in images. These models consider significant points of faces from several images and study them in order to obtain the mean shape of the object and the main modes of variation. Active Shape Models and Active Appearance Models use the Point Distribution Model to segment the modelled object in new images. In this paper these models, their automatically building and some application examples from objects like faces represented in images are describe.

1 INTRODUCTION

One of the most recent areas of interest in Computational Vision is image analysis based on flexible models. In this field, the use of statistical methods for object modelling has proved to be suitable to deal with problems in which the objects have variable shapes.

This work is mainly concerned with the employment of Point Distribution Models (PDMs) in the modelling of objects represented in images (Cootes et al. 1992). These models are obtained by analysing the statistics of the co-ordinates of the landmarks that represent the deformable object in study: after aligning the object shapes, a Principal Component Analysis is made and the mean shape of the object and the main modes of its variation are obtained. Grey levels of the objects can also be modelled and used to build Active Shape Models (ASMs) and Active Appearance Models (AAMs), in order to segment (identify) the modelled object in new images.

These statistical models have been very useful for image analysis in different applications of Computational Vision. For instance, they can be used on areas like: medicine, for locating bones and organs in medical images; industry, for industrial inspection; and security, for face recognition.

Usually, because is manually made, the determination of the landmark points of the objects to be modelled is the most time consuming step of the construction of PDMs, and so of ASMs and AAMs as well. Consequently, some authors, like (Hill &

Taylor 1994, Baker & Matthews 2002, Hicks et al. 2002, Angelopoulou & Psarrou 2004, Carvalho & Tavares 2005, Vasconcelos 2005), have been developing methodologies to fully automate this stage. In this work, we present three methodologies to automatically extract significant points from faces represented in images.

The main goals of the present work are: the introduction to the Point Distribution Models and its variants, namely ASMs and AAMs; the building of these models for faces represented in images using fully automatic procedures; and its application, namely, for its automatic segmentation in new images.

This paper is organized as follows: in the next section, the models considered are presented; in section 3, are described our methods to automatically extract landmark points of the faces to be modelled, using the models previous presented; in section 4, some experimental results are presented; finally, in the last section, some conclusions and perspectives of future work are addressed.

2 POINT DISTRIBUTION MODEL

(Cootes et al. 1992) describe how to build flexible shape models for objects called *Point Distribution Models*. These models are generated from examples of shapes of the object to be modelled, where each shape is represented by a set of labelled landmark points. The

Figure 1. Training image, landmarks and an image labelled with the landmark points (from left to right).

landmarks can represent the boundary or significant internal locations of the object (Fig. 1).

In this modelling method, all the training examples are aligned into a standard co-ordinate frame and a Principal Component Analysis is applied to the co-ordinates of the landmark points. This produces the mean position for each landmark, and a description of the main ways in which these points tend to move together. The equation below represents the *Point Distribution Model* or *Shape Model* and can be used to generate new shapes:

$$x = \bar{x} + P_s b_s,\qquad(1)$$

where x represents the n points of the shape:

$$x = \left(x_0, y_0, x_1, y_1, \ldots, x_{n-1}, y_{n-1}\right)^T,$$

(x_k, y_k) the position of point k, $\bar{x}$ the mean position of the points, $P_s = (p_{s1} p_{s2} \ldots p_{st})$ the matrix of the first t modes of variation, p_{si}, corresponding to the most significant eigenvectors in a Principal Component Analysis of the position variables, and $b_s = (b_{s1} b_{s2} \ldots b_{st})^T$ a vector of weights for each mode.

If the shape parameters b are chosen inside suitable limits (derived from the training set), then the shapes generated by equation (1) will be similar to those given in the original training set.

The local grey-level environment about each landmark point can also be considered in the modelling of an object represented in images. Thus, statistical information is obtained about the mean and covariance of the grey values of the pixels around each landmark point. This information is used in the PDMs variations: to evaluate the match between landmark points in Active Shape Models and to construct the appearance models in Active Appearance Models, as we explain next.

2.1 Active shape model

After build the PDM and the grey level profiles for each landmark point of an object, we can segment that object in new images using the Active Shape Models, an iterative technique for fitting flexible models to objects represented in images (Cootes & Taylor 1992a).

The referred technique is an iterative optimisation scheme for PDMs allowing initial estimates of pose, scale and shape of an object to be refined in a new image. The used approach can be summarized on the following steps: 1) at each landmark point of the models is calculated the necessary movement to displace that point to a better position; 2) changes in the overall position, orientation and scale of the model which best satisfy the displacements are calculated; 3) finally, any residual differences are used to deform the shape of the model by calculating the required adjustments to the shape parameters.

In (Cootes et al. 1994) is presented an improvement for the active shape models, which uses multiresolution. Thus, initially the method used constructs a multiresolution pyramid of the images to be consider, by applying a Gaussian mask, and then study the grey level profiles on the various levels of the pyramid built, making this away active models faster and reliable.

2.2 Active appearance model

This approach was presented in (Cootes et al. 1998) and allow the building of texture and appearance models. These models are generated by combining a model of shape variation (a geometric model), with a model of the appearance variations in a shape-normalized frame. The used statistical model of the shape it is also described by equation (1). To build a statistical model of the grey level appearance, we deform each example image so that its landmark points match the mean shape of the object, by using a triangulation algorithm. We then sample the grey level information, g_{im} from the shape-normalized image over the region covered by the mean shape. To minimize the effect of global light variation, we normalize this vector, obtaining g. By applying a Principal Component Analysis to this data, we obtain a linear model, the texture model:

$$g = \bar{g} + P_g b_g,\qquad(2)$$

where $\bar{g}$ is the mean normalised grey level vector, P_g is a set of orthogonal modes of grey level variation and b_g is a set of grey level model parameters.

Therefore, the shape and appearance of any example of the object modelled can be defined by vectors b_s and b_g.

Since there may be some correlation between the shape and grey levels variations, we apply a further Principal Component Analysis to the data of the models. Thus, for each training example we generate the concatenated vector:

$$b = \begin{pmatrix} W_s b_s \\ b_g \end{pmatrix} = \begin{pmatrix} W_s P_s^T \left(x - \bar{x}\right) \\ P_g^T \left(g - \bar{g}\right) \end{pmatrix},\qquad(3)$$

where W_s is a diagonal matrix of weights for each shape parameter, allowing the adequate balance between the shape and the grey models. Then, we apply a Principal Component Analysis on these vectors, giving a further model:

$$b = Qc , \qquad (4)$$

where Q are the eigenvectors of b, and c is the vector of appearance parameters controlling both the shape and the grey levels of the model. Thus, an example object can be synthesized for a given c by generating the shape-free grey level object, from the vector g, and deforming it using the landmark points described by x.

3   AUTOMATIC EXTRACTION OF LANDMARK POINTS

3.1   Face contour extraction

This method extracts significant points of faces represented in images; namely, on chin, eyes, eyebrows and mouth.

The first step of our method uses a skin detection algorithm to localize the face region. This algorithm uses a skin representative model, built with skin samples of the individual in study. Studies like (Jones & Rehg 1999, Tien et al. 2004, Zheng et al. 2004, Carvalho & Tavares 2005) show that the skin colour have usually the same luminance range and with the study of the skin chromatic colours it is possible to build a probability function for skin regions.

Studies like (Campadelli et al. 2003) show that the use of chrominance maps are useful for eyebrows and eyes localization in images. Chromatic colours can be obtained from the RGB colour space using the transformation:

$$\begin{cases} Cr = \dfrac{R}{R+G+B} \\ Cb = \dfrac{B}{R+G+B} \end{cases} . \qquad (5)$$

Usually, eyes are characterized in $CbCr$ plane by low values on the red component, Cr, and high values on the blue component, Cb, so the chrominance map for eyes can be defined by the following equation:

$$EyeMap = \frac{1}{3}\left\{ \left(Cb^2\right) + \left(\hat{C}r\right)^2 + \left(\frac{Cb}{Cr}\right) \right\} , \qquad (6)$$

where Cb^2, $\hat{C}r^2$ and Cb/Cr are normalized to the range [0,255] and $\hat{C}r$ is the negative of Cr (ie, $\hat{C}r = 255 - Cr$). In our work, the $EyeMap$ is used also to identify the eyebrows region with good results.

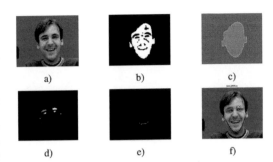

a)   b)   c)

d)   e)   f)

Figure 2.   a) Training image, b) segmentation result using the skin algorithm, c) face contour extracted, d) eyebrows and eyes found, e) mouth identified, and f) final contours obtained.

In other hand, in our method, the mouth region is identified using the HSV space, where H, S, V represent hue, saturation and value, respectively; where mouth is habitually characterized by having high values on the saturation component.

By congregating the contours of the face, eyebrows, eyes and mouth is possible to extract landmark points from each of these zones. Considering that the zone of the chin is the most important segment of the face contour, we only use the inferior part between ears.

In Figure 2, are present some results in a training image example using our method to automatically extract landmark points of faces represented in images.

3.2   Face regular mesh

The second method developed for automatically extraction of the landmark points of faces represented in images is based on the worked presented in (Baker & Matthews 2004) that, to construct active appearance models, consider the landmark points as the nodes of a mesh defined on the object to model.

Our method starts to identify face and eye regions like described in the last section, and adjust a regular rectangular mesh to the face region detected, rotating it according to the angle given by the eye's centroids. The nodes of the mesh obtained are then considered as landmark points of the object and used to build active appearance models for the same one.

Figure 3, shows the face mesh result obtained in a training image example using this method to automatically extract landmark points of faces represented in images.

3.3   Face adaptative multiresolution mesh

Finally, our third method combines the philosophy of the first method described, using face, eyes and mouth localization, and of the last method, considering the landmark points as the nodes of the defined meshes.

a)   b)

Figure 3. a) Training image, b) face regular mesh (red points) adapted to the face region (face contour in blue) and rotated according to the eyes direction (yellow).

a)   b)

Figure 4. a) Training image and b) example of an adaptative multiresolution mesh obtained for a face.

So, our new method builds a multiresolution mesh considering the face, eyes and mouth positions.

After localizing the face, eyes and mouth regions in the input image like described before, this new method constructs adaptive meshes, in the eye and mouth regions detected according to their localization; and then adds additional nodes, in the large mesh (that contains the face region), defined by the external edges and the bounds of the sub-meshes used in the regions of eyes and the mouth.

One example of the resulting final mesh using our third method for faces represented in images is presented in Figure 4.

In all implementation developed for our methods presented for automatically extract landmark points of faces represented in images, we can choose the parameters that define the resulting contour or mesh; that is, the number of landmark points defined in each interesting zone of the object to be modelled.

4 RESULTS

The methods described in this paper were used in this work to automatically build active shape and active appearance models for objects like faces represented in images.

During this work we developed an application in MATLAB to build shape models, using the *Active Shape Models software* (Hamarneh 1999). For the appearance models, we used the *Modelling and Search*

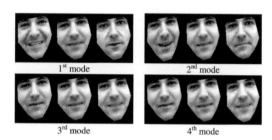

1st mode   2nd mode

3rd mode   4th mode

Figure 5. First four modes of appearance variation for the contour face model built ($\pm 2\,sd$).

Software available in (Cootes 2004). The images used in this paper are available in (Cootes 2004a).

For modelling faces represented in images, we used a training set of 22 images and other 4 images were used just for testing purpose. The active shape model was build using the first method presented in this paper for automatic extraction landmark points of faces represented in images and the other two methods presented from the same purpose were used to build active appearance models.

We present results for active models using the three approaches proposed for extracting landmark points: the face contour method extracted 44 landmark points, we extract 49 landmark points with the regular mesh approach and the third method extracted 54 and 75 landmark points respectively.

To the active shape model built, using 44 landmark points, the first 10 modes of variation could explain 90% of all the shape variance of the object modelled.

For the first face shape model trained (face contour), it was found that for 95% of the shape variance could be explained only by the first 13 modes of variation. By other hand, for the texture model it was found that 95% of the variance could be explained by the first 15 modes of variation. Finally, the appearance model needs only 12 modes of variation to explain 95% of the observed variance.

The first four modes of appearance variation are shown in Figure 5.

For the model trained using an adaptative multireso-lution face mesh, it was found that 95% of the variance of the object modelled could be explained only by the first 3 modes of variation. In the other hand, for the tex-ture model, it was found that 95% of the variance of the same object could be explained by the first 14 modes of variation. In last, the appearance model needs only 8 modes of variation to explain 95% of the observed variance of the object modelled.

The first four modes of variation of the texture and appearance models built are shown in Figure 6.

In Figures 7, 8 and 9 are presented some segmen-tation results obtained in a test image using the active appearance models built using the face contour model,

438

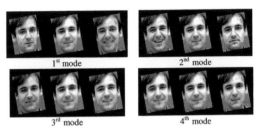

1st mode   2nd mode

3rd mode   4th mode

Figure 6. First four modes of appearance variation for the adaptive multiresolution face mesh model considered ($\pm 2\,sd$).

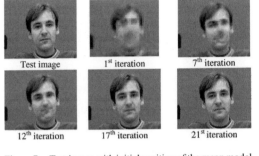

Test image   1st iteration   7th iteration

12th iteration   17th iteration   21st iteration

Figure 7. Test image with initial position of the mean model overlapped, and after the 1st, 7th, 12th, 17th and 21st iteration of the search with the active appearance model built for the face contour model.

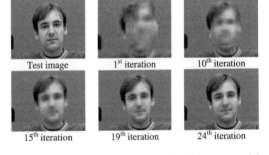

Test image   1st iteration   10th iteration

15th iteration   19th iteration   24th iteration

Figure 8. Test image with initial position of the mean model overlapped, and after the 1st, 10th, 15th, 19th and 24th iteration of the search with the active appearance model built for the regular face mesh model.

regular face mesh model and adaptive face mesh model.

In the active appearance search process, 5 levels of resolution were used and a maximum of 5 iterations were allowed per level. The active shape models built using he alignment process that consider the variance of landmark points, retained 95% of the variance of the object modelled and used the grey level profile of 7 or 15 pixels long, were the ones that obtained the

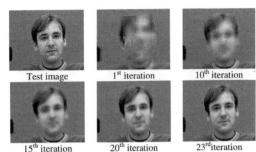

Test image   1st iteration   10th iteration

15th iteration   20th iteration   23rd iteration

Figure 9. Test image with initial position of the mean model overlapped, and after the 1st, 10th, 15th, 20th and 23rd iteration of the search with the active appearance model built for the adaptive face mesh model.

best segmentation results. For the active appearance models, the models built that obtained best segmentation results considered 99% of variance and 50000 pixels for the texture model.

In the face models, the mean error for segmentation was between 6.2 and 15.5 pixels for the active shape model, between 4.1 and 6.1 pixels using the face contour extraction method, 1.3 and 4.9 pixels using the face regular mesh method, and 1.5 and 3.7 pixels using the face adaptive multiresolution method, for the active appearance models. The mean error calculated for each test image consists in the Euclidean distance between the landmark points obtained by the model used and the object to be segmented.

5 CONCLUSIONS AND FUTURE WORK

A methodology to automatic build flexible models was presented, using a statistical approach, for deformable objects represented in images, namely for faces objects.

The methods developed to automatically extract landmark points from faces represented in images showed to be reliable and that allow the building of active shape models and active appearance models in a fully automatic way.

The segmentation results obtained in this work showed that the active appearance models built with the regular face mesh model and the adaptive face mesh model present better results than using the face contour model.

In general, active appearance models allows the construction of a robust model using relatively few landmark points compared to active shape models; so the first one is preferred in problems in which the landmark points extraction is not a easy process.

For future work, the use of previous knowledge about physical proprieties of the objects to be modelled can be considered in the building of its statistical models. Other interesting work can be the study of the influence in the models built of the number of training images used.

ACKNOWLEDGMENTS

This work was partially done in the scope of the project "Segmentation, Tracking and Motion Analysis of Deformable (2D/3D) Objects using Physical Principles", with reference POSC/EEA-SRI/55386/2004, financially supported by *FCT – Fundação para a Ciência e a Tecnologia* from Portugal.

REFERENCES

Angelopoulou, A. N. and A. Psarrou (2004). Evaluating Statistical Shape Models for Automatic Landmark Generation on a Class of Human Hands. International Archives of the Photogrammetry, Remote Sensing and Spatial Information Sciences, Istanbul.

Baker, S. and I. Matthews 2002. Automatic Construction of Active Appearance Models as an Image Coding Problem IEEE Transactions on Pattern Analysis and Machine Intelligence 26: 1380–1384.

Baker, S. and I. Matthews 2004. Automatic Construction of Active Appearance Models as an Image Coding Problem IEEE Transactions on Pattern Analysis and Machine Intelligence 26: 1380–1384.

Campadelli, P., et al. (2003). A color based method for face detection. International Symposium on Telecomunications, Isfahan, Iran.

Carvalho, F. J. S. and J. M. R. S. Tavares (2005). Metodologias para identificação de faces em imagens: Introdução e exemplos de resultados. Congresso de Métodos Numéricos en Ingeniería 2005, Granada, Espanha.

Cootes, T. F. (2004). Build_aam. http://www.wiau.man.ac.uk/~bim/software/am_tools_doc/download_win.html.

Cootes, T. F. (2004a). Talking Face. http://www.isbe.man.ac.uk/~bim/data/talking_face/talking_face.html.

Cootes, T. F., et al. (1998). Active Appearance Models. Proceedings of European Conference on Computer Vision, Springer.

Cootes, T. F. and C. J. Taylor (1992a). Active Shape Models – 'Smart Snakes'. Proceedings of the British Machine Vision Conference, Leeds.

Cootes, T. F., et al. (1992). Training Models of Shape from Sets of Examples. Proceedings of the British Machine Vision Conference, Leeds.

Cootes, T. F., et al. (1994). Active Shape Models: Evaluation of a Multi-Resolution Method for Improving Image Search. British Machine Vision Conference, BMVA.

Hamarneh, G. (1999). ASM (MATLAB). http://www.cs.sfu.ca/~hamarneh/software/code/asm.zip.

Hicks, Y., et al. 2002. Automatic Landmarking for Building Biological Shape Models International Conference of Image Processing, Rochester, USA 2: 801–804.

Hill, A. and C. J. Taylor (1994). Automatic Landmark Generation for Point Distribution Models. Fifth British Machine Vision Conference, England, York, BMVA Press.

Jones, M. J. and J. M. Rehg (1999). Statistical Color Models with application to skin detection. IEEE Conference on Computer Vision and Pattern Recognition, Ft. Collins, CO, USA.

Tien, F.-C., et al. 2004. Automated visual inspection for microdrills in printed circuit board production International Journal of Production Research 42, n° 12: 2477–2495.

Vasconcelos, M. J. 2005. MSc Thesis: Modelos Pontuais de Distribuição em Visão Computacional: Estudo, Desenvolvimento e Aplicação. Estatística Aplicada e Modelação, Universidade do Porto.

Zheng, H., et al. 2004. Blocking Adult Images Based on Statistical Skin Detection Electronic Letters on Computer Vision and Image Analysis 4: 1–14.

Computational Modelling of Objects Represented in Images –
João Manuel R.S. Tavares & R.M. Natal Jorge (eds)
© 2007 Taylor & Francis Group, London, ISBN 978-0-415-43349-5

Locally adaptive smoothing method based on B-splines

G. Vidal-Cassanya
Physics and Applied Mathematics Department, University of Navarra, Pamplona, Spain

A. Muñoz-Barrutia
Oncology Division, Center for Applied Medical Research, Pamplona, Spain
Electrical, Electronics and Automotive Department, University of Navarra, San Sebastián, Spain

M. Unser
Biomedical Imaging Group, Swiss Federal Institute of Technology (EPFL), Lausanne, Switzerland

ABSTRACT: This paper presents a novel method for the edge-preserving smoothing of biomedical images. It is based on the convolution of the image with scaled B-splines. The size of the spline convolution kernel at each image position is adaptive and matched to the underlying image characteristics; i.e., wide splines for smooth regions and narrow ones for pixels belonging to edges. Consequently, the algorithm reduces image noise in homogeneous areas while, at the same time, preserving image structures such as edges or corners. We argue that the proposed adaptive filtering strategy provides a good balance between the improvement in the Signal to Noise Ratio (SNR) and perceptual quality. Our algorithm takes advantage of the unique convolution and factorization properties of B-splines. Specifically, the input signal is expressed in a B-spline basis; the inner product with a B-spline of arbitrary size is then computed by using an adequate combination of 1D integrations (preprocessing) and rescaled finite differences. The method is computationally efficient with a cost per pixel that is fixed and independent upon the scaling factor.

Keywords: *B*-spline, scale-variant smoothing, perceptual metrics.

1 INTRODUCTION

Biomedical images are often degraded by noise. This tends to produce artifacts that affect many image processing tasks such as segmentation, registration, visual rendition and feature extraction. Noise reduction is therefore of considerable interest in many applications.

The simplest and most widely-used method for denoising is to low-pass filter the image; for example, by moving averaging, Gaussian filtering or, possibly, Wiener filtering. These classical scale-invariant restoration techniques improve the SNR, but they have the drawback of introducing a significant amount of blur.

To overcome this limitation, a variety of local image feature-dependent adaptive filtering strategies have been developed during the past two decades. One of the first methods that appeared in the literature is the gradient inverse weighted smoothing described in (1).

In this paper, we present a generalization of Wang's method that use B-splines as smoothing kernels. The key component of the algorithm is the convolution at each location with a smoothing B-spline whose scale

depends on the local characteristics of the image in the pixel neighborhood. In our case, the masks storing the scale values are calculated from the inverse image gradient and the more sophisticated Noise Visibility Function (NVF) (2).

We choose local convolution kernels that are rescaled (normalized) B- splines (with a unit integral); by tuning the spline degree, we are able to switch from a moving average to a weighted Gaussian-like smoothing. We also adopt a continuous domain formulation by interpolating the input image and expressing it into a B-spline basis. Thanks to the convolution properties of B-splines, we are then able to derive an exact and efficient scale-variant filter implementation; it uses a combination of moving sums and size-adjustable finite differences that are implemented efficiently by means of a look-up table.

2 ADAPTIVE FILTERING ALGORITHM

The continuous input signal $h(x)$ is represented by its spline interpolant which is in a one-to-one relation with the discrete input samples $h(k)$. Thus, we have $h(x) = c * \beta^{n_1}(x) = \sum_{k \in Z} c[k]\beta^{n_1}(x - k)$ where the sequence of interpolation coefficients $c[k]$ is calculated as shown in (3).

The output smoothed signal $f(x)$ at position x is calculated as the convolution of the input signal $h(x)$ with a B-spline kernel at scale a denoted by $\beta\left(\frac{x}{a}\right)$ which is given by

$$f(x) = c * \beta^{n_1} * \beta^{n_2}\left(\frac{x}{a}\right). \tag{1}$$

For details about the derivation of the algorithm, see (4).

The first step of the algorithm that provides the exact evaluation of the convolution given in equation (1) is to calculate the $(n_2 + 1)$-fold integral of the interpolation coefficients c_k

$$g = \Delta^{-(n_2+1)} * c, \tag{2}$$

where Δ^{-1} is the inverse finite-differences operator defined as $\Delta^{-1}(x) = \sum_{n \geq 0} \delta(x - n)$. The second step is to compute the inner products

$$f(b) = \sum_{k=0}^{n_2+1} \sum_{p=0}^{n_1+n_2+1} g(p + p_0)w_a(k,p) \tag{3}$$

where

- $w_a(k,p) = \frac{1}{a^{n_2}} q(k)\beta^{n_1+n_2+1}(\tau - ak - p - p_0)$ is a filter mask which we can store in a look-up table.
- $q(k) = \binom{n_2+1}{k}(-1)^k$ is the kth weighting coefficient of the $(n_2 + 1)$-finite difference at scale a.
- $p_0 = \left\lceil \frac{(a-2)(n_2+1)}{2} - ak - \frac{n_1+1}{2} \right\rceil + b$ is the first meaningful index in the sum over p that is computed using the compact-support property of B-splines.
- $\tau = \frac{(a-1)(n_2+1)}{2}$ is a time shift.

The operation described by equation (3) is equivalent to a discrete convolution with a modified 'a trous' filter. The scale-variant smoothing computation at each position b consists in filtering the coefficients $g(p)$ with $(n_2 + 2)$ 'clusters' of length $(n_1 + n_2 + 2)$, each 'cluster' being separated from its neighbors by a distance a as shown in Figure 1.

A box diagram for a fast implementation of this scale-variant smoothing algorithm is shown in Figure 2. For each of the N scales in which we choose to quantize the mask, we compute the weights w_a and store them in a 3D look-up-take of dimensions $(n_2 + 2) \times (n_1 + n_2 + 2) \times N$. In the initialization step, the B-spline expansion coefficients c_k of the sampled signal $h(x)$ are calculated and the running sum operator Δ^{-1} is applied $(n_2 + 1)$-times. The intermediate result $g(p)$ does not depend on the scale a. The filtered

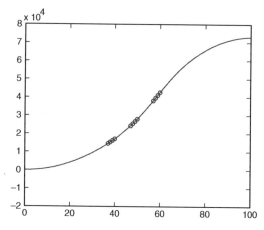

Figure 1. Spatial structure of the filter w_a to calculate $f(b)$ with $b = 50$. The filter weights form $(n_2 + 2) = 3$ 'clusters' of length $(n_1 + n_2 + 2) = 4$, with an inter-cluster distance of $a = 10$. ('-' integral function g and 'o' coefficients of g ponderated by the filter weights).

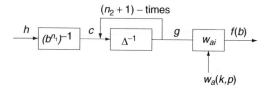

Figure 2. Schematic representation of the locally adaptive smoothing algorithm. $(b^{n_1})^{-1}$: Computation of the interpolation coefficients c. $\Delta^{-(n_2+1)}$: Calculation of the $(n_2 + 1)$-fold integral of c. $w_{a_i}(k,p)$: Look-up table calculation where $k \in [0, n_2 + 1]$, $p \in [0, n_1 + n_2 + 1]$ and $a_i \in [a_1, a_N]$ with N the number of scales. w_{a_i}: Filtering with the weights calculated for each scale. $f(b)$: Smoothing output f at position b.

output $f(b)$ at position b is calculated with the modified 'a trous' filter whose coefficients are stored in the ith plane w_a of the 3D look-up-table. The values w_a and the intercluster distance for the filtering depend on a, but the computational complexity is constant and does not depend on a. Not taking into account the initialization step, we need $(n_2 + 2)(n_1 + n_2 + 2)$ multiplications and $(n_2 + 1)(n_1 + n_2 + 1)$ additions per point, corresponding to the filtering of g with the weights w_a.

We obtain a higher dimensional scheme by successive 1D processing along the various dimensions of the data. For the rest of the paper, we will restrict ourselves to 2D processing.

3 EXPERIMENTAL RESULTS

We propose now to realize a quantitative comparison of the performance of the scale-invariant denoising

442

Figure 3. Original Lena image.

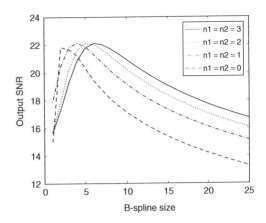

Figure 4. For a noisy Lena with SNR = 15 dB: SNR vs. B-spline size.

filtering using different B-spline kernels. We add Gaussian noise to Lena as to have a SNR = 15 dB (see Figure 8(a)). In order to be able to evaluate the scale-invariant denoising with respect to a ground truth, i.e., original Lena image (see Figure 3). We perform the evaluation in terms of noise reduction (using the Signal to Noise Ratio (SNR) measure) and perceptual quality (using the Average Edge Width (AEW) and the Michelson contrast (MC) measure). The metrics above are defined as:

- $SNR = 10 \cdot \log 10\left(\sum_{i=1}^{n_x} \sum_{j=1}^{n_y} \frac{in(i,j)^2}{(in(i,j)-out(i,j))^2}\right)$ where in is the input image, out the denoised image and n_x and n_y the x- and y-dimensions.
- $AEW = \frac{TotBM}{NbEdges}$ where TotBM is the total number of pixels in the image that belongs to a edge and NbEdges is the total number of edges in the image. For more information in how to compute this quantity we refer to (5).
- $MC = \frac{\max(out) - \min(out)}{\max(out) + \min(out)}$ where $\max(out)(\min(out))$ is the maximum (minimum) value over all the pixels of the image out.

The SNR of the scale-invariant smoothed image versus the B-spline size is shown in Figure 4. For the correspondence between scale factor and B-spline size note that the support of a B-spline of degree n and scaled by a factor a is $a(n + 1)$. The peak SNR obtained for the best compromise between denoising and edge blurring is 22 dB independently of the B-spline degree but it is reached for a higher B-spline size as the degree increases. From the slower falling of the cubic B-spline SNR curve with respect to the rest we can deduce that cubic B-splines offers a good compromise between noise reduction and boundary sharpness conservation for a scale-invariant smoothing with a kernel of a given size.

The AEW of the scale-invariant smoothed image versus the B-spline size is shown in Figure 5 (a). We observe a linear increment of the AEW measure with

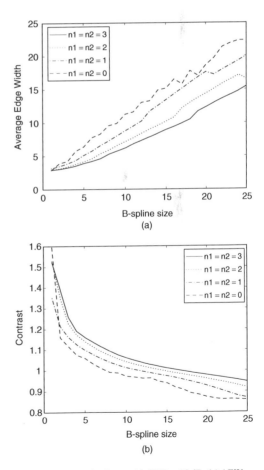

Figure 5. For a noisy Lena with SNR = 15 dB: (a) AEW vs. B-spline size. (b) MC vs. B-spline size.

443

Table 1. SNR, AEW and MC measures for the smoothed noisy Lena with SNR = 15 dB.

	Original	Uniform 2.0	Gradient [0.5, 2.0]	NVF [0.5, 2.0]
SNR (dBs)	21.737	14.987	21.949	21.843
AEW (pixels)	5.245	2.857	4.868	4.708
MC	1.594	1.094	1.354	1.354

Table 2. SNR, AEW and MC measures for the smoothed noisy centromeres image.

	Original	Uniform 3.0	Gradient [0.5, 3.0]	NVF [0.5, 3.0]	NVF [0.5, 1.0, 3.0]
SNR (dBs)	11.912	Inf	17.098	24.544	19.868
AEW (pixels)	12.061	6.286	8.658	6.660	7.333
MC	0.98443	0.98096	0.98377	0.98624	0.98432

respect to the B-spline size for all the B-spline kernels. The cubic B-spline curve has a smaller slope that the rest of the considered kernels. Thus, cubic B-splines results in less blurring than their counterparts for a given B-spline size.

The MC of the scale-invariant smoothed image versus the B-spline size is shown in Figure 5(b). The contrast decreases steeply for the convolution with a B-spline of a smaller size than the corresponding to the peak SNR output. It seems that the pronounced contrast loss is due to the dramatic noise reduction caused by the convolution with incremental size kernels. The linear decrement in contrast observed after that point seems to be caused by the linear increment of the blurring with the B-spline size. Moreover, the cubic B-spline curve keeps itself over the rest for all the sizes. So, we observe from another point of view how the cubic B-splines results in images of higher quality for a given B-spline size.

As the three quality measures agree in indicating that the use of cubic B-splines results in images of higher quality for a given B-spline size, we decide to use cubic B-splines for the interpolation and convolution in the rest of the experiments.

Next, we will compare the performance of the scale-independent and scale-dependent smoothing implemented as described in Section 2. We have chosen to use the following masks: the binarized inverted gradient and the binarized inverted NVF function. For a detailed description of the NVF function consult (2). The inverted gradient and NVF function are binarized using the unimodal background symmetry method. The masks computed for the noisy Lena are shown in Figure 8(b) and (c). We can observe from them how the size of the spline convolution kernel adapt somehow to the underlying image characteristics. Wider splines are applied to smooth regions and narrower ones to edges.

We show in Table 1 the quantitative comparison of the denoising of Lena with SNR = 15 dB (see Figure 8(a)) using the scale-invariant filter with a scale 2.0 and the gradient and NVF masks with scales [0.5, 2.0]. We observe that the scale-variant filtering approach (for both masks) have a gain of 0.2 dB in SNR, 0.345 in MC and a AEW 0.5 pixels smaller that the corresponding scale-invariant filtering. Figure 8 shows the visual result. We observe that the scale-dependent smoothing reduces image noise in homogeneous areas, while preserving edges much better than the scale-invariant smoothing.

To shown an example of the performance of the approach on a real biomedical image, we have applied our algorithm on a fluorescent microscopy image of chromosome centromeres (see Figure 9). The quantitative comparison between the three approaches (scale-invariant filter with a scale 3.0, and the gradient and NVF filter with scales [0.5, 3.0]) is shown in Table 2. In this case, we measure the SNR with respect to the original noisy centromeres image. It seems from the data collected in the table that the NVF method outperforms the other two: a higher SNR and MC, and a lower AEW value. The truth is that the NVF mask separates the background from the cells but does not distinguish the spots. In consequence, the NVF filtering does not clean very much from the noise in the inter-spot regions inside the nucleus. On the other hand, the gradient mask clearly segments the spots. So, the gradient method gives a better compromise between denoising and spot conservation. As expected, the scale-invariant filtering introduces much more blurring than the other two methods.

The key point here is that the masks we have implemented indicate where the contours and the fine structures are in the image but they do not give any clue on what is the size of the smooth regions. High values in the magnitude of the gradient refers mainly

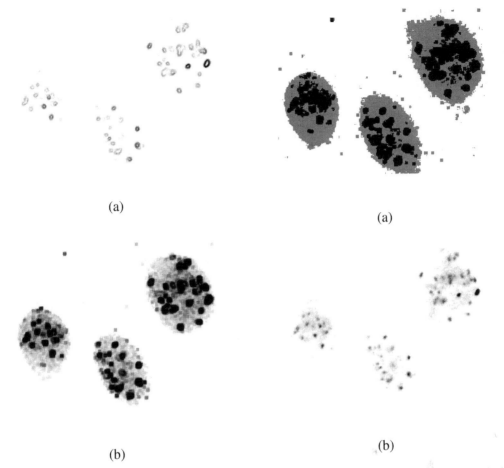

(a)

(a)

(b)

(b)

Figure 6. For the centromeres image: (a) Inverted gradient magnitude. (b) Inverted NVF function.

Figure 7. For the centromeres image: (a) Inverted tri-scale NVF mask. (b) Inverted NVF smoothed image with [0.5, 1.0, 3.0]. We have use cubic B-spline interpolation and convolution.

to the contours of the image. High values in the NVF function correspond not only to the contours but also to areas of the images with high to medium activity. The differences can be observed in Figure 6. We use this fact to point out the interest to introduce an estimation of local scale (equivalently, spatial frequency) into the method. For example, we have constructed a NVF mask that separates the three differentiated scale regions in the centromeres images: the background, the nucleus and the spots (see Figure 7 (a)). We have computed the corresponding NVF denoised centromeres image with scales [0.5, 1.0, 3.0] which is included in Figure 7 (b) to facilitate a visual comparison with the other methods. We observe sharp spots and clean smooth regions. The quantitative comparison of the denoising with the tri-scaled NVF with respect to the binarized gradient method shows a gain of 1.8 dB in SNR, 0.0025 in MC, and a AEW 1.3 pixels smaller as collected in Table 2.

In consequence, we observe the convenience to introduce a measure of local scale into the method and incorporate such an information to the mask to be able to smartly control the degree of smoothing that is done in different regions of the image. Another weak point of the method is that the separable filtering we propose although fast is not quite equivalent to dilating the B-spline in 2D (only if the x- dilation is the same over the region covered by the y-dilation). For that reason we plan to extend the method to work for a non-separable tensor spline kernels.

4   CONCLUSIONS

In summary, we have presented a novel adaptive B-spline based smoothing algorithm that is capable of

445

Figure 8. (a) Noisy Lena with a SNR $= 15$ dB. (b) Inverted gradient mask. (c) Inverted NVF mask. (d) Space independent smoothed image for $a_2 = 2.0$. (e) Inverted gradient smoothed image with $[a_1, a_2] = [0.5, 2.0]$. (f) NVF smoothed image with $[0.5, 2.0]$. We have use cubic B-spline interpolation and convolution.

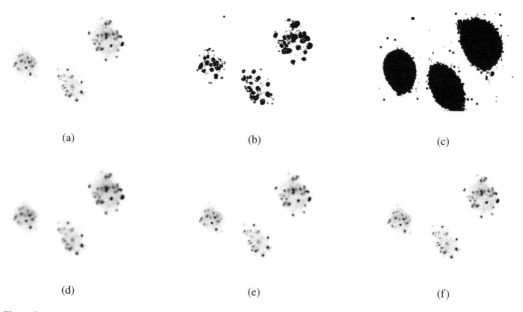

Figure 9. (a) Inverted original noisy centromeres image. (b) Inverted gradient mask. (c) Inverted NVF mask. (d) Inverted space independent smoothed image for $a_2 = 3.0$. (e) Inverted gradient smoothed image with $[a_1, a_2] = [0.5, 3.0]$. (f) Inverted NVF smoothed image with $[0.5, 3.0]$. We have use cubic B-spline interpolation and convolution.

reducing noise without degrading edges. The proposed scheme is general and flexible. It works for all spline degrees which allows one to easily switch from moving average to Gaussian-like filters of arbitrary sizes. It is also extendable for more sophisticated masks that are linear combination of B-splines and that can be rescaled in a signal-adaptive fashion. The overall operation count only depends on the degrees of the chosen B-splines but it is independent of the value of the local scale. The results we show are promising and a further improvement of the method could be obtained in a future work by the introduction of a measure of local scale and by its extension to use 2D non-separable spline kernels.

ACKNOWLEDGEMENTS

A. Muñoz-Barrutia is funded by a Ramon y Cajal fellowship of the Spanish Ministry of Science and Technology.

REFERENCES

[1] D. C. C. Wang, A. H. Vagnucci, and C. C. Li, "Gradient Inverse Weighted Smoothing Scheme and the Evaluation of its Performance", in *Computer Graphics Image Proc.*, vol. 15, pp. 167–181, 1981.

[2] S. Voloshynovskiy, A. Herrigel, N. Baumgaertner, and T. Pun, "A Stochastic Approach to Content Adaptive Digital Image Watermarking", *in Int. Workshop on Inf. Hiding, Lecture Notes in Computer Science*, vol. 1768, pp. 212–236, Oct. 1999.

[3] M. Unser, "Splines, A Perfect Fit for Signal and Image Processing", in *IEEE Signal Proc. Mag.*, vol. 16, no. 6, pp. 22–38, Nov. 1999.

[4] A. Muñoz-Barrutia, R. Ertlé, M. Unser, "Continuous Wavelet Transform with Arbitrary Scales and $O(N)$ Complexity", *Signal Proc.*, vol. 82, no. 5, pp 749–757, May 2002.

[5] P. Marziliano, F. Dufaux, S. Winkler, and T. Ebrahimi, "A No-reference Perceptual Blur Metric", *in Proc. of the Int. Conf. on Image Proc.*, vol. 3, pp 57–60, Sept. 22–25, 2002.

Computational Modelling of Objects Represented in Images –
João Manuel R.S. Tavares & R.M. Natal Jorge (eds)
© *2007 Taylor & Francis Group, London, ISBN 978-0-415-43349-5*

Acquisition and filtering of digital video in real time using Matlab® GUIs

J.M. Vilardy, J.E. Calderon, C.O. Torres & L. Mattos
Optics and Computer Science Laboratory, University Popular of the Cesar, Valledupar, Cesar, Colombia

ABSTRACT: An interactive graphical user interface (GUI) was developed to capture, save, and filter digital video in real time, with the Matlab platform. The system developed supports different types of video capturing devices, like: Matrox Inc, Coreco Imaging Inc, Data Translation Inc, DCAM (IEEE 1394, FireWire), Windows Driver Model (WDM), or Video for Windows (VFW) drivers (USB e IEEE 1394 Web Cameras, Digital Video Camcorders and TV tuner card), the frames are treated in matrix manner for their real-time visualization and filtering through spatial convolution with the different filters implemented. This solution to video acquisition may be used to treat digital video efficiently, given that the frames are in matrix form, thus enabling the application of many treatment techniques of images and digital video.

1 INTRODUCTION

For the treatment of the digital video signal in determined applications, it is necessary to conduct operations as fast as possible in order to make on-the-spot decisions by analyzing the sequences of images treated. In these applications, it is necessary to acquire the digital video signal and bring it to the environment where it is to be treated.

At the Optics and Computer Science Laboratory, a real-time digital video acquisition and filtering system was developed, using commercial cameras, the Matlab® technical programming platform, and the Matlab Graphical User Interface (GUI). It is worth noting that, in this work, real time is defined as the time needed to acquire, process, and visualize the digital video signal according to the persistence of the human eye to observe moving images.

2 SPATIAL FILTERING

Spatial filtering is the operation applied to the raster video frames (bit map) to improve or suppress spatial details in order to improve visual interpretation. Common examples include the application of filters to improve details at frame boundaries, or to reduce or eliminate noise patterns. Spatial filtering is a "local" operation in the sense that it modifies the value of each pixel according to the values of nearby pixels and the convolution kernel used (generally, this kernel is a

square, uneven-sized matrix mask), using the Discrete Convolution operation, thus:

$$I_{m,n} = \sum_{i=-r}^{r}\sum_{j=-r}^{r} H_{i,j} G_{m-i,n-j} = \sum_{i=-r}^{r}\sum_{j=-r}^{r} H_{-i,-j} G_{m+i,n+j} \quad (1)$$

From the previous equation we have: $G_{m,n}$ as the frame to be filtered of m rows and n columns, $H_{i,j}$ is the convolution kernel or the mask, which is assumed of uneven size (for the purpose of placing the pixel to be changed in the center of the mask, and conducting the filtering process as previously mentioned) with $(2r+1)(2r+1)$ coefficients, that is $2r+1$ rows and $2r+1$ columns, and $I_{m,n}$ is the result of the filtering operation.

The filters used in this article yield, as a result, the softening and extraction of video frame borders through the operation carried out in equation (1). Hereinafter, we show the different characteristics of the filters used and their corresponding kernel.

2.1 *Average filter*

It is a filter applied directly to the spatial domain, obtaining a softened video frame, whose intensity for each pixel is reached by averaging the intensity values of the pixels included in the nearby environment predefined by the kernel for every pixel. In addition to softening the frame, this pixel suppresses noise that can appear due to sampling, quantization, and transmission, or by system perturbations such as dust particles on the optical system, obtaining homogeneity in the filtered frame. The kernel for this filter is:

$$H = \frac{1}{9}\begin{bmatrix} 1 & 1 & 1 \\ 1 & 1 & 1 \\ 1 & 1 & 1 \end{bmatrix} \quad (2)$$

2.2 Sobel filter in both directions

This type of filter emphasizes the borders surrounding an object within a frame in all directions (both horizontal and vertical) utilizing the gradient operator. This filter is also known as the Sobel operator and, besides emphasizing the borders, it has the added feature of softening the filtered frame, eliminating part of the noise and, hence, minimizing the appearance of false borders due to the effect of noise magnification derived from the operators. For the Sobel operator in both directions, the kernel is given by (Jahne et al. 2003):

$$H = \frac{1}{8}\begin{bmatrix} 0 & -1 & -2 \\ 1 & 0 & -1 \\ 2 & 1 & 0 \end{bmatrix} \quad (3)$$

2.3 Gaussian Laplacian filter

The Gaussian Laplacian filter is a second-derivative operator isotropic capable of detecting borders in all possible orientations, reducing the effect of noise when softening the frame through the Gaussian effect. This second derivate operator, as indicated by its name, is constituted by a bi-dimensional Gaussian function to which the Laplacian operator is applied; its mask or kernel is defined by (De La Cruz et al. 2004):

$$H = \begin{bmatrix} 0.0448 & 0.0468 & 0.0564 & 0.0468 & 0.0448 \\ 0.0468 & 0.3167 & 0.7146 & 0.3167 & 0.0468 \\ 0.0564 & 0.7146 & -4.9048 & 0.7146 & 0.0564 \\ 0.0468 & 0.3167 & 0.7146 & 0.3167 & 0.0468 \\ 0.0448 & 0.0468 & 0.0564 & 0.0468 & 0.0448 \end{bmatrix} \quad (4)$$

2.4 Sharpen filter

Unlike the two previous border-extraction filters (Two-directional Sobel and Gaussian Laplacian), this filter highlights the borders without creating a dark background and without suppressing colors. This type of filter is a variation of discrete highpass filters (which improve the fine details of an image), since these manage to offer better contrast and visual sharpness to the

Figure 1. GUI of the application presentation.

processed frame. The kernel of the Sharpen filter is represented by (MathWorks[1] et al. 2005):

$$H = \begin{bmatrix} -0.1667 & -0.6667 & -0.1667 \\ -0.6667 & 4.3333 & -0.6667 \\ -0.1667 & -0.6667 & -0.1667 \end{bmatrix} \quad (5)$$

3 DEVELOPMENT OF THE SYSTEM OF ACQUISITION AND FILTERING OF DIGITAL VIDEO IN REAL TIME

The graphical user interface implemented is constituted by three windows, which are: The window of the application presentation, which has three buttons called Acquisition, Filter, and Exit – the first two buttons lead to the other two remaining windows; that is, if users only need to acquire but not filter, they should press the Acquisition button and the corresponding window is deployed for video acquisition. If users wish to acquire and filter, they should at the same time press the Filter button to open the window that runs said operations. To conduct the acquisition and/or filtering of digital video, we used Matlab® version 7.1, with the Image Acquisition Toolbox (IMAQ), GUIDE (Graphical User Interface Development Environment), and the Image Processing Toolbox, a T.V. Tuner Card to import video and bring it to the Matlab® WorkSpace, analog cameras (SONY Handycam Vision CCD-TRV108 NTSC) or USB, and an IBM PC with a 2.26 GHz processor and 256 MBytes of RAM memory. Figure 1 shows the GUI of the application presentation.

3.1 Acquisition

In this GUI we have the following objects:
The function of the Popup Menu is that of providing information to the user about the adaptors and devices connected to the computer system, the adaptor is the

450

Table 1. Type of data & objects of the acquisition GUI

Description	Type of data	GUI object
Adaptor	Character	Static text
Names of adaptors	Character cell	Popup menu
Device	Character	Static text
Name of devices	Character cell	Popup menu
Reset	–	Push button
Preview	–	Push button
Acquisition	–	Push button
Stop	–	Push button
Save	–	Push button
Exit	–	Push button

Figure 2. Video acquisition GUI.

software that the toolbox (Image Acquisition) uses to communicate with the device that video acquires through the driver of said device; hence, depending on the number of adaptors for the system supports, the user could establish communication with a device supported by any adaptor. Upon initiating this GUI, a reset is applied to the IMAQ, information is requested from existing adaptors supported by Matlab®, and this information updates the popup menu that contains the names of the adaptors. Depending on the adaptor selected and the video capturing devices (cameras) connected, the other popup menu is updated, containing the names of the cameras. The push buttons execute actions related to their name:

1. Previews: Shows the video captured by the camera selected.
2. Acquisition: Creates a video object in Matlab®, starts it, acquires the frames of the captured video in the manner specified by the object of the video (MathWorks[2] et al. 2005), and, lastly, these frames are written onto a hard-disk file in AVI format.
3. Stop: The stop button to stop the Preview or the video Acquisition.
4. Save: This determines whether or not to save the video that was acquired and taken to a hard-disk file.
5. Reset: Returns the IMAQ to a known state, erasing all the video objects.
6. Close: Determines if the user wishes to exit the application or not.

The GUI of video Acquisition is shown in Figure 2.

3.2 Acquisition and filtering

For this GUI we have the same objects as from the previous GUI, but here the acquisition button is now the Filter button and three Radix buttons were added, called: Sharpen, Sobel, and Laplacian, the names of the filters to apply and they are contained in an iu-button group (MathWorks[3] et al. 2005); these radix buttons are excluding – meaning that if one is not selected the rest are not either. Depending on the radix button selected, when the Filter button is pressed, the effect of the filter selected will be visualized at once through programming events carried out in the following manner: When the video object is started, such is immediately triggered; thus, the images captured by the camera selected are saved in a buffer of the PC memory, after initiating and triggering the video object when the time specified by the Timer Period has elapsed (this is the time in seconds that specifies the time between events, as long as the video object is initiated), the specified function by the Timer-Fcn is executed (this is a feature of the video object (MathWorks[2] et al. 2005)), in the function called upon to be executed, the procedures conducted are:

1. Acquisition of a frame from the PC memory buffer to be bring to the Matlab® WorkSpace.
2. Erasing of the frame acquired from the PC memory buffer.
3. Visualization of the frame acquired in the GUI.
4. Filtering in the spatial domain of the frame acquired, utilizing the appropriate kernel for the filter selected.
5. Visualization of the filtered frame in the same GUI, where the acquired frame was acquired.
6. The acquired and filtered frames are saved in two video files (in AVI format), respectively.

Once these steps are executed, the IMAQ toolbox measures the time established by the Timer Period and when such has elapsed, the prior procedures are repeated and so on. The Timer Period established in this work was of 1/25 seconds for a digital video resolution of 320 × 240 pixels, that is that the application implemented can acquire, visualize, and filter 25 frames per second, for the digital video resolution mentioned (visualization shows the acquired frame, as well as the filtered frame). The equipment (computer) used to carry out the acquisition, filtering, and visualization, as was already mentioned, has a 2.26 GHz

451

Figure 3. GUI of video acquisition and filtering.

processor and a RAM memory of 256 MBytes, occupying approximately 100% of the processor and 50% RAM memory when the Acquisition and Filtering GUI is acquiring, processing, and visualizing digital video, hence if we seek to save the acquired and filtered video, we would additionally need a faster processor compared to the one used in this research work (at least a 3.0 GHz processor is needed), since the operation of writing the frame onto a hard-disk file demands a good percentage of processor use, adding more execution time to the programming conducted of the event. The GUI of video Acquisition and Filtering is found in Figure 3.

4 CONCLUSIONS

Within this work, we presented the acquisition and filtering of digital video in real time by way of the implementation of a digital video system controlled by an coded program in the Matlab® technical programming platform, with the aid of the image acquisition toolbox, and the GUI development environment, revealing how the frames are matrix treated for their filtering in the spatial domain and visualization. This made it attractive to acquire video through said tool, given that any frame could be modified in any manner desired. Since what is being treated is video, the speed at which the IMAQ toolbox can process the different frames depends upon the speed of the processor, the complexity of the algorithm, and the speed at which the capturing device acquires the frames; therefore, given a rather quick processor, a simple algorithm, and a speed of frame acquisition adjusted to the speed of frame acquisition of the capturing device, the IMAQ toolbox can process the frames very rapidly.

REFERENCES

De La Cruz, J. & Pajares, G. 2004. Visón por Computadora. In Alfaomega (eds), 65 – 179. Colombia.
Jahne, B. 2003. Digital Image Processing. In Springer-Verlag (eds). Alemania.
MathWorks[1]. "MATLAB Image Processing Toolbox help". http://www.mathworks.com/access/helpdesk/help/pdf_doc/images/images_tb.pdf, September 2005.
MathWorks[2]. "MATLAB Image Acquisition Toolbox help". http://www.mathworks.com/access/helpdesk/help/pdf_doc/ imaq/imaq_print.pdf, September 2005.
MathWorks[3]. "MATLAB Creating Graphical User Interfaces help". http://www.mathworks.com/access/helpdesk/help/pdf_doc/matlab/buildgui.pdf, September 2005.

Computational Modelling of Objects Represented in Images –
João Manuel R.S. Tavares & R.M. Natal Jorge (eds)
© 2007 Taylor & Francis Group, London, ISBN 978-0-415-43349-5

Integration of skin segmentation methods using ANNs

A. Viloria-Lanero & V. Cardeñoso-Payo

Departamento de Informática, Universidad de Valladolid, Spain

ABSTRACT: In this paper, we provide a thorough review of color based skin segmentation methods. A selection of the best documented methods found in the state of the art has been made in order to get a consensus-like decision making system which has been used as the reference skin classification system to train a multilayer perceptron (MLP) artificial neural network (ANN) in order to map color information into skin/non-skin class information. Results prove that this integration clearly provides a lower error rate for color based skin segmentation than any of the individual methods, thus offering a promising alternative for fast, accurate and robust skin segmentation ideal both for real time image processing environments and adaptive skin tracking solutions.

Keywords: Face detection, Objects segmentation, tracking and analysis of movement, Computational vision.

1   INTRODUCTION

Skin color segmentation methods (SCS) are becoming widespread as a means to locate or to track faces or other nude parts of the body, both in static photographs and in video sequences. Many applications of skin detection and tracking have been recently reported: surveillance environments, adult content detection, videophone tasks, biometric identification, or leisure video games, among others. Identification of the Areas of Interest (AOI) inside an image is a crucial step before proceeding to any deeper analysis on reduced and normalized regions over the input image or video frame. Skin Color Segmentation provides a fast and robust alternative for the segmentation of the image into AOIs, although inaccurate results usually require a post-processing stage. Color based skin segmentation methods have well recognized advantages and bear some known limitations (J. Brand 2000; Son Lam Phung 2005). Pixel level segmentation is a fast way to locate areas containing skin patterns inside images and dramatically reduces the search space as a previous step to higher level image classification. However, these methods usually give poor results under uncontrolled illumination conditions, suffer from region discontinuities due to shadows or occlusions, pose difficulties to separate neighbor regions having similar color schemes, introduce extra noise because of skin-like pixels in the background, provide a medium to high degree of inaccuracy which requires additional processing and are completely useless for gray level

images. Nevertheless, SCS provide a good means to get an initial bounding region for AOIs within an image before any specialized identification step and are thus an ongoing and increasing focus of interest within the image recognition community.

2   CONCEPTUAL MODEL

As already pointed out in (Son Lam Phung 2003), a number of fundamental prerequisites have to be considered when designing any skin segmentation method: a very low rate of False Negatives when False Positives rate is low, a good detection rate for different skin tones, a good disambiguation capability, and good enough robustness to cope with changing lightning conditions. Here, we provide a solution which minimizes the limitations of Skin Color Segmentation Methods enumerated in the introduction and, at the same time, complies with these prerequisites.

In this work, we propose to merge in a consensus-like classifier the decisions provided by the best documented skin color segmentation methods found in the literature. The experts votes provide a reference measure on the likelihood that a given pixel value is to be labeled with a skin value or not. If we take this probability measure as the desired output of a highly nonlinear and complex mapping from pixel color coordinates to image class, an artificial neural network could be trained to provide this mapping. This ANN based mapping would provide an integrated skin classifier over the color space which has the classification power of each method, given that the output of that method is consistent with the one of a given majority of them.

This work has been partially supported by the Spanish Ministry of Education, project n° TIC2003-08382-C05-03.

Through integration, we are able to obtain: better focusing on more plausible color values of skin; a reasonable smoothing which provides a more robust labeling than under any specific individual technique; and a skin detection rate which will, at least, be as good as the one of the best methods. A careful selection of the neural network model, an appropriate conditioning of the color space training data and the right quality assessment measure will provide the desired operation point of the system, balancing the False Negatives and False Positives rates.

3 CHOOSING THE SCS EXPERTS

As input for the decision making process, we have considered 46 methods, chosen from the many ones found in the literature. Although the list could be extended and the ANN easily trained to new conditions, the key selection criterion has been the availability of enough data as to implement the segmentation method proposed by the authors.

All SCS methods classify a given pixel of the color space into one of two classes: skin and non-skin. Most of them use a set of explicit geometrical rules in order to define the color regions corresponding to these classes. Just a few of them fully document parametric methods to describe the same regions.

When using explicit geometric rules, these are expressed in term either of thresholds, planes, parabolas or ellipses. Thresholds provide a computational efficient and easy way to define a cluster just comparing values of color components. Authors that used this strategy were (C. Wang 1999; Loyola 2003; Gomez 2002; M.M. Fleck 1996; Kapur 1997; K. Sobottka 1998; S. Tsekeridou 1998; D. Chai 1999; T. Sawangsri 2005; Ing-Sheen Hsieh 2002; Kyung-Min Cho 2001; F. Dadgostar 2006; Ahlberg 1999; Jamie Sherrah 2001). Other authors delimit the cluster using linear combinations of the color components (J. Kovac 2003; Christophe Garcia 1999; G. Gomez 2002b; Hwei-Jen Lin 2005; Wang Xian-bao 2005; Martinkauppi 2002; B. Martinkauppi 2005; G. Gomez 2002a). Another way to define the cluster is trying to fit it between two parabolas, as in (J. Fritsch 2002; Ming-Chieh Chi 2006; M. Soriano 2000a; M. Soriano 2000b). Finally, in several color spaces, the skin cluster looks compact enough as to be delimited by ellipses, as showed in (J. Kovac 2003; R.-L. Hsu 2002).

All the papers using parametric methods applied statistics analysis over a big set of skin samples to get the statistical cluster parameters, but very few of them report the set of final parameters which were used. (P. Kranthi Kiran 2004) showed the parameters to define a gaussian over the distribution of skin samples individually on each component of normalized-RGB color scheme. (M.J. Jones 1999) is a most referenced

paper in which a mixture of gaussians was used to parameterize the clusters.

In Table 1, we show a comparison of all the skin color segmentation methods which were implemented in this work, in terms of the quality measurements which will be described later.

4 SEGMENTATION USING ANNs

As already pointed out, all the reference SCS methods were implemented and applied to obtain a skin classification answer over all the corresponding color space. The likelihood of a pixel x being labeled as skin is estimated as a weighted sum of the individual expert labeling:

$$P_{ref}(x = skin) = \frac{\sum_i \alpha_i \cdot vote_i(x)}{\sum_i \alpha_i} \qquad (1)$$

where $vote_i(x)$ is 1 if pixel x corresponds to skin under the criterion of expert i, and 0 otherwise. We chose $\forall i : \alpha_i = 1$, but another criteria could easily be applied if one wants, for example, to favor some experts which provide better labeling, both in a given color region or over all color space.

A Multilayer Perceptron (MLP) artificial neural network was selected to build the mapping from pixel values x into the $P_{ref}(x)$ provided by the SCS experts. Several 3D color spaces have been considered for the input pixel representation: RGB color components (in scale [0,1]), normalized-RGB (rgb), Hue-Saturation-Value (HSV), YCbCr (as an instance of an independent chromatic channels color space), and Tint-Saturation-Lightness (TSL). We have also considered 2D-color spaces: rg, CbCr, HS and TS. Finally, RGB ratios (R/G, R/B, G/R, G/B, B/R and B/G) were tested in some cases, as extra input dimensions to the input pixel representation, although we couldn't find scarcely any improvement and, thus, will not report any results here.

The MLP output $P_{out}(x)$ represented either a continuous probability measure of $x = skin$ or two valued function $0 : non - skin/1 : skin$, depending on the nature of the output activation function. When a discrete 0/1 version was needed, we applied the typical criterion $P_{out,ref}(x) > 0.5$ to map a pixel x into skin. In any case, the training procedure for the MLP produces the set of weights which minimizes the distance measure between the output and the reference objective function over the training set.

4.1 MLP configurations

The size of the input layer was 3 neurons for 3D color schemes and 2 neurons for 2D color spaces (except

Table 1. Comparative results for the 46 SCS methods implemented in this work.

Method reference	Color space	Succ (%)	Sens (%)	Spec (%)	Score (%)
(J. Kovac 2003)	RGB	84,01	95,38	83,48	75,13
(J. Kovac 2003)	YCbCr	95,48	94,79	95,51	90,85
(J. Kovac 2003)	YCbCr	97,28	78,61	98,14	85,97
(J. Fritsch 2002)	rg	94,56	69,84	95,69	78,26
(Christophe Garcia 1999)	YCbCr	97,23	74,42	98,28	83,96
(Christophe Garcia 1999)	HSV	98,11	78,30	99,03	86,99
(P. Kranthi Kiran 2004)	rgb	97,63	82,51	98,33	88,27
(J. Brand 2000)	RGB	86,73	86,34	86,75	75,06
(C. Wang 1999)	YI'Q'	84,44	53,68	85,86	58,92
(Loyola 2003)	YI'Q'	69,56	95,63	68,36	57,04
(G. Gomez 2002b)	RGB	90,30	80,81	90,73	77,45
(G. Gomez 2002b)	RGB, HSV	95,01	69,00	96,21	78,48
(G. Gomez 2002b)	RGB, HSV	93,94	58,90	95,55	72,55
(Gomez 2002)	H-GY-Wr	97,17	39,56	99,83	67,73
(R.-L. Hsu 2002)	YCbCr	97,99	75,35	99,04	85,45
(D.A. Forsyth 1996)	IRgBy	91,55	53,59	93,30	67,24
(Kapur 1997)	IRgBy	93,09	77,62	93,80	79,78
(Ming-Chieh Chi 2006)	YCbCr	96,89	43,87	99,33	69,37
(Hwei-Jen Lin 2005)	RGB, rg, HSV	97,74	95,60	97,84	94,54
(Wang Xianbao 2005)	RGB	95,41	12,25	99,25	53,19
(K. Sobottka 1998)	HS	96,70	77,22	97,59	84,52
(S. Tsekeridou 1998)	HSV	95,94	51,00	98,01	71,48
(D. Chai 1999)	CbCr	95,26	95,53	95,25	90,87
(T. Sawangsri 2005)	YCbCr, HSV	94,58	83,48	95,10	84,45
(T. Sawangsri 2005)	YCbCr, HSV	97,06	57,29	98,89	75,79
(M. Soriano 2000a)	rg	94,82	64,15	96,24	76,04
(M. Soriano 2000b)	rg	91,72	90,35	91,78	83,52
(Martinkauppi 2002)	rg	84,49	99,51	83,80	77,43
(B. Martinkauppi 2005)	rg	90,66	89,87	90,69	81,85
(B. Martinkauppi 2005)	rg	86,62	57,75	87,96	63,11
(G. Gomez 2002a)	RGB	95,91	50,06	98,02	71,01
(G. Gomez 2002a)	RGB	96,88	30,50	99,94	63,18
(G. Gomez 2002a)	RGB	97,20	47,19	99,50	71,29
(G. Gomez 2002a)	RGB	97,21	49,57	99,41	72,41
(G. Gomez 2002a)	RGB	97,21	46,23	99,56	70,86
(G. Gomez 2002a)	RGB	96,87	37,44	99,61	66,38
(G. Gomez 2002a)	RGB	96,67	38,04	99,37	66,41
(G. Gomez 2002a)	RGB	96,53	21,58	99,98	58,67
(M.J. Jones 1999)	RGB	87,36	98,71	86,83	81,04
(Ing-Sheen Hsieh 2002)	HSI	84,99	98,95	84,34	77,89
(Kyung-Min Cho 2001)	HSV	90,07	86,00	90,26	79,39
(F. Dadgostar 2006)	HSV	87,06	81,31	87,33	73,41
(Ahlberg 1999)	CbCr	96,45	41,69	98,98	67,84
(Jamie Sherrah 2001)	HSV	91,70	78,51	92,30	78,31
(Jinfeng Yang 2004a)	Polar-rgb	94,55	96,87	94,44	90,44
(Jinfeng Yang 2004b)	Polar-rgb	93,86	98,47	93,64	90,15
Best result		98,11	99,51	99,98	94,54
Mean variance		5,95	33,75	6,72	13,01

when testing the effect of RGB ratios addition). MLP has only one hidden layer, which size was initially chosen following multiples of the input size (9, 12, 15 and 18 in 3D color spaces). For better joint comparison with 2D color spaces the hidden layer size was finally fixed to 8, 10, 12 and 16 elements.

Two different activation functions have been tested for the hidden layer: logistic and hyperbolic tangent. As

for the output layer, we tried three different activation function: linear, logistic and SoftMax.

Linear activation function in the output layer (1 neuron) was first tried to approximate the output value to the probability of the pixel to be classified as skin (given by the ratio of methods giving that output). Results were discouraging in terms of stability and additional experiments using several different input and output contexts gave no extra success. Next, we tried a logistic activation function (1 neuron), representing only information about the binarization of the color space into two classes (skin and non-skin). This gave better results, although instabilities in the learning process arose.

Finally, SoftMax activation function was tested for the output layer (2 neurons). As recommended by the Neural Network Group (Sarle 1997), this should be the right output activation function when, as in our case, the output represents the probability that a given input belongs to a class. This activation function gave the best results overall and was then selected as the final activation function of the output layer for the rest of the experiments.

If x_k is the output value of the k-th neuron in the output layer, the SoftMax activation function is given by:

$$SoftMax(x_i) = \frac{e^{x_i}}{\sum_j e^{x_j}} \qquad (2)$$

When the Softmax activation function is used in the output layer, the MLP output value represents an estimated probability of the input color to be classified as skin. In this case, the Cross-Entropy function

$$E_c = \sum_{x \in \mathbf{C}} P_{ref}(x) * \ln P_{out}(x) \qquad (3)$$

was used as the objective function when training, where $\mathbf{C}$ represents the training Color Space. The maximization of E_c gives the best fit of the network output to the consensus decision of the 46 experts.

4.2 Training and evaluation

Backpropagation with momentum was used as the MLP training method in all cases to get the maximum Cross-Entropy or the minimum Mean Square Error, depending on the kind of output activation function.

Since just 4.4% of color space cells are labeled as skin after the consensus of at least 50% of the methods, any system providing an 'always non-skin' output would provide a 95.6% classification rate. Thus, a totally dumb classifier would give a very high absolute classification quality although a totally useless one. To cope with the effects of this negative sample biasing, we had to devise a training set selection mechanism and a revision of the evaluation measurements.

An importance sampling is carried out over the full color space, so that the number of skin and non-skin pixels is identical. The non skin pixels were chosen completely at random and the MLP was trained using this balanced data set. For testing, the full color space was used in all cases.

To provide a reasonable quality assessment measure which avoids the no-skin unbalancing problem and which provides a robust comparison and analysis of the results, we propose the function $Score$, which weights the straight success and failure indicators by means of an specificity and sensitivity components. The quality scoring function is defined as:

$$Score = Succ * (\alpha * Sens + \beta * Spec) \qquad (4)$$

Different combinations of this weights have been tested and the results did not show any special trends, so that α and β were set to 0.5.

The success and failure rates are obtained from the number of True Positives (TP), True Negatives (TN), False Positives (FP), and False Negatives (FN):

$$Succ = \frac{TP + TN}{TP + TN + FP + FN} \qquad (5)$$

$$Fail = \frac{FP + FN}{TP + TN + FP + FN} \qquad (6)$$

Sensitivity ($Sens$) and Specificity ($Spec$) give, respectively, measures of the fraction of type I (false rejection) and type II (false accepting) errors.

$$Sens = \frac{TP}{TP + FN} \qquad (7)$$

$$Spec = \frac{TN}{TN + FP} \qquad (8)$$

This $Score$ corrects out the success rate $Succ$ by means of a weighted average of $Sens$ and $Spec$, so that pathological cases related to the unbalanced nature of the data set are discarded in favor of those cases were $Sens$ and $Spec$ are at the same time similar and with high enough values. So, we fulfill one of the fundamental requisites discussed in (Son Lam Phung 2003).

Since the SoftMax activation function provides a continuous likelihood estimation value, a likelihood scoring threshold has to be applied before labeling. This threshold was evaluated over the training set applying the usual Equal Error Rate (EER) criterion. As expected from the configuration of the training set, this threshold was always very close to 0.5.

5 RESULTS AND DISCUSSION

All the 46 SCS methods have been implemented and compared in terms of the $Succ$ and $Score$ measures, as shown in Table 1. Clearly, the highest success rate is not always related to a higher quality (big $Score$) behavior. In particular, the best $Succ$ result has a $Score$ of just 86.99. Also, the scoring we propose shows a greater

Table 2. Comparative results for several hidden and output layer activation functions.

Hidden-Output	#Hidden	Succ (%)	Sens (%)	Spec (%)	Score (%)
Logistic-Linear	H16	98.78	86.09	99.37	91.60
Logistic-Sigmoid	H16	99.65	95.86	99.82	97.50
Logistic-Tanh	H16	95.75	47.62	97.97	69.70
Logistic-SoftMax	H16	99.39	92.98	99.69	95.75
Tanh-Linear	H8	95.07	42.18	97.50	66.39
Tanh-Tanh	H16	99.25	91.01	99.63	94.61
Tanh-SoftMax	H10	94.55	38.13	97.15	63.95
Best result		99.65	95.86	99.82	97.50
Mean variance		2.32	37.43	1.16	18.43

mean variance among all the results than expected just looking at the reported success rates. Comparing the integration results with the individual experts showed that the cluster proposed in (Christophe Garcia 1999) (HSV color space) has the best success rate (98.11%), and the cluster proposed in (Hwei-Jen Lin 2005) (using RGB, rg and HSV color spaces) has the best score (94.54%).

Table 2 shows the evaluation results of the different hidden-output activation functions tested in this work. Although the *Score* is better for the Logistic-Sigmoid combination, this MLP configuration showed a highly instable behavior along the training epochs and thus, we discarded it in benefit of the SoftMax configuration which, moreover, outputs a continuous likelihood estimation instead of a binary skin/non-skin direct classification, which eases the tuning of the operation point of the classifier to the particularities of the training data in a second stage.

Figure 1 shows the evaluation of the influence of the choosing of the input color space both on the *Succ* and the *Score* measurements, for different numbers of neurons in the hidden layer. While in terms of success rate, only an improvement around 1% is achieved when using a 3D-color space, a notorious improvement was obtained in terms of total 'score' compared to 2D-color spaces. Anyway, the color space selection did not show to be a determinant factor when using a 3D-color scheme: RGB, HSV and TSL have the better results, so that RGB might be chosen in terms of computation costs. Table 3 shows the best results in terms of *Score* for every color space. As can be seen, the number of neurons in the hidden layer is usually the highest available, although is not determinant above 8. This could suggest additional experimentation with higher values (although this would badly affect computational efficiency).

Best results correspond to a HSV color space, 16 neurons in the hidden layer, a Logistic function in the hidden layer, and a SoftMax activation function in the output layer (*Succ* = 99.47% and *Score* = 96.30%). The cross-entropy value was around 0.09 in the best cases. Of course, all these results correspond to an

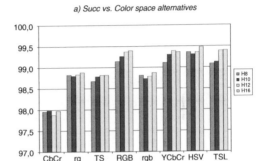

a) Succ vs. Color space alternatives

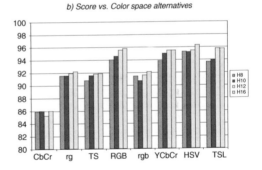

b) Score vs. Color space alternatives

Figure 1. Evaluation of the influence of the selection of color space parametrization. The activation functions for the hidden-output layers of the MLP correspond Logistic-SoftMax in all cases.

EER output likelihood thresholding over the training set (the set of points in color space which got the vote of the majority of the experts); we haven't experimented the influence of the majority qualification on this yet.

Finally, the ANNs can be easily trained to new conditions and have a small enough cpu time footprint as to provide real time skin tracking to the face tracking and recognition platform we are developing. Figure 2 shows a screen capture of the skin segmentation monitoring filter designed using the best ANN: the image

Table 3. Comparative results for the SoftMax function for different color spaces.

#Hidden	Color space	Succ (%)	Sens (%)	Spec (%)	Score (%)
H16	CbCr	97,97	76,64	98,95	86,01
H16	rg	98,88	87,02	99,42	92,18
H16	TS	98,81	86,51	99,38	91,84
H16	RGB	99,39	92,98	99,69	95,75
H16	rgb	98,86	86,93	99,41	92,12
H16	YCbCr	99,35	92,46	99,67	95,45
H16	HSV	99,47	93,88	99,73	96,30
H12	TSL	99,38	92,97	99,68	95,73
Best result		99,47	93,88	99,73	96,30
Mean variance		0,51	6,52	0,26	3,70

Figure 2. Screen capture illustrating the SCS provided by our ANNs.

on the left shows the real scene with a bounding rectangle framing the detected face region and the image on the right a B/W representation of pixels inside the detected skin regions (W).

6 CONCLUSIONS

An integration of relevant Skin Color Segmentation Methods has been provided using an MLP. Different configurations and input color spaces have been tested in order to obtain the best approximation. 3D-color space show always better results than 2D. Small differences between RGB and YCbCr integrations can be found. A Logistic-SoftMax activation function for the hidden-output layers has been proposed and several others have been evaluated.

The color skin segmentation method we propose here performs better than any other studied method and adequately fulfills the main requirements of a high-end skin segmentation method while it still can run in real time.

REFERENCES

Ahlberg, J. (1999). A system for face localization and facial feature extraction. Technical report, Linkping University, Sweden.

B. Martinkauppi, M. Soriano, M. P. (2005). Comparison of skin color detection and tracking methods under varying illumination. *Journal of Electronic Imaging 14*(4).

C. Wang, M. B. (1999). Multi-source face tracking with audio and visual data. In *IEEE 3rd Workshop on Multimedia Signal Processing*, pp. 169–174.

Christophe Garcia, G. T. (1999). Face detection using quantized skin color regions merging and wavelet packet analysis. *IEEE Trans. on Multimedia 1*(3), 264–277.

D. Chai, K. N. (1999). Face segmentation using skin-color map in videophone applications. In *Circuits and Systems for Video Technology*, Volume 9, pp. 551–564.

D. A. Forsyth, M. F. (1996). Identifying nude pictures. In *3rd IEEE Workshop on Applications of Computer Vision. WACV '96*, pp. 103–108.

F. Dadgostar, A. S. (2006). An adaptive real-time skin detector based on hue thresholding: A comparison on two motion tracking methods. In Elsevier (Ed.), *Pattern Recognition Letters*, Volume 27, pp. 1342–1352.

G. Gomez, E. M. (2002a). Automatic feature construction and a simple rule induction algorithm for skin detection. In *ICML Workshop on Machine Learning in Computer Vision*, pp. 31–38.

G. Gomez, M. Sanchez, L. S. (2002b). On selecting an appropriate colour space for skin detection. In *MICAI'02: Proceedings of the Second Mexican International Conference on Artificial Intelligence*, London, UK, pp. 69–78. Springer-Verlag.

Gomez, G. (2002). On selecting colour components for skin detection. In *16th International Conference on Pattern Recognition*, Volume 2, pp. 961–964.

Hwei-Jen Lin, Shu-Yi Wang, S.-H. Y. Y.-T. K. (2005). Face detection based on skin color segmentation and neural network. In *International Conference on Neural Networks and Brain*, Volume 2, pp. 1144–1149.

Ing-Sheen Hsieh, Kuo-Chin Fan, C. L. (2002). A statistic approach to the detection of human faces in color nature scene. *Pattern Recognition 35*(7), 1583–1596.

J. Brand, J. M. (2000). A comparative assessment of three approaches to pixel-level human skin-detection. In *15th International Conference on Pattern Recognition*, Volume 1, pp. 1056–1059.

J. Fritsch, S. Lang, M. K. G. F. G. S. (2002). Improving adaptive skin color segmentation by incorporating results from face detection. In *IEEE Int. Workshop on Robot and Human Interactive Communication (ROMAN)*, pp. 337–343.

J. Kovac, P. Peer, F. S. (2003). Human skin color clustering for face detection. In *The IEEE Region 8 EUROCON 2003. Computer as a Tool*, Volume 2, pp. 144–148.

Jamie Sherrah, S. G. (2001). Skin color analysis. http://www.dai.ed.ac.uk/CVonline/LOCAL_COPIES/GONG1/cvOnline-skinColourAnalysis.html.

Jinfeng Yang, Zhouyu Fu, T. T. W. H. (2004a). Skin color detection using multiple cues. In *17th International Conference on Pattern Recognition. ICPR 2004*, Volume 1, pp. 632–635.

Jinfeng Yang, Tieniu Tan, W. H. (2004b). Face contour construction with multiple information. In *Sixth IEEE International Conference on Automatic Face and Gesture Recognition*, pp. 451–456.

K. Sobottka, I. P. (1998). A novel method for automatic face segmentation, facial feature extraction and tracking. In *Signal Processing: Image Comm.*, Volume 12, pp. 263–281.

Kapur, J. P. (1997). Face detection in color images. http://www.geocities.com/jaykapur/face.html. EE499 Capstone Design Project Spring 1997.

Kyung-Min Cho, Jeong-Hun Jang, K.-S. H. (2001). Adaptive skin-color filter. *Pattern Recognition 34*(5), 1067–1073.

Loyola, R. C. (2003). *Implementación de una herramienta gráfica para modelar el rostro humano a través de mallas de triángulos Delaunay en 2D*. Ph. D. thesis, Universidad del Bío-Bío. Concepción. Chile.

M. Soriano, B. Martinkauppi, S. H. M. L. (2000a). Skin detection in video under changing illumination conditions. In *15th International Conference on Pattern Recognition (ICPR'00)*, Volume 1, pp. 839–842.

M. Soriano, B. Martinkauppi, S. H. M. L. (2000b). Using the skin locus to cope with changing illumination conditions in color-based face tracking. In *IEEE Nordic Signal Processing Symposium (NORSIG 2000)*, pp. 383–386.

Martinkauppi, B. (2002). Face colour under varying illumination – analysis and applications. Technical report, Department of Electrical and Information Engineering and Infotech Oulu. University of Oulu.

Ming-Chieh Chi, Jyong-An Jhu, M.-J. C. (2006). H.263+ region-of-interest video coding with efficient skin-color extraction. In *International Conference on Consumer Electronics. ICCE '06*, pp. 381–382.

M. J. Jones, J. R. (1999). Statistical color models with application to skin detection. In *IEEE Conf. Computer Vision and Pattern Recognition*, pp. 274–280.

M. M. Fleck, D. A. Forsyth, C. B. (1996). Finding naked people. In *European Conference on Computer Vision*, Volume 2, pp. 592–602.

P. Kranthi Kiran, T. L. (2004). Face recognition using distance based features. In *National Conference on Communications*.

R.-L. Hsu, M. Abdel-Mottaleb, A. J. (2002). Face detection in color images. *IEEE Trans. on Pattern Analysis and Machine Intelligence 24*(5), 696–706.

S. Tsekeridou, I. P. (1998). Facial feature extraction in frontal views using biometric analogies. In *IX European Signal Processing Conference*, Volume 1, pp. 315–318.

Sarle, W. (1997). Neural network faq. ftp://ftp.sas.com/pub/neural/FAQ.html.

Son Lam Phung, D. Chai, A. B. (2003). Adaptive skin segmentation in color images. In *IEEE International Conference on Acoustics, Speech, and Signal Processing*, Volume 3, pp. 353–356.

Son Lam Phung, A. Bouzerdoum, D. C. (2005). Skin segmentation using color pixel classification: Analysis and comparison. *IEEE Trans. on Pattern Analysis and Machine Intelligence 27*(1), 148–154.

T. Sawangsri, V. Patanavijit, S. J. (2005). Face segmentation based on hue-cr components and morphological technique. In *IEEE International Symposium on Circuits and Systems. ISCAS 2005*, Volume 6, pp. 5401–5404.

Wang Xianbao, Cao Wenming, L. G. (2005). A method of face location in complex background. In *International Conference on Neural Networks and Brain*, Volume 3, pp. 1507–1510.

459

Computational Modelling of Objects Represented in Images –
João Manuel R.S. Tavares & R.M. Natal Jorge (eds)
© 2007 Taylor & Francis Group, London, ISBN 978-0-415-43349-5

Author index

Printed and bound by CPI Group (UK) Ltd, Croydon, CR0 4YY

05/11/2024

01784060-0001